PRACTICAL FORESTRY
FOR THE AGENT AND SURVEYOR

CYRIL HART

By the same author

Royal Forest: A History of Dean's Woods as Producers of Timber (1966).
Clarendon Press, Oxford.
British Trees in Colour (1973). Michael Joseph, London.
*Forestry in Europe: A Report for the Committee on Agriculture of the
Council of Europe* (1979). Strasbourg.
Effect of Taxation on Forest Management and Roundwood Supply (1980).
A Report for United Nations, ECE and FAO. Geneva.
Private Woodlands: A Guide to British Timber Prices and Forestry Costings (1987).
Taxation of Woodlands (1988).
Trees: A brief Introduction to Forestry and Arboriculture (1988).

First published 1991
Reprinted with amendments, 1994
Reprinted 1998, 2002, 2005
This edition published 2022

The History Press
97 St George's Place, Cheltenham,
Gloucestershire, GL50 3QB
www.thehistorypress.co.uk

© Cyril Hart, 1991, 1994, 1998, 2002, 2005, 2022

The right of Cyril Hart to be identified as the Author
of this work has been asserted in accordance with the
Copyright, Designs and Patents Act 1988.

All rights reserved. No part of this book may be reprinted
or reproduced or utilised in any form or by any electronic,
mechanical or other means, now known or hereafter invented,
including photocopying and recording, or in any information
storage or retrieval system, without the permission in writing
from the Publishers.

British Library Cataloguing in Publication Data.
A catalogue record for this book is available from the British Library.

ISBN 978 0 7509 9941 0

Typesetting and origination by The History Press
Printed and bound in Great Britain by TJ Books Limited, Padstow, Cornwall.

 Trees for L🌱fe

Jo
D., J. and *A.*

'In the woods is perpetual youth. Within these plantations of God a decorum and sanctity reign, a perennial festival is dressed, and the guest sees not how he should tire of them in a thousand years. In the woods we return to reason and faith . . . The waving of the boughs in the storm is new to me and old. It takes me by surprise, and yet is not unknown. Its effect is like that of a higher thought or a better emotion coming over me when I deemed I was thinking justly or doing right.'

RALPH WALDO EMERSON

'A man of eighty, *planting!*
To *build* at such an age might be no harm
Argued three youngsters from a neighbouring farm,
But to *plant trees!* Th' old boy was plainly wanting.
"For what, in Heaven's name," said one of them,
"Can possibly reward your pains,
Unless you live to be Methusalem?
Why tax what little of your life remains
To serve a future you will never see?"

"Is it so?" said he.

"My children's children, when my trees are grown,
Will bless me for their kindly shade:
What then? Has any law forbade
The Wise to toil for pleasure not his own?
To picture *theirs* is my reward *today*,
Perhaps tomorrow also: who shall say?"'

JEAN DE LA FONTAINE

The last major revision of this text was in 1991 (with minor amendments in 1994); therefore, the reader needs to bear in mind the following:

1. Prices of timber and woodland products may have changed.
2. Some operational costs may have changed.
3. Many grants from the Forestry Commission have increased.
4. Some incidents of taxation may have changed.

CONTENTS

Foreword by Dr Julian Evans	xviii
Preface	xix
List of Colour Plates	xxvi
List of Figures	xxvii
List of Tables	xxviii
Abbreviations	xxx

INTRODUCTION: PRIVATE FORESTRY IN THE BRITISH CONTEXT	**1**
Changes in support of private forestry	4
Ownership of private woodlands	5
Today	8
The future	9

Chapter One

TREES, TIMBERS AND FOREST MEASUREMENT	**11**
1. *Britain's woodland heritage*	11
Native (indigenous) trees	12
Exotic (introduced) trees	13
2. *Identification of trees*	15
Winter twigs (or young shoots) and buds	15
Leaves/needles	16
Flowers	17
Fruits and seeds	18
Bark	19
Form and outline	19
3. *Identification of timbers*	20
Conifers (Softwoods)	26
Broadleaves (Hardwoods)	28
4. *Forest measurement*	31
Timber measurement — general	32
Measuring the volume of a standing tree	34
Measuring the volume of a stand	35
Measuring the volume of felled timber	37
Weight measurements	38
Hoppus measure	39
References	39

Chapter Two
NURSERY PRACTICE, SEED AND FOREST GENETICS — 41

1. *Nursery practice* — 41
   - Whether to start a nursery — 41
   - Labour requirement — 42
   - Choosing the site — 43
   - Preparing the site — 44
   - Seed supply — 44
   - Seedbeds — 44
   - Sowing density — 45
   - Transplanting — 46
   - Precision-sowing with undercutting — 48
   - Protection of nursery stock — 48
   - Container grown seedlings — 49
   - Genetically-improved Sitka spruce — 49
   - Poplar production — 50
   - Willow production — 51
   - Production of ornamental trees and shrubs — 51
   - Vegetative propagation — 52
   - Nursery herbicides — 52
   - Weed control in seedbeds — 52
   - Weed control in transplant lines — 53
   - Weed control on fallow ground — 53
   - Nursery fungal diseases — 53
   - Nursery pesticides — 55
   - Nursery insect pests — 55
   - Cold storage — 56
2. *Tree seed* — 57
   - Seed sources — 58
   - Collection of seed — 60
   - Extraction and storage of seed — 60
   - Seed pretreatment — 61
   - Insects attacking seed — 62
3. *Genetics in silviculture: tree improvement by selection and breeding* — 62

References — 66

Chapter Three
FACTORS OF THE FOREST SITE — 68

1. *Climate* — 69
   - Topography — 69
   - Temperature (warmth/cold; drought/frost) — 69
   - Rainfall (wetness/dryness): precipitation and evaporation — 71
   - Water catchment areas — 73
   - Water quality — 74
   - Snow and ice — 74
   - Lightning — 75

		Light	75
		Nutrient requirements	76
		Air pollution and plant health	76
		Global warming – the 'greenhouse effect'	79
		Wind	80
	2.	*Ground vegetation as indicating site factors*	87
	3.	*Forest soils*	90
		Mycorrhizae	91
		Nitrogen fixing	91
		Groups of forest soils	91
		Soil and site classifications for forestry	93
		Maps of land capability for forestry	95
		Scotland	95
		England and Wales	97
		Soil, geological and tree relationships	97
		Modification of soils	98
	4.	*The forest ecosystem*	99
	References		100

Chapter Four

SILVICULTURAL CHARACTERISTICS AND SPECIES CHOICE   103

	1.	*Silvicultural characteristics*	103
		Conifers	105
		Broadleaves	111
	2.	*Species choice*	120
		Objectives	120
		Timber production	122
		Site factors influencing species choice	123
		Origin/provenance	126
		Conifers	130
		Broadleaves	135
	3.	*Pure crops or mixtures?*	138
		Pure crops	138
		Mixtures	138
		Nurse species	139
		The pattern of a mixture	140
		Types of mixtures	141
		Components of a mixture	141
	4.	*Broadleaves in Britain*	143
		The broadleaved resource	143
		Forestry Commission broadleaved policy	145
		The Government's measures for broadleaves	146
		Ancient semi-natural woodland	146
	References		147

Chapter Five
ESTABLISHING PLANTATIONS AND SPECIAL TREE CROPS — 149

1. *Establishing plantations* — 149
    - Preparation of the site — 150
    - Ground preparation by cultivation — 151
    - Ground preparation by chemical treatment — 152
    - Ploughing — 152
    - Scarifying — 154
    - Drainage — 154
    - Fencing — 156
    - Spacing: narrow (close) versus wide planting — 162
    - Plants — 165
    - Plant prices — 167
    - Planting — 168
    - Treeshelters — 171
    - Weeding — 172
    - Beating up (Gapping) — 178
    - Initial fertilizing — 179
    - Afforestation in uplands (Plantation silviculture) — 180
    - Restocking conifers in uplands — 182
    - Broadleaves in uplands — 185
    - Failures in establishing — 185
    - Costs of establishing plantations — 186
    - Operational costs of establishing plantations — 188

2. *Establishing special tree crops* — 190
    - Decorative quality broadleaved timber — 190
    - Ornamental (decorative) foliage — 191
    - Poplars — 191
        - Growth and yield — 193
    - Christmas trees — 198
    - Cricket-bat willow — 202
    - Shelterbelts — 204
    - Hedgerows and hedgerow trees — 206
    - Land reclamation and forestry — 208

References — 210

Chapter Six
AFTER-CARE (TENDING) OF PLANTATIONS — 213

1. *Cleaning, rack-cutting, brashing and pruning* — 214
    - Cleaning — 214
    - Rack-cutting — 215
    - Brashing — 215
    - Pruning — 217

2.	*Fertilizing established plantations (top-dressing)*	219
	Top-dressing of conifers	219
	Aerial fertilizing	222
	Top-dressing of broadleaves	223
3.	*Protection against insect pests and diseases*	223
	Insect pests	224
	Insects and storm-damaged trees	227
	Diseases	228
	General importance	229
	Conifers	229
	Broadleaves	230
4.	*Protection against animals and birds*	231
	Woodland animals	231
	Woodland birds	238
	Assessment of wildlife damage in forests	239
5.	*Fire risk and hazard*	239
6.	*Maintenance of woodlands*	242
	Insurance	242
7.	*Costs of plantations to the thicket-stage*	243
	Conifers	243
	Broadleaves	243
	Operational costs of after-care (tending) of plantations	244
	References	245

Chapter Seven

SILVICULTURAL SYSTEMS, THE 'NORMAL' FOREST, NATURAL REGENERATION AND ENRICHMENT — 248

1.	*Silvicultural systems*	248
	High forest systems	250
	Regular (even-aged) silviculture	250
	Irregular (uneven-aged) silviculture	251
	Shelterwood systems	251
	Group regeneration	252
	The selection system	252
	Accessory systems	255
	The history of irregular silviculture in Britain	255
	Garfitt's contribution to irregular silviculture	257
	Bradford-Hutt Plan for continuous canopy forestry	258
	Economics of irregular silviculture	262
	Coppice systems	264
	Conversion of silvicultural systems	273
2.	*The 'normal' forest and sustained yield*	273
3.	*Natural regeneration*	275
	Natural regeneration of broadleaves	276
	Preparing for natural regeneration	276

		Managing the overstorey	278
		Tending of young natural regeneration	279
		Examples of broadleaved natural regeneration	279
		Natural regeneration of broadleaves in the uplands	281
		Natural regeneration of conifers	281
		An example of natural regeneration of conifers	283
		Economics of natural regeneration	284
	4.	*Enrichment*	285
	5.	*Basket willow coppice*	286
		References	287

Chapter Eight

THINNING 291

	1.	*Thinning classes and regimes*	291
		Thinning regime	292
	2.	*Traditional thinning chiefly in lowland stands*	296
		Conifers	298
		Broadleaves	300
		Epicormic branches and knots	303
		'Free growth'	304
	3.	*Thinning of broadleaves and mixtures chiefly in lowland stands*	304
		Mixtures of broadleaves and conifers	305
		Broadleaved mixtures	306
	4.	*Current thinning of conifers chiefly in uplands*	307
		Timing of first thinning	307
		Systematic thinning	308
	5.	*Economic aspects of thinning*	310
	6.	*Thinning practice*	312
		Forestry Commission	312
		To thin or not to thin?	313
	7.	*Thinning control*	315
	8.	*Respacing of upland conifers*	316
		Methods of respacing	316
		Oceanic silviculture	317
		Chemical thinning in the pole-stage	318
		Self-thinning mixtures	319
	9.	*Harvesting of thinnings*	319
		References	320

Chapter Nine

BRITAIN'S FOREST RESOURCE AND TIMBER INDUSTRY 321

	1.	*Britain's forest resource*	321
		The case for national forestry	322

2.	*Britain's timber industry*	324
	Timber demand	326
	Timber supply	326
	Future timber supply and demand	327
	Conifer sawmilling	328
	Broadleaved sawmilling	328
	Small diameter roundwood processing	328
	Future timber production	329
	Forecasting and achieving production of roundwood	331
3.	*Properties of British-grown timbers*	332
	Chemistry of wood	334
	Properties of the main British-grown timbers	335
	Conifers (softwoods)	335
	Broadleaves (hardwoods)	342
	Determinants of wood quality	347
	Improving quality of British conifers	347
	Sawn softwood: stress grading and strength classes	348
	Softwood sawlog grading	349
4.	*Utilization of British-grown roundwood*	351
	Broadleaved utilization categories	355
	Roundwood (Broadleaves)	355
	Sawnwood (Hardwood)	359
	Conifer utilization categories	360
	Roundwood (Conifer)	360
	Sawnwood (Softwood)	363
	Wood processing industries using small diameter roundwood	367
	Paper and paperboard	368
	The export pulpwood market	369
	Wood-based panel board	369
	Particleboard (chipboard)	370
	Fibreboard	370
	Cement-bonded board, moulded-wood, and wood wool	370
	References	371

Chapter Ten

HARVESTING 373

1.	*The rotation length*	373
2.	*Techniques of harvesting*	376
	Felling	378
	Harvesting windthrown trees	380
	Whole tree harvesting: forest biomass	382
	Extracting	384
	Combined costs of harvesting	387
	Transporting	387
	Combined costs of harvesting and transporting	388

3.	*Forestry roads*	389
	Terrain	390
	When to construct roads	390
	Road considerations in regard to harvesting system	391
	Road specifications	392
	Road planning and construction	393
	Timber traffic on highways	393
4.	*Interaction of harvesting and restocking*	394
5.	*Environmental considerations in harvesting*	395
6.	*Conversion; and estate sawmilling*	397
	Conversion in woodlands	397
	Peeling and pointing by machine	399
	Conversion in wood-yard or depot	401
	Estate sawmilling	402
7.	*Preservation of wood*	405
	Decay in service	405
	Preservative treatment	407
	References	409

Chapter Eleven
MARKETING AND PRICES   411

1.	*Timber and other woodland produce*	411
	Determinants of sale value	411
	Seeking the markets	412
	Markets for conifers	414
	Markets for broadleaves	415
	Markets for coppice	417
2.	*Forestry Commission marketing policy*	418
3.	*Points of sale*	419
	Standing sales	419
	Sales of felled trees	420
	Sales of converted products	420
4.	*Grading of roundwood*	421
	Broadleaves	421
	Conifers	422
5.	*Methods of marketing timber*	422
	Particulars and Conditions of Sale	423
	Contracts	424
	Charges for marketing	424
6.	*World prices of timber*	425
	Outlook for timber prices	426
7.	*Timber price-size relationship*	427
	Price-size curves for roundwood	427

8.	*Prices of roundwood*	430
	Conifer	432
	Broadleaved	433
	Standing parcels of broadleaves	435
	Wood processing industries using small diameter roundwood	436
	Pulpmills using conifers	436
	Pulpmills using broadleaves	436
	Boardmills	437
	Wood wool and Moulded-wood mills	439
9.	*Prices of converted roundwood products*	439
	Minor forest products	439
	References	443

Chapter Twelve

FOREST PLANNING — 445

1.	*The yield class system*	445
	Yield Class	445
	Volume increment	446
	Maximum mean annual increment	447
	Classification by yield classes	447
	General Yield Class	447
	Estimating yield class of unplanted land	448
	Production Class	448
	Local Yield Class	451
	The effect of variations in growth rate	451
	The effect of different treatments	451
2.	*Yield models for forest management*	451
	Using yield models	452
	Choosing the right yield model	454
	Using yield models to forecast production	455
3.	*Data collection and use*	455
	Sub-compartment notes	456
	Prescriptions of work	458
	Planning, organizing and controlling wood production	458
	Forecasting production	458
	References	459

Chapter Thirteen

FOREST MANAGEMENT — 461

1.	*Choice of forestry intensity*	461
2.	*The management scheme*	463
	The Plan of Operations	464
	The Organization Plan	464
	The Financial Plan	466
	Charges for management	466

3.	Management primarily for wood production	467
4.	Management to include non-wood benefits	471
	Mixed woods	472
	Guidelines for broadleaved management	473
	Guidelines for management of ancient semi-natural woodlands	473
	General management considerations	474
	Progress of the Broadleaved Policy	475
5.	Management for landscape	476
	Coniferous woodlands in the landscape	476
	Broadleaved woodlands in the landscape	477
	Forest landscape design	477
	Afforestation: landscape principles and guidelines	479
6.	Management for recreation	484
7.	Management for wildlife conservation	485
	Woodland animals	487
	Effect of silvicultural systems	487
	Conservation of habitat	489
	Managing wildlife in woodlands	494
	The approach to nature conservation	495
	Operational guidelines	497
	Woodland birds	499
8.	Management for water	502
9.	Management for sporting	504
	Gamebirds	505
	Wildfowl, fishing and foxhunting	507
	Deer management	509
	Economics of game management	509
10.	Urban woods, community forests, and new national forests	510
	Urban and urban fringe woods	510
	Community forests	512
	New national forests	514
11.	Farm forestry	514
	Integrating forestry and farming	515
	Agroforestry	516
	Short rotation coppice crops	517
	Farm woodland	517
	Advice on farm forestry	520
12.	Amenity woods	520
	References	521

Chapter Fourteen

MANAGERIAL ECONOMICS OF PRIVATE FORESTRY — 524

1.	Introduction	524
	Decision-making	525
	The effect of time-scales on financial calculations	526

	Discounting	526
	Non-discounting criteria	528
2.	*Measures of profitability*	528
	Inflation	531
	Risk and uncertainty	532
3.	*Economic appraisal for management decisions*	533
4.	*Principles of economic appraisal*	534
	Non-wood benefits and costs	536
5.	*Data and formation of assumptions for economic appraisal*	537
	Costs	537
	Revenues	537
	Yields	538
	Grants	538
	Taxation	538
6.	*Applications of economic appraisal*	539
	Land Expectation Value	539
	Options in establishment	540
	Options in after-care (tending)	542
	Options in thinning	544
	Options in clear-felling and restocking	544
	Options in roading and extraction	546
	Options in machinery	547
	Conclusion of the appraisal	547
7.	*Profitablility of private forestry*	548
	Scarcity of financial data	549
	Factors which influence profitability	550
	Comparison with State forestry profitability	550
	Profitability of conifers	551
	Profitability of broadleaves	553
	Forestry compared with agriculture	558
8.	*Forests and the social good*	559
	Social cost and benefit	560
	References	561

Chapter Fifteen
SUPPORT FOR PRIVATE FORESTRY   563

1.	*State support and incentives*	563
2.	*Grants and other subsidies*	565
	Woodland Grant Scheme	565
	Storm damage restocking supplement	566
	Grants for new native pinewoods	567
	Woodland Management Grants	567
	Grants for farmland planting	569
	Set-aside Scheme	570
	Grants for amenity planting	570

3.	*Taxation of woodlands*	570
	Income tax	571
	Capital gains tax on land following the disposal of woodlands	573
	Inheritance tax on woodlands	573
	Estate duty	577
	Sporting in woodlands	577
	Recreation in woodlands	577
	Estate forest tree nurseries	577
	Christmas trees	577
	Estate sawmills	578
	Grants (taxation incidence)	578
	Rates	578
	Value Added Tax	578
4.	*Controls and restraints*	579
	Environmental aspects	579
	Environmental assessment of afforestation projects	579
	Indicative forestry strategies	580
	Consultation with statutory authorities	580
	Control of tree felling	580
	Tree Preservation Orders	581
5.	*Other aid and encouragement*	582
	Education and training	582
	Health and safety at work	584
6.	*Forest research and development*	585
	References	590

Chapter Sixteen

WOODLAND ACCOUNTS AND FORESTRY VALUATIONS — 592

1.	*Woodland accounts*	592
	Financial accounting	592
	Cost accounting	592
2.	*Woodland records*	594
3.	*Forestry costings*	594
	Labour	594
	Forest machinery	595
	Work Study	595
	Scale economies/diseconomies	596
4.	*The growing-stock*	597
5.	*Computers in forestry*	598
6.	*Forestry valuations*	600
	Valuation of woodlands	600
	Types of forestry valuations	602
	Methods of forestry valuations	604
	Young plantations up to 5 years old	604
	Thicket-stage plantations 8 to 15 years old	605

Pole-stage to mid-age crops	607
Nearly or quite mature crops	608
Hazel underwood	608
Coppice	608
Forestry land	609
Forest properties	609
Effect of grants, taxation and other personal circumstances	609
Valuation of amenity	609
Valuation of non-wood benefits	610
Charges for valuation	610
References	610

Chapter Seventeen

INVESTMENT IN PRIVATE FORESTRY   612

1. *The case for investment*   612
    - Past investments   613
    - Selection of a forestry investment   616
    - Types of forestry investment   617
    - The commercial case for investment in private forestry   619
    - Specific investment objectives   620
    - Management of forestry investments   621
2. *The market for woodlands and plantable land*   622
    - Forestry Commission disposals of woodlands and plantable land: 1981–90   622
    - Private sector sales of woodlands: 1981–90   624
    - Effects of the 1988 Budget on woodland sales   624
    - Market values of woodlands and plantable land   625
        - Plantable land   626
        - Stocked woodlands   626
    - Types of investors in private woodlands   628
    - The future   630
3. *Loans for forestry*   631
4. *Forestry in the European Community*   632
5. *European forestry and timber trends*   635

References   638

Index   639

FOREWORD

Cyril Hart's *Practical Forestry for the Agent and Surveyor* is a modern classic. Indeed, I used the second edition, published in 1967, when reading forestry at Bangor and even then it was already ten years since the book first saw the light of day. The bringing out of a third edition, rather like Matthews' complete revision in 1989 of Troup's famous *Silvicultural Systems,* once again restores to British forestry an important work of reference.

Practical Forestry for the Agent and Surveyor is, first and foremost, for reference and aimed primarily at the private grower. It complements the Forestry Commission's summary of silviculture and forest management *Forestry Practice,* known to many as Bulletin 14 but published in 1991 as Handbook 6, but goes into greater depth and embraces more topics. Indeed, the range of material covered in *Practical Forestry for the Agent and Surveyor* is breathtaking: from silviculture to marketing, from law to accounting and, uniquely, it addresses authoritatively the whole question of valuation. I stress this breadth of treatment because herein lies the book's great strength; it is a real introduction to the whole of forestry in Britain. This has only been made possible by the author's enormous experience. Not only has he written more than ten books covering local history as well as forestry, but, as this book reveals, he is familiar with numerous estates and woodlands throughout Britain. The skill displayed in these pages is his use of this knowledge to effect in conveying sound principles and practices.

Forestry is often described as both an art and a science, the timescale of forestry simply precludes experimental proof of every precept. Thus, a necessary handmaiden to factual knowledge is the art of wise application of acquired skill and experience. This is not whimsical since from the time of Evelyn and his *Sylva* such a blend has been evident in the books designed to help British foresters achieve success in their endeavours. Cyril Hart's *Practical Forestry for the Agent and Surveyor* continues admirably this tradition.

Julian Evans
Chief Research Officer (South)
Forestry Commission

PREFACE

Great Britain's 2.3 million hectares of woodlands and forests (about 10% of its land area) – roughly 40% State and 60% private sector – are complex renewable resources which provide about 13% of the country's wood requirements as well as invaluable wildlife habitats, attractive countryside and many other non-wood benefits.

Private forestry has experienced many changes since the writing of the second edition (1967). To be welcomed is the continued partnership of private woodlands and State forests (through the Forestry Commission to whom appropriate tribute is given in the Introduction as well as in chapters 9 and 15). The government currently expect the private sector to undertake the large majority of afforestation. Its aim is for forestry to increase economic, social and environmental benefits.

In recent years the forest industry has seen advances both in the fortunes of the wood processing sector and in the setting of new standards of environmental awareness and forest practice. Techniques of silviculture, harvesting and marketing have significantly improved, and increased attention given to business methods, among them work study; all leading to greater productivity. The public perception of forestry has received a dramatic reappraisal, with increasing public recognition of the many environmental and other benefits provided by woodlands and forests. Steps have been taken to build environmental objectives into forest policy. Advances are being made in the area of broadleaved trees planted and in the management of broadleaved woodlands. Multi-purpose woodland management is being pursued to aid wildlife conservation and recreation, along with wood production. Reshaping of forests and hillside woodlands is being undertaken for landscape reasons as they are harvested and restocked. More thought is being given to uneven-aged silvicultural systems. Plans are laid to create new woodlands on the fringes of towns and cities, and extensive new forests are envisaged close to major centres of population.

The case for an expansion of forestry becomes compelling when its benefits are fully appreciated, along with the need to find alternative uses for agricultural land and with the recognition that trees lock up carbon from CO_2 in the atmosphere that would otherwise contribute to the 'greenhouse effect.' The location and type of new planting is now changing; there is emphasis on a higher proportion of broadleaved trees and on planting on better quality land 'down the hill' and on lowland farms – all encouraged by appropriate incentives. The higher environmental standards for forestry are not meant to be a constraint, but the very considerable costs of 'conservation forestry' can never be ignored by the private woodland owner.

The sudden change made by the government to the tax status of private forestry in 1988, albeit soon alleviated in part by increased grant-aid, did not signify a change in government policy towards private forestry, but only in the mechanism through which successive governments previously sought to implement it; capital taxation concessions remained. However, the manner of the changes caused a temporary crisis of confidence in the private sectors of growing, harvesting and processing, and some of the effects, including a fall in the level of annual planting, were not fully foreseen. The forest nursery and management companies were harmed.

Important fiscal incentives still remain. In particular, investors enjoy freedom from income and corporation taxation on the growing tree crop and the harvesting of timber; also inheritance tax relief. The value of these reliefs is supported by grant-aid under various schemes. Moreover, fiscal policy plays a notable part in the 'financial protection' of estates through tax exemptions and reliefs accorded to heritage property such as land of outstanding scenic interest – important in succession and tax plans. The government is continually seeking and implementing an appropriate balance of grant-aid and other measures of support. (Forestry Commission grant-aid to private forestry during the year ended 31 March 1990 amounted to about £14 million. Other support, such as research and advice, was also substantial.)

The Countryside Commissions, English Nature (NCCE) and the Countryside Council for Wales (noted in the Introduction) have had an increased influence on forest policy; so too have other bodies, as well as individual conservationists, ecologists and economists. Conflicts between foresters and conservationists are decreasing as better understandings are reached. There is no inherent difficulty in striking a balance between forestry for economic and for social/environmental reasons. The economic use of forest resources and the conservation of the land and of its associated wildlife, are interdependent and not contradictory objectives.

Much of the fresh thinking about private forestry is oriented towards reducing agricultural surpluses and to enhancing environmental features of woodlands – amenity, landscape, recreation and wildlife conservation (concurrently many woodlands continue to provide much sought-after sporting values). Provision of social benefits is encouraged by nontaxable planting and management grants and by capital taxation concessions; and in the case of farmland by annual and other potentially taxable payments. In most areas, lowland and upland, there is now much greater emphasis on establishing and tending broadleaves. However, in almost all forestry, timber production continues to be a major objective, with the trade for sawlogs usually being adequate, and that for small diameter roundwood being subject to peaks and troughs caused by storm damage (home or abroad) and fluctuations in the economic climate.

Forestry requires a wide variety of skills whether directed to wood production or to non-wood benefits or to a combination of them. It draws on knowledge and experience from many disciplines and professions. Foresters have to be managers of resources and people, businessmen alert to the needs of the market and the demands of the public. Often they can claim to be conservationists, in practice if not by designation. Their task is to achieve all the many objectives of management set by the wide variety of owners and organizations for which they work. Foresters in the private sector have for long been employed by landed estates, management companies, consultancy firms and contractors; those employers are now supplemented by county, regional and district councils, universities, the Agricultural Development and Advisory Service, Scottish Agricultural College, the National Trust and a wide range of conservational and educational organizations. A matter for debate is: the range of skill foresters now need to encompass in these changing circumstances, especially in view of the range of other professions that are impinging on forest and tree management. They cannot be expected to display expertise in everything, and it is difficult to discern where the professional boundaries between foresters, land agents, agriculturalists, conservationists and arboriculturists lie; their interests often overlap. There is an increasing number of 'environmental foresters' due particularly to development of urban, community and new national forests, as well as to the policies of the Forestry Commission, the two Countryside Commissions, English Nature and the Countryside Council for Wales – all regarding woodlands in terms of the environment.

Throughout this book, as the title implies, the term *forester* refers chiefly to any agent/surveyor who has a direct role in forest management requiring adequate knowledge and experience; but it also refers to professional foresters in any of the organizations mentioned above, as well as to any owner who competently manages woodlands. The book is chiefly for them as well as for students of forestry and associated disciplines, whatever their eventual calling. Certainly it contains information that I would have welcomed access to within a single volume when preparing for a vocation in forestry and land economy. This book is intended to be a helpful addition to the many invaluable texts which have been published recently, notably: Savill and Evans *Plantation Silviculture in Temperate Regions* (1986), Price *The Theory and Application of Forest Economics* (1989), Matthews *Silvicultural Systems* (1989) and Aaron and Richards *British Woodland Produce* (1990). To be added are the numerous relevant publications of the Forestry Commission, among them those of Evans, Hibberd (ed.), Insley (ed.), Edwards, Christie and Hamilton – all noted in my text. Research is, of course, continually leading to improvements in the practice of forestry.

Forestry in this book means simply the growing of forest trees (i.e. silviculture) for whatever objective, as well as harvesting and marketing. *Restocking* is replanting following felling, and *afforestation* is new planting of bare land. *Forest* is an extensive wooded area, generally owned by the Forestry Commission or managed by forestry companies, or by firms of chartered surveyors and chartered foresters. *Wood* and *woodland* relate to large or small areas, mainly in the private sector. *Lowland* (below c. 250 m) concerns particularly mid and southern parts of England and Wales, though planting 'lower down the hill' in Scotland can be called lowland. *Upland* relates mainly to Scotland, northern England and North Wales. There is a decrease in the amount of broadleaves from the south to the north of Britain.

Much difference lies between the type of work/skills required of foresters managing extensive upland plantations – chiefly comprising monocultures of even-aged conifers – and those managing traditional lowland mixed wooded estates or small isolated woodlands. In the uplands, where practically all forestry is a commercial venture which has been developed on a large scale, foresters are expected to seek profit, but their management needs to be consistent with accepted environmental standards and their own professionalism. By contrast, lowland foresters have to seek a balance between many objectives – commerce, amenity, wildlife conservation and sporting. Both types of management may also be involved in landscape and recreation, and public access will be a pertinent consideration.

The structure of the book. The Introduction briefly sets the scene of forest policy and practice in Britain and the contributions of State forests (through the Forestry Commission) and private woodlands. Attention is drawn to the comparatively recent changes in taxation of woodlands, to grant-aid, and to the main targets of forest policy, particularly as to the environment and the role of broadleaves. The types of ownership of private woodlands are noted. Comments are made on the current economic recession in Britain and its effect on woodland investment; it appears likely to be only a temporary cyclical recession.

Chapter 1 deals with the main native and exotic trees, and with concepts of forestry (particularly Britain's sylvan heritage), identification of trees and their timbers, and forest measurement. Chapter 2 relates to tree seed, the raising of planting stock in nurseries, polytunnels and glasshouses, and genetics in silviculture. The question is posed as to whether or not a woodland estate should set up its own forest tree nursery.

Chapter 3 deals with the forest ecosystem, factors of the forest site, particularly climate and soils, and ground vegetation as indicating site factors. The information is intended to help in species choice, discussed in chapter 4, which same chapter lists the main silvicultural characteristics of the most commonly planted forest trees, and discusses the pros and cons of pure crops and mixtures. Broadleaves, because of their increasing importance, are then treated more extensively.

Chapter 5 discusses the establishment and costs of plantations, from preparation of the site, through all the relevant operations to completion. Chapter 6 deals first with the after-care (tending) of plantations – cleaning, rack-cutting, brashing and pruning. Fertilizing (top-dressing) of established plantations is then examined, followed by protection against insect pests, diseases, animals, birds and fire. These subjects could not have been adequately discussed without the help of experienced practitioners noted in my text. There follows information on the maintenance of plantations (roads, fencing and the like), and indications of costs of operations and of tending to the thicket-stage.

Chapter 7 treats of the silvicultural systems used in Britain, particularly the various high forest systems (giving some notable examples in practice) and the coppice systems. This chapter was only accomplished after study of renowned texts on the subject. Natural regeneration, the *normal* forest and sustained yield are dealt with. Chapter 8 concerns the important subject of thinning in all its aspects, treating of past and present methods and regimes, and recommendations. The important question 'to thin or not to thin' is discussed.

Chapter 9 deals with Britain's forest resource, the timber industry, and supply and demand. The properties of home-grown timbers are set out along with determinants of wood quality and methods of grading. The utilization of British grown timber is discussed, and can be considered along with markets and prices included in chapter 11. Chapter 10 considers rotation length. Techniques of harvesting are explained, together with information on forestry roads. The interaction of harvesting and restocking are dealt with, as well as the environmental considerations in harvesting. Then follows information on conversion and preservative treatment of wood. Chapter 11 concerns, first, marketing of timber and other woodland produce, including points of sale, grading of roundwood and methods of marketing. The remainder of the chapter relates to prices of timber and price-size relationship. Substantial information is provided of timber used by pulpmills, boardmills and the like, together with indications of prices. Prices of veneer logs and minor forest products are also provided.

Chapter 12 discusses forest planning, explaining the yield class system, yield models for forest management, and data collection and use. Chapter 13, after touching on the intensity of forestry practised, proceeds to consider management for wood production, followed by management for non-wood benefits – landscape, recreation, wildlife conservation, water quality and sporting. There follows information on forests for the community and farmland forestry. The sections relating to management for environmental benefits could not have been written without study of the erudite publications of experienced practitioners noted in my text.

Chapter 14, briefly introducing management economics of private forestry, owes much to renowned forest economists named in my text. The measures of profitability are explained. Investment appraisal is then examined. The chapter proceeds to discuss the profitability of private forestry and the basics of private woodland investment. Briefly touched upon are the wider political economy of forest development and social well-being. Peripheral information on forest economics is included in parts of other chapters. The main aim of this chapter is to provide an introduction to the subject. Brief comments relating to the 'greenhouse effect' include references to evaluation of the relevant benefits of forestry.

Chapter 15 outlines State support for private forestry, the controls and restraints upon it, and the grant-aids and fiscal concessions, as well as noting encouragements relating to education and training, health and safety at work, and forest research and development. Chapter 16 provides a very brief introduction to woodland accounts and records – each owner of a woodland property uses a chosen peculiar system which meets requirements. Forestry valuations are treated in much greater detail, being a subject on which relatively little has hitherto been written. Finally chapter 17 outlines the case for investment in private forestry and indicates the market for stocked woodlands and plantable land, as well as the loans available for forestry. It then touches on forestry in the European Community and on European timber trends – matters of increasing importance particularly in view of the forthcoming European Single Market.

Readers are reminded that constituents of herbicides, pesticides, fungicides and fertilizers – and the regulations relating thereto – can change, and hence, when contemplating their use, relevant up-to-date information must be obtained.

Species of trees are referred to by their English names; scientific names are given in the Index. Physical measurements are in metric units. Indications of costs and prices are provided, but as they vary greatly through time and conditions it has been difficult to summarize them. The processing industries, being the growers' main market, are appropriately referred to. Useful addresses are included as an Appendix.

Reasons for the appearance of this third edition – so long after the second – include the hope that thereby I am able to make a contribution to private forestry that might assist the agent/surveyor throughout Great Britain; and concurrently to attempt to keep abreast of the many changes witnessed in forestry conditions, objectives and practice during the last two decades.

The book's content, as the title implies, relates specifically to the private forestry sector (wherein my calling lies). Its writing is the result of practical forestry experience therein, supplemented by knowledge gained from: (i) the Forestry Commission's numerous fine examples, notable research and development, and publications, along with splendid co-operation received from its staff – acknowledgement for which is gratefully recorded below; (ii) many private estates, their owners and their foresters; (iii) nurserymen, sawmillers, wood processing executives; and (iv) forest economists, educationists and consultants.

Acknowledgements. In this wide-ranging semi-technical book on private forestry it has been difficult to treat all components comprehensively or always with appropriate balance and originality. Site conditions and many other factors under which silviculture is practised in the lowlands are different to those met in the uplands, and hence it has not always been possible to write adequately of both sets of conditions, particularly as to establishing, harvesting, wildlife conservation and sporting. I have not gained first-hand experience of all aspects of forestry but, following good fortune I can to this day unroll a personal map of forestry memories, and on it flag visits to numerous woodlands, forests, sawmills, pulpmills, board mills, nurseries, research stations, colleges and universities where the opportunities have been taken to widen my experience throughout Great Britain and in parts of Europe, Scandinavia and the west coasts of America and Canada.

The many references given at the end of each chapter indicate my profound indebtedness to numerous foresters, harvesters, wood processors, conservationists and economists, all expert in their particular field, many possessing also a remarkably wide knowledge of associated subjects. Those now in retirement or semi-retirement following distinguished careers include: Arnold Grayson CBE, John Matthews CBE, David Johnston, Roger Lines OBE, John Workman OBE, Alan Mitchell, Dick Steele, John Brazier, Jack Aaron, John Christie, Roy Faulkner, John Aldhous, Jim MacGregor, Michael Williams, John Morgan (BRE), Dick Stumbles, Kenneth Rankin, Phil Hutt MBE, Ian Falconer, Bert Dover and Ken Stott.

From within the Forestry Commission, invaluable information and assistance were provided by the following (named in alphabetical order): Eddie Arthurs, Don Bardy, Simon Bell, Calvin Booth, Richard Britton, Richard Broadhurst, Roger Busby, Ken Buswell, Fred Currie, Jim Dewar, Peter Edwards, Julian Evans, Hugh Evans, John Everard, Moira Foote, Peter Freer-Smith, John Gibbs, Peter Gosling, Douglas Green, Graham Hamilton, Wilma Harper, Roger Hay, Brian Hibberd, Simon Hodge, David Hughes, Richard Illingworth, John Innes, Hugh Insley, Steve Lee, Mike Lofthouse, Alan Low, Bill Mason, Bob McIntosh, Keith Miller, Andy Moffat, Andy Neustein, Tom Nisbet, Steve O'Neill, Derek Patch, David Patterson, Harry Pepper, Steve Petty, Carolyn Potter, Mark Pritchard, Graham Pyatt, Steve Quigley, Chris Quine, Phil Ratcliffe, Tim Rollinson, Robert Strouts, Charles Taylor, Paul Tabbush, Donald Thompson, Richard Toleman, Graham Tuley, Chris Walker, John Wallace, Roger Warn, Joan Webber, Adrian Whiteman, David Williamson, Tim Winter and Mark Yorke.

From within a wide spectrum of the private sector, splendid help in diverse ways was received from: Adrian Baird (CLA), Michael Bax, Arnold Beaton, Chris Burd, Peter Chantler, Graham Colborne, David Cooper, Mark Crichton-Maitland, Angus Crow, Huw Davies (ADAS), Vernon Daykin, Roy Dyer, John Eadie, Geoff Elliott, Garth Evans, Ted Garfitt, Andy Gordon, David Goss, Phil Hare (ETSU), Anthony Hart, Dick Hartnup, Rodney Helliwell, Mike Henderson, Harry Hindle, Chris Hood, Barrie Hudson, Chris Hughes, Keith Kirby (NCCE), Mike Kirby (CC), Peter Lambert, Douglas Lamont, Mike Lane, Arthur Lloyd, Hanslip Long, Roy Lorrain-Smith, Geoff Machin, Philip Mackenzie, Douglas Malcolm, John McHardy, Ian McCall, Paul Mitchell, Keith Openshaw, Alan Parks, David Pearce, Mark Potter, Colin Price, Richard Prior, Tony Richardson, Peter Savill, Bob Smith, Ronnie Smith, Mike Spilsbury, Robin Sym, David Strang Steel, David Taylor, Barrie Wellington, Guy Watt, David Wood and Geoff Yates.

From the continent of Europe information was provided by Tim Peck (ECE/FAO, Geneva) and Jeremy Wall (EC, Brussels), co-operation deeply appreciated.

To the above-named helpers – as well as to any inadvertently not recorded – I express profound thanks and appreciation. Any errors in the text must of course be attributed to me alone.

I also accept this opportunity to place on record my appreciation of certain distinguished foresters additional to those already mentioned through whom in one way or another I have widened my experience of forestry, namely: Charles Ackers OBE, Tom Alexander, Mark Anderson MC, Alan Bayliss, Bill Binns, Charles Chavet, Herbert Edlin MBE, Gwyn Francis CB, Wilfrid Hiley CBE, George Holmes CB, Melville Hyett, Jimmy James OBE, John Jobling, George Lane, Russell Meiggs, C. Syrach Larsen, Alec Lockhart, Henri Naudet CLHCG, Judith Rowe, George Ryle CBE and A.D.C. Le Sueur. Without having known them, my forestry would have been much less interesting and rewarding.

Grateful thanks are expressed to the Forestry Commission for the splendid co-operation received from so many of its staff, and for permission to include several extracts, tables, figures and photographs – all acknowledged in the text where they now appear. For Figures 2 and 3 I am indebted to TRADA and the Buckinghamshire College; for Figure 21 to the Fraser of Allander Institute; and for Figure 16 to John Matthews and Clarendon Press, Oxford.

As to colour illustrations, it has been neither necessary nor possible to include sufficient photographs adequately to illustrate all the subjects in the text. However, a selection has been made, chiefly comprising views of some of the woodlands where alone or in close association with friends in forestry, I have been privileged to act as consultant or manager, or as agent either for vendors or purchasers. It is to be hoped that the choice, in which I have benefited from much appreciated help given by Chris Hughes, appropriately enhances the text. Credits for individual photographs are given in the captions and in the List of Colour Plates.

Finally, in writing extensively on the silviculture of broadleaves in Britain I have of necessity had to rely substantially on the notable research and literature of Julian Evans – particularly in Forestry Commission Bulletins 62 and 78. Permission to base on them some sections of my relevant chapters has been invaluable. Furthermore, I feel greatly honoured by, and deeply appreciate, Julian's writing of the Foreword to accompany this contribution to private forestry in Britain.

Cyril Hart, April 1991

LIST OF COLOUR PLATES

1 (a,b,c,d)		Botanical features of four commonly planted commercial broadleaves [C. Raymond]
2 (a,b,c,d)		Botanical features of four commonly planted commercial conifers [C. Raymond]
3		Ancient semi-natural woodland [M. Fitchett]
4		Westonbirt Arboretum [Forestry Commission]
5 (a,b)		Upland cultivation [a: Forestry Commission; b: K. Taylor]
6		Tavistock Woodland Estate [T. Burrows]
7		Workmans Wood National Nature Reserve [J. Workman]
8 (a,b)		Longleat Estate Woodlands [J. McHardy]
9 (a,b)		Longleat Estate Woodlands [J. McHardy]
10		Caledonian Scots pine [G.A. Dey]
11		Springtime in Ashridge Park [W. Frances]
12		Thorpe Wood Local Nature Reserve [P. Wakeley]
13		Sparrowhawk nesting in Norway Spruce [R. Wilmhurst]
14		Pheasant shooting [J. Edenbrow]
15		Red deer [Forestry Commission]
16		Talymaes Woodlands [C. Hughes]
17		Broom Hill Wood [C. Hughes]
18		Woodlands at Huntley [C. Hughes]
19		Blaisdon Wood [J.P. Birch]
20		Coast redwood at Leighton [Anon.]
21 (a,b)		Dawn redwood at Leighton [a: M. Howells; b: H. Hindle]
22		Newtonhill Woodlands [T. Baker]
23		Minnygryle Forest [A. Crow]
24		Carsphairn Forest [A. Crow]
25		Margree Forest [R.J. Smith]
26		Coed Craig Ruperra [C. Hughes]
27		Ramsden Coppice and Widow's Wood [C. Hughes]
28		Llandegla Forest [C. Hughes]
29		Park Wood near Pen-y-Clawdd [C. Hughes]
30		Kilcot Wood [C. Hughes]
31		Hadnock Court Woods [C. Hughes]
32		Wyastone Leys Woods [Anon.]

LIST OF FIGURES

		page
1	Structure of timber; (a) hardwoods, (b) softwoods	21
2	Timber structures	22
3	Choice of seed trees	64
4	Average annual rainfall in the British Isles	72
5	Wind zonation of Great Britain	84
6	Regions of seed provenance within Great Britain	127
7	Seed collection zones in Western North America	128
8	Fencing terms	158
9	Rabbit fence specification	160
10	Stock fence specification	160
11	Deer fence – light specification for roe	161
12	Deer fence – heavy specification for fallow, red and sika	161
13	Pruning regime for fast-growing plantation poplar	194
14	Poplar: Top height/age curves for estimating General Yield Class; and Production Class Curves	196
15	The selection silvicultural system	253
16	Bradford-Hutt Plan (Tavistock): Schematic representation	259
17	Bradford-Hutt Plan (Tavistock): Schematic representation	261
18	Bradford-Hutt Plan (Tavistock): Schematic representation	261
19	Thinning: Classification of the types of trees found in a crop, based on canopy classes	293
20	The structure of the forestry industry	325
21	Locations of the major wood processing mills using small diameter roundwood (1991)	369
22	Patterns of volume increment in an even-aged stand	446
23	Sitka spruce: Top height/age curves giving General Yield Class; and Production Class Curves	450
24	Oak: Top height/age curves giving General Yield Class Curves	450
25	Yield Model: Sitka spruce, YC 14	453
26	Yield Model: Oak, YC 6	453
27	Forests and landscape	482
28(a)	Forests and conservation: I	492
28(b)	Forests and conservation: II	493
29	Forests and water	503
30	Forestry Commission Research Division Organization	587

LIST OF TABLES

		page
1	Basal areas	33
2	Density of fresh felled timber	34
3	Estimation of tree volume from diameter	34
4	Overbark to underbark volume conversion factors for conifer sawlogs	37
5	Volume/weight ratios, by species	38
6	Spacing in transplant lines	46
7	Average critical heights at which the onset of windthrow can be expected	86
8	The main mineral and shallow peaty soils (peat less than 50 cm)	94
9	Peatland soil types (peat 50 cm or more)	94
10	Various soils	95
11	Indications of costs of establishing conifer plantations to the fifth year	186
12	Indications of costs of establishing broadleaved plantations using treeshelters to the fifth year	187
13	Indications of operational costs of establishing plantations	187
14	Yields of unthinned stands of poplar YC 14 at spacings of 4.6 m and 7.3 m	195
15	Poplar: An indication of cash flow	197
16	Christmas trees (Norway spruce): An indication of cash flow	200
17	Cricket bat willow: An indication of cash flow	203
18	Indications of insurance premium rates for woodlands	242
19	Indications of costs of conifer plantations to the thicket-stage	243
20	Indications of operational costs of after-care (tending) of plantations	244
21	Sweet chestnut coppice: Yield table based on age	267
22	Sweet chestnut coppice: Yield table based on mean diameters	268
23	Threshold basal areas for fully stocked stands	295
24	Standard ages of first thinning	308
25	Productive private woodlands at 31 March 1990	322
26	Volumes of UK (a) imports, (b) production and (c) apparent consumption of wood and wood products during calendar year 1989	327
27	Volumes of deliveries of British-grown roundwood to wood processing industries during calendar year 1989	327
28	Forecasts of roundwood production in Britain	329
29	Forecasts of demand for small diameter roundwood and wood residues, excluding pitprops, fencing, export, firewood and miscellaneous	331
30	Sawn softwood visual grades	349
31	Forestry Commission conifer sawlog categories	350
32	Transmission pole classes	361
33	Forecast for market share development of sawn softwood by end uses 1993–97, compared with 1983–87	364
34	Felling ages for conifers	374
35	Broadleaves: minimum rotation lengths (years)	376
36	Indications of combined costs of harvesting	387
37	Indications of costs of road haulage	388
38	An example of costs of harvesting and transporting whereby the standing value of the produce can be determined	388
39	Natural durability of the heartwood of some British-grown timber species	406
40	Price-size curve: Conifers: England and Wales: at 1990–91 prices	428

List of Tables xxix

		page
41	Price-size curve: Conifers: Scotland: at 1990–91 prices	429
42	Price-size curve: Broadleaves: at 1990–91 prices	430
43	Prices of conifer roundwood	432
44	Prices of broadleaved roundwood	433
45	Prices of veneer logs	434
46	Values of standing parcels of broadleaved timber	436
47	Some representative discounting and compounding factors at 3%, 5% and 7%	527
48	An indication of operational input and planting grant for one hectare of lowland coniferous planting throughout the first ten years	551
49	A cash flow and calculation of NPV using a discount rate of 3%	552
50	An indication of operational input and planting grant for one hectare of lowland broadleaved planting throughout the first ten years	554
51	Land Expectation Values at 3% in £/ha (at 1986 values) and excluding the then income tax relief and planting grants	556
52	Land use: International comparisons	633

ABBREVIATIONS

Economics

BCR	benefit-cost ratio
CBA	cost-benefit analysis
DE	discounted expenditure
DR	discounted revenue
IRR	internal rate of return
LEV	land expectation value
NPV	net present value
YC	yield class
GYC	General Yield Class

Mensuration

dbh	diameter at breast height
MAI	mean annual increment (volume)
ob	overbark
td	top diameter
ub	underbark

Organizations

ADAS	Agricultural Development and Advisory Service
APF	Association of Professional Foresters
BRE	Building Research Establishment
CC	Countryside Commission
CCS	Countryside Commission, Scotland
CPRE	Council for the Protection of Rural England
DAFS	Department of Agriculture and Food, Scotland
EC	European Community
ECE	Economic Commission for Europe
FAO	Food and Agriculture Organization
FC	Forestry Commission
FICGB	Forestry Industry Committee of Great Britain
FSC	Forestry Safety Council
FTC	Forestry Training Council
HSE	Health and Safety Executive
ICF	Institute of Chartered Foresters
MAFF	Ministry of Agriculture, Fisheries and Food
NCC	Nature Conservancy Council
RICS	Royal Institution of Chartered Surveyors
RSPB	Royal Society for the Protection of Birds
SAC	Scottish Agricultural Colleges
TGUK	Timber Growers United Kingdom
TRADA	Timber Research and Development Association

Various

AONB	Area of Outstanding Natural Beauty
BS	British Standard
ESA	Environmentally Sensitive Area
FISG	Forest Industrial Safety Guide
LFA	Less Favoured Area
RGP	root growth potential
SDA	Severely Disadvantaged Area
SSSI	Site of Special Scientific Interest
TPO	Tree Preservation Order

INTRODUCTION: PRIVATE FORESTRY IN THE BRITISH CONTEXT

Forests and woodlands cover one-third of the world's surface but they are being destroyed at an alarming rate – both the temperate forests in the northern hemisphere and the hardwood forests and groves of the tropics. Forests are vital to man in numerous ways. Their wanton destruction provides not only a great threat to man, and to the ecology of this planet, but is one of the 'engines' driving the 'greenhouse effect'. Deforestation accounts for about one-quarter of annual CO_2 emissions.

Of Britain's land area about 10% carries trees, a land use which contributes considerably to the rural economy. However, Britain is one of the least forested countries in Europe – the European average is 25%. Her timber production provides only about 12% of her requirements of wood and wood products, yet it substantially helps to reduce an annual import bill currently amounting to over £7 billion. Britain's population uses annually at least 50 million tonnes of timber. More than three-quarters is conifers (softwood); over half is for paper and paper products. Upwards of 40,000 people owe their jobs to forests. Forests provide people with havens of peace and quiet in which to relax; moreover, to countless millions, forests enhance the landscape, and they provide important and sometimes unique habitats for flora and fauna.

Ownership of productive woodland in Britain is shared about 40:60 by the State (0.89 million hectares) and the private sector (1.25 million hectares). The Forestry Commission, established in 1919, is charged with the general duties of promoting the interests of British forestry, the establishment and maintenance of adequate reserves of trees and the production and supply of timber.[1] Since 1985, the Commission has also had a statutory duty to seek to achieve a reasonable balance, in carrying out its functions, between the needs of forestry and the environment.

The Forestry Commission is legally and practically a government department, reporting directly to Forestry ministers. The lead minister is the Secretary of State for Scotland, but the Commission equally is responsible for advice on the implementation of forest policy to the Minister of Agriculture, Fisheries and Food, and the Secretary of State for Wales. The Commission is, however, different from the usual departments of State in that there is a statutorily-appointed chairman and board of commissioners with prescribed duties and powers. The current (1991) chairman is J.R. Johnstone CBE, and the deputy chairman and director general, T.R. Cutler. For the private forestry sector the commissioner is R.T. Bradley, the

---

1 The Forestry Commission as *Forestry Enterprise* develops its forests for the production of wood for industry, manages its estate economically, protects and enhances the environment, provides recreational facilities, stimulates and supports employment and the local economy in rural areas by the development of forests and the wood-using industry, and fosters a harmonious relationship between forestry and other land use interests, including agriculture.

head of the division for private forestry and environment being A.H.A. Scott. The Commission's headquarters, and main offices are indicated in Figure 1. Some changes in organization are imminent.

The Commission as *The Forestry Authority* advances knowledge and understanding of forestry and trees in the countryside, develops and ensures the best use of the country's forest resources, promotes the development of the wood-using industry, endeavours to achieve a reasonable balance between the interests of forestry and those of the environment, undertakes forest research, combats forest pests and diseases, and advises and assists with plant health, and safety and training in forestry. Finally it encourages good forestry practice in private woodlands by administering grant-aid and felling controls, as well as through forest research and advice – functions explained in chapter 15.

The Commission's immense timber production, and its ever increasing contribution to landscape, recreation and nature conservation are widely known. It welcomes the public on foot and cycles to all its forests, provided this does not conflict with forestry operations and that there are no legal or other constraints on public access.

The role of the private sector has been supplementary to that of the Forestry Commission. It provided an immense contribution of timber during two world wars. In particular, many long established well-wooded estates (those of Bath, Bathurst, Bradford, Buccleuch, Clinton, Devonshire and Dulverton come readily to mind) have continued their traditional forestry[2] producing timber and contributing to landscape and wildlife conservation. In addition are the forestry enterprises of numerous individual owners of woodlands, small and large. Following the concentration on rehabilitation of woodlands felled during the second World War, from the mid-1960s private forestry management companies, operating within a favourable fiscal climate, planted extensive areas of mainly monoculture even-aged conifers chiefly throughout the uplands, for individuals, syndicates and financial institutions.

The 1980s formed a period of unprecedented innovation in British forest policy, with a wide range of developments that marked a turning point in growing and processing. Previously the main aims of government policy were the expansion of a basic resource to supply the markets for wood products, diversification of the economic base of communities in remote rural areas and, particularly for State forests, provision of opportunities for outdoor recreation. The private forestry sector continued to play an important role. This stage culminated in the early 1980s when the Forestry Commission worked closely with industrialists to stimulate new investment in sawmills, pulpmills, board factories and other processing industries, with the aim of using the new resource to supply domestic markets, until then mainly dependent on imports.

In 1985 forest policy developed further with the decision of the government to introduce a policy specifically directed at nature conservation and expansion of broadleaved forestry, supported by a new planting grant scheme offering higher grants, and with a detailed set of guidelines to protect and increase the country's broadleaved woodlands. The focus began to move towards conservation of existing

---

2   The large, hereditary, traditional, well-wooded estates were recently ably commented upon by the Duke of Buccleuch (1990) – particularly as to the value of their role in wildlife conservation. The total extent of such estates has fallen by two-thirds during this century. ('Landed with a Duty to Nature'. *Weekend Telegraph,* 17 November 1990, II.)

woodlands, and the planting of a wide range of broadleaved species, mainly native, including some that will only flourish on the more fertile soils.

Changes in support of private forestry. In 1988 the government reformed the incentives for private sector afforestation and restocking, shifting the emphasis from income tax and corporation tax relief (see chapter 15) to non-taxable planting grants – while continuing the capital taxation concessions. The chancellor of the exchequer had responded to a substantial media campaign for the reform of forestry taxation (especially the Schedule D/B 'switch'). The media had sought to link two main issues. First, that commercial conifer afforestation over extensive areas of Scotland and elsewhere was unnecessary and damaging to the environment. Second, that forestry was a very inviting 'income tax shelter' which enabled investors to set off the cost of planting (not the cost of land) against their other taxable income. So far as the environmental aspects were concerned, critics implied that the choice of a site for afforestation, followed by species choice and management, were being determined by tax saving considerations.

The government at once re-confirmed that the tax and grant changes did not mean that there had been any fundamental change in its policy of encouraging forestry, in an environmentally acceptable way. It remained of the view that a healthy forestry industry was in the national interest and that long-term confidence in both forestry and wood processing industries in this country was fully justified. As far as the substance of the changes was concerned, it was generally expected that a grant-aid system would provide a more widely acceptable system for private forestry. As a result of the switch to all-grant support, occupiers of rural land who were not high-rate tax payers were now somewhat loath to contemplate planting. (Incidentally, the reduction of the higher rate of income tax from 60% to 40% would itself have reduced the attractiveness of afforestation under the previous support regime.)

The level of grant support proved inadequate. Most forestry investment suffered – not only afforestation but also some of the maintenance of (particularly) lowland woods. Much of the motivation for afforestation was halted, i.e. the tax shelter accorded to, especially, 'absentee investors' (which had created the stability and continuity of afforesting which the forestry management companies in particular needed). Under the pre-1988 system of support, inward investment into forestry was essentially motivated by considerations of income (i.e. through income tax relief) to convert income into capital. The new system is mainly capital based (i.e. the investment of existing capital in bare land or stocked woodlands) with the support of planting aid (and, more recently, the promise of conditional management grants). Incentives therefore have had to be evaluated by investors on this basis and in comparison with available alternative opportunities.

Various interpretations of the results have been made. Commercial afforestation declined substantially during 1988–90, with a major decrease in acquisition of bare land, the market for which was depressed, although prices asked remained high. In Scotland forestry management companies continued some afforestation, along with the compulsory restocking. Furthermore, some Scottish landowners continued planting, at initial costs probably halved by the new grant-aid. Minimal planting was undertaken in England and Wales, together with some planting under farm schemes. Other planting was undertaken by occupiers benefiting from the transitional Schedule D benefits (available until 5 April 1993) though ineligible for the higher rates of grant – likely to ensure completion of their new planting, restocking and early tending by that date.

The result in the medium to long term may be to the benefit of the forestry industry, particularly when forestry properties come to the stage of yielding net income. Following a period of adjustment to the effects of the 1988 changes (to be improved by the forthcoming conditional grants towards management), forestry may be placed on a more even keel, enabling afforestation targets, if realistic, to be pursued. Forestry is no longer a tax-driven investment, and rightly this has caused potential investors to consider more sharply the commercial returns available (supplemented by the non-wood benefits and satisfactions). The importance of keeping all costs under prudent control is recognized.

The changes brought greater control and surveillance by the Forestry Commission and more specific targeting recognizing six objectives of policy: wood production, employment, landscape, recreation, nature conservation and alternative uses of excess agricultural land. The main new planting was to be undertaken by the private sector. However, the major changes resulted in a fall in new planting levels and a period of adjustment while forestry interests assessed the new situation. A surplus of planting stock was reported by the nursery trade. Against this background the government reaffirmed national annual planting targets, mainly under the Woodland Grant Scheme, and made available potentially taxable payments for farmers under the Farm Woodland and Set-aside Schemes, designed to reduce surpluses of arable crops. The emphasis on farmland planting and on broadleaves is significantly concerned with the environment. The market response to all the changes and incentives has not yet been fully enthusiastic, but there are encouraging signs of an improvement.

In January 1990 the House of Commons Agriculture Committee published its second Report 'Land Use for Forestry',[3] with comments and recommendations on surplus agricultural land and forestry policy. The response by the government was largely the retention of the status quo of current policy for the State and private sectors.[4]

A vast array of statutes and regulations now exists covering almost every issue on the environmental agenda: this accumulation of powers is beginning to point to a policy for the economic, social and environmental benefits that forests and woodlands provide.

Ownership of private woodlands. Britain's private woodlands are remarkably fragmented. So too is ownership, with about 90% of owners holding 30% of woodlands in blocks of less than 50 ha. Financial institutions during the 1970s and 1980s acquired UK plantations, partly in response to inflationary pressures, and now own some 30,000 ha. Commercial forestry management companies manage over 200,000 ha, achieving economies of scale in employment of labour and machinery and in marketing. The services offered by companies, firms, associations and co-operatives vary from complete management and marketing to arranging contracts for single silvicultural and harvesting operations.

Fragmented woods involve longer travelling distances for workers and machinery, longer fence lines per unit of area enclosed, increased protection worries and

---

[3] House of Commons Agriculture Committee: Second Report: 'Land Use and Forestry'. Session 1989–90. HC16–1. Vol. I. 10 January 1990, HMSO.
[4] Response by the Government to the Second Report from the House of Commons Agriculture Committee, 'Land Use and Forestry'. Session 1989–90. HC16–1. 9 May 1990, HMSO.

less remunerative timber sales. By contrast, the exploitation of soils for maximum volume of food production and wood and the pleasing landscape pattern thus created, provide some counter-balancing benefits: small blocks of woodland make an immense contribution to the environment, although public access to them is often denied.

The whole private forestry industry – growers, harvesters and processors – receives the co-operation, help and encouragement of the Forestry Commission both as the Forestry Enterprise and the Forestry Authority. The sector forms a complex interdependent industry involving many individual trades and organizations. Their interests are promoted by the Forestry Industry Committee of Great Britain (FICGB)[5] supported by various forestry associations, societies and wood processors. Many changes of ownership have occurred in recent years. Acquisitions have been made by individuals new to forestry, charitable trusts and financial institutions; and, since 1981, the Commission has sold to the private sector about 10% of its holdings, mainly isolated/fragmented woods. In summary, current types of woodland owners include:

1. Many thousands of traditional owners of large wooded estates in lowland England and Wales, the woodlands tending to be on the better quality land, and to comprise most of the nation's broadleaves;
2. Several hundreds of traditional owners of large wooded estates, chiefly coniferous, in Scotland;
3. 'Absentee' individual investors (mainly in upland conifers);
4. Numerous small woodland owners, including an increasing number of farmers;
5. Financial institutions;
6. Several charities, societies, conservation trusts, universities, National Trust, Ministry of Defence, water companies, and many local and national authorities;
7. A few wood processors – sawmills, pulpmills and board mills; and
8. Overseas investors, chiefly from Scandinavia and Europe.

The interests of most of the foregoing, especially the growers, are promoted by FICGB (as noted earlier) and Timber Growers United Kingdom Ltd, the Royal Forestry Society of England, Wales and Northern Ireland, the Royal Scottish Forestry Society, Country Eandowners' Association, Scottish Landowners' Federation, and the Farmers' Unions. Consultation and management is encouraged by the Institute of Chartered Foresters, the Association of Professional Foresters, the Royal Institution of Chartered Surveyors, co-operatives and ADAS. Additional to the above are the contractors for establishing, tending, harvesting and marketing,

---

5. The task of FICGB is to influence the economic climate within which the forestry industry can operate and expand – through political lobbying, public relations, information dissemination and education. In December 1987 FICGB published *Beyond 2000: The Forestry Industry of Great Britain*. It sets out the view of the private sector of the industry as a whole, providing a comprehensive picture of the various components of the industry, emphasizing their interaction with each other and describing the policies and priorities which the Committee saw as being necessary to sustain the industry's growth and competitiveness. The publication was most useful in focusing attention on the important contribution of forestry to trade, employment, the rural economy and the countryside. In December 1989 FICGB published *Options for British Forestry 1989–1990*. This set out FICGB's perceptions of the way in which the forest industry is developing, and the future opportunities and challenges that it faces. It outlined the decisions and actions necessary to realize the industry's potential as an important and influential force for growth and progress in Britain beyond 2000. Some wood processors are supporting FICGB with a levy on their supplies of raw materials.

road hauliers, roundwood merchants, specialists in veneer logs, transmission poles, turnery and mining timber and bark products, as well as individuals producing firewood, fencing materials, tent pegs, rustic work, hurdles, charcoal and foliage products. There are of course labour and management unions. Education and training are attended to by universities, colleges, the Forestry Commission, the Forestry Training Council and the Forestry Safety Council.

Conservation is encouraged by the two Countryside Commissions,[6] English Nature[7], the Countryside Council for Wales and local authorities.[8] Many of the above interests overlap.[9] The government proposes to integrate the NCC in Scotland and the CCS into an agency to be called Scottish Natural Heritage (SNH).

6   The Countryside Commission (CC) and the Countryside Commission for Scotland (CCS) have a statutory duty to promote the conservation of the countryside. Their particular interest lies in the contribution which woodlands can make to the rural landscape. Financial assistance for both establishment and maintenance of amenity woodland is provided directly and through the agency of local authorities and national park authorities; it is also available for the appointment of advisers or project officers and to support schemes such as the Demonstration Farms Project in England and Wales and the Central Scotland Woodlands Project in Scotland.
7   English Nature (NCCE) and the NCC for Scotland have overall responsibility for providing scientific advice about wildlife conservation, and has particular responsibility for establishing and managing nature reserves and for safeguarding Sites of Special Scientific Interest (SSSI). The ability to provide direct financial assistance for woodland management is very limited but under the terms of the Wildlife and Countryside Act 1981 payments may be made, in certain circumstances, for revenue forgone in SSSIs in the interests of nature conservation.
8   Local authorities are concerned about the environment and take an interest in both landscape and wildlife conservation. The aim of maintaining the traditional character of the landscape is featured in many structure plans. However, this principle if applied too inflexibly in relation to harvesting and restocking through the imposition of Tree Preservation Orders (TPOs) can frustrate sound woodland management. The main expertise on woodland amenity frequently resides in the county or regional level but arrangements usually exist for it to be made available to other authorities on request. In addition to full-time or part-time forestry officers a number of local authorities provide specialist staff for specific projects orientated towards woodland amenity management. Often they are also able to harness the resources of the voluntary organizations for this type of project.
9   The conservation movement is represented by many organizations including the Royal Society for the Protection of Birds, the Royal Society for Nature Conservation, county naturalist trusts, the Scottish Wildlife Trust and the Woodland Trust. The latter is specifically concerned with the conservation of broadleaved woodlands. Others such as the Council for the Protection of Rural England and the Ramblers' Association are equally concerned with amenity. The National Trust and National Trust for Scotland are much concerned with conservation in its broadest sense and both have large areas of woodland under their direct control. Various other bodies with a particular interest in woodlands include the Institute of Biology, the Landscape Institute and sporting organizations. The conservation movement, beyond its role in forming and influencing public opinion on such matters as the management of woodlands, has at its disposal considerable resources of finance and voluntary labour which can be mobilized to acquire, lease or manage areas of woodland, particularly broadleaved, and there is ample scope for further development of this important role. The voluntary groups within the Farming, Forestry and Wildlife Advisory Groups (FFWAGs) bring together farming and conservation interests with the aim of assisting farmers to manage their land, including woodlands, in such a way that positive action is taken to promote wildlife conservation. Most of the public sector organizations mentioned above are represented on local FFWAGs which are now established in every county in England and Wales and are rapidly gaining ground in Scotland.

Of wood processors, the paper and particleboard sections are represented partly by the UKWPA, and partly by the UK and Ireland Panel/Chipboard Association. The particular interests of sawmills are attended to in England and Wales by the British Timber Merchants' Association, and in Scotland by the United Kingdom Softwood Sawmillers' Association. There are of course many smaller utilizers of wood – turneries, fencing manufacturers and mines. All can receive encouragement from many of the above sources, as well as from the Timber Research and Development Association (TRADA) and the Building Research Establishment (BRE). Another component of the private sector is forest tree nurseries, their interests being undertaken by the Horticultural Trades Association and the Cell Grown Plant Producers Association. The combined influence of all relevant private sector interests has a considerable bearing on the extent to which forestry policies are carried through in practice.

Today. Britain currently is experiencing an economic recession. Several sectors of industry are depressed. This in part is due to the moves against inflation and to changes in the policies and economics of other countries. The current situation has significantly curtailed private forest investment, which in any case was reduced following the changes made in 1988 in forestry taxation reliefs despite their substitution by increased grant-aid. Consequently, the levels of private sector new planting, woodland tending, roading and employment are low relative to previous years.

A sharp reduction has occurred in the amount of plantable land coming on to the market. Land prices prior to 1988 averaged about £750 ha, soon fell to around £350 ha, and currently average about £500 ha. Farmers and other landowners are reluctant to sell plantable land at today's prices, and even so there are relatively few potential investors under the prevailing economic conditions.

The market for stocked woodland is non-uniform throughout Britain in two main market types – (i) commercial/investment, and (ii) conservation/sporting (here the financial rate of return is generally not significant, though the location, structure and age are important). The demand for the investment type is rather opportunist than active: most potential buyers are seeking commercial bargains. Young plantations in particular are relatively cheap, some owners being forced to sell as the cash position of their forest investment becomes overstretched.

However, there is a brighter side. The forestry industry as a whole demonstrates its underlying resilience by maintaining and, in some sectors, increasing its share of the domestic wood market despite a fall in demand and increased competition from abroad (particularly in price and by 'dumping'). About 13% of Britain's wood requirement is met by home-grown production. Nonetheless, difficulties are evident due to the down-turn in housing and other construction, furniture, and home improvements activity, adversely affecting particularly boardmills and, to a lesser extent, sawmills, and in consequence somewhat reducing the growers' market and prices.

Foresters hope that as the economic situation shows signs of improvement, and as woodland owners continue to adjust to the withdrawal of part of the forestry taxation relief, there will be an increase in new planting and other forest investment. Much will depend on the level and conditions of grant-aid given for the expected multi-purpose woodland management. Forestry is an increasingly important part of the rural economy. It is changing to meet public demands, and to face the uncertainties arising from the problem of agriculture surpluses and the search for alternative land uses. The employment it generates, though comparatively small in national terms, is vital to the maintenance of a healthy local infrastructure notably in many of the remote parts of Britain. Afforestation is continuing and forestry activity is increasing, particularly harvesting and restocking as the major plantings of the 1950s and 1960s yield large timber. The need to promote visual amenity, wildlife conservation and recreation – complementing timber production – is leading to a renewed interest in development of at least a modest transformation from uniform plantations to uneven-aged (irregular) woodlands.

Desirable though non-wood objectives are, it is essential to remember that the overwhelming demand continues to be for softwood timber for all sorts of uses from paper, through panelboard to construction timber. Conifers are essential to meet the needs that people have in their daily lives for materials made from wood, a fact not often appreciated by those who enjoy the countryside and see mainly the attractions of broadleaved species for amenity or possibly for conservation reasons. There is of course a place for all. The extra benefits ('spin-off') of forestry are clearly set out in *The Impact of Forestry in the United Kingdom* sponsored in 1989 by the Fraser of Allander Institute (see chapter 9). There is a strong case for afforestation both nationally (see chapter 9) and in the private sector (see chapter 17) – the planting of more trees 'of the right species in the right place for the right reasons' – an investment for the future. Site conditions for it are ideal in most parts of the country. In all the above, the private sector continues in a major role; a main thrust of current government policy is directed towards encouraging private investment. The Forestry Commission's valuable commercial and silvicultural experience has enabled it successfully to stimulate investment in the wood processing industries. The private sector values being able to deal with State staff who have practical experience of day-to-day forestry operations and who are able to give examples of high standards of multi-purpose forestry. The State's disposal policy for about 10% of its woodlands is discussed in chapter 17.

Britain's State forest parks, first established in 1935, continue today the tradition of foresters providing for the public to enter the forests and enjoy the facilities. From 8,000 km of waymarked walks all over the country, to major visitor centres in popular tourist areas and special facilities for car rallies and orienteering, millions rely on the State forests for their leisure. The Forestry Commission's forty-six forest nature reserves represent only the cream of the country's wildlife sites; several thousand sites are managed to conserve forest flora and fauna.

The large majority of afforestation proposals are now approved by the Forestry Commission without giving rise to controversy. Forestry, like agriculture, is not subject to planning control, but Indicative Forestry Strategies[10] are considered to perform two useful functions in relation to planting grant applications: (i) providing a framework for responses by planning authorities when consulted, and (ii) providing an indication to landowners and other forestry investors of the opportunities for further forestry development, of the degrees of sensitivity of areas of land to afforestation, and of the extent of consultation likely to be required. The strategy, to be successful in influencing the location of future afforestation, takes due account of the requirements of the forestry sector. Preferred areas are not

---

10 Indicative Forestry Strategies at their best are effectively regional plans drawn up by local authorities, now either in place or in an advanced state of preparedness in Scotland. They have not yet been extended to England and Wales.

simply the residue left over from a constraints exercise; they contain land which is genuinely attractive for afforestation.

The future. The range of demands placed upon woodlands is intensifying and expanding. There is increasing public pressure for recreation and for the conservation of wildlife to take a more prominent place in the management of the sylvan resource. These demands supplement a widespread awareness of and expectation concerning the quality of the environment and the management of woodlands for multi-purposes in which timber production may be seen as only one of the aims and not necessarily the dominant one. Woodlands, more than most other land uses, can be managed to generate a great variety of benefits and are increasingly managed with multi-purpose objectives.

Afforestation in Britain when on a large scale is likely to be in the northern and western uplands where land prices are relatively low, soils poor, the climate wet, and agriculture not highly profitable. Some planting will take place 'lower down the hill'. Lowland farmland, particularly in England, will also receive some small-scale afforestation. The availability of better quality land for planting significantly widens foresters' options – permitting a greater diversity of species, a generally smaller scale of operations, and a more intimate allocation of land between agriculture and forestry. However, there is a general presumption against large-scale, predominantly conifer, afforestation on unimproved land in the English uplands. Attempts are planned to establish new national forests, community forests and urban woods on the outskirts of towns and cities by joint initiatives of the Forestry Commission and the Countryside Commissions. From 1992 Woodland Management Grants will be available (see chapter 15), the aims focused particularly on good silvicultural practice after establishment, on conservation of ancient and natural woodlands, on environmental improvement of other woodlands and on conservation of all as a national resource. Broadleaved forestry is particularly promoted.

A balance continues to be sought between forestry and other interests in the countryside environment – pursuing multi-purpose forestry, for the commercial potential and the many non-wood benefits that it can provide. More trees are needed in urban areas and countryside alike to diversify wildlife habitats and provide shelter in open areas, enhance landscapes, supplement poor natural regeneration, replace hedgerow trees which have died or been destroyed, provide recreation and reduce Britain's imports from tropical forests and elsewhere. Trees bring nature into or near the heart of towns and cities, and make them more pleasant places in which to reside and work. Inner city areas, derelict industrial sites and urban fringe areas can all be transformed by planting trees and creating woods.

The major share of afforestation remains to be undertaken by the private sector. Private foresters are responding in a positive way to the changing environmental perceptions – ensuring that planting is fully compatible with landscape, wildlife, water and archaeological needs; at the same time securing for Britain a measure of self-dependency in a substantial invaluable resource.

Chapter One

TREES, TIMBERS AND FOREST MEASUREMENT

1. Britain's woodland heritage

Britain has a rich heritage of trees – some native, others introduced (exotics) – some evergreen, others deciduous – in a wide range of beauty, interest, growth and utility. Most notably, untold numbers clothe national and private forests and woodlands. Others enrich her gardens, parks, roadsides, hedgerows and watercourses: and many stand invitingly in arboreta, pinetums and forest gardens.

Conifers ('softwoods') have for long been extensively planted throughout Britain's now sylvan lands – chiefly for economic reasons. This change from the traditional broadleaved ('hardwood') trees is regretted by some people, but even constructive critics should concede that broadleaves, though possessing individual beauty, in mass during winter often present a sombre leafless picture, relieved only by the evergreen of holly, yew, box and climbing ivy, the stems of silver birch, along with the faded brown leaves clinging to low branches and young trees of oak, beech and hornbeam. Conifers add significantly to a landscape that, by natural causes, has been deprived of them, and when not in mass, add body and shape, and a variety of textures and colours. They provide what much of Britain's woods so signally lack – winter colour, high cover and shelter as well as the second form of tree-growth, the spire form (Mitchell 1980). Conifers also add structure and can provide the ultimate in trees – the groves of giants and monarchs. Since the introduction of conifers in mass much of Britain's sylvan scene has a patchwork cloak of many colours throughout all seasons. Her winters are relieved by the evergreen conifers and the hues of bare larches, all of which supplement the carpets of decaying leaves and needles.

In February, early spring is heralded by the yellow lambs' tails of hazel, the flower-haze of the elms and the silver and yellow catkins of sallow; while the buds and catkins of the alders continue to display a purple sheen. Later appear the changing greens of the hawthorn, the larches and birch. April and May bring the flower-haze of ash, the fruit-haze of elms and the bud-haze of beech and oak, all followed by the first flushes and deepening foliage of these and most other broadleaved trees, supplemented by the white blossoms of wild cherry (gean) and blackthorn.

Throughout summer, Britain's trees are crowned with a mantle of leaves and needles of many shades of green, sometimes contrasted by faintly blue or silver sheens. Here and there the scene is enriched with the blossoms of hawthorn, rowan, whitebeam, crab apple and guelder rose. When, during this season, the broadleaves are in such mass that, particularly from afar, their greens are difficult to distinguish, the secondary flush of foliage (for example, of some oaks) provides a little contrast and relief to the scene. Thereafter, colour and interest are added by resplendent berries, and by fruits which vary from the pink seed-pods of spindle trees to clusters of red-tinged wings of sycamore and maples.

Before summer is forgotten, and as some compensation for the shortening days, the glory of autumn falls on the deciduous trees – a rich mingling of browns, russet-reds and golden-yellows – heralding the fall of fruits, cones, seeds and spent foliage. Ceaselessly the rhythm of the beauty of woods and trees recurs as the seasons wax and wane and, as Emerson noted, 'a perennial festival is dressed'.

Year by year, many of the vast coniferous forests planted this century pass from unprepossessing youth and adolescence to handsome maturity and great utility. Although some people who remember Britain's old woodlands, predominantly broadleaved, are sad at their partial passing, on the generation that knows only the newer trees as they mature, the magic of the sylvan scene casts its spell as powerfully as on their forebears. Thus Britain's trees, both broadleaved and coniferous, are a delight. People depend on trees; they *expect* them to be there. But in a modern industrial country trees are dependent on people. They cannot be taken for granted: they live long but are not immortal. Unless there is afforestation, restocking after felling and prudent management, coming generations will not enjoy the treasure-house of beauty, shelter and utility passed down by the people of the eighteenth, nineteenth and twentieth centuries. Only foresight and care will perpetuate and enhance Britain's sylvan heritage.

The trees noted herein are grouped into two sections which correspond with two natural divisions of all trees found in Britain. The first, the angiosperms, are represented here by the dicotyledons; the form of their leaves gives the class one name, broadleaved trees, the structure of their wood another, hardwoods – though a few are soft-wooded (lime, willow). The second division is that of gymnosperms, comprised of the class broadly termed conifers; the structure of their wood gives rise to their general classification as softwoods – though yew has a very hard wood. Within the above two subdivisions come the families signified by the termination *aceae;* for example, *Fagaceae,* the beech family, which includes six genera, among them the oak genus Quercus, and sweet chestnut, *Castanea.*

The final unit of classification is the species, one or more of which form the genus; thus there is *Quercus robur,* English oak, and several other species of *Quercus.* The specific names have varying meanings; for example, *alba,* white, *glutinous,* covered with a sticky exudation, and *grandis,* large. Occasionally there are variations that are sufficiently distinct from the typical form of the species to warrant separate description; for example, *Populus nigra* 'Italica', Lombardy poplar. Two hybrids, the offspring of distinct parents, have arisen by chance in Britain, namely hybrid larch and Leyland cypress. Many hybrid poplar have been produced by design and now play a useful role in British forestry. Each tree (see Index) bears the name of a genus, followed by the species and where relevant the varietal name. A suffix, usually abbreviated – for example, L. (for Linnaeus), refers to the botanist (Carl Linnaeus of Sweden, 1707–79) who first published the name used. Alternative acceptable names or synonyms of the trees are sometimes in use.

Native (indigenous) trees. These are trees that colonized Britain unaided by man's effort. When the last ice-sheet was retreating from Britain 9–10,000 years ago it left a land devoid of trees. The combined effect of that ice age and the formation of the English Channel led to Britain having a poor native tree flora of 32 species only – 29 broadleaves and 3 conifers (Mitchell 1981).

Broadleaves. The main native broadleaves are listed below, modified from Mitchell (1981) and Evans (1984), in an approximate order of arrival. (The scientific names are given in the Index.)

1	Downy birch		16	Hawthorn
2	Silver birch		17	Crack willow
3	Aspen		18	Black poplar
4	Bay willow		19	Whitebeam
5	Common alder		20	Midland thorn
6	Hazel		21	Crab apple
7	Small-leaved lime		22	Wild cherry (Gean)
8	Bird cherry		23	White willow
9	Sallow (Goat willow)		24	Field maple
10	Wych elm		25	Wild service tree
11	Rowan		26	Large-leaved lime
12	Sessile oak		27	Beech
13	Ash		28	Hornbeam
14	Holly		29	Box
15	Pedunculate oak			

Conifers. Britain has only three native conifers – Scots pine, yew and the miniature tree, common juniper.

Exotic (introduced) broadleaves. By the year 1600 man was responsible for the introduction of many additional broadleaved species, including several dating back to Roman times. The great influx of park and garden trees began after 1600. Some of the commoner introduced broadleaved species are listed below, modified from Mitchell (1981) and Evans (1984).

Broadleaves
Pre-1600

White poplar	} very early	Sweet chestnut	Roman
Grey poplar		Swedish whitebeam	
English elm	} pre-Roman	Sycamore	
Smooth-leaved elm		Holm oak	} c. 1500
Common walnut		Oriental plane	

1600–1800

Common lime	c. 1600	Turkey oak	1735
Horse chestnut	1616	Serotina poplar	1750
Black walnut	pre-1656	Lombardy poplar	1758
Norway maple	pre-1683	Grey alder	1780
London plane	c. 1685	Cricket-bat willow	1780
Red oak	1724		

Relatively recent introductions

Italian alder	1820	Red alder	pre-1880
Southern beeches	since 1830	Black cottonwood (poplar)	1892
(Nothofagus obliqua	1902)	Robusta poplar	1895
(Nothofagus procera	1913)	Poplar 'Balsam spire'	1948
Eucalypts	since 1846		

Exotic (introduced) conifers. After 1600 the great number of conifers introduced included a flood of widely varied species from the eastern USA. Among the important timber trees of North America, the Mediterranean and the Far East that arrived between 1600 and 1800, were European larch and Corsican pine. The first of the Western North American conifers arrived in 1827; others followed in 1831–33. Coast redwood arrived in 1843 and Wellingtonia in 1853. A major influx of plants from Japan arrived in 1861 with a second wave between 1878 and 1890. Four trees came from China before 1800 and a few in 1850–54; many additions came in 1901 and 1907–08. Most of the commoner introduced conifers are listed below, modified from Mitchell (1981).

Conifers

c. 1500

Maritime pine
Norway spruce

1600–1800

Common silver fir	1603
European larch	1625
Cedar of Lebanon	1639
Swamp cypress	1640
Corsican pine	1759
Monkey-puzzle	1797

Relatively recent introductions

Deodar cedar	c. 1822	Wellingtonia	1853
Douglas fir	1827	Atlas cedar	1854
Grand fir		Western red cedar	1854
Noble fir	1831	Lawson cypress	1854
Sitka spruce	–1833	Lodgepole pine	1855
Monterey pine		Japanese larch	1861
Monterey cypress	1838	Austrian pine	1890
Coast redwood	1843	Caucasian fir	1898
Western hemlock	1851		

Dawn redwood or 'Fossil tree' discovered in Central China in 1941 was received here in January 1948.

Indigenous tree species are not necessarily native everywhere in Britain. Beech for example, is native in southern England and south-east Wales because it occupied these areas naturally. It is not native in northern Britain, although it is common enough as a planted tree, and as self-sown individuals springing from seed produced ultimately by a planted tree. Conversely, Scots pine was once native in England and Wales but was naturally replaced by broadleaved species, and is now regarded as native only in the Scottish Highlands.

Literature relative to Britain's sylvan heritage has been listed recently by Marren (1990) and Watkins (1990). Mitchell *et al.* (1990) have listed by genus and species the 'champion trees' in the British Isles – recording the tallest and the stoutest. Dimensions used are metric: the height in metres, the diameter in centimetres derived from the measurement of girth at 1.5 m (one inch of girth is 0.8085 cm diameter and one foot is 9.7 cm diameter).

2. Identification of trees

A prerequisite of successful forestry is a sound knowledge of the relevant materials, machinery and equipment, their uses, cost, measurement and value. The sylvan materials of forestry are trees, the main components of which are timber, bark and foliage. Foresters will, of course, readily identify trees both when growing throughout all seasons and when felled, and even when peeled and converted. They will also be well acquainted with their silvicultural characteristics, noted in chapter 4.

Broadleaved trees (hardwoods) are mostly deciduous, i.e. they shed their leaves in winter; notable exceptions are evergreen oak, holly and box. Conifers (softwoods) are chiefly evergreen (deciduous species are larches, dawn redwood and swamp cypress). The foliage of a tree termed 'evergreen' may be some other colour than green, e.g. golden, grey or blue. The needles or leaves of evergreens remain alive for only two, three or a few more years, and then fade to brown or yellow before they unobtrusively fall.

In identifying a tree one must look to its general form, the bark, twigs (or young shoots), buds, leaf scars, leaves (or needles), flowers and cone or fruit. Trees vary widely in botanical features throughout the seasons in their progress through the production of bud to foliage, flower and fruit. Aids to identification likewise vary throughout the year – in some species only the bark and the form are constant.

Thus in winter, a typical deciduous broadleaved tree has bare branches, with only form, bark, twig, bud and perhaps persistent fruits to reveal its identity. In spring, the same tree will display bursting leaf-buds, but the first flush of leaves may be smaller than or of a colour different from those typical of later months. In summer, identification is relatively easy, because the foliage is usually supplemented by flowers and fruits. Through autumn the leaves, although changing in colour, and the maturing fruits or their remnants in varying forms, aid identification. The time of leafing, flowering and fruiting varies according to species, location and climate.

Most common conifers are relatively easy to identify, particularly the evergreens with their almost constant foliage. Seasonal changes occur through the production of shoots, flowers and the developing and maturing cones. The few deciduous conifers are bereft of foliage in winter, yet other features readily reveal their identity.

The important features to aid identification are given below. Description of colour is usually an approximation – shoot and bud, sometimes bark, are generally much more brightly pigmented on the surfaces that are exposed to the sun; and the greens of the foliage are so varied as sometimes to defy exact description.

Winter twigs (or young shoots) and buds. Winter twigs of broadleaved trees are of varying form – from round to angular – from straight to zigzag – and they may be hairy or otherwise, slender (birch), stout and black (ash), pith chambered (elder), or pith laminated (walnut). Some trees, for example poplar, rowan and larches, produce 'short shoots' or 'spurs', which develop only a fraction of a centimetre each year. Compared with the long shoots, they are thick and wrinkled, and later generally bear a tuft of leaves or flowers at the end. The arrangement of the winter (resting) growth buds may be opposite (ash), alternate (beech), or spiral (oak); they may be round (lime), squat (hazel), conical (beech), appressed (hornbeam, willow), hairy (hazel), or large and sticky (horse chestnut). They have a wide range of colour – for example, green (hazel, sycamore), brown (beech, horse chestnut), purple (alder), red (maple, lime) and black (ash). The number and arrangement of the bud scales are pertinent. So too are the shape and markings of the scar left by the fallen leaf-stalk, which have many different forms – for example, large and conspicuous in horse chestnut and walnut, and small and narrow in birch. The flower buds, as distinct from the leaf-buds, are not always easy to differentiate until development proceeds. Important in conifers is the bud shape, and whether or not the buds are resin-coated.

Leaves/needles. In broadleaves the arrangement of the leaves follows that of the bud. The leaves vary in shape whether in simple or compound form. Simple leaves may be linear or lanceolate (some willows), oval (beech), cordate (lime), obcordate (alder), triangular (birch, most poplars) or asymmetric (elm). Compound leaves may be digitate (horse chestnut) or pinnate (ash, rowan, elder, walnut). Palmately lobed leaves include those of sycamore, maple and plane. The length, width, venation and number of leaflets are diverse; so too are the margins, which may be entire, wavy, toothed (serrated) or lobed (oak). If the leaf is stalkless it is termed sessile. The surface of the leaf may be smooth (without hairs, i.e. glabrous) or hairy (pubescent, as in elm), and in colour may vary between the upper and under surface (whitebeam). The colours of most leaves change between flushing and falling; for example, some poplars, oak and walnut open with a light khaki or brown colour and develop through varying shades of green to their autumn hues. Beech leaves, which on opening lay a carpet of light red bud-coverings, are light translucent green in spring while in autumn they are golden or russet. The leaves of hornbeam turn a rich yellow, while those of whitebeam, green above and grey-white underneath, fall in autumn to provide a wrinkled carpet of purple-grey hue. Leaves of broadleaves low down (particularly leaves on suckers) are often larger than the typical ones above. Leaves in the seedling stage differ, on most trees, from those typical of the species; each kind of tree has its particular germination pattern. Oak, for example, always has its two cotyledons (seedleaves) in the soil, whereas beech expands its two broad green seed-leaves above ground. The cotyledons are followed by the primary ('juvenile') foliage which in general is different from the secondary or 'adult' kind usually developed in the second year of growth. Thus, a seedling ash has four kinds of leaf: first the long oval seed-leaf, then an individual juvenile leaf, next a compound leaf with only three leaflets, and finally the adult form of compound-pinnate leaf, having many leaflets.

In conifers, from the foliage point of view the main aid to identification is the needles (more correctly termed leaves). In pines they appear grouped in basal sheaths, when their number (two, three or five), and length and colour are important. Some conifers have short needles in rosettes or in horizontal ranks. Others have scale-like leaves either in fern-like sprays or in cord-like growths.

By their 'adult' needles, the main conifers can be divided for the purpose of identification into the following four simple categories:

Conifers

1. With long needles in bundles. *Pinus* (all evergreen)

In twos	In threes	In fives
Scots pine	Monterey pine	Weymouth pine
Corsican pine		
Lodgepole pine		

2. With numerous short needles in rosettes or 'bunches'

Deciduous	Evergreen	
The larches	True cedars	General branching habit
	Atlas	ascending
	Deodar	drooping
	Lebanon	horizontal

3. With short needles spirally arranged, sometimes seemingly in horizontal ranks

Deciduous	Evergreen
Dawn redwood ('Fossil tree')	Spruces
Swamp cypress	Grand fir
	Douglas fir
	Western hemlock
	Coast redwood
	Yew
	Noble fir (needles massed on top of twig)
	Caucasian fir

4. With scale-like leaves (all evergreen)

In fern-like sprays	In cord-like growths
Lawson cypress	Wellingtonia (Giant sequoia)
Western red cedar	
Leyland cypress	

When most conifers are in the seedling stage they have a number of seed-needles in the form of a little whorl or rosette; solitary needles almost always follow and often the typical needle formation does not develop until the second year. 'Juvenile' foliage is found on all terminal shoots in larches, dawn redwood and true cedars.

Flowers. A fascinating botanical feature of trees is their flowers. In broadleaves, male and female flowers are in some species borne on separate trees *(dioecious,* e.g. willow, poplar), and in other species on the same tree *(monoecious,* and sometimes *hermaphrodite,* i.e. both sexes in one flower, e.g. rowan, elm). In conifers, both sexes are borne separately, though usually on the same tree (yew is one of many exceptions). The variety of flowers is great: they range in the broadleaves from the inconspicuous green and yellow 'catkins' of oak and the tiny red female flowers of hazel, to the purple-tinted tufts of ash or dark purplish-red tufts of elms, and to the large spiked inflorescence of horse chestnut. Some broadleaves flower before the leaves appear, for example, elm, sallow, hazel, alder, hornbeam, blackthorn, ash and Norway maple. The flowers of conifers are usually female conelets and male pollen-bearing catkins which are simply shortlived clusters of stamens. They are generally small and set somewhat inconspicuously amid the foliage; thus they often escape notice. Some are very attractive – for example, in European larch the male flowers are small bright clusters of anthers bearing showers of golden pollen, while the female 'larch roses' are rich pink in colour and have the form – in miniature – of the later cone; in noble fir the clusters of deep purple male catkins are a delight to behold. The age at which trees begin to produce flowers and fruits is very variable (see chapters 2 and 7); youngish trees probably have neither flowers nor fruits/cones. The shedding of pollen generally takes place in spring or summer; in the true cedars it is liberated in autumn. (Never has the beauty of 'flowers of the forest' been so excellently portrayed as in the twenty-four water-colour botanical drawings by the late John Patten which he presented to the Royal Agricultural College, Cirencester in 1961.)

Fruits and seeds. The arrangement of the female flowers, whether in broadleaves or conifers, obviously decides the arrangement of the subsequent fruit or cone and seeds. For example, female flowers of elm and ash being in clusters, it follows that the fruits (respectively, soft transparent roundish wings, and hard elongated 'keys') are found likewise. Fruits of broadleaves are in wide variety – from the small nuts of beech and hazel to large chestnuts and walnuts, and from the small berries of hawthorn, blackthorn, spindle and dogwood to the double-samaras of sycamore and maples. There are also the female woody cone-like fruit body and winged seeds of alder, the acorn, the cherry, the 'bobble' of plane, the hair-tufted seeds of poplar and willow, and the fruits with bracts – lime and hornbeam.

In conifers the fruit in the advanced family *Pinaceae* is always a hard scaly cone, bearing one- or two-winged seeds at the base of each scale; by contrast, yew has a fleshy fruit – a seed surrounded by an aril. The cones vary from the small round structure of Lawson cypress and the small 'Grecian urn' of western red cedar, to the pendulous large and scaly cone of Weymouth pine and the erect 'barrels' of true cedars and silver firs. Among many other kinds, there are the 'whiskered' cone of Douglas fir and the crinkled-scale cone of Sitka spruce. Most cones take one year to ripen; almost all pines take two years. Some disintegrate on the tree leaving only a central spike (true cedars and some firs); other cones fall entire (spruces and pines), but sometimes their seeds have already been dispersed. The seeds of conifers show an interesting range of form and size, an idea of which can be obtained by comparison with the numbers needed to make up one kilogram by weight – from almost two thousand to more than two hundred thousand. Often a handful of seed may represent a hectare or more of potential forest. Seed dispersal is effected by several methods. Dissemination by wind applies chiefly to light and winged seeds, as in many pines and other conifers as well as broadleaved species like ash, birch, sycamore and maples. Less commonly, water (notably for alder), animals or birds (notably for acorns) may be the agents of dispersal.

Bark. Trees in youth usually have a thin, smooth and green bark which only later changes to a typical form and colour. Thus a single tree may show at different stages of its growth, quite distinct and even contrasting patterns, colours and thicknesses. A few trees (for example, beech) retain a thin and smooth bark, but others when older may develop bark of sometimes greater thickness, which is scaly (Sitka spruce), patchy (plane), or horizontally ribboned (birch, cherry). Other trees in later life develop bark which is furrowed (Douglas fir), or ridged and patterned – for example, oak (cubed), walnut (diamond-shaped) and sweet chestnut (spiral). Corky bark may be found on young elm and field maple, while the bark of coast redwood and Wellingtonia is very thick, spongy and fibrous. The colour of some bark changes as the age increases, and may differ according to the direction of exposure – oak, beech, sycamore and larch sometimes have in part a pink or purplish sheen. The bark of beech and hornbeam has a distinct metallic appearance, while they (and the bark of other trees) may be coloured or discoloured by varying shades of lichen. Resin blisters usually develop in young Douglas fir, grand fir and noble fir.

Form and outline. Many trees from a distance can be identified at sight by their form, outline or the disposition of their branches, coupled with their general colour. Much will depend on whether the tree is growing healthily in open conditions, confined in a woodland, or standing in a wind-swept location. An isolated oak is likely to have a broad-spreading crown and a short, thick bole, whereas one of the same age growing in a wood will have a much narrower crown and a taller and branchless trunk. A young Scots pine, wherever grown, will have a roughly conical outline and may be clothed with foliage from ground level; whereas an old tree will usually bear a rugged crown springing from the top of a branchless orange-coloured trunk.

Broadleaved trees generally have a less formal growth than conifers and a more oval crown of branches, and are thus more static in shape and outline. Conifers tend to have a pointed shape and a more formal habit of growth. The features of the top shoot (leader) of a conifer are often helpful, particularly in Lawson cypress (drooping), western red cedar (erect) and western hemlock (whiplike, pendent). European larch branches sweep down; Japanese larch has strong horizontal branches. Pines, grand fir and spruces have their branches in distinct whorls. Turkey oak has ascending giraffe-like branches. The form of a tree must be interpreted with regard to all these features.

Competence in identification of trees is achieved only by appropriate practice and observation, aided by an adequate book of keys to species; for example, such publications as those of Hadfield (1957), Edlin (1971), Hart (1973), Mitchell (1972, 1974, 1985), Edlin and Mitchell (1985), Mitchell and Jobling (1984) and Mitchell et al. (1990). The study can be pursued in many ways and in increasing detail until all the species met in British forestry can be recognized at all stages of growth and throughout all seasons. Such knowledge can later be complemented by the recognition of other less common trees which appear within or near woods and forests. The process of learning to identify trees is a progressive one – without end if one moves into the sphere of studies requiring comprehensive botanical manuals, of which Bean's five volumes (1914–88) are splendid examples.

A knowledge of wild flowers, herbs and grasses is also rewarding; many of these, noted in chapter 3, convey ecological and vegetational information which sometimes assist species choice in afforestation and restocking. Covert plants are another group which ought to be readily recognized; they are named in chapter 13. Easily recognized are honeysuckle and wild clematis, attractive in appearance throughout some months, but likely to be harmful particularly in the thicket-stage, the former species being a strangler and the latter a smotherer of untended trees. The supervision of weeding young plantations calls for knowledge of, and acquaintance with, such weeds as rose-bay willow herb, bracken, bramble, broom, gorse and foxglove. This knowledge helps in choosing the best time for weeding, as well as in the fixing of piece-work rates if the operation is done manually.

Some foresters will be concerned with planting and tending an avenue or park trees, and possibly with maintenance and surgery of specimen trees. Thus identification of a wide variety of ornamental trees and shrubs is desirable. Likewise, knowledge may be essential of canker-resistant poplars, of the cricketbat willow and of osiers where foresters manage sites suited to these crops. Poplars are readily identified by family, but many species, cultivars and hybrids are difficult to identify; the best source of information is Jobling (1990).

3. Identification of timbers

Foresters should complement their knowledge of silviculture, harvesting and marketing with the ability to identify the common British timbers, and to be familiar with their more important properties and uses. It is unnecessary to have a profound knowledge of the biology and chemistry of wood but some understanding of its structure and properties helps in the exploitation of its good qualities and the accommodation of its lesser ones. The timber will be at various stages of conversion: (i) in the round as full length stems, or cross-cut into sawlogs, veneer logs, pitprops, pulpwood, boardwood, posts, stakes, rails and the like, with or without bark; (ii) sawn-through or cleft, lengthwise; or (iii) fully converted, i.e. squared, quartered or otherwise-sawn, and perhaps planed.

Wood is a complex, versatile material with a variability that gives it utility and beauty but also imposes limitations on its use. The relationships between growth and structure, and between structure and properties, have been adequately described by Johnston (1979, 1983). A simplified introduction to timber structure is given hereunder. Figures 2 and 3 are relevant. (Properties and uses of timber are noted in chapter 9, along with a few comments on the chemistry of wood.)

Just below the bark and forming a complete sheath around the stem and branches is a thin layer of tissue known as the *cambium* which produces *wood cells* towards the centre of the tree and soft tissue beneath the bark towards the outside. The wood cells of the stem provide mechanical support for the crown, conduct water and nutrients from the roots to the leaves, and store food made in the leaves by photosynthesis which involves the assimilation of carbon dioxide. Water conduction and food storage occur in the *sapwood* – an outermost zone of wood which, depending upon the species, age and condition of growth of the tree, is usually about 12–75 mm wide. Inside the sapwood is the *heartwood* which, after drying, can be more durable than the sapwood, contains less moisture and is usually darker in colour although in some species the sapwood and heartwood are visually indistinguishable.

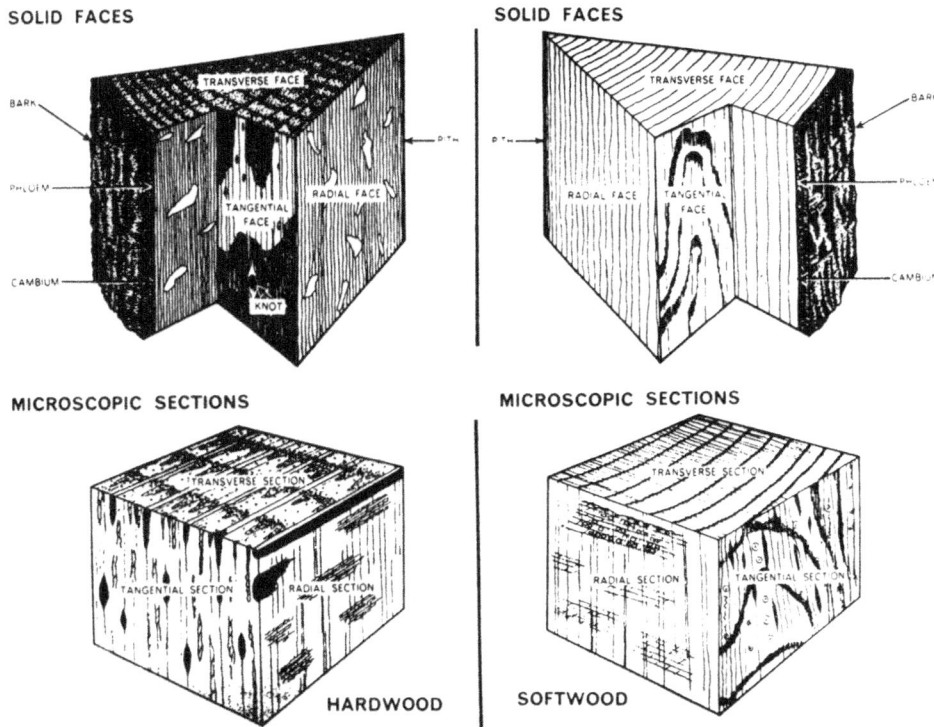

Figure 1. Structure of timber: hardwoods (left), softwoods (right)
Source. TRADA and The Buckinghamshire College.

The *annual growth rings*[1] in all temperate species are formed almost entirely in the spring and summer. In most conifers and some broadleaved species, such as oak and ash, the wood cells produced in the spring are larger and thinner walled than those produced in the summer and there is a very conspicuous annual growth ring made up of the usually softer and paler *springwood* (earlywood) and the harder *summerwood* (latewood). In other species, such as beech and sycamore, there is far less contrast in the cell structure, and growth rings, although present, are less obvious. The boundary between the spring and summerwood may be abrupt as in larch, Douglas fir and most pines, or gradual as in the firs and spruces. On a flat sawn surface the rings appear as stripes and more or less irregular loops; on a quarter-sawn surface the rings appear as parallel stripes.

As the stem of a tree grows it encloses the branches which form knots in the wood. While the branch is still living the enclosed knot is, structurally, part of the stem and is known as a live knot. But after a branch has died the dead wood enclosed by the stem has no structural connection with the surrounding tissue and is known as a dead knot. In most conifers, especially those with thick bark, dead knots often fall out after the tree has been felled and sawn, while in broadleaves

---

1   Dendrochronology – tree-ring dating (particularly for archaeology) – is a subject in which foresters have shown little interest. The best introduction is probably that of Trenard (1982).

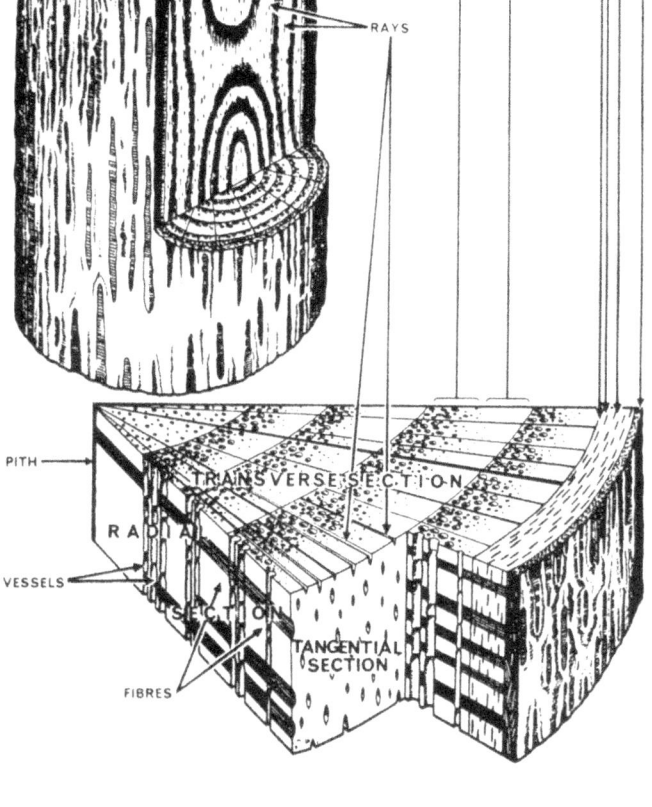

Figure 2. Timber structures
Source: TRADA and The Buckinghamshire College.

they are less common and tend to rot rather than fall out. Flat sawn surfaces exhibit circular knots, while quarter-sawn surfaces show splay knots.

For any given species the character of the wood can, to a considerable extent, be influenced by foresters who can select for vigour, straightness and branching habit in their tree breeding programme, and who can influence growth rate by their choice of planting site and by the use of fertilizers. Diameter growth and branchiness, but not to any significant extent height growth, can be influenced by spacing so that trees planted wider apart, or given more space by thinning, become larger in girth and more tapered and have thicker branches, and hence larger knots, than trees grown more closely together.

The new cells generally grow from the cambium in a direction more or less parallel with the long axis of the stem. Sometimes, however, they grow at an angle, thus forming a spiral round the tree. *Spiral grain* is more common in some species than others and frequently occurs, for example, in sweet chestnut. Wood from near the centre of the tree, known as *juvenile wood,* is less dense, less strong and more knotty than the more mature wood and in conifers tends to have more spiral grain. The juvenile characteristics are only pronounced in the ten to twenty annual rings in the centre of the stem. All wood is composed of small cells the principal structural material of which is *cellulose,* the principal constitutent of the cell walls. This is reinforced with *hemicellulose* and *lignin* which add strength and rigidity. Various other chemicals occur in different species and these affect the colour, smell and natural durability of the wood.

There are significant differences in structure between conifers (softwoods) and broadleaves (hardwoods).

In conifers there are two types of cell: (i) mainly *trachieds* which serve both to conduct water and nutrients and to provide mechanical strength; (ii) food storage cells (or *parenchyma*) which occur in very small quantities. For any given species density, which is closely correlated with strength, depends upon cell diameter and cell-wall thickness and upon the proportions of springwood and summerwood. Fast-growing trees have wider annual rings and tend to have larger and thinnerwalled, and therefore weaker, trachieds than slow-growing trees. Fast growth does not necessarily, however, result in weak wood of low density. On some favourable sites and under good growing conditions, trees continue to grow until late in the season and produce a relatively large proportion of strong, thick-walled, latewood trachieds. The storage cells, which are inconspicuous in conifers, occur among the trachieds or in the form of *rays* extending radially through the wood. The rays are typically only one cell wide but are several cells deep. Usually they are barely visible to the naked eye. Many conifers contain *resin canals* or resin ducts which give the characteristic resin smell to the wood and which, on occasions, make the wood sticky to work.

In broadleaves the wood cells are generally shorter than those in conifers and there is a difference between the cells which conduct water and nutrients and those which provide strength and rigidity. The storage cells are similar in structure and function to those of conifers but they are more abundant and more variable.

The main conducting cells, which are called *vessels,* range in diameter from 0.02 to 0.5 mm so that in some species they are large enough to be seen easily by the naked eye. They always occur in vertical series which form tube-like structures extending vertically for some distance in the tree. In most species there is little contrast in the size or pattern of the vessels across the growth ring and these are known as *diffuse-porous* species. In some species, however, the transition from large springwood vessels to smaller summerwood vessels is abrupt and the springwood vessels form a series of distinct rings which are clearly visible in the end grain of the wood. These are the *ring-porous* species which include oak, elm and sweet chestnut. In oak, for example, the diameter of the springwood vessels can be as much as ten times that of those in the summerwood. Fast growth accentuates larger vessels and produces fewer fibres; and certainly leads to coarser texture.

The thickness of the fibre wall and its physical and chemical properties are major factors in determining the strength and working properties of the wood.

The storage tissue occurs in various patterns amongst the vessels and fibres, sometimes forming distinct bands, and sometimes distributed throughout the wood. It also forms radial rays which are generally larger and more conspicuous than those in conifers. The rays are usually several cells wide and in some woods, such as oak, are many millimetres deep forming a major feature on radial surfaces.

Grain, texture and *figure* are components concerning more the worker of timber, not least the craft furniture-maker. Figure depends to a considerable extent upon the direction in which wood is veneered (rotary or sliced), or cleft. These components are comprehensively discussed by Johnston (1979, 1983).

The timbers of most common broadleaves are fairly easy to distinguish from those of conifers, but the kinds of timber within these two classes differ in many features. As noted in section 2 *supra* under 'Identification of trees', the term *hardwood* is used for the timber of broadleaved trees, and the term *softwood* for that of conifers. But some broadleaved trees, e.g. willow, poplar, lime and alder, furnish a comparatively soft wood, particularly when dry; and the timber of yew, a conifer, is indeed hard.

Most timbers display an inner zone of heartwood which may be darker and, in some species but by no means all, can be more durable than the outer zone of sapwood. Knowledge of this is necessary for an appreciation of natural durability as well as the need or otherwise for, and the possibility of, preservative treatment. Sweet chestnut and larch have an exceptionally narrow sapwood; in Corsican pine and Turkey oak it is comparatively wide. Some timbers show little or no difference in the colour of their sapwood and heartwood; among them are ash, beech, sycamore, holly and spruce. Where the decorative value of a timber depends on the dark colour of the heartwood, as in oak and walnut, the sapwood in every log is treated as a defect. By contrast, the white sapwood of willow is preferred for most grades of cricket bats. From the point of view of strength there is usually no difference between heartwood and sapwood.

The general colour, and sometimes the discoloration, of timbers is an aid to identification. Sycamore is white, beech is pale brown, yet if their seasoning is delayed they are both apt to turn grey. Good quality ash, which is nearly white or light yellowish-brown, sometimes turns pink when freshly cut. On exposure alder becomes darker and western red cedar changes from red-brown to grey.

Weight, when the timber is dry, can sometimes be another aid: yew, elm, beech, oak and holly are heavy; poplar, willow, lime, western red cedar, western hemlock, Lawson cypress, grand fir, spruces, coast redwood and Wellingtonia are relatively light. The total weight of water in freshly felled 'green' timber may amount to more than 100% of the dry weight of its wood substance; the moisture content of air-dry timber may range between 14 and 23% according to the season of the year and the species of timber concerned.

Douglas fir sometimes has the smell of fresh apples, and western red cedar a strong but pleasant aroma. Pines have a resinous smell and elm an unpleasant pungent one. True cedars, too, have a distinct pleasant scent.

If the timber still carries its bark, identification is usually greatly eased. As noted under 'Identification of trees', some species may have smooth, rugged or scaly bark when they are of substantial size and age; but in youth most have smooth bark which changes to a typical form only later on. A single tree shows, at different stages of its growth, quite distinct and even contrasting patterns, a change that develops as the years go by. Scots pine is deeply fissured and dark brown at the base and smooth and red higher up. Incidental to identification, bark is an indicator of growth (it may show splits caused by fast growth), health, disease and irregular grain (which may raise, or lower, the commercial value of the timber).

Several useful publications of the Building Research Establishment set out the salient points for the identification of most home-grown timbers. Likewise of great use are books by Schwankl (1956), Johnston (1979, 1983) and a bulletin by Brazier (1990). By studying these and applying the information they provide on every possible occasion foresters will become proficient in identifying most of the timbers of British woodlands. The knowledge they require is not of cell-structure and the like, but rather a practical knowledge which will enable them readily to identify timbers by their outward appearance without the use of a lens or microscope. A combination of characteristics – colour, ring-structure, type of knots, presence or absence of heartwood, smell and bark if available – helps more than any printed 'key'.

Complementary to the above, foresters should be able to recognize the defects in round or sawn timber. These include: knots, rot, shakes, fluting, ingrown bark, pitch or resin pockets, woodpeckers' holes, wane, spiral grain (e.g. in sweet chestnut and Sitka spruce), blue stain, green sap stain in sycamore, excessive sapwood (as in young oak and walnut, and in Turkey oak and Corsican pine) and seasoning defects (checks and splits and distortion, for example, bow, spring, twist and cup). Foresters should be able to assess the degrade in value due to the defects. They should also recognize in timbers the characters that are not serious technical defects. These include: (i) 'Brown oak' (believed to owe its colour to an attack of 'beefsteak fungus' *Fistulina hepatica*: strength is not affected); (ii) black-hearted ash (quite sound and equal in strength to normal white ash); and (iii) red-hearted beech (not much affected as to its technical properties). However, some such characteristics may have an effect on the commercial value of the timber.

Identification of timber depends largely upon whether the main visible portion of the wood is an end radial section, or a vertical or horizontal flat surface. The bark if still attached is a distinct aid as noted in section 2 *supra* under 'Identification of trees'. Brief descriptions to aid identification are given hereunder of each main timber's superficial colour and other features – when peeled, cleft or sawn.

CONIFERS (SOFTWOODS)

Species (Scientific names are given in the Index)	Colours and other features of timbers (Properties and uses are noted in Chapter 9)
	The colours indicated relate chiefly to the heartwood. However, individual timbers show variation in colour as well as changing colour in use. Timber exposed to light will change colour, and unprotected timber exposed to the weather eventually will become silver-grey.
Scots pine	The pale yellowish-brown to reddish-brown heartwood, when dry, contrasts with the lighter-coloured sapwood usually 5–100 mm wide. Knots are in pronounced 'whorls' (i.e. a distinct line of knots in flat surfaces), and sometimes large. Annual rings clearly marked by the contrasting light-coloured springwood and the dark summerwood zones. Resinous.
Corsican pine	Pale yellowish-brown. Similar to Scots pine in appearance, but with a wider sapwood and is typically of more vigorous growth, particularly in juvenile wood, giving larger knots and more pronounced knot 'whorls' (i.e. a distinct line of knots in flat surfaces). Resinous.
Lodgepole pine	Pale, straw colour. Heartwood is difficult to distinguish from the comparatively narrow band of sapwood. Straight-grained, and little contrast between the springwood and the summerwood in the annual rings. Resinous.

Larch
  European
  Japanese
  Hybrid
	The pale reddish-brown to brick-red heartwood contrasts sharply with the narrow pale brown sapwood. The annual rings are clearly marked by the light springwood and dark summerwood zones. Typically has many fairly small but often dead knots. Resinous. Relatively heavy in weight. Occasional cankers on European larch.

Douglas fir	The light reddish-yellow to brown heartwood is clearly distinguished from the comparatively narrow band of pale coloured sapwood. Marked 'flame-like' growth ring figure on rotary cut veneers. Generally straight-grained. The marked contrast between springwood and summerwood in the annual rings gives rise to a prominent growth ring figure. Resinous. Smell of fresh apples.

Species (Scientific names are given in the Index)	Colours and other features of timbers (Properties and uses are noted in Chapter 9)
Norway spruce	White to yellowish-brown throughout. Heartwood not readily distinguished from the sapwood. Slightly resinous. A natural lustre.
Sitka spruce	Light colour, of lustrous appearance, sometimes with a very pale pink or pinkishbrown core. When dry, there is no real distinction between heartwood and sapwood. Low resin content.
Lawson cypress	Heartwood grey to dark brown. Sapwood yellowish-white. Relatively light in weight.
Western red cedar	Heartwood at first bright orange-brown, weathering to silver-grey. Sapwood pale yellowish. Strong pleasant aroma. Relatively light in weight. An acid timber which may corrode metals under damp conditions and cause iron staining. Non-resinous.
Western hemlock	Heartwood dark. Sapwood yellowish-brown. Clearly marked by moderately dense summerwood. Practically non-resinous. Relatively light in weight.
Yew	Heartwood deep golden or reddish-brown (even orange-brown to purplish-brown). Sapwood narrow and white. Hard and relatively heavy in weight.

True cedars
  Atlas
  Deodar
  Lebanon
	Heartwood light-brown. Sapwood narrow and white. Pungent (fragrant) cedar odour.

Coast redwood Wellingtonia (Giant sequoia)	Heartwood reddish-brown. Sapwood, thin zone, pale yellow. Relatively light in weight when dry.

Firs
  Grand
  Noble
	White or pale cream, with no marked heartwood. Knots in distinct lines. Relatively light in weight.

BROADLEAVES (HARDWOODS)

Species (Scientific names are given in the Index)	Colours and other features of timbers (Properties and uses are noted in Chapter 9)
	The colours indicated relate chiefly to the heartwood. However, individual timbers show variation in colour as well as changing colour in use. Timber exposed to light will change colour, and unprotected timber exposed to the weather eventually will become silver-grey.

Oak
  Pedunculate
  Sessile
	Heartwood warm rich deep-brown or golden-brown. Sapwood white to pale fawn. Ring-porous. Wide, easily visible radiating silver-coloured rays giving an attractive figure on radial surfaces. May contain 'shakes' in older trees. Relatively heavy in weight. 'Brown oak' sometimes found.

Beech	Whitish to pale brown; pinkish-red when steamed. Diffuse-porous. Distinctly visible rays which show as pale to mid-brown flecks on tangential surfaces.
Ash	Pale white to light-brown. Whitish to yellowish or greyish-white, sometimes with a pale brown heartwood. Ring-porous. Wellmarked annual rings. Straight, attractive grain.
Sycamore	Very pale, almost white sheen, sometimes yellowish-white. Diffuse-porous. No wellmarked figure or grain; but figured ('rippled') sycamore has an attractive pattern of light and dark shades in the radial grain. Odourless.
Sweet chestnut	Yellowish-brown. Ring-porous. Little sapwood, much heartwood. Fine, virtually invisible radiating rays. Often spiral-grained. May contain 'shakes' in older trees.

Alder
  Common, Grey
  Italian, Red
	When stripped of bark, a brilliant-orange colour develops. When dry, light reddishbrown without a distinguished heartwood. Diffuse-porous.

Birch
  Silver
  Downy
	Almost white to pale fawn, with a dull surface. Diffuse-porous.

Species (Scientific names are given in the Index)	Colours and other features of timbers (Properties and uses are noted in Chapter 9)
Wild cherry (Gean)	Heartwood warm rich reddish-brown, tinged with gold and green. Pale brown. Diffuse-porous. An intricate attractive figure.
Red oak	Yellow to dull-brown with slight pinkish tinge. Ring-porous. Less character than English oak.
Holm oak	Curiously patterned with dark brown and light brown, and shows silver grain along its rays. Diffuse-porous. Relatively heavy in weight.
Turkey oak	Wide sapwood. Relatively heavy in weight. Ring-porous. Relatively difficult to season and unstable in service.
Horse chestnut	Nearly white to pale yellowish-brown. Diffuse-porous. Virtually all sapwood. Relatively light in weight.
Southern beeches (*Nothofagus* spp.)	Uniform pinkish-brown. Sometimes pale salmon. Superficially similar to beech. Sapwood narrow and paler in colour. Diffuse-porous. Usually straight grain, fine texture and medium weight.
Hornbeam	Whitish. Diffuse-porous. Relatively heavy in weight.
Norway maple	White-grey. Diffuse-porous. Resembles sycamore.
Plane	Fawn to pinkish-brown, without any distinct heartwood. Mottled red-brown. Diffuse-porous. Conspicuous rays.
Walnut	Heartwood rich greyish-brown, often with patches of dark brown, black or paler brown that give it a lively attractive figure. Irregular dark streaks. Sapwood broad and pale greyish-brown. Semi-ring-porous. Black walnut: rich dark brown. Common walnut: grey-brown with dark streak. Often handsomely marked. Relatively heavy in weight.
Elm	Heartwood dull brown or light to dark reddish-brown. Sapwood thin band of paleyellow; or medium-brown. Ring-porous. Interlocked (crossed) grain. Relatively heavy in weight. Unpleasant pungent smell.

Species (Scientific names are given in the Index)	Colours and other features of timbers (Properties and uses are noted in Chapter 9)
Holly	Ivory or pale creamy-white, with faint greenish tinge. Diffuse-porous. Extremely small pores. Fine even texture and somewhat irregular grain. Relatively heavy in weight.
Lime	Pale white to yellowish-green. Diffuseporous. Even-grained. Relatively light in weight.
Poplar	Generally uniformly white, yellowish-white, pale brown or greyish when freshly felled. Heartwood and sapwood are usually the same colour, though the former is sometimes slightly deeper in shade. Diffuseporous. Relatively light in weight. See Jobling (1990).
Willow	Pale brown to white pinkish-brown. Heartwood pinkish. Sapwood white to greyish-brown. Diffuse-porous. Relatively light in weight.
Crab apple	Red-brown. Old trees can have dark streaks. Diffuse-porous. Fine grained.
Box	Creamy-white to bright orange-brown. Relatively heavy in weight. Britain's heaviest timber. Diffuse-porous. Fine even texture.

In addition to identifying trees at all seasons, and timbers in all common forms, would-be foresters should be able to recognize fasciation, witches broom and the like, and should train themselves to identify the more common insects, cankers and diseases which are harmful to forestry, as noted in chapter 6.

Identification of trees and timbers as outlined in this chapter comprises an interesting and rewarding study. With such knowledge foresters are better able to practise sound silviculture, collect good seed, manage a nursery, sawmill or wood-yard, provide good conditions for a shoot, choose species for planting of amenity areas, avenues, parks, screens, shelterbelts and hedges, and perhaps attend to the maintenance and surgery of specimen trees.

4. Forest measurement

Measurement of timber is necessary for several purposes. The most obvious is to estimate forest produce quantities for sale, but measurement is also needed for management purposes such as payment for labour, machinery and other services, inventory, planning, control of resources, and the measurement of yield class for production forecasting (see chapter 12). There are many different ways in which timber can be measured, some being more appropriate for one purpose than others. In addition the choice of method may be influenced by the size and quantity of produce, its value and possibly by the local conditions and facilities available. The main units used in forestry are:

Linear measurements
cm   centimetre
m   metre   0.01 metre
km   kilometre   1,000 metres

Area measurements
m^2 square metre
ha   hectare   10,000 sq m; 100 ares
    are   100 sq m

Volume measurement
m^3 cubic metre   1,000 litres

Weight measurement
    tonne   1,000 kilogrammes

Constants π is 3.1415927
1 m^3 of water weighs 1 tonne

Forest measurement procedures are comprehensively described by Hamilton (1975). Edwards (1983) provides all the mensuration information which may be needed *in the forest* for measuring both standing trees and felled timber. The two sources are the basis of this section. A useful additional source is Philip (1983).

Timber quantity can be expressed in terms of solid volume, stacked volume, green weight, dry weight, length, diameter, number of pieces, etc. Traditionally, timber quantity has mostly been expressed in terms of solid volume (m^3). Measurements may also be expressed in weight (for example, for energy crops where the final product is chipped and yield expressed in tonnes).

In a sale or purchase the choice of whether the timber should be measured standing or felled depends on who is responsible for the felling and extraction. Where this is done by the buyer the tendency is to measure standing, but if the grower is responsible, felled measure is more usual. However, with small lots of timber, it may be prudent to provide a rough and cheap estimate of standing volume in order to attract the interest of the prospective buyer, but to agree to sell on the basis of an appropriate felled measure.

Timber measurement – general

Measuring length and height. The *length* of a piece of timber should be measured in metres (m), using a tape following the curvature of the log. Lengths are conventionally rounded down to the nearest tenth of a metre for lengths up to 10 m, and to the nearest whole metre for lengths greater than 10 m. The *total height* of a standing tree is the vertical distance from the base of the tree to the uppermost point (tip). The total height of felled trees is the straight line distance from the base to the tip. The total height of young standing trees can be measured with graduated poles; that of other trees should be measured with a hypsometer or clinometer. The distance of the observation point from the tree should be about 1 to 1.5 times the height of the tree. The *timber height* of a tree is the distance from the base of the tree to the lowest point on the main stem where the diameter is 7 cm overbark. In broadleaves and occasionally in conifers this point may alternatively be the 'spring of the crown', i.e. the lowest point at which no main stem is distinguishable.

The *top height* of a stand is the expected value of the average of the total heights of a number of 'top height trees' in the stand, where a 'top height tree' is the tree of largest breast height diameter in a 0.01 ha sample plot. This is not necessarily the tallest tree. The *mean height* of a stand is the average total height of all the trees in the stand. It can be estimated from the top height of the stand but the relationship varies with the type of thinning.

Measuring diameter. All *diameters* should be measured in centimetres (cm) using a girthing tape marked in cm diameter, which is placed round the circumference of the tree or log. They are conventionally rounded down to the nearest whole centimetre. Diameters can also be measured with calipers, while the diameters of the ends of logs are usually measured with a ruler. The *breast height point* is the point on a tree which is 1.3 m above ground level. The *dbh* is the diameter at the breast height point. Trees with a dbh of less than 7 cm are assumed to have no volume and so are conventionally classified as 'unmeasurable'. The *mean diameter* of a stand is the diameter of the tree of mean basal area, which is the same as the quadratic mean of the dbhs of all the trees. The mean diameter can be calculated by using the basal area table (Table 1); or a calculator which has a square root key; or a local volume table or a tariff number. Or for a quick check, diameter can be estimated by Weise's 40% rule. This rule, as well as full information on basal area, are provided by Edwards (1983).

Measuring basal area. The *basal area* of an individual tree is the cross-sectional area in m^2 of the tree at its breast height point. The formula to be used is: BA = $\varpi d^2/40,000$, where BA = total area in m^2, and d = dbh in cm. The *basal area of a stand* (expressed in m^2 per hectare) is the sum of the basal area of all the trees in the stand. It can be estimated in two ways: (i) using a relascope,[2] choosing a suitable number of sample points and sweeps; (ii) using sampling plots, choosing a plot size which includes from seven to twenty measurable trees. Tables of numbers of relascope sweeps, minimum distances of sample point, circular, square and rectangular plot sizes,

---

2 A *relascope* is a simple instrument which can be used to estimate the basal area of stands without the need for laying down plots. This instrument is alternatively described as an angle gauge (Bitterlich 1984). A glass or plastic wedge prism relascope with a factor of 2 is the relascope in current common use.

and number of sample plots required are provided by Hamilton (1975). Conversion of dbh or diameter to basal area is possible from Table 1.

Table 1. Basal areas.

dbh or diameter (cm)	Basal area or cross-sectional area (sq m)	dbh or diameter (cm)	Basal area or cross-sectional area (sq m)
7	0.0038	34	0.091
8	0.0050	35	0.096
9	0.0064	36	0.102
10	0.0079	37	0.108
11	0.0095	38	0.113
12	0.0113	39	0.119
13	0.0133	40	0.126
14	0.0154	41	0.132
15	0.018	42	0.139
16	0.020	43	0.145
17	0.023	44	0.152
18	0.025	45	0.159
19	0.028	46	0.166
20	0.031	47	0.173
21	0.035	48	0.181
22	0.038	49	0.189
23	0.042	50	0.196
24	0.045	51	0.204
25	0.049	52	0.212
26	0.053	53	0.221
27	0.057	54	0.229
28	0.062	55	0.238
29	0.066	56	0.246
30	0.071	57	0.255
31	0.075	58	0.264
32	0.080	59	0.273
33	0.086	60	0.283

Source: Edwards (1983). FC Booklet 49.

Measuring weight. All weights should be measured in tonnes. The weight of a load of timber is usually measured by subtracting the weight of the empty lorry (tare weight) from the weight of the lorry when loaded (gross weight). The weight of fuel, oil, water, bolsters and chains, etc., must be the same at each weighing.

The weight of any stack of timber will decrease as the timber dries out, and the rate of drying depends on the weather and the way the timber is handled. In particular, if the bark is removed the timber will dry out much more quickly. Coniferous timber with the bark on is likely to lose about 2% of its weight each week during the summer and about 0.5% weekly during the winter. Table 2 gives an estimate of the density of *fresh felled* timber in tonnes per cubic metre. The figures are only averages and will vary with the time of year, the site and the age of the tree. The measurement of weight in order to estimate the volume of a load is described later.

Table 2. Density of fresh felled timber.

Species	Tonnes per m^3	Species	Tonnes per m^3
SP	1.02	WRC/LC	0.89
CP	1.00	GF	0.85
LP	0.95	NF	0.93
SS	0.92	Oak	1.06
NS	0.96	Beech	1.03
EL	0.90	Ash	0.78
JL/HL	0.83	Sycamore	0.83
DF	0.87	Birch	0.93
WH	0.93	Elm	1.03

Measuring the volume of a standing tree

All volumes should be recorded in cubic metres (m^3), measured overbark (ob), except for individual trees of high value. The conventional top diameter limit for volume is 7 cm overbark, or the point at which no main stem is distinguishable, whichever comes first. So trees with a diameter at breast height of less than 7 cm are normally ignored when estimating volume.

In estimating the volume of a standing tree, measure the dbh of the tree; measure its total height if it is a conifer, or its timber height if it is a broadleaved tree; use the appropriate single tree tariff chart (discussed later) in Hamilton (1975) to find the tariff number of the tree; and find the volume from the tariff chart in Edwards (1983) or from the tariff tables in Hamilton (1975). If the single tree tariff charts are not available or the tree is outside the range of the charts, two alternative methods for estimating the volume of a standing tree are to be found in Edwards (1983). None of the foregoing methods include branchwood.

A rough estimate of the volume of a standing tree can be derived from the following method, based on form factors (related to taper[3]): measure the dbh in cm; calculate the tree's basal area in m^2; measure the total height in m; estimate the form factor (0.5 is appropriate for most mature conifers in plantations, and 0.4 for open grown conifers, while a lower form factor, down to 0.35, is more appropriate for younger trees); and multiply the total height by the basal area and by the form factor to give the estimated total volume in m^3.

An alternative very rough estimate of the volume can be obtained from tree diameter using Table 3.

---

[3] *Taper* is conventionally expressed as a height to diameter ratio. Ratios increase as competition between trees increases and so they rise with age. Closely spaced stands or conventionally thinned stands often lead to ratios of well over a hundred. The height to diameter ratio decreases with increased spacing.

Table 3. Estimation of tree volume from diameter.

Tree diameter at breast height (cm)	Estimated tree volume (m³, overbark)	
	Conifers	Broadleaves
5	0.01	0.01
10	0.04	0.04
15	0.1	0.1
20	0.25	0.25
25	0.45	0.4
30	0.7	0.6
35	1.0	0.9
40	1.5	1.2
45	2.0	1.5
50	2.5	1.9
60	3.6	2.6
70	4.6	3.3
80		4.2
100		6.0

Measuring the volume of a stand

There are several ways of estimating the volume of a stand, and the following simplified key should be used to choose the best method for any given situation (when timber is being offered for sale, the seller must make clear exactly which method has been used):

Value of stand

Very high: Sell on the basis of felled timber, measured by mid diameter. If in place of this, an estimate of the volume is required before felling, either measure the volume of each tree or tariff measure the stand (explained overleaf).

High: Sell on tariff measure.

Average or low★: If to be marked, sell on tariff measure. If not to be marked, derive the volume estimate using either abbreviated tariff or the basal area/form height method (also used for inventory, or assessing piece-work rates).

★ A stand may be of low value because the trees are small or of poor quality, or because the total volume is small, because access is difficult, or for some other similar reason.

Tariffing. This method should be used when measuring, for sale, stands of high value, and stands of average value when the trees are being marked. It is essential that a forester engaged on tariffing is fully familiar with the whole tariffing procedure as described in Hamilton (1975). The basic three steps in estimating the volume of a parcel of trees using the tariff system are to be found therein. Abbreviated tariffing and the basal area/form height method are conveniently described by Edwards (1983).

A frequently used method of estimating the volume of an even-aged, single species stand is set out below. (In a stand containing a mixture of species or ages, each component of the mixture should be considered separately.)

1. Choose a convenient sample plot size which will contain between 7 and 20 trees from the following list of radius or side lengths for various plot sizes:

Length (m) for plot size (ha):	0.005	0.01	0.02	0.05	0.10	0.20	0.50
Circular (radius):	4.0	5.6	8.0	12.6	17.8	25.2	39.9
Square (sides):	7.1	10.0	14.1	22.4	31.6	44.7	70.7

2. Decide on the number of plots to be measured from the following:

Area of stand (ha)	Number of sample plots according to crop	
	Uniform	Variable
0.5–2	6	8
2–10	8	12
Over 10	10	16

3. Select the plots at random, and in each plot measure and record (a) the dbh of all trees greater than 7 cm dbh, and (b) the total height of the tree of largest dbh within a 5.6 m radius of the plot centre.

4. Calculate the average height of all the samples taken. This gives an estimate of the top height of the stand.

5. Estimate the tariff number of the stand using a table for top height/tariff numbers.

6. Calculate the mean dbh using the measurements taken in all plots together. This can be done using a calculator with a square root key as follows: square each dbh, add all the squared values together, divide by the number of trees, and calculate the square root, which is the mean dbh.

7. Estimate the mean volume from the mean dbh and tariff numbers which have already been calculated, by using a Tariff Number Chart.

8. Work out the average number of trees in the plots and divide by the individual plot area to give the estimated number of trees per hectare.

9. Multiply the mean volume by the number of trees per hectare to give the total volume per hectare.

10 Multiply the volume per hectare by the stocked area of the stand to obtain the total volume of the stand.

Foresters intent on using tariffing will of course need to acquire the appropriate Forestry Commission tariff tables.

Measuring the volume of felled timber

Overbark and underbark measurement. The volume of felled timber can be measured either overbark, i.e. including the bark, or underbark, i.e. excluding the bark. When timber is being offered for sale, and the details include an estimated volume, the description must state clearly whether the volume is overbark or underbark. Timber can be measured overbark by any of the methods already noted in this chapter. If the top diameter method of measurement or the end diameter method is being used, the timber can be measured either overbark or underbark. It is possible to convert overbark volumes to underbark volumes and vice versa, but this is rarely necessary except when measuring long sawlogs which are normally measured mid-diameter overbark, but the volumes are usually quoted underbark. The overbark volumes should be converted to underbark volumes by multiplying them by the factor given in Table 4 which is suitable for average sawlogs 8–16 m long.

Table 4. Overbark to underbark volume conversion factors for conifer sawlogs.

Species	Conversion factor	Species	Conversion factor
SP, LP	0.87	EL	0.82
CP	0.83	JL, HL	0.85
SS, NS, GF, NF	0.92	DF	0.88
		WH, WRC, LC	0.90

Mid-diameter method. This is the traditional method for estimating the volume of sawlogs. Although it is a very accurate method, it becomes less accurate as the length increases. If the log is longer than 15 m it should be measured in two or more sections. The measurements required are: the length of each log in metres, and the diameter of the mid-point of each log in cm. The mid-point must be found by measuring half the rounded down length from the *butt end* of the log. The volume (m^3) is calculated using the following formula: V = ϖd^{2}L/40,000, where V = volume in m^3, d = mid-diameter in cm, and L = length in m. In practice the calculation is made by using the mid-diameter volume tables given in Forestry Commission Field Book 11 (1990).

Top diameter method. The top diameter method for estimating the volume of logs is usually confined to softwood logs. The method is less time consuming and so cheaper than the more traditional mid-diameter method and is best suited to batches of uniform length logs. Top diameters, normally measured underbark, should be measured across the smallest diameter. The description of logs in terms of top diameter and length provides a useful basis for size classifications. A standard taper rate of 1:120 (i.e. 1 cm diameter in 1.2 m length) has been used in constructing the tables. This rate refers to the taper between the top and the mid-point of the log. The taper rate of individual logs can be expected to differ significantly from the standard rate, so volume estimates derived by this method are likely to be subject to greater errors than by the mid-diameter method. The top diameter method is not recommended for estimating the volume of logs with lengths longer than 8.3 m. The tables, assuming a taper rate of 1:120 and applying to the softwoods, are available as Forestry Commission Field Book 1, *Top Diameter Sawlog Tables,* 1987.

Stacked timber. Estimating the *solid volume* of timber (i.e. the volume of timber, excluding air spaces) can only be undertaken with stacks of timber of *uniform length*. It is easiest with straight logs which are stacked neatly. The main necessity is to estimate an appropriate *solid/stacked conversion factor*. This must be estimated separately for different timber specifications, for different methods of stacking and for different types of tree. The conversion factor will vary with many factors, including the lengths of the logs, their straightness and smoothness, their taper, their diameter and the method of stacking. Conversion factors might be expected to be about 0.55–0.65 for broadleaved timber; 0.65–0.75 for average quality coniferous timber; and 0.75–0.85 for short, straight good quality coniferous timber. The measured stacked volume should be multiplied by the *worked out* solid/stacked conversion factor to give the solid volume of timber.

The volume of branchwood in *mature* broadleaves may be considerable, 20–30% of stem volume for woodland trees and up to 100% for open-grown trees. Such material is normally cut to a standard length – the traditional stack being the 'cord' where 4 ft lengths are built into a stack of height 4 ft and length 8 ft (128 ft^3 stacked volume, approximately 80 ft^3 solid volume).

Weight measurements

The volume of a load can be estimated by multiplying its weight by its volume/weight ratio. The weight should be multiplied by the *worked out* volume/weight conversion factor to give the volume of timber. An indication of volume/weight ratios (m^3/tonne) of green density, fresh felled is given in Table 5.

Table 5. Volume/Weight ratios, by species.

Conifers				Broadleaves	
SP	0.98	JL/HL	1.20	Oak	0.94
CP	1.00	DF	1.15	Beech	0.97
LP	1.05	WH	1.07	Ash	1.28

	Conifers			Broadleaves	
SS	1.08	WRC/LC	1.12	Sycamore	1.20
NS	1.04	GF	1.17	Birch	1.07
EL	1.11	NF	1.07	Elm	0.97

Source: Hamilton (1975). FC Booklet 39.

Hoppus measure

For hardwood sawlogs and veneer logs (and occasionally for softwood sawlogs) some roundwood merchants, sawmillers and contractors continue to use imperial units in which volume is expressed in hoppus feet (hft), derived as follows:

$$\text{Hoppus Volume (hft)} = \frac{(\text{Mid Quarter Girth in ins.})^2}{144} \times \text{Length (ft)}$$

The girth tape used is calibrated in quarter girth (QG) in inches, and the cross-sectional area is approximated by squaring the QG and dividing by 144. One Hoppus foot = 1.273 true ft^3; 27.74 h ft = 1 m^3; and 1 h ft = 0.03605 m^3. A Hoppus foot is about 21% short of a true ft^3 – the reduction helping to compensate the processor for his loss in converting roundwood to sawnwood. The following additional conversion factors should be noted:

1 cm diameter = 0.31 ins. QG;
1 hectare (ha) = 2.47 acres (ac); 1 ac = 0.405 ha;
1 m^3/ha = 11.2 h ft/ac; and 1 h ft/ac = 0.089 m^3 ha.

References

Bean, W.J. (various editions: 1914–88), *Trees and Shrubs Hardy in The British Isles*. John Murray, London.
Bitterlich, W. (1984), *The Relascope Idea: Relative Measurements in Forestry*.
Brazier, J.D. (1990), 'The Timbers of Farm Woodland Trees'. FC Bulletin 91.
Edlin, H.L. (1971), *Wayside and Woodland Trees*. A revision of the original by Edward Step. Warne, London.
Edlin, H.L. and Mitchell, A.F. (1985), 'Broadleaves'. FC Paperback.
Edwards, P.N. (1983), 'Timber Measurement: A Field Guide'. FC Booklet 49.
Evans, J. (1984), 'Silviculture of Broadleaved Woodland'. FC Booklet 62.
Hadfield, Miles (1957), *British Trees: A Guide for Everyman*. Dent, London.
Hamilton, G.J. (1975), 'Forest mensuration handbook'. FC Booklet 39. Slightly ammended 1985.
Hart, C.E. (1973), *British Trees in Colour*. Michael Joseph, London.
Jobling, J. (1990), 'Poplars for Wood Production and Amenity', FC Bulletin 92.
Johnston, D.R. (1979), *The Craft of Furniture Making*. Batsford, London.
—— (1983), *Wood Handbook for Craftsmen*. Batsford, London.
Kerr, G. and Evans, J. (1993), 'Growing Broadleaves for Timber'. FC Handbook 9.
Marren, P. (1990), *Woodland Heritage*. David & Charles, Newton Abbot.

Mitchell, A.F. (1972), 'Conifers in the British Isles'. FC Booklet 33.
—— (1974), *Trees of Britain and Northern Europe*. Collins, London.
—— (1980), 'Conifers in the Landscape'. *Country Life,* 6.11.1980, pp. 1646, 1647.
—— (1981), 'The Native and Exotic Trees in Britain'. FC Arboriculture Research Note 29/81/SILS.
—— (1985), 'Conifers'. FC Paperback.
Mitchell, A.F., Hallett, V.E. and White, J.E.J. (1990), 'Champion Trees in the British Isles'. FC Field Book 10.
Mitchell, A.F. and Jobling, J. (1984), *Decorative Trees for Country, Town and Garden*. HMSO, London.
Philip, M.S. (1983), *Measuring Trees and Forests*. Aberdeen University Press.
Schwankl, A. (1956), *What Wood is That?* Trans. and ed. by Edlin, H. L. Thames & Hudson, London.
Trenard, Y. (1982), *Making Wood Speak: An Introduction to Dendrochronology. Forestry Abstracts* 43 (12) Commonwealth Forestry Bureau, Oxford.
Watkins, C. (1990), *Woodland Management and Conservation*. David & Charles, Newton Abbot.

Chapter Two

NURSERY PRACTICE, SEED AND FOREST GENETICS

1. Nursery practice

Whether to start a nursery. The principal objective of a forest tree nursery is to produce at acceptable cost good quality sturdy healthy plants with a high root:shoot ratio, and a well-developed fibrous root system. Achievement is generally by adequate ground preparation, careful sowing of good quality seed, predator control and correct use of irrigation, fertilizers and herbicides. The techniques maximize seed germination, plant survival and healthy growth. Plants should conform to BS 3936, Part 4: Nursery stock, which may be revised in the future (Aldhous 1989). In considering whether to start a nursery foresters have three options: to produce both seedlings and transplants, to purchase seedlings and transplant them, or to purchase all the plants required. The number of hectares of woodland to be planted annually will have a great bearing on the decision. The important new developments of growing in containers, and of vegetative propagation are noted later.

There may be particular advantages in some estates having their own nursery. Plants may be lifted as required, they may be planted without delay, and should have a better chance of survival than those brought from afar. The cost of plants to the estate may be lower, and packing and transport is not required. However, the disadvantages of an estate nursery have to be taken into account. Skilled and keen supervision, efficient labour, a suitable site and a sustained demand for plants are necessary for success. The estate has to bear all risks of production, including failures, culls and any unsaleable surplus.

Efficient trade nurseries can supply all the plants required for private woodlands. Healthy competition between them keeps prices at a fair level, possible only by the mechanization of most operations and by producing in large quantities. Unless the estate annual planting programme is substantial (see *infra*), the best and most economical course is to purchase the needed plants; the whole cost is then at once known and the risks entailed in production remain the concern of the trade.

The production of seedlings can be difficult and hazardous; it requires skilled direction and supervision by someone who has an adequate nursery experience. A newcomer to nursery work has generally to 'buy' as well as to 'acquire' his experience, a process that can be long and expensive. However, it provides work of engrossing interest throughout the year and affords, in addition to the production of forest trees, the possibility of raising whips, covert plants and ornamental trees and shrubs for use on the estate, and possibly for sale. But unless the operations of the nursery are carefully costed (and this task, particularly by species, is rare) it is difficult to know whether the stocks produced are cheap or expensive.

For foresters intending to raise their own plants, it can be noted that one hectare of nursery (gross area, including paths and fallow) is required to produce 100,000–250,000 plants for forest planting. For every unit area cultivated approximately 60% is cropped, generally 60–90% of this area can be taken as the area under transplants. Between 25% and 35% of the cultivated area is likely to be fallow or green-cropped at any one time.

Preparing a new nursery may cost £1,000–£3,000 per hectare, exclusive of overheads and the cost of land and any buildings. The cost of seed can be readily ascertained (see section 2 *infra*) but other costs are variable. Transplanting (lining-out) manually may cost £5 or more per 1,000 plants and by machine at least £3. To these figures should be added labour and machinery oncosts and overheads. Other costs include sterilants, grit, fertilizers, insecticides, fungicides, herbicides, weeding, irrigation and protection. Foresters may well consider that nursery practice is a task best left to the trade. However, most of this Section is based on the assumption that, after due consideration of both the advantages and disadvantages, it has been decided to establish a reasonably sized and comprehensive estate nursery which may or may not be permanent. What follows is not a treatise on nursery practice but an indication of some of the chief factors encountered, yet obviously far short of what is met in a substantial trade nursery. For much greater detail recourse should be made to Aldhous (1994, new edition), supplemented by Gordon and Rowe (1982), Gordon (ed.) (1991), and information obtainable from the Forestry Commission Research Stations.

Briefly, seed given appropriate pre-sowing treatment is sown densely in prepared seedbeds, designed to ensure maximum germination, and good growth of the seedlings in their first growing season. The seedlings (1 + 0) are then lifted and transplanted (lined out) where they have enough space to grow into sturdy well-shaped plants fit to go out into the woodland at the end of a further growing season (as 1 + 1 transplants). There are modifications. The less climatically favoured nurseries or the slower growing species may not produce usable plants in two years, and longer periods may have to be allowed in seedbeds and/or lines. Sometimes the seedlings are left in the beds for a second year (becoming 2 + 0) during which they are undercut, and later lifted (as lul) or lined out to become 2+1 transplants; or 2 + 2 if left for a fourth year. Summer lining out is a widespread practice, producing 1½ + 1½ transplants, and often is a practical solution to the problem of having to make a choice between small 1 + 0 and oversize 2 + 0 seedlings. An alternative is to produce precision-drilled undercut and wrenched plants. Depending on species, usable plants are normally produced and sold as ½u½ or lul. Irrigation facilities must be available. The ½u½ option is becoming increasingly popular, largely for economic reasons, and particularly for broadleaved species. Larches and pines can also with skill be grown as ½u½ in southern England.

Labour requirement. In a non-mechanical nursery the labour required, over a year, may average two workers (about 500 working days) to about one hectare. If a substantial amount of mechanization is introduced the requirement may average only the equivalent of one worker (about 250 working days) to one hectare or more. The labour required may not be evenly distributed during the year: at some periods all that is necessary are occasional inspection and protection. The ideal is to have workers reasonably experienced in nursery practice available to be brought from the woods or other departments as occasion requires. The busiest time will be during seedbed preparation, sowing, weeding, transplanting and lifting. If piecework rates can be used for some of the operations this should reduce the amount of labour required. Some nurseries are able to draw on a pool of local seasonal workers for lining out, lifting, packing and despatching.

Choosing the site. If, as is assumed, skilled supervision and labour are available, the choice of a suitable site is the next requirement. It needs to have a light freely drained soil capable of being cultivated at most times of the year. The soil should have a moderately acid reaction, between pH 4.5 and 5.5 for conifers, pH 6 for most broadleaves, and pH 5.0–6.5 for poplar. Soil-texture is important. Well drained sandy loams, usually free of many stones, are the most suitable. They are friable and thus permit work to start in early spring and continue into late autumn; they also minimize the risk of frost-lift. Some good nurseries lie on sandy soils. Clay soils should be avoided for obvious reasons. It may be difficult to find a soil in a limited area which suits all the principal species of both conifers and broadleaves. Freedom from persistent weeds is necessary; a heavy weed-infestation is costly to remove or control and will increase costs and substantially reduce the quantity and profitability of the plants produced. Although shading by tall trees close to the site should be avoided, the nursery needs to be sheltered from drying wind. The ideal is perhaps to have a fairly large wood 'in the background' on the side or sides of the prevailing wind. It should not be too open to early morning sun as some seedlings may be harmed by a rapid thaw after a frosty night. In the south of England, a nursery may be best facing south-east; in more northerly districts perhaps facing south. Sites are usually improved by being on a slight slope to facilitate drainage of water and cold air. The site should not be subject to late spring frosts, and an elevation of 75 to 150 m may be best. Local rainfall should be 600–1000 mm, well distributed over the seasons. It is desirable to have a nearby supply of suitable water. Elaborate irrigation equipment is only worthwhile (cost-wise) in a large nursery, but some watering facility is always advisable. In any case, when irrigation is being considered, contact should be made at the outset with the local water authority, to enquire if there are any limitations on the amount or timing of water abstraction. The site should be convenient to most of the estate woodlands and have easy access for wheeled vehicles in all weathers. It should not be too distant from the living accommodation of a skilled worker who, during or after usual working hours, can undertake shading of newly emerged seedlings from frost/sun, or watering, and can be aware of possible damage at an early stage.

The size of the nursery, allowing for fallows, should be approximately onetwentieth to one twenty-fifth of that of the annual planting area. About one-fifth to one-tenth of the cropped area would be taken up as seedbeds. If poplar and a few ornamental trees and shrubs are also to be produced, the size of the nursery should be increased accordingly; for such stock a heavier loam may be best. Foresters must decide what is the minimum economic size of a nursery; this will vary according to circumstances. An estate with a planting programme of 5–10 hectares would need only a half hectare nursery but this is rarely an economic size; there are certain minimum costs and one hectare (the minimum) will probably not cost double that for a half hectare. In practice, the situation is confused by the role of production of non-forestry species. Some foresters may plan for a surplus if there is a known market. The shape of the nursery should be rectangular to facilitate cultivation; the nearer it is to a square, the smaller will be the amount of protective fencing required to enclose it. Possibilities of mechanizing some of the operations must be constantly borne in mind; long uninterrupted runs of 75–100 m for the machines are desirable, though it is rare for an estate nursery to be of sufficient size to permit this.

Preparing the site. The site for a the nursery should be cleared before it is enclosed, and the soil subsoiled, ploughed, disced and harrowed, perhaps more than once to remove roots and other woody material, care being taken to avoid bringing subsoil to the surface. After these initial cultivations, fencing and gating against rabbits, perhaps against deer and livestock, should be completed and water led to an internal tank. The pH value of the water should first be tested. If it is unsuitable, plans must be made to collect rainfall. Sheds, roomy enough for machines, tools, pesticides, grading, packing and some wet-weather work need to be provided. If the site is exposed to strong winds, hedges of such as hornbeam, hawthorn or Lawson cypress could be planted where required along the interior of the perimeter; beech or western red cedar may be used provided plants of these species are not to be produced (because of pests and disease). Elsewhere within the nursery hedges may be a hindrance to mechanization. The use of hedges may be unwise, particularly in view of the fact that a site may have to be changed after a period of years. They are costly to maintain and harbour pests, and can cause the build-up of frost, particularly at the bottom of a slope. Temporary breaks in lieu of hedges can be provided by tying lath screens on end to stakes; these can be rolled up when necessary to admit machines. The use of wattle hurdles is another possibility.

Perennial weeds are best controlled using recommended herbicides. Nutrient deficiencies can be revealed by foliage- or soil-analysis and corrected by fertilizing, but different species have different requirements. The value of hop-waste, particularly as a retainer of moisture, can be remembered; likewise the fact that nurseries require plenty of humus. Farmyard manure often brings more trouble in weeds and diseases than it is worth. The real value of organic fertilizers is in physical properties; they add nutrients, retain moisture and encourage fibrous root development.

Seed supply. Information on seed supply is provided in section 2 *infra*. For any woodland planting programme the time for obtaining the necessary amount of seed will be determined by the age at which the plants eventually are to leave the nursery. Thus planning may have to be from one to three years ahead. The calculated amount of seed can either be collected by the estate staff or ordered through the trade or the Forestry Commission. Indications of current prices of seed are given in section 2 *infra*. An increase of 0.5% in volume of timber produced justifies an increase in expenditure on seed of from 50 to 420%. Thus it is essential to use certified seed of recommended origin/provenance (discussed in section 3 *infra* and in chapter 4, section 2), of good quality and of acceptable germination percentage.

Seedbeds. The number of square metres of seedbed required for a planned production can be calculated from tables referred to hereafter. The site for the seedbeds should be ploughed in the autumn and any necessary bulky organic manures worked in. Only on heavy ground need the area be ridged up over winter. No additional work may then be required until the early spring, unless broadleaves are to be sown immediately on collection. The possibility of pre-sterilization of the bed is discussed *infra*; it will control weeds and disease, and will enhance seedling growth. At least four weeks before sowing in the autumn or in the spring, the beds must be thrown up, across a slope if storm damage is possible. The width of the beds should be 1–1.3 m, with 0.5 m allowed for alleys between them. The height of the beds will vary according to the soil and climate; but never more than 15 cm, particularly if machines are to be used. The beds should be cultivated to a good tilth and levelled; at the same time any necessary inorganic fertilizer should be applied. Raking and rolling are then necessary. Consolidation is very important where soils are light and liable to dry out, and especially if sowing is late and is followed by dry windy weather. The need for, and use of, fertilizer of whatever kind must be carefully considered. A true assessment of the requirements can usually be made only from experience, but potassic super-phosphate is necessary for seedbeds on most sites. It is sometimes beneficial for conifers to be grown on soils containing *mycorrhizae* (see chapter 3).

Sowing density. Planning the production of seedlings requires a knowledge of densities of sowing. Tables for conifers are provided by Aldhous (1972) and for broadleaves by Gordon and Rowe (1982). The density depends on the species, the viability, and the strategy, i.e. whether the seedlings will be raised for transplanting or grown on in the seedbed and undercut. The normal aim for conifers is to produce 300–800 usable 1 + 0 seedlings per square metre, and for broadleaves 100–300, according to species.

To remove the need for nursery managers to calculate sowing densities separately, the Forestry Commission includes them at the bottom of its Seed Supplier's Certificate (Gosling 1990). Trade seedsmen do likewise. Suggested sowing densities are presented as the number of square metres of seedbed into which the despatched quality of seed should be sown to produce either 1 + 0 or 2 + 0 seedlings. Accuracy in sowing to the correct density comes with practice. It must be remembered that not every viable seed will produce a usable seedling. Mechanical and broadcasting machines should be properly serviced and calibrated before use. To ease hand sowing of small seed 'Waxoline red' or talc can be used as a dressing. Seed colourants should not be applied to pretreated seed.

The best time for sowing, except when broadleaved seed is sown in the autumn (when it must be well protected), will probably be, in the south, at any time after mid-February, and in the north, in April; it is usually determined by the soil conditions and the climate. Only by trial over several seasons will the nursery manager discover the periods most suitable for each species. 'Know-how' plays an important part, as it also does in the choice of the methods of sowing, by drill, broadcast or band. Drilling is more often used when there is to be no mechanization and spraying or when the labour is inexperienced. The depth and type of covering calls for care and will vary. The covering must be of non-calcareous grit, as this lessens the risk of 'caking' which can check germination; it also reduces damage by frost-lift. The main purposes are to retain adequate moisture around the seeds, reduce surface crust formation and generally provide a micro-environment more conducive to the seed germination and seedling establishment than soil.

Seedbeds need close attention. Attacks by birds, insect pests and fungal and bacterial diseases (all discussed later) must be attended to. Irrigation, little and often, is necessary during the early germination phase when seedlings are often at their most vulnerable. Some protection may be necessary against frost. The length of time for covering against frost will vary: for beech it may be only six weeks after germination; for silver firs only up to the primary needle stage. At least one large commercial nursery has a network of pipelines and sprinklers (covering species susceptible to frost damage especially in the chancy spring) which can be activated when the night temperature falls below the accepted level. The plants are covered with water which quickly freezes, giving latent heat of fusion to the plants without damage. Frost-lift is perhaps more prevalent in northern parts of Britain where there is a high frequency of frost alternating with mild conditions; it occurs more in heavy soils when they are wet than light ones. Protection against birds is discussed *infra*. Weeding will necessitate carefully timed pre-emergence or post-emergence spraying, or hand-weeding; usually a combination of them is required. Techniques of weeding are noted *infra*. Undercutting is necessary for most species of broadleaves, especially oak, ash and sweet chestnut, if 1 + 0 seedlings are to be produced or if they remain in the seedbeds for more than one year. The same applies to Corsican pine and Douglas fir. More root fibre is thereby induced.

Opinions differ on what height and age seedlings ought to attain before they are ready to be either transplanted or planted direct into the woods. Oak and sweet chestnut are sometimes ready to go out into the woods as 1 + 0 seedlings. For lining-out, seedlings of most species should be at least 6 cm in height. Some nurserymen prefer to line-out one-year seedlings but others prefer to wait until they are two years old, according to local rates of growth. Lifting and grading, as well as bundling, packing and despatching must be done with great care. Special coextruded black/white polythene bags are usually used. Cold storage, discussed later, is also being resorted to, enabling plants to be set out safely very late in the spring or early in the summer.

Transplanting. The land on which to line-out must be clean and workable; it should have been ploughed or rotavated during the previous late summer or autumn and left to weather. It should be re-cultivated before lining-out begins; and required fertilizers should be worked in at the same time. When preparing the lining-out programme the number of square metres of land required must first be calculated. An approximate range of density is 100–150 per square metre. Species and kinds of seedlings require different spacings; their size has to be taken into account and whether they are to remain for one or for two years in the transplant lines. Generally speaking a 1 + 1 or a lul is best if the plant is big enough to go out into the woods. It is the most balanced plant with a good ratio of root to shoot. The next best may be a 2 + 1, but a 2 + 2 is often necessary because of extensive weed growth in plantations; in Scotland, because of slower growth rates in nurseries, one cannot in many cases procure a usable plant unless 2 + 1 or 2 + 2. A standard spacing between the rows is needed for the mechanized control of weeds. Alleys between lining-out strips must be allowed for. Table 6 sets out the general recommendations for distances for mechanical lining-out in fertile nurseries of good quality. On less fertile sites slightly reduced espacements can be adopted.

Table 6. Spacing in transplant lines.

Expected average height when lifted from the lines	Spacing at lining-out: rows 20 cm apart	No. of plants per 100 sq metres (approximate)
	Spacing in rows	
Less than 20 cm	3 cm	16,500
20 to 40 cm	5 cm	10,000
Over 40 cm	7 cm	7,500

The 20cm spacing between rows is that most widely used. In theory, the spacing between rows should be adjusted to suit the size of plants being raised; but most nurserymen have some cultivating, spraying or lining-out equipment adjusted to a particular row spacing. In practice it is best to adhere to the spacings shown for all commonly required sizes of plant. Row spacings of 18 cm are also common. One spacing should be used throughout any particular nursery.

Correct spacing in and between the lines is important to allow plants to develop to the required size, but the success and cost of this operation is very dependent on the workability of the soil. Light sandy soils are both easy to work and quicker to dry out in the spring. The lining-out process can be carried out manually, using boards, but is usually partly or wholly mechanized. Partial mechanization involves the use of special ploughs which place the soil against the plant roots, at the same time cutting the face for the next line of seedlings. Full mechanization depends on five or six row machines that place the seedlings in position and firm soil round their roots, but even these still require the seedlings to be fed into the planting mechanism by hand; five or six planters are needed, and often another person has to follow behind the machine to ensure appropriate firmness of the plants. Mechanization requires greater space to turn the machines at the end of beds, and hence long lines to obtain maximum output. This is one reason why the maximum benefit of mechanization will only be obtained in larger nurseries.

In general, the more trees lined-out in the autumn the better, but it really does not matter which months from October to April are chosen for transplanting, if periods of hard frost, snow or drought are avoided. Small seedlings are best lined-out in the spring to avoid frost-lift, while the autumn is favoured for beech. As the larches are the earliest to flush they should be lined out before the middle of March. Alternatively dormant seedlings held in cold store can be lined-out during late April, May and early June, thus extending the lining-out season. Irrigation facilities must be available. Usually autumn-lined seedlings will be significantly larger at the end of the following growing season than spring-lined stock.

Summer lining-out can sometimes be highly successful and advantageous. In Scotland, Scots and lodgepole pines, Douglas fir and even Japanese larch can be lined-out in July; the outstanding success is with spruces. The theory is that from the end of June to some time in August root activity is intense, particularly with spruces whose shoot growth at this time has slowed up. The seedlings are lifted while they are rising $2 + 0$ so that instead of producing a $2 + 1$, a truer age category would be $1½ + 1½$. This system, however, involves a quicker turnover in land use and presents difficulties in preparing ground in time unless there is plenty to spare.

Methods of transplanting vary. It must be decided whether to work with long or short lining-out boards and whether ploughs or machine transplanters can be used. The machines currently available necessitate large sections of reasonably flat stone-free ground; they are rarely of use in small estate nurseries. Good, simple rules for the work are: the trench must be sufficiently deep to allow the roots to be spread out freely and straight-backed to avoid bending the roots; the plants should be set at the depth they had in the seedbeds and well-firmed. During lining-out, fine roots must on no account be allowed to dry. They must always look damp. Weeding can be by hand, using dutch hoes or wheeled hoes, by mechanization, or by spraying of chemicals as noted *infra*. Shading is rarely required but in very dry conditions spraying with water may be necessary. Attacks by insect pests and fungal and bacterial diseases (all discussed *infra*) must be attended to. Lifting, grading (including culling), bundling and packing of transplants can be done from October to April whenever they are required and the weather is suitable. Care taken in these operations will help to ensure greater success in the woods. The shorter the time the plants are out of the soil the better.

Precision-sowing with undercutting. This nursery system is relatively new in Britain, the accent being on the production of uniform stock of conifers of a specified standard – planned to be accompanied with improved site preparation and defined standards for plant handling and planting (Mason 1986). Good quality seed is essential: impurities foul the machinery; poor germination wastes space. The results are good for all conifers tested to date. Precision-sowing involves the use of a drill sower which spaces out the seed regularly in each line, providing seedlings with sufficient room to grow on in the seedbed until they are ready for lifting and despatch as two-year-old plants. This method is designed to replace the traditional technique of lifting and transplanting after one year's growth in the seedbed which is more costly and can sometimes lead to considerable plant losses. Plants grown by the system need careful side-cutting and undercutting with special machinery, combined with close control of irrigation and nutrient supply to keep root and shoot growth in balance. With good quality seed and stone-free soils, the technique offers a substantial advance on conventional methods in terms of cost saving, plant yields and quality.

There are, however, advantages and disadvantages. Of advantages, the need to organize and execute lining-out is avoided, thus creating considerable scope for cost saving as well as a simpler system to manage; research evidence to date indicates that undercut plants are more likely to survive planting out than transplanted stock (particularly in the case of the more sensitive species such as Douglas fir, Corsican pine and the larches); and it is probable that plant yield per germinating seedling is higher. Of disadvantages, drought after sowing can cause problems because of the need to irrigate a much larger area of seedbeds than with conventional methods; root development along the rows, which cannot be controlled by undercutting or side-cutting, can make lifting difficult; and intensive management, with frequent undercutting, is necessary if height growth is to be adequately controlled in the second year.

Protection of nursery stock. The fencing which may be required against wildlife and livestock has already been referred to. Moles, mice and voles should be controlled by trapping or poisoning. Rooks, wood pigeons and pheasants must not be allowed to take seed from beds particularly of oak, beech and sweet chestnut; nor small birds, such as finches and linnets to extract small conifer seed or to use the beds for 'dusting'. The use of polythene netting is effective. Plastic small-mesh netting, e.g. Netlon, supported on wire hoops is most frequently used. Sheltering from excessive sun and frost has already been mentioned. Erosion of soil and beds by rain and flood must be guarded against; raising the beds and perhaps siting them across a slope, and providing channels to enable storm-water to escape, may be the best plan. The worst pests in the nursery are insects inhabiting the soil, sap-suckers and leaf-eating insects. These, together with fungal diseases – and their control – are noted *infra*.

Container grown seedlings. By contrast to bare-rooted plants raised in open ground nurseries, there are various types of containers for producing tree seedlings (Mason and Jinks 1990). Many millions of seedlings are produced in containers each year. Most have been conifer species, but even more broadleaves are now being produced. There are seedlings raised in greenhouses or polyhouses in small containers such as 'Japanese Paper Pots', 'Ecopots' and 'Rigipots'. These container grown seedlings are planted with the roots surrounded by the growing medium which reduces planting shock. The 'JPP' plants are suitable for lowland use only, because the pots do not degrade rapidly enough in the colder upland soils and can restrict root development. The 'Ecopots' are suitable for universal use. Most conifer species have been Corsican pine, Douglas fir and Sitka spruce. The first-named has been used extensively in East Anglia and on other heathland soils in lowland Britain, particularly for restocking; their use fits in well with mechanized planting.

A more recent innovation is 'Rootrainers' – hinged plastic 'books' for producing seedlings in plugs of growing medium, with accelerated and protected techniques. They are a modern container system for the propagation of seeds and cuttings. The claimed advantages over other forms of container are the hinged opening enabling the plants to be inspected or removed, with minimum risk of damage or 'shock', the internal grooves and greater depth promote optimum root formation and sturdy plants, and handling and planting are simpler. The containers are lightweight, can be flat-packed, and are re-usable.

A crucial point about containers is the size (i.e. the volume of rootable medium) and in some circumstances the container density (i.e. the number of plugs per square metre of the container unit – higher densities of containers per square metre can lead to plants being drawn up more than is desirable, but do make better use of greenhouse/polyhouse space thus leading to lower costs). Beyond the foregoing the other main consideration is price: different container types can influence growing costs, because of their varying purchase price.

A target specification for container grown seedling planting stock for upland conifers is 20–30 cm tall with a root collar diameter of 3–4 mm, produced on a six to twelve month production cycle (Mason and Hollingsworth 1989; Hollingsworth and Mason 1989). Technical information on container stock is provided by Mason and Jinks (1990). From the nursery viewpoint, there are attractions in using controlled conditions: container nurseries are not confined to sites with a suitable soil, and there is a lowered risk of a build up of weeds, diseases and insect problems. However, a high capital investment is required for facilities such as greenhouses or polyhouses, compost mixtures and filling machines, irrigation and liquid feed systems. There is a short period of production (a few months to a year) and a high output rate per person. The price obtainable is up to double that of bare-rooted seedlings.

Genetically-improved Sitka spruce. Increasing numbers of cuttings of genetically improved (C½ + 1½) Sitka spruce are available (Mason, Biggin and McCavish 1989). The Forestry Commission charge £200 per lot of 2,500 seeds. It obtains seed by hand pollination of selected and tested parent trees. Because this high quality seed is available only in small quantities, the seedlings produced from it are used as a source of cuttings for vegetative propagation in order to increase the number of plants from each seed. The seedlings are first raised as stock plants in pots inside greenhouses or polyhouses for two years and then divided up to produce 60–100 cuttings each. After two years' growth, the young plants are again used to produce a second cycle of cuttings which are rooted and transplanted into the nursery, to be despatched as two-year-old plants. In this way, each seed will produce 400–700 usable plants of identical genetic composition. Trees grown from these selected sources are expected to provide an increase in timber yield of some 15% above that of unimproved Sitka spruce (see section 3 *infra*). The price obtainable is about double that of unimproved stock.

Poplar production. Most poplars are propagated vegetatively. They are easily raised from cuttings from dormant shoots; about 90% of such cuttings normally grow to become usable plants (Adhous 1972). The site for raising of poplar should be moist, fertile, sheltered and easily cultivated (Jobling 1990). The soil needs to be of a pH between 5.0 and 6.5, capable of working easily and remain reasonably moist during the growing season. Clay soils and sands are normally to be avoided. All poplars are heavy feeders and soil fertility must be maintained by regular addition both of organic matter such as hop-waste and of inorganic fertilizers.

Cuttings should be 20–25 cm in length, taken from straight, well-ripened one-year-old shoots with strong well-formed dormant buds (Aldhous 1972), the top end of the cutting being cut just above a bud. Material can be prepared at any time between leaf-fall and the end of March. In early spring, the ground must be well-cultivated to a depth of at least 20 cm and the cuttings pushed vertically into the soil shortly afterwards until the upper end is level with the soil surface. The cuttings should be spaced about 40 cm between cuttings and 50 cm between rows. When buds break, two or more shoots may be thrown up by each cutting. After the risk of late-spring frost has passed, and the shoots are 15–20 cm tall, the strongest and straightest of them are selected as the leading shoot and others removed. Thereafter, the ground should be hoed or sprayed to control attacks of weevil or beetles. At the end of the season, one-year rooted cuttings (C1 + 0) are likely to be from 1.4–2 m tall. If they are sturdy and well-branched they are suitable for planting out. The price obtainable ranges from £0.95 to £1.20 each.

Those one-year-old plants which are too small for planting should be lifted and replanted at 0.6–0.9 m spacing in ground prepared in the same way as for cuttings. After planting, the main stems are cut back to within 2 cm of the top of the original cutting. The shoots so removed can be used for cuttings and this is an important way of increasing stocks. The stumped back transplants will often make as much growth in one year as would have been made by uncut plants in two years; stumped plants are also usually sturdier. In the early part of the season, the shoots from a stumped plant should be singled, favouring the strongest and straightest shoot, in the same way as for cuttings. After one year in the transplant lines the plants, having a two-year-old root system and a one-year-old shoot (C1 + S1), will almost always be suitable for planting out and will often be over 2.3 m tall. They are as easily established as well-grown one-year-old plants, but are bulky to transport, more time-consuming to plant and may require staking. Plants may also be grown from cuttings for two years without disturbance, but are likely to be taller than stumped transplants and may require more careful handling and deeper planting.

Poplar stool bed. By cutting back one-year-old cuttings each year, most growers will have sufficient cutting material to meet their needs. For larger-scale propagation, however, a regular supply of cuttings is most conveniently obtained from stool beds. The young shoots should be spaced 1.2 x 1.2 m and cut back after planting and at the end of each successive season. The shoots produced in successive years from stools will increase in number and quality until, by the fourth season, each stool will be producing 20–60 cuttings. Like the cutting and transplant beds, the stool bed must be on good ground. With regular manuring, stools should remain vigorous until they are six or seven years old at least.

Production of poplar sets. Very occasionally long cuttings or sets (i.e. without roots) can be established satisfactorily on the most favoured planting sites, but their survival rate is usually lower than that of rooted plants and their growth is appreciably slower for one or two years after planting. One year sets, 1.8–2.3 metres in height, are usually large enough for most sites although longer material can be used provided the sets are straight and are planted to even greater depth. If one year's growth in the nursery produces shoots less than 1.5 metres long these may be grown on for a second year to produce sets of a good size. Deep planting (not less than 50 cm) and care in ensuring soil contact with the whole length of the buried portion are essential to success.

Recommended timber-producing poplars. Recommended poplars are noted in chapter 5. The material should be resistant to bacterial canker and should be from officially approved reproductive sources recorded in the National Register of Basic Material. The stock should conform to BS 3936 Part 6:1985.

Insect pests and diseases. Insects attacking poplars in nurseries and stool beds include (Bevan 1987) the Brassy willow beetle *Phyllodecta vitellinae* and Blue willow beetle *P. vulgatissima*. Both attack during May to September. Control is by sprays of Gamma HCH (under 'long-term off-label arrangements'). Various diseases to be expected are noted by Phillips and Burdekin (1982).

Willow production. Traditionally, willows have been planted as sets. However, rooted plants are being used to ensure good growth and straight stems. Rooted plants can be raised both from cuttings and sets. Plants are raised from 20–25 cm cuttings, 15–25 mm thick, in the same way as for poplar. Sets are grown on stools 90 cm apart. Shoots 2.5–3 m should be cut and planted in the nursery at 30–60 x 60–90 cm, remaining for one or two years according to size and growth. White and Patch (1990) give details of propagation of lowland willows.

Production of ornamental trees and shrubs. Foresters having mastered the techniques of efficient forest-tree nursery work, can undertake the growing of a few plants for the estate's coverts and hedges and of a few ornamental trees and shrubs. But this is really the job of the experienced trade nurseryman. The horticultural trade is a competitive one and foresters may find it cheaper to purchase stock from commercial nurseries growing ornamental trees. However, it is possible to produce a few covert plants, some hedge plants such as western red cedar, Lawson cypress and beech and perhaps a modest number of trees for the estate parks and gardens. Some foresters may, in addition, and if only as a hobby, attempt a measure of propagation, including budding, grafting, layering and the raising from cuttings of such plants as Leyland cypress and coast redwood. This will provide at least an introduction to some of the problems of a plant propagator.

Vegetative propagation. Few forest trees can be grown easily from cuttings, the exceptions being poplar, willow and elm, and amongst the conifers Sitka spruce, larch, western red cedar, coast redwood and Leyland cypress (Samuel and Mason 1988; Hollingsworth and Mason 1991). The technique is particularly valuable when seed is not readily available, as in the case of particular clones and cultivars. Difficult subjects can be raised under glass in a heated frame or greenhouse, or in a polythene tunnel. Further references to vegetative propagation are included in section 3 *infra*. In-vitro propagation as it relates to forest trees, is only in the experimental stage.

Vegetative propagation is also seen as the way forward for hybrid larch production, given the difficulties in obtaining sufficient quantities of good quality seed. The Forestry Commission is about to embark on large scale production (up to 1.5 million per annum) of hybrid larch container stock using cuttings taken from larch hedges grown from seed taken from phenotypically superior British parents. Suitable seed for establishing hedges will also be made available to the private sector.

Nursery herbicides. Foresters having to control nursery weeds should consult Williamson and Mason (1989) to identify suitable herbicides, determine the approval status of herbicides before using them and conduct limited trials of any new herbicide before adopting it on a wide scale. Herbicide efficacy is influenced by weather conditions, e.g. residual herbicides should be applied to moist soil to aid incorporation and activate the chemicals; and herbicides should not be applied in hot, sunny conditions because of the risk of crop damage. Some herbicides have both contact and residual properties but usually their activity is biased towards one of these modes of actions. *Soil acting (residual) herbicides* must be applied to moist soil which has a fine tilth and is free from clods. They require moisture in the soil to be activated and may need to be incorporated immediately after application by irrigation with 10–15 mm depth of water. The length of time residual herbicides remain effective varies depending on the chemical in use, soil type and climate – particularly rainfall, sunlight and temperature. *Contact and translocated herbicides* are absorbed through the point of contact on the leaf and stem and are independent of the condition of the soil. Most give better results when the target weed species is actively growing and at the appropriate growth stage for the herbicide in use. The same herbicide should not be used repeatedly over a number of years on the same area of ground: this can lead to development of resistant strains of particular weed species, or to the development of unwanted soil microflora which can rapidly degrade the chemical active ingredient.

Weed control in seedbeds. In *pre-sowing treatments, soil sterilization* usually takes place in late summer-early autumn prior to sowing, and controls many soil pests (nematodes, fungi) as well as weeds and weed seeds. The growth of the seedlings will usually be improved. The favourite soil sterilant, Dazomet (approved product Basamid), must be incorporated by rotavation and sealed in by rolling or polythene sheeting. At least four weeks must elapse before cultivating the soil to release gas; if applied in the autumn it is usual to wait until the spring. It is important to aerate the soil and release all the sterilant residues prior to sowing. An alternative to sterilization is a *stale seedbed technique* carried out either before or after seedbeds have been formed. A fine tilth is created which allows the germination of weeds; these are then destroyed by further cultivation or by herbicides (usually glyphosate: approved product Roundup, or paraquat: approved product Gramoxone).

Seeds of small-seeded broadleaves and conifers are normally sown on to the surface of raised seedbeds and covered with 2–3 mm depth of grit. *Pre-emergence pesticides* can then be applied, depending on species, immediately after sowing before crop germination. Large-seeded broadleaves such as oak, beech and sweet chestnut are usually drilled into seedbeds and then covered with at least 25 mm of soil. Such species are therefore usually more tolerant of pre-emergence herbicides. There is a wide choice of herbicides, dependent on the range of weeds expected, and the tolerance of the seed species. A similar wide choice is available for *post-emergence pesticides,* but great care has to be taken with the timing and method of application; and their use is no substitute for pre-emergent treatments.

Weed control in transplant lines. Residual herbicides are widely used in transplant lines, being normally applied immediately after lining-out and repeated as necessary. Many of the residual herbicides available do not control germinated weeds and therefore must be applied before weeds emerge. Undercut stock can generally be treated in the same way as transplants, provided herbicide application does not occur until soil has settled after undercutting/wrenching (this is because of the risk of herbicides coming into direct contact with tree roots). There is a wide choice of herbicides, dependent on the range of weeds present or expected.

Weed control on fallow ground. The fallow period in a nursery rotation provides an opportunity for controlling deep-rooted perennial weeds by a combination of cultivation and chemical control. Repeat applications may be necessary. The usual herbicides are glyphosate or paraquat. Before these or any other herbicides are used the label should be read: it carries full instructions for use and for the protection of the operator and the environment. Further advice is obtainable from the Forestry Commission at Alice Holt Lodge.

Nursery fungal diseases. Several diseases have adverse effects on nursery seedlings and plants (Hibberd (ed.) 1991; Phillips and Burdekin 1982). Various soilinhabiting fungi cause losses at a very early stage of germination (pre-emergence *damping-off*), and they also infect the roots and cause the death of young seedlings (post-emergence *damping-off*). The fungi concerned include *Pythium, Rhizoctonia* and *Fusarium.* Post-emergence killing usually occurs before the end of June, and is most serious in dense seedbeds under warm, moist conditions. Infected seedlings may either be scattered throughout the seedbed, or in irregularly distributed groups. On light or medium loam soils, soil sterilization is usually with Dazomet (approved product Basamid), before sowing, but treatment with fungicides after damage appears is unlikely to prevent further development of the disease.

Botrytis cinerea on foliage, recognizable by a typical 'grey mould' that develops on killed tissues of conifer seedlings, occurs usually in late summer, autumn and early winter. Spraying at about ten-day intervals with an approved product based on thiram, captan or benomyl (under 'long-term off-label arrangements') gives protective control, and should be a routine measure with seedbeds of *Sequoia, Sequoiadendron, Cryptomeria* and *Cupressus* species, as these are very susceptible. Other conifers, particularly Japanese larch, Sitka spruce, western hemlock, Douglas fir and lodgepole pine, may be attacked under dense, humid seedbed conditions, when the fungus colonizes shaded needles and then infects the lower stems, causing death of entire plants. The disease may also become established on parts damaged by frost. Dense stocking of seedbeds must be avoided, and spraying should be instituted at the first sign of infection.

Leaf cast of larch, caused by the fungus *Meria laricis,* infects young needles, causing them to turn yellow and wither. European and hybrid larch are susceptible, and Japanese resistant. Infection proceeds up the shoot, the top of which often remains green. The disease rarely kills plants, but it may seriously weaken them and greatly increase the proportion of culls. The fungus persists in needles which have fallen from or have been killed and remain attached to the infected plants. These are the main source of new infections, so to lessen the risk of an outbreak in the second year the plants are best distanced from the fallen needles by transplanting and also sprayed with a suitable fungicide. The risk of the disease breaking out is also reduced if young plants are raised well away from older larch trees as these are probably a source of the fungus. In susceptible nurseries, spraying with colloidal sulphur or zineb at the time of flushing and at intervals of three or four weeks until the end of July gives good control.

Needle cast of pine *Lophodermium seditiosum* is severely damaging. Current needles become infected in late summer or in autumn. The needles turn brown and die and fall during winter, and by spring defoliation can be complete (i.e. 100% loss). If transplanted, the stress of the defoliation on top of the stress of the transplanting (especially outplanting) can result in the death of plants. Severe defoliation on its own may kill first year seedlings. Older plants usually survive if left undisturbed or carefully transplanted. Infection in the following year, from fallen needles, can be guarded against by applying an appropriate approved fungicide to the new needles during summer and autumn. Older pine trees usually harbour the fungus so are best removed where close to pine nurseries, or it may be necessary to routine spray the young plants.

Keithia disease in western red cedar is caused by the fungus *Didymascella (Keithia) thujina.* Infection causes browning of individual scattered leaflets and where infection is severe, death of whole shoots. The fructifications – small, round or oval structures covered by a hinged lid produced on leaflets – are slightly swollen and olive-brown when mature, but when moribund appear as blackened cavities in dead leaflets. The disease can cause heavy losses in nursery stock between two and five years old. Infection is rarely seen on first-year seedlings, and after about five years plants acquire a considerable degree of resistance. The recommended control is by the fungicide Octave (active ingredient prochloraz) under 'long-term off-label arrangements'. Depending on the prevailing weather, infection can occur from late April until November, so the fungicide should be applied according to the instructions on the label at fortnightly intervals during the period.

Oak mildew is caused by the fungus *Microsphaera alphitoides,* which grows mainly on the outside of succulent leaves and shoots, covering them with a white bloom, leading to distortion, poor growth and dieback of Lammas shoots rather than complete death. The fungus overwinters between bud scales, and the emerging shoots are the first affected in spring. Later, the disease is spread by wind-dispersed spores. Spraying with an approved product based on dinocap or benomyl at the start of flushing, and later at intervals of two or three weeks if secondary infections appear, gives good control. However, benomyl alone should not be relied upon as the fungus can become resistant.

Phytophthora root rot kills plant roots but does not decay them. It is caused by various soil-inhabiting species of the genus *Phytophthora,* and can lead to heavy losses among various species, including Lawson cypress, Douglas fir, sweet chestnut and beech. The rot is often introduced in badly-made farmyard manure. Roots are infected by spores that are motile in water, and the development of the disease is dependent on wet soil conditions in summer. Regular treatment with various fungicides is claimed to prevent the disease developing in nursery plants. However, these do not eradicate the fungus, and there would seem to be a risk that the disease could develop once the plants are sold and treatment ceases. The fungus is readily transported in soil and on infected plants, which may be symptomless. Care should therefore be taken to avoid purchasing plants from infected nurseries. *Phytophthora* may also be harboured in streams, ponds and dirty water tanks and spread to plants during irrigation. Information on obtaining clean water supplies can be obtained from the British Effluent and Water Association, ADAS or DAFS. There should be no risk if mains or artesian water is used so long as storage tanks are kept clean and covered.

Nursery pesticides. The Forestry Commission have published a series of eight guidelines and two checklists for the use of pesticides. *Provisional code of practice for the use of pesticides in forestry* (Occasional Paper 21, 1989) conforms with the Control of Pesticide Regulations 1986. Pesticide includes herbicide, fungicide and insecticide. The guidelines cover general principles, product approvals, competence of users, operator protection, storage and handling, equipment and reduced volume application. The checklists give salient points on the decision to use pesticides and daily operation. In effect, the Regulations disallow the use of pesticides for any purpose or crop unless specifically approved for the same, but for the time being 'long-term off-label arrangements' relax this situation. It is the specific product that is approved, not the ingredient. Before any pesticide is used the label must be read. It carries full instructions for use and for the protection of the operator and the environment. Further advice on control of diseases in nurseries is obtainable from the Forestry Commission at Alice Holt Lodge.

Nursery insect pests. The most important nursery insect pests are soil-inhabiting insects and sap-suckers. Leaf-eating insects are not usually troublesome in the nursery, but occasionally some moth and sawfly caterpillars and species of leaf beetle damage broadleaves (Hibberd (ed.) 1991; Bevan 1987). These pests can be controlled with insecticides applied at the rates recommended in normal horticultural practice.

Cutworms are the caterpillars of various species of noctuid moths which remain in the soil during daytime and emerge at night to feed upon the seedlings, gnawing at the root collar region and usually resulting in the young plants being severed at about soil level. The caterpillars are dirty grey-green in colour and measure about 25 mm in length. Advice on their control can be obtained from ADAS.

Chafer grubs are white, curved and wrinkled and measure up to 40 mm in length when full grown. They are the larvae of various species of the scarabid beetles of which the best known is the large May bug *Melolontha melolontha*. A smaller species, *Serica brunnea,* is common in the north. The grubs live in the soil for from one to four years and during this period feed on the roots of seedlings and transplants. The roots are either stripped of bark or chewed through. The first obvious symptom of attack is browning of the foliage, usually followed by the death of the plant. Control is with an approved product Lindane (based on Gamma HCH) either worked into the top 5 cm of the soil between the rows or incorporated during the last cultivation before sowing or planting.

Springtail *Bourletiella hortensis,* a soil dweller, attacks germinating and emerging seedlings, eating growing buds and causing multiple shoots. The tiny jumping wingless insects, when found, can be controlled by spraying with an approved product based on malathion (under 'long-term off-label arrangements').

Sap-sucking insects – aphids and adelgids – are fairly common in nurseries, and their attacks may check and stunt the growth of plants. Adelgids are restricted to coniferous trees and their presence can be detected by patches of white wool which they produce to cover themselves. The most common and sometimes damaging are *Adelges cooleyi* on Douglas fir, *A. laricis* on larch and *Pineus pini* on Scots pine. *Cinara pilicornis* feeds on the new shoots of spruce. *Phyllaphis fagi* on beech and *Myzus cerasi* on cherry feed on the leaves causing leaf curl and stunted growth. Most of the aphid species are controlled by the use of an approved product based on malathion or Lindane (Gamma HCH) under 'long-term off-label arrangements'. With adelgid species, which protect themselves under wool and have spring and summer egg stages, some difficulty may be experienced in reducing damage to an acceptable level. Careful timing of treatment and a suitable prevailing temperature are critical factors in adelgid control; a warm period in late autumn, winter or early spring provides optimum conditions for spraying. The Conifer spinning mite *Oligonychus ununguis* can be a serious pest of young conifers, particularly the spruces in rather dry growing conditions. This tiny relation of the spiders suck the sap from the needles causing them to turn a dirty brown colour and leaving them netted with fine silk. Control can be obtained through the use of Kelthane (dicofol) under 'long-term off-label arrangements'. Small weevils such as *Otiorrhynchus, Phyllobius,* and *Barypithes* occasionally cause damage in the nursery by feeding upon bark and leaves. These insects can be controlled by an approved product based on Gamma HCH, usually Lindane. Dipping or spraying against *Hylobius* and *Hylastes* spp. is referred to in chapter 5.

Advice on control of insect pests in nurseries is obtainable from the Forestry Commission at Alice Holt Lodge. Information on pesticides and their use can be obtained from the Pesticides Information Division, while ADAS also give advice on relevant husbandry.

Cold storage. Cold stores offer nurseries and foresters great flexibility (Tabbush 1988). In nurseries, seedlings for lining-out can be lifted and placed in cold store, particularly to avoid later inclement weather conditions, and be withdrawn as and when required. For foresters, planting stock can be safely held far beyond the end of the normal planting season. Once plants have become fully 'storable' in late autumn they may be lifted to suit nursery soil conditions, and the availability of labour, and put into store unconstrained by weather conditions or programme considerations in the woodland. Plants can be held in store where they will keep a high root growth potential (RGP) and inactive buds well beyond the time in spring when freshly lifted stock has active buds and RGP has decreased. Thus trees from a cold store with a high survival potential can be planted in late spring or early summer when woodland soil conditions are favourable. However, serious damage can be inflicted if plants are put into store before they become inactive in autumn, or after they become active in spring.

Most cold stores are directly refrigerated, i.e. the cooling unit and fan are within the storage chamber. This has a desiccating effect on the atmosphere inside the store, and plants must therefore be held inside sealed waterproof containers. By contrast, in a humidified store, air is blown across cold water and the resultant cold, moist air is ducted to all parts of the storage chamber. The temperature cannot be reduced to 0°C or below. In directly refrigerated stores the temperature should be selected within the range +2°C to − 2°C and maintained within 1°C of the selected figure. It should not be allowed to fluctuate above and below 0°C. In humidified stores the temperature should not exceed +2°C, and the relative humidity should be maintained above 95% (Tabbush 1988). Guidance on the monitoring and maintenance of temperature and humidity for overwinter storage of planting stock in directly, humidified and indirectly refrigerated cold stores is provided by Tabbush and Gregory (1987).[1]

The nursery production difficulties and hazards mentioned will not all be met in one season or year; some may never be encountered. The provision of an estate nursery can be both challenging and interesting as well as a profitable adjunct to woodland management. Careful attention to the site, seed, cultivation and protection are the chief guarantees of high and good quality yields. Profitability demands skill and ability in the supervision, experience in the workers, attention to costings, planned progression of the work, adequate stocking and full use of the ground, kindness of Mother Nature and the weather, and finally, sound marketing of any surplus. The planned revision of Aldhous (1972) – the main general guide to forest tree nursery practice for two decades – will provide up-to-date information enabling a decision to be made regarding the enterprise of producing planting stock. Meanwhile, Hibberd (ed.) (1991) – chapter 3 by Williamson and Mason – provides useful guidance.

2. Tree seed

The production of trees with good vigour, health, stem-form and crown-habit depends upon two interacting factors; the genetic constitution of the seed and the environment in which the plants raised from it is planted. Accordingly it is necessary to ensure that the seed used is of satisfactory genetic quality and that the plants are planted in the environment to which they are best suited. To raise high quality plantations, foresters require healthy seeds capable of producing plants which have an ability to grow well on the chosen forest sites when given sound silvicultural treatment. To produce high quality wood for the market intended, it is essential that trees have good stem-form and crown-habit (defined by the Forestry Commission as 'straight unforked circular stems with regularly spaced small-sized branches coupled with good natural pruning, horizontal branching and freedom from spiral grain or fluting'), and good health with resistance to diseases and insects. Foresters can meet these requirements by carefully selecting their seed sources. Useful guidance on seed matters is given in Hibberd (ed.) (1991) – chapter 2 by Gosling.

Seed of the thirteen 'EC species' referred to in the Forest Reproductive Material Regulations (noted *infra*) must be sold only with a current test certificate, issued by an official seed-testing station. The three official stations in this country are the

---

1   As a note of caution regarding cold storage of plants, whilst QCI origin Sitka spruce can be stored safely (i.e. without significant deterioration) for several months if lifted at the optimum time, this is not true for all species. Douglas fir and larches can deteriorate rapidly in cold store and foresters contemplating cold storage of plants for more than the few days between lifting and dispatch are advised to seek advice from the Forestry Commission Research Division at the Northern Research Division, Roslin in Midlothian.

Ministry of Agriculture seed-testing station at Cambridge, the Department of Agriculture seed-testing station at East Craigs, near Edinburgh, and the Forestry Commission at Alice Holt Lodge. Certificates issued by other official testing stations in the EC are also valid. However, particularly for home-collected seed, the bulk of forest tree seed of the thirteen 'EC species', both private and state, is tested at Alice Holt Lodge. For 'non-EC species' there is no statutory requirement for testing although reputable seed merchants normally provide details of a recent test result with any seed supplied.

Seed sources. The importance of origin/provenance of seed is noted in chapter 4 section 2. These are comprehensively dealt with by Lines (1987); and Fletcher (1990) gives information on the seed sources available from the Forestry Commission during 1990–91. Samuel (1990) provides an explanation of the Commission's seed identity number scheme and of the more important numbers for countries/ states from which seed might be imported. A comprehensive account is given in Gordon (ed.) (1991). Foresters will secure such seed that will produce plants suited to their woodland sites and possibly for the sites of local purchasers. Seed is obtainable from home-collection, from the Forestry Commission[2] or from Forestart.[3] Indications of current (1991) prices of seed are:

	Range £/kg according to origin/ provenance, quality and quantity
Conifer seed	
Scots pine	120–400
Corsican pine	85–95
Lodgepole pine	250–275
Norway spruce	90–160
Sitka spruce	80–135
Douglas fir	85–145
European larch	250–260
Japanese larch	140–165
Hybrid larch	300–700
Western hemlock	195–220
Western red cedar	80–120
Grand fir	60–150
Noble fir	50–100
Lawson cypress	50–60
Broadleaved seed	1–2
Oak, pedunculate	2–3
sessile	1–2
red	10–20
Beech	50–200
Alder	7–8
Ash	1–2
Sweet chestnut	11–20
Wild cherry (Gean)	7–8
Sycamore	

---

2 Forestry Commission (Seed Section), Alice Holt Lodge, Wrecclesham, Farnham, Surrey GU10 4LH (Tel. 0420 22255).
3 Forestart, Church Farm, Hadnall, Shrewsbury, Shropshire SY4 4AQ (Tel. 09397 638).

Home-collected seed. The genetic quality of seed is dependent upon both the mother tree and the source of the pollen. For this reason, home-collected seed must be taken from the very best *seed stands* in the region where it is intended to plant. The stands should be well separated from low quality stands of the same species, where cross-pollination may occur. The stands must be uniformly good and any trees of poor form, or suffering from disease or some other weakness, should not be used as seed parents. Some young conifer stands of particularly good quality are selected and managed purely for seed production. Early and heavy thinnings are aimed at retaining as much live-crown, and therefore as much potential seed-bearing surface area, as possible on the remaining trees: only the best trees are kept. All this assumes that phenotypically good stands are genetically superior, which they may not be. Origin is more important than phenotype.

The *Forest Reproductive Material Regulations* made under EC legislation, require the Forestry Commission to maintain a National Register of seed stands, seed orchards and poplar stool beds which have been inspected and found to meet minimum quality standards. The National Register covers thirteen species and one genus *(Populus)* all common in European forestry: European silver fir, European larch, Japanese larch, Norway spruce, Sitka spruce, Austrian and Corsican pine, Weymouth pine, Douglas fir, beech, Sessile oak, Pedunculate oak and Red oak. Other commonly used forest species are on a voluntary basis – currently including ash, sycamore, lodgepole pine, western red cedar and a small number of minor species. As far as seed originating from sources in Great Britain is concerned (except for seed marketed under 'Derogation') only that coming from stands or orchards included in the National Register may be marketed in the United Kingdom and other EC countries. It must have been tested at the Forestry Commission Official Testing Station at Alice Holt Lodge. An owner wishing to have a stand of one of the listed species or a poplar stool bed considered for the National Register should contact the nearest Conservator of Forests. An inspection fee will be charged by the Commission whether or not the stand meets the criteria prescribed by law. For species not covered by EC regulations, seed from recommended sources is obviously to be preferred to seed from doubtful sources. There are desirable and less desirable origins, and higher prices can be expected for those in the first category.

Seed, dependent on species, is a comparatively small proportion of the total cost of producing a plant; and seed cost is commonly less than 0.1% of the total sum needed for establishing a plantation. Thus the use of seed from sources which give even small increases in growth-rate, or improved timber quality, is a good investment. Persons marketing forest tree seed (likewise, plants) of the thirteen EC species are required to furnish the buyer with a Supplier's Certificate. An explanatory booklet *The marketing of Forest Tree Seed and Plants within the Economic Community* is available from the Forestry Commission. For species not covered by the regulations, seed from recommended sources is still to be preferred to seed from 'unknown' sources but only the normal consumer legislation protects the purchaser.

Pests and diseases can be carried in plants and seeds. If a nursery intends to export such material it must comply with the importing country's plant health regulations. Guides for exporters are given in Forestry Commission 'Plant Health Leaflet No. 6' (1990). Guides for importers are noted in chapter 6.

Seed orchards. The Forestry Commission has established 'clonal' and 'seedling' seed orchards of some species (Faulkner 1975) as discussed in section 3 *infra*. They are composed of individuals selected from parents with highly desirable attributes which by testing have been shown to pass these attributes on to their offspring. They are grown either by grafting or by raising seed derived from controlled parental crosses. They are well isolated to minimize contaminant pollination from other sources, and are sited and especially managed to produce frequent, abundant, easily collected seed. Such seed is preferred to that which is collected from a registered seed stand.

Imported seed. When only imported seed is available, it is most important to obtain it from origins known to grow well in the region in which the plants are to be planted. Seed from unsuitable sources must obviously be avoided. Within the EC, the parent sources of imported seed of the listed species must have been approved by a recognized authority within the country of origin. Comprehensive information on origins/provenances has been made available by Lines (1987) as discussed in chapter 4, section 2. Detailed advice on suitability can be obtained from the Forestry Commission.

Collection of seed. Foresters wishing to collect their own seed can find relevant information in various publications of the Forestry Commission. Seed collection necessitates a knowledge of the ages at which trees start to produce sound seed, the times and occurrence of masts and the times of ripening of cones (Matthews 1963). The frequency of full seeding in the common species is from two to eleven years. Times for the collection of conifers and broadleaves are given by Aldhous (1972) and by Gordon (ed.) (1991). Collecting seed off the ground is sometimes inadvisable and collecting cones from standing trees can be dangerous for workers without the necessary skill. Seed of broadleaves should be collected during the period between embryo maturation and seed dispersal. For species of many genera, this period is short (e.g. poplar and willow), but for others the seeds may remain on the tree for months (e.g. ash). Alternatively, the seeds may lie beneath the tree for some time (e.g. oak). A table of information on seed production for the more widely used forest species is provided by Hibberd (ed.) (1991) and Gordon (ed.) (1991), indicating for each species the ages of first good seed crop and of maximum production, average interval between good seed years, and recommended time for seed collection.

Heavy seed crops often follow a year with a prolonged hot, dry, early summer, and collection costs are reduced in heavy crop years. It is common practice to collect several years' supply of seed for those species the seed of which can be satisfactorily stored for long periods. Large seeds such as acorns, beech-nuts and chestnuts, are collected when fallen, and to facilitate this the ground must be cleared of vegetation and debris beforehand. Ripe cones are collected using short ladders or, in mature seed stands, from felled trees to provide a mixture of genetically diverse material. When trees are climbed (and this is becoming rare) the climbers must be fully trained for the task and provided with approved and tested safety equipment. Any climbing equipment should be covered by British Standards Specification.

Extraction and storage of seed. Collections should only be made in dry weather since moist cones and fruit frequently heat up or go mouldy during temporary storage. Bags of cones or fruits must be kept well ventilated. The bags, which must be porous and not impervious, should be stored where they are protected from the rain, and any damp cones should be spread out on the floor of a well-ventilated shed and turned until dry. Most conifer species of seed are storable, but in broadleaved species only oak, sweet chestnut and sycamore. There is a tendency in some nurseries to sow seeds of oak, beech, sycamore, maple and sweet chestnut immediately after collection. This avoids the cost and hazards of storage. The facilities of the Forestry Commission's Alice Holt Lodge cold store are available for conifer seed at a charge of £2/kg per annum. Official tests of seed quality are made there on payment of fees ranging from £49-£72 each full test, £15 for 'advisory' tests and £7.50 for a test of moisture content.

Conifers. The ease of extracting seeds from cones varies according to species. Cones of noble fir and grand fir break up soon after collection. Western hemlock, western red cedar and Lawson cypress cones can be opened by spreading them in shallow boxes or trays in a cool well-ventilated building. Cones of most other species are normally dried in a special kiln (like a tumble-drier) in which the temperature of the air-stream is progressively raised as drying proceeds. After drying, the seeds are separated by shaking through coarse sieves or by rotating them in a wire-mesh drum, and then collected in a tray. Seeds with detachable wings are de-winged by gentle rubbing between cloth sheets or in a machine especially designed for the purpose. Pines and spruces are most easily de-winged if moistened. Wings, cone scales, empty seed and other impurities are separated from the seed by winnowing. The seed is then ready for testing for purity percentage, germination percentage and moisture content, and for storage. Seed can be satisfactorily stored in this condition until the following spring, provided it is kept cool in sealed containers. However, seed to be stored beyond six months should be dried in warm dry air (radiant heat should not be used) until the moisture content is between 6 and 8% (8 and 10% for true firs). The dried seed is placed in airtight containers and stored in a cold store. For long-term storage of seed (i.e. for more than three years) and for even short-term storage of some species (e.g. Douglas fir and noble fir) storage should be at $-10°C$. Short-term storage of other species (e.g. Sitka spruce and Scots pine) should be at $+2°C$ for three to five years, provided the moisture content is suitably adjusted.

Broadleaves. Broadleaved seeds and fruit vary in size and shape, ranging from those in husks, such as chestnut or beech, or berries such as wild cherry, to those which are dry and may or may not be winged. In general, seed should be separated from its husk if dry, or if a berry from its soft coat. Broadleave seed like birch with small wings has to be sown with the wings remaining on the seed coat. Pulpy fruits should first be macerated by squashing or gentle mashing and mixing with water; pulp and skins can usually be separated from the seed by washing through sieves to remove unwanted heavy material and by flotation in water to remove unwanted light material.

Seed pretreatment. Mature seeds of most woody plant species will either not germinate at all, or at least not germinate promptly when placed under conditions which are normally regarded as suitable for germination. Such seeds are said to be dormant. They can be induced to germinate within a reasonable time if certain predetermined conditions are satisfied – chiefly pre-chilling under controlled conditions or stratification in a stratification pit (for broadleaves see Gordon and Rowe 1982). Pre-chilling is usually recommended for Douglas fir, grand fir, noble fir, lodgepole pine, ash and sycamore.

Moist pre-chilling is being recommended for many seedlots, often because the treatment enhances the seeds' ability to germinate over a range of conditions, rather than just the optimum. Putting seed into pretreatment in these circumstances is equivalent to sowing, so that the nominal date of sowing for dry untreated seed should be deferred by up to three weeks for pre-treated seed. The only draw-back is that pre-treated seed must not be allowed to dry out after sowing; if drought follows sowing, watering will be essential to avoid either total loss of seed, or reversion to a dormant state. The main distinction between pre-chilling and stratification is that the former is done in a cold store, and the latter outdoors in a stratification pit. Little conifer seed is now stratified; most is pre-chilled. Stratification is most widely practised for ash and lime which require more than the three to six weeks pretreatment required by most conifers.

Insects attacking seed. Many insects may live in the developing seeds, and cones of forest trees and their attacks can sometimes result in appreciable losses (Bevan 1987). The Douglas fir seed wasp *Megastigmus spermotrophus*, whose larvae hollow out the seeds, is particularly damaging. As infestations by this insect are sometimes very heavy and can cause near total loss of the seed crop, it is advisable to make a pilot assessment of the seed to determine its soundness before large scale cone-collecting is undertaken. Other species of *Megastigmus* infest noble and grand fir, larch and Norway spruce seeds. The caterpillars of a number of moth species such as *Dioryctria abietella* and *Cydia* spp., the larvae of the pine cone weevil *Pissodes validirostris,* and maggots of some dipterous (two-winged) flies, feed upon and destroy the seeds of various conifers. Their attacks, however, are not often serious. The Knopper gall wasp *Andricus quercuscalicis* can destroy large numbers of acorns by transforming them into gross galls. The grubs of the weevils of the genus *Curculio* attack and hollow out acorns, while beech-nuts are similarly infested by the caterpillars of the moth *Cydia fagiglandana*. Again the attacks are not usually of a serious nature, but they may on some occasions affect the success of natural regeneration schemes or the economics of seed collection. Other types of insect infestation can produce indirect effect on seed production; for example, the defoliation of oak by the oak leaf roller moth *Tortrix viridana,* or of the winter moth *Operophtera brumata,* may result in a marked reduction in acorn yield.

3. Genetics in silviculture: tree improvement by selection and breeding

'The large size and long life cycle of most forest trees have made it difficult to unravel the effects of genotype and environment on the phenotype – the tree in the forest – but since 1950 provenance research and tree breeding have advanced sufficiently to place forest genetics alongside tree physiology and forest ecology as one of the foundations of silviculture' (Matthews 1989, p. 190). The Forestry Commission was one of the first national forest services to try to increase the economic value of its planting stock through selection and breeding. There are many current developments, particularly in the fields of flowering induction, vegetative propagation, tissue culture techniques, and early progeny and clone testing procedures. Increasing quantities of seed are becoming available from the Commission's seed orchards (Faulkner 1975), and for Sitka spruce from controlled pollinations between genetically superior females and pollen mixtures ('polymixes') from proven superior males to provide seed termed 'Bulked Family Mixtures'. Ultimately seed will be obtained from fullsib crosses (single pair-matings) of high specific-combining-ability (SCA), i.e. families having higher than expected breeding values. There have also been developments in the production of lodgepole pine seed from special seed plantations based on seedlings derived from superior parents of known origin, or of F1 (first generation) hybrids between parents of different origins, e.g. coast BC x Skeena River BC.

While tree breeding is a relatively young discipline and few tree improvement programmes have progressed beyond their first generation of selection, the private forestry sector will benefit from knowledge of the developments, particularly as improved quality seed is beginning to become available. The gains from selecting the correct origin/provenance are substantial and easily won if available advice is taken (Lines 1987). This will be emphasized in chapter 4, section 2, when discussing 'Species choice'. Once the best origin is known, further improvements can be made by the tree breeder, as explained hereafter.

The object of a tree breeding programme is to increase the quality, adaptability, yield and therefore economic return from a forest system by exploiting the genetic variation that exists within it (Lee 1986). Tree breeding programmes start by selecting superior trees within the best adapted origins (see Figure 4). Further improvement within an origin involves selecting individuals of the highest genetic quality on the basis of progeny tests, and bringing them together to intermate. The seed produced should be of a better quality than that available before selection for whatever characteristics (syn. traits), for example, volume, form and height, it was selected. The amount of improvement depends on how intensively the trees were selected and the heritability of the trait. In the next generation the best individuals from the best families are again selected, allowed to intermate and again generate seed even better than that from the previous generation.

The theories and practices as they relate to tree breeding in Britain are outlined by Lee (1986). Some terms frequently used by breeders are explained hereafter. Phenotype (P) describes the physical appearance of or a given value of a trait; this value is an expression of the combined genetic and environmental influence upon it. Genotype (G) is the genetic constitution of an individual tree. The job of the breeder is to determine the good genotypes (trees of good genetic quality) from those phenotypes which are good merely because of the good micro-environment in which they happen to be, since phenotype = genotype + environment or P = G + E. Within a population this is expressed as: phenotype variation = genotype variation + environmental variation. Generic variation is in turn made up of additive genetic variation and non-additive genetic variation. Additive genetic variation can be 'added' to by selecting and breeding in each successive generation. Non-additive genetic variation is peculiar to that generation, e.g. two clones may combine to yield a particularly good cross that could not be predicted based on the value of the individual clones. Non-additive genetic variation cannot be passed on to the next generation although it will manifest itself again in new, unpredictable and often beneficial ways.

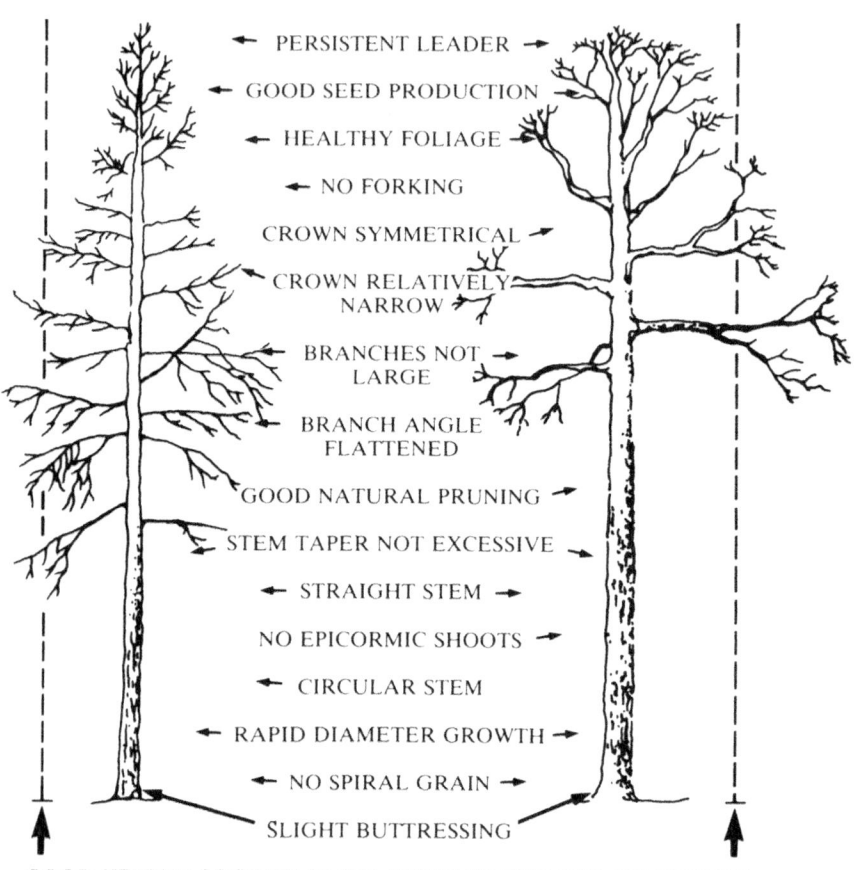

Figure 3. Choice of seed trees
Source: A.W. Coram and J.D. Matthews.

Progress in the Forestry Commission's breeding programme is controlled by how quickly reliable estimates of future growth can be made in progeny tests and, more importantly, how early in life, how frequently and how profusely potential breeding trees can be made to bear flowers. Breeding strategies have long been established for five species or hybrids of conifers: Sitka spruce, lodgepole pine, Scots pine, Corsican pine and hybrid larch. In 1989 strategies were planned for oak, ash and sycamore among broadleaved species. Methods of selecting superior oak have been described by Harmer (1989).

The strategy of the Commission's tree improvement programme is basically the same for each species. The first steps are usually the selection of 'Seed stands' and 'Plus' trees. Seed stands are those of high phenotypic value – they have shown themselves to be well adapted to a particular sort of site and usually have been thinned to leave the best phenotypes to intermate. The average genetic quality of the seed from such stands is assumed to be better than unselected stock. Since the genetic value of the genotypes is unknown at the time of selection, it is assumed that a good phenotype normally reflects a good genotype (although it is by no means always the case). Some stem and branching characteristics are clearly controlled genetically when selected without the influence of competition. Seed stands are interim sources of seed; they are generally phased out as seed of better genetic quality becomes available later in the programme. However, if a full breeding programme cannot be justified for a species, seed stands remain a cheap way of slightly raising the quality and adaptability of the planting stock. 'Plus' trees are the best individuals (best phenotypes) from the best stands on a variety of site types throughout the country; they combine high vigour, form, branch quality and freedom from disease or insect pests. Many of the early years of the Commission's tree breeding team were spent carrying out this intensive 'Plus' tree selection process. In the 1960s progeny testing began on a large scale in order to determine the genetic quality of the 'Plus' tree when combined with pollen from other trees in general. Trees of the very best phenotype will be selected from within such tests and used for future breeding and in second generation seed orchards. Around 140 ha of seed orchards have been established of which about one-quarter are of Sitka spruce. About 10 ha of new orchard planting is achieved each year.

Initially the Forestry Commission's Genetics Branch (now the Tree Improvement Branch) placed most effort on Scots pine so that much of the Scots pine seed used in Britain now comes from orchards. With lodgepole pine it has proved impossible to find a single origin that combines acceptable vigour and stem qualities and origin hybrids are being created to combine the two. With larch the objective is to produce F1 (first generation) hybrids between European and Japanese larch, producing hybrid vigour or 'heterosis' and having straight stems, relatively fine branches and resistance to canker.

A tree breeding programme must produce seed of increased economic value to justify its existence. So often, however, twenty or more years elapse between selecting 'Plus' trees and obtaining improved seed from a tested orchard. However, seed from Sitka spruce orchards based on tested clones is now available from the Commission in annually increasing quantity (Faulkner 1987) although still less than 5% of UK normal needs for Sitka spruce for open nursery production. The increased genetic gain relative to unimproved imported Queen Charlotte Islands (QCI) material should be around 15% for final rotation volume; equivalent to one full Yield Class (Lee 1990, 1991). When planted on a site of Yield Class 14 the financial gain (discounted to year 0 at 5%) is likely to be in the region of £410/ha.

Three areas have the greatest potential for future advances in tree breeding: exploitation of the non-additive genetic variation, shortening of generation times, and use of vegetative propagation – all of which overlap to a certain extent. The lengthy period before the full genetic potential of a tested clone can be confirmed is frustrating; so too is waiting for a selected tree to flower so that a controlled cross can be made and the programme proceed. However, some trees may be made to flower at a younger age by using treatments such as stem girdling, root pruning, or by injecting minute amounts of a mixture of gibberellins 4 and 7 (naturally occurring chemical compounds found in many plants) although such trees cannot be used for selection work on growth characteristics.

Vegetative propagation is used for exploiting non-additive genetic variance and also for shortening the time taken to realize genetic gain. The potential exists to double the gain achieved to date. Only through vegetative propagation can the *full* genetic value, additive and non-additive, be passed on to the next generation. Currently, vegetative propagation (in addition to hybrid larch, noted in section 1 *supra*) is only commercially feasible for Sitka spruce and even so only when used on one-to-six year old seedlings since 'rootability' rapidly declines with increasing age. The opportunity of getting genetic improvement more quickly into commerce than via the orchard route is being exploited by collecting pollen from a number of different 'tested' and proven superior genotypes and applying this pollen mixture to female flowers on other 'tested' and proven superior mother trees represented in clone banks of grafted trees. Seed is collected later the same year and the small amounts are then bulked-up by growing large stock plants under greenhouse conditions and multiplying these vegetatively by rooted cuttings ('Bulked Family Mixtures'). Over two cycles, up to 500 healthy rooted cuttings per seed can be obtained by this method. Seed showing 15% improvement in final rotation volume is now available to the public and private sector. Any extra cost of the seed or planting stock relative to improved material is more than justified in terms of increased growth-rate, a shorter rotation length, form, and proportion of final crop making sawlogs.

If methods can be achieved to 'rejuvenate' trees older than six years then rooted cuttings could be produced from trees which have proved themselves well adapted to particular situations, for example, resistant to insect pests and diseases, 'acid rain', or capable of growing in nutrient deficient soils, and would form the basis of clonal forestry – forests in which foresters could decide how many different genotypes they should plant. Clonal forestry has been practised in Japan for many centuries with Japanese red cedar and in many European countries with clones of *Populus* spp. Great genetic gains are attained when only the very best clones are used but at a high risk if the number of clones used is small and their adaptability narrow. If a single clone or narrow range of clones is used for forestry, any unpredictable pathogen attack could be much more devastating than on trees of orchard origin. However, if the crop remains healthy, just a few of the very best clones could give the highest returns; for obvious reasons the risks are incalculable. Since breeding will always be needed in order to recombine genes in successive generations, only through these recombinations and further reselection will further genetic advances be possible. Furthermore, as trees have greater longevity than agricultural crops, the planting stock will always require a broader genetic base in order satisfactorily to grow on a diverse range of sites and under the varied climatic conditions experienced during the rotation; hence the practice of using as breeding parents only those trees from families which have shown superior growth on a range of site types. The number of genetic combinations and therefore the number of genotypes produced by cross-pollination in a seed orchard is vast and the risk of catastrophic losses from harmful agencies minimal (Faulkner 1990).

References

Aldhous, J.R. (1989), 'Standards for Assessing Plants for Forestry in the United Kingdom'. *Forestry Supplement 62*, pp. 13–19.
Aldhous, J. and Mason, B. (1994), 'Forest Nursery Practice'. FC Bulletin 111.

Bevan, D. (1987), 'Forest Insects'. FC Handbook 1.
Faulkner, R. (1975), 'Seed Orchards'. FC Bulletin 54.
—— (1987), 'Genetics and Breeding of Sitka Spruce'. Proceedings of the Royal Society of Edinburgh, 93B, pp. 41–50.
—— (1990), personal communication.
Fletcher, A.M. (1990), 'Seed Origins and Provenances'. FC seed catalogue.
Gordon, A.G. (ed.) (1991), 'Seed Manual for Commercial Forestry Species'. FC Bulletin 83. In preparation.
Gordon, A.G. and Rowe, D.C.F. (1982), 'Seed Manual for Ornamental Trees and Shrubs'. FC Bulletin 59.
Gosling, P. (1990), 'Notes on Seed and Seed Handling'. FC seed catalogue.
Harmer, R. (1989), 'Selection of Superior Oak'. FC Research Information Note 149.
Hibberd, B.G. (ed.) (1991), 'Forestry Practice'. FC Handbook 6.
Hollingsworth, M.K. and Mason, W.L. (1989), 'Provisional Regimes for Growing Containerized Douglas Fir and Sitka Spruce'. FC Research Information Note 141.
—— (1991), 'Vegetative Propagation of Aspen'. FC Research Information Note 200.
Jobling, J. (1990), 'Poplars for Wood Production and Amenity'. FC Bulletin 92.
Lee, S.J. (1986), 'Tree Breeding in Britain'. *Forestry and British Timber.* June, July, August 1986.
—— (1990), 'Potential Gains from Genetically-improved Sitka Spruce'. FC Research Information Note 190.
—— (1991), 'Likely Increases in Volume and Profitability from Planting Genetically Improved Sitka Spruce'. FC Bulletin on proceedings of 'Super Sitka for the 1990s'. (In press 1991).
Lines, R. (1987), 'Choice of Seed Origins for the Main Forest Species in Britain'. FC Bulletin 66.
Mason, W.L. (1986), 'Precision Sowing and Undercutting of Conifers'. FC Research Information Note 105/86/SILN.
Mason, W.L., Biggin, P. and McCavish, W.J. (1989), 'Early Forest Performance of Sitka Spruce Planting Stock raised from Cuttings'. FC Research Information Note 143.
Mason, W.L. and Jinks, R.L. (1990), 'The Use of Containers as a Method for raising Tree Seedlings'. FC Research Information Note 179.
—— (1990), 'Notes on Nursery Practice'. FC seed catalogue.
Mason, W.L. and Hollingworth, M.K. (1989), 'Use of Containerized Conifer Seedlings in Upland Forestry'. FC Research Information Note 142.
Matthews, J.D. (1963), 'Factors affecting the Production of Seed by Forest Trees'. *Forestry Abstracts, 24* (1), pp. i–xiii.
—— (1989), *Silvicultural Systems.* Clarendon Press, Oxford.
Phillips, D.H. and Burdekin, D.A. (1982). *Diseases of Forest and Ornamental Trees.* MacMillan, London.
Samuel, C.J.A. (1990), 'An Explanation of the Forestry Commission Seed Identity System'. FC seed catalogue.
Samuel, C.J.A. and Mason, W.L. (1988), 'Identity and Nomenclature of Vegetatively Propagated Conifers used for Forestry Purposes'. FC Research Information Note 135.
Tabbush, P.M. (1988), 'Silvicultural Principles for Upland Restocking'. FC Bulletin 76.
Tabbush, P.M., and Gregory, S.C. (1987), 'Guidelines for Monitoring and Management of Cold Stores'. FC Research Information Note 115/87/SILN.
White, J.J., and Patch, D. (1990), 'Propagation of Lowland Willows by Winter Cuttings'. FC Arboriculture Research Note 85/90/SILS.
Williamson, D.R., and Mason, W.L. (1989), 'Forest Nursery Herbicides'. FC Occasional Paper 22.

Chapter Three

FACTORS OF THE FOREST SITE

Britain's climate, topography and soil generally provide almost ideal conditions for tree growth. Conifer growth rates are probably over three times those in Sweden and five times those in Finland. After about fifty years, the time when many of our conifers are felled, they are of comparable size, or larger, than old more slowly-grown Scandinavian trees of the same species. Our broadleaved growth rates compare favourably with those in Europe.

Each species (while recognizing that origins/provenances within species differ) has inborn characteristics which have suited it for its natural habitat, i.e. where it is naturally indigenous. Some have only a limited extent of natural habitat, e.g. Monterey pine from a small region in California. Others have a wide range, e.g. Norway spruce and Scots pine throughout most of Scandinavia and northern Europe; and Douglas fir, Sitka spruce and associated conifers extending from California up to Alaska. Occasionally a species grows much quicker in a new habitat, e.g. Monterey pine grows remarkably faster in New Zealand and South Africa. When seed of exotics is imported, the Forestry Commission select origins/provenances that will appropriately suit most parts of Britain (see chapters 2 and 4).

Site factors are important. Those most readily apparent that influence a tree's growth are geology, soils, climate, altitude, exposure and aspect, along with the ground vegetation. Therefore, in silviculture, consideration has to be given to climatic factors especially in relation to temperature (warmth/cold), rainfall (wetness/dryness) and wind (exposure), as well as to soil (including rootable depth, and nutrient and water availability), vegetation competition and terrain. These site factors impose varying degrees of limitation on silviculture and management techniques. Some, such as difficult terrain or a drought-prone soil, are permanent while others such as waterlogging, infertility or incidence of luxuriant weed growth can be overcome to some extent by drainage and the application of fertilizers and herbicides.

Site classifications (in terms which include soil, ground vegetation and terrain; or land capability for forestry) are used for estimating future yields, and for helping to decide species choice and treatments such as drainage and fertilizing. Site productivity is expressed in terms of the maximum mean annual increment, i.e. the basis of the yield class system (discussed in chapter 12). Such an index of growth has limited reference, however, where the attributes of unplanted ground must be evaluated, when assessments must be made on the basis of features with an integrated character such as soil and ground vegetation. Complementary to the foregoing site classifications is that of windthrow hazard (discussed *infra*). Terrain (discussed in chapter 9) takes into account factors such as the steepness of slopes, ground roughness, and the occurrence of boulders. All have considerable practical importance to foresters. Difficult terrain may present limitations to the use of some kinds of machinery, and also affect the output of the labour force. Ground vegetation classifications are noted in section 2 *infra*. Of prime importance are soil classifications, dealt with in section 3 *infra*.

1. Climate

The climate in Britain is favourable to tree growth particularly because of relatively mild winters and adequate rainfall distributed throughout the year. Seasonal frosts and summer droughts may have a severe effect on newly planted trees but usually have less affect after the establishment phase. However, climate is the one factor of the site environment over which foresters have practically no control; of all the physical factors bearing on silviculture it is probably the most pervasive and influential. Climate generally determines the type of woodland which exists or can conceivably exist in a particular region. (The broad pattern in Britain is: broadleaves in the lowlands, conifers in the uplands, pines in the east and north; and spruces in the west and north.) There are relatively large climatic differences from north to south and more particularly from east to west. Very broadly, climate determines the rate of growth of trees and hence to a large extent the relative attractiveness of different sites for forest investment. It also affects the practicability and ease of managing forests, silviculture being a labour-intensive activity. Rainfall – the annual amount and the spread of the precipitation throughout the year – is of great significance. Micro-climate shows wide variations, and its incidence is extremely important to foresters, over which they have modest control.

Topography influences and interacts with climate and is particularly important because forests are often associated with hills and mountains, partly as a result of such areas having become available because they are rarely favourable to intensive agriculture, and not always economically viable for sheep grazing. Increasing altitude reduces the length of the growing season; and the potential responses of tree crops to soil improvement by fertilization or cultivation treatment will often be greater at lower altitudes. However, the greatest effect of altitude relates to exposure (windiness), discussed *infra*. In general as altitude increases yields decrease; in some species this decrease is so large that the species would not produce an economic crop of timber. Exposed outside 'edge trees' in a woodland are often shorter, more heavily branched and produce lower quality timber than trees inside the woodland which benefit from mutual shelter. Our western coasts have high average wind speeds; which are the main limiting factor to tree growth and survival at high altitudes. There is a marked contrast between the vegetation of north- and south-facing aspects at the same elevation. Most trees will grow below 300 m; above this, the choice is more limited, and the economic 'tree line' may be about 450 m though where sheltered by other high areas the 'line' may sometimes be pushed up a further 150 m or so. The effects of topography on the mechanized operations necessary for establishment and harvesting and the design and construction of forest roads are discussed in chapters 5 and 10.

Temperature (warmth/cold; drought/frost). Climatic warmth has an important effect on tree growth. Usually there is an optimum range for each species (for example sweet chestnut needs warmth). Temperature depends on altitude, exposure, longitude and latitude; yet a tree's crown, stem and roots live in entirely different temperatures. Since the temperatures, even in winter, are generally above freezing point, it is possible to cultivate the ground and therefore carry out nursery-work and tree planting from autumn through winter to spring; but this favourable circumstance cannot be entirely relied on, for work is apt to be stopped by a spell of frosty weather that may last for several weeks – and hence there is a need to formulate plans for 'wet weather work'. The growing season is normally considered to start when an average daily temperature of 6°C is reached and maintained, though this varies both between and within species. The growing season varies in duration from nine months or more in southwestern coastal areas, seven or eight months in lowland areas, five or six months in most of the uplands and as little as four months in the highest parts of the Grampians and the west Highlands of Scotland. The period between leaf expansion and leaf fall in deciduous woodlands in lowland England may vary between 190 and 240 days. In very mild areas the critical temperature which initiates flushes of species able to take advantage of a long growing season occurs in early February, while in north Scotland it is not reached until well into April. The effect of increasing elevation is more drastic.

In summer the heat of the sun sometimes scorches unshaded seedlings in nurseries, causing them to collapse and wither. It can also harm young plants, especially where the ground cracks around the collar; and can kill patches of bark on stems of thin-barked species such as beech if they are suddenly exposed to full light after having been shaded. Drought may cause wilting and dieback of trees, especially on sands, gravels and very shallow soils over rock; and may cause cracks in the stems of fast-growing conifer dominants, especially grand fir and Sitka spruce. Prolonged drought will slow the growth of trees of all ages and can be a contributory factor to the death of larger specimens already subject to other stress, e.g. root decay.

Damage can arise from extremely low temperatures in winter, and persistent cold drying winds. Many species become hardy to low temperatures in autumn, and quite low midwinter temperatures may do no harm. Susceptibility to damage rises until the end of May or early June from when it remains more or less constant until October or November, though differences of timing and degree occur between species, provenances and locations.

Seasonal frosts, which depend very much on the local topography, are a prime source of injury to trees. Besides the harm done in nurseries (see chapter 2), and in flowering and seeding, in winter the foliage of evergreens may be damaged when there are sudden changes from mild to freezing conditions, or when long periods of frost with bright sunshine cause repeated and fairly rapid freezing and thawing. Early frosts in autumn may damage shoots that have not hardened off, but the most troublesome damage is that caused by late frosts in the spring, when the tender developing shoots and sometimes the bark and cambium of young trees are particularly liable to injury. Early and late frosts occur mainly on clear still nights, when air in contact with surfaces rendered cold flows down slopes to collect in valleys and hollows or on shelves of hillsides; frost injury is liable to occur in such situations. On level ground, small plants are more likely to be damaged where there is a grass mat or similar cover than where there is bare soil. Heat loss by radiation from the bare soil surface is replaced by heat from the underlying soil, provided it is moist, whereas grass insulates the soil and has no large store of heat to replace that lost from its surfaces by radiation.

On sites where late frosts are likely to be frequent and severe, hardy species should be planted in preference to frost-tender species (listed in chapter 4). Where suitable woody growth is present, however, under- or inter-planting with shade tolerant species can be used to gain shelter from frost; they should grow safely after they get their heads above the normal level of spring frosts. On level ground, ploughing for planting, with its concomitant reduction of surface vegetation and exposure of bare soil, appreciably reduces the severity of late frosts. Sometimes, pockets in which cold air collects can be counteracted by felling trees and scrub on any adjacent lower ground, and sometimes by drainage.

Rainfall (wetness/dryness): precipitation and evaporation. The amount of precipitation, its form, distribution throughout the year and the degree to which moisture is retained in the soil, have a particular influence on the growth of trees. Too much moisture, or too little, is harmful to tree growth. An excess of water, as in boggy ground, severely restricts soil aeration resulting in anoxia and root death (Nisbet et al. 1989). Cultivation such as ploughing and drainage can lower soil water, improving conditions for tree growth in wet soils. Soil compaction, e.g. by machinery or the trampling of man or animals, can impair soil drainage and aeration. The growth of a forest crop will lower the soil water, particularly if the species are poplar, willow or alder. The drying effect of lodgepole pine in peaty soils can result in irreversible cracking of the soil (Pyatt et al. 1987). Clear-felling of a crop is likely to result in a re-wetted site.

The average annual rainfall in the British Isles is indicated in Figure 5. In England, the minimum rainfall is over 500 mm and is fairly evenly distributed throughout the year. Norfolk and Suffolk are areas of low rainfall. Medium rainfall of 700–900 mm occurs in the central and southern counties, except Devon and Cornwall where it is higher. In the west Midlands and the South-West there is a fall of 1,000–1,150 mm; the same applies to Wales, but heavier falls, 1,500 mm and more, occur in the upland areas, particularly near the west coast. In Scotland the south and central portions have a fall of 1,100 to 1,150 mm, and there are areas of low rainfall in places such as Nairn and the Black Isle; on its west coast the precipitation is at its highest, varying from 2,000–2,500 mm. The eastern lowlands of Scotland have 750 mm or less.

Trees, like all plants, require water in order to survive and grow; a mature tree needs a great deal and its very size and longevity not only affect the way rain reaches and enters the soil but also impose marked and persistent soil-drying patterns (Binns 1980). An ample supply of water to a plant's roots is necessary for maximum growth. The amount which a soil can hold against gravity and which is available to plants varies widely, depending on the soil texture. Evaporation of water, wherever it occurs, is controlled by the weather: sunshine has the greatest influence, but windspeed and, to a lesser extent, air humidity also affect evaporation. The plant uses the sun's energy to convert carbon dioxide from the air and water from the soil into sugars by the process of photosynthesis. The carbon dioxide enters the leaf and oxygen and water escape from it through stomata, this diffusive loss of water being called transpiration.

In the lower rainfall areas droughts and potential water deficits occur in summer and these can be particularly severe in the lowland areas. Though the frequency of rainfall is normally sufficient to prevent drought damage, especially in deep-rooted trees, some species are particularly sensitive to water stress and benefit greatly when young from good weed growth control which reduces competition for water as well as nutrients. Drought damage can occur on overmature or stressed trees, particularly beech on chalk soils. Beech, birch, western hemlock and the larches were all badly affected by the severe drought in 1976. Much of the reduction in growth potential can be accounted for by the effects of high vapour pressure deficits in closing stomata and halting photosynthesis.

Figure 4. Average annual rainfall in the British Isles. The figures indicate the precipitation in inches. One inch = 25.4mm

Different species have different water requirements. Douglas fir, spruces, western hemlock and western red cedar flourish in the heavier rainfall areas; they along with Japanese larch and poplars, do best in areas with a rainfall of 1,150–1,500 mm. Although all these species will grow reasonably well in districts where the precipitation is over 750 mm they are really at home in the wetter areas. Annual precipitation of between 700 and 800 mm was shown to depress yields of Sitka spruce growing on freely draining soils in NE Scotland (Jarvis and Mullins 1987). In areas of 400–500 mm rainfall, such as Norfolk and Suffolk, Scots and Corsican pines are the best choice. Most broadleaves are tolerant of low rainfall: sweet chestnut does well in the south on extremely dry soils, while beech, sycamore and ash appear quite satisfied in areas of 600–700 mm, but depending on the moisture retentive quality of the soil. Much depends on how the rainfall is distributed throughout the year.

Water intercepted by the forest is evaporated much faster than that by short vegetation, because the canopy, or surface of the forest, is 'rough' and better ventilated; and in coniferous forest the evaporation rate of intercepted water can be higher than the transpiration rate. Binns (1980) suggests that in Britain, evergreen forest intercepts about one-third of the annual rainfall and deciduous forest about one-quarter, though in any year the actual proportion will depend on the pattern of rainfall.

Precipitation increases with altitude, and most afforested uplands are areas of major water surplus. Only in unusually dry summer spells and on south-facing slopes with freely draining soils, may there be water stress during the growing season as rainfall is generally well distributed in all months. In winter there is usually an excess of soil moistures everywhere.

Water catchment areas. Many upland forested watersheds serve as important catchment areas for water supplies. They often contain reservoirs (notable examples include Lake Vyrnwy in North Wales and Thirlmere in the Lake District). The question is sometimes raised as to how much extra rain in such areas is evaporated back into the atmosphere as a result of the forest cover. Forests and forestry operations have an effect on both the volume and timing of run-off. The nature and extent of any effect will depend on a number of site characteristics, specifically climate and soil type. Although the ploughing and drainage of peaty sites can increase the volume of run-off for a number of years after planting, doubling base flows (Robinson 1986), the subsequent development of the forest canopy will reduce run-off as a result of the increased interception of rainfall. Closed canopy stands of conifers are generally believed to reduce the volume of run-off from a given upland moorland catchment by between 1 and 2% for every 10% that is forested. Pyatt (1984) explains a simple method of predicting the effect of afforestation on annual run-off. The effect of forestry on run-off is likely to be less where it replaces other types of vegetation, e.g. bracken, which intercept more rainfall than moorland grasses. Studies by Robinson (1986) and Nicholson *et al.* (1989) have shown forest ploughing and drainage to result in an increase in peak flow (20–30%) and a decrease in time to peak (up to 30%). Such effects are likely to decrease with increased forest growth and revegetation. (Silviculture in watersheds is constrained by controls on the use of fertilizers and herbicides.)

Although run-off in lowland forests will differ from other vegetation in the same way as in the uplands, the actual differences in total water use are likely to be smaller, because more agricultural land is managed with vigorous crops transpiring freely, often with irrigation to make up any water deficits, and because there is less rain anyway, usually under 1,000 mm. Mature broadleaved trees probably use more water than agricultural crops at all times during the growing season because they intercept more but transpire about as much. By contrast, conifers, which have high stomatal resistances, transpire less during a drought so that grass may, for a short time, use water at a higher rate than coniferous forest. During a prolonged drought, however, shallow-rooted grasses soon exhaust the available water and, in any event, as soon as light showers return the greater interception of the trees will more than restore the difference.

Where either the water itself or hydroelectricity is sold, resultant net revenue gained or lost can be computed. Barrow *et al.* (1986) have shown that the opportunity cost of lost water yield may be greater than net timber value on hydroelectric power catchments. Forestry may impose financial costs: a technique due to Collett (1970) evaluates adverse hydrological effects (Price 1989).

It is popularly supposed that trees attract rain but there is little evidence to support this view. It does seem, however, that mists, which might otherwise pass by without much effect, can collect on foliage and branches of trees and fall onto the ground – an effect most common along coasts, where sea fogs are frequent, and in high-lying forests in the north and west, but is unimportant in the rest of the lowlands.

Water quality. Forest management has profound effects on water catchments, reservoirs, watercourses, lakes and lochs. Well-designed drainage systems improve conditions for tree growth and protect forest roads and other works. Drainage entering the natural watercourses influence the rich variety of wildlife both within and alongside streams. Good water management consists not only simply of avoiding harm, whether to water running through the forest or to downstream users, but also involves taking active measures to keep the right environmental balance and to conserve streamside flora and fauna (Mills 1980; Forestry Commission 1988). Poor water management can lead to flooding, increased soil and stream erosion, greater turbidity and sedimentation. These can cause damage to roads, fisheries and wildlife generally and can increase costs of treatment for drinking water. The primary cause of acidification of water is atmospheric pollution.

Acid deposition from the atmosphere within susceptible areas of Britain has affected fresh water flora and fauna, causing the decline and in some instances the complete loss of fish population. Currently there is a debate about whether the presence of forests has increased the acidity of surface waters and contributed to the observed decline (Nisbet 1990). The evidence for the significance and scale of such a forest effect is by no means clear and only limited conclusions can be drawn from studies undertaken so far. Long-term monitoring of streamwater chemistry within industrial catchments is being undertaken. Management of forests for water is discussed in chapter 13, section 8.

Snow and ice. Mean temperatures in the British Isles, even in the coldest months, are above freezing. However, heavy snowstorms are fairly frequent, particularly in the eastern and northern parts of the country, but in winter snow seldom lies for long below 600 m. (In parts of Scotland severe snow-break of conifers can occur.) Clinging wet snow or hoar frost is occasionally severe, causing damage to branches and foliage; broadleaved stands are unlikely to be devastated. Snow can cause harm by drawing down bramble over young plants, and may damage rabbit and deer fences. Therefore foresters, following snow, should inspect their fences, release bent-over plants, and where necessary stake disturbed young trees. A combination of snow, rain and wind should always be looked upon as a possible recipe for disaster. The weight of ice (rime), as notably in January 1940, can cause severe breakage of trees of all species. Snow and frozen ground precludes a number of silvicultural operations such as ground preparation and planting. Haulage during a thaw can have disastrous consequences for roads and rides.

Lightning. Tall, isolated trees are particularly at risk to lightning strikes; groups of trees in plantations, hedgerows and screens are also frequently struck (Rose 1990). Damaged trees may lead to the entry of decay-causing fungi; and usually their timber value is lowered. Oak, poplar, Scots pine and Wellingtonia appear to be the species most frequently damaged. Strikes appear to be most frequent in the Thames Valley, to the north of London, in the central Midlands, south of the Severn estuary, along the Welsh Marches, and in the Lake District. Lightning conductors can be fitted to isolated highly valuable trees where the risk of a strike is high (British Standards Institution, 1985, The Protection of Structures against Lightning, *Code of Practice,* 6651).

Light. Light is necessary for tree growth, acting as a growth-producing and form-determining agent. Formation of tissues is dependent upon assimilation; the more light that is utilized the more volume is produced. Light can be direct or diffused ('dappled shade'), and trees can adapt themselves to the light available (ash will tend to grow towards a gap in the canopy; leaves can turn 90° or more to the sun's rays). Almost all trees grow better in full light (exceptions may include western hemlock and noble fir). Top light draws trees up, whilst side light is relatively unhelpful, prolonging branching and allowing epicormic shoots to develop (particularly in oak, poplar and sweet chestnut – see chapter 4). According to the degree of light and temperature depends the degree of photosynthesis, respiration and transpiration. Different parts of the British Isles have different lengths of growing season. Some trees can tolerate less light than others, as indicated below:

Light demanders		Moderate shade bearers		Shade bearers	
Broadleaves	Conifers	Broadleaves	Conifers	Broadleaves	Conifers
Oak	Larches	Sycamore	Douglas	Beech	Western
Ash★	Pines	Lime	fir†	Hornbeam	red cedar††
Elm		Norway	Coast		Western
Sweet		maple	redwood		hemlock‡
chestnut★		Hazel	Leyland		Lawson
Poplar			cypress		cypress
Birch					Grand fir
Alder★					Noble fir
					Yew

★ Will stand some shade when young or as coppice   † Side shade only   ‡ Almost shade demanders when young

Shade-tolerance plays an important part in natural regeneration, the natural alternation of tree species and the structure of mixed stands. Shade bearers can be grown pure, or in mixture, or as an understorey; they tend to create a bare forest floor. A stand of light demanders is more open than one of shade bearers; they have thinner crowns, cast a light shade, encourage a richer ground flora and lose lower branches earlier. A stand of light demanders can be underplanted successfully by shade bearers, but not vice versa; the result is a two-storey stand.

Nutrient requirements. Plant growth can be substantially reduced if there is an insufficient supply of any of at least twelve nutrient elements (N, P, K, Ca, Mg, S, Mn, Fe, Zn, Cu, B, Mo). These exclude carbon, hydrogen and oxygen, which are derived from the air and from water, and others such as sodium, silicon and cobalt, which may be beneficial under some circumstances and for some species. Different species of trees have different requirements of soil nutrients (see chapters 4 and 5). On limestone and chalk soils (see section 3 *infra*), where the high lime content prevents the absorption by roots of certain other minerals, induced nutrient deficiency is liable to cause yellowing, dieback and death of some trees, particularly Scots pine, Japanese larch, Douglas fir and sweet chestnut. Symptoms usually do not appear until trees start to close canopy or even later. An abundance of nitrogenous matter in very calcareous soils appears to intensify lime-induced chlorosis. The best choice on chalk and limestone soils is likely to be Corsican pine, European larch, western red cedar, beech, ash, sycamore, wild cherry and Norway maple.

In conifers, nutrient deficiencies are common (see chapter 5). In broadleaves, nutrient deficiencies are unlikely to occur on most soils – except that lime-induced chlorosis frequently occurs on high pH soils with free calcium carbonate, especially on old arable land. Sites with a long history of coppice working may have depleted phosphate reserves. Nutrient deficiencies can be corrected by application of fertilizers.

There is, however, a nitrogen problem. All plants need nitrogen to grow, but some trees may be receiving more nitrogen than is good for them. Nitrogen reaches forests in a number of forms: it falls as nitrate in 'acid rain' *infra*, and it reaches trees in a number of gaseous forms and they seem to be able to take it up through the leaves or needles. Most importantly, it reaches forests in the form of ammonia – a gas released from a variety of sources, with farmyard manure being one of the most important. Nitrogen can cause a number of problems to trees; for example, it makes trees grow faster and they may run out of other essential nutrients such as magnesium or manganese. Some insects, particularly aphids, seem to favour trees with high nitrogen levels; nitrogen enrichment may therefore result in increased levels of insect attack. While these problems have been recorded elsewhere in Europe, there is currently little evidence of them in Britain (Forestry Commission Leaflet 'Environmental Threats to Forests', 1990).

Air pollution and plant health. Industrial and domestic fumes can cause a wide variety of stresses to, and sometimes kill, trees. Point sources of atmospheric pollution include the combustion of various fuels (by aluminium smelters, brickworks and the like). Aerial pollutants such as fluorine may cause damage to conifers. Industrial smoke reduces sunlight; the fall-out of sooty matter on to the foliage of trees further reduces the light available for photosynthesis and clogs the stomata. The gaseous fraction of smoke also causes damage. Trees in some industrial areas may suffer badly. Conifers (with the possible exception of the larches and Corsican pine) are more susceptible to pollution injury than broadleaves. Particularly tolerant among the latter are sycamore, oak, elm, beech and poplar. Pollution by salt spray blown from the sea during summer causes leaf-burn (conspicuous browning of foliage) for up to several miles inland. Salt spray from roads may cause leaf-burn on trees in their immediate vicinity. Where acute injury occurs after severe exposure to fumes in the vicinity of point sources of pollution, the cause is fairly obvious, but the extent to which trees are harmed by relatively low concentrations of pollutants well away from sources is problematical. In the latter situation, damage and poor growth caused by other agencies is quite often wrongly attributed to pollution.

'Acid rain'. Evidence from the European mainland strongly suggests that some forms of air pollution (generally referred to as 'acid rain') are involved in the 'decline of forests' that has been observed there over the last decade (Innes 1987). 'Acid rain' refers to rainfall that is more acidic than it should be, but the term has also been used to cover all forms of air pollution. Rainfall is naturally acidic. Some of the acidity comes from carbon dioxide dissolved in rain water and additional acidity comes from natural sources of sulphur such as volcanoes and the ocean. 'Acid rain' occurs when certain types of pollutants are dissolved in rain. The proportion of man-made and natural acidity is variable throughout Britain. In the west of Scotland, about half the acidity is probably man-made. Moving eastwards, the proportion of man-made acidity steadily increases, and on the east coast 90% of the acidity may be from pollution.

The burning of fossil fuels, such as coal and oil, is the biggest source of man-made acidity in the air. The acidity is derived from two gases produced during burning: sulphur dioxide and nitric oxide (which quickly turns to nitrogen dioxide). These react with other substances in the air to form sulphuric acid and nitric acid, the most important man-made acids in 'acid rain'. Most of the sulphur produced in Britain comes from power stations, but car exhausts are an important source of the oxides of nitrogen. The gases and particles released during burning can fall to the ground in the dry form (known as dry deposition) where there may be a resulting increase in the acidity of the soil. Recent research suggests that the acidity of cloudwater, fog and mist can be even greater than rainfall and when these contact plant surfaces, damage may occur.

The increase in the acidity of some freshwaters is mainly due to 'acid rain'. Some forest soils have become more acidic. This may affect the trees but, so far, no adverse effects on trees have been associated with this form of acidification, but it is suspected that other plants within the forest ecosystems, such as lichens, mosses and fungi may have been affected. The direct effects on trees are much less clear. Experiments indicate that rain at the sort of acidity level that is normally found in Britain does not directly damage the leaves or needles of trees. However, fog and mist can sometimes be very acidic and this may cause direct damage.

Sulphur dioxide, ozone and other gases. A number of gases affect plants directly. Close to sources, sulphur dioxide may be a problem. This gas is mainly derived from burning fossil fuels. The concentrations of sulphur dioxide in Britain are currently decreasing but the concentrations of other gases, particularly ozone, may be increasing. Ozone occurs at ground level and at very high altitudes within the atmosphere, about 20 to 30 km above the Earth's surface. The higher band acts as a protective layer, preventing the sun's harmful ultraviolet rays from reaching the surface of the Earth. Pollutants, mainly chlorofluorocarbons (CFCs), are damaging this layer, resulting in a thinning of the protective layer, one sign of which is the so-called 'ozone hole' over Antarctica – with serious implications for the whole world. As it is a serious problem action is being taken to stop the thinning from spreading or getting any worse. The Montreal Protocol of 1988 sought to contain CFC production and use. The Protocol was strengthened substantially in June 1990 in London. CFCs are, effectively, to be phased out by 2000, or before. Ozone at ground level currently presents a more direct problem. Ozone concentration high enough to damage trees may have occurred in Britain, though no damage caused by ozone has been detected.

'Forest decline' on the continent caused concern throughout the 1980s. However, since 1985 there has been a steady improvement in the condition of trees in many areas. The causes of the decline are still problematical, but one of the most likely explanations is a multiple stress hypothesis involving air pollution and adverse soil conditions as predisposing stresses, and short-term climatic factors as inciting stresses. Although adverse weather conditions may have triggered the decline, it seems probable that this decline would not have occurred if the trees had not already been weakened by air pollution or some other stress. If the multiple stress hypothesis is going to be accepted, it is misleading to talk about forest decline as a single phenomenon. It appears that there are a number of areas where a genuine forest decline is occurring. However, over much of Europe, there does not appear to be a long-term problem. The feeling among many scientists is that there are series of local or regional problems which all have different explanations. In each case, a combination of stresses is seen as being important but the actual composition of this combination may change between sites. In some cases, air pollution is clearly involved, in others, the evidence for its involvement is much less certain. Innes (1987) reviewed information available on the interaction between air pollution and forests, dealing mainly with long-range pollution and its possible regional-scale effects. Research undertaken by the Forestry Commission has been updated by Innes (1990).

The Forestry Commission is investigating the effect of pollution relative to clean air using open-top chambers. These chambers can also be used to determine the effects of added pollutants. One series of experiments has been concerned to assess the effects of ozone events on the biomass distribution within tree saplings and the seasonal variations of plant sensitivity; likewise the dependence upon phenological condition of some conifers and the variations between individuals within populations (Taylor *et al.* 1989). The monitoring of forest condition by the Forestry Commission suggests that climate is the main factor affecting the crown density of Scots pine, beech and oak. Associations with pollution levels are generally positive: crown density *increases* with levels of most pollutants (Innes and Boswell 1990).

The Forestry Commission believe there is still little to suggest that British trees are being significantly damaged by pollution, but they are responding to pollution. For example, the sulphur contents of spruce and pine needles are higher in areas with high levels of sulphur dioxide, but the sulphur contents appear to be unrelated to any form of damage. Similar patterns have been found with nitrogen contents and the oxides of nitrogen. Filtration experiments suggest that trees may grow better in cleaner air, although rising levels of carbon dioxide and nitrogen deposition may also result in increased growth.

In the east of Britain the rain is more acidic than in the west, but as there is less of it, less acid reaches the soil. In the west, any effects will be from the total amount of acid reaching the soil, and because rocks and soils in the west tend to be more acidic anyway, damage is more likely than in the east (this is why most reports of acid lakes and streams come from the west). Most of the gases come from industrial or power-generating centres and concentrations are higher in these areas. The most polluted part of Britain in this regard is Nottinghamshire and Yorkshire, with the north-west of Scotland being the cleanest. Ozone tends to build up where there are a lot of oxides of nitrogen, and when the weather is still and sunny. These conditions occur most frequently in the south of England and this is where the highest concentrations of ozone have been recorded. There is much more to learn about trees. The Forestry Commission continues to pay great attention to all relevant aspects of pollution (Innes and Boswell 1990; Innes 1990).

Global warming – the 'greenhouse effect'. Besides ozone and ammonia a number of gases are increasing in the atmosphere. One of the most important of these is carbon dioxide which, together with certain other gases, plays an important role in controlling the temperature of the Earth's atmosphere. Trees need carbon dioxide to grow and they absorb a lot during their lifetime. When they die, the carbon dioxide is gradually released as the wood breaks down. Under natural conditions, the carbon dioxide would then be taken up by new trees. This cycle has been upset by Man. Tropical rain forests are being cut down rapidly and burnt. Few forests are being replaced and the carbon dioxide is not being put back. The soils from these former forests can also break down and release carbon dioxide. The destruction of the tropical forests is therefore an important cause of the 'greenhouse effect'. The average temperature of the Earth has increased by about half a degree (Centigrade) over the past eighty years. The UN Intergovernmental Panel on Climate Change (IPCC) suggests that warming will increase at a rate of 0.1°C–0.3°C per decade. By 2050, therefore, mean temperatures would be 0.6°C–1.8°C higher than today, with significant regional variations. A change of this size could have major effects on Britain's trees. However, the predictions about how the weather will change in Britain are still uncertain – wetter or drier, warmer or cooler – so it is unknown what effects there will be on trees. Nor is it known what effects the increasing carbon dioxide levels will have on trees.

Assuming current industrial practices, a continued rise in the carbon dioxide concentration of the atmosphere is likely and will, in combination with other greenhouse gases, cause a general warming of the atmosphere. Forest destruction, by contributing carbon dioxide, is part of the problem; forest creation can be part of its solution. Trees represent the most efficient land-based biological system for locking up carbon, and more planting of fast-growing trees and the maintenance of a high average growing stock could provide part of the solution to global warming. Grayson (1989) indicates that on good sites in Britain, Douglas fir (Yield Class 16 if grown on a rotation of maximum mean annual increment – see chapter 12) can fix an average of 4 tonnes of carbon per hectare per year, and beech (Yield Class 8) 2.5 tonnes. In the long run, trees either die or are harvested. With death, saprophytes cause the release of carbon to the air. When wood is harvested for fuel, burning clearly causes an even more immediate release to the atmosphere. Pulping to make paper and board staves off the release a few years but may lead, if the used paper is buried in landfill, to the production of methane which is, molecule for molecule, many times more powerful as greenhouse gas than CO_2. Use of wood in panel products and sawnwood, however, especially when these are treated with preservatives, leads to an extension of the period of locking up by several decades at least.

Forests provide a most effective defence against the 'greenhouse effect' as they lock up carbon dioxide for tens or even hundreds of years (Thompson and Matthews 1989;

Cannell *et al.* 1989; Freer-Smith 1990). Conversely, their wanton destruction provides a great threat to the climatic and ecological stability of this planet. Of a variety of actions needed, those influencing forestry include: (i) understanding climatic changes that will occur, in particular those affecting Britain; (ii) assessing the various impacts on the growth of trees, including new species, and of trees relative to agricultural crops; (iii) evaluating the contribution of an increased forest area to carbon fixing; and (iv) helping to slow the high rate of deforestation in tropical countries. It is too early to say categorically what effects climatic change will have on our forests or vice versa. Current thinking suggests, perhaps surprisingly, that the effects are likely to be beneficial.

Pearce (1990) has discussed the economic value of afforestation in terms of its carbon-fixing capability. The value of forests as fixers of carbon is an example of an unmarketed benefit – no one buys or sells this service. Nonetheless, it is a real economic benefit since forest growth can be thought as a means of avoiding expensive CO_2 abatement measures or of avoiding the damage done by CO_2 emissions. Pearce (1990) estimates the value of the forest with the second procedure – the 'damage avoided' approach – and shows that the carbon-fixing values of new forests are at least equal to their recreational values. Anderson, D. (1990) adopts the 'avoided abatement cost' approach and derives even higher 'carbon credits' for forests. Both approaches are legitimate and offer foresters a novel economic defence for growing trees. The economics of the subject is further noted in chapter 14, section 5.

Wind. Parts of the British Isles are amongst the windiest parts of the world where commercial forestry is practised. In regions where wind damage is a serious threat to the viability of forestry, the risk of damage to tree crops from windthrow is a constant concern of foresters and a major constraint on silviculture. Special forms of silviculture developed to minimize the risks of damage, include respacing, altering thinning regimes, attention to the general structure and composition of stands, and shortening rotations.

In general, the prevailing winds are warm, moisture-laden, south-westerly ones; but cold, dry, north-easterly winds may blow for several successive weeks, particularly in spring. Average wind speeds range from about 16 km per hour in the Midlands of England to 24 km per hour on the west coast of Scotland. Gales with wind speeds up to 130 km per hour occur quite frequently. During an average winter most parts of the country experience two or three severe gales, usually from the south-west but occasionally from the north-west or north-east. In some districts of Scotland, wind is the most limiting climatic factor in silviculture; and in many parts of Britain windthrow is likely to be a serious risk, especially for spruces, Douglas fir and larches, on shallow rooting clay soils, and for most species on shallow peat soils, particularly on high exposed ground. Besides windthrow, the risks include wind-break and wind distortion. Wind-loosening of trees severely retards growth, and from an early stage can result in basal bowing, notably of lodgepole pine. There are also the risks in nurseries and young plantations of plants drying out due to excessive transpiration or to movement around their collar.

In windthrow, there is a clear relationship between spacing, thinning regime and stability. The most serious risk occurs after thinning and when the ground is waterlogged; and is greatly increased when thinning has started after a long period of delay. In these situations it is generally advisable to thin early and selectively, or not at all (see chapter 8). Line thinning attracts a greater risk of windthrow than selective thinning. Windthrow can, of course, shorten rotations, resulting in lower returns from smaller, less valuable crops, and also a truncated yield class because the period of maximum mean annual volume increment has not been attained.

Wind behaviour is dependent upon geographical position, topography, altitude, aspect, and degree of slope; also upon the structure of the stand both inside and on its periphery. Points for foresters to remember include correct choice of species, appropriate cultivation and drainage and a sensible thinning regime. Before clearfelling, consideration should be given to the problems of restocking: for example, it may be prudent to select and retain windfirm boundaries in old crops as a shelter.

In broadleaved stands the problem of wind damage, especially windthrow, is rarely serious. The sites generally planted are modestly exposed and the soils are usually deeply rootable; the deciduous habit in autumn and winter reduces the sail area during the time of year when storms are most frequent; and on exposed sites broadleaves generally do not grow very tall and maintain a smaller height-root ratio. As a result some broadleaves make useful wind-breaks; beech and sycamore are often seen in this role in south-west England and Scotland. On exposed sites, persistent wind leads to crown deformation and poor growth. Much damage can occur in autumnal storms with branch break and uprooting of old over-mature trees, particularly if the roots are diseased, but usually little damage occurs to younger stands. Long neglected broadleaved stands may suffer wind damage when thinning is eventually undertaken because many trees will be tall and whippy.

Windthrow in conifer plantations has become an increasingly serious problem in many parts of the uplands of Scotland, northern England and mid and north Wales. Miller (1985) writes comprehensively on the problems. Much of the large-scale afforestation undertaken by the Forestry Commission since 1919, and some by the private sector during the past twenty years, is located on exposed upland areas in the north and west, where the wet gley soils often have impeded drainage and other characteristics which restrict the development of windfirm root systems. Rooting of the trees has often been restricted to the top 25–45 cm of the soils by waterlogging, and anchorage of the trees has often proved inadequate in strong winds. As the trees grow taller, they become increasingly susceptible to windthrow, which frequently begins well before plantations have reached the economically desirable rotation age, and the yield of valuable large-diameter timber may be considerably reduced. Selection of appropriate silvicultural techniques, and effective planning of timber harvesting and marketing, must take into account the possibility of windthrow damage.

Wind damage can be classified as catastrophic or endemic. The former results from severe storms and can be locally disastrous. The latter starts as small pockets of windthrown trees spreading progressively through a stand over a period of years. (This is through the inherent weakness of the trees to withstand ordinary high winds; it is therefore regarded as somewhat inevitable, and in that sense it is not wholly unpredictable.)

Catastrophic windthrow arises as a result of storm conditions of unusual severity. These have long recurrence periods, and on average affect some part of the United Kingdom about once in every fifteen years. Such damage is caused by exceptional climatic conditions, e.g. in east Scotland in January 1953, west and central Scotland in January 1968, Wales, central England and East Anglia in January 1976, south-east England in

October 1987[1] and south-west England and Wales in January 1990.[2] The extremely high wind speeds involved can cause serious damage to plantations both on unstable and windfirm sites, and the degree of damage is influenced by soil conditions or silvicultural practices. Catastrophic damage produces serious harvesting problems in affected forest areas, often exacerbated by a substantial proportion of trees being broken rather than uprooted. It is not possible to predict where future catastrophic damage will occur, nor in general to mitigate significantly the effects of such storms by silvicultural means, other than by increasing the diversity of stand ages and heights within a forest and so reducing the proportion of the total forest area at risk at any one time.

Endemic windthrow is of greater economic importance than catastrophic windthrow, and arises as a result of normal winter gales with a relatively moderate mean speed of approximately 72 km gusting to 108 km per hour. Most upland forest areas experience gales of this type several times each year, and a common result is the occurrence of fresh wind damage in the less stable parts of semi-mature plantations. Very little stem breakage occurs, uprooting of trees on wetter soils being the predominant effect. Damage is often sporadic, but is influenced strongly by site conditions and silvicultural practice. This offers a means of predicting the occurrence of endemic windthrow damage, as well as the prospect of selecting silvicultural treatments likely to delay or restrict the incidence and extent of such damage.

Factors influencing windthrow. Numerous plantations in the uplands become increasingly susceptible to windthrow with increasing stand height, and the *top height* of the stand (based on the mean height of the 100 trees of largest diameter at breast height per hectare) when windthrow damage begins, is termed the *critical height* (defined as 3% windthrow). When the progression of damage within the stand reaches 40%, the stand top height at this stage is termed the *terminal height,* and under normal circumstances clear-felling is desirable to enable recovery (harvesting) of the fully productive capacity of the site. Windthrow in forests involves complex aero-mechanical interactions between turbulent wind passing over the forest and the

---

1 The great storm in the early hours of 16 October 1987 in south-east England (to the east and south of a line from Bournemouth to King's Lynn) had windspeeds of up to 90 knots (Hill 1988). Some 15 million trees (about 15,000 ha of woodland) were blown down, more than 50% being broadleaves, probably 70% (by volume) in private woodlands, 25% in Forestry Commission woodlands, and 5% in hedgerows, parks and gardens. The 'Forest Windblow Action Committee' (see chapter 10), together with, especially, the Forestry Commission, quickly issued sound guidance on the harvesting of the wind damaged trees and restocking. The windthrow was remarkable for the high proportion of broadleaved timber blown. Ownership of the private woodlands was widely dispersed in many hundreds of estates. Not all these woodlands had been managed with timber production as a primary objective; significant areas had been managed for shooting and amenity and some had not been managed at all. Impact and responses to the storm are related by Grayson (ed. 1989), and a special edition of *Weather* (March 1988, Vol. 43, No. 3) is devoted to the catastrophe.
2 Gibbs (1990) gives guidance on the clearance of trees after the gales in January and February 1990, and Evans and Paterson (1990) on restocking after windthrow. Grants and supplementary aid of up to £400/ha were available for restocking (see chapter 15).

dynamic response of the stand, comprising trees of various shapes and sizes anchored imperfectly by the roots and soil. It is impossible to predict the location, severity and duration of potentially damaging gales with any accuracy, and the great variation in soils, topography and forest structures increases the difficulty of windthrow prediction. However, windthrow hazard is closely related to the following four site features, which can objectively be quantified to some extent (Miller 1985):

1. Windiness of regional climate. The north and west of Britain experience strong winds more frequently, and at greater strength than other parts of the country; and coastal areas are also associated with more frequent gales than inland areas. The country has been zoned (Figure 6) according to the incidence and severity of strong wind conditions.

2. Elevation. Mean wind speed increases with elevation, as does gale frequency, and forests at higher elevation are therefore generally more prone to wind damage than lowland sites within any geographic area. Rainfall also increases, leading to wetter soil conditions with a lowering of the rooting capacity and a possible reduction in soil strength at higher elevation.

3. Topography. The effects of increasing elevation on the wind exposure of a site are modified by the influence of surrounding topography. The sheltering effect of adjacent high ground can be particularly important in reducing local wind speeds. The incidence of windthrow is also related to topographical shelter. A simple, objective characterization of the topographical shelter of a site can be obtained by *topex* assessment (noted *infra*). This involves measurement on site of the angle of inclination to the horizon, at the eight major compass points; by summing the eight angular measurements a topex value for the assessment site is obtained. This topex value is a useful relative indicator of general site exposure. In practice, topex assessments are often combined with soil surveys.

4. Soil conditions. Root morphology is strongly affected by soil type, with the depth and strength of roots being influenced by soil moisture and aeration, and by physical conditions within the soil profile. Soil strength is also related to soil type, with soil moisture status being particularly important. The method of site preparation used for the establishment of plantations also influences root architecture and soil strength, with consequent effects on plantation stability later in the rotation.

Forestry Commission research on tree stability is currently targeted at predicting the likelihood of damage based on site, crop and wind characteristics, and also at improving understanding of the interaction between trees and airflow, and the relationship between topography and airflow (Quine and Reynard 1990). The influence of cultivation type and position on root architecture is also studied.

Windthrow Hazard Classification. The Forestry Commission system of classifying wind damage susceptibility by windthrow hazard classes for coniferous forests has been in use for several years. Its practical value as a management aid has been clearly demonstrated by experience gained since their introduction (Booth 1977) and subsequent revision and refinement. Miller (1985) explains the basics of the classification, which is a practical approximation intended for the broad zonation of forest areas of the order of 500 ha or more in extent. (Windthrow hazard is of course only one of the several factors to be taken into account in the appraisal of silvicultural options.)

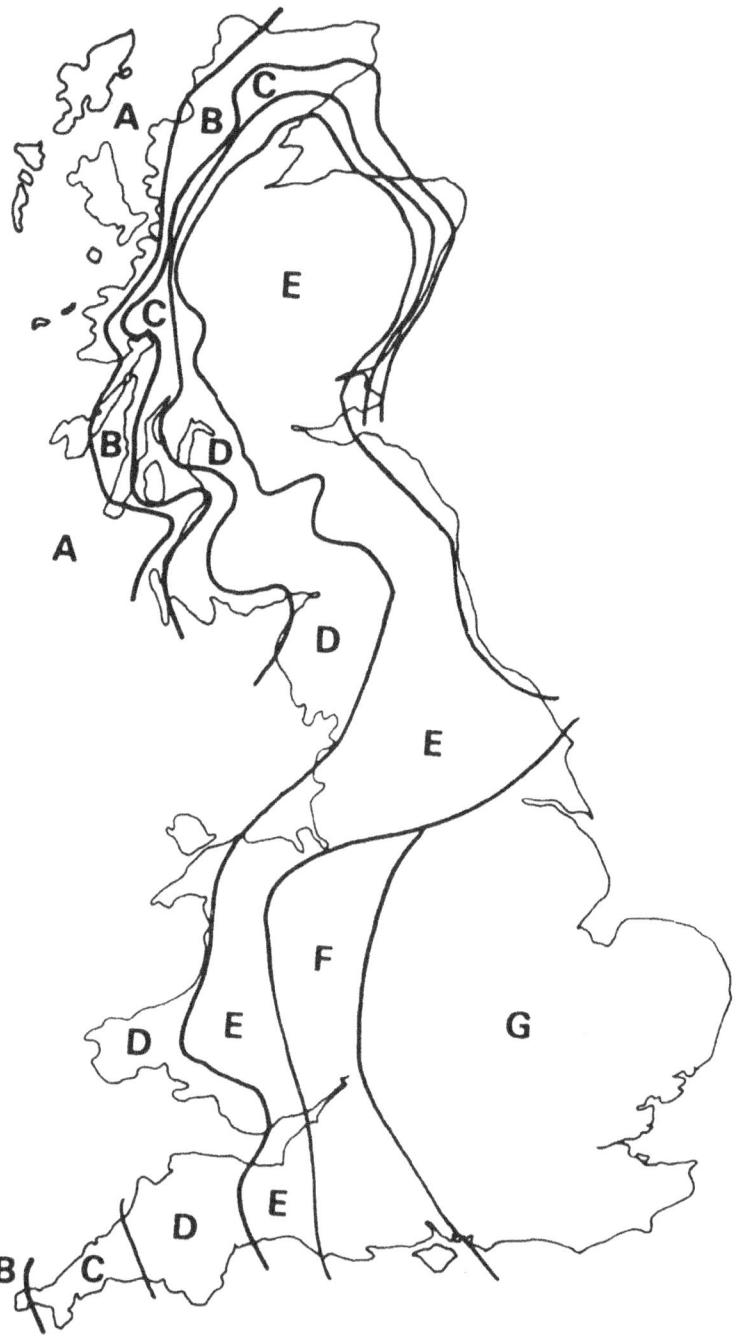

Figure 5. Wind zonation of Great Britain, based on a combination of wind speed data and exposure flag surveys. (Gradations of wind exposure from high – A, to low – G)

Source: Miller (1985). FC Leaflet 85.

Assessment of windthrow hazard. The windthrow hazard class of a site is derived from scoring assessments of the four site factors, wind zone, elevation, exposure (topex), and soil, as noted below. The total score for the four factors indicates the hazard class (Miller 1985):

1. Wind zone. The relevant wind zone (of seven) for a forest area is determined from the wind zonation map (Figure 6). The appropriate score for wind zone is then obtained thus: wind zone A (13.0), B (11.0), C (9.5), D (7.5), E (2.5), F (0.5), G (0).

2. Elevation. The elevation above sea level of the assessment site can be obtained from Ordnance Survey maps; or a barometric altimeter can be used. The score for the relevant band is then noted thus: elevation (m) 541+ (10), 466–540 (9), 406–465 (8), 361–405 (7), 316–360 (6), 286–315 (5), 256–285 (4), 226–255 (3), 191–225 (2), 141–190 (1), 61–140 (0.5), 0–60 (0).

3. Exposure (topex). The assessed degree (°) of exposure as influenced by topographic features surrounding the site, is scored thus: topex value (total °) 0–9 (10), 10–15 (9), 16–17 (8), 18–19 (7), 20–22 (6), 23–24 (5), 25–27 (4), 28–40 (3), 41–70 (2), 71–200 (1), 101+ (0). Alternatively, where topex-based exposure maps have already been prepared, scores can be obtained thus: severely exposed (10), very exposed (7), moderately exposed (3), moderately sheltered (1), very sheltered (0). In the absence of topex data or exposure maps, the influence of local topography on site exposure can be assessed subjectively by using the foregoing five exposure classes.

4. Soil (stabilizing effect). Where detailed soil information is not available, or where observation of root development on blown trees enable an objective assessment of the restriction to root development, scores can be based on root development and broad soil groups, thus:

Root development	Soil group	Score
Unrestricted (in excess of 50 cm)	Brown earths, podzols, intergrades to ironpan	0
Restricted (but in excess of 25 cm)	Deep peats, loamy gleys	5
Very restricted (under 25 cm)	Peaty gleys, surface water gleys, shallow indurated soils, waterlogged soils	10

With detailed soil maps or soil surveys a more sophisticated soil scoring is available.

Derivation of Windthrow Hazard Class. The scores for wind zone, elevation, exposure (topex), and soil are summed to give a total score for windthrow hazard. The classes are: 1 low (score range up to 8.0), 2 (8.0–13.5), 3 (14.0–19.0), 4 (19.5–24.5), 5 (25.0–30.0), 6 high (over 30.5).

Management applications of Windthrow Hazard Classification. The value of the classification to foresters lies primarily in the field of production-planning by assisting decisions on the location and extent of future thinning and felling operations. (It has an effect on investment market values of both bare land and plantations – chapter 16.) In upland conifer forests subject to high risk of premature windthrow, an important management objective will be to achieve the maximum possible crop rotation length on susceptible parts of the forest, thus increasing the yield of higher value sawlogs from these stands. It is often the case that a relatively small increase in rotation length can substantially improve the value of a stand at time of clear-felling, due to an increase in average tree size. The six Windthrow Hazard Classifications are associated with a range of *critical heights* at which the onset of windthrow can be expected. The average critical heights (m) and estimated terminal heights (m) for three generalized thinning regimes for upland Sitka spruce plantations are shown in Table 7.

Table 7. Average critical heights at which the onset of windthrow can be expected.

Windthrow hazard class	Critical heights (m) (onset of windthrow)			Estimated terminal heights (m) (40% stand area blown)		
	Non-thin	Selective thin	Systematic thin	Nonthin	Selective thin	Systematic thin
1	---------- Unconstrained by windthrow ----------					
2	25.0	22.0	21.0	31.0	28.0	28.0
3	22.0	19.0	17.0	27.0	25.0	23.5
4	19.0	16.0	14.0	24.0	21.5	18.5
5	16.0	13.0	12.0	19.5	17.5	15.5
6	13.0	10.0	9.0	15.5	13.5	11.5

Source: Hibberd (ed.) (1991). FC Handbook 6.

Table 7 shows that thinning generally leads to windthrow, starting at lower top heights (and earlier age) than in unthinned stands, and systematic thinning (line or chevron patterns) is worse than selective thinning. In addition, the timing and intensity of thinning also influence windthrow. It is important to appreciate that the actual critical and terminal heights which are observed in any forest may deviate widely from the figures shown in Table 7. The *critical heights* shown are quite well supported by field validation surveys, but the *terminal heights* given are only to an approximation, based on casual observation, and the rate of spread of windthrow is highly variable. As a general rule, stands in the highly susceptible windthrow hazard classes 5 and 6 should be left unthinned. For those in the most stable hazard classes 1 and 2 foresters are free to choose any normal thinning technique, and stands can usually be grown to full economic rotation without difficulty. In the intermediate hazard classes 3 and 4, thinning options are more limited, and care must be taken over the timing, pattern and intensity of thinning to avoid precipitating the onset of serious windthrow.

The Windthrow Hazard Classification and derived critical and terminal heights are very broad approximations, and cannot be applied safely to individual stands in a forest. The main purpose of the classification is the zonation of extensive forest areas according to windthrow susceptibility, as an aid to general decisions on thinning policy and to production forecasting. Decisions on felling or thinning of individual stands are subject to a wide range of additional constraints and objectives. The clearance of windthrow pockets and normal clear-felling operations can also carry an increased risk of windthrow spreading in surrounding forest, by exposing unstable edges of plantations to the full force of the wind. It is recommended that in the clearance of windthrown pockets, leaning edges should be left intact, and clearfelling operations should be planned to avoid leaving susceptible forest margins, by cutting to existing rides, roadlines and species or age class boundaries.

Forestry Commission economic appraisals have led to the decision not to thin crops in areas of high windthrow hazard. Analysis of data published in yield class tables shows that although the final crop of a thinned stand will be made up of larger trees than an unthinned stand, the total yield of sawlogs to say, 18 cm top diameter over bark may not be significantly greater. For normal planting spacings of about 1.7–2.0 m, sawlog production over a conventional rotation of between forty and sixty years is not markedly different in a thinned and unthinned crop. In areas of high windthrow hazard, unthinned crops can, as a result of their longer rotation, produce substantially higher yields of sawlogs than thinned crops.

Reducing risk of windthrow. Site preparation and soil type are important influences on the incidence and extent of windthrow. In particular, spaced furrow ploughing can result in restrictions to root spread, with the main structural roots tending to align along the plough ridge, and being unable to cross the open furrow. Double mouldboard ploughing is likely to produce less serious root restriction than single mouldboard ploughing, since there are fewer open furrows per hectare, and the more gently sloping sides of the ridges and furrows may encourage increased development of structural rooting at these points. On the drier podzols and ironpan soils complete ploughing may promote improved root architecture, and on the wet gley soils subsurface drainage by mole ploughing with shallow surface cultivation, or mounding, are attractive alternatives to spaced furrow ploughing for improved rooting and stability (Hibberd (ed.) 1991).

Spacing or respacing of plantations can be expected to influence windthrow susceptibility to some extent, mainly through their effect on tree shape. The use of wide spacing at planting or subsequent respacing treatment will encourage development of heavily tapered individual tree stems which may resist windloading more effectively than close grown stems with lower taper. However, any benefits of improved individual resistance to windloading from more tapered stems may well be counteracted by the increased canopy roughness which will generate higher turbulence above the forest. The deeper tree crowns in widely spaced stands will also permit deeper wind penetration and, in combination with increased turbulence, the windloading on individual stems will increase. In addition, widely planted stands will suffer from a reduction in timber quality (due to increased branch size, branch retention and large juvenile core) which reduces the attractiveness of this option even further. Other silvicultural options such as precommercial thinning, or chemical thinning of plantations have potential application on high windthrow hazard class sites where conventional thinning is undesirable (see chapter 8). Provided the operations are carried out well before critical height, stability should not be impaired and timber quality is unlikely to be prejudiced.

2. Ground vegetation as indicating site factors

The vegetation of a site is usually indicative of the soil, moisture and other conditions, and hence sometimes assists species choice (though other factors of the site and relevant conditions need to be considered). Tree growth and the composition of the lower vegetation are to a large extent determined by the same basic variables of temperature, light, moisture, soil aeration and fertility. Where there is enough natural or semi-natural vegetation, it can be a sensitive indicator of remaining site variability in areas of uniform climate, soil and land surface.

Classifications based on vegetation are commonly used in Northern Europe and elsewhere. The main problem is that vegetation is often highly complex and difficult to sample objectively. The basis of all systems is that various indicator plants or plant communities are used to give a guide to the productive potential of the site and other features of interest, such as drainage status or fertility. (One of the oldest and best known classifications is that of Cajander, 1921, in Finland who classified natural and thinned stands of Scots pine and Norway spruce, as well as treeless sites, using the dominant plant species in the ground vegetation. It enabled reasonably accurate predictions of site productivity to be made and the need for drainage is based on this classification today.) On suitable sites in Britain a ground classification is quite widely used as an aid to species selection based on the work of Anderson, M.L. (1961).

Vegetation, apart from its use in the classification of sites, is also of interest as a potential nuisance as weeds, for its rarity value as something which requires conservation, or for its potential in the management of domestic or wild animals as a source of food or shelter. In Britain, most soils have been influenced by man, and hence ground vegetation gives only a general indication of the potential of a site for tree growth. But usually this general indication makes it possible to distinguish, for example, (i) a fertile from an infertile soil, (ii) an acid from an alkaline soil and (iii) a well-drained from an impeded site. This may help to suggest species choice and probable yield class. But past and present treatment often dominates the site (sheep-grazing, ploughing, draining, liming, burning) and hence ground vegetation may be misleading. (A point of interest is that some seeds can survive in the soil for so long that they can outlive forest plantations established over them. Seeds of several woodland plants, including foxglove and bramble, can survive for forty years or more; and hence seeds could fall from plants growing amid infant trees and still be viable when the trees are felled as mature timber.) Soils where their nature and potential are indicated by plants include:

1. Damp, wet soils (usually requiring drainage)
    Rush *Juncus* spp., sedge, cotton grass *Eriophorum vaginatum*, *Sphagnum* moss, purple moor grass *Molinia* spp. See also 'Peat' *infra*.

2. Heavy, moist clays
    Ferns, horsetails *Equisetum* spp., tufted hair grass *Deschampsia caespitosa*.

3. Limestone and chalk

    Dog's mercury *Mercurialis perennis*★
    Rock rose *Helanthemum* spp.
    Wild strawberry *Fragaria vesca*
    Wild thyme *Thymus* spp.
    Wild garlic or Ransoms *Allium ursinum*: moist★
    Travellers' Joy or Old Man's Beard *Clematis vitalba*
    The grasses *Brachypodium* spp., and *Bromus erectus*

    Planting of beech is usually indicated, with sycamore and ash as alternatives on the better and deeper soils. Corsican pine, western red cedar and European larch are the best of the conifers.

    ★Dog's mercury and wild garlic, along with nettle *infra*, are sometimes indicators for ash, but Dog's mercury is abundant on many soils which are too dry or shallow to support good ash.

4   Fertile soils
    Nettle *Urtica dioica:* indicates a soil rich in nitrogen.
    Bramble *Rubus fruticosus:* a soil improver.
    Bracken *Pteridium aquilinum:* when luxuriant indicates good soil aeration and depth.
    Rosebay willow-herb *Epilobium angustifolium*
    Foxglove *Digitalis purpurea*
    Bluebell *Endymion non-scriptus*
    } Rich humus typical of old woodland sites.

5   Sand dunes
    Marram grass *Psamma arenaria:* suitable only for pines.

6   Infertile soils
    Heather Calluna vulgaris: possibly avoid planting spruces.
    Cross-leaved heather Erica tetralix: damper site; acid.
    Bell-heather Erica cinerea: drier site.
    Bilberry Vaccinium myrtillus.
    Mat grass Nardus stricta.
    Gorse *Ulex europaeus:* sometimes indicates dry, fairly free-drained soil.

7   Peat soils
    Bog myrtle, cotton grass, rush, sedge, *Sphagnum* moss and heather.

Certain plants if left to nature's own devices will give rise to particular types of woodland; thus grassland on clay soil often tends to change to oakwood and on chalk to yew, whitebeam and beechwood. It is not always easy to decide how far our natural ground vegetation indicates conditions suitable to *exotic* trees.

Soil, rather than ground vegetation, classifications provide the most useful form of classifications for silviculture in many places where bare land is being afforested, for three reasons:

1   A large number of sites have been so extensively modified by agricultural practices and burning, that vegetation is not always a reliable indicator of site conditions. (This is particularly true on sites dominated by heather and bracken.);

2   In establishing tree crops, many sites are so modified microclimatically and in terms of drainage, cultivation and nutrition that pre-existing semi-natural vegetation is profoundly changed; and

3   Within a few years of planting the exotic species commonly used in plantations, there is virtually no ground vegetation left. Most is shaded out and destroyed though the seeds of some species can live for decades in the soil and lead to rapid recolonization of a site late in the rotation and at restocking.

3. Forest soils

Soils are the foundation of tree life. Broadly they can be classed as fertile or infertile, mineral or organic, poorly or well drained, deep or shallow – all qualities inimical or agreeable to a particular tree's regime of tolerance. A 'poor' soil can be improved by cultivation, application of fertilizer, or drainage; whereas a 'good' soil can be abused by compaction or erosion.

Soils are derived from a 'parent material', which can be hard rock weathering *in situ*, soft rock weathering *in situ*, or drift – generally consolidated material transported and laid down at a site by glaciers, rivers, sea, wind or simply downslope mass movement. Often the distinction between soft rock weathering and drift as soil parent material is unclear. The soil itself comprises a sequence of vertical layers termed horizons. Below a dark topsoil lie browner (or redder, greyer or even greener) subsoil layers which pass either sharply or gradually to the parent material. Sometimes the subsoil, or several distinct layers making up the subsoil, may be missing and the topsoil rests directly on rock – especially common in mountainous (hard rock) areas.

In situ rocks and drift have been mapped by the British Geological Survey. 1:250,000 scale soil maps and accompanying bulletins of the Soil Survey of England and Wales and the Soil Survey of Scotland describe the soils of mainland Britain. A standard text on the properties of soils is that of Pritchett and Fisher (1989). Valentine (1986) provides much detailed information on soil surveys. Nonetheless, despite the extensive reference maps and texts, foresters on many sites still find that a spade or soil auger is necessary to show the various layers. In excavating they expose or pierce through a series of layers; and their understanding of the nature of the soil will depend much on the study of this layered structure or horizons.

The physical conditions of depth, porosity and water content are important. (The major structural roots of a tree provide most of its stability, while the fine roots take up most of the water.) Depth of soil needs to be sufficient to allow the roots to make strong anchorage and to provide extensive access to water and nutrients. Depth requirements vary: a species like oak, for instance, needs greater depth than spruce, whose roots naturally tend to spread out in the upper soil layers. Trees that will grow in shallow soil, say 30–45 cm, include Norway spruce, beech and birch; those requiring deeper soils, 75 cm or more, include oak, ash, sweet chestnut, lime and the larches. Compact soil layers, whether created by man or natural in origin, pose problems. They are particularly common in glaciated upland soils, also in some lowland soils, and in disturbed soils. Some soils contain ironpans which also form a barrier to tree roots. (These hard layers can usually be dealt with by a tine plough, with or without a mouldboard.) Reclaimed soils, and soils in urban areas are often compacted, and this feature must not be overlooked when trees are planted on these sites.

Soils vary enormously in fertility. Foresters, except in some lowlands, are unlikely to be concerned with sites at the more fertile end of the range. Their lot is most likely to be degraded upland grazing, wet upland moors, and lowland sandy heaths. All these sites are likely to be low in phosphate; and organic soils (peaty gleys and deep peats) may also be deficient in potash. Chalk and limestone soils require special care, but no fertilizers. In general, the more favourable soils in Britain have been cleared for agriculture, and foresters are usually concerned with the less fertile or less tractable soil types, or with steep slopes.

Mycorrhizae, upon which most plants rely for their nutrient uptake, have been briefly mentioned in chapter 2. Mycorrhiza (from two Greek words, literally meaning fungus-root), are the result of symbiosis between certain soil fungi and the roots of plants. The fungus receives food materials (sugar and possibly plant hormones) from the plant. The plant is supplied with water and dissolved nutrients (particularly P) from the fungus. A kind of mycorrhiza (ectomycorrhiza) can easily be seen on many conifers and broadleaved trees used in Britain forestry. Other trees have mycorrhiza of a kind (endomycorrhiza) that cannot be seen without special treatment of the roots.

There are several types of mycorrhiza, but the two of most interest to the silviculturist (and also to the aboriculturist and horticulturist) are the alreadymentioned ectomycorrhiza, and the vesicular-arbuscular mycorrhiza (usually shortened to VAM). In conifers, the former are found particularly with pines, spruces, larches, firs and Douglas fir, and in broadleaves, especially with oaks, beeches and birches. The latter, a kind of endomycorrhiza, is found with such trees as maples, ash, cypresses, and coast redwood. Some trees, e.g., alders, willows and poplars may have both kinds, sometimes on the same root. Pines can have a third type of mycorrhiza, the ect-endomycorrhiza, particularly in nurseries. They seem to act in the same way as ectomycorrhizas, but may be more adapted to high-nutrient soils, so are not so commonly found in the forest.

Mycorrhiza are generally thought to be mutally beneficient relationships. The plants and fungi become inseparably linked and, in some instances, each may be totally dependent on the other. Indeed, this type of symbiosis is so common that, under natural conditions, roots of most plants do not exist independently. Often, therefore, nutrient and water uptake is mainly by way of mycorrhiza, not, as is commonly believed, directly by roots.

Nitrogen fixing. Alders *Alnus* spp. have a symbiotic relationship with a bacterial-like micro-organism of the genus *Frankia* which forms nodules on the roots. These micro-organisms are able to extract (fix) nitrogen from the air. For this reason alders are important species for reclamation work. As reclaimed land is usually low in available nitrogen, it is essential that alder plants destined for these sites should possess nodules before outplanting (McNeill *et al.* 1990), see chapter 5, section 8.

Groups of forest soils. Several systems of soil classification have been devised and are discussed briefly later in this section. They should not be thought of as distinct sharply defined classes, but rather as soils that may merge in various ways. Nor should they always be regarded as occupying large homogeneous areas (other than in upland moorlands and the like), for within short distances quite different soils may be traversed, even in a district where the geological map shows only one main rock formation. Soil differences are the result of many locally variable factors, including slope, plant cover, climate and particularly past treatment, man's management. In general, six groups of forest soils with which foresters should be acquainted are outlined below:

1 Brown forest soils. These fertile and desirable soils, eagerly sought for agriculture, will support a very wide range of tree species. Typically beneath a woodland cover, they have a thin carpet of leaf litter or *mull,* which is always being decomposed by the activities of earthworms, mites, fungi and micro-organisms to become incorporated into a rich brown fertile loam, several centimetres deep, merging gradually into the underlying drift

sor weathered bedrock. The whole structure, though reasonably retentive of moisture, is free-draining; and deep rooting is usually possible. Fertilizers are seldom needed. Vigorous weed growth can be a major problem in crop establishment.

2. Rendzinas (shallow calcareous soils). This group forms over calcareous (chalk and limestone) rocks, e.g. on the Chilterns, the Cotswolds and the Downs. They have a very thin but fertile and freely draining layer of good soil often rich in organic matter, black in colour, declining abruptly into the calcareous rock below. Although these soils will support most kinds of tree in their younger stages, only those that are tolerant of lime, e.g. beech, ash, sycamore, wild cherry, Norway maple, Corsican pine, European larch and western red cedar are likely to grow to full stature. Lime-induced chlorosis (indicated by yellowing of the foliage) often occurs in conifers and some broadleaves. Not every site over chalk or limestone has a rendzina soil; e.g. the calcareous bedrock may be overlaid by a good depth of sand, clay or gravel. The absolute prohibition of any soil inversion is imperative; deep ripping with a tine eases planting and aids establishment, but bringing the solid bedrock up from below and burying the topsoil is a recipe for disaster. There is little benefit from applying fertilizer.

3. Podzols. This soil type, in its uncultivated state, is very unfavourable to tree growth. The only trees it can support successfully are Scots pine, Corsican pine and birch. Podzols are found on most sandy heathlands up and down Britain in areas of low rainfall; they are common around Aldershot, in parts of the New Forest, in north-east Yorkshire and on the east side of Scotland. The soil consists of three major layers:

   i   A surface layer of dark brown or black acid organic matter or *mor,* formed of decayed heather, stems and roots.
   ii A pale grey zone from which many of the minerals have been leached by organic acids from the top layer.
   iii A rust-red, dark brown, black or orange-yellow zone (usually 20–40 cm below the surface) in which iron-rich minerals have been deposited, and which can be quite hard and compact. In extreme cases an *ironpan* may be formed which is usually impervious to water and tree roots. Below the pan lies little altered subsoil. This soil is termed the ironpan podzol.

Forestry Commission literature separates podzols from ironpan soils. Podzols are usually freely drained, whereas ironpan soils are imperfectly drained. Ironpan soils usually develop in wet uplands. They start with a peat layer overlying as a waterloggedanaerobic horizon. An ironpan forms at the interface between this and the freely drained subsoil. Podzols can develop this way too in wetter eastern uplands, e.g. North Yorkshire moors. Ironpan is a barrier to drainage and tree roots. The horizon above the pan is waterlogged and anaerobic (likewise where lower subsoil is indurated and impermeable). An organic surface horizon is common. The soils are strongly acidic. Considerable improvement is possible if the pan is disrupted by tining and cultivation giving good conditions for growth, similar to podzols. Application of phosphate is often necessary. Ploughing to beneath the ironpan is usually necessary for afforestation of sites with ironpan soils; and then most pines (and Sitka spruce in wetter climates) will succeed. (Heather, if not controlled with herbicides, inhibits growth of some tree species, especially spruces; and no broadleaved species will make an economic crop.)

4. Gleys. These soils are found where there is impeded drainage or a fluctuating water-table. The soils are dominantly grey in colour, often with prominent ochreous (yellowish, brownish or rusty red) mottling throughout the subsoil. They develop when vertical drainage is impeded by the fine-textured, dense and structureless lower subsoils or where these are indurated and impermeable. An organic surface horizon is common in cool wet uplands, i.e. the well-known peaty gley. The gleys are often strongly acidic on the surface, but the pH rises with depth. Perched water-table leads to shallow rooting

and instability in crops even after drainage for removal of excess water. Fertility levels vary widely but crops frequently benefit from phosphate on peaty gleys. All gleys require drainage where the mottling occurs within the normal tree rooting zone, and if this is provided these heavy soils are quite fertile for growth particularly of spruces, especially the *Juncus*/grassy non-peaty surface water gleys.

5. Deep peats. Conventionally peats are classified as such if deeper than 50 cm (peaty gleys merge into this as peat exceeds 50 cm). They consist of partially decayed organic matter which accumulates under moist conditions, and are found in upland areas of high rainfall and low temperatures. Rooting is shallow, even after essential intensive drainage. Most upland peats are very infertile, and require high inputs of nutrients. The species choice is usually very limited. Those areas of peat used for afforestation are usually the moorland peats. When these very acid soils can be ploughed or turf-planted, they usually support satisfactory coniferous crops, particularly of Sitka spruce, with an admixture of lodgepole pine where the presence of heather (if not controlled with herbicides) precludes a pure spruce crop. The trees need an application of phosphate to start them off. Pyatt (1990) updates information on long term prospects for forests on peatland; and Dutch *et al.* (1990) indicate the notable benefits of application of potassium fertilizer to Sitka spruce on deep peat.

6. Undeveloped soils. These include skeletal soils, sand dunes and man-made soils, most of which have problems associated with shallowness and infertility. Among them are those of colliery spoil heaps and opencast mining sites. Sand dunes, if first stabilized either by thatching with brushwood or establishing marram grass, can be successfully planted with Scots and Corsican pines, but protection from exposure is needed.

Soil and site classifications for forestry. These are commonly used to indicate the potential of forest sites since soil types often express nutrient status, or water-holding capacity and reflect many aspects of climate. Unless nutrient conditions are particularly difficult, as on many peats and sands, soil physical properties are more important than chemical ones for forestry classifications. Emphasis is placed upon soil properties which affect forest yield and which, though they may be modified by cultural operations, impose relatively permanent limitations on the use of soils. These include the natural drainage, available depth for rooting, the presence of compact or cemented layers, the texture and general level of acidity or alkalinity and the occurrence of peat.

During the past three decades the Forestry Commission has developed and used (and encouraged the private sector to use) a specialized classification of soils as a necessary basis for efficient silvicultural practices. The classification, designed for upland Britain (Carboniferous and older rocks), also assists in making a forecast of tree performance in areas not yet planted. The classification forms the basis for the delineation of types of ground for each of which a distinct form of silviculture may be appropriate. From it, specifications can be given which aid selection of species and indicate the need for cultivation, drainage and nutrition. When exposure of the site has been estimated, it allows the assessment of windthrow hazard (Miller 1985). On the basis of the classification, several silvicultural guides have been produced for the twelve or so regions of upland Britain (Busby 1974) within each of which there is a narrow range of lithological types, a characteristic range of terrain types, soil types, climate, and growth rates (Toleman and Pyatt 1974). The classification has complemented studies by Pyatt *et al.* (1969), Pyatt (1970) and Busby (1974). Pyatt (1982) has made clearer the structure of the classification, part of which is given in Tables, 8, 9 and 10 (see overleaf).

Table 8. The main mineral and shallow peaty soils (peat less than 50 cm)

	Soil Group	Soil Type
Soils with well-aerated subsoil	1 Brown earths	Typical brown earth Basic brown earth Upland brown earth Podzolic brown earth
	3 Podzols	Typical podzol
	4 Ironpan soils	Intergrade ironpan soil Ironpan soil Podzolic ironpan soil
Soils with poorly-aerated subsoil	5 Ground-water gley soils	Ground-water gley
	6 Peaty gley soils	Peaty gley Peaty podzolic gley
	7 Surface-water gley soils	Surface-water gley Brown gley Podzolic gley

Source: Pyatt (1982). FC RIN 68/82/SSN.

Table 9. Peatland soil types (peat 50 cm or more)

	Soil Group	Soil Type
Flushed peatlands	8 *Juncus* bogs (Basin bogs)	*Phragmites* bog *Juncus articulatus* or *acutiflorus* bog *Juncus effusus* bog *Carex* bog
	9 *Molinia* bogs (Flushed blanket bogs)	*Molinia, Myrica, Salix* bog Tussocky *molinia* bog; *Molinia, Calluna* bog Tussocky *Molinia, Eriophorum vaginatum* bog Non-tussocky *Molinia, Eriophorum vaginatum, Trichophorum* bog *Trichophorum, Calluna, Eriophorum, Molinia* bog (weakly flushed blanket bog)
Unflushed peatlands	10 *Sphagnum* bogs (Flat or raised bogs)	Lowland *Sphagnum* bog Upland *Sphagnum* bog
	11 *Calluna, Eriophorum, Trichophorum* bogs (Unflushed blanket bogs)	*Calluna* blanket bog *Calluna, Eriophorum vaginatum* blanket bog *Trichophorum, Calluna* blanket bog *Eriophorum* blanket bog

	Soil Group	Soil Type
	14 Eroded bogs	Eroded (shallow hagging) Deeply hagged bog Pooled bog

Source: Pyatt (1982). FC RIN 68/82/SSN.

Table 10. Various soils

2 Man-made soils	Mining spoil, stony or coarse textured Mining spoil, shaly or fine textured
12 Calcareous soils (soils on limestone rock)	Rendzina (shallow soil) Calcareous brown earth Argillic brown earth (clayey subsoil)
13 Rankers and Skeletal soils (Rankers = shallow soils to bedrock Skeletal = excessively stony)	Brown ranker Gley ranker Peaty ranker Rock Scree Podzolic ranker
15 Littoral soils (Coastal sand and gravel)	Shingle Dunes Excessively drained sand Sand with moderately deep water table Sand with shallow water table Sand with very shallow water table

Source: Pyatt (1982). FC RIN 68/82/SSN.

Insley (ed.) (1988) classifies soils according to the main groupings used by farmers and then follows the classification to describe the species suitable and operations required to establish and grow tree crops of these species. Helliwell in 1987 developed an interesting point-score system to assess land potential for tree growth.

Maps of land capability for forestry in Scotland. Five maps (scale 1:250 000) in the Forestry Commission's Field Book series, and their accompanying illustrated booklets, were commissioned in 1989 from the Macaulay Land Use Research Institute in Aberdeen. They identify and describe the land area of Scotland according to its suitability for different forest tree species. Seven land classes are used based on an assessment of physical features developed by Bibby *et al.* (1988). The classification for forestry is based on an assessment of the degree of limitation imposed by the physical factors of soil, topography and climate on the growth of trees and on silvicultural practices. The principal tree species considered are those conifers and broadleaves commonly grown in Britain, and the classification assumes a skilled management level that will include cultivation, drainage, fertilizer application and weed control where these are necessary. The classification is based on seven types of limitation, these being climate, windthrow, nutrients, topography, droughtiness, wetness and soil. The seven classes ('flexibility' referring to the growth and management of tree crops) are:

F1 Land with excellent flexibility. The soils are deep and well supplied with moisture, and neither climate nor site factors seriously restrict the growth of the main tree species used in Britain. A wide range of coniferous and broadleaved species can be planted.

F2 Land with very good flexibility. The soils have no or only limited periods of seasonal waterlogging, but some mineral gleys may be included if, with drainage, the water-table can be controlled at depths which prevent serious waterlogging of the root system. Minor areas of shallower or wetter soils are acceptable but should not exceed 10% in total. Minor restrictions on cultivation and harvesting due to slopes or minor climatic restraints are also acceptable. Both broadleaved and coniferous species may be planted but choice is more restricted than in Class F1. In areas where available water is limited, those species with high water demand are unsuitable; in areas with water surplus soil drainage may be necessary.

F3 Land with good flexibility. The soil range extends to include mineral gleys with sandy or loamy textures and flushed gleys with humose topsoils. Drainage is necessary on gley soils. Windthrow risk is not high and land management is primarily concerned with limitations imposed by drainage, sloping land or patterns of variable soils. The land is suitable for a wide range of conifers and for a restricted range of broadleaved species.

F4 Land with moderate flexibility. The soils include the more fertile peaty soils and the problem mineral soils, e.g. gleys with clayey textures or soils with calcareous horizons. Ploughing difficulty may be encountered due to stony or shallow soils but this should not be more than 20% of the area. There is a risk of small areas of windthrow which should not be sufficiently severe to reduce rotation lengths or influence management practices. The land is suitable for many coniferous species and in places for the less demanding broadleaves.

F5 Land with limited flexibility. The soils are primarily podzols, peaty gleys and peat, but where limitations are sufficiently severe to limit species selections, other soils may be included. Ploughing is possible but may be more difficult than in the previous classes. Sites in which the risk of windthrow affects management by modifying the thinning practice fall within this class. In the uplands species choice is limited to conifers, such as spruces, larches and pines, and to birch, alder or other hardy broadleaves.

F6 Land with very limited flexibility. The principal limitations are adverse climate and poor soil conditions. The soils include podzols, peaty gleys and peats, and soils affected by toxicities. Sites on which the risk of windthrow effectively prevents thinning and seriously curtails the rotation length and sites with very severe surface terrain which imposes great difficulty in ploughing or extraction fall within this class. Species choice is limited to lodgepole pine and Sitka spruce and to amenity broadleaves such as birch and alder.

F7   Land unsuitable for producing tree crops. Land is considered unplantable if its physical characteristics preclude the growth or establishment of tree crops by normal methods. These characters include extremes of climate, wetness, rockiness and extreme slopes.

The maps cover five separate areas of Scotland – north, west, east, south-west and south-east – and each of the accompanying booklets describes the areas in more detail. The five maps (£6.00 each) and the accompanying illustrated booklets (£2.50 each) are obtainable from the Forestry Commission or the Macaulay Institute, Craigiebuckler, Aberdeen. They clearly demonstrate the considerable potential for forestry in Scotland and are of value not only to foresters in deciding species choice, but also to the Regional Authorities in devising indicative strategies for forestry (see Introduction, page 9).

Maps of land capability for forestry in England and Wales. There is no series of maps and booklets for England and Wales on the lines of those noted above for Scotland. However, bulletins produced by the former Soil Survey of England and Wales for some regions (1984) have sections on land suitability for forestry which when used in conjunction with the regional maps can give quite valuable information on the potential for growth, expected problems and best suited species for quite small sites. The Soil Survey is now known as the Soil Survey and Land Research Centre (SSLRC) based at the Cranfield Institute of Technology, Silsoe, Bedfordshire, and is no longer a Government agency.

SSLRC has undertaken maps of land capability for forestry of parts of the north Pennines and the Cambrian Mountains under contract to the Countryside Commission, and is undertaking research to develop methods of yield prediciton. A Soil Survey database provides information on soil and climate from which the yield potential for different tree species can be assessed. The SSLRC has a system of linking the database with another containing information about the characteristics and soil/climate preferences of the main trees grown in England and Wales. This system, being marketed under the name of 'Treefit', provides useful information for landowners who are considering the economic and environmental benefits of growing trees. Many other individual firms and consultants offer independent and reliable advice on problems related to soil survey and land suitability for tree growth.

Soil, geological and tree relationships. Obviously soil composition has an enormous effect on tree growth. Old Red Sandstones of the Devonian Period, except at high altitudes, include some of the best forest soils; they provide strong, fertile loams ideal for most forest trees and produce prime ash wherever there is depth and moderate shelter. Silurian shales and limestones also provide numerous fertile forest sites. Limestones of the Carboniferous Period afford sites of varying fertility; good beech can usually be grown, and so can ash and sycamore on the lower slopes of the valleys where the ground is moist. Beech is the most suitable broadleaved species for chalk and limestone soils, with ash, sycamore and Norway maple where the soil is deep; appropriate conifers include Corsican pine, western red cedar and European larch, if mixed with beech. By contrast, on such soils the planting of Douglas fir, Japanese larch and sweet chestnut is usually avoided. Questions of lime tolerance do not arise everywhere over these alkaline formations. Clay with flints, for example, also plateau drift and terrace gravels, occur over chalk and provide conditions for trees other than those tolerant of lime. On the Cretaceous Greensands, subject to safeguards from exposure, and with adequate rainfall, excellent crops of almost any of the broadleaves and conifers of higher yield can be grown. Scots pine regenerates freely on the Bagshot sands and gravels where Corsican also does well. Clays from rocks of several geological Periods in lowland England produce excellent oak, e.g. in parts of the New Forest. On blown dune sands, as those in East Anglia, pines do best, especially Corsican, which are relatively resistant to salt spray.

Modification of soils. Forests as they grow modify soils and ground vegetation. Anderson, M.A. (1983) has discussed the effects of tree species on vegetation and nutrient supply in lowland Britain, pointing out that both nutrient supply and plant communities vary most under different tree species 20–25 years after planting at about the first thinning stage. Thereafter, major nutrient concentrations and plant communities become progressively more similar under a variety of tree species.

The reciprocal influences of soil on forest and of forest on soil are not easily disentangled, partly because of the long periods over which these influences act (Matthews 1989). The effect of different kinds of trees on the long term fertility of the soil is problematical (Helliwell 1982). It is generally accepted that most broadleaved species improve the soil or at least maintain its fertility. By contrast it is often assumed that conifers cause the soil to deteriorate, a belief not held by all silviculturists; those who doubt it can instance natural conifer stands hundreds of years old. Some assert that it is only on poor soils that there is a likelihood of such deterioration. The soil is, of course, influenced not only by the trees but also by the ground flora and other undergrowth. Some broadleaves appear better than others (e.g. birch better than oak or beech), some conifers better than others (e.g. western red cedar and Douglas fir better than western hemlock and spruce). There is certainly little risk of the soil deteriorating under larch stands if, on thinning sufficiently heavily, bramble or broadleaved woody growth such as elder invade; if, on the other hand, thinning admits heather, bilberry or dense mats of undesirable grasses to develop, deterioration is more likely.

Problematical too is whether the effects of uneven-aged stands are better or worse than even-aged (see chapter 7). Probably on soils of low nutrient status trees can affect nutrient levels. Under certain conditions mixtures of broadleaves with conifers stimulate microbial activity. Coppicing may cause long-term soil degradation (Pryor and Savill 1986). Large quantities of nutrient rich bark are removed, and the bare-ground state, with accompanied leaching and acidification, occurs much more regularly than with high forest systems. Much of the sessile oak coppice of south-west England is believed to have reduced soil phosphate levels, probably due to a combination of nutrient removal (particularly tan-barking) and leaching. Standards with the coppice would reduce the leaching, but not the removal in bark. The continuity of forest cover has a 'conservative' effect on the soil. In a wet climate, the clear-felling of a stand eliminates interception by the canopy and leads to a rise in water content and wetter site conditions.

The foregoing information on soils is used in chapter 4 when noting the silvicultural characteristics of individual species and when discussing species choice.

4. The forest ecosystem

A forest is an ecosystem – the product of climate, geology, terrain, soil and time, and of the trees, shrubs, animals, fungi, and other organisms living and interacting together on the site (Matthews 1989). Natural forest ecosystems vary greatly in their composition and structure. They may comprise one or several tree species which are similar in age or size, or they may differ in these respects. Consequently, two important aspects of forest ecology are whether managed forests should comprise stands that are pure or are mixed in species composition and whether, in structure, they should be regular (even-aged) or irregular (uneven-aged) as discussed in chapter 7. The woodland community is composed of many kinds of living organisms which differ greatly in appearance and abundance (Ovington 1965; Minckler 1975; Helliwell 1982; Packham and Harding 1982; Marren 1990; Watkins 1990). The interaction between living things in woodlands is complex and dynamic and the activities of one group may greatly affect others.

In a forest, the canopy of leaves and branches supported by the tree stems intercepts incoming radiation and precipitation in the form of rain, snow and mist. The canopy also affects the flow of air over and through the upper parts of the forest, making it turbulent; wind speeds in the lower parts of the forest are one-quarter to one-half of those in the open. The humidity of the air inside is higher than outside. The amount of light reaching the ground when the full canopy is present is between 1 and 15% of incident light. When the microclimate within a forest stand is compared with that of a nearby site without trees, the former is more equitable.

On clear-felling, the overhead and side shelter provided by the canopy is lost and the microclimate becomes similar to that of a nearby site without trees. The range of temperature widens and the desiccative effects of wind have full play. The hydrological cycle is broken, interception with evaporation and transpiration cease, and the water content of the soil rises. Surface run-off also increases. Wet soils become wetter and dry soils drier. The nutrient cycle is disrupted. The litter and humus layer is exposed and rapidly broken down.

If the forest ecosystem is treated without due care and a cleared site cannot readily be restocked, the disturbance caused by harvesting is more severe and long lived. The site may have become less productive. The most common symptoms of site deterioration are: soil compaction, excessive wetness or dryness of the surface and upper horizons of the soil, loss of soil due to surface run-off and erosion, and rank growth of rushes and other weeds which flourish in disturbed soils.

As the management of forests and woodlands becomes more intensive, and as ever more of the tree – stump, roots and bark, together with the mineral nutrients which it contains – is removed in forest products, so foresters require an increasing knowledge of the ecology of woodland processes (Packham and Harding 1982). The same is true of foresters endeavouring to conserve woodlands of aesthetic and scientific value. Because of their complex structure and dynamic nature, forest ecosystems will always pose management problems and provide a challenge to foresters. They will have a certain reverence for the forest ecosystem, trying to understand it, to learn how to modify it for man's needs, and to keep it healthy while being used, observed, or enjoyed; and will bear the ecosystem in mind whenever they ponder the choice of silvicultural system, method of establishing, tending, harvesting and restocking.

A large amount of research is being done into the processes that link soil and trees in forest ecosystems, into methods of site improvement, and into raising the genetic potential for growth and yield of the trees themselves – all with the object of maintaining and increasing the productivity of sites and stands (Matthews 1989).

Conservationists generally seek 'naturalness' in the forest ecosystems or wildlife habitats which they value most – 'naturalness' being that which has evolved without interference by man, i.e. the opposite of artificial (Ratcliffe 1990). But, almost all British ecosystems and habitats have been modified by man's activities, so usually sought is some degree of naturalness. Semi-natural is used to describe ecosystems which consist mainly of native species in an assemblage close to that which probably occurred naturally: they often contain the widest range of species possible within the physical and chemical limitations of the site. A discussion of ancient seminatural woodland is included in chapter 4, section 4.

Present management objectives will affect the forest ecosystem and influence the future wildlife value of all sites – matters discussed in chapter 4, section 4 and chapter 13, section 7.

References

Anderson, D. (1990), 'Carbon fixing from an economic perspective'. Draft for comment 4 September 1990, Department of Economics, University College London. Study on 'The Forestry Industry and the Greenhouse Effect' for the Scottish Forestry Trust and Forestry Commission.

Anderson, M.A. (1983), 'The Effects of Tree Species on Vegetation and Nutrient Supply in Lowland Britain'. FC Arboriculture Research Note 44/83/SSS.

Anderson, M.L. (1961), *The Selection of Tree Species* (2nd edn.). Oliver and Boyd, London.

Barrow, P., Hinsley, A.P. and Price, C. (1986), 'The Effect of Afforestation on Hydroelectricity Generation: a quantitative assessment'. *Land Use Policy,* 3, pp. 141–5.

Bibby, J.S., Heslop, R.E.F. and Hartnup, R. (1988), 'Land Capability Classification for Forestry in Britain. Soil Survey Monograph'. The Macaulay Land Use Research Institute, Aberdeen.

Binns, W.O. (1980), 'Trees and Water'. FC Arboricultural Leaflet 6.

Booth, T.C. (1977), 'Windthrow Hazard Classification'. FC Research Information Note 22/77/SILN.

Busby, R.J.N. (1974), 'Forest Site Yield Guide to Upland Britain'. FC Forest Record 97.

Cajander, A.K. (1921), 'Über Waldtypen in allgemeinen'. *Acta Forestalia Fennica, 20,* pp. 1–77.

Cannell, M.G.R., Grace, J. and Booth, A. (1989), 'Possible Impacts of Climatic Warming on Trees and Forests in the United Kingdom: a Review'. *Forestry, 62,* (4), pp. 338–64.

Collett, M.E.W. (1970), 'External Costs arising from the Effects of Forests upon Streamflow in Britain'. *Forestry, 43,* pp. 87–93.

Dutch, J.C., Taylor, C.M.A. and Worrell, R. (1990), 'Potassium Fertilizer – Effects of different Rates, Types and Times of Application on Height Growth of Sitka Spruce on Deep Peat'. FC Research Information Note 188.

Evans, J. and Paterson D.B. (1990), 'Restocking after Windthrow in Southern Britain'. FC Research Information Note 175.

Forestry Commission (1988), 'Forests and Water Guidelines'.

Freer-Smith, P.H. (1990a), 'Forests and Climate Change'. *Timber Grower,* Autumn 1990, pp. 17, 19.

—— (1990b), 'Climate Change: the Contribution of Forestry to Response Strategies'. FC Research Information Note 189.

Gibbs, J.N. (1990), 'The Clearance of Trees after the Gale of 25 January 1990'. FC Research Information Note 172.

Grayson, A.J. (1989), 'Carbon Dioxide, Global Warming and Forestry'. FC Research Information Note 146.
—— (ed.) (1989), 'The 1987 Storm: Impacts and Responses'. FC Bulletin 87.
Helliwell, D.R. (1982), *Options in Forestry.* Packard, Chichester.
Hibberd, B.G. (ed.) (1991), 'Forestry Practice'. FC Handbook 6.
Hill, G. (1988), *Hurricane Force: The Story of the Storm of October 1987.* Collins, London.
Innes, J.L. (1987), 'Air Pollution and Forestry'. FC Bulletin 70.
—— (1990), 'Assessment of Tree Condition'. FC Field Book 12.
Innes, J.L. and Boswell, R.C. (1990), 'Monitoring of Forest Condition in Great Britain 1989'. FC Bulletin 94.
Insley, H. (ed.) (1988), 'Farm Woodland Planning'. FC Bulletin 80.
Jarvis, N.J. and Mullins, C.E. (1987), 'Modelling the Effects of Drought on the Growth of Sitka Spruce in Scotland'. *Forestry, 60,* pp. 13–30.
Marren, P. (1990), *Woodland Heritage.* David & Charles, Newton Abbot.
Matthews, J.D. (1989), *Silvicultural Systems.* Clarendon Press, Oxford.
McNeill, J.D. *et al.* (1990), 'Inoculation of Alder Seedlings to Improve Seedling Growth and Field Performance'. FC Arboricultural Research Note 88/90/SILN.
Miller, K.F. (1985), 'Windthrow Hazard Classification'. FC Leaflet 85.
Mills, D.H. (1980), 'The Management of Forest Streams'. FC Leaflet 78.
Minckler, L.S. (1975), *Woodland Ecology.* Syracuse University Press, New York.
Nicholson, I.A., Robertson, R.A. and Robinson, M. (1989), 'The Effects of Drainage on the Hydrology of a Peat Bog'. *International Peat Journal, 3,* pp. 59–83.
Nisbet, T.R. (1990), 'Forests and Surface Water Acidification'. FC Bulletin 86.
Nisbet, T.R., Mullins, C.E. and Macleod, D.A. (1989), 'The Variation of Soil Water Regime, Oxygen Status and Rooting Pattern with Soil Type under Sitka Spruce'. *Journal of Soil Science, 40,* pp. 183–97.
Ovington, J.D. (1965), *Woodlands.* English Universities Press, London.
Packham, J.R. and Harding, D.J.L. (1982), *Ecology of Woodland Processes.* Arnold, London.
Pearce, D.W. (1990), 'Assessing the Returns to Society and the Economy from Afforestation', *mimeo,* Department of Economics, University College London. October 1990. Paper presented at the AGM of the FICGB, 22 May 1990, London. *Timber Grower,* Autumn 1990.
—— (1990), personal communication. University College London.
Price, C. (1989), *The Theory and Application of Forest Economics.* Basil Blackwell, Oxford.
Pritchett, W.L. and Fisher, R.F. (1989) (2nd edn.), *Properties and Management of Forest Soils.* John Wiley & Sons, New York.
Pryor, S.N. and Savill, P.S. (1986), 'Silvicultural Systems for Broadleaved Woodland in Britain'. Oxford Forestry Institute Occasional Paper No. 34.
Pyatt, D.G., Harrison, D. and Ford, A.S. (1969), 'Guide to Site Types in Forests of North and mid-Wales'. FC Forest Record 69.
Pyatt, D.G. (1970), 'Soil Groups of Upland Forests'. FC Forest Record 71.
—— (1982), 'Soil Classification'. FC Research Information Note 68/82/SSN.
—— (1984), 'The Effect of Afforestation on the Quantity of Water Run-off'. FC Research Information Note 83/84/SSN.
—— (1990), 'Long-term Prospects for Forests on Peatland'. *Scottish Forestry, 44* (1), pp. 19–25.
Pyatt, D.G., Anderson, A.R. and Ray, D. (1987), 'Deep Peats'. FC Report on Forest Research.
Quine, C.P. and Reynard, B.R. (1990), 'A New Series of Windthrow Monitoring Areas in Upland Britain'. FC Occasional Paper 25.
Ratcliffe, P.R. (1990), 'Naturalness in Woodland Management'. *Timber Grower,* Summer 1990, p. 13.
Robinson, M. (1986), 'Changes in Catchment Run-off following Drainage and Afforestation'. *Journal of Hydrology, 86,* pp. 71–84.

Rose, D.R. (1990), 'Lightning Damage to Trees in Britain'. FC Arboriculture Research Note 68/90/PATH.

Soil Survey of England and Wales Regional Bulletins (1984), 'Soils and their Use in northern England, Wales, south-west England, and south-east England'.

Taylor, G., Dobson, M.C., Freer-Smith, P.H. and Davies, W.J. (1989), 'Tree Physiology and Air Pollution in southern Britain'. FC Research Information Note 145.

Thompson, D. A. and Matthews R.W. (1989), 'The Storage of Carbon in Trees and Timber'. FC Research Information Note 160.

Toleman, R.D.L. and Pyatt, D.G. (1974), 'Site Classification as an Aid to Silviculture in the Forestry Commission of Great Britain'. Paper for 10th Commonwealth Forestry Commission, UK.

Valentine, K.W.G. (1986), *Soil Resource Surveys for Forestry*. Oxford Science Publications, Clarendon Press, Oxford.

Watkins, C. (1990), *Woodland Management and Conservation,* David & Charles, Newton Abbot.

(a) Sessile oak (b) Sweet chestnut

(c) Ash (d) Silver birch

Plate 1. Botanical features of four commonly planted commercial broadleaves

[C. Raymond]

(a) European larch

(b) Douglas fir

(c) Sitka spruce

(d) Western hemlock

Plate 2. Botanical features of four commonly planted commercial conifers

[C. Raymond]

Plate 3. Ancient semi-natural woodland. This aerial view, though not of good quality, is important, being of woodland surrounding the 'Norman Castle of Dene' – the origin of the name 'Forest of Dean', in Gloucestershire. The woodland and earthwork castle were presented to the Dean Heritage Museum Trust by a private forester in 1987. The main species are sweet chestnut, of a provenance brought to the locality in Roman times

[M. Fitchett]

Plate 4. Westonbirt Arboretum near Tetbury, in Gloucestershire. Started by Robert Holland in 1829 on open agricultural land and acquired in 1956 by the Forestry Commission, the arboretum has significantly assisted the training in tree identification of land agents/surveyors at the Royal Agricultural College, Cirencester

[Forestry Commission]

(a) Torrachilty Forest in the Highland region of Scotland

[Forestry Commission]

(b) Eskdalemuir Forest in Dumfries and Galloway, Scotland

[K. Taylor]

Plate 5. Upland cultivation. Typical scenes of deep ploughing – a necessary cultivation for afforestation – transforming the Scottish upland landscape. Both sites now carry thriving conifer plantations. Afforestation inevitably creates dramatic changes in the landscape and the local ecosystems, but later there are many compensating benefits. An ebb and flow of mammals, birds and plants occurs as plantations are created and change in structure with time

Plate 6. Tavistock Woodland Estate in Devon. One of the older plots of the Bradford-Hutt Plan for continuous cover silviculture (see chapter 7). The older trees are P10 western red cedar. Below are younger western red cedar and western hemlock. The mixed species, uneven-aged, continuous cover selection silvicultural system is based on a geographical pattern

[T. Burrows]

Plate 7. Workmans Wood National Nature Reserve ('Poppets Corner') in Gloucestershire. Autumn colour in fine 110-year-old beech with well-established natural regeneration. Prudent silvicultural thinning over a long period with appropriate control of rabbits and deer have produced a receptive ground cover. Now almost half of the 325 hectares of the nature reserve are filled with natural regeneration, some of it up to forty years old (see chapter 7)

[J. Workman]

(a) 'Warminster Plain': Mixed conifers and broadleaves one to twenty years old under forty-year-old highpruned Japanese larch

(b) 'Redway Plain': Douglas fir one to twenty years old with forty-five-year-old Scots pine in the background

Plate 8. Longleat Estate Woodlands in Wiltshire. Fine examples of natural regeneration (see chapter 7)
[J. McHardy]

(a) 'Haucombe North Hanging': Douglas fir one to fifteen years old under sixty-seven-year-old Douglas fir

(b) Young Japanese larch and grand fir under forty-year-old high-pruned Japanese larch, with fifteen-year-old Lawson cypress in the background

Plate 9. Longleat Estate Woodlands in Wiltshire. Fine examples of natural regeneration (see chapter 7)
[J. McHardy]

Chapter Four

SILVICULTURAL CHARACTERISTICS AND SPECIES CHOICE

1. Silvicultural characteristics

Botanical features of the main forest and associated trees have been noted in chapter 1, section 2 under 'Identification of trees'. Silvicultural characteristics are now introduced. Some foresters stress the distinction between silviculture and silvicultural characteristics – the nouns are silvics and silviculture. The former refers to the characteristics of individual species, in some ways paralleling autoecology.[1] The term silviculture they confine to the handling of a group or stand of trees, including cutting systems and interventions in the development of the crop. Silvicultural features include, where appropriate:

- Native or exotic; conifer or broadleave; evergreen or deciduous
- Root system, stem-form and crown
- Branch habit, size, whorls, angle, persistency and apical dominance
- Climate, soil, moisture and nutrient requirements
- Type of planting stock, and espacement at planting
- Light requirements; tolerance to shade; the shade cast
- Ability to coppice, naturally regenerate, or sucker
- Sensitivity to drought, sun-scorch, frost, wind, exposure, pollution, salt-spray, disease, fungi and insect attack
- Susceptibility to deer browsing and bark stripping; and to damage by grey squirrels and rabbits
- Volume, quality, cleavability, natural durability, amenability to preservative treatment, and utility of thinnings and final crop
- Rate of growth, vigour, longevity, rotation length, ultimate form, height and size attained, and response to pruning and thinning
- (As relates to conservation, amenity and landscape) general appearance, flowers, fruit, and colour of foliage and bark throughout the four seasons, partially discussed in chapter 1.

1 *Silvics* is defined as 'the study of the life history and general characteristics of forest trees and stands, with particular reference to locality factors, as a basis for the practice of silviculture' (Ford-Robertson 1971). Silvics deals primarily with ecological relationships of trees in natural stands.

Additional to the foregoing are such features as reproductive maturity (Matthews 1963; Gordon and Rowe 1982; Evans 1988) – minimum seed-bearing age, interval between large seed crops, age after which seed production begins to decline, and time of seed-fall or seed dispersal. For the main commercially-important species, the basic timber properties, including density, strength, shrinkage, movement, natural durability, amenability to preservative treatment and working properties are noted in chapter 9.

Silvicultural characteristics in this chapter are dealt with as a conventionally familiar term – briefly the particular features in the growth of trees, their potentialities and their value within the forest.

The tree's root-system and crown are equally important, the former having a bearing on both the depth of soil the tree requires and on its ability to resist wind, while the crown has a direct effect on the shape and volume of its stem and branches. A species which naturally forms a deep main root (e.g. oak) will not develop to its best unless it has sufficient depth of soil; one which tends to form large spreading branches may be less suitable for the growth of sawlogs than one of pyramidal shape with short, slender branches, though this to a great extent can be controlled by thinning.

A tree's need for light is likewise of importance (see chapter 3). Light-demanding species are thinly foliaged and their lower branches usually die readily in shade. They, and pioneer species, accumulate volume rapidly at first, and reach a peak early, after which increment declines, and hence annual percentage increase in volume falls dramatically. Shade bearers are heavily foliaged, and their lower branches are persistent and are not so much affected by the shade in which they grow. This last characteristic is pertinent, for species whose branches die very slowly (e.g. western red cedar and western hemlock) must be kept specially close together, particularly in early and middle life, in order to encourage natural pruning to produce knot-free timber; however, if manual pruning is employed close espacement may not be vital. Shade-bearing species, and climax species, grow slowly at first and only gradually build up to peak annual increment; and hence increment percent is upheld longer than for a light-demanding species. The effect of light on epicormic shoots has been noted in chapter 3.

A tree's requirements for soil, moisture, climate and location are all crucial (see chapter 3). So too is the rate of growth, both in girth and height, as well as the quantity and quality of the thinnings and mature trees produced. Rate of growth is determined by the species and the site factors; and, in respect of girth, by the thinning regime. These factors will sometimes determine the length of the rotation. Within the limits usually practised, height and total yield are unaffected by thinning regime. But this statement regarding height holds up less well for some of the slower-growing broadleaves; for example, oak at wide spacings (3 m) has been found to grow somewhat slower in height than at close spacings (less than 2 m) or when in mixture with conifers. Also, in regard to yield, biomass experiments indicate that on very short rotations (less than five years) spacing is a principal determinant of yield.

Some species (e.g. poplar) grow at a fairly uniform rate throughout their development, others (e.g. beech, grand and noble firs) usually grow slowly in early life and more rapidly later. Light demanders tend to grow rapidly when young and thereby attain early much of their required height; shade bearers, in contrast, grow slowly in height at first and more rapidly later on and may eventually catch up with or even overtake light demanders. The volume produced in thinnings and final crops and the time taken to produce it varies with species, sites, and management: factors which affect yield class and, along with markets, profitability.

Sensitivity to damage by wind, frost, insects and fungal diseases is important (see chapters 3 and 6). Trees of robust stock, in a suitable environment and tended with care, on the whole are less harmed or more resistant to these injuries. Although most pests attack certain species only, the two most serious fungal diseases, root and butt rot *Heterobasidion annosum* and Honey fungus *Armillaria mellea,* are almost catholic in their tastes. Some agencies of damage can be guarded against, e.g. the chemical protection of young planting stock – either bare-rooted or raised in containers – against Large pine weevil *Hylobius abietis* and the Black pine and spruce beetles *Hylastes* spp. which are frequently destructive on felled conifer sites.

Notes on the main silvicultural characteristics of the most commonly planted forest trees, set out hereafter, will assist in species choice (see section 2 *infra*) and in deciding the appropriate type of planting stock and the espacement for forestry. (A table of the number of trees required to plant per hectare is given in chapter 5.)

Main silvicultural characteristics of conifers

For conditions justifying selection for timber growing, for generally unsuitable conditions and for some recommended provenances see 'Species choice' in section 2 *infra*. The following notes show those conifers most commonly planted for timber. All are evergreen except the larches. For thinning recommendations see chapter 8. For timber properties and uses, see chapter 10.

Species (Scientific names are given in the Index) Planting stock and approximate spacing[2]	Abbreviations used: N: Native; E: Exotic; PB: Persistent branches; LD: Light demander; SB: Shade bearer; MSB: Moderate shade bearer; WF: Windfirm; WW: Windweak; MWF: Moderately windfirm; FH: Frost hardy; FT: Frost tender; MFH: Moderately frost hardy (this relates to spring frosts in the growing season, not to winter cold). YC = Yield class (with average) as discussed in chapter 12.
Scots pine N, LD, WF, FH, PB 1u1, 1+1 or 2+1 1.5 x 1.5 m	Hardy, stands exposure, good wind-break. Accommodating as to most soils. Low demands on nutrients and water. Tolerates poor soils and harsh microclimate. Easy to establish. Early growth fast, thereafter moderate. Volume production not high compared with more exacting species. Withstands competition from grass and heather. Good pioneer and nurse species. Occasionally regenerates freely. First thinnings usually of moderate quality. Sometimes grown for Christmas trees at 1.0 x 1.0 m spacing, or wider. Fire danger on sandy heaths. High conservation and landscape value in older ages. Scotland's native pinewoods are among the least modified woodland areas in Britain and as such are extremely important for conservation. Pests: Pine shoot beetle and Pine shoot moth; also *Lophodermium* fungi and *Peridermium pini* rust, particularly at Thetford. Blue stain in the timber can be a problem. YC 4–14(8).

---

2   The number of trees required to plant per hectare is given in chapter 5.

Species (Scientific names are given in the Index) Planting stock and approximate spacing	Abbreviations used: N: Native; E: Exotic; PB: Persistent branches; LD: Light demander; SB: Shade bearer; MSB: Moderate shade bearer; WF: Windfirm; WW: Windweak; MWF: Moderately windfirm; FH: Frost hardy; FT: Frost tender; MFH: Moderately frost hardy (this relates to spring frosts in the growing season, not to winter cold). YC = Yield class (with average) as discussed in chapter 12.
Corsican pine E, LD, WF, FH, PB 1u1 or 1+1 or container grown seedlings 2.0 x 2.0 m	Hardy, stands exposure, good wind-break. Tolerates dry areas and alkaline conditions. More difficult to establish than Scots pine because of tap root. Best planted in February–March. Potentially fast grower. First thinnings usually of moderate quality. Produces timber faster than Scots pine, gives greater volume, but is less tolerant of cold and poor soil conditions. The higher yield than Scots pine is due largely to its more rigorous growth in the years following establishment. Good stem-form, often better than Scots pine, but has somewhat larger branches in more conspicuous whorls. A profitable crop. Increasingly established as container grown seedling planting stock. Relatively resistant to air pollution. Susceptible to the fungus *Gremmeniella abietina* on wetter upland soils. YC 6–20(11).
Lodgepole pine E, LD, FH, WF, PB 1u1 or 1+1 or 2+1 2.0 x 2.0m	Excellent on peat, pure as a pioneer species, or as a nurse to Sitka spruce. Low nutrient requirements. Rapid early growth. Modest growth potential. Seeds at an early age. Grows better than Scots pine on grass, heather or peat. Stands cold and exposure better than spruce. Occasionally crippled by Pine Beauty moth *Panolis flammea* in north Scotland, and by *Ramichloridium pini* in Wales. Planted on a large scale in Scotland, Wales and Ireland. Tends to grow very coarsely on moist fertile soils. Very susceptible to deer damage especially bark stripping by red deer. Fairly tolerant of air pollution. For optimum results the choice of correct provenance is important (see section 2, 'Species choice'); those provenances available differ in vigour, flushing, form and branching habit. YC 4–14(7).
European larch E, LD, MFH, MWF 1u1 or 1+1 or 2+1 1.5 x 1.5 m	Deciduous. Branches sweep down. Flushes earlier than JL. Moderately hardy. Best adapted to the drier parts of the country. Reasonably fast-growing in early life but yield not exceptionally high. Good quality first and subsequent thinnings. Can be heavily thinned and underplanted. Suitable nurse species to oak and beech. A useful nurse for Sitka spruce on poor sites. Rapid early growth gives short economic rotation on better freely-drained sites; but relatively low yield potential. Easy to brash and prune. Lower branches die early and, unless removed, give rise to the small encased knots which are a feature of much larch timber. Not a high yield species but is usually a profitable crop. Canker and dieback are problems. The provenance is very important. Relatively susceptible to *Heterobasidion annosum*. High amenity value for landscaping and conservation purposes. Grows well at high altitudes but commonly responds to prevailing winds by producing a swept lower stem which causes problems on conversion, with the wood tending to spring from the saw. YC 4–16(8).

Species (Scientific names are given in the Index) Planting stock and approximate spacing	Abbreviations used: N: Native; E: Exotic; PB: Persistent branches; LD: Light demander; SB: Shade bearer; MSB: Moderate shade bearer; WF: Windfirm; WW: Windweak; MWF: Moderately windfirm; FH: Frost hardy; FT: Frost tender; MFH: Moderately frost hardy (this relates to spring frosts in the growing season, not to winter cold). YC = Yield class (with average) as discussed in chapter 12.
Japanese larch E, LD, MFH, MWF 1u1 or 1+1 2.0 x 2.0 m	Deciduous. Strong horizontal branching. More accommodating as to soil than EL; but more subject to damage by drought. Likes plenty of rainfall. Fairly hardy. Fast starter. Tolerant to air pollution. Faster growing than EL, with higher yield to middle age. Good quality first and subsequent thinnings. A profitable crop. Wind-sway and spiral growth on rich soil. Chlorosis generally on alkaline soil. Relatively susceptible to *Heterobasidion annosum*. Poor form in exposed conditions. Usually planted pure. Needs heavy thinning. Ideal for fire-break, after hazardous early years. Can be heavily thinned and underplanted. Usually canker-free. Better adapted than EL for growth in the higher rainfall areas of the west and north and on less fertile soils. Greater resistance to larch dieback compared with EL. High amenity value for landscaping and conservation purposes. YC 4–16(8).
Hybrid larch N, Hybrid, LD, MFH, MWF 1u1 or 1+1 2.0 x 2.0 m	Deciduous. A natural hybrid. Characteristics between those of EL and JL, but depend on the qualities of both parents. Quick starter. On good sites can grow even more quickly than JL. A profitable crop. Preferable to JL when seed available, particularly on low fertility sites. Needs heavy thinning. Outgrows either parent on all but the very best larch sites and its superiority becomes very marked under conditions that are marginal for growth of larch. Hardier and more resistant to pests than EL and JL. Relatively susceptible to *Heterobasidion annosum*. High amenity value for landscaping and conservation purposes. YC 4–16(8).
Douglas fir E, MSB, FT, WW, PB 1u1, 1+1 or 2+1 2.0 x 2.0 m	High growth potential on better freely drained sites at low elevations. Unsuitable in wet and exposed areas being shallow rooted and has a heavy crown hence leading to instability in unsheltered sites, i.e. very windthrow susceptible. Dislikes dry areas, chalk and limestone. On deep, moist and well drained soils it grows rapidly and produces a high volume of excellent timber. Choice of provenance is important (see section 2, 'Species choice'). Best seed origins are now available. Nowhere common but is widely planted in western Britain, particularly in south-west England and Wales, also on the lower slopes of some of the western Scottish glens. Yields can exceed those from Sitka spruce. Tolerates side but not top shade. Best as a pure crop. Thinning at too late a date may lead to windthrow. Very susceptible to deer damage. Yields fine strong

Species (Scientific names are given in the Index) Planting stock and approximate spacing	Abbreviations used: N: Native; E: Exotic; PB: Persistent branches; LD: Light demander; SB: Shade bearer; MSB: Moderate shade bearer; WF: Windfirm; WW: Windweak; MWF: Moderately windfirm; FH: Frost hardy; FT: Frost tender; MFH: Moderately frost hardy (this relates to spring frosts in the growing season, not to winter cold). YC = Yield class (with average) as discussed in chapter 12.
	timber, but end uses more specialized than for spruces. Relatively resistant to *Heterobasidion annosum* and Honey fungus. Pests include woolly aphid *Adelges cooleyi*. A profitable crop. Increased use of containerized seedling planting stock. YC 8–24(14).
Norway spruce E, MSB, MFH, WW, PB 1u1 or 2+1 or 2+2 1.5 x 1.5 m	Moderately hardy. Shallow rooted. Accommodating as to soil if fertile and not too dry. Better adapted than Sitka spruce for growth in the drier, eastern parts of Britain. Rarely as vigorous as well-grown Sitka spruce but is of good stem-form and modest branch size. High growth potential on fertile sites. Checked by heather. Good drainage desirable. Sensitive to wind exposure. Susceptible to salt-laden winds and browning of winter foliage on exposed sites. Fairly high volume producer. Some origins of very late flushing are suitable for frost hollows. Sudden exposure by thinning can lead to ill-health and die-back. More susceptible than Sitka spruce to deer and sheep damage. Pole stage and older stands susceptible to bark stripping by red deer. Relatively susceptible to Honey fungus and *Heterobasidion annosum*. Often grown for Christmas trees at 1.0 x 1.0 m spacing, or closer. YC 6–22(12).
Sitka spruce E, LD, FT, MWF, PB 1u1 or 1+1 or 2+2 2.0 x 2.0 m	High growth potential. Faster-growing than Norway spruce and larger volume. Has a good stem-form. Accommodating as to soil except dry or alkaline. Deeper-rooting than Norway spruce. Relatively resistant to wind exposure, but doubtful stability on sites with high water-table. The most commonly planted tree in British forests, mainly of QCI origin (see section 2, 'Species choice'). Tolerant of a range of site conditions but grows best in the higher rainfall areas of south and west Scotland, north and south-west England, and Wales. Best in high annual rainfall over 1,000 mm. Frost susceptibility depends on provenance (QCI origin is susceptible to late frost; Alaskan is not but is less productive. In frost hollows Norway spruce is necessary instead of Sitka spruce). Readily established, and grows well. Increased use of container grown seedling planting stock, as well as planting genetically improved stock raised from cuttings. Suffers defoliation by Green spruce aphid *Elatobium abietinum*, often cyclically, but survives. Relatively susceptible to *Heterobasidion annosum*. A profitable crop. Tolerates substantial deer pressures: much less bark stripping by red deer than Norway spruce. Generally disease resistant. Can be severely checked by heather. YC 6–24+(12).

Species (Scientific names are given in the Index) Planting stock and approximate spacing	Abbreviations used: N: Native; E: Exotic; PB: Persistent branches; LD: Light demander; SB: Shade bearer; MSB: Moderate shade bearer; WF: Windfirm; WW: Windweak; MWF: Moderately windfirm; FH: Frost hardy; FT: Frost tender; MFH: Moderately frost hardy (this relates to spring frosts in the growing season, not to winter cold). YC = Yield class (with average) as discussed in chapter 12.
Western hemlock E, SB, FT, MWF, PB 1u1 or 2+1 or 2+2 1.5 x 1.5 m	Whip-like, pendent leader. Fast growth, good volume. High growth potential even on sites rather dry for Sitka spruce. Tolerates shade; establishment is easier under a light shade. Casts dense shade. Acidic litter. Excellent underplant (for example, of birch). Susceptible to bole-fluting; also to *Heterobasidion annosum* on the better soils. Sensitive to frost and exposure. Very susceptible to deer pressures. Foliage accepted by florists. YC 12–24+(14). Can be very 'aggressive' and dominant.
Western red cedar E, SB, MFH, MWF, PB 1u1 or 2+1 or 2+2 1.5 x 1.5 m	Erect leader. Tolerates alkaline conditions. Reasonably fast growth and good volume. Needs to be kept dense. Casts dense shade. Useful in mixture with or as an underplant of oak or larch. Prone to *Didymascella thujina* disease in nurseries, but appears immune later in life. Relatively resistant to *Heterobasidion annosum*. Good hedge plant. Foliage accepted by florists. YC 6–24(12).
Lawson cypress E, SB, FH, WW, PB 1u1 or 2+1 or 2+2 1.5 x 1.5 m	Drooping leader. Of limited forestry value. Slow-growing. Tendency to fork. Liable to suffer from snow. Useful as a nurse for oak. Accommodating as to soil; best in medium to high fertility. Likes fair to good rainfall. Dislikes heather land and dry infertile soils. Keep dense, and single the forks until inaccessible. Good hedge plant. Diseases: *Heterobasidion annosum* and Honey fungus. Foliage accepted by florists. YC 8–20(12).
Grand fir E, SB, FT, MWF, PB 1u1 or 2+1 or 2+2 1.5 x 1.5m or 2.0 x 2.0 m	Tallest tree species in Britain: c. 63 m in 1990. Fast, very heavy volume producer of perhaps second-rate timber (though this depends on the end use). Stem-form good. Best planted pure or in mixture with Douglas fir or Japanese larch. Sensitive to exposure. Good underplant, being shade tolerant. Slow early growth, then fast. Prone to drought-crack on dry and clay soils, and to windbreak. Sensitive to deer pressures. Relatively resistant to *Heterobasidion annosum* and Honey fungus. YC 8–34+(14).

Species (Scientific names are given in the Index) Planting stock and approximate spacing	Abbreviations used: N: Native; E: Exotic; PB: Persistent branches; LD: Light demander; SB: Shade bearer; MSB: Moderate shade bearer; WF: Windfirm; WW: Windweak; MWF: Moderately windfirm; FH: Frost hardy; FT: Frost tender; MFH: Moderately frost hardy (this relates to spring frosts in the growing season, not to winter cold). YC = Yield class (with average) as discussed in chapter 12.
Noble fir E, MSB, MFH, WF, PB 1u1 or 2+1 or 2+2 1.5 x 1.5 m or 2.0 x 2.0 m	Hardy, stands exposure. Useful shelterbelt under certain west coast conditions. Slow early growth and then moderately fast. High volume. Prone to drought-crack. Relatively resistant to *Heterobasidion annosum*. Often grown as Christmas trees at 1.0 x 1.0 m spacing, or wider. YC 8–34(14).
Austrian pine E, LD, FH, WF, PB 1+1 or 2+1 1.5 x 1.5 m	Hardy, stands exposure. Good wind-break. Slower-growing and heavier- and coarser-branching than Corsican pine. More resistant to lime-induced chlorosis than Scots pine when planted on chalk sites. Tolerates air pollution better than other conifers except perhaps Japanese larch. Timber inferior to Scots and Corsican pines in most properties. Mainly planted as a shelterbelt. Preferred origins are EC registered stands. On thin soils over chalk or limestone, Austrian seedlots may be more healthy and more resistant to chlorosis than those from Corsica, but with poorer volume production and stem-form (Lines 1987).
Coast redwood E, MSB, FT, MWF, PB 1+1 or 2+1 2.0 x 2.0 m	Shade bearer. Frost tender. Likes deep, moist, fertile soils and mists. Fast growth. Successful crops, e.g. at Leighton, Huntley, Longleat, and Forest of Dean. Usually slow to establish. Will coppice. Best planted under tall cover. Good specimen landscape tree. One of the tallest species in the world: up to 112 m in California.
Serbian spruce E, LD, FH, WF, PB 2+1 1.5 x 1.5 m	A slender tower of dark blue. Slower-growing than Scots pine but tolerates poorer soils and is more frost hardy. Useful in frost hollows where other conifers may fail. Very susceptible to Honey fungus. Timber similar to Norway spruce.

Conifers additional to the foregoing are important in parks and substantial gardens. Occasionally their timber comes onto the market, particularly in their older ages and larger sizes. A selection of species is given hereunder. Their ultimate height is largely dependent on such factors as soil, aspect, and local weather conditions, but indications are: S (Small) 4.5 to 9 metres, M (Medium) 10 to 18 metres, and L (Large) over 18 metres. All are light demanders, and all are evergreen except Dawn redwood. For timber properties and uses see chapter 10.

Wellingtonia (Giant Sequoia) (L)	A tall tower of dark green. Hardy. Likes deep, moist soils, snow and high rainfall. Good specimen tree in landscape. Successful small plots, e.g. at Leighton, Bedgebury and Westonbirt. Largest species in the world: up to 1,490 nr bole volume and 3,200 years of age in California; largest measurement: 83 m x 9.7 m.
Yew (M)	A well-known native dark-green spreading tree which can exist for 750 years or more. Slow-growing. Not to be planted where cattle might browse it and be poisoned. Can grow on chalk, on clay or on sands. A fine veneer timber, when not shaken.
Leyland cypress (L)	A natural cross between Alaska yellow cedar and Monterey cypress. Propagated by cuttings. Many clones, some quite unsuitable. Upright pyramidal growth. Poor windfirmness. Rather dullgreen appearance. An often planted hedge plant; can be clipped, but its vigorous growth sometimes causes problems. There are several golden, grey and blue varieties. *Coryneum cardinale* canker may become a problem.
Japanese red cedar (L)	A tall neat tree with columnar conic crown. Brown-red stripping bole. Foliage turns bronze in autumn. Chief timber tree in Japan. Related to the Sequoias.
Atlas cedar (L)	A broad conic tree with generally ascending branches. Silvery bark. Male catkins in autumn. A blue-grey form, *glauca,* is particularly attractive. Requires adequate space.
Deodar cedar (L)	Horizontal branches with pendulous tips. Blackish bark. Male catkins in autumn. Fast grower. Requires adequate space.
Cedar of Lebanon (L)	Hugely spreading. Mostly horizontal branches; some ascending. Male catkins in autumn. Requires adequate space. Susceptible to snow-break.
Dawn redwood (Fossil tree) (L)	Deciduous. Sensational discovery in central China in 1941. Arrived in Britain in 1948. Vigorous growth. Early into leaf. In late autumn foliage changes from fresh green to pink, deep red or red-brown. Susceptible to bole-fluting.

Main silvicultural characteristics of broadleaves

For conditions justifying selection for timber growing, and for generally unsuitable conditions, see 'Species choice' in section 2 *infra*. The following notes show those broadleaves most commonly planted for timber. All are deciduous except Eucalypts. For comprehensive information on broadleaves see Evans (1984). For thinning recommendations see chapter 8. For timber properties and uses, see chapter 10.

Species (Scientific names are given in the Index) Planting stock and approximate spacing[3]	Abbreviations used: N: Native; E: Exotic; PB: Persistent branches; LD: Light demander; SB: Shade bearer; MSB: Moderate shade bearer; WF: Windfirm; WW: Windweak; MWF: Moderately windfirm; FH: Frost hardy; FT: Frost tender; MFH: Moderately frost hardy (this relates to spring frosts in the growing season, not to winter cold). YC = Yield class (with average) as discussed in chapter 12.
Pedunculate oak Sessile oak N, LD, FT, WF 1+0 or 2+0 seedlings or ½u½ or 1+1 or 1u1 1.5 x 1.5 m, preferably in groups	Native oaks. Evans (1984) details the features distinguishing the two. Relatively slow-growing. Accommodating as to soils but need adequate rooting-depth. Dislike exposure and frost hollows. Strong recovery from frost. Regenerate and coppice freely. Appreciate a nurse species, e.g. Norway spruce, European larch, Lawson cypress, western red cedar and beech. Very high conservation value. The most common standard tree in the coppice with standards silvicultural system. Early thinnings poor in quality. Epicormic shoots a problem, and hence pruning of selected stems may be worthwhile. Pedunculate is best on heaviest clays. Sessile is hardier, has a more shapely crown and tolerates less rich soils; also is more resistant to mildew and *Tortrix viridana* moth. Stem-rots and shake may be prevalent on light soils. Good timber is produced on heavier soils where trees are less prone to shake. Both species may hybridize freely. As to provenances, see section 2, 'Species choice'. 'Pests' include: rot (pipe, brown cubical); Oak leaf roller moth and Winter moth, Spangle galls and Knopper galls, and aphids and beetles. YC 2–8(4).
Beech N, SB, FT, WF 1+1 or 2+1 1.5 x 1.5 m, preferably in groups	Shade bearer, and casts a dense shade. Tolerates calcareous soils provided there is a thin acid layer. Likes well-drained soils. Stands exposure. Best as a final crop in a mixture. As to provenances, see section 2, 'Species choice'. Good underplant being shade tolerant. Slow early growth. Keep dense when young. Regenerates freely, but as coppice is poor. Early thinnings of poor quality. Rarely suffers from epicormic branching. Limited conservation value (often creates a bare forest floor), but superb autumn colour. Suffers disease problems both in the pole stage and in the overmature phase. 'Pests': grey squirrel, rabbits, Beech bark disease and Felted beech coccus insect. Frost damage and sun-scorch may be prevalent. Sometimes grows poorly in treeshelters. Much used for hedging due to persistent leaves. There are several varieties of beech of arboricultural interest, most notably Copper beech, the fastigiate Dawyck beech, and fernleaf beech. YC 4–10(6).

3 The recommended planting stock and spacing relates to conventional forest planting. Where treeshelters are used, small stock – 20–40 cm – is usually advisable. In contrast, where whips are planted, the appropriate size is 90–120 cm. The number of trees required to plant per hectare is given in chapter 5.

Species (Scientific names are given in the Index) Planting stock and approximate spacing	Abbreviations used: N: Native; E: Exotic; PB: Persistent branches; LD: Light demander; SB: Shade bearer; MSB: Moderate shade bearer; WF: Windfirm; WW: Windweak; MWF: Moderately windfirm; FH: Frost hardy; FT: Frost tender; MFH: Moderately frost hardy (this relates to spring frosts in the growing season, not to winter cold). YC = Yield class (with average) as discussed in chapter 12.
Ash N, LD, FT, WF 1+1 or 2+1 2.0 x 2.0 m	Mostly confined to the lowlands but is found at higher altitudes than oak. Frost tender. Regenerates and coppices freely. Attracts the best timber markets when rapidly grown. Best in mixture. Tendency to fork due to Ash bud moth and frost. Individual trees may bear wholly male, female, or hermaphrodite flowers; those with male flowers often exhibit the best stem-form because female flowers are terminal. Late to flush in spring. Tests' include: bacterial canker and some fungi. 'Ash dieback' is widespread in some regions but mainly confined to hedgerow trees owing to adjacent agricultural practice. YC 4–10(5).
Sycamore* E, MSB, MFH, WF 1+1 or 2+1 2.0 x 2.0 m * Called plane in Scotland	Accommodating as to soil. Moderate shade bearer. Vigorous growth, clean stem, natural pruning. Can be pure or in mixture. Tolerates exposure, air pollution and salt-spray. Useful wind-break. Resistant to lime-induced chlorosis. Regenerates profusely and coppices freely. Concern is sometimes expressed about its place in nature conservation on account of its invasive regeneration swamping other species; however, it is one of the better trees for epiphytes. Pests include grey squirrel. YC 4–12(5).
Sweet chestnut E, LD, FT, WF 1+0 or 2+0 seedlings or 1+1 1.5 x 1.5 m	Likes warm, sunny, sandy-loam slopes. Dislikes chalk, limestone, frosty and wet sites. Fast growth as coppice, and is a profitable crop. If left to reach larger size, deterioration through shake and spiral grain can be problems, especially if growth rate slows down. Foliage yellow and russet in autumn. YC 4–10(6).
Cherry, Wild (Gean) N, LD, FH 1+0 or 2+0 seedlings or 1+1 2.0 x 2.0 m, preferably in groups	One of the few tree species to produce both good timber and showy blossom. Often found naturally established. A highly recommended tree for amenity as well as for veneer and high class joinery timber. Regenerates and suckers freely. Rapid to establish. Shallow rooting. Useful in mixture with ash, beech or oak. Appears to be immune from grey squirrel damage, but very susceptible to browsing and fraying by deer. Pruning in the summer and heavy regular thinning are recommended. Height and diameter growth is rapid; volume yield is high, and rotations of 65–75 years are feasible. Rapid decay when overmature. Can suffer from rot fungi. For a comprehensive treatise on wild cherry, see Pryor (1988). YC 4–10(6).

Species (Scientific names are given in the Index) Planting stock and approximate spacing	Abbreviations used: N: Native; E: Exotic; PB: Persistent branches; LD: Light demander; SB: Shade bearer; MSB: Moderate shade bearer; WF: Windfirm; WW: Windweak; MWF: Moderately windfirm; FH: Frost hardy; FT: Frost tender; MFH: Moderately frost hardy (this relates to spring frosts in the growing season, not to winter cold). YC = Yield class (with average) as discussed in chapter 12.

Alder
  Common (N)
  Grey (E)
  Italian (E)
  Red (E)
 	
LD, FH, WF
 	
lul or 1 + 1
 	
1.5 x 1.5 m
	Very hardy. Like moisture and can tolerate wet sites where drainage is poor, flooding and exposed sites. Will grow in conditions of wetness of soil where no tree, other than willow, will thrive. (Grey and Italian alder are also tolerant of dry sites.) Very useful for reclamation work. A good coppice crop, except Italian alder. Good as fire-break; also as wind-break. A suitable crop to grow as coppice for hardwood pulp. Offer the prospect of high yielding crops on short rotations. Naturally regenerate freely on freshly exposed soils in wet locations including banks of watercourses. Have the ability to extract nitrogen from the air through their symbiotic relationship with a bacterial-like micro-organism of the genus *Frankia* which forms nodules on the roots. Generally lack autumn colour.

Birch
  Silver
  Downy
 	
N, LD, FH, WF
 	
1 + 0 or 2 + 0 seedlings or 1 + 1
 	
1.5 x 1.5 m
	Silver birch is graceful, slender and pendulous, usually with attractive white bark. Downy birch has twisted non-pendulous shoots and therefore poorer form; but is the best for poorly-drained soils. The two native birches occur both in highland and lowland birch woods. They produce seed very freely, frequently and from an early age. Readily established naturally where there is bare mineral soil of moderate phosphate status. Accommodating as to soil. Excellent pioneer and nurse species, but the very rapid initial growth must not be allowed to swamp the main crop species. Natural growth often worth keeping as shelter for a new species crop. Coppices freely when young. Reputed to be a soil improver. Very frost hardy. Rapid grower. Boles of large trees are often fluted. Striking yellow foliage in autumn. An important landscape tree, with increasing potential for planting in Scotland. For planting, use seed sources appropriate for the intended site, collected from British stands of good appearance taken from similar or slightly more southerly latitude than the planting site. If unobtainable, then use seed from western Europe at similar latitudes in Britain. Avoid northern provenances from Scandinavia, and those from southern Europe; also those from elevations above 500 m.

Red oak E, LD 1+0 or 2+0 seedlings or 1+1 1.5 x 1.5 m	A native of North America. The most widely planted exotic oak. Rich autumn colouring after a dry summer. Growth rates and yields comparable to native oaks on fertile sites, but tends to be superior on poor, especially sandy soils. Likes fertile sandy soils but does well even on poor dry sites. Not widely planted for timber, but valuable for amenity. A general purpose hardwood timber, inferior to the native oaks, for which it is not a substitute. Not durable but takes preservative well. Timber used mainly for cheaper furniture and flooring.

Species (Scientific names are given in the Index) Planting stock and approximate spacing	Abbreviations used: N: Native; E: Exotic; PB: Persistent branches; LD: Light demander; SB: Shade bearer; MSB: Moderate shade bearer; WF: Windfirm; WW: Windweak; MWF: Moderately windfirm; FH: Frost hardy; FT: Frost tender; MFH: Moderately frost hardy (this relates to spring frosts in the growing season, not to winter cold). YC = Yield class (with average) as discussed in chapter 12.
Turkey oak E, LD 1+1 1.5 x 1.5 m	Grows more rapidly, with better stem-form (but lanky in branching) than the native oaks. Ascending giraffe-like branches. Good on chalky soils. Widely naturalized in southern Britain. Forms hybrids with native oaks. Relatively poor timber, rarely accepted for sawing.
Norway maple E, LD, FH 1+0 or 2+0 seedlings or 1+1 1.5 x 1.5 m	Frost hardy. Regenerates freely. Similar to sycamore. Useful for amenity, particularly as an attractive belt around other tree crops. Foliage gold to orange in autumn. Best on moist, deep, fertile soils over chalk and limestone but tolerates shallow soils over these rocks. Tests' include grey squirrel. Timber not so hard or strong as sycamore; good for turnery.
Hornbeam N, SB, FH 1+1 or 2+1 1.5 x 1.5 m	Native to south-east England and parts of Somerset and Gwent. Strongly shade-bearing and casts a dense shade. Useful on damp heavy clay and in frost hollows where beech will not flourish. Thrives on both acid brown earths and soils derived from chalk and limestone. Perhaps best on sandy or gravelly soils. Often coppiced or pollarded. Irregular slow growth. Not usually damaged by grey squirrel.
Southern beeches (*Nothofagus* spp.) E, FT, LD 1+0 seedlings or 1+1 2.0 x 2.0 m	Fast-growing species, usually from Chile. Frost tender as they flush early; susceptible to damage by extreme cold but become winter hardy later in the autumn. Dislike frost hollows. Regenerate and coppice well. Can suffer from poor survival and indifferent growth rates. Prefer deep, freely-drained soils but will tolerate a wide range of conditions apart from very wet soils and peats. *N. procera* appears to be more suited to the wetter west and to the better soils, with *N. obliqua* growing more successfully in the south-east and on poor, gravelly soils. Sometimes a substitute for common beech on fertile heavy clays. Rarely attacked by grey squirrel but are attractive to deer. Susceptible to 'climatic stem canker'. The best single origin of *N. procera* for frost resistance appears to be from Puesco in Cautin Province (Chile), and this is also the tallest in north Britain experiments (Lines 1987). (Argentinian seed probably has significantly great cold hardiness.) Seedlots from the forests around the towns of Victoria, Temuco and Cuneo have grown well. Seedlots from north of latitude 38°S should be avoided.

Species (Scientific names are given in the Index)	Abbreviations used: N: Native; E: Exotic; PB: Persistent branches; LD: Light demander; SB: Shade bearer; MSB: Moderate shade bearer; WF: Windfirm; WW: Windweak; MWF: Moderately windfirm; FH: Frost hardy; FT: Frost tender; MFH: Moderately frost hardy (this relates to spring frosts in the growing season, not to winter cold). YC = Yield class (with average) as discussed in chapter 12.
	Imported *N. obliqua* shows less seed origin variation. Seed from Llanquihue, Malleco and Cautin provinces (Chile) are recommended in that order. Both species cannot be planted anywhere in Britain without some risk of frost damage, but many seed origins appear to make a good recovery after the shoots have been frosted. For comprehensive information on Southern beeches, see Potter (1987a, 1987b) and Tuley (1980).
Eucalypts E, LD, some FT, some FH	Evergreen. Amongst the fastest-growing trees. Found throughout Australia but only species growing in the colder mountainous regions of New South Wales, Victoria, and central Tasmania may be hardy enough to grow in Britain, where many species have been tried. They have great variety in leaf form and colour (juvenile and adult foliage) and several are planted as garden ornamentals. Nearly all species grow rapidly. Some coppice freely. Many are frost tender, some are quite hardy (e.g. *E. gunnii*). For suitable sites, provenances, planting, tending and uses, see Brooker and Evans (1983).
Willow N, LD, some FT, some FH	Generally associated with damp conditions, though sallow *S. caprea* tolerates drier sites than other willows. Cricket-bat willow is the only species grown for timber (see chapter 5). Many willows are very vigorous, and some when grown on a three to five year cycle, are capable of yielding up to 10 dry tonnes ha (see chapter 5); tracts of fertile land are necessary. Structural damage to houses can occur under certain site conditions. Osier-growing, for willow furniture hand-made by craftsmen, is discussed in chapter 5.
Poplar LD, some FT, some FH, E (other than *P. nigra* and aspen)	The best known poplars are horticultural cultivars propagated clonally by vegetative means. Some are grown solely to produce merchantable timber, while others are planted for screening, shelter and ornament (Jobling 1990). A few cultivars are used both for timber production and for landscaping, as well as to protect horticultural and agricultural crops. Most poplar are fast-growing, and need wide spacing. Some are hardy to frost damage, some are rost tender. Some have striking autumn colours (e.g. Aspen). Lombardy poplar is well known for its narrow columnar habit. Poplar are light demanders, and are generally windfirm. They are often susceptible to a bacterial canker and rust; only clones generally resistant should be planted. Poplars are very exacting;

Silvicultural Characteristics and Species Choice

Species (Scientific names are given in the Index)	Abbreviations used: N: Native; E: Exotic; PB: Persistent branches; LD: Light demander; SB: Shade bearer; MSB: Moderate shade bearer; WF: Windfirm; WW: Windweak; MWF: Moderately windfirm; FH: Frost hardy; FT: Frost tender; MFH: Moderately frost hardy (this relates to spring frosts in the growing season, not to winter cold). YC = Yield class (with average) as discussed in chapter 12.
	suitable sites are limited. The ideal site is a base rich loamy or alluvial or fen soil in a sheltered situation, well-drained and wellwatered. Unsuitable sites are those with poor drainage, pH values of less than 5.0 and with a soil depth of 50 cm or less. All upland, northern and western sites should be avoided; also exposed sites and shallow soils. Stagnant water is fatal, but occasional winter floods do not harm. The Balsam poplar withstand slightly more acid soils than the Black hybrids and are more suited to the cooler and wetter parts of Britain. The spacing for poplar should be about 8 x 8 m (156 ha). Yield classes range from 4–14. Either rooted plants or sets should be used. Regular pruning is required (discussed in chapter 5). Poplar can be quite profitable under suitable conditions. They are fast-growing and their wide spacing allows inter-row agricultural cropping for several years followed by grass seeding and summer grazing for most of the remaining poplar rotation. Several poplar cultivars appear capable of yielding 10–12 dry tonnes ha/yr when grown as an energy/fuel crop on a three to five year coppice rotation. Tracts of fertile land are necessary. For further information on poplar see chapter 5 and Jobling (1990).

Broadleaves additional to the foregoing are important on farms, in parks and substantial gardens and sometimes in woodlands. Some can produce merchantable timber, particularly in their older ages and larger sizes, as well as having other attributes. A selection of species is given hereunder. Their ultimate height is chiefly dependent on such factors as soil, aspect and local weather conditions; but indications are: S (Small) 4.5 to 9 metres, M (Medium) 10 to 18 metres, and L (Large) over 18 metres. All are light demanders except Holm oak, holly and box. For timber properties and uses, see chapter 10.

Holm oak (M)	Evergreen. Bushy-crowned. Slow-growing. Not hardy but stands exposure well. Dense shadecaster. Useful for coastal planting. Rather dull foliage. Rarely planted for its timber.
London plane (L)	Useful for cities, parks and large gardens. Tolerates air pollution. Can be pollarded.

Walnut (L) Common Black	Generally restricted to central and southern Britain. Frost tender. Usually grown as individuals, not in clumps. Needs shelter when young and plenty of space later. Likes fertile well-drained deep light soils. Dislikes exposed, frost prevalent and dry sites. Black walnut is usually the more vigorous. Deserves to be planted more often for veneer and high class joinery.
Horse chestnut (L)	An ornamental, not a timber tree. Showy white or pink flowers. Adequate space required. No forestry potential.
Lime (L) Small-leaved Large-leaved	Frost hardy and windfirm. Some good natural crops, particularly of coppice (especially small-leaved). Boles of large trees often fluted, and base surrounded by shoots. Honeydew from aphids creates a nuisance below the tree. High conservation interest in areas where native. For the ecology and silviculture of limes, see Pigott (1988) and Helliwell (1989).
Elm, English (L)	Because of Dutch elm disease, woodland planting of elm cannot yet be encouraged. However, many hedgerows show elm regrowth and this can be expected to survive for about 20 years and then succumb, to be followed by further regeneration. A hybrid elm 'Sapporo Autumn Gold' *Ulmus pumila* x *japonica* is reputed to be resistant to the disease. For comprehensive information on elm, see Greig (1988). Wych elm is a broadly domed, spreading tree. Moderately tolerant of exposure. Needs damp but well-drained soils. Striking yellow autumn colour.
Hazel (S)	An ingredient of many broadleaved areas. Widely occurring underwood species and component of hedgerows. Often grown as coppice in south England. Conspicuous catkins in February/March; and nuts in autumn. Accommodating as to soils and will grow on even heavy clay. Hardy. A small area on an estate or farm is useful. Ideal pheasant cover if regularly coppiced. Can have high conservation value. Usually a small bushy clump especially when coppiced. Many people regret the gradual passing of a lovely form of hazel woodland, which incidentally often harbours primrose, wood anemone, violet, bluebell, campion and daffodil.
Rowan, Mountain ash (S)	Not a timber tree but has high conservation value. Hardy. Resistant to exposure. Small erect crown. Conspicuous white flowers; fruits regularly – large scarlet berries. Striking coloured autumn foliage. Coppices strongly. Good on chalky soils or acid peats. Found at higher altitudes than almost any other broadleaved species. Useful amenity tree in the uplands.
Spindle tree (S)	A small shrubby tree, common in wayside and hedgerow. Small white flowers followed by four-lobed seed pod turning to fuchsia pink. Likes chalk and limestone. Is the alternate host of the Broad bean aphid.

Whitebeam (M)
  Common
  Swedish
	A small compact tree. Vigorous early growth. Likes soils derived from chalk and limestone. Underside of leaves conspicuously light-coloured. White flowers, followed by large scarlet berries. Coppices freely. Relatively short lived. An attractive component of mixed broadleaved woodland.

Wild service tree (M)	'Chequers tree.' White flowers, followed by brown speckled berries. Strong autumn colour. Coppices and produces suckers freely. Generally found on clays and on soils derived from chalk and limestone. Can endure some shade. An attractive component of mixed broadleaved woodland. Generally an indicator of an ancient woodland site.
Wayfaring tree (S)	A bush shrub masquerading as a tree! Flowers creamy-white, followed by oblong berries slowly maturing from red to black. Found in some waysides and hedgerows. Grows well on chalky soils.
Guelder rose (S)	A small shrubby tree. Flowers of flattened corymbs surrounded by showy white ray florets, followed by bunches of glistening red, translucent berries. Common in wayside and hedgerow. Prefers moist soils.
Dogwood (S)	A small shrubby tree, common in wayside and hedgerow on calcareous soils. White flowers. Fruit: a black berry.
Elderberry (S)	Large shrub or small tree. Grows wild almost anywhere. Very accommodating as to soil but prefers high level of fertility. Likes chalk and clay soils. White flowers in large loose clusters, followed by purplish-black berries. Mainly found as an understorey.
Blackthorn (Sloe) (S)	A small bushy shrubby tree often forming dense thickets. Relatively short lived. Its root suckers can create problems. Conspicuous white flowers, followed by black fruits (sloes). Produces spines. Frost hardy. Good autumn yellow colour. Best suited for group planting.
Crab apple (M)	Frequent to rare in hedgerows and copses. Flowers white, faintly pink, followed by small apples.
Field maple (M)	An attractive bushy tree of wayside and hedgerow, with a splendid autumn golden colour. High conservation and landscape value. Exposure resistant. Best on heavy calcareous soils. Coppices strongly. Rarely of timber size.
Bird cherry (S)	Conspicuous white scented flowers, followed by small 'cherries'. Prefers alkaline soils but tolerant of more acid sandy soils. Relatively short lived. Coppices and produces root suckers. Good autumn yellow colour. High conservation value in north and west Britain.

Holly (M)	Evergreen. Slow-growing tree well known for its small white flowers, conspicuous red berries and glossy green spiked leaves. Only females produce berries. A common understorey species in oak and beech woods.
Hawthorn (S)	A traditional small component of hedgerows, and when not laid or trimmed forms so much a part of the British landscape. High conservation value. Relatively short lived. Conspicuous white flowers ('May'), followed by red berries. Produces spines. Good autumn foliage colour. Common component of upland 'scrub' woodland with birch and rowan. Tolerates wide soil range, but dislikes very dry or wet conditions. Frost hardy.
Box (S)	Evergreen. A small native tree found on chalk and limestone in southern England. Slow-growing but producing a fine even texture, creamy-white timber much valued in turnery and craft work.

2. Species choice

The selection of species to achieve objectives and to match the site and other factors, is probably the single most important decision to be made for success in forestry. If the selection is unsound, no matter how carefully the crop is established and subsequently tended, the interim and final results are likely to be financially and otherwise disappointing. Many species when young grow fairly well on sites where they cannot grow into sound large timber (e.g. oak on sandy soils, Scots pine on chalky soils); after two or three decades, perhaps earlier, they fail to exhibit the characters desired or may even become unhealthy. The error of judgement is not always apparent until many years have elapsed.

Four main considerations determine which species are planted: (i) the objectives; (ii) site factors, mainly climate, rainfall, frost, wind, nutrients, topography and soil (see chapter 3); (iii) any biological constraints (see chapter 6); and (iv) restraints imposed by the woodland owner.

Afforestation during this century, because of the land market and the need for fast volume production, has been confined mostly to the wetter uplands, in Scotland, Wales and northern England, and to the inferior soil types where agricultural crops and grazing have not been competitive and land prices have been lower. The planting has almost always been of conifers. The current grant-aid schemes now encourage also the planting on more fertile land in comparatively sheltered lowland situations of a wide range of species, both broadleaved and coniferous. In particular, the emphasis placed on broadleaved species has led to a greater increase in their planting. A wide species choice is available: some trees grow relatively fast, and some grow more slowly; unit values of some are high, and of some are low. A major consideration is rotation length, as most conifers and a few faster-growing broadleaves offer the prospect of a final crop within the lifetime of the planter, whereas most traditional broadleaves imply longer rotations and planting largely for the benefit of future generations.

Objectives. The objectives of forestry planting need to be determined; failure to make them clear from the outset will create difficulties and mistakes in subsequent operations. Most large scale planting will have as its main objective the production of timber; but often, particularly in broadleaved woodland and some mixed woodland, there will be subsidiary objectives – landscape, amenity, shelter, recreation, wildlife conservation and sporting. These objectives are likely to result in some reduction in timber production. They may influence species choice (likewise will the varying levels of grant-aid and the expected high percentage of broadleaved species). Current grant-aid requires timber production to be at least one of the objectives. A woodland established with one major objective in mind invariably provides a range of benefits; and the objectives set and pursued may be changed by a new owner (for example, a woodland established for commercial timber production, when acquired by a conservation trust, may be assigned new objectives).

However, as timber production provides the major source of woodland revenue it is likely to have the greatest influence on management. Wherever other objectives are overriding (e.g. sporting), the production of timber need seldom be abandoned. It may not be produced in the same quantity and over the same rotation as might have been possible had management been primarily for wood production, nevertheless marketable timber can be grown. Conflicts over silvicultural systems, species choice and rotations have to be overcome: there are abundant precedents for successful reconciliation (notably in the Cotswolds, the Chilterns, the Forest of Dean and the New Forest). Some woodlands because of their value as wildlife habitats may have conservation as the primary objective; in which circumstances it will be necessary to strike a balance and determine priorities for the secondary objectives.

Objectives of maintenance or enhancement of landscape, wildlife conservation, recreation and sporting are discussed comprehensively in chapter 13. *Landscape* may occasionally be an overriding objective (policy planting is one example). In recent years Britain's predominantly urban population has become increasingly interested in the appearance of the countryside and there have been objections to unsympathetic changes in landscape caused by afforestation or clear-felling. Changes due to forestry activities can usually be blended into the landscape by paying appropriate attention to the shape of plantation edges, the location of different species and the retention of any existing broadleaves; but effects on the landscape within the forest may be more difficult to handle. *Wildlife conservation* will often have great influence on species choice and be important when setting objectives. The level of priority necessary will vary greatly, from the site of unique conservation value (e.g. ancient semi-natural woodland), to simple conservation measures involving little or no sacrifice to timber potential. A detailed guide to appropriate native species of trees and shrubs, together with a specification for uses compiled by the Nature Conservancy Council, is given in the Appendix to Forestry Commission 'Forest Nature Conservation Guidelines' (1990) noted in chapter 13, section 7. A need to cater for *recreation,* present or potential, should have a defined place in the objectives. *Sporting* (game management), *shelter* and other objectives should likewise be planned. Species choice and decisions on planting layout can take the needs into account.

For many of the objectives noted above, given suitable site factors and locations, native broadleaves will often be the popular choice. In the lowlands, for optimum growth, the principal broadleaved species require sheltered sites on deep fertile soils which are moist but freely draining, in other words good agricultural land. In the uplands, the main site limitations for broadleaves are lower temperature, wetness and acidity of the soils and wind exposure. Although they may grow under these conditions they will tend to form poor crops and produce low quality timber. Nevertheless there are some sheltered locations where they will thrive and attention should be concentrated on these areas (see chapter 5).

A deterrent to the successful growth of broadleaved plantations of some species is the stripping of bark by grey squirrels (see chapter 6). Methods of controlling populations are available and have been shown to be effective in preventing damage if properly applied and carried out at the right time although it has to be recognized that many woodland owners and farmers neglect to take the necessary action. Uncontrolled grazing by stock is a serious threat to the future of many woodlands and a severe constraint on successful regeneration.

The woodland owner's likes, dislikes and prejudices may impose constraints on foresters' species choice. They may relate to objectives other than timber production, or directly to species. There may be desires for timber and fuelwood for the use and enterprises of a landed estate or farm; and perhaps for early returns (e.g. thinnings from larches, Christmas trees, fast-growing poplar and sweet chestnut coppice). The requirement to use timber on the estate as firewood probably favours relatively fast-growing broadleaves which coppice well and produce reasonable fuel wood on a fairly flexible rotation (sweet chestnut, ash, sycamore and alder). If fencing materials are required, then sweet chestnut is the principal broadleaved species; conifers capable of producing timber that can be used without preservative treatment (larches and western red cedar) also may be favoured. Yet species choice ought still to be within the framework of economics and suitabllility of species for the site. Thus questions which may be posed in species choice include: which are eligible on both economic (financial) and ecological grounds?; are early returns required, and long rotations undesirable?; are there any other constraints?; what markets, local or not too distant, are likely to be available? The species choice must match and complement these factors as well as those of the site.

Timber production. This is likely to remain the chief objective in most forestry planting (see chapter 13). It is necessary to consider the type of timber required, the silvicultural system under which it should be grown (see chapter 7), the length of time it will take to produce and the potential markets. With all this in mind, it is possible to select species which will give the best yield for the site concerned. In most cases both quantity and quality of timber produced will be important. Problems will arise if substantial quantities of low grade material are produced, unless there is a suitable market not too distant: sites with poor access are generally useless for low value produce. Good quality sawlogs, whether coniferous or broadleaved, will find a satisfactory market. In upland areas and on infertile lowland sites, use of coniferous species will almost invariably give the best financial return. Good quality broadleaved timber can only be produced on the more fertile lowland sites and if grown properly in reasonably well-stocked stands. Coppice, a major influence on broadleaved woodlands in the past, is still practised extensively with sweet chestnut.

Species which can be expected to grow equally well on a given site may show very different financial results. The most favourable species will be the one that shows the highest ratio of discounted revenue to discounted expenditure at appropriate rotations, using a chosen rate of interest (see chapter 14). Two different species may give very similar rates of return, but require quite different inputs of capital.

Foresters should concentrate on planting species which they expect to grow well, but must also give prudent thought to future markets. Trees have not always served those end-uses for which they were originally planted (oak for ships, poplar for match-splints), but it is likely that there will continue to be a market for large-sized clean-grown broadleaves, conifer sawlogs, and small diameter roundwood for pulp, boardwood and box-making. Well-conducted market research taken in conjunction with a forecast of possible production can assist foresters in determining whether a species is likely to be commercially viable.

The relationship between the user and the grower is less apparent in practice because the user is thinking of the present or a few years ahead, whilst foresters think of the distant future. Growers therefore have to take a view of what they expect the user to require in perhaps thirty or more years' time. From the user's point of view, the principal factors of interest which, to a greater or lesser extent, are under the control of the grower are: species, size, form, wood characteristics (strength, working properties, seasoning properties) and the level and regularity of production. An important consideration for the grower is versatility: not to produce a highly-specialized product (exceptions include sawlogs and decorative veneers) which may not mature for many decades and for which a market may not exist in the distant future. Yet, a market is likely to be assured for sizeable quantities of a consistent produce with good inherent qualities and no obvious defects. The distance from markets will often be another factor for consideration; for instance, the growing on short rotations of broadleaves, chiefly as coppice, may help to determine species selection in regions within economical distance of a hardwood pulpwood or boardwood mill or a turnery.

Site factors influencing species choice. Britain's climate, topography and soil generally provide good conditions for tree growth, and are the main factors that determine whether a particular species has the potential to grow well and produce good timber on a given site. Matching species to site factors is fundamental to sound silviculture. The alternative of modifying site conditions to suit a species is sometimes feasible, but the possibilities are limited and mainly involve: (i) improving soil physical conditions by cultivation or draining; (ii) raising nutritional status by adding fertilizer; and (iii) modifying the microclimate, to reduce frost damage or moderate windspeeds by leaving side shelter or a light overstorey, or by using a nurse species in mixture. However, some sites have a number of permanent limitations which restrict the planting of certain species to the local climate (Douglas fir, ash and poplar have quite specific requirements). For broadleaves, Evans (1984) provides a method for selecting species based on the potential of different soils according to pH, texture, rooting depth, drainage and fertility. The physical condition and fertility of the soil must be maintained and if possible enhanced.

Lowlands. For lowland areas a simple subdivision of soils into those derived from clays and loams, sands, and chalk and limestone gives a good basis for species choice. On clays and loams a wide range of species will grow well, namely: of conifers, Douglas fir, Corsican pine, Norway spruce, larches, western red cedar and firs; of broadleaves, oak, ash, beech, sycamore, sweet chestnut, cherry, poplar and southern beech. On sandy soils (unless they are deep and not too acid) the choice is restricted to species capable of surviving with high water stress and frequently in acidic conditions (e.g. heathlands): of conifers, Scots and Corsican pines and larches; of broadleaves, sweet chestnut and birch. On chalk and limestone, alkaline-tolerant species must be used, among them: of conifers, Corsican pine, western red cedar and European larch (Japanese larch and Douglas fir are inappropriate if lime-induced chlorosis is expected in later life); of broadleaves, ash, beech, sycamore, cherry and Norway maple (sweet chestnut particularly is unsuitable).

The main climatic and topographical influences (see chapter 3) on species choice are rainfall, altitude and exposure, seasonal frosts and winter cold. Rainfall has a limited effect on choice of broadleaves but a considerable effect on conifers, the annual amount, as well as the distribution of the precipitation throughout the year, being of great importance. The possibility of summer drought must be borne in mind. In heavier rainfall areas, the choice may lie in spruces, Douglas fir, lodgepole pine, western hemlock, western red cedar, grand fir, Japanese or Hybrid larches, and (sometimes) poplar. In low rainfall areas the likely species are Scots and Corsican pines, beech, sycamore, ash and sweet chestnut.

Variation in mean temperature over the British Isles has remarkably little influence on species choice among the broadleaves species, though sweet chestnut, poplar and walnut appear to have high sunshine and warmth requirements for good growth. The only recommendations based on gross climatic criteria are to confine planting of sweet chestnut, black hybrid poplar and walnut to south and south-east England and not to plant southern beeches in eastern regions. Most important are local factors of exposure, incidence of frost and other microclimate effects which mostly reflect conditions of local topography. Because of exposure and generally poorer soil conditions, attempting to establish productive broadleaved high forest at altitudes above 300 m will rarely be worthwhile. (Small woods of birch, rowan, sycamore and alder do occur above this altitude, and occasionally of ash, oak and beech, but the trees are mostly slow-growing and poorly-formed.)

Altitude and exposure have similar effects and, as they increase, yields generally decrease and the proportion of conifers planted rises. Wind can be a serious risk particularly for spruces, Douglas fir and larches. (Windthrow Hazard Classification has been noted in chapter 3.) Frost and winter cold are particular problems in the early establishment of many species of conifers and broadleaves. On sites where late frosts are likely to be frequent and severe, hardy species should be planted: alder, most poplar, birch, hornbeam, hazel, cherry, sycamore, lime, Scots and Corsican pines, and Norway spruce. Frost tender broadleaves when young include ash, oak, beech, sweet chestnut and walnut; and among conifers, Sitka spruce, Douglas fir, western hemlock, coast redwood and larches. A traditional way of minimizing frost damage to young broadleaves, especially on exposed sites, is to plant them in mixture with conifers. Frost and winter cold can at a later stage contribute to stem cracking and failure of plantations.

Trees' demand for light has been noted in chapter 3 relating to light demanders, moderate shade bearers, and shade bearers – all listed in section 1 of this chapter under 'Silvicultural characteristics'. Nutrient requirements have been noted in chapter 3; deficiencies can be corrected by fertilizer application. Ground vegetation (as noted in chapter 3) can be indicative of the soil, moisture and other site conditions – and hence sometimes assists in species choice – but factors such as frost need to be taken into account.

Trees are subject to damage by several other agencies, and hence an assessment is necessary of the predisposition of a species to damage. The presence of deer may influence species choice: broadleaves in particular, along with some conifers, are favoured by deer for browsing or bark stripping, and fencing against them is costly. The risk from sheep, goats, cattle and rabbits can be obviated by fencing or by the use of treeshelters or guards. The prevalence of grey squirrel damage to beech, oak, sycamore and Norway maple will influence species choice; cherry and ash are less susceptible. Atmospheric pollution of various kinds can be damaging and restrictive. The risks from insects, fungi, viruses, and bacteria must also be considered. Attack by *Hylobius abietis* and *Hylastes* spp. on young conifer plants has to be guarded against as appropriate, especially after recently felled conifer crops. Following *Heterobasidion annosum,* usually found on old conifer woodland sites, particularly on soils with a relatively high pH value, Douglas fir, grand fir, and most broadleaves may be prudent restocking choices. Honey fungus, usually found on old broadleaved woodland sites, may again favour Douglas fir and grand fir; all provided other factors and objectives are favourable.

In the lowlands the selection of species will not always be easy. The relatively recent availability of better land for planting considerably extends the range of species likely to produce an acceptable return from timber growing. The wide range of possible objectives of management, the varying rates of grant-aid for different species and the possibility of producing top quality broadleaved timber also bear on the choice. Although many combinations of previously minor species may be planted, the bulk of lowland planting will continue to be of well tried and tested species (often in mixture) which have proved their worth. The larches, Douglas fir, Norway spruce and Corsican pine will be the major coniferous species; while oak, ash, sycamore, cherry, beech and sweet chestnut are likely to be the favoured broadleaves. This does not rule out the selective planting on fertile sites of a few walnut for decorative timber.

Uplands. In upland areas – typically peatland and moorland – the species choice in general will be Sitka spruce, lodgepole pine, Norway spruce, Scots pine and larches; along with birch, sycamore, common alder and red alder as the broadleaves included mainly for environmental reasons, rarely for expectation of profitable timber crops. When restocking in the uplands (see chapter 5), foresters will often be faced by many of the conditions which affected species choice in the first rotation, but there are also likely to be new opportunities as well as new limitations (Low (ed.) 1985, 1986; Tabbush 1988). As a guide, foresters will usually have the actual long-term performance of several species on different site types within the locality, though this information must be used with caution as the origin/provenance, ground preparation and later treatment may not have been optimal. The original ground vegetation (e.g. *Molinia* or *Calluna*) may have been markedly changed by the influence of the first crop. Depending on the size of felling coupes, there may now be greater side shelter than at the start of the first rotation. However, some restocking sites may also have an increased risk of frost caused by the effect of surrounding stands on cold air movement. If stumps of the first rotation are heavily affected by *Heterobasidion annosum,* some species are unsuitable. There may also be a need to take action against *Hylobius abietis* and *Hylastes* spp. Damage by deer (both roe and red) on many sites is likely to be far greater than at afforestation, and may be one of the main factors determining species choice. (Whereas Norway spruce and Douglas fir are very susceptible to deer browsing and bark stripping, Sitka spruce withstands substantial browsing pressure.) Sitka spruce is likely to be the first choice conifer for restocking because of its high growth potential, adaptability to a wide range of site conditions, tolerance of deer browsing, and general purpose timber. However, the use of other conifers may be suggested by site factors, established or expected timber markets, a desire for greater species diversity, and considerations of wildlife conservation and landscape. Information for and against various alternative coniferous species is summarized by Low (ed.) (1985).

Upland site conditions rarely favour the use of broadleaves for commercial timber production. However, native species such as alder, birch, oak, rowan and willow have an important role in enhancing landscapes and enriching wildlife habitats. Yet, in comparison with coniferous species on similar sites, young broadleaves are generally more demanding of moisture and nutrients, more sensitive to early competition and more susceptible to climatic and deer damage. The use of large plants, and planting at close spacing in compact groups, may facilitate establishment, protection measures and any necessary tending. Complementary native broadleaves will often seed naturally into an area, if given protection from browsing and fire (Low (ed.) 1985, 1986).

Origin/provenance. In species choice it is essential to consider the origin and provenance of the plants which may best suit the site. Tree species which occur over wide geographic areas develop sub-populations with slightly different characteristics which may, for example, be related to altitude and day length. These populations are usually not physically distinguishable from one another but each one is just a little better suited to its own particular environment. Seed from these populations or ecotypes is usually better adapted to grow in a similar environment to that in which its ancestors grew. In order to realize a tree's maximum potential growth it is clearly wise to use plants on a site to which they are best suited.

Provenance is the place in which any stand of trees, whether native or exotic, is growing; *origin* is the place in which a stand of native trees is growing, or the place from which a non-native stand was originally introduced. In general forestry use, the term 'provenance' is commonly extended to mean not only the geographical source of the seed, but also the trees raised from it. The definition of 'origin' means, for example, that if seed was imported from Vancouver Island to form a stand in Perthshire the *origin* is still 'Vancouver Island' but the *provenance* is 'Perthshire'. Further comments on origins/provenances have been made in chapter 2, section 3 under 'Genetics in silviculture', and later in this section.

Afforestation and much of the restocking in Britain is based predominantly on wide-ranging exotic species – almost wholly coniferous – and hence choice of seed origin/provenance is a crucial constituent of success. Choice of provenance can be a very critical stage in forest management, in some cases resulting either in a flourishing plantation or in a failure. The problem is confounded by the large site differences which exist in Britain which may require selection of a different origin in, say, parts of Scotland from that used in southern England, or even on different site types within a region.

The cost of seed is a relatively insignificant part of the cost of establishing a forest: Faulkner (1962) has shown that for some species an increase of only 0.5% in the value of the final crop justifies a 50% increase in the price of seed.

Lines (1987), for more than three decades responsible for Forestry Commission origin/provenance investigations, succinctly guides foresters to appropriate seed origins for the site conditions and objectives, and equally importantly warns against unsuitable sources. He points out that in broad terms site types in Britain may be linked with the European north temperate forest zonation to give two main types: a northern zone similar to the boreal coniferous forests and a southern zone similar

Figure 6. Regions of seed provenance within Great Britain
Source: Forestry Commission.

to the west European mixed conifer and broadleaved species. In Figure 7, the former type would broadly equate with Regions of Provenances 10 and 20 and the latter with Regions 30 and 40. The map emphasizes that Britain is split also into western and eastern sectors, which for some site factors may be just as important as the north/south division.

There are a few basic factors governing the choice of seed origin (Lines, 1987, provides a detailed discussion): (i) broad climatic matching is generally required between the origin and the site on which it is to be planted, but the principle should not be too strongly emphasized; (ii) day-length at the place of origin usually exerts a strong influence on flushing and growth cessation, but there may be some compensating effect of elevation on this effect; and (iii) edaphic factors may have some influence.

Lines (1987) has collated origin/provenance data on the species of most importance in British forests: Scots pine, Corsican pine, lodgepole pine, Sitka spruce, Norway spruce, European larch, Japanese larch, Douglas fir, western hemlock, western red cedar, grand fir, noble fir, oak, beech, birch and Southern beech. Foresters wishing to investigate seed origin in more detail can consult the references Lines has provided as well as the forthcoming publication by Gordon (ed.) (1991). As many of the species come from western North America, Lines included maps of the Tree Seed Zones for Washington, Oregon and British Columbia, together with the Forestry Commission map of seed collection zones in Western North America (see Figure 8): this has formed the basis of the seed identity code since 1956. The Commission and the privately owned seed company Forestart in their annual seed catalogues include notes on the particular

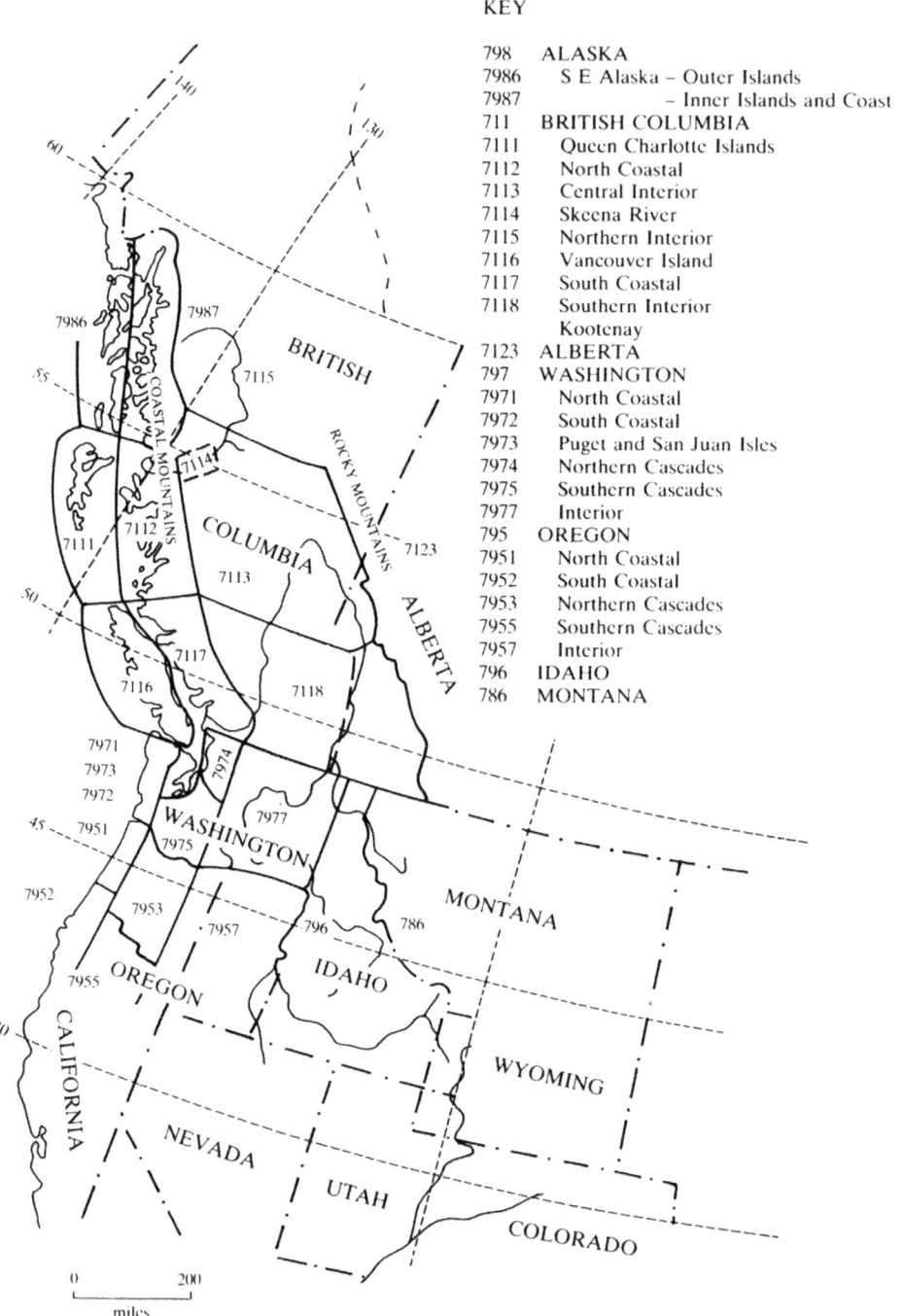

Figure 7. Seed collection zones in Western North America
Source: Forestry Commission.

seed origins/provenances offered for the main species. Recommendations are noted later in this section under 'Species choice – conditions justifying selection'.

In species choice, foresters' first and almost inevitable choice is between broadleaves and conifers. For long there has been a traditional 'obligation' to plant some broadleaves; the English and Welsh peoples may prefer the appearance of them; yet in Scotland conifers have long been accepted as part of the traditional scene. For most broadleaves, the more fertile soils are required if they are to have a chance of being grown profitably; yet conifers generally prove far more profitable on the same soils. This presents the case for planting broadleaves where their visual and other attributes are more gratifying than their profitability as timber. Usually the growth of conifers is far quicker, their thinnings more easy to sell, and their conventional rotation much shorter. Many conifer stands are a source of income by the time they are twenty years old; and some quickly repay the cost of establishment and early maintenance. Most broadleaves on the other hand, except where firewood is profitable, will continue on the debit side; yet at the end of the rotation may result in fine timber and a financial profit.

Partly for these reasons, and partly because Britain's climate is favourable for tree growth, there is a wide choice, and foresters may be tempted to grow too many species. Enthusiasm for variety should be tempered by considerations of economics and marketing. Many prudent foresters have concentrated on 'bread-and-butter' conifers (spruces, larches, Douglas fir and pines) and have regarded other species as relatively minor (for example, western red cedar, western hemlock, grand fir and noble fir). Markets and demands favour substantial quantities of well-known timbers rather than small amounts of several different kinds; and marketing is easier if, in the main, the crop is limited to a few kinds well adapted to local requirements. However, there will still be reasons for having stands of minor species; and many are particularly useful for underplanting and enriching inadequately stocked stands. There will always be a place for some broadleaves, grown pure or in mixture with conifers, especially where environmental considerations are important (see section 4 *infra*).

Forestry Commission/Macaulay Institute land capability for forestry maps for Scotland (1989), noted in chapter 3, are of help in deciding species choice. Likewise are the soil and forestry survey services offered for England and Wales by firms and individuals. Mindful of the silvicultural characteristics of the individual tree species set out in section 1 *supra,* and of the factors of the forest site (especially climate, topography, soil, wind, ground vegetation) and light, nutrient and moisture requirements, discussed in chapter 3, it is possible to consult the following pages as an aid to species choice for the main species planted for timber. First are set out conditions justifying selection and, second, generally unsuitable conditions. The opportunity is taken to include the main recommended origins/provenances for the conifers (Lines 1987) and for oak and beech (Evans 1984; Lines 1987; Gordon 1982). The choices suggested in these notes are based almost wholly on commercial silvicultural principles. Where consideration has to be given to landscape, recreation or wildlife conservation (see chapter 13) any required adjustment can be made to species choice, but usually they are unnecessary.

Species choice for conifers

Conifers most commonly planted for timber. All are evergreen except the larches.

Conditions justifying selection; and the main recommended origins/provenances. (To be considered in conjunction with silvicultural characteristics noted in section 1 *supra*.)	Generally unsuitable conditions.
Scots pine Succeeds over a wide range of conditions. Along with lodgepole pine is the easiest and often the only species to plant on dry heath sites. Can be grown almost anywhere but is most successful in central and northeast Scotland, East Anglia and south-east England. Thrives on light and sandy soils but growth is relatively slow compared with that of spruces. Does well in low rainfall areas. Useful for nursing Sitka spruce on drier nutrient-deficient sites. *Origin/Provenance:* Seed from orchards should normally be the first choice. Otherwise use seed from Registered EC stands in the Vosges Mountains or West Germany at 50°-52° N, in south and west England or seed collected in local stands of good form. In Scotland, local seed is the second choice to orchard seed. Forrest (1990) gives recommendations for restocking and regeneration of native Caledonian pinewoods.	Avoid (a) smoky areas, soft ground and sites exposed to sea wind, wet coastal uplands and heavy soils; and (b) heavy wet mineral soils and shallow calcareous soils, especially chalky rendzinas. Not easy to establish on moorland under high rainfall. On wet coastal uplands may suffer needle loss and snow damage. (Native Scots pine in Scotland calls for special treatment and management, see chapter 15, 'Grants'.)
Corsican pine Low elevations, particularly sandy areas near the sea. Best growth is on sandy soils in areas of low summer rainfall and correspondingly high summer temperature. Light sandy soils, also heavy clays in the Midlands, south and east of England. Low rainfall areas. Also on the north-east coast of Scotland, notably at Culbin. More successful on chalky soils than Scots pine. Tolerates air pollution better than other conifers. Useful in mixture with beech. *Origin/Provenance:* For southern sites in low rainfall areas, home collections from registered ('select') stands of Corsica origin are best. The next choice is seed from registered EC stands in Corsica. For sites north of the Midlands, the two sources above will also grow well, though for higher volume production, seed from second generation stands of Calabrian origin, e.g., Koekelare in Belgium, or Les Barres in France or, if available, seed from healthy stands of Corsican type in northern Britain, are suggested.	Avoid high elevations. Unsuitable for the northern and western uplands of Britain and wetter upland areas where there is a likelihood of dieback caused by *Gremmeniella abietina*.

Conditions justifying selection; and the main recommended origins/provenances. (To be considered in conjunction with silvicultural characteristics noted in section 1 *supra.*)	Generally unsuitable conditions.
Lodgepole pine Flourishes on poorest heaths and peats, where few other trees will grow, after appropriate ground cultivation and fertilization. Planting is often on grassheaths because the yield is higher than that of Scots pine. A useful pioneer species, especially in the west and north. Alaskan origins useful for nursing Sitka spruce on poor sites. *Origin/Provenance:* Seed origin choice is complex, as often a compromise must be reached between e.g., growth rate and stem-form or exposure tolerance. Lines (1987) provides a table which allows origins to be selected with whichever characteristics are considered to be of greater importance. For use in mixture with Sitka spruce, Alaskan or North Coastal origins are most likely to produce a sound nursing effect and have a good chance of resulting in a self-thinning mixture leading to a final crop of pure spruce. If faster-growing origins, e.g., Skeena River or Southern Interior or British Columbia (BC) are used, then a higher proportion of the pine is thinned or otherwise treated to control its growth. For a combination of good form and volume production, on many sites the choice will lie between Skeena River, Vancouver Island and Southern Interior of BC origins. Vancouver Island sources should not be used on very exposed sites; and seed from the Southern Interior region should be from the Wet Belt. If good form is of more importance than volume production, then Central Interior of BC, e.g. Prince George or Bulkley River origins, are indicated. Avoid South Coastal Washington and Oregon provenances.	Tends to grow very coarsely on most fertile sites. On some poor, deep peat sites in Scotland there is a risk of epidemic attack by the Pine Beauty moth (see discussion by Leather 1987).
European larch Site requirements exacting. Must have a frost-free site. Best on moist but well-drained moderately fertile loams. A good nurse species. Some tolerance of air pollution. The most desired characteristics are: growth rate, stem-form, resistance to canker and dieback, and the date of flushing. Lines (1987) has ranked the main groups of origins/provenances for each of these characteristics. *Origin/Provenance:* Registered Seed Stands in Britain should be the first choice. The next choice are imports from the Sudetan region of Czechoslovakia, or from low elevation stands outside the native range in Germany (e.g., Schlitz) and Austria (e.g., Wrenerwald). Seed from high elevations in Swiss, French and Italian Alps must be avoided.	Avoid damp, badly-drained or very dry sites, frosty places, shallow soils over chalk, poor sands, peat soils, severely leached soils, exposed sites at high elevations or near the sea, and sites carrying dense growth of heather or grass.

Conditions justifying selection; and the main recommended origins/provenances. (To be considered in conjunction with silvicultural characteristics noted in section 1 *supra*.)	Generally unsuitable conditions.
Japanese larch Thrives over a wide range of conditions of the west and north. Suitable for upland sites including grassy and heathery slopes. Likes better freely-drained sites. Useful for checking dense weed growth, including coppice shoots on former woodland. Some tolerance of air pollution. Best in hilly districts in the milder, wetter regions; especially on well-drained, moderately fertile soils which are not too heavy. *Origin/Provenance:* The best source is likely to be seed from selected British stands, as these will have been thinned to remove individuals with poor stem and crown features. This seed may contain a proportion of natural hybrids with European larch. For Japanese imported seed, the Suwa region of Nagano is the first choice. Acceptable origins are the Nikko region, Tochogi Prefecture and stands on Hokkaido. Seed collected above 1800 m and from outlying areas in Honshu should be avoided.	Avoid (a) dry and low (or too high) fertility sites and areas where rainfall is under 750 mm; (b) badly-drained sites, frost hollows and very exposed situations.
Hybrid larch Special value on sites at limits for use of European or Japanese larch. Likes humidity. First generation hybrid from selected parents outstanding; second generation also good, but third generation often very inferior to European and Japanese larch.	In general, as for Japanese larch.
Douglas fir Well-drained soils of good depth, moderate fertility and valley slopes. Particular care needed in site selection. Best on wetter, western regions; though will grow satisfactorily in lower areas of south and south-east England and foothills of the Grampians. Good for middle valley slopes. *Origin/Provenance:* The most vigorous seed origins come from a U-shaped area along the western foothills of the Cascade Mountains in Washington, USA, westwards along the 46th parallel to the coastal range and north as far as Forks in Clallam County. Trees from Coquille, Oregon grow well on a range of sites, but the more southerly Oregon sources should be avoided in Scotland. Some Vancouver Island sources, e.g., Jeune Landing, grow well in Scotland and quite satisfactorily in England. Little seed is now available from the Lower Fraser River area, which supplied much of the seed 50–60 years ago, as the stands have now been felled. Seed from registered British stands dating from this era are a good choice anywhere in Britain, but supplies are very limited. Origins from the interior of British Columbia north of 50° should be avoided.	Avoid (a) exposed or low fertility situations, heather ground, dry areas, and wet, shallow and alkaline soils; (b) rich sites with poor drainage; and low rainfall areas. Liable to windthrow on loose ground, especially when young.

Conditions justifying selection; and the main recommended origins/provenances. (To be considered in conjunction with silvicultural characteristics noted in section 1 *supra.*)	Generally unsuitable conditions.
Norway spruce Moist grassy or rush sites, and shallow but less acid peats. Succeeds on old woodland sites and most soils of moderate fertility including heavy clays. *Origin/ Provenance:* East European origins from Romania, Poland and Czechoslovakia combine fast growth with late flushing times. Some southern Austrian sources also have these characteristics. German origins are rather variable and only the Harz region appears generally suitable. Most French, Alpine and Scandinavian origins are not recommended either because of slow growth, early flushing or both. Provenances for Christmas tree production are noted in chapter 5.	Avoid (a) uncontrolled heather land and dry sites, particularly on the east side of Britain; (b) sites exposed to wind; and (c) sites of low fertility. Often checked by frost in hollows unless very late flushing origins are used.
Sitka spruce Successful on a wide range of sites. Damp sites, generally, including exposed high land. Requires high rainfall. Can withstand a much greater degree of leader browsing by deer than Douglas fir, Japanese larch or broadleaves. Planting stock derived from proven superior families raised from cuttings ('C½+C1½') is now available. *Origin/Provenance:* Most seed collected from wellgrown stands in Britain will be of Queen Charlotte Islands (QCI) origin – general-purpose seed source which is reasonably frost hardy, very resistant to exposure and well-proven as a producer of acceptable timber. It can be reliably used throughout Britain, and is also the best provenance for very exposed sites. Where sites are less exposed, and particularly on sites in south-west England and Wales, there is a choice between QCI and more southerly origins with the latter offering a slightly increased risk of autumn or winter frost damage (particularly in the nursery), but greater timber production. On these more favourable sites north Oregon seedlots grow best, while those from west Washington and some Vancouver Island origins grow faster than QCI. These southern sources also grow well on sites close to the sea as far as 57° N in Scotland. Use of these very fast-growing origins carries a risk that the timber density of some trees may be unacceptably low. There is no case for using Alaskan origins in Britain or those from the upper Skeena River region, where introgressive hybridization with White spruce *Picea glauca* occurs.	Avoid (a) frosty or dry sites, areas of high water table or heather ground; and (b) low rainfall districts. Requires nutrient input or species mixture on poorest sites. Honey fungus risk in scrub and coppice.

Conditions justifying selection; and the main recommended origins/provenances. (To be considered in conjunction with silvicultural characteristics noted in section 1 *supra*.)	Generally unsuitable conditions.
Western hemlock Acid mineral soils and better peats. Less fertile, moist but well-drained sites, especially slopes of valleys. May have to be grown on short rotations due to *Heterobasidion annosum*, unless stumps of any previous crop were thoroughly treated on felling. Use for underplanting is important. *Origin/Provenance:* Vancouver Island sources, e.g., Courtenay, are good general purpose origins, while on dry sheltered sites in the south and east of England, Camano Island, Washington grows very well. For sheltered sites in northern Britain, Alaskan origins are the hardiest, but may be less vigorous than those from QCI.	Avoid dry sites, sands, sandy soils, limestones, sites prevalent to *Heterobasidion annosum*, and frost areas. Difficult to establish pure on exposed ground, so need a nurse species. Slow to establish on heaths.
Western red cedar High rainfall, and deep, moist loams. Moderately fertile soils, even if rather shallow, and fairly heavy clays. Succeeds on chalk and limestone. Good as a nurse species on chalky soils. *Origin/Provenance:* First choice is the north slope of the Olympic Mountains, Washington below 150 m in elevation. Some Vancouver Island seed sources, e.g., Ladysmith, show promise. The interior origin from Shuswap lake has grown well, except in the high rainfall area of Benmore in Scotland.	Avoid poor or very acid soils and exposed or frosty sites.
Grand fir Well-drained, moist, deep soils of moderate fertility. Tolerates drier sites than Sitka spruce. Likes abundant rainfall. Useful as an underplant, being shade tolerant. *Origin/Provenance:* Origins from the Olympic Peninsula, Washington and the Puget Sound have grown uniformly well on all sites; individual seedlots from Elwha, Louella and Sequim are outstanding. Excellent height growth has characterized the two seedlots from the S. Oregon coast; however, these flush early and have suffered some dieback on exposed sites. Vancouver Island seed origins are somewhat slower-growing and they are also rather early in flushing; however, they have the advantage that their use is well proven in Britain, e.g., a stand of YC 34 at Dunkeld, Perthshire, is from Campbell Lake, Vancouver Island.	Avoid frost hollows, exposed situations, and poor soils, particularly really acid ones.
Noble fir Flourishes on well-drained, deep, moist soils. Tolerates fairly acid soils and high elevations. Best on low ground on the western seaboard. Likes abundant rainfall.	Avoid poor soils and dry sites.

Conditions justifying selection; and the main recommended origins/provenances. (To be considered in conjunction with silvicultural characteristics noted in section 1 *supra*.)	Generally unsuitable conditions.
Origin/Provenance: The most promising source appears to be Larch Mountain, Oregon, east of Portland. Areas to avoid are the introgression zone south of latitude 44° N and perhaps Mary's Peak. However, the Laurel Mountain origin, which is also in the Coast Range, has grown well.	

Species choice for broadleaves

Broadleaves most commonly planted for timber. All are deciduous.

Conditions justifying selection; and the main recommended origins/provenances. (To be considered in conjunction with silvicultural characteristics noted in section 1 *supra*.)	Generally unsuitable conditions.
Oak: Pedunculate; Sessile Well-aerated, deep, fertile loams. Grow fairly well on fertile, heavy soils and marls. Pedunculate oak may be used on heavier soils but in general Sessile oak should be chosen. Sessile oak tolerates less rich soils than does Pedunculate. The best chance of a profitable crop may be on fertile soils in mixture with conifers; pruning and concentration on production of veneer quality trees may also be profitable. Respond well in treeshelters. *Origin/Provenance:* In Britain (1991) there were 14 Registered Seed Stands of Sessile oak and 33 of Pedunculate oak. These currently appear to offer the best prospects for establishing satisfactory oak plantations. An alternative scheme being investigated by the National Hardwoods Project team at Oxford University is to use scions from selected 'plus' trees, grafted on to stock plants to form a multiclonal tree bank. Trees in this bank will be regularly pruned to produce suitable material for the bulk production of cuttings for rooting. Evans (1984) has provided a Table to enable selection of whichever oak species is more appropriate for the given site. The first choice of seed would be from Registered Seed Stands in Britain; if unobtainable, then from recommended seed from Registered Seed Stands in EC countries such as France or Germany. A last option is to obtain seed from trees of good form in a local stand.	Avoid (a) shallow, ill-drained or infertile soils, frost hollows and exposed sites; (b) strongly calcareous soils; and (c) because of the association with shake, light, very freely-drained sandy soils, especially with fluctuating water-tables.

Conditions justifying selection; and the main recommended origins/provenances. (To be considered in conjunction with silvicultural characteristics noted in section 1 *supra*.)	Generally unsuitable conditions.
Beech Chalk and limestone soils. Good loams of all types; if heavy, should be well-drained. Likes a mild sunny climate. Best on deep, well-drained loams to clay loams, including those over chalk and limestone with acid surface, as well as generally on deep, sandy loams to loams of moderate acidity. The best chance of a profitable crop may be in mixture with conifers. Benefits from a nurse species, e.g., Scots pine, on exposed sites. Essential to control grey squirrel. *Origin/ Provenance:* As a first choice, seed from the Forêt de Soignes, Belgium is to be preferred (often in short supply). The next choice is seed from a Registered Seed Stand in the region; if unobtainable, then recommended is seed from Registered Seed Stands in Belgium, Holland, northern France and Germany. A last option is to obtain seed from trees of good form in a local stand.	Avoid (a) frost hollows, valley bottoms, soils on badly-drained sites, and leached soils, including heathlands; (b) poorly-drained sites with clay or silty clay texture soils; and (c) very shallow soils over impermeable substratas which are liable to excessive drying out.
Ash An exacting species demanding good soil conditions. Likes sheltered situations and deep calcareous loams, moist but well-drained. Thrives on chalk and limestone where the soil is deep. Likes nitrogen-rich alluvial and non-alluvial soils; also valley bottoms if free of frost. Best on deep, moist, freely-draining and fertile soils of about neutral pH reaction. Favour fertile pockets of soil conducive to sports-quality and veneer growth. Vegetation indicators of potentially good ash soils are Dog's mercury, wild garlic and wild angelica.	Unsuitable for extensive planting or exposed sites. Avoid (a) dry or shallow soils, grassland, heath or moorland, compacted soils, ill-drained ground and heavy clays; (b) frost hollows; and (c) planting adjacent to arable fields due to tendency to die-back.
Sycamore Fertile, rich, alkaline soils, with adequate moisture. A useful tree as a windfirm mixture component with conifers in shelterbelts. Tolerates air pollution and salt-spray. Best on deep, well-drained soils over chalk and limestone and on acid brown earths. Essential to control grey squirrel.	Generally as for ash *supra* but tolerates exposure and is frost hardy. Avoid acid peats, and infertile sandy soils or those of low phosphate.
Sweet chestnut Needs deep, acid soils. Best in a mild climate, and on south-facing slopes with plenty of sun and no frost. If large timber, not coppice, is to be grown, sites expected not to produce shake should be chosen; sites on Old Red Sandstone appear to be good in this respect.	Avoid (a) less fertile sites, frosty or exposed sites, badly-drained ground, heavy clays and alkaline soils; (b) areas north of a line from Liverpool to the Wash; and (c) sandy soils where shake may be expected.

Conditions justifying selection; and the main recommended origins/provenances. (To be considered in conjunction with silvicultural characteristics noted in section 1 *supra*.)	Generally unsuitable conditions.
Wild Cherry (Gean) Likes fertile soils, deep loams over chalk, and welldrained sites. Because relatively short-lived, best results are achieved when grown to merchantable size as fast as possible. Best planted in small groups or in mixture with ash, beech or oak. Especially good on the Downs of south-east England and the Chilterns and Cotswolds. Pruning is essential to produce veneer quality timber.	Avoid (a) infertile strongly acid soils, compacted soils, heavy clays, exposed sites and depressions where waterlogging may occur; and (b) shallow, dry and poorly-drained soils.
Poplar Likes base-rich loamy soil in a sheltered situation, with the water-table 1–1.5 m below the surface in summer. Also suitable are the heavy clay soils in the south Midlands and Thames valley. A soil with a pH in the range of 5.0–6.5 may be expected to support satisfactory growth of all species and cultivars.	Avoid (a) sphagnum peats, (b) light sand and gravel heath soils of south and east England, and (c) wet, badly-drained or waterlogged sites (Jobling 1990).

Evans (1984) recommends for species choice adherence to the following procedure: (i) define the objective or purpose(s) of growing trees; (ii) list objectives in order of priority, preferably into main, secondary, and minor; (iii) list species which satisfy the main objective; (iv) eliminate species ill-suited to the site, unavailable, or biologically undesirable (e.g. elm); (v) eliminate species not compatible with the secondary objective; and (vi) choose species least in conflict with minor objectives. The procedure is usually followed subconsciously but species' selection merits orderly and logical considerations since the one chosen may occupy the site for 50 to 100 years or more. The consequences of a wrong choice can be costly and disappointing.

The process of matching species with site and other factors may become over-complicated, and maximizing yield through use of a wide range of species may introduce management and marketing problems. Choosing the species or mixture of them can be performed confidently only with experience, especially that acquired locally. Experience and far-sighted common sense should be brought to play on the task. Most foresters, with a fund of knowledge of the locality, will make their species choice with little apparent difficulty; even so, seemingly without effort, they will have carefully considered site and other factors, as well as objectives.

Species choice, including the best origin/provenance, is of no avail unless followed by the use of appropriate planting stock, carefully handled, and adequately planted and tended – matters discussed in chapter 5.

3. Pure crops or mixtures?

For each planting site foresters have to decide whether to grow a pure crop of one species or instead a mixture of at least two (i.e. two conifers, or two broadleaves, or a conifer with a broadleave). Mixtures though sometimes more difficult to manage have many silvicultural, biological and aesthetic attributes. There are several species which consort well together, i.e. without any component being suppressed or harmed. However, in a mixture, one component must not be much fastergrowing than the other. Sometimes mixtures are more successful when one component is shade tolerant. Foresters' decisions can be aided by considering the advantages and disadvantages of each option. (Some of them are more theoretical than practical.)

Pure crops. Advantages of pure crops compared with mixtures include their simplicity, and ease and lower cost of establishment. They are generally of more even growth, size, grade and quality, easier to manage, thinning less arduous, and small diameter roundwood and sawlogs more easily harvested and marketed. Later in the rotation, should natural regeneration of a single species be an objective, it can be attained more easily from a pure overstorey. (In practice, however, other tree species often tend to invade, particularly sycamore, ash and birch.) Wherever extensive plantation silviculture is practised, pure even-aged crops tend to predominate due to their economic advantages, i.e. their simpler establishment, management, harvesting and marketing.

Disadvantages of pure crops are evident. Soil and other site factors do not always suit a single species throughout a full rotation. Pure crops of most broadleaves cannot easily be grown at a profit (exceptions may include poplar and sweet chestnut coppice). The first thinnings of some species (e.g. oak and beech) are generally of low value, and a financial return is long delayed. Local markets may demand more than one species, and hence it may be unwise to depend on one only, as a wrong choice in species could be disappointing. Damage by the elements, insects or fungal diseases are sometimes more harmful in a monoculture. Deterioration of soil *may* occur under a pure crop of conifers (but one cannot be dogmatic about this; management has a great impact on soil fertility). Flora and fauna are generally less varied. Pure crops of some conifers, particularly when in large even-aged blocks, usually have low amenity, landscape, recreational and sporting attributes – dependent on age and structure.

Mixtures. Advantages and disadvantages of mixtures are generally in reverse to those stated for pure crops. The main problem is that to be successful mixtures generally require extra skill in management. Mixed crops, although more costly to establish and manage, often have economic advantages; for instance, they may be more flexible in meeting market demands, and mistakes in species choice are less serious. Broadleave/conifer mixtures provide earlier financial returns (from the conifer): owing to long rotations and slow growth of most broadleaves, a matrix of a faster-growing conifer which matures sooner and hence brings earlier income should improve overall profitability. Mixed crops, especially of broadleaves and particularly if uneven-aged, have certain aesthetic attributes: their amenity, landscape, recreational and sporting attributes are generally greater; and their flora and fauna are more diverse and abundant. They may exploit a site more fully, i.e. the components root in different zones. When at least one component is broadleaved, they *may* diminish the risk of soil deterioration.

Mixtures adapt better to damaging events. Intimate mixtures might give maximum risk insurance. Transmission of narrow-spectrum pests and disease is retarded. Broadleaves in mixture with conifers appear to suffer less from grey squirrel damage than do wholly broadleaved stands. Beech bark disease is probably less prevalent in beech mixed with pine than when pure. Canker on European larch may be less widespread when in mixture (though canker is probably due to unsuitable provenance and worsened by frost damage).

Windthrow might be lessened when a relatively deep-rooting species is mixed with one which is shallow-rooting. There may be a reduced risk of damage by snow. Strategically-located lines of less combustible species may help to limit the spread of fire. Species interaction may be beneficial in mixtures, as when one component is a nitrogen-fixer (e.g. alder on impoverished sites), or is a water-tolerant species whose roots ameliorate the site for a more productive but less tolerant partner. Differences in vigour may be deliberately exploited in *selfthinning mixtures* of conifers; they are in use on sites of high windthrow hazard where conventional thinning may be unwise (see chapter 8). Such mixtures may be of Sitka spruce and lodgepole pine; or be achieved by planting of one species having two different growth rates, e.g. in Sitka spruce, the fast-growing Queen Charlotte Island provenance with the slow Alaskan provenance (Lines 1981, 1987).

Management problems may be presented by intimate mixtures throughout the rotation: they may not develop exactly as planned. Different growth rates may cause premature suppression of one component. Thinning regimes may differ between species. Harvesting and marketing may be rendered more costly by the need to separate products by species as well as size. However, the challenge of the possibilities should be met: the extra skill required for a soundly chosen mixture is well worth while from the silvicultural, harvesting and marketing points of view, as well as from the enhanced aesthetic values and satisfactions usually provided.

Methods of creating mixtures include planting, natural regeneration or enrichment. Uneven-aged mixtures have been encouraged, e.g. oak underplanted with beech, with the intention of maintaining or enhancing fertility and to help control development of epicormic shoots in the former, which degrades its timber. (For example, the Forest of Dean has fine 1910 natural oak underplanted with beech in 1956.) Mixtures frequently include *pioneer* species, i.e. those which are early colonizers of bare or newly-planted land. Birch, the most common example, is useful for its provision of shelter and protection from sun, wind and frost. (Nature herself uses pioneer species, particularly following fire or windthrow.)

Nurse species can form a component of mixtures, performing one or more of three functions: (i) rapidly suppressing competing vegetation (this may have an adverse effect on desirable ground flora); (ii) modifying the microclimate, to protect tender species from frost and exposure during their early years; and (iii) benefiting the nutrition for another species. On some sites, growth of broadleaves in mixture with conifers may be superior to that of a pure crop and will often produce taller, straighter stems on the broadleaves. Beech, western hemlock and grand fir may grow better initially if mixed with a faster-growing nurse or if planted under the canopy of a near-mature crop. Often the nurse species will provide early returns.

When deciding upon a nurse species, it is necessary to know what site factors are required to be changed: these may include temperature, evaporation, competing vegetation and the elements of the soil. In the uplands, lodgepole pine, a well used species on difficult moorland when peat is prevalent, aids especially the establishment of Sitka spruce which suffers chronic nitrogen deficiency in many infertile sites. On such sites the once common practice of planting conifers in mixtures is again being undertaken and expanded onto the deep peats of the north – previously the domain of lodgepole pine (Taylor 1985). On upland heathland, a special type of nurse mixture has been where admixture with suitable origins of Scots pine, lodgepole pine or Japanese larch has improved the growth of Sitka spruce due to the ability of these species to suppress the heather – to nurse the Sitka spruce through 'heather check' (discussed in chapter 5).

A decision has to be made as to whether one species is to act only temporarily as a nurse or whether the whole crop is to remain mixed for a long period, and perhaps throughout a whole rotation. Thinning can control the composition and development of the mixture: in one containing a temporary nurse species, for instance larch, the first thinning would be confined chiefly to the removal of most of the nursing trees, eventually resulting in a pure crop of the second species. Another mixture might comprise European larch with oak or beech, later changed to a pure crop of broadleaves by removing the larch. Difficulties in management of conifers mixed with broadleaves mainly arise where the conifers are allowed critically to outgrow the broadleaves.

The pattern of a mixture. The pattern of the lay-out of the components of a mixture must be decided. Planting of alternate species in single rows, or of alternating rows of each of two species is usually unsatisfactory. Rarely do two species grow at the same rate or have the same shade-tolerance; thus the slower becomes suppressed, resulting in a pure crop of the more vigorous. However, if three or more rows of each species alternate (e.g. a three/three row of oak and European larch), the conifer row next to the broadleaves should be harvested at the first thinning. Many of the trees in the middle row of each can remain for a long period, with little undue competition between them.

Block planting, which creates a 'chequer-board' matrix, is sometimes adopted; both establishment and management are relatively complicated, progress is not easy to observe and control, and thinnings may be difficult to harvest. An alternative method is to plant groups (5, 9 or 25 in number) of one broadleaved species in a matrix of such a conifer as is unlikely to suppress it, the spacing of the groups being commensurate with the required espacement of the expected broadleaved final crop. This pattern is sometimes followed when oak or beech or cherry is mixed with European larch; the mixture provides some nursing effect and makes for economy in the use of the more costly broadleave planting stock. The intention is likely to ensure that one outstanding broadleave for the final crop arises from each group; meanwhile all broadleaves in the group are helping to achieve straight stems and induce natural pruning, most thinnings yielding small but merchantable proportions. Yet, the mixture may not always develop in the neat geometrical pattern set at planting.

Patterns of rows or groups in undulating countryside can create landscape problems by creating undesirable 'pyjama-stripes' or 'chequer-board' effects; in a level scene the problems do not arise. Belts of broadleaves around conifers in hill country can create undesirable landscape effects; in a level scene the practice is effective (a surround of cherry, red oak or Norway maple can be scenically attractive – but careful design is needed).

Types of mixtures. Mixtures of oak or beech or ash with one species of conifer (e.g. European larch or Scots pine or Norway spruce) give satisfactory results, provided the oak or ash, being slow-growing and (by contrast to the beech) intolerant of shade, are not suppressed by the conifer. In such mixtures, whether in rows, or in broadleaved groups, appropriate early brashing and thinning of some of the conifer are desirable.

In a mixture of conifers, several species consort well throughout their rotation; but there are others which are only suitable components for part of their economic life. Where Scots pine and European larch are mixed, the larch tend to outgrow the pine. Douglas fir and larch stand well together for a while, but the larch usually have to be removed early to avoid being suppressed. The site factors are critical; for instance, Scots pine will outgrow Norway spruce on light soils in East Anglia (wrong conditions for spruce in any case) while the result is usually reversed (because the choice is correct) on heavier soils in the north and west of Britain. In all mixtures great importance attaches to the thinning regime and the encouraging of any one component. By timely action a mixture can be swung according to the objective; especially is this so in such a mixture as Japanese larch with Douglas fir or Sitka spruce or western red cedar. How mixtures of exotic conifers will develop is often uncertain. (For instance, in the Forest of Dean, in an experimental mixture comprising four north-west American species, western red cedar and western hemlock for a long period competed well with Douglas fir and Sitka spruce.)

Certain mixtures are considered to be reasonably sound under most conditions. When comprised of two light demanders, recommendations are oak with European larch , and oak or ash with Norway spruce. When of two shade bearers, satisfactory are western red cedar with beech or very occasionally Douglas fir, though the cedar may suppress the fir if the latter (usually through wrong species choice) is seriously checked by woolly aphid *Adelges cooleyi*. Mixtures comprising a shade bearer and a light demander may comprise: beech with ash, cherry, Corsican pine or European larch, and occasionally oak; western red cedar with European larch or oak; and grand fir with hybrid larch. Mixing a light demander with a shade bearer to maturity may incur difficulties unless the latter is the slower growing of the two and is sufficiently shade-tolerant to survive when over-topped by the light demander; a notable exception is oak with beech.

Components of a mixture. The compatibility of components of a mixture will vary with species, site factors, spacing and management. Matching of species both to each other and to the site is important. Mixtures which generally work well on suitable sites include:

Mixtures of conifers:

- Sitka spruce with European larch, though there is a tendency for the larch to outgrow the spruce, especially if the stand is underthinned.

- Scots pine with Norway spruce, depending on site conditions: on light soils the pine may suppress the spruce; on heavier soils the tendency may be the reverse.
- Douglas fir with European larch, though this mixture can seldom be maintained to an advanced age: the larch must be removed in early thinnings.
- Japanese or hybrid larch with Douglas fir or Sitka spruce or western red cedar or grand fir. The mixture may tend to become a pure crop of either species, and an early decision must be made as to the objective; generally the larch should be removed. The mixture can be swung either way according to the objective. Japanese larch can also consort well with Scots pine or Corsican pine given suitable sites.

Mixtures of broadleaves with conifers:

- Sweet chestnut with European larch.
- Oak with European larch (especially on lighter loams and alkaline clays); and occasionally with Norway spruce (on heavy acid clays); or with Scots pine (especially on freely-draining soils); or with western red cedar.
- Beech with western red cedar or Scots pine or Corsican pine and occa-sionally with European larch or Norway spruce.
- Ash with European larch (on the drier sites) or with Norway spruce (on the wetter sites) subject to vigorous thinning to provide the ash with ample headroom.

As a guide to compatibility, when deciding on a broadleaved with conifer mixture, generally the expected yield class of the conifer should never be more than double that for the broadleaved component. However, on some high pH sites, the broadleaved component may dominate and suppress the conifer.

Mixtures of broadleaves:

- Oak with ash or cherry (especially on brown earths or clays over chalk); also with sweet chestnut or beech.
- Ash with oak or cherry or sycamore or sweet chestnut.

In mixtures of broadleaves, compatability between the species is less of a problem than between broadleaves and conifers; and mixture layout is relatively unimportant. Planting formal mixtures of broadleaves is uncommon, but inclusion of a small proportion of a second broadleaved species (e.g. cherry) can benefit both silviculture and interests such as amenity and landscape. (On some sites, especially recently felled woodland, mixtures of broadleaves will often arise by way of natural regeneration among the restocked trees; the colonization is likely to include sycamore, ash, birch, alder and sallow.) Some broadleaved species, notably oak, require fairly dense stand conditions to ensure the development of a reasonable formed stem and upward growth.

A recommended pure broadleave mixture, given suitable site factors, is one comprising oak, ash and cherry with a view to the two latter species maturing early. (Cherry grows rapidly in the early years and has rotation lengths of only five or ten more years longer than those of most conifers.) Regular thinning should be undertaken to ensure that the oak and ash are not suppressed by the cherry. Shortly following the removal of the cherry, the ash component should be removed leaving a stand of oak to grow to maturity.

Whatever the mixture chosen, the intention for establishing it should be recorded, thus obviating different aims being sought during the various thinning stages, i.e. one component being favoured during one thinning and the other subsequently, in which case neither may yield its best. Foresters must continually bear in mind what they are hoping to achieve in any mixture they are managing.

4. Broadleaves in Britain

Broadleaves because of their many attributes are receiving significant attention in Britain, playing an important role as a vital and cherished part of the countryside, whether as woodlands, hedgerows, single specimens or clumps of trees. They are an integral feature of the landscape heritage, and of great environmental importance. Broadleaved woodlands are used extensively for recreation, from the simple pleasure of walking to more specialized pursuits, and provide cover for country sports notably shooting and hunting. The diversity of the structure – particularly in medium and older ages – gives them great value for wildlife conservation. Their importance goes well beyond their visual appearance and other aesthetic attributes (Steele 1972; Rackham 1980; Peterken 1981; Malcolm *et al.* 1982; Evans 1984; Forestry Commission 1984; 1985a, 1985b; Potter (ed.) 1988; Marren, 1990, Watkins, 1990). The foregoing features and attributes, though sometimes preventing the maximizing of timber production, can be turned to advantage of both the woodland owner and the public. The woodlands' many roles bring interest, complexity and challenge in managment (see chapter 13).

Broadleaved tree species are mainly native (see chapter 1). There are a few other species, notably sycamore, English elm and sweet chestnut, which for long have been treated as 'traditional' trees. Native woodlands can support a wide variety of flora – spring flowers such as primrose, wood anemone and bluebell, and many others throughout summer – as well as lichen, fungi, moss, fern, bird, insect and other fauna. (Exotic tree species likewise support wildlife, but woodlands of native broadleaves usually provide the richest habitats for both flora and fauna.)

The broadleaved resource. Broadleaves in Britain comprise around 615,000 ha or 29% of total productive woodland area; in England alone the extent is about 55%. The bulk of broadleaved high forest and coppice is in England, and 70% occurs in the mid and southern lowlands (thus, silviculturally, compared with much coniferous forest, broadleaves generally experience more sheltered conditions in sites of generally superior fertility but subject to greater environmental constraints).

The private sector owns just over 90% of the broadleaved area; 39% (almost 300,000 ha) of all such woodland consists of small woods, predominantly on farms, of less than 10 ha. Small extent tends to increase costs and depress returns from growing timber, but, if anything, enhance environmental importance as a landscape feature. Such woods are often important components of a lay-out of farmland for game bird management.

Of the total volume of timber in Britain's woodlands, broadleaves account for around half: this is because of their older average age than conifers and because most isolated trees and small clumps outside the woodland areas are broadleaved. They supply half of the country's consumption of hardwood (about one-fifth of annual wood production) and could yield more if actively managed. Besides trees comprising broadleaved woodland, Britain has a further 90 million broadleaved trees occurring as isolated specimens, or in groups and hedgerows. The overall situation is much the same as almost fifty years ago, but its composition has changed and significantly there has been a loss of ancient semi-natural woodland.

In 1980, of 752,000 ha of broadleaved woodland, 564,000 were high forest, 12,000 coppice with standards, 28,000 coppice and 140,000 scrub. The volume of the whole growing stock was 91 million m^3 ob. The age-class distribution of broadleaved high forest is remarkably well-balanced, percentages by species being: oak 30.7, beech 13.2, sycamore 8.8, ash 12.4, birch 12.2, poplar 2.4, sweet chestnut 1.8, elm 1.7, other broadleaves 5.2, and mixed broadleaves 11.6. In the very long term, maintenance of the current rate of new planting and restocking of broadleaves in woodland, on farmland, and as park and hedgerow trees will be sufficient to sustain an annual yield of about 900,000 m^3 ob.

The post-war afforestation programme led to a substantial increase in coniferous high forest but it is noteworthy that the area of broadleaved woodland has also increased, part resulting from new planting and part from natural colonization by broadleaved species. The significant change in the composition of broadleaved woodlands is due to the decline in coppice working and the natural conversion of scrub woodland into broadleaved high forest. As a consequence, productive broadleaved woodland (i.e. the total of high forest, coppice and coppice with standards) has increased by some 15% since 1947.

In recent times there has been an increasing awareness of the contribution which broadleaved woodland and trees make to the British, and particularly to the English, lowland landscape. Their future has become a matter of keen interest: any apparent threat to them produces an understandable outcry. Public interest arises from an appreciation of their richness, diversity, visual beauty, the variety of wildlife habitats which they provide and their value to the national heritage. There is much concern about their conversion to coniferous woodland, or to agricultural use, and about the prospects for the ageing stock of hedgerow and park trees almost all of which are broadleaved. The disappearance of the elm from many areas as a result of Dutch elm disease, and of ash dieback, has also heightened appreciation of the risk to broadleaved trees in both woodland and farmland.

The current areas of broadleaves are best sustained and expanded through the use of natural regeneration, suckering or coppicing; or through the planting of stock raised by way of seed from the same woodland. (Yet many areas of the broadleaved woodland which is now valued so highly were created by means of some of the conifer/broadleaved mixtures that are in use today.)

Forestry Commission broadleaved policy. Management is essential for broadleaved woodlands if they are not to deteriorate. The Government's Broadleaved Policy, announced in 1985, is based on this premise and emphasizes the importance of their silviculture. The effect of the policy on forestry in Britain is dynamic and impelling, aimed at encouraging positive and sympathetic management to arrest depletion, to increase the quality of timber and to expand this valuable national resource to meet the various supplementary objectives. Important to the success of these new policy aims is that they should be pursued with some flexibility within the framework of an essentially voluntary approach which harnesses the interest and goodwill of landowners, farmers, and conservation organizations.

More specifically, the government's aims for broadleaves are: (i) to enhance the broadleaved character of the well-wooded areas of the country; (ii) to promote the planting of broadleaved woodlands where they are scarce, including areas on the periphery of towns and cities; (iii) to encourage the maintenance and greater use of broadleaves in the uplands, particularly where they will enhance the landscape and the wildlife interest – supplementing the extensive conifer plantations; (iv) to encourage greater use of all types of broadleaved woodland for conservation, recreation, landscape and sport, in conjunction with timber production; and (v) to ensure that the special interest of ancient semi-natural woodland (discussed *infra*) is recognized and maintained.

Part of the aim of this policy is the need to increase the quality and value of timber produced by broadleaved woodlands. Priorities vary with local circumstances and objectives. Woodlands designated as Sites of Special Scientific Interest require special management attention, as do ancient semi-natural woodlands and those of high landscape value. Even then, however, timber production will feature in the management plan and help to finance other costs. (Practical management is discussed in chapter 13. Planting and management grants are set out in chapter 15.)

Some farmers are coming to realize that well-managed woodlands can be assets of considerable value, and that it is essential for the future of the large number of small broadleaved woodlands that they are brought more closely into the overall economy of the farm. Farmers are being given grant-aid, professional advice, and practical demonstrations of what can be done, the effort required and the rewards of good woodland management. There is scope on farms for more planting of woodlands, copses and hedgerows, and a concerted campaign to encourage farmers to step up their efforts in this direction is being led by Government agencies (see chapters 13 and 15) particularly as a means of taking surplus land out of agriculture.

In the uplands the scope for expansion of broadleaved woodland is very much less because of harsh conditions, unsuitable soils and deer pressure. However, on suitable sites there should be a much greater use of broadleaves, particularly where they offer greatest environmental benefit or can relieve the appearance of the predominantly coniferous, timber-producing plantations. Particular attention should be paid to the possibility of introducing broadleaves when restocking felled woodlands; site conditions may then have improved. Native broadleaves will be preferred where conservation objectives dominate, but other species compatible with the landscape may be used elsewhere.

The future for broadleaved trees outside woodlands seem fairly well assured, thanks to private and public planting. The grants for planting small groups of trees paid by the Countryside Commissions are effective. Tree Preservation Orders give local authorities power to protect endangered trees.

The policy represents only the latest stage in an essentially evolutionary process, the extent of success in which will depend on the goodwill and co-operation of all concerned with the countryside.

The Government's measures for broadleaves. These measures include woodland planting and management grants (see chapter 15) offering significantly higher rates for the planting and managing of broadleaved species, for restocking by natural regeneration and for the rehabilitation of neglected woodlands. The controls on felling are adequate, and there is greater insistence on restocking broadleaved woodland with broadleaved species unless the site has proved to be totally unsuitable. The Forestry Commission has drawn up management guidelines (discussed in chapter 13) for different types of woodlands in collaboration with relevant Government agencies, forestry organizations and voluntary bodies. The Commission has given appropriate training to its own staff and those of the Agriculture Departments and other interested organizations in the various aspects of broadleaved management, including landscape design and wildlife conservation. Local grey squirrel control groups have been encouraged in vulnerable broadleaved woodland areas.

Ancient semi-natural woodland is particularly valued. Most is broadleaved, the main exceptions being the native pinewoods in Scotland. Ancient woodlands, defined as those which have had a continuous woodland cover since at least AD 1600, may be 'primary', i.e. on land considered always to have been woodland, or 'secondary', i.e. on land formerly farmland or moorland within historical times but before 1600 (Marren 1990; Watkins 1990). Ancient woodlands in general are ecologically richer than other types but, if their full wildlife value is to be preserved, they usually have restricted timber producing potential.

Identification of woodland of high conservation value is greatly helped by the Register of Ancient Semi-Natural Woodland prepared by the Nature Conservancy Council (Inventories, 1989). This national inventory lists, on a county by county (or equivalent Scottish districts) basis, all probably ancient semi-natural woodlands, also other woodlands, such as plantations, on ancient woodland sites. The register, continually being checked and verified, also contains some recent semi-natural woodlands which are known or believed to be important sites for nature conservation. Copies of the register are held by the Forestry Commission offices, by regional secretaries of TGUK and of the Country Landowners' Association, and by county councils. There is not a formal mechanism for appeal against inclusion of a site on the inventory but if owners believe a site to be wrongly classified they should present the relevant information to the Forestry Commission. Registration has an effect on management (see chapter 13). A proportion of ancient semi-natural woodlands occur now as the result of skilled management involving the use of conifer components of mixtures. However, most originate from coppice or regeneration and it is unlikely that a conifer mixture was involved in their establishment. Certainly some high forest examples may have included a conifer component – the Chiltern beech woods are frequently quoted in this respect – but they comprise no more than about a third of ancient semi-natural sites. Flexibility in the application of Forestry Commission guidelines for management is essential to take account of these silvicultural skills.

No woodland in Britain can be regarded as entirely natural – such is the influence of man on the environment – but woodland on sites continuously wooded for a great number of years and having a tree and shrub layer composed of species native to the site, are the closest approximation to natural broadleaved forest to be found in Britain. In their strictest form, they consist of trees which have regenerated naturally from seed or suckers, or of coppice re-growth from trees which were themselves natural seedlings. It is not possible to make firm distinctions, however, between truly semi-natural stands and long-established plantations of native species. Many ancient woodland sites, though continuously wooded, are no longer semi-natural; they now carry plantations of broadleaves or conifers, either pure or in mixture. Such woodland often retains a variety of native ground and shrub flora, particularly in glades and rides, which can be developed by sensitive management.

Broadleaved woodlands not on ancient woodland sites, whether planted or naturally derived, may also have significant environmental value. They may lack the varied ecology of ancient woodland, but they may well have greater habitat diversity than the open ground they replaced. They are readily colonized by some woodland plants and animals, but colonization by more specialized woodland species takes place more gradually and may never be complete. Naturally regenerated woods of recent origin are particularly important for nature conservation in upland districts. The best and most representative examples of ancient semi-natural woodland, and some woods which are not ancient or semi-natural but which have identifiable special features, are designated as Sites of Special Scientific Interest or National Nature Reserves. In these selected woodlands, nature conservation objectives (along with landscape objectives, and scientific interest if present) take precedence. Such woodlands are within the scope of Forestry Commission guidelines. In most cases it is possible to integrate nature conservation and timber production. Woodlands of high landscape or recreational value are often identified in National Park Plans or in local authority plans. Individual local authorities have detailed knowledge of important landscape woodlands; and owners of such woodlands are usually aware of their importance.

Management of broadleaves under Forestry Commission guidelines is discussed in chapter 13 concerning five specific objectives – wood production, landscape, recreation, wildlife conservation and sporting. The progress of the Government's Broadleaved Policy is noted in chapter 13; and planting and management grants for furthering it are set out in chapter 15, section 2. The Commission intend to publish a review of its Broadleaved Policy in June 1991; this will relate specifically to the Woodland Management Grant, and enable applications to be made from July 1991 (see chapter 15).

References

Brooker, M.I.H. and Evans, J. (1983), 'A Key to Eucalypts in Britain and Ireland; with Notes on Growing Eucalypts in Britain'. FC Booklet 50.
Evans, J. (1984), 'Silviculture of Broadleaved Woodland'. FC Bulletin 62.
—— (1988), 'Natural Regeneration of Broadleaves'. FC Bulletin 78.
Faulkner, R. (1962), 'Seed Stands and their Management'. *Quarterly Journal of Forestry,* 56 (1), pp. 8–22.
Ford-Robertson, F.C. (ed.) (1971), 'Terminology of Forest Science, Technology, Practice and Products'. Society of American Foresters, Washington DC, USA.
Forestry Commission (1984), 'Broadleaves in Britain: a Consultative Paper'.
—— (1985a). 'Guidelines for the Management of Broadleaved Woodland'. Reprinted 1986.
—— (1985b), 'The Policy for Broadleaved Woodland'. FC Policy Paper No. 5. Reprinted 1989.

Forrest, I. (1990), 'Fingerprints of the Caledonian Pine'. FC *Forest Life* (7).
Gordon, A.G. (ed.) (1991), 'Seed Manual for Commercial Forestry Species'. FC Bulletin 83. In preparation.
Gordon, A.G. and Rowe, D.C.F. (1982), 'Seed Manual for Ornamental Trees and Shrubs'. FC Bulletin 59.
Greig, B.J.W. (1988), 'English Elm Regeneration'. FC Arboriculture Research Note 13/88/PATH.
Helliwell, D.R. (1989), 'Lime Trees in Britain'. *Arboricultural Journal, 13* (2), pp. 119–23.
Jobling, J. (1990), 'Poplars for Wood Production and Amenity'. FC Bulletin 92.
Leather, S.R. (1987), 'Lodgepole Pine and Seed Origin and the Pine Beauty Moth'. FC Research Information Note 125.
Lines, R. (1981), 'Self-thinning Mixtures'. FC Report on Forest Research 1981.
—— (1987), 'Choice of Seed Origins for the Main Forest Species in Britain'. FC Bulletin 66.
Low, A.J. (ed.) (1985), 'Guide to Upland Restocking Practice'. FC Leaflet 84.
—— (1986), 'Use of Broadleaved Species in Upland Forests'. FC Leaflet 88.
Malcolm, D.C., Evans, J. and Edwards, P.N. (eds.) (1982), *Broadleaves in Britain*. Symposium, Loughborough, Leics. 1982. Institute of Chartered Foresters, Edinburgh.
Marren, P. (1990), *Woodland Heritage*. David & Charles, Newton Abbot.
Matthews, J.D. (1963), 'Factors affecting the Production of Seed by Forest Trees'. *Forestry Abstracts 24* (i).
Peterken, G. (1981), *Woodland Conservation and Management*. Chapman & Hall Ltd., London.
Pigott, C.D. (1988), 'The Ecology and Silviculture of Lime *(Tilia* spp)'. Oxford Forestry Institute Occasional Papers No. 37.
Potter, M.J. (1987a), Provenance Selection in *Nothofagus procera* and *Nothofagus obliqua*'. FC Research Information Note 114/87/SILS.
—— (1987b), 'What Future for *Nothofagus* in British Forestry?' *Forestry and British Timber,* June.
—— (ed.) (1988), 'Broadleaves – Changing Horizons'. Proceedings of a Discussion Meeting, Institute of Chartered Foresters, Edinburgh.
Pryor, S.N. (1988), 'The Silviculture and Yield of Wild Cherry'. FC Bulletin 75.
Rackham, O. (1980), *Ancient Woodland*. Edward Arnold, London.
Steele, R.C. (1972), 'Wildlife Conservation in Woodlands'. FC Booklet 29.
Tabbush, P.M. (1988), 'Silvicultural Principles for Upland Restocking'. FC Bulletin 76.
Taylor, C.M.A. (1985), 'The Return of Nursing Mixtures'. *Forestry and British Timber, 14* (5), pp. 18–19.
Tuley, G. (1980), '*Nothofagus* in Britain'. FC Forest Record 122.
Watkins, C. (1990), *Woodland Management and Conservation*. David & Charles, Newton Abbot.

Chapter Five

ESTABLISHING PLANTATIONS AND SPECIAL TREE CROPS

1. Establishing plantations

The species choice having been made, it is necessary to plan and undertake the operations involved in establishing – one of the labour-intensive phases of silviculture. Much will depend on the type of site involved, invariably one of the following:

'New to woodland' (being planted for the first time, i.e. afforestation)
- (i) bare land in uplands
- (ii) lowland grassland or arable
- (iii) reclaimed industrial sites and mineral workings

'Recently felled' (requiring restocking)
- (i) in uplands, chiefly after conifers
- (ii) in lowlands, after either broadleaves or conifers

'Derelict scrub/woodland' (requiring special treatment, possibly enrichment)

Mainly in lowlands, but also on some upland sites; chiefly neglected broadleaves.

Methods and costs of establishment depend largely on the site conditions and other factors, not least on the vegetation and debris to be encountered. Consideration must be given to the age, type and size of the plants, their appropriate spacing, and the technique and pattern of planting. The number of plants required needs to be reserved with a nursery, and a date fixed by which they have to be available. Approval for grant-aid must be obtained before planting which means applying in good time and certainly before any operations are started.

Thought has to be given as to whether site preparation is necessary. In the uplands particularly there may be a need for ploughing or scarifying. Attention has to be given to drainage and appropriate protection, including vermin control. An initial application of fertilizer may be necessary. Relevant aesthetic and sporting interests should be borne in mind. Furthermore, there must be the scheduling of the approximate dates of planting, enabling a plan to be formulated for completing prior operations. Labour, tools and machinery have to be provided, and fencing materials assembled. The location and lay-out of roads, rides, turning places, gates and stiles must be settled. Provision for roads may have to be started. Due attention must be given to constraints such as water-pipes, gas-pipes, electricity or telephone lines, bridges, watercourses, marshes/bogs and conservation sites. Perhaps an adjacent stand will serve as a wind-break, or a stretch of old crop may have been retained to provide shelter for the new crop. Alternatively it may be decided to include in the species to be planted a potentially more stable species to create a wind-break on the windward side. Fire-breaks may be provided either as purposefully unplanted land or by planting of a less vulnerable species. An amenity belt, if thought desirable, may be planned.

Forest establishment has traditionally been accomplished by raising trees in nurseries and transferring the best quality plants to the forest site. The nursery techniques maximize seed germination, plant survival and healthy growth. By contrast, *direct sowing* is the artificial sowing of the seed on to the final site for the proposed crop. It is frequently preceded by intense ground preparation and followed by some years of aftercare; work in the establishment phase is extended beyond that normally expected. Results from it during this century have been extremely variable: seed and seedling losses are difficult and expensive to control. The unpredictable results obtained in Forestry Commission experiments lead to the conclusion that traditional establishment techniques remain preferable to direct seeding (Stevens *et al.* 1990). Few sites are now established by direct sowing – very occasionally Scots pine, less frequently Corsican pine, on heathland sites in East England. It can perhaps complement conventional planting and natural regeneration when the needs, sites and species are appropriate (Matthews 1989). Direct sowing of acorns has occasionally resulted in successful establishment. However, the technique of direct sowing has not generally proved reliable for reasons which include high losses from predation by mice, grey squirrels, insects and birds, the large quantities required to achieve reasonable stocking, and the patchy stocking comprised of bunches of plants and gaps.

The establishment phase – usually extending over three to five years – is the most critical time in the life of a plantation. (More than one third of the total cost of a rotation may be spent in the first year alone, and if the land purchase is included the proportion rises to half.) Equally important, many fundamental decisions are made in the first year which are irrevocable and the result of which, for good or ill, may not be apparent for a decade or more. As well as species choice, ground preparation and planting density are all vital decisions; but the most important factor is the site in terms of location, terrain and quality.

For each operation, methods need to be critically appraised in order to ensure that the right intensity and timing are chosen. Output Guides and Standard Time-tables for most of the operations are available from the Forestry Commission Work Study Branch (see chapter 16). Coniferous plantations are the least costly to establish. Broadleaved plantations, and this is mainly in the lowlands, are costly – often by a factor of at least two – and individual sites tend to be smaller and suffer from diseconomies of scale. Consequently where size and shape of a site would make conventional fencing prohibitively expensive it is becoming common to provide each plant with its own individual guard, either of plastic mesh or translucent plastic treeshelters which provide protection and give a reduction in weeding costs as well as providing benefit of a 'greenhouse' growth effect. Broadleaved plants are more costly than coniferous and usually grow more slowly. Whereas upland coniferous plantations are usually safe from weeds after the first year, lowland broadleaves have often to be weeded for several years, occasionally twice in a year.

Preparation of the site. The need for ground preparation generally depends on whether the land has recently carried a tree crop. The site may comprise:

(i) previously unplanted land (which probably needs either cultivation or chemical treatment to destroy vegetation); (ii) a recently felled area (which may or may not need treatment); or (iii) 'derelict scrub/woodland' (which requires special treatment). The aim of ground preparation is to make the land suitable for planting and to encourage rapid establishment and early growth of the crop to be planted. Where pre-clearing of vegetation or debris is necessary, managerial decisions are directed mainly at the selection of the cheapest effective method.

'Derelict scrub/woodland' is rather a broad term for woodland which has been clear-felled or very heavily depleted and subsequently neglected (a condition now much less common due to imposed conditions under felling licences). Dense re-growth is more common in the southern half of Britain particularly where run-down coppice and stools are more widespread.

On felled areas to be restocked, if the lop and top (in the case of broadleaves) or brashwood/slash (in the case of conifers) has been removed as part of harvesting, no pre-planting ground preparation is needed. If not removed, or the site has been neglected and become overgrown, the costs of preparation may be high, particularly where the debris is mixed with coppice shoots and other woody vegetation. Occasionally it is possible to plant through the debris; or consideration can be given to partial clearing, in narrow lanes or strips. In broadleaves, selected seedlings may be worth retaining to form a part of the next crop; alternatively, some or all may be retained temporarily as nurses to trees, chiefly shade bearers, to be planted beneath or amidst them. These methods, considered in chapter 7 under 'Enrichment', have the advantages of preventing the baring of soil, of nursing the planted trees from frost and wind and of retaining or enhancing amenity, habitats and sporting.

Manual or mechanical clearing may be replaced by chemical spraying. However, clearing is usually undertaken by machine cutting, swiping, crushing or chipping, depending in each case on the regeneration, the estimated cost and the effectiveness of the method. On old coppice sites no longer required as such, the cost of clearing is extremely heavy (the re-growth after planting may be dense and persistent). On clear-felled conifer sites, the pushing up of brashwood/slash into windrows by tractor, and then burning the heaps, can be costly. On clear-felled broadleaved sites the cost may be less, provided there is no heavy scrub, bramble, briar or rhododendron. In 'derelict scrub/woodland' the presence of a heavy stocking of hazel, sallow, elder and thorn, makes clearing extremely expensive, and may necessitate the use of large machines. If the site for restocking is covered by tree stumps and stools but is clean of debris, no ground preparation is necessary, but planting should be done before re-growth of woody and herbaceous weeds.

A particular problem in restocking is the re-wetting of ground (through the rise in soil moisture following felling), the effects of which are particularly serious on inhospitable clay soils, often prone to frost. Another problem is soil compaction by harvesting machines, or where planting is likely to be arduous in hard ground, e.g. shallow soil over chalk. Here cultivation, particularly ripping, may be necessary.

Ground preparation by cultivation. Cultivation of previously unplanted reasonably clean land is generally by ploughing. The purpose is to enable the tree root to be planted on cultivated mineral soil, also to reduce costs of planting and weeding. On lowland sites where no weed growth is present either following an agricultural crop or because the site has been chemically treated, there will be no need to plough; but on grassland and organic soils ploughing is usually essential to bury weeds and mix the organic material with the mineral soil.

Ploughing small areas is generally by a single furrow agricultural plough with furrows pulled at the intended plant spacing, usually 2 m. On larger areas, a double mouldboard plough can be used pulled by a 80 bhp tracked tractor, with the furrows spaced at 2 m intervals. Alternatively, ploughing can be done by a single mouldboard plough pulled by a tracked or 4xwd tractor again with the furrows spaced at 2 m intervals; however, the risk of windthrow of maturing trees may be greater. On heavy wet or organic soils, mounding may be required, using a machine to dig out dollops of soil or peat and deposit them in the intended planting positions. A rarely used technique on wet sites is to plant in upturned turves. Where a *pan* has to be broken up, ripping is necessary, e.g. by a mounted double ripper pulled by a 4xwd tractor, usually at 2 m spacing. Scarifying (noted *infra*) is advantageous on some sites. Nature conservation considerations when preparing ground are noted in chapter 13, section 7.

Ground preparation by chemical treatment. An alternative to burying weeds by cultivation is preparation of ground by chemical treatment. Methods and choice of chemical will depend on type and amount of vegetation. Grass/herbaceous weeds can be treated with glyphosate (Roundup) for overall weed kill either with a tractor-mounted boom sprayer or forestry spot gun. Treatment should be given far enough in advance of planting to allow time for the treated vegetation to die and the chemical to become broken down. Overall bracken control can be either by glyphosate or Asulox (asulam), the former being preferred where bracken is growing with other weeds, and the latter for pure bracken crops, usually with grass below. Application can be by spraying by boom or helicopter. Heather is best dealt with before planting by burning or swiping followed by ploughing, but where this is not possible chemical treatment may be used – 2, 4-D ester, or glyphosate applied by knapsack sprayer. Woody weeds can be killed by glyphosate as a foliar spray. Gorse, broom and rhododendron[1] are best controlled by triclopyr (Timbrel) as a foliar spray, the cost varying according to intensity of infestation. In all the foregoing chemical treatments prescription for the use of herbicides is complicated, and guidance should be obtained from Williamson and Lane (1989). The method and rates of application are important, as well as the timing.

Pine sites, before restocking, require the removal of brashwood/slash and any ground vegetation to help discourage the Large pine weevil. In pine stands of East Anglia destruction of lop and top by heavy-duty brush chopper has been carried out partly to facilitate use of a planting machine which makes a shallow furrow.

Ploughing is the most widespread method of cultivation in the uplands. It is used to make a cultivated, raised and weed-free planting position, particularly on wetter,

---

1 *Rhododendron ponticum,* an introduced species, is a formidable weed now commonplace on forested slopes in the west of Britain, difficult to control and its continued presence and rapid spread leads to an impoverishment of the local ecosystem. Tabbush and Williamson (1987) describe the biological characteristics and recommend both physical and chemical control measures. Their advice on control strategy concludes with a summary of recommended herbicides, with related warnings on their use. Clearance is generally accomplished through a combination of physical (bulldozing, winching, chopping or cutting) and chemical means (foliage sprays or stump treatments). Where there is a scattered growth beneath a mature tree crop, it is worth applying herbicide in advance of felling. Stables and Nelson (1990) give an update of control experiments.

heavier ground, and is generally in spaced furrows which form the planting line. On some soils the operation is to break through an impervious layer. It should be a means of reducing costs of drainage, planting and weeding. Ploughing can virtually eliminate serious weed competition; the vegetation is buried under the spaced inverted plough ridges, and by the time the weeds begin to recover the tree crop is well on the way to occupying the site and has a lead which may make further weed control unnecessary.

A primary objective of ploughing is to improve rooting conditions for the young trees. Increased soil aeration, drainage, soil temperature (reducing risk of frost damage), and nutrition should encourage root growth. Improved root development should lead to increases in shoot growth, yield and resistance to wind loosening or windthrow of young trees. Three kinds of physical change can occur in ploughing: aeration by drainage, aeration by fragmentation, and soil mixing (Thompson 1984). The furrow, whether empty of soil or filled with loose soil, acts as an outlet for excess soil water (or as a sump creating a hydraulic gradient down which soil water tends to move): this is aeration by drainage. Ploughing causes aggregates to lift, shear and fragment so that their average size is reduced. This effect is very limited on peats and in wet gleys and most marked in clayey soils, but provided waterlogged conditions do not prevail the volume of pore space is increased. Pores produced in this way are mainly large and will drain quickly under gravity: this is aeration by fragmentation. Forest ploughs are designed to move soil material both vertically and laterally – movements which cause soil layers to become realigned relative to each other. This mixing of soil horizons by ploughing alters the natural distribution of bacteria and fungi in the soil, leading to changes in nutrient cycling. It promotes the breakdown of accumulated surface organic matter and stimulates a rapid release of nutrients which can be taken up by plant roots. All three physical changes to the soil occur to some degree with ploughing but it is possible by using different types of plough to emphasize any one aspect.

The effectiveness of ploughing depends largely on the soil conditions of the site. On freely-drained soils (brown earths and podzols) without induration, ploughing improves tree establishment primarily by reducing competition from ground vegetation but is unlikely to result in any long-term site improvement. On ironpan soils, there is a close correlation between growth rate and the volume per unit area of soil disturbed by ploughing and particularly with the degree to which any pan or induration has been broken. In the long run the most effective treatment may be deep complete ploughing (instead of spaced furrow ploughing). On wet sites (deep peats, peaty gleys and surface-water gleys) early growth of the crop is positively correlated with the size of the plough ridges, and depth and interval of ploughing interact in providing a drainage effect. Maintenance of satisfactory growth on such wet sites in the long term depends on the provision of a satisfactory system of adequately maintained deep cross drains.

In ploughing, it is important to select the most cost-effective combinations of equipment to meet the silvicultural objectives for the site (Thompson 1984). Ploughing costs are high, and are incurred at the beginning of a lengthy rotation, and hence must be justified by the firm expectation of considerable improvement in establishment costs and tree growth rates. The anticipated influence of ploughing on tree stability in the future must also be considered. Before ploughing, consideration should be given to the provisional road alignments and natural main water channels. The ideal time for ploughing when planting is to be done in the spring, is in late August to October of the preceding year. It is usually done up and down hill – the chosen spacing between plants dictating the distance between furrows. It is essential to provide cross drains at appropriate places, as noted under 'Drainage' *infra*.

The resultant ribbon patterns of spaced furrow planting can restrict the lateral spread of roots, increasing vulnerability to premature windthrow. Alternative site preparation methods are directed at promoting effective establishment and improved rooting patterns. One method is *subsoiling* (Miller and Coutts 1986), which may give better root development, and may be preferable on uniform surface-water gley soils, particularly on sites with a high yield potential. There appear to be some circumstances where the adoption of subsoiling will give a reasonable prospect for improvement over conventional ploughing, but these cases depend very much on the degree of soil structure modification achieved and the extent to which increased stability is assumed to result. It is important to appreciate that subsoiling carries with it a greater degree of uncertainty in controlling waterlogging on most upland soils, and will also incur greater establishment costs and some growth penalties compared with ploughing. Afforestation sites may prove more responsive to subsoiling than restocking sites, but care must be taken in selection of suitable soil types for treatment, and in the timing of subsoiling operations. On peaty gleys and other soil types in the uplands, standard double mouldboard with tine ploughing is the preferred treatment. Forestry Commission research is continuing to define further the development of roots on sites prepared by subsoiling and other methods with potential for improving anchorage and long-term stand stability.

Scarifying, an alternative to ploughing, is being increasingly adopted for mineral soil sites in the lowlands where full ploughing is unnecessary (or impractical) and usually there is no drainage requirement. Scarifiers and mounders offer opportunities to sacrifice some site amelioration for reduced risk of sediment laden water reaching streams. Some machines may be capable of giving brash-free planting sites by rake-and-release action on restocking areas. *Ripping* – using a ripper or a ripper scarifier – instead of ploughing of heather-covered sites is claimed to have advantages which include enhanced crop stability, less restriction on the movement of vehicles, ease of walking for shooting and a much reduced impact on local visual amenity. However, usually there is a slower early rate of growth and increased costs of planting and weeding. A range of scarifying equipment is available (Chadwick 1989; Nelson and Quine 1990).

There are economic interactions between operations in establishing, e.g. cultivation may reduce costs of drainage, plants, planting, weeding and beating up. Furthermore, the site improvements may be undertaken not just for the current crop but for its successors, so benefits in perpetuity must be included in economic calculations (Price 1989). The usual way of evaluating investments in operations is to calculate whether the expected growth response, along with other factors, justifies the expenditure. Ploughing before planting should only be done if the costs of establishing are less than the costs of establishing on unploughed land; or if it is expected that growth rate (yield class) will be significantly better on cultivated ground.

Drainage. Much of the foregoing information relating to cultivation has a direct correlation with drainage. Trees require a considerable amount of soil moisture but they cannot thrive where the ground is waterlogged, particularly if the roots are submerged for more than a short period. They are unable to achieve adequate anchorage in soft wet ground, and in consequence windthrow may occur. Some species can tolerate waterlogging better than others (see 'Species choice' in chapter 4).

In the lowlands it is not possible always to determine the wetness of the soil and to know adequately what moisture conditions are best for specific tree species on a particular site. Poor drainage is easily recognized, but the appropriate degree of artificial drainage required will depend on soil texture, slope and rainfall. Any sudden alteration in the soil conditions may damage standing crops. When in doubt, it may be wise to under-drain as a start, but experience will usually indicate when and what drainage is required; attention to a spring may sometimes suffice. After the clear-felling of a substantial crop the soil moisture usually rises (re-wetting). In appraising the need for drainage it may be necessary to observe the degree of moisture in the site throughout a full year; inspecting a site at a particularly wet time or during a dry period may not provide a reliable indication.

Drainage of sites usually comprises the cutting of open drains (i.e. ditches) to remove excess soil water. Possible reasons for drainage are: to improve tree growth, particularly in the first ten years; and to increase the depth of rooting and so reduce the risk of windthrow. Drainage may improve crop stability on wet sites but it is unlikely consistently to improve growth, and minimizing soil erosion and sediment transport to streams requires careful planning of the drainage system (Pyatt and Low 1986; Pyatt 1990).

In the uplands, one of the main problems on almost all soils being afforested is excessive soil water. The main soil types which inherently require drainage are the clayey gleys, the sandy or loamy gleys, peaty gleys, the deep peats and the wet coarser types. In addition, ploughed areas of other soils, particularly ironpan soils, will usually require drains for the control of erosion. Heavy clay gley soils are very difficult to drain because of limited lateral water movement, and the maintenance of a drainage system with drain intensity greater than 100 m/ha is recommended.

In the uplands, drains should be aligned towards the head of a valley. Gradients should not exceed 2°; gentler gradients are advised for the lower stretch of the drain and for particularly sensitive catchments. The effective depth of drains should be 60 cm; this may necessitate an initial depth of 90 cm to allow for accumulation of sediment. Cut-off drains should be planned first, followed by collecting drains to give an average spacing of 40 m, except for loamy gleys and deep peats on slopes of less than 3° where the spacing should be reduced to 20 m average. Drains should end 15–30 m from main watercourses or 5–10 m from lesser streams.

In the lowlands, drains should normally be no deeper than is necessary to intercept the water from the highest seepage layer, identified by digging holes in late winter. Normal drains are 40–60 cm deep, with a width of 60–70 cm at the top and 25–30 cm at the bottom. The depth will sometimes be that at which the mineral soil is reached. The drain will be in mineral soil on most sites, the only peat being in the fens or Somerset levels. The direction and pattern will normally conform to the contours of the site, and the amount of drainage required will dictate length and distance apart. Drainage of lowland sites for broadleaved species is rare, though wet pockets may need to be planted with alder or willow. Poplar is a possibility, but although it requires plenty of soil moisture it dislikes being waterlogged and may still need drainage.

Initial drainage will normally be carried out as an adjunct to ploughing on heavy wet or organic soils. Some manual work will be required to tidy up drain ends and where furrows cross. Up to 100 m/ha may be required for well-drained loamy soils, and 250 m/ha for heavy wet soils. Drainage equipment information and recommendations have been provided by Thompson (1979), Pyatt and Low (1986) and Pyatt (1990). Ploughs are much cheaper to operate than excavators for the larger new planting sites and can operate well in most soils but are generally unsuitable for small areas, restocking sites and maintenance operations. Excavators can be used for a wide range of site conditions but are slow and expensive and therefore most suitable for small areas of afforestation or restocking. Rotary machines give much greater outputs than excavators in peaty sites, and more robust versions are being developed for peaty gleys.

Once established, trees' roots effectively withdraw some moisture from peats and clays, and this can result in a drying and cracking process which is not fully reversed in winter. Drain maintenance should be limited to the clearance of gross blockages and the stoppage of overflows. Particular attention is required following wet periods, the fall of needles and leaves in autumn/winter and after each thinning. Remedial drainage should only be considered for plantations less than 5 m tall and then only where windthrow risk is likely to be exceptionally high, e.g. where the plough furrows contain standing water. In such delayed drainage, the costs increase with the age of the crop. Continuing erosion in plough furrows or drains, especially if persisting after canopy closure, will justify modification to the drainage system.

When restocking, particularly in upland wet areas on soft soils, the original drainage system inherited from the previous rotation is likely to have been seriously disrupted by the passage of harvesting machinery, and by brashwood/slash accumulations. The provision of temporary culverts, using lightweight pipes, would have reduced damage to main drains. Site damage due to rutting and soil compaction would have been minimized if harvesting operations on wet sites had been carried out during the drier seasons. However, it is inevitable that on wet sites substantial work will be necessary to re-establish a suitable drainage system. Especially on impeded soils, the old root channels will provide a certain amount of drainage, notably where the roots have penetrated the ironpan or induration. Under these circumstances new drains should only be put in to tap wet hollows and spring lines. On impervious soils the site will also have been modified by the previous rotation, and water movement may be much enhanced, particularly on peat soils. Nature conservation considerations in drainage are noted in chapter 13, section 7. (Drainage concerns the tree-grower in other ways – for instance, where poplar roots may block underground drains of buildings and where erosion of road surfaces needs to be prevented.)

Fencing. Young plantations can be seriously damaged by horses, cattle, sheep and goats, as well as by wild animals, particularly rabbits, hares and deer. Fencing is an efficient method of protection, but it can be extremely costly. The increasing demand to minimize costs of establishment and protection makes imperative the need to investigate ways of economizing. Improvements to fence design and construction are constantly under review (Pepper and Tee 1986). There have been new materials, and costs have been reduced by introducing improved labour-saving tools and methods of working. Any savings obtained can be wasted if the initial planning has not been thorough. Many forest fences are a compromise between expense and effectiveness.

Protection of the young plantations may not be the only consideration. Owners of farm animals are required by law to keep them under control and to fence against their own stock, but many conveyances include clauses which require a woodland owner to maintain a stock fence on the boundary. There may be other legal constraints such as rights of way, both public and private, though usually these may be mitigated by gates, stiles or cattle grids at appropriate places.

There will be sites where fencing is unnecessary because no damaging animals are present or because their numbers can be reduced to levels causing little or tolerable damage; rabbits and deer are cases in point. There will also be instances when the cost of fencing would be so high that it is cheaper to provide individual plastic mesh tree guards or treeshelters for each tree than to fence the whole site. The choice between fencing and individual protection is governed by the area to be protected and the number of trees per hectare planted. Individual tree protection is normally cheaper than fencing on areas less than 2–5 ha. Each case will vary depending on the area, the length and type of fence involved, and the number and cost of individual tree protection. But guards may have many attributes, discussed in a later section, relating to growth rate and protection. There are few planting sites where some fencing is not required, and it will be obvious what is to be fenced against, what the type of fencing, guard or treeshelter shall be and the minimum life looked for in it.

How to minimize the length of fencing will usually depend on the area and shape of the site. The cost per hectare of fencing decreases as the area enclosed increases. A square hectare needs 400 m, a site of four hectares square only 200 m to each hectare, and one of sixteen hectares square only 100 m to each hectare. For square sites the length of fencing per hectare is reduced by half each time the area is quadrupled. The nearer the site is to being a circle, or at next best a square, the less is the fencing needed per hectare; but a circular fence is costly because of the extra number of stakes, posts and struts required. Conversely, the longer and narrower, or more irregular, a site, the more will be the fencing per hectare. Thought should therefore be given to how irregular boundaries may be evened out and to whether a site can have other ground enclosed with it in order that the maximum area can be fenced at one time; or conversely an area can be excluded from the fencing and trees protected individually.

Fence specifications. Spring steel wire is recommended for all forest fences. Its advantage is that once tensioned in a fence it has the ability to accept further accidental tensioning from animals or humans or fallen trees without deforming; consequently it returns to its original position and tension once the accidental load is removed. Multiple wires are not used in spring steel fences but instead netting of various sizes, depending on the animals constrained. The spacing of the supporting fence posts can be increased substantially, thus reducing both material and erection costs. Pepper and Tee (1986) provide details of fence components and associated tools, fence construction and maintenance, as well as fencing principles and specifications. The specification of the fence will largely depend upon its purpose and desired duration; it may include gates and stiles, and possibly badger gates. Figure 9 indicates the main fencing terms.

The British Standard for strained wire spring steel fences comes into BS 1722: Part 2: specification for rectangular wire mesh and horizontal wire netting fences, currently being revised. Wire netting remains the most durable and economic type of netting available despite the introduction of various types of synthetic netting. Polythene and nylon fibres have been used to make traditional knotted netting for fencing purposes. Both materials are degraded by ultra-violet radiation. The advantage of polythene netting is that it is light and easy to handle. However, as

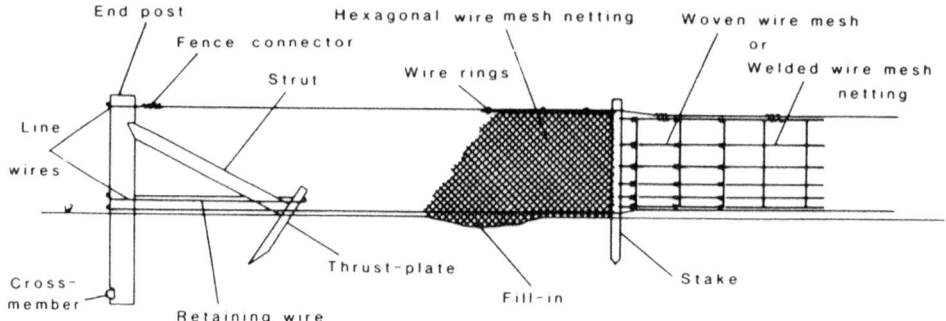

Figure 8. Fencing terms
Source: Pepper and Tee (1986). FC Leaflet 87.

animals can easily become entangled in it, such netting should not be used in any fence, in any circumstances. Plastics technology may lead to the development of more acceptable types of netting.

The strength and durability of all the components in a fence should match the strength of fence required and the period of time for which the fence is needed. However strong and durable the metal components, the fence will cease to be effective if the wood components fail. Conversely, it is equally wasteful to erect wooden posts and stakes that will outlive their metal counterparts by many years. The classification of the natural durability of the heartwood of most timbers is noted in chapter 10. Fencing wood containing a proportion of sapwood can be classed as perishable and will only have an average life of five years. If the fence is required for a period longer than the normal service life of the untreated timber, then it is necessary to treat it with preservative after debarking and seasoning.

Treated round fencing material lasts longer than sawn material of the same species. The absorbent sapwood provides a protective barrier of treated timber. In general, preservative treated hardwoods do not last as long as similarly treated softwoods. However, the remarkable durability of sweet chestnut and heartwood of oak, whether round, cleft or sawn, is noteworthy. It is impossible to erect a fence without cutting into some of the treated timber and when doing this it is important to renew the protective layer of preservative; but any retreatment is inferior to the original treatment. The length of posts, struts and stakes required vary; first, according to the height of the fence to be erected, which is dependent on the species of animal to be excluded, and, second, according to the depth the post is to be in the ground – this is dependent on the soil texture. The main sizes are:

	End posts		Struts		Stakes	
	Length (m)	Top Diam. (cm)	Length (m)	Top Diam. (cm)	Length (m)	Top Diam (cm)
Rabbit	2 or 2.3	10–13	2	8–10	1.7	5–8
Rabbit and stock	2.3	10–13	2	8–10	1.7	8–10
Sheep	2.3	10–13	2	8–10	1.7	8–10
Cattle	2.3	10–13	2	8–10	1.8	8–10
Roe deer	2.8	10–13	2.5	8–10	2.5	5–8
Fallow, red, sika deer	2.8	12–18	2.5	10–13	2.6	8–10

The line a fence may be required to take is often rigidly defined by law, or by the geography of the area, allowing for little subsequent variation. Where this is not so, it may be possible to make worthwhile savings by straightening out the line to eliminate one or more corner posts even at the expense of excluding some land. A fence should not be used simply to mark an irregular boundary where there is an option to straighten the line. There may be instances where, to economize in fencing, a combination of a straight fence line coupled with a number of tree guards or treeshelters outside may be suitable and cheapest. Provision must be made for any public ways, and requirements for shooting and hunting must be heeded.

The ease of digging-in and firming straining posts should be considered, and waterlogged areas and shallow soils over rock avoided. Where possible, avoid fencing over excessively undulating ground where it may be difficult to prevent the fence lifting off the ground, or through hollows which may become snow-filled. By choosing a relatively level line fewer stakes are required and 'filling-in' can be avoided. Care must be taken either to remove banks and tree stumps outside the fence which reduce the effective height of it, or to raise the height of the fence by adding an 'apron' of netting. Straining posts are needed particularly at every point of definite change in line direction and sudden change in gradient.

Methods of erection vary, but foresters will bear in mind the use of 'Drivalls', 'post-hole diggers' and strainers. A two man gang is generally the most efficient size of a fencing team. When it is not possible to distribute the material along the fence line with a vehicle, carefully sited dumps of materials should be placed within easy reach of the fence line to reduce the distance the material has to be carried. If possible, a tractor-mounted post driver should be used for intermediate stakes and a tractor-mounted post-hole borer for posts.

The fencing specification must match in height and strength the animals it is wished to constrain. Four examples follow.

Rabbit fences. The basic specification for rabbit fencing is shown in Figure 10. The height should be 0.75 or 0.9 m, using 31 mm hexagonal mesh netting. Additional modifications are required when it is necessary to exclude domestic stock in addition to rabbits: these involve slightly larger and more robust posts and an extra line wire. Generally the netting is 1,050 mm wide, lapped out horizontally on the surface 150 mm towards the rabbits, and held down with pegs and/or turfs. Fencing over tree stumps and rock outcrops should be avoided as these can be difficult to make rabbit-proof. The number of gates in a fence should be kept to a minimum because it is difficult to make and maintain rabbit-proof gates. Badger gates must be provided wherever a fence crosses an established badger run. Culverts must be netted where rabbits are present. Suggestions for improved specifications for rabbit fencing for tree protection are made by McKillop *et al.* (1988). Costs per hectare vary according to specification, size and shape of area, and terrain.

Stock fences. The basic specification for excluding sheep and cattle is detailed in Figure 11. When fencing against cattle, many farmers require a barbed wire on the top of the fence in preference to the plain line wire shown. However, where deer, especially fallow deer, are present the barbed wire should not be used. Costs per hectare vary as stated for rabbit fences *supra*.

Deer fences. The light specification is only suitable for excluding roe deer or a combination of roe with rabbits and sheep (Figure 12). The heavy specification

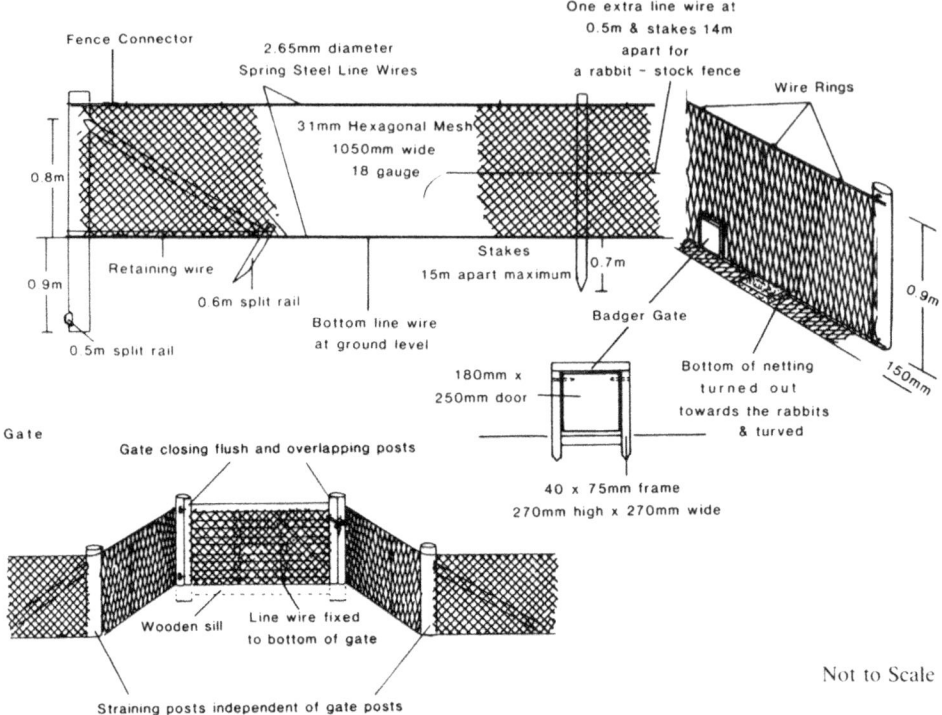

Figure 9. Rabbit fence specification
Source: Pepper and Tee (1986). FC Leaflet 87.

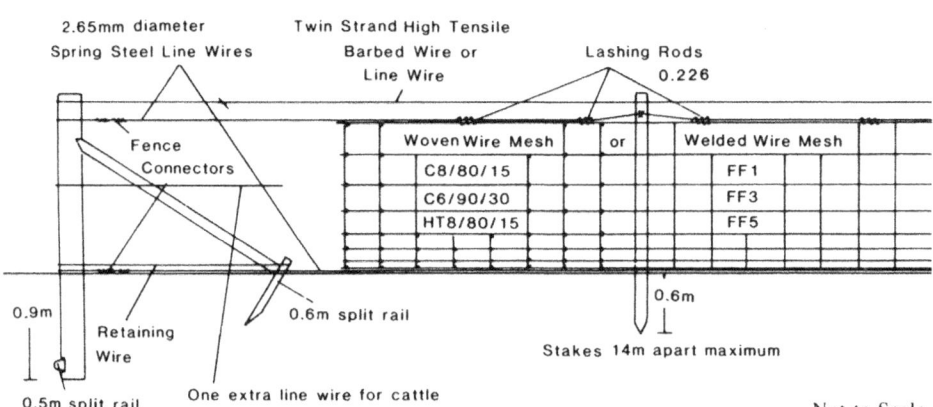

Figure 10. Stock fence specification
Source: Pepper and Tee (1986). FC Leaflet 87.

Establishing Plantations and Special Tree Crops

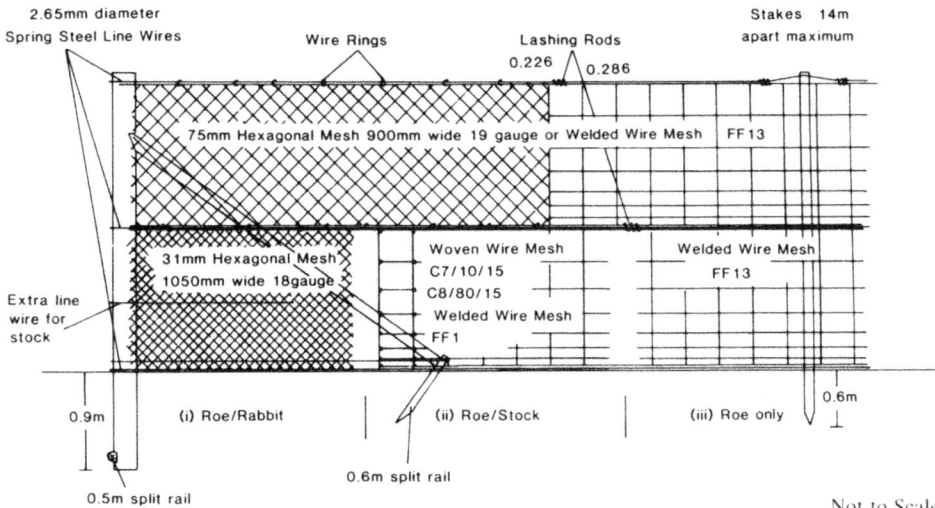

Figure 11. Deer fence – light specification for roe
Source: Pepper and Tee (1986). FC Leaflet 87.

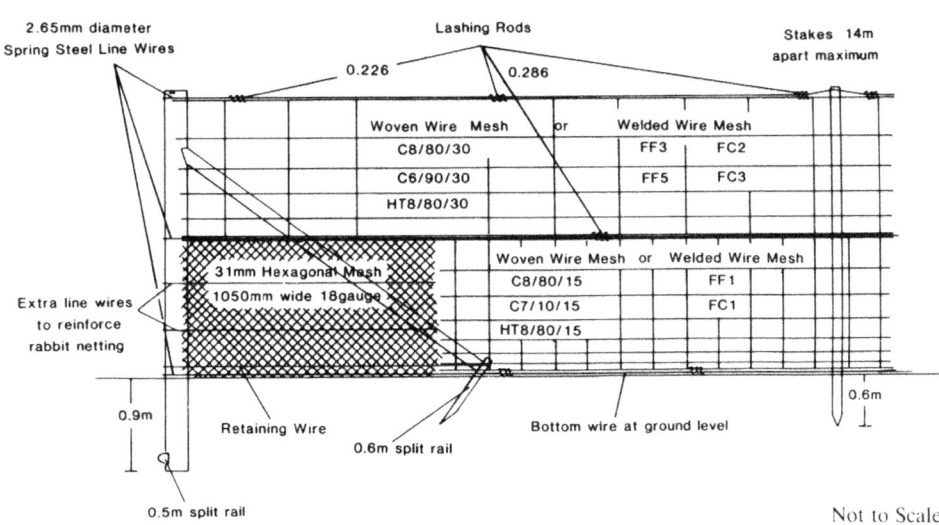

Figure 12. Deer fence – heavy specification for fallow, red and sika
Source: Pepper and Tee (1986). FC Leaflet 87.

should be adopted when fencing out fallow, red or sika deer or any combination of these with roe deer, domestic stock or rabbits (Figure 13). The fencing should be 1.8 or 2 m high with spring steel wires and a combination of two nets will provide a cost-effective barrier. The choice of netting to use will depend upon the deer species to be excluded. In uplands the minimum specification as shown must be increased to compensate for the combined effect of severe weather, difficult soil conditions and heavy deer pressure. Deer jumps should be incorporated into the fence to provide an exit for animals that have broken into the plantation. Costs per hectare vary according to specification, size and shape of area, and terrain.

The foregoing comparatively brief introduction to fencing is not an adequate substitute for the comprehensive information supplied by Pepper and Tee (1986) which will repay study.

Spacing: narrow (close) versus wide planting. The more space a tree has the more it will grow in volume, although very similarly in height. Deciding on the spacing of planting stock at establishment is important (Hamilton and Christie 1974). Spacing affects the growth of individual trees and the yield of products from the crop. It ultimately influences timber properties through its effect on wood density, number and size of knots, taper and the size of juvenile core, as well as affecting establishment and harvesting costs and marketing factors. However, spacing is irrelevant to cost of fencing, and of aerial application of fertilizers or herbicides. Foresters must appreciate the *pros* and *cons* of narrow and wide planting. In general, advantages of narrow (close) planting include:

- Beating up may be less essential.
- More rapid closure of canopy. This is important where weed growth, e.g. on heathland and grassland, is capable of serious competition with planted trees. Costs of weeding and cleaning might thereby be reduced.
- Encouragement of fine branching which may lead to trees of better form. A relatively large number of stems are available for removal at the first thinning.
- Prevention of large knots by the early suppression and thereby natural pruning of side branches. This should increase the value of the timber.
- Many stems of good form should be available from which to choose the final crop.

Disadvantages of narrow (close) planting include:

- Initial costs of establishing and tending are higher.
- Individual size of maturing trees is smaller and may be less in value. This is important for the effect on the profitability of first thinning at times when the timber market is depressed. If not thinned, the final crop at felling or windthrow will have a smaller average tree size than from wide spaced stands.
- Each tree has less space for its roots even though they interlace, and thus the risk of windthrow may be greater.
- Diseases may spread.
- Ground flora of conservation interest may be suppressed earlier.

Assessing the factors in another way, wide planting is generally advocated when:

- The species is a fast grower.
- The site is good in both quality and location.
- Reducing the costs of initial establishing and tending is both desirable and prudent.

- The aim is for increases in individual size of trees, and thereby increased prices.
- Ground flora is of conservation interest – but ultimately it will be suppressed.

Initial plant spacing affects the rate of development of mutual shelter within a plantation, but the importance of this will vary considerably with site and location. On sheltered sites, e.g. in valleys, on lee slopes or in clear-felled small sites surrounded by standing crops, the development of mutual shelter has relatively little influence on tree growth. By contrast, on severe climatic and exposed upland sites (usually relating to Sitka spruce and lodgepole pine) the development of mutual shelter generally improves tree growth by providing a more favourable microclimate and by reducing the incidence of leader breakage and damage to foliage. While some broadleaves rarely show quick early upward growth unless planted close, height growth in conifers is not affected by spacing, except on much exposed or weedy sites where close planting has been found to be beneficial.

Widely spaced trees grow faster and hence produce merchantable volume more quickly than closely spaced trees, but this is offset by losses in total volume production as the available growing space is not fully utilized. Spacing affects not only the total volume production and the distribution of individual tree sizes within a crop, but also tree quality (Brazier 1977). In areas of high wind risk, wide spacings may have advantages. The mean tree will be greater so it is likely that more sawlog-sized material will be produced before the trees are windthrown. The trees will also have larger root plates which may increase stability (conversely, the original exposed structure may reduce stability). On the other hand, early canopy closure and weed suppression is important in the establishment of crops in the uplands. (In these cases manipulation by respacing – see chapter 8 – may be a suitable means of altering spacing without increasing total establishment costs.) The effects of different treatment in thinning and pruning superimposed on, or in conjunction with, spacing are noted in chapter 8.

Planting distances. When considering the most commonly planted species – Sitka spruce – it is notable that the yield of timber suitable for load-bearing structural use declines as spacing increases between 0.9 x 0.9 m and 2.4 x 2.4 m spacing, in such a way that spacing beyond 2 x 2 m cannot be recommended given current prospects for marketing. The decline is thought to be more important for the inherently low density spruce than for the higher density pines, larches and Douglas fir. Recommended *established* stocking densities are therefore 2.1 x 2.1 m spacing (2,267/ha) for most conifers; and 2 x 2 m spacing (2,500/ha) for spruces.

For conifers, the present practice is to plant the slower-growing, such as Norway spruce and the pines, at 1.8 x 1.8 m (say 3,000/ha) and the faster-growing at 2 x 2m (2,5000/ha). Broadleaves need to be planted relatively close, 1.5 x 1.5 m to 1.8 x 1.8 m being preferred. Currently (1991), for grant-aid the maximum spacings are 2.1 x 2.1 m (2,267/ha) for conifers and 3 x 3m (1,111/ha) for broadleaves. There are conventional planting distances for the different species (generally relatively close for most broadleaves and wider for conifers) – see chapter 4 – but these spacings may vary according to quality of site, exposure and objective. The number of plants required per hectare, at various spacings when planted 'on the square' is given below (allowance should be made for land to be left unplanted e.g. along boundaries and rides):

Spacing/m	plants/ha	
1.0 x 1.0	10,000	Norway spruce for Christmas trees (wider for *Abies* spp. and Scots pine)
1.5 x 1.5	4,000	Slow-growing conifers and broadleaves
1.8 x 1.8	3,086	⎫
2.0 x 2.0	2,500	⎬ Fast-growing conifers
2.1 x 2.1	2,267	⎭
3.0 x 3.0	1,111	Broadleaves in treeshelters
8.0 x 8.0	156	Poplar. See range of spacings *infra*.
12 x12	70	Cricket-bat willow

The number of plants required per hectare for standard metric square spacings is given below:

spacing/m	plants/ha	spacing/m	plants/ha	spacing/m	plants/ha
1.0	10,000	3.1	1,041	5.4	343
1.1	8,264	3.2	977	5.5	331
1.2	6,944	3.3	918	5.6	319
1.3	5,917	3.4	865	5.8	297
1.4	5,102	3.5	816	6.0	278
1.5	4,444	3.6	772	6.2	260
1.6	3,906	3.7	730	6.4	244
1.7	3,460	3.8	693	6.5	237
1.8	3,086	3.9	657	6.6	230
1.9	2,770	4.0	625	6.8	216
2.0	2,500	4.1	595	7.0	204
2.1	2,268	4.2	567	7.5	178
2.2	2,066	4.3	541	8.0	156
2.3	1,890	4.4	517	8.5	138
2.4	1,736	4.5	494	9.0	123
2.5	1,600	4.6	473	9.5	111
2.6	1,479	4.7	453	10.0	100
2.7	1,373	4.8	434	10.5	91
2.8	1,276	4.9	416	11.0	83
2.9	1,189	5.0	400	11.5	76
3.0	1,111	5.2	370	12.0	69

If planting is 'on the triangle', 15.5% more plants are required, i.e. the number 'on the square' is multiplied by 1.155. Such a pattern may be used for poplar or cricket-bat willow. Spacing for trees in treeshelters is noted *infra*. Spacings for poplar, according to species, cultivars and hybrids range from 5–9 m (Jobling 1990). The spacing adopted must be a matter of judgement on the part of foresters after taking into account the species, the factors of the site and the intended crop.

The Forestry Commission continues to experiment on various spacings and how trees are affected by thinning and pruning. Judgements have been formed of the most suitable and profitable spacings up to and including the early thinnings, and also how closer spacing *within* rows, accompanied by wider spacing *between* rows, might facilitate mechanical weeding where this is a problem, and, later, aid both thinning and extraction.

Plants. The species and desired origin/provenance having been selected, foresters must consider the age, type and size of plants best suited for the microsite; and decide whether to use bare-rooted seedlings, transplants, 'undercuts', or 'whips'; or, instead, container grown seedling planting stock, i.e. quickly-raised seedlings with roots undisturbed in a plug of peat. In the case of willow and poplar either rooted plants or (occasionally) unrooted sets are used.

All plants must be in accord with the soil, aspect, exposure and conditions of the site. Origin/provenance have been noted in chapter 4, section 2. All should conform to BS 3936: Specification for Nursery Stock (Part 4: Forest trees; Part 5: Poplars and Willows). Overgrown, leftover plants should not be accepted. The plants should desirably be inspected in the nursery where they are growing, and a sample retained for comparison with the bulk supply when it is collected or delivered; this will prevent or decide a possible dispute. Ages and types of suitable plants have been suggested in chapter 4, but in practice the size, root system and leader are better criteria than age and type on their own (White 1990). The roots of oak, ash and sweet chestnut, and more rarely those of Corsican pine and Douglas fir, if they have not been undercut in the nursery, may need to be shortened before planting. Following Forestry Commission extensive research and practice, invaluable information has been provided by Low (ed.) (1985) and Tabbush (1988) on plant quality, handling, storage and timing of planting. Their publications are again referred to under 'Restocking conifers in uplands' *infra*. For broadleaved species much credit is also due to the Department of the Environment, contracting with the Commission.

Morphological factors are readily observable and are in general use for culling, grading and marketing plants (Tabbush 1988). They include shoot factors – length, sturdiness, foliage colour (related to nutrition), degree of damage (insects, fungal infection, wilting), visual indications of 'hardness' (lignification), and condition and size of the terminal bud. In addition are root factors – total root length, 'branchiness' of the root system, visual assessment of root:shoot ratio, and root collar diameter (Mason 1991).

A number of physiological factors give an indication of forest performance – root:shoot ratio (on a dry weight or volume basis), root moisture content and root growth potential (RGP).[2] The latter is the most useful index of physiological status or vitality given favourable site conditions. Initial survival of planted stock depends in large measure on their readiness to produce new roots and thereby establish intimate contact with the soil, and RGP is a key factor in determining early survival and growth. Soil contact is broken on lifting and only poorly re-established even with careful planting; new stock growth is required to prevent the development of severe water stress. Site conditions have an effect on the expression of RGP; in particular, dry soil, low soil temperature and soil compaction reduce early root growth. RGP does not by itself give an accurate prediction of forest performances; rather plants of a high RGP are able to succeed under harsher conditions than plants with a low RGP.

2 RGP is generally measured by growing sample plants in moist peat or aerated water for 7–28 days under long days' temperatures of 20–25°C, and fairly high light intensity, i.e. nearly ideal conditions for root growth. Root development is then measured or scored. Relativities in RGP are maintained within a wide range of test temperatures. RGP is usually assessed on a sampling basis.

Bare-rooted plants. The techniques of forest tree nursery work were discussed in chapter 2. In forest planting, the tradition is to use bare-rooted plants, either transplants which typically spend one to two years in seedbeds and are then transplanted (lined-out) to produce 1 + 1, 1+2, or 2+1, or 'undercut' seedlings one to two years old (undercut to produce 1u0 or 1u1). The effect of either transplanting or undercutting is to stimulate the production of a compact fibrous root. Typical nursery grown plants range from about 20–90 cm in height (there are wide variations depending on species) with a root system in balance with the crown and a sturdy root collar diameter. A height of 20–40 cm is generally ideal for planting on exposed sites (or for broadleaves within treeshelters), while larger stock – 30–45 cm or 45–60 cm – are preferred in more fertile sites where exposure is less and weed growth is greater. In the case of alders, wild cherry and southern beech, 45–90 cm plants are likely to be appropriate because of their rapid first-year seedbed growth. On the market are precision-sown undercut 1u1 seedling conifers, a uniform stock of a specified standard planned to be accompanied with improved site preparation, and defined standards for plant handling and planting (Mason 1986). Also available are increasing numbers of Sitka spruce raised from cuttings of genetically-improved (C½ +1½) stock, as noted in chapter 2. Trees grown from these selected sources are expected to provide an increase in timber yield of some 10% above that of unimproved Sitka spruce (Mason *et al.* 1989). The price of the plants is about double that of unimproved plants.

Container grown plants. By contrast to bare-rooted plants raised in open ground, there are various types of container grown seedlings (noted in chapter 2), particularly 'Japanese Paper Pots' and 'Rootrainers'. The price of the plants is up to double that of bare-rooted seedlings. Their use makes possible extension of the planting season well beyond the normal limits for bare-rooted stock – and hence is of value to foresters especially when labour is short. Main advantages claimed for container or cell-grown seedlings is reduced transplanting shock because roots are not damaged and safer handling. Claims about superior survival of container plants compared with bare-rooted plants are often based on anecdotal evidence (Mason and Jinks 1990). Forestry Commission experiments have provided little evidence to show that container grown plants are superior to good quality bare-rooted stock; reported superiority often reflects special situations such as late season planting (Mason and Hollingsworth 1989; Hollingsworth and Mason 1989). Container grown seedlings seem unlikely to replace bare-rooted plants for standard forest use in the UK. They will find wider use in the uplands if nurseries consistently produce seedlings which are both biologically robust and cost-competitive with bare-rooted plants. The two systems are more likely to supplement each other with the flexibility of container growing being employed to grow seedlings of 'sensitive' species or from seedlots of highly genetic quality, and the bare-rooted techniques used to mass-produce the standard species and genotypes (Mason and Biggin 1988).

Plant handling. Careful handling of bare-rooted plants is of particular importance. From lifting in the nursery to planting in the forest, plants undergo a series of handling, storage and transport shocks, almost all of which result in some loss of vitality. Damage may occur through desiccation, mechanical shock, tearing or bruising, extremes of temperature (heating or drying) or chemical toxicity, the latter usually the result of insecticide treatment. The utmost care is needed in lifting, grading, culling, packing, storing, transporting and unpacking. Fine roots are fragile and are easily broken or damaged by desiccation. A high level of supervision is required to prevent the loss of plant potential from reaching critical levels.

Plant storage. It is often necessary to cold store bare-rooted plants for short periods between handling and treatment operations (e.g. culling, grading or treatment with insecticide), or for longer periods between lifting and planting. Container grown seedlings, provided they are properly hardened, can be cold stored like bare-rooted stock. The flow of plants through the system must be organized so as to reduce the risk of the types of handling damage noted above.

Cold stores (see chapter 2) offer nurseries and foresters greater flexibility. In the past, 'heeling in' ('sheughing') or stacking in 'beehives' have been commonplace for bare-rooted stock, but are best avoided since this involves a good deal of plant disturbance. Plants can be stored short term in coextruded black/white polythene bags, in a cool, shaded environment; they eliminate the overheating likely in clear polythene bags. Roots which retain a covering of nursery soil dry out more slowly, but excess nursery soil in long storage can lead to problems including fungal infection.

Dipping roots in water-retentive substances such as sodium alginate protects against drying, provided it is applied at lifting, but can be messy, and inevitably leads to a hiatus in the plant handling process, which can result in further rough handling (Tabbush 1987). Under some conditions dipping is slightly beneficial but under others it can be slightly harmful (Davies 1988; Hodge and Walmsley 1990).

Treatment of plants against insect pests. Planting stock of any conifer or broadleaved species used to replace recently felled conifer crops is at risk from attack by adult Large pine weevil *Hylobius abietis* feeding on thin bark above ground; young conifers are also at risk from adult *Hylastes* spp. feeding on the root system. Plants may be protected against the insects by application of insecticides at pre-planting and/or post-planting, as noted in chapter 6.

Plant prices. Prices of plants to BS 3936 depend on species, origin/provenance, type, size, quality and quantity. Indications £/1,000, 'collected' ('EC species' are indicated by ★) are:

Bare-rooted: transplants or undercut seedlings: £/1,000:

	15–30 cm	30–45 cm	45–60 cm	60–90 cm
Conifers:				
Fir, Douglas★	75	90	100	—
Caucasian	200	300	—	—
Grand	100	130	—	—
Noble	125	200	…	—
Lawson cypress	90	125	200	—
Larch, European★	80	100	110	—
Japanese★	75	90	100	—
Hybrid	85	120	130	—
Spruce, Norway★	75	100	110	—
Serbian	135	—	…	—
Sitka★	70	90	100	—

	15–30 cm	30–45 cm	45–60 cm	60–90 cm
Pine, Austrian	80	95	–	–
Corsican★	100	100	–	–
Lodgepole	70	100	–	–
Scots★	75	95	–	–
Coast redwood	200	–	–	–
Wellingtonia	200	–	–	–
Western hemlock	125	200	–	–
Western red cedar	105	175	200	–

Broadleaves:

	15–30 cm	30–45 cm	45–60 cm	60–90 cm
Alder (4 main species)	100	130	150	175
Ash	85	125	150	175
Beech★	150	200	275	300
Birch (2 main species)	100	145	200	225
Cherry (Gean)	125	150	170	200
Bird	150	200	250	275
Chestnut, Sweet	140	175	200	225
Hazel	150	200	250	275
Lime (small-leaved)	200	250	300	325
Maple, Norway	100	125	170	200
Oak, Pedunculate★	150	200	250	275
Sessile★	165	215	265	300
Red★	215	250	275	300
Poplar★ (C+1)	–	–	300–1,000	–
Rowan (Mountain ash)	–	175	200	225
Southern beech (2 main species)	160	180	200	250
Sycamore	100	150	200	225
Walnut (2 main species)	250	300	350	400

Broadleave whips: Prices according to species: £/1,000:

90–120 cm	120–150 cm
500–1,000	1,000–3,000

Container grown plants: JPPs: Prices according to species.

Rootrainers: Prices according to species: £/1,000:

	15–30 cm	30–45 cm	45–60 cm
Main broadleaves:	250–300	300–350	350–450

Planting. The selected species having been ordered, the planting can be organized. Correct planting is an art, and the need for care to be applied to it cannot be over-emphasized; all the good work of the nurseryman can be spoiled in a short time. The expense and care of ensuring that trees are correctly planted is always amply repaid in later years.

Planting, under a wide variety of conditions and locations, must be done in a manner which will give the plants the best chance of survival and vigorous early growth. The roots should be kept moist and spread out naturally; and the trees planted upright, at the right depth and with appropriate firmness. To achieve all this is difficult when speed of planting is necessary for economy's sake, and hence the speed of planting should be restricted to that at which sound planting can be ensured.

The most usual method is notching with a planting spade or mattock, the plants being carried by the planter in a shoulder bag. The form of notch, L or T, and which shape of spade (the straight-backed garden type or, now rarely used, the tapered Schlich and Mansfield types) may be according to the workman's preference, and to the soil, size of plants and terrain. The first cut is vertical, then using the spade to hinge back the soil or turf, allowing the plant to be inserted uprightly in the slot at an appropriate depth and with the roots evenly spread. Pit-planting though costly may be advisable where large rooted or expensive plants are used. Planting of poplar and cricket-bat willow, a different technique, is referred to later in this chapter.

The depth of planting is crucial. On upland peats and gleys (where the watertable can be within 10 cm of the surface for most of the year, and with anaerobic conditions toxic to roots below this level) the root system must be planted shallowly in a raised, locally drained position on a mound or ridge created by cultivation equipment (see 'Cultivation' *supra*) or close to a stump or root plate on uncultivated ground. On well-drained or freely-draining mineral soils, all that is necessary is to notch the plants directly into the ground after screefing away or chemically killing any competing grass or other vegetation. On very dry sites, the plants must be put in more deeply, with the root collar perhaps 5 cm below the surface. Generally it is better to plant a little on the deep side rather than too shallow.

Broadleaves sometimes need 'stumping back' – the practice of cutting back top growth at or a few years after planting to stimulate a vigorous initial shoot. It is the traditional method of improving poorly-formed oak, ash and sweet chestnut; also to help sweet chestnut to form coppice. Generally the technique is carried out in late winter, and the new growth singled to one shoot in the summer.

The direction of the planting lines must be planned. Where ploughing has formed spaced furrows the lines will have been decided. Where there is no such cultivation, the direction will usually be determined by the angle required with the main line of future extraction. On slopes, the general aim should be to plant in lines running up and down, with deviations only to ensure that the planting position takes advantage of the most favourable microsite conditions. The first line to be planted, other than on spaced furrows, can be set out by alignment of sticks or thick twine. The subsequent lines will be found by moving the sticks or twine after each row is completed. When spacing the trees in the rows, the planters judge the required distance by the length of their planting tool or by stepping. The planting of more than one species creates some difficulty, but only in the case of a (usually inadvisable) intricate mixture (e.g. of three or more species) or of a group planting (e.g. oak in groups of say nine to twenty-five amidst conifers) are additional indicators and thought required. In awkward corners trees frequently have to be 'dodged-in'. Skilled planters will know how far they must plant away from tree stumps and stools which may throw up damaging re-growth, and will not waste a tree where they encounter particularly stony ground.

Concentrations of logging residues (especially brashwood/slash of conifers) can lead to inadequate planting. Here a single vertical notch is all that may be possible, and great care must be exercised to close the notch completely by reinsertion of the spade alongside it. It can be very difficult to firm the plants and this may result in the creation of air pockets below the ground causing the roots to dry out. Mattock planting may be appropriate for hard, dry sites, especially on sloping ground, using the adze blade to screef away the vegetation, cultivating the patch with the pick, and finally opening a planting hole with the adze. On peaty ploughed soils, it may be possible to use the semi-circular spade technique for planting in the centre of the spoil ridge.

Careful handling of plants on every occasion is very important. Before and during planting, root drying-out should be avoided. The plants must either be kept in appropriate bags, or (least recommended) heeled-in, i.e. the bundles are placed in a trench with their roots covered by soil, or placed at the bottom of a moist furrow with their roots packed close together and not exposed. If necessary, such 'sheughs' must be irrigated. In the actual planting, particularly during dry, windy or hot days, only one plant at a time should be removed from the waterproof planting bag. Chemical-proof gloves should be worn when handling treated plants and when planting on recently treated sites.

Mechanized planting by a tractor-mounted machine has its limitations. Most machines are designed for clean flat terrain or completely ploughed sites; they have been found suitable for planting Christmas trees at close spacing on level, clean land, free of stumps and ditches. This method of planting is proving useful on ex-arable land under the Farm Woodlands Scheme.

Much of the foregoing information on planting relates also to container grown seedlings. Specialist planting tools are available, e.g. 'Pottiputki' for use with Japanese Paper Pots.

Timing of planting. The physiological condition of the plant varies as the season progresses. Dormancy is induced in autumn in response to declining day length and temperature. Full dormancy is attained in October–December, depending on the season, and from then on, dormancy is released progressively as the 'chilling requirement' is met by accumulated winter cold. During this period RGP rises steeply.

Planting dates should be chosen to ensure that roots will grow rapidly after planting and before budburst; planting after the buds begin to develop makes the plant vulnerable to decay. Lifting, handling and planting should be undertaken when RGP is high, but coupled with reasonably high soil temperature in the site. Seasons for plant handling are more restricted in the south and at lower elevations where the growing season is longer; for example, storage of bags in shaded environment should normally finish by the end of March in Wales and southern England. Planting can take place earlier in spring at warmer locations. Autumn planting is most appropriate where soils are likely to remain warm into November.

Plants may be lifted for autumn planting from early October until the end of November provided shoots have become fully lignified ('hardened'). From then on, cold soil will limit establishment success. Sitka spruce, because it tends to be planted on cooler sites, is generally less suitable for autumn planting than Douglas fir. In general, for bare-rooted stock, planting after the end of March (April-May in Scotland) should use cold stored plants.

In the lowlands, the planting season is generally October to early April. In the uplands, particularly in Scotland, planting may continue through May and June. Trees protected by cold storage, as well as small container grown plants, can extend the planting season. Adverse weather conditions should be avoided, particularly periods of frost, snow, severe cold winds, and dry periods. Whether to plant during the autumn or to wait until February, March or later is debatable; but in general broadleaves may prove more successful if planted in the late autumn/early winter and conifers towards early spring. A difficulty about waiting to plant in spring is that if it becomes unduly delayed or coincides with a prolonged dry spell losses can be high. Bare-rooted stock should not be planted later than early May, even if flushing has been delayed by cold storage, unless the weather is very wet and the plants are still dormant.

Where planting conditions are difficult, this should be recognized in the method of organizing and paying the planters so that they are encouraged not to sacrifice care for speed. Allocation of individual work-areas and payment on a unit area basis will ensure that the standard of work of each planter can be properly assessed. Inspection should involve more than a count of the number of plants per hectare; it is necessary to spot-check for uprightness, depth and firmness. Further inspection is advisable following a drought, thaw, snowfall or strong winds, in order to rectify any lifting or other undesirable movement. Firming by use of the heel of one's boot is generally sufficient.

Treeshelters. The technique of using treeshelters – vertical translucent or transparent tubes – around newly-planted or naturally regenerated trees is widely practised in the lowlands and in some well-sheltered uplands. This has eased the problems of foresters facing small planting programmes. They can be an expensive investment with little return if used incorrectly and not maintained. When correctly used, they are a reliable and relatively inexpensive means of protection in the early years of tree life, and can lessen deer browsing and rabbit damage. Trees inside shelters, because of the 'greenhouse benefit', will grow faster than non-sheltered trees, oak in particular. Shelters also readily identify where the tree is planted and permit easier weeding with herbicides. Most are 1.2 m tall, sufficient to give protection against roe and muntjac deer; and 1.8 m against red and sika deer. For cattle, 2 m is recommended; for sheep, 1.5 m. Against rabbits 0.6 m is adequate, and for hares, 0.75 m, provided the ground is level and reasonably even (although deep snow will reduce the effective height of protection). Stakes and fixing methods must be robust to avoid movement by animals or wind. There are a variety of types of treeshelter, either square or circular in a range of colours, and they include ultra-violet inhibitors to slow the rate of breakdown in sunlight.

Treeshelters should not be used on compacted, waterlogged, or exposed sites. Before using, the planting position should be cleared of grass and weeds by screefing, and roots of bramble and bracken dug up. It is more prudent to choose the best site than avoid the worst.

Treeshelters can be expected to improve planting survival and height growth rates during the critical first three years. 'Fertility' and yield class of the site and crop are not changed thereby – the trees grow more rapidly only through the expensive establishment phase. A treeshelter life of at least five years is desirable so that the tree can grow out of the top and produce an adequate stem and early branching. Ideally it should provide 'greenhouse' conditions ('treeshelter microclimate') for the first two or three years and then continue to give support and protection for another few years. The shelter should remain around the tree until it disintegrates, which should be between five and ten years.

Most broadleaved species show improved height growth when inside treeshelters. The best suited are oak, birch, cherry, lime and hawthorn; the response is less in beech, sweet chestnut, walnut and southern beech. Of conifers, pines, spruces, larch and Douglas fir have done well in experiments, but problems have arisen with very shade-tolerant species such as *Abies* and *Tsuga*. Plants should be small but sturdy transplants (15–40 cm), with a good terminal bud, well-furnished with roots and at least a 6 mm diameter at the root collar. Treeshelters can be applied at least equally successfully to naturally regenerated seedlings (though driving of the stake must not be allowed to damage the plant).

Treeshelters cost around 55p depending on dimension and quantity purchased. The stake and clip cost about 25p. The combined cost of plant, planting, shelter and fixing lies between 90p and £1.50, dependent largely on the site and the number planted. The maximum 'acceptable spacing' for grant-aid is 3.0 x 3.0 m (1,111 ha); for wider spacing the grant is pro rata. The current (1991) erected gross cost per hectare, including plants and planting, is likely to be in the £1,250-£2,000 region. Treeshelters should not be seen as an encouragement towards wider spacing, particularly with oak and beech, for which most foresters would view a 3.0 x 3.0 m spacing as unacceptably wide.

On small areas treeshelters are likely to be a more economical means of offering protection than fencing, though the point at which the financial balance swings in favour of shelters is debatable. Foresters must make their own calculations in the light of their individual circumstances and priorities. The comparison should take into account the expected improvement in survival, the increased speed and the convenience of weed control, as well as the differences in the costs of shelters and fencing. If a shelter fails one tree is exposed, whereas if a fence is breached all trees are at risk. There may be instances where, to economize in fencing, a combination of a straight fence line coupled with a few treeshelters outside may be suitable and cost-effective.

Treeshelters do not obviate the need for effective weed control during the establishment period. For optimum growth the trees benefit from at least 1.0 m diameter free from weeds. A medium volume herbicide application from a knapsack sprayer can be applied safely to weeds outside the shelter when the trees and weeds are most susceptible. Inspection and maintenance operations should not be neglected.

Some doubts about treeshelters have been raised as to: (i) their stability in wind and cattle/sheep grazing (they may cost up to £50/ha to maintain); (ii) the weed growth inside the shelter; (iii) the support of the tree after the shelter has degraded; and (iv) the aesthetic disadvantage. There is also the problem of small birds entering shelters and not being able to escape (probably a small gap at the base may be desirable). Treeshelters appear very artificial, particularly in large numbers. They should be positioned in irregular fashion, not in straight lines or intrusive geometrical patterns. In landscape sensitive areas, an unobtrusive colour should be used to blend with surroundings; russet browns or olive greens are usually best, and white or garish greens should be avoided. Hence there is need for due consideration being given to the nature of the site and the managerial implications before treeshelters are employed (Tuley 1989; Potter 1989, 1991).

Weeding. Grass, broadleaved weeds, gorse, broom, bramble, bracken, heather and woody re-growth, can greatly harm young plantations. They reduce the survival and early growth of newly planted trees by competing for light, soil moisture and nutrients, and by doing physical damage by excluding air and light through smothering, by collapsing in autumn/winter especially during heavy rain or snowfall, or by harbouring bark-gnawing rodents such as voles. The main objectives of weeding are to eliminate or suppress competing vegetation during the early life of a tree crop, and to secure rapid early growth and the even and successful establishment of the plantation. Root competition reduces survival and early tree growth especially on dry sites or in dry seasons, being most detrimental when plants have already been stressed by poor handling or planting. Competition between the trees and weeds will to some extent be determined by the conditions of site at planting; it is much reduced in cultivated sites or where screefing or herbicide application has taken place. Weeding is more often required on richer lowland than harsh upland sites. Mowing or cutting weeds reduce their above ground harmful effects, but does not reduce root competition.

In lowland conditions, weeding poses problems such as: when and how to weed, when it can be safe and prudent not to weed, or when it is desirable to retain the weed growth. Regular inspection of plantations will reveal which sites need weeding. For most effect, weed control should take place before competition is obvious. Assessment should be made of the kind of weed on the site, their densities and physical strengths, the species and height of the tree planted and the progress of the plantation. Some weeds, particularly bracken, grow particularly fast and a week or two may make much difference. Heavy rank growth may be expected on fertile soils. The effects of seasonal vagaries are also apt to make weeding more onerous, particularly during wet periods. Even if the operation has been undertaken in early summer, secondary summer and early autumnal growth may be lush and necessitate another weeding. A particular danger is prevalent in the autumn on areas carrying bracken, bramble and coarse grasses; these when dying are usually relatively heavy and wet, and some collapse on small trees, forming dense mats over them. A late weeding may then be necessary even if it is only to rake, push or scuffle the decayed vegetation away from the trees. Heavy snowfalls make the risk of damage from collapsing even more serious. The time of weeding should never be dictated by the gamekeeper; but every possible co-operation should be extended to him, particularly during nesting times (see chapter 13).

Weed growth if kept within reasonable bounds can sometimes be beneficial, in protecting soil structure, preventing erosion, and acting as a temporary 'reservoir' for nutrients in young crops. Savill and Evans (1986) have tabulated 'ideal' weed characteristics. A moderate amount of robust growing weed affords temporary shelter from sun, frost and drying wind; it may assist in keeping the ground cool or moist and encouraging the trees by their nursing effect. Some tree species, such as western hemlock, western red cedar, coast redwood, and *Abies* spp., benefit from moderate shelter, and the weeding treatment of these and other shade bearers can sometimes be different to that needed by light demanders. Low vegetation is particularly beneficial for protection on shallow and dry soils, such as those of much of the Cotswolds, provided one square metre is kept weed-free around each tree. However, frost damage sometimes appears to increase where the grasses predominate among the ground flora.

The need for weeding having been decided, the desired intensity of it must be considered. Seldom is it necessary to weed a plantation completely unless machines are used for the work, when it may be cheaper to do it reasonably fully. All that is necessary is to ensure that the leader of each tree has adequate freedom and light. On other occasions it is sufficient to weed selectively around each tree, the central weed growth being ignored if it is unlikely, when alive or dead, to topple over the young trees. An alternative may be lane weeding. In certain circumstances there is the benefit of shelter from intelligent manipulation of weed growth, but no suspicion of smothering or overcrowding can be tolerated. The weeding of

Christmas trees needs to be more intensive, and hence is more costly, because even the lowest branches must be kept free to prevent them decaying or dying.

Exposing trees too suddenly or too soon in summer may induce sun-scorch and wilting; if left too late their leaders may become so weak and tender that dieback sets in. On dry sites early treatment may be necessary in order to condition the trees to the heat of the sun; if the first weeding is unduly delayed the result can be serious. Young larches are difficult to locate in early weeding because usually they do not add growth in the leader until August; the easiest to find as the weeders work along the rows are the shade-bearing conifers and the spruces and firs. When working among such as Sitka spruce, Lawson cypress and ash, any double leaders should be singled (formative pruning). Climbing or trailing plants encountered should be removed and their roots pulled up, unless conservation interests dictate otherwise.

Natural seedlings, suckers and stool re-growth pose a special problem during weeding. Prudently tended they can afford protection and nursing for some species, or even take the place of a failed or inferior planted tree. Experienced weeders can decide in all cases whether to cut or to encourage. In general, such natural growth needs to be removed, especially where the planted survival rate is high.

Most plantations will require some weeding in early life. Some may need it twice in a season. Some will require weeding for one year only; others for two or more years. Thus the total cost per hectare can be substantial. Constantly, as in other woodland operations, every prudent effort should be made to reduce expenditure. Foresters who have undertaken weeding find it of assistance to decide the method to be followed, what tools are appropriate, and perhaps what piece-work rate to fix. They are the better able to make known their desires as to the extent of weeding required and to give instructions as to the saving of any natural seedlings. Even if contractors are engaged foresters' practical experience will stand them in good stead.

Nowadays the main methods of weed control are by pre-planting cultivation and/or chemical treatment. Chemical weeding is most used, but before fully discussing it, other methods are noted below.

Manual weeding. This method is expensive and ineffective in fully removing weed competition. It involves the forest worker using an edge tool, cutting back the weeds from around each tree. The technique is simple to learn but the timing is restricted to the growing season and with some weed species may have to be repeated more than once in a season. Thus weeding by hand can create a peak demand for labour. Its advantages are that the equipment and training are minimal and it is very versatile both in weed type and terrain. A disadvantage is that it is the least effective means of reducing competition, and on grass sites may actually increase it. In some circumstances (e.g. following up chemical-treated bracken, or beside watercourses where chemical must not be used) manual weeding is necessary.

The amount and kinds of the weeds and the experience of those undertaking the work will decide the tools used. In the early growth of bracken a stick may be used to swipe off the tops of the fronds; this may have to be repeated during a season. For other growths a short-handled sickle or reaping hook is the normal tool, but experienced workers may prefer a long-handled hook. Great care must be exercised in avoiding the cutting of trees (workers refer to it as 'Sheffield blight'!). The tree should be found using a hooked stick or crook before a cutting stroke is made towards it. The weeders must be made aware of the spacing of the trees. Trampling is occasionally helpful instead of cutting but this will depend on the softness or brittleness of the weed. The species of weed will determine too the power needed behind the cutting stroke; weeds may be either soft or tough, thus an experienced operator will constantly adjust the strength of the stroke according to the species and will know whether to ease cutting by first bending the weeds. A billhook or slasher is usually needed for such as stool re-growth, gorse and broom. The two latter should always be cut before they seed. All cutting tools must be frequently sharpened; they blunt more quickly on some growths, especially tough grasses. Safety precautions in working are paramount. In general, weeding is no longer necessary when the trees reach above the weeds and are growing vigorously.

Mechanical weeding. This method, although apparently cheaper than manual weeding, is ineffective likewise in removing competition. There are several mechanical cutting techniques employed, among them: rotating saw blades, flails and chains (swipes), reciprocating cutter and crushing roller. These can be employed on several different types of machines: tractor, pedestrian-controlled machine and hand-held machine. The larger, more powerful of these (the tractor) fitted with the appropriate tool has the capacity to deal with substantial woody growth or has a high output and low cost on easier weed types. But terrain is a limiting factor and tractors cannot operate in the weeding role on steep slopes or rough terrain. The pedestrian-controlled machines (for instance the auto-scythe) have slightly better versatility as far as terrain is concerned but outputs are low and costs high. The hand-held machines like the brushcutter can be used on steep and rough terrain and are effective on woody weeds but costs are high. In general, machine methods can be most effective and less costly on easy terrain. Capital expenditure is involved and training in operation and maintenance is essential. In some cases, supplemental hand weeding around individual trees may be necessary.

Chemical weeding. The control of weeds by the application of herbicides is now the most widely accepted and cost-effective technique. Weedkillers, when responsibly used, are of great assistance in containing the costs of afforesting or restocking. Application of herbicide is the preferred method except where amenity and environmental consideration are a constraint against such use. On environmental grounds, those herbicides recommended for forestry are well tested and when applied in accordance with manufacturers' recommendations the risks are minimal; but misuse can have serious consequences.

Detailed information and guidance on chemical weeding are provided by Williamson and Lane (1989) whose recommendations conform to present legislation under the Control of Pesticides Regulations 1986, the Control of Substances Hazardous to Health Regulations 1988, and Forestry Commission *Provisional Code of Practice for the use of Pesticides in Forestry,* 1989. (Pesticide includes herbicide, fungicide and insecticide.) The guidelines cover general principles, product approvals, competence of users, operator protection, environmental protection, storage and handling, equipment and reduced volume application. The checklists give salient points on the decision to use herbicides and daily operation. The herbicides appropriate for use against a range of forest weed species are set out in a wallchart available from the Forestry Commission. Recommendations are linked to proprietary products which have the necessary approval. Supplementary information on herbicides has been provided by Nelson and Williamson (1990); and on the pesticide regulations by Williamson (1990). The method, rates of application, and timing are important. Before any herbicide is used the label should be noted – it carries full instructions for use and for the protection of the operator and the environment.

Herbicides act either through leaves and only kill the part of the plant touched, or by leaf translocation to kill whole plants. Their application requires skill and care to avoid problems of wrong dosages and spray drift. The timing of applications and area treated around each tree are important attributes of success. There are a number of factors that make the effective use of herbicides rather difficult. Most are selective, killing some weeds and only weakening others and often leaving some unharmed. Many can be used safely in relation to the trees only over a relatively limited period, few can be used in the period May–July: this means that weed control by herbicides must be planned well in advance. The number of possible chemicals is large, and successful treatment depends on the application of the appropriate chemical to the weed types and trees at the right time at the correct dosage. Methods of application continue to develop rapidly, generally making the work easier, improving output and reducing costs. A decision must be made as to which application pattern suits the weeding requirements among the options of complete (incremental), band (over-row and inter-row) or spot (with or without tree guards or treeshelters). The applicator will depend upon the nature of the herbicide (granular or liquid) and on the application pattern required. Operators should check their liability under the Control of Pesticides Regulations 1986 in relation to the need for certification.

Grasses compete vigorously for light, nutrients and, especially, for water; effective control of them is therefore usually essential for successful crop establishment and growth. (Peat soils being water-retentive, moisture competition is not so severe; and the need for weeding is generally less than on mineral soils.) As forest weeds, grasses can be grouped into two categories: coarse grasses which are generally tall, bulky, rank, stiff, often rhizomatous and tussocky and others which, in contrast, are known as soft grasses. The latter are generally more susceptible to herbicides while coarse grasses usually show a somewhat greater resistance. Herbicides recommended include atrazine, cyanazine, glyphosate (Roundup), hexazinone and propyzamide (Kerb).

Bracken competes strongly with young trees for light during the later part of the growing season. Towards the end of the year it collapses and can smother and flatten small trees with its weight, increasingly so if snow lies on top of them both. Pre-cultivation, generally ploughing, does give some control of bracken for the first season but on sites where bracken is vigorous, the fronds on either side of the plough ridge can overcome conifers. If a crop of trees is present, chemical control must be followed at least one month later by hand cutting before the fronds collapse on the trees and cause damage. Whenever possible, herbicide should be applied pre-planting to avoid this problem. The herbicides used for bracken control are glyphosate (Roundup) or asulam (Asulox).

Where the dominant vegetation is *heather*, generally on sites where the availability of mineral nitrogen limits tree growth, nitrogen deficiency may develop in certain species of conifer. This may need to be alleviated by complete spraying to kill the heather. (The need to control heather can often be avoided by planting a non-susceptible species, burning the heather before ploughing, restocking felled areas before the heather has time to invade, or planting spruce in mixture with suitable origin/provenance Scots pine, lodgepole pine or Japanese larch.) Herbicides recommended for heather control are 2,4-D ester or glyphosate (Roundup).

The *woody weed* group contains a wide range of species including bramble, climbers and shrubs of all types. Species such as birch in some circumstances are part of the desired crop and in others are unwanted and threaten it (situations which require a precise definition of management objectives and constraints, such as the present and future amenity effects of broadleaved components of a stand). Herbicides recommended are glyphosate (Roundup), triclopyr (Timbrel) or ammonium sulphate (Amcide). *Gorse* and *broom,* both evergreen shrubs, occur either separately or together and locally may present a major weed problem. The recommended herbicide is triclopyr (Timbrel). *Rhododendron ponticum* and *laurel* are to be found on acid sites, mainly in the wetter western half of the country, in all phases of colonization from a light scatter of small seedlings to impenetrable thickets 2–5 m in height. The early stages can be treated with herbicide but bushes more than about 0.5 m high must be manually or mechanically cleared to allow the stumps and the more susceptible regrowth to be sprayed, preferably before re-growth is more than 1.0 m tall. Recommended herbicides are as for the woody weed group *supra*. Stables and Nelson (1990) give an update of experiments on control of rhododendron.

On upland afforestation sites, much weeding can be avoided by good management practice. Cultivation by ploughing or scarifying is the main method of weed control and, if planting is prompt and successful, weeding is usually unnecessary. Many upland restocking sites are initially weed-free and are not cultivated, and the need to weed can still be avoided by planting before weeds have re-invaded. However, herbicide application is necessary where the conditions are hostile and grasses are abundant.

Most lowland sites, even when cultivated, will require chemical weed control. Killing weeds before afforesting or restocking reduces the need for subsequent weeding. It also enables non-selective herbicides to be used without danger to trees. Where bramble, bracken or woody re-growth are expected to invade rapidly following clear-felling it may be better to delay restocking for one growing season to allow the flush of weeds which arises following the influx of light to be killed off easily and effectively by herbicides. Restocking can take place in the following winter or spring. On dry sites spring-planted trees suffer from water-stress immediately after planting, and it is important to ensure that this is not exacerbated by grass competition.

The choice of which herbicide to use after planting is determined not only by the weed species to be controlled, but also by the tree species which are present. Very few herbicides are tolerated by trees in active growth and many broadleaved tree species will not tolerate overall applications of some herbicides even during the dormant season.

Optimum benefits from weed control are obtained where either residual herbicides or mulch mats (noted *infra*) are applied in the winter/spring in anticipation of weed competition. The application of herbicides at this time of year means that many broadleaved species are more tolerant of herbicide application. Residual herbicides also have the added advantage of being less demanding in terms of weather condition and the date of application. The optimum cost/benefit can be achieved by keeping an area of one square metre around each tree free from weeds. The post-plant application of herbicide is therefore usually limited to either spot or band application. The former is applied by knapsack sprayer, forestry spot gun, pepper pot or weed wiper; the latter by knapsack sprayer, micron-herbi, or tractor-mounted sprayer adjusted for band application. When applying herbicide around a small tree it is difficult if not impossible to treat the weeds immediately around the base of the stem or foliage of the trees. The benefits from killing these last few weeds must be weighed against possible damage to trees; a steady reduction in tree increment can be expected as the size of the untreated clump of weeds increases.

Mulching. This method of weed control acts by smothering weeds or preventing them from germinating. It has the additional benefit of reducing or eliminating evaporation from the soil surface, keeping nutrient-rich upper soil layers moist enough to nourish tree roots. On waterlogged sites suffering from anaerobic soil conditions, the reduced evaporation from mulched surfaces can exacerbate the problem. The effectiveness of mulching varies with the material used. Black polythene with an ultra-violet light inhibitor and a thickness of 125 microns is recommended; either laid on bare ground or on ground where the weeds have been killed by herbicide. Clods of earth can be placed on the mulch mats to hold them in position; this deters rodents entering below them, and prevents the mats being dislodged by wind and weeds.

Costs of weeding. The cost of weeding is related to the method, terrain, cultivation, species, spacing and number of trees to be found and cleared per unit area, the type and intensity of weed growth and the number of weedings necessary. Costs in the lowlands are much higher than in the uplands, because of more vigorous and prolific weed growth on the more fertile soils. Broadleaves grow successfully chiefly on sites likely to produce vigorous weed growth, and because of their slower growth the trees require weeding for a longer time than conifers. The direct cost of maintaining one square metre around a tree for three years using herbicides may be about one-fifth that of using mulch mats; but when additional travelling and supervisory costs are taken into consideration there may not be a great difference between the two methods of weed control over a three year establishment period (Williamson and Lane 1989).

The most desirable strategy of weed control, which can sometimes eliminate or greatly reduce the need for additional measures, is to ensure rapid establishment to harness the competitive ability of the trees themselves (Savill and Evans 1986).

Beating up (Gapping). This operation is the replacement of plants which have failed on a recently planted site. The ideal is to achieve a fully stocked plantation, but beating up is expensive and may be unnecessary if not more than a scattered 20% of plants have failed in a close spacing, or 10–15% in a wide spacing; but if the failures are in groups some beating up will always be desirable. Failures noticed before the end of the first planting season may be replaced without any delay to growth; but usually beating up is undertaken during the autumn or winter following the original planting (unless container grown plants are used). Rarely is it again necessary. The new plants need to be healthy and well-balanced; preferably sturdy ones slightly larger than average. Scots pine is a particularly successful beating up species. Good quality natural seedlings, suckers or coppice shoots can occasionally be accepted in place of a dead plant. The unit cost of beating up is usually about 15–25% above that of the original planting, because of the need to locate gaps. The unit cost will be greater the smaller the proportion of failed trees being replaced. The best course is to eliminate or minimize the need for beating up by initially adopting high standards of plant handling and planting.

Initial fertilizing. Fertilizing accelerates crop growth in both height and diameter, and in some afforesting is vital. Likewise it can increase weed growth!

Fertilizing conifers at planting. Particularly in upland areas, trees require an adequate supply of several major nutrients and trace elements for satisfactory growth. Fertilizers are used on infertile sites to augment the natural supply of plant nutrients from the soils in order to forestall or correct deficiency (Mayhead 1976; Binns 1980; McIntosh 1984; Taylor 1986b; Taylor and Tabbush 1990). Binns (1975) and King *et al.* (1983) provide a list of fertilizer materials commonly used in forestry, the three main types being: P as ground rock phosphate, P as superphosphate granules, and P and K together in granular form. The need for nitrogen fertilizer in spruce plantations is noted later under ' "Check" of spruce in heather'. Urea and ammonium nitrate are the most commonly used nitrogen fertilizers, both being high in nitrogen content and water soluble. The benefits of application of potassium (K) to Sitka spruce on deep peat are noted by Dutch *et al.* (1990).

Fertilizing of conifers at planting is largely with the aim of reaching canopy closure without serious delay due to nutrient deficiencies. Each plant can be treated individually by manual application, but it is costly. Broadcast application by tractor-mounted spreader can be satisfactory. Application by helicopter is only suited to large-scale areas, and is usually done by a specialist contractor. The technique of aerial application is discussed in chapter 6 under 'Top-dressing of conifers'. Where P and K are applied it is normal to make separate flight runs for each because of the difficulty of mixing and achieving an even spread of the two together.

The main soils requiring fertilizing are the upland podzols, ironpans, peaty gleys and deep peats, and the lowland heaths. Brown earths, surface-water gleys and inter-grade soils do not usually require treatment. At planting in southern England west of the New Forest, Everard (1974) recommends that 75 kg P/ha should be applied on heathland or former heathland soils. In Wales most upland deep peat sites should receive 50 kg P/ha; but most other sites do not require fertilizer. Recommendations relative to upland conifers in afforestation and restocking are noted in later sections. Possible application of fertilizer as a top-dressing is discussed in chapter 6, along with foliar analysis as a diagnostic technique of nutrient content. Taylor (1991) has updated relevant information.

Fertilizing broadleaves at planting. Most broadleaved planting will be on relatively fertile soils and normally will not require addition of fertilizers. Nutrients are unlikely to be a limiting factor to growth. However, on some kinds of sites nutrient deficiencies may occur: lowland heaths – phosphorus; chalk downland – nitrogen and potassium; woodland where coppice working has long been practised – phosphorus; and restored, man-made substrates – nitrogen. Analysis of soils to assess the likely fertilizer needs of trees is rarely undertaken, but it is prudent to check the soil pH before a site is planted, since either alkaline or very acid soils will rule out some species. Taylor (1991) has updated relevant information. Recommendations relative to upland broadleaved afforestation are noted in a later section.

Afforestation in uplands (Plantation silviculture). The major expansion of forestry – by way of afforestation – has been essentially of conifers (mainly exotics) in the uplands on land sub-marginal for agriculture. Afforestation is usually associated with irregular, hilly terrain, difficult ground conditions and harsh environments; the soils involved often have serious physical limitations to growth, and very rarely will have been cultivated previously. One of the main problems overcome has been excessive soil water. On many moorland and heathland soils worthwhile establishment would not have been possible without initial cultivation, cross drainage and the application of fertilizer, particularly phosphates. The cultivations, including the breaking up of *pans* in podzols, and other impervious soil conditions, have been standard practice. (The need for fertilizer application is noted under ' "Check" of spruce in heather' *infra*.) Pure even-aged crops tend to predominate everywhere that plantation forestry is practised due to their economic advantages – the simpler silviculture, harvesting and marketing.

Afforestation has enabled foresters to select the species, in contrast to being dependent on existing local natural forest types. It has been possible to ensure that the whole site is fully stocked for the whole rotation (subject to windthrow) with the kind of trees most wanted; thus full use is made of the available land. Three factors in this conventional plantation forestry have led to great uniformity of end product: use of one or few species (almost always coniferous), raising a crop to form an even-aged stand, and applying the same silvicultural treatment over a whole stand. As a result of these factors the productivity of plantation forests is almost always much greater than that of natural woodland (Savill and Evans 1986). The intensive nature demands high levels of commitment, skill and resources. The uniformity of plantation crops brings possible biological risks as well as other disadvantages, particularly aesthetic, and where semi-natural vegetation on a site is superseded by trees, possible loss in conservation value. But, where timber production is important, such losses are more than outweighed in most cases by the opportunities plantation silviculture afford of greater production of timber.

The attainment of high yields compared with those in traditional forests is usually one of the aims of plantation silviculture. How these may be achieved and what risks may be involved in the longterm are among the most fundamental considerations (Savill and Evans 1986). Where production of industrial wood is an important objective, the most common plantation forestry strategy involves establishing one or two fast-growing species on a large cultivated bare site, growing the crop until approximately the age of maximum mean annual increment and then clear-felling the whole (followed by restocking). This ensures high productivity and is simple to carry out. The system has the economic advantages of being relatively cheap to establish, manage, harvest and market. Whatever the biological arguments in favour of stands with a diversity of age classes and species, they may carry much less short-term economic weight. For the six most widely-planted conifers in Britain, maximum current annual production of stemwood lies in the range 6–20 m^3/ha; means are about 4–14 m^3/ha. (In the cooler, more continental temperate forests, production may be one-quarter of the British maximum and in warm temperate forests about 12–24 m^3/ha.)

A matter for debate is whether the superior productivity of first rotation monocultures, especially of coniferous exotics (Sitka spruce, lodgepole pine), can be maintained without site-degradation, disease, insect pests and serious loss of yields in subsequent rotations. (Virtual monocultures are found often in nature; but plantation trees often have a much narrower genetical base than the wild populations of natural monocultures from which they were selected.) It is frequently argued that the introduction of some diversity into monocultures would result in the better use of the soil and less risk of pests and diseases. Provided management is sufficiently intensive and careful, serious site deterioration does not usually occur in coniferous monocultures, and declining yields in second rotations will be the exception. A question is: whether mineralization of nutrients can keep pace with their removal from the site through harvesting. One of the more serious threats to continued productivity results from the compaction of soils by heavy harvesting machinery. The avoidance of this, or amelioration by cultivation can be important. Much has still to be learned about the biological stability and potential problems associated with monocultures in the uplands.

'Check' of spruce in heather. Growth 'check', more usually in Sitka spruce, on moorland and heathland soils, is indicated by short and yellowing foliage (and in any case can be determined by foliar analysis – see chapter 6, section 2). It is a situation where the spruce are growing extremely slowly or hardly at all – often fairly patchy in extent, which makes remedial action very difficult, as areas in check cannot usually be singled out for treatment. Checked crops, untreated, will remain in that condition for an indefinite period: on the more fertile heathland sites, it is possible that the crop would emerge from check after a lengthy period – unlikely to be less than ten years. Such delays would generally be financially unacceptable.

Until the 1970s, check was thought to be due solely to competition from heather. It can be alleviated by means of appropriately timed applications of 2,4-D herbicide (Mackenzie *et al.* 1976). However, increased planting of Sitka spruce on very nutrient-poor soils revealed that, even after removal of heather by herbicide treatment, growth was still limited by low availability of nitrogen. This can be caused by limited soil nitrogen capital and/or slow rate of nitrogen mineralization (Taylor and Tabbush 1990). Applications of nitrogen fertilizer can overcome this deficiency although several applications may be required to achieve full canopy closure. Once this stage is reached demand for nutrients is reduced due to shading of competing vegetation, improved nutrient cycling and capture of atmospheric nutrients, and further inputs of nitrogen should not be required (Miller 1981). The major difficulty faced in determining the treatment of a nitrogen-deficient stand is deciding whether heather control, application of nitrogen fertilizer, or a combination of both, will yield the most cost-effective response on a given site. Taylor and Tabbush (1990) explain the background to the problem, categorize the range of site types involved, and advise on available treatment.

In summary, nitrogen deficiency in young Sitka spruce plantations can be caused by heather competition, low rates of mineralization or a combination of the two. Where heather competition is the major limiting factor, then herbicide control will normally prove to be the most cost-effective treatment. However, as sites become increasingly low in available nitrogen, heather control is less effective and applications of nitrogen fertilizer (urea or ammonium nitrate) are more appropriate (Taylor and Tabbush 1990). Alternatively, it is possible to avoid the need for remedial treatment by planting suitable origins/provenances of Scots pine (on heathlands), lodgepole pine (on unflushed moorlands) or Japanese larch with the Sitka spruce in a nursing mixture when afforesting or restocking. The nurse species not only suppress competing heather but also have the ability to increase the amount of nitrogen available to the spruce.

Mechanical application of fertilizer is unlikely to be much used because vehicle access will be restricted by tree size at the stage when nitrogen treatment is required. *Manual application* can be very accurate, although distribution can be time consuming. Care needs to be exercised in the placement of fertilizer, and the base of the tree must not be used as a target. (High concentrations of nitrogen fertilizer around the root collar can be very damaging and can even lead to tree death; and foliage can be scorched by high concentrations of fertilizer lodging on the needles.) The best manual application technique is to broadcast the fertilizer over the whole site, avoiding the furrows, which not only avoids tree damage but is just as efficient for crop response. *Aerial application* by helicopter is the most common method on substantial areas. (The technique is noted in chapter 6 under 'Top-dressing of conifers'.) The evenness of the distribution can be visually assessed a few months after application, due to the colour response in nitrogendeficient Sitka spruce. (The following spring should be awaited in the case of autumn application.) Contracts should include provision for re-application to untreated areas.

Restocking conifers in uplands. Restocking (second rotation silviculture) has increased in importance in Britain as the post-war plantings, chiefly of State owned conifers in the uplands, have become mature and been harvested. It has introduced new silvicultural problems, along with opportunities, in addition to those associated with the first rotation. Forestry Commission restocking practice was reviewed by Low (ed.) (1985). Subsequent research with an integrated approach has led to an even greater understanding of the establishment process, and Tabbush (1988) drew together knowledge as a basis for the design of improved systems for upland restocking.

The cost of silvicultural restocking operations and quality of establishment achieved are strongly influenced by the harvesting operation. Costs of both should be optimised by careful planning. To make the second rotation economically sound it is important that it is established to specified standards of stocking and uniformity at minimum cost. Excessive reductions in expenditure on planting stock, site preparation and planting can result in severe cost penalties in beating up, additional weeding, prolonged protection, loss of uniformity, lower timber quality and ultimately delayed harvesting. Foresters have now been set standards of establishment and informed of the minimum outputs necessary to achieve them. They have needed to acquire a sound understanding of the biological processes involved (Tabbush 1988).

Nelson and Quine (1990) discuss the silvicultural advantages of cultivation on restocking, review the suitability of a range of cultivation machinery for different site types and give potential outputs. They further discuss the characteristics of the required microsites for planting, and show the recommended planting positions on these. Cultivation is particularly beneficial for plants which have an inherently low RGP, e.g. Douglas fir.

Since clear-felled sites in the uplands are generally more hostile than newly cultivated grazing land in terms of tree establishment, somewhat higher standards of plant quality are required than those for afforestation. Tabbush (1988) gives recommended size standards for culling and grading plants intended for restocking, based on stem length ranges and corresponding minimum root collar diameters.

The quality of planting stock at the time of planting is the result of nursery techniques (see chapter 2) and the effects of lifting, handling, storage and transport (all noted earlier in this chapter). The performance of plants of a given quality will depend on site conditions, and clear-felled sites generally offer a more testing environment than reclaimed grazing land, because stumps and brashwood/slash make site preparation expensive and planting difficult, and because of damage from insects and mammals. Good quality planting stock is a prerequisite of successful restocking. Although large plants generally perform better than small ones given adequate root:shoot ratio, small shoot length can confer some advantage on a site where exposure is the limiting factor; and small plants may also be appropriate where weed competition is minimal.

Site conditions both above and below ground are likely to have been modified in various ways by the growth of the first crop and by the presence of adjacent unfelled stands. The chief problem will be the logging residues presenting a physical barrier to cultivation, drainage and planting. Other problems may be the Large pine weevil *Hylobius abietis* and Black pine beetles *Hylastes* spp. developing substantial populations on restock sites, browsing by mammals, particularly deer, and pathological problems such as *Heterobasidion annosum* and Honey fungus *Armillaria* spp. – but rarely troublesome in woods.

In restocking, there is an opportunity to change the species or origins/provenances distribution in order to improve yield, landscaping or wildlife conservation. This is because of changes in soil structure, shelter from neighbouring plantations, and lessons learned from the previous or neighbouring crops. In forest design (see chapter 13), a compromise is needed between the demands of landscape, wildlife conservation, harvesting, restocking and management. The balancing factor should be cost, related to the quality and visibility of the site. For landscape design, clear-felling and restocking provide important opportunities for improving the appearance of the forest by the correction of previous bad design and the introduction of more varied stand appearance and tree size. For wildlife conservation diversification of forest structure and, to some extent, of coniferous tree species, will benefit wildlife as well as landscape.

Fertilizing requirements of second rotation upland sites need adequate consideration. Recommendations are given by Low (ed.) (1985) based on results from experiments in Sitka spruce. That species does not require or benefit from, application of fertilizer at time of restocking, deriving adequate nutrients from the breakdown of the litter layer and brash (Taylor 1990). Later top-dressing is unlikely to be required on sites that did not require fertilizer in the first rotation, but there are some indications that it may be required on sites low in available nitrogen or phosphate. Pure Sitka spruce crops will become nitrogen-deficient on poor heathland sites where, if used, it should be planted in mixture with suitable origins/provenances of Scots pine, lodgepole pine or Japanese larch.

For a given site, nutrient deficiency problems will, at worst, be similar to those encountered in the first rotation and will frequently be considerably reduced. The nutrient relations of the site may be affected by the previous crop in three ways: (i) there is normally a considerable litter layer on the site, decomposition of which may provide a valuable source of nutrients in the first few years of the successor crop's life; (ii) originally present weed species like heather may have been shaded out and so will not form part of the recolonization vegetation following felling; and (iii) on peat soils, particularly following a lodgepole pine crop, considerable shrinkage and cracking of the peat may have occurred, creating conditions which may be more favourable for mineralization of the organic matter. (An explanation of shrinkage and cracking on blanket peatland is given by Pyatt, 1987.) The effect of the three factors is generally to create conditions leading to improved nutrient status of the successor crop. There may be nutrient leaching in the late thicket-stage – when a crop would be making maximum demand on the site and the litter layer would have largely decomposed. Low (ed.) (1985) and Taylor (1986) have used four main categories of soil types when relating to the likely nutrient (and fertilizer application) requirements in restocking:

1. Fertile mineral soils (brown earth, surface-water gley and ironpan soils) which originally carried a fairly rich natural vegetation of fine grasses, herbs and bracken. These soils supported a high-yielding first rotation crop (chiefly of Sitka spruce and Douglas fir) with no fertilizer input so it is unlikely that the second rotation crop will either need or benefit from fertilizer application.

2. Heathland mineral soils (ironpan, podzols and gley soils) dominated by heather which often carried pine or larch crops in the first rotation but latterly have been planted with Sitka spruce. The higher-yielding spruce required phosphate fertilizer with heather control and/or nitrogen inputs to perform well on these sites in the first rotation. In the second rotation there is no requirement for phosphate at restocking and the re-invasion of heather may be delayed if shaded out by the previous crop. However, on the poorest sites it is likely that nitrogen deficiency will be encountered again if pure spruce is planted. (See recommendations noted earlier as to nitrogen application and the use of nurse species for Sitka spruce.)

3. Peaty gleys. Second rotation crops on flushed peaty gleys should not require fertilizer inputs for satisfactory growth, but on the more infertile peaty gleys there may be a requirement for phosphate fertilizing, with additional potassium where peat depth exceeds 30 cm when the crop reaches the thicket-stage. However, even on these poorer sites there is no requirement for fertilizer at time of restocking.

4. Deep peats. The most remarkable change has occurred on peats over about 50 cm depth where the first rotation crop, particularly lodgepole pine, has caused considerable cracking and shrinkage of the peat. This irreversible drying provides favourable conditions for mineralization of nutrients which are available to the second rotation crop. Again, there is no need to apply fertilizer in the early stages, unlike the first rotation where phosphate and potassium inputs were essential. On those sites which are low in available nitrogen (such as the poorer heathland soils) pure spruce will require nitrogen fertilizer applications for satisfactory growth. However, when spruce is planted in mixture with Scots pine, lodgepole pine, or Japanese larch on such sites, nitrogen deficiency does not develop. The processes involved in this 'mixture effect' are not yet fully understood but do appear to be associated with an inherent adaptation of pine and larch to obtain nitrogen on poor sites. The spruce are probably then able to tap this nitrogen, either via the nutrient cycle of the mixed crop or by mycorrhizal associations – symbiotic root fungi – and thus maintain satisfactory nitrogen levels. Therefore there is considerable benefit to be gained by planting such mixtures when restocking sites of low nitrogen status. Pyatt (1990) updates information on long term prospects for forests on peatland. The benefits of application of potassium (K) to Sitka spruce on deep peat are noted by Dutch *et al.* (1990).

It appears to be evident that in general replacement crops do not require fertilizer at restocking, in contrast to the first rotation, and inputs of fertilizer over the whole rotation may well be substantially reduced. However, the need for application of nitrogen fertilizer on moorland and heathland soils should be considered (Taylor and Tabbush 1990). Taylor (1991) has updated relevant information.

Broadleaves in uplands. The silviculture of broadleaves in the lowlands has been comprehensively discussed by Evans (1984). In the uplands, although climatic and soil conditions generally necessitate the planting of conifers if acceptable timber yields are to be obtained, even limited use of broadleaved species can do much to diversify wildlife habitats and landscape in commercial forests (Lines and Brown 1982). Furthermore, past planting of broadleaves in Scotland has been more widespread than is generally known (Anderson, M.L. 1952). Low (1986) has given practical guidance on how best to select and establish broadleaved tree species within upland areas. The guidance applies both to afforestation and to restocking but does not relate to the establishment or management of broadleaves aimed primarily at timber production on unusually fertile well-sheltered sites in upland forests – where management of such crops is essentially similar to that for broadleaves on lowland sites.

Locations for broadleaved planting in upland sites will normally be selected as having above-average nutrient status. In consequence, the fertilizer regime applied to a neighbouring conifer crop will usually be adequate, even though broadleaved species generally have higher soil nutrient requirements than the commonly used conifers. Sites to be planted with broadleaves should be treated with broadcast application of phosphate or phosphate/potash mixture as required for the main conifer crop, and should also be included in any subsequent top-dressing operations. Even on poor peatland, where the choice of broadleaved species is restricted to birch and possibly rowan, willow and alder, there is little evidence to suggest that extra fertilizer applications (including additional nitrogen) will produce a worthwhile response in terms of vigour, or extend the range of possible species.

Broadleaves are likely to form only a small proportion of upland forests but their extent is expected gradually to increase, contributing to timber production, wildlife diversification and landscape enhancement.

Failures in establishing. Failures in establishing may be due to many factors: inadequate ground preparation, inappropriate types or qualities of plants, bad silviculture (in handling, planting, weeding and beating up) and lack of protection. Despite much successful practical work in afforestation and restocking there have been far too many failures. In initial planting, to obtain no more than 90% survival is expensive in beating up, weeding, understocking and delayed revenues. Poor survival on clear-felled sites is a problem where significant factors contributing to the low stocking levels are related to the size and type of plants used and the way in which they were handled and planted (Tabbush 1988a).

Good quality plants correctly handled are a prerequisite of successful planting. The behaviour of the root system after planting is strongly influenced by soil temperature, and this largely depends on the method of site preparation. Freedom from compaction too is likely to have a direct effect on survival. RGP integrates many aspects of plant quality, in particular the absence of damage, the size and configuration of the root system and its readiness to develop. High survival rates will be achieved when plants with a high RGP are planted on microsites with favourable conditions for root growth.

One of the most common causes of planting failure is the development of an imbalance between the amount of water lost through the foliage in transpiration and the amount that can be taken up through the roots, leading to excessive moisture stress. There are several ways in which this imbalance can arise, for example as a result of root or top damage during the various stages of plant handling, or of competition from grass and weeds for available moisture. The choice of plant type, size and handling method is guided for the most part by the need to minimize moisture stress after planting.

The remedy is largely in improved plant quality, i.e. the readiness of the plants to survive and grow after planting. Research in this area promises high returns. The risk of failure can be further reduced by appropriate site preparation, principally to provide microsites favourable for early root development. The attendant reduction in planting costs is an added benefit.

In summary, success in planting is generally obtained by preventing loss of establishment potential from reaching levels critical for the particular microsite into which planting will take place. If plants to a desired specification are carefully handled at each stage, and adequately planted, weeded and protected, success rates should be high.

Foresters welcome the time when their young plantations can be regarded as adequately stocked and no longer in danger from weed growth. This may be in the second to fourth years in conifers (earliest on cultivated sites) and in the third to fifth years in broadleaves, largely dependent on species, site factors and management. The height of conifers may then be about 1.0–1.5 m, and of broadleaves 0.75–1.0 m.

For all the silvicultural operations noted in this chapter, it is prudent that records should be made not only of what has been done but also the purpose. Few foresters will serve one estate throughout a long rotation, and managerial changes should not interrupt the course of perhaps a half century or more forestry if the enterprises are to yield their best and the objectives attained.

Costs of establishing plantations. The costs of achieving establishment (by the fifth year) may be ascertained on the pattern in Table 11 which indicates current (1991) average costs, planting being at 2 m (2,500/ha). Contractors' charges would include their profit; their costs of labour, machinery and materials usually have the benefit of scale.

Table 11. Indications of costs of establishing **conifer** plantations to the fifth year.

Operation	Year(s) of operation*	Cost £/hectare (Labour, machinery, materials, and labour oncosts; not overheads)	
		Small lowland sites	Extensive upland sites in England, Wales and Scotland
Ground preparation	0	150	100
Drainage	0	50	150
Fencing	0	450	250
Plants (2,500)	0	250	175
Planting	0	300	150
Initial fertilizing	0	–	125
Initial establishment		1,200	950

Operation	Year(s) of operation*	Cost £/hectare (Labour, machinery, materials, and labour oncosts; not overheads)	
		Small lowland sites	Extensive upland sites in England, Wales and Scotland
Beating up (plants and planting)	1	150	100
Weeding	0–4	400	100
Protection and Maintenance†	1–4	50	25
Total establishment		1,800	1,175

* The year convention adopted in planting models is that the year of planting is always described as year 0, the year before planting as -1, 2 years after planting as year 2, and so on.
† Maintenance of fences and drains, pest control, fire protection, management and insurance.
The above costs exclude overheads, land, roading and road maintenance, and interest on capital. If roads were included the extra cost might be £135-£250/ha. Grant-aid, where relevant (see chapter 15), would reduce the total costs.

Where large-scale plantations are being established, savings in fencing, and other economies of scale, may reduce average costs per hectare by 10 to 20%; conversely, costs for small woods may be 25% higher per hectare established.

Broadleaves (other than poplar and cricket-bat willow). If all or most of the trees in Table 11 under 'Small lowland sites' were broadleaved, the total establishment costs could be £300/ha-£500/ha higher than those for conifers. Establishment of broadleaves is more expensive and usually takes longer. However, the grant-aid may be higher, dependent on the scheme and band involved.

Broadleaves in treeshelters. If broadleaves were established in treeshelters, instead of being fenced, and spaced at 3 m (about 1,100/ha), the costs on 'Small lowland sites' might be as indicated in Table 12.

Table 12. Indications of costs of establishing **broadleaved** plantations using treeshelters to the fifth year.

Operation	Year(s) of operation	Cost £/hectare (Labour, machinery, materials, and labour oncosts; not overheads)
Ground preparation	0	150
Drainage	0	50
Plants (1,100)	0	165

Operation	Year(s) of operation	Cost £/hectare (Labour, machinery, materials, and labour oncosts; not overheads)
Treeshelters 1.2 m (1,100)	0	825
Planting, plus erection of treeshelters	0	375
Initial establishment		1,565
Weeding around treeshelters	0–4	225
Protection and maintenance	1–4	50
Total establishment		1,840

For accurate costing of establishment in Tables 11 and 12, it would be necessary to take into account the years in which the items of expenditure and grant-aid (where relevant) occur: the main limitation in the two Tables is the exclusion of compound interest. (An example of a calculation embodying expenditure and income over time, and compound interest, is given in chapter 14.)

Operational costs of establishing plantations. Costs indicated in Table 13 are generalized and will vary between scale, and individual sites and weather conditions. Contractors' charges may lie within the ranges shown but would include profit; their costs of machinery and materials usually have the benefit of scale. More detailed indications of operational costs, as well as output guide times, are given by Insley (ed.) (1988).

Table 13. Indications of operational costs of establishing plantations.

Operation (Output Guides and Standard Time tables for most of the operations are available from the Forestry Commission Work Study Branch – see chapter 16)	£/hectare (Labour, machinery, materials, and labour oncosts; not overheads)

Ground preparation.

(a) Cultivation. According to terrain, site conditions and method.

Ploughing, spaced furrows at intended planting spacing (usually 2 m) according to whether single mouldboard plough or double mouldboard plough:	60–120
Ripping to break 'pan'	150–160
Scarifying	40–50
Mounding	300–400

(b) Chemical. According to terrain, site conditions, method, and type and amount of vegetation:

Grass/herbaceous weeds (lowest is spot spraying)	50–75
Bracken	80–100
Heather	100–120

Operation (Output Guides and Standard Time tables for most of the operations are available from the Forestry Commission Work Study Branch – see chapter 16)			£/hectare (Labour, machinery, materials, and labour oncosts; not overheads)
Woody weeds			125–175
Gorse, broom, rhododendron			300–400
Drainage. According to terrain, site conditions and method.			60–150
Fencing. According to terrain, site conditions, size and shape of planting area and specification:			
(a) Netting, including erection:		per metre	
Rabbit		£2.00–2.50	
Stock		£2.50–3.50	
Deer		£4.00–5.50	
(b) Treeshelters 1.2 m, including stake and erection: (1,100)			750–1,500
Plants. According to species, provenance, type, size, quality and quantity:			
(a) Container grown conifer seedlings (2,500)			200–450
(b) Container grown broadleaved seedlings (2,500)			300–650
(c) Conifer transplants (2,500)			200–350
(d) Broadleaved seedlings or transplants (2,500)			300–700
(e) Broadleaved seedlings or transplants for treeshelters (1,100)			150–300
Prices of plants by species have been given earlier in this section.			
Planting. According to terrain, species, type and size of plant, site conditions and method. Labour only.			
(a) Container grown conifer seedlings (2,500)			125–150
(b) Container grown broadleaved seedlings (2,500)			150–200
(c) Conifer transplants (2,500)			125–200
(d) Broadleaved transplants (2,500)			225–300
(e) Broadleaved whips 90–120 cm (1,100)			300–500
(f) Broadleaved seedlings or transplants for treeshelters (1,100)			150–300
Initial fertilizing (P or PK). According to terrain and method. Cost of the fertilizer depends on analysis, size of order, and delivery charge:			
(a) Manual application			125–150
(b) Tractor application			120–135
(c) Helicopter application. According to terrain, distance involved, extent and weight			100–110

Operation (Output Guides and Standard Time tables for most of the operations are available from the Forestry Commission Work Study Branch – see chapter 16)	£/hectare (Labour, machinery, materials, and labour oncosts; not overheads)
Weeding. According to terrain, method, species, type and size of plant, type and amount of vegetation, and number of times undertaken:	
(a) Manual application — Each weeding	100–200
(b) Mechanical application — Each weeding	50–100
(c) Chemical application — Each weeding	40–180
Beating up. According to terrain, gaps, species, type and size of plant, site conditions and method. Plants and planting (15%–25% of original cost).	75–200
Protection and Maintenance: Annually	5–10
Insurance: Storm and fire: according to location, species and age: Annually	4–6

2. Establishing special tree crops

In addition to establishing plantations – by way of afforestation or restocking – there are other needs for establishing which may fall within foresters' sphere of management. They are discussed below.

Decorative quality broadleaved timber

This enterprise usually refers to veneer and high quality joinery timber (discussed in chapter 10). Exceptional butts of most broadleaves are suitable for these markets but some, notably cherry and walnut, can be open-grown specially for decorative timber. Here it is the individual tree eventually attaining large size that is valuable. However, decorative quality timber occurs in all the major broadleaved species – ash, oak, sycamore and, to a lesser extent, beech. Identification and separate marketing of such trees at harvesting is always worth while. Some turners, cabinet makers and fine craftsmen will seek to feature the unusual in their product, and an ill-shapen, gnarled tree, can occasionally be sold for this purpose.

Specifications of butts for decorative quality timber are listed by Evans (1984). They are exacting, but the very high prices commanded – up to several £100/m^3 – mean that this market should never be ignored. Since the supply of high quality tropical hardwoods is gradually diminishing, a firm market can be expected to continue for all home-grown good quality hardwoods. Not all the desirable characteristics which dictate high quality are visible externally and selection of superior butts after felling must follow the initial identification of good stems. Owing to the great natural variability among broadleaved trees only a very small proportion make the top grade, though by pruning and care the proportion of good stems in a stand can be increased. Encouragement of open growth to obtain fast diameter increment and well-developed crowns may be desirable. This is usually achieved by planting at wide (6 to 8 m) initial spacings to avoid competition, or, where trees have been closely spaced at planting, by frequent thinning to favour suitable quality stems.

Ornamental (decorative) foliage

Several evergreen forest trees have foliage and branches that are saleable to florists, mainly for wreaths and decoration. The required conifers are mainly silver and blue firs (particularly noble fir and caucasian fir), western hemlock, western red cedar and Lawson cypress; of broadleaves, laurel, holly and eucalyptus (blue gums and white gums). The eucalyptus (mainly *E. gunnii*) are valued for their attractive sprays of glaucous or bluish juvenile foliage. The firs are much in demand, particularly those origins/provenances with specially selected characteristics such as glaucousness and greenness. The market in Britain is rather fragmented and unco-ordinated, but there is a continuing demand for export of firs to Denmark and Germany.

Trees for the purpose can be grown as single species blocks, as coppice (only broadleaves) to keep the height down and maximize foliage production, or as part of a mixed planting with the intention of removing the foliage component before its quality is reduced when the timber tree canopies start to form.

In some locations, purchasers of foliage will trim conifers (in effect brashing and low pruning) and this is a benefit to the grower if undertaken efficiently and carefully; in addition a sum per tree, per bundle or per hectare is received, the amount depending on species, quality, undergrowth and access.

Poplars

A resurgence of interest in poplar has arisen from the very fast growths and early returns obtainable when these trees are grown on agricultural soils. The best known poplars are cultivars propagated clonally by vegetative means. Some are raised only for screening, shelter and ornament, while others are grown for timber production, as well as to protect horticultural and agricultural crops. The most informative guide to poplar is Jobling (1990).

Poplar are fast growers and need full daylight, and wide spacing if grown for timber. They are very exacting and require high fertility, ample well-oxygenated soil, moisture and little competition from weeds; and a pH above 5.0 (not above 7 for balsam poplars). The site must have available moisture at rooting-depth throughout even the driest summer. Suitable sites for poplars are limited; they do best on loamy soils in sheltered situations, rich alluvial or fen soils, both well-drained and well-watered, and banks of watercourses. To be avoided are high elevation exposed sites, shallow soils, acid peats and heathlands. Stagnant water is fatal but occasional winter floods do no harm. In the north of Britain, poplars should be limited to the most favourable sites. Successful plantations in Scotland are rare except for *P. trichocarpa* hybrids. New clones selected for northerly conditions could have great potential. Balsam poplars withstand slightly more acid soils than the Black poplar hybrids and are more suited to the cooler and wetter parts of lowland Britain.

Because of the importance of canker resistance only a relatively few poplar are in the Forestry Commission 'approved list' for timber production. The main Black hybrids, *Populus* x *euramericana (P. deltoides* x *nigra)* are: 'Casale 78', *P.* 'Eugenei', *P.* 'Gelrica', *P.* 'Heidemij', *P.* 'Robusta', *P.* 'Serotina'. Approved Balsam poplars are *P. trichocarpa* 'Fritzi Pauley' and 'Scott Pauley', and *P. tachamahaca* x *trichocarpa* 'Balsam Spire'. Belgian hybrids which are in the 'approved list' are *P.* x *euramericana* 'Primo', 'Ghoy', 'Gaver' and 'Gibecq'; and *P.* x *interamericana* 'Beaupre' and 'Boelare' (Potter *et al.* 1990). The 'list' is no longer for grant-aid purposes but for commercial propagation for plantation purposes, i.e. the Forestry Commission can register commercial stoolbeds of those clones. Approval for grant-aid requires a clone to come from a registered source *and* to be suitable for the site. The 'list' is likely to be amended fairly frequently as new clones are introduced and old ones deleted.

The raising of poplars as rooted plants or sets (long cuttings) or 25 cm cuttings by vegetative propagation, has been described in chapter 2. Under Common Market regulations those used for timber growing must be raised from officially approved reproductive sources recorded in the National Register of Basic Material.

To form plantations, one- or two-year-old plants, sets about 1.5–2.0 m tall, or 25 cm cuttings are planted. Spacing for fastest growth of sawlogs or veneer logs needs to be very wide, usually about 7 x 7 m (204/ha) or 8 x 8 m (156/ha); sometimes a spacing 'on the triangle' instead of 'on the square' is followed. Closer-spaced plantations (i.e. 3 x 3 m) are likely to be established using unrooted cuttings into suitable cultivated ground. The maximum planting distance eligible for grant-aid is 8 x 8 m, though the Woodland Grant Scheme rates will be reduced pro rata from those payable at the 3 x 3 m maximum permitted spacing for broadleaves. At the recommended spacings, most trees in the plantation may be expected to reach veneer log dimensions before competition seriously inhibits diameter growth, so obviating the need for thinning.

Rooted plants require pit-planting to accommodate the roots; sometimes tractordriven soil augers are used to make planting holes. Sets are easier to plant, being placed in a crow-bar hole ensuring that there is no gap between soil and set. On sites that remain wet for much of the planting season, planting can be on mounds of soil. Depth of planting for sets should be sufficient to ensure permanently-moist soil for the developing root-system and for stability during the first two seasons. The minimum should be 60 cm. When 25 cm cuttings are used, they should be inserted almost to ground level. Young poplars, other than hybrids with very large leaves, especially on exposed sites, rarely need to be staked. Usually they have to be protected by a suitable perimeter fence or individually guarded against livestock, rabbits and deer, using sleeves of netting or treeshelters (which, however, may create problems of abrasion or stability due to fast crown growth).

Weed competition seriously reduces the growth of newly-planted poplars; and hence all weeds around young plants and sets, to a diameter of at least 1 m, should be rigorously removed preferably by herbicide application. Mulches of locally cut vegetation, straw or shredded bark, are frequently used to check weed growth but need replacement. An alternative method is to place a sheet of an opaque, durable material such as black polythene around the tree immediately after planting. Agricultural and horticultural cultivators can be employed to till the soil to obtain weed-free conditions; and repeated shallow cultivation may be needed. As an alternative, residual herbicides can be applied overall after planting, in the dormant season.

Much the most serious disease of poplars is bacterial canker caused by *Xanthomonas (Aplanobacter) populi*. The principle control measure is to use canker resistant clones, named above. Leaf rusts, leaf spot diseases and mosaic virus can do harm. Usually there are no serious insect pests but Poplar leaf beetles and White Satin moth occasionally cause harm; so do the wood boring larvae of some moths.

Production of veneer quality stems is dependent upon regular pruning, which may be undertaken throughout the dormant season and again in mid- to late summer. (April to early June should be avoided: the bark tissue tears too easily.) To minimize epicormic growth, pruning should never be excessive (see Figure 14). Annual pruning of stems is necessary from about the third year according to vigour and growth. Sometimes every other year will suffice. To be effective, pruning should remove just one main whorl and branches up to the base of the next on each occasion. Start in the third year and leave two whorls, then continue pruning one whorl in each of two out of three years. Probably four or five operations are required to reach a minimum of 6 m of pruned stem. Pruning should be completed by about the eighth or ninth year in fast-growing plantations. Final pruned height is usually about 6m, achieved by using pole-saws, without recourse to special tools or climbing equipment.

Poplars grow rapidly if the variety is sound and the quality of the site is good. Beaton (1987) describes the establishment by Bryant and May (Forestry) Ltd during 1965–78 of some 1,600 hectares of poplar plantations in Herefordshire, Worcestershire, Bedfordshire and East Anglia. It was of hybrid poplars using both rooted plants and unrooted sets, spaced triangularly at about 8 x 8 m (185/ha) for match-splint production. The spacing allowed inter-row cropping, with cereals for the first eight years, followed by grass seeding and summer grazing for most of the remaining poplar rotation. Following the demise of the home match-splint industry in 1978 the poplar enterprise was discontinued.

Growth and yield. Favourably located *P. trichocarpa* cultivars and hybrids with *P. deltoides* can achieve height increments of 1–2 m in the first season after planting, and can continue to grow at this rate until the end of the twelfth season to reach 25 m or more, and some attain over 30 m in only eighteen years from planting. Some of the new hybrids grow even faster. *P. x euramericana* cultivars are generally less vigorous and seldom grow more than 1.5 m in a season even during the first ten years.

As to yield, Evans (1984) records that on fertile sites with a stocking of 200–250 stems/ha, volumes of 250 to 300 m^3/ha in twenty five-years (Yield Class 10–12) are achieved. At lower stockings, 150–200 stems/ha (trees 8 m and 7 m apart respectively) yields are 220 to 270 m^3/ha at twenty-five years (Yield Class 10). On sites of below average quality, yields are 200 to 230 m^3/ha at thirty to thirty-five years (Yield Class 6–8). When trees in the stand are high pruned to about 6 m, volume to pruned height (veneer quality wood) is more than 80% of total volume. Fast-grown trees begin to produce logs of veneer size at twelve to fifteen years of age (the minimum top diameter specification for veneer logs is usually 25.4 cm,

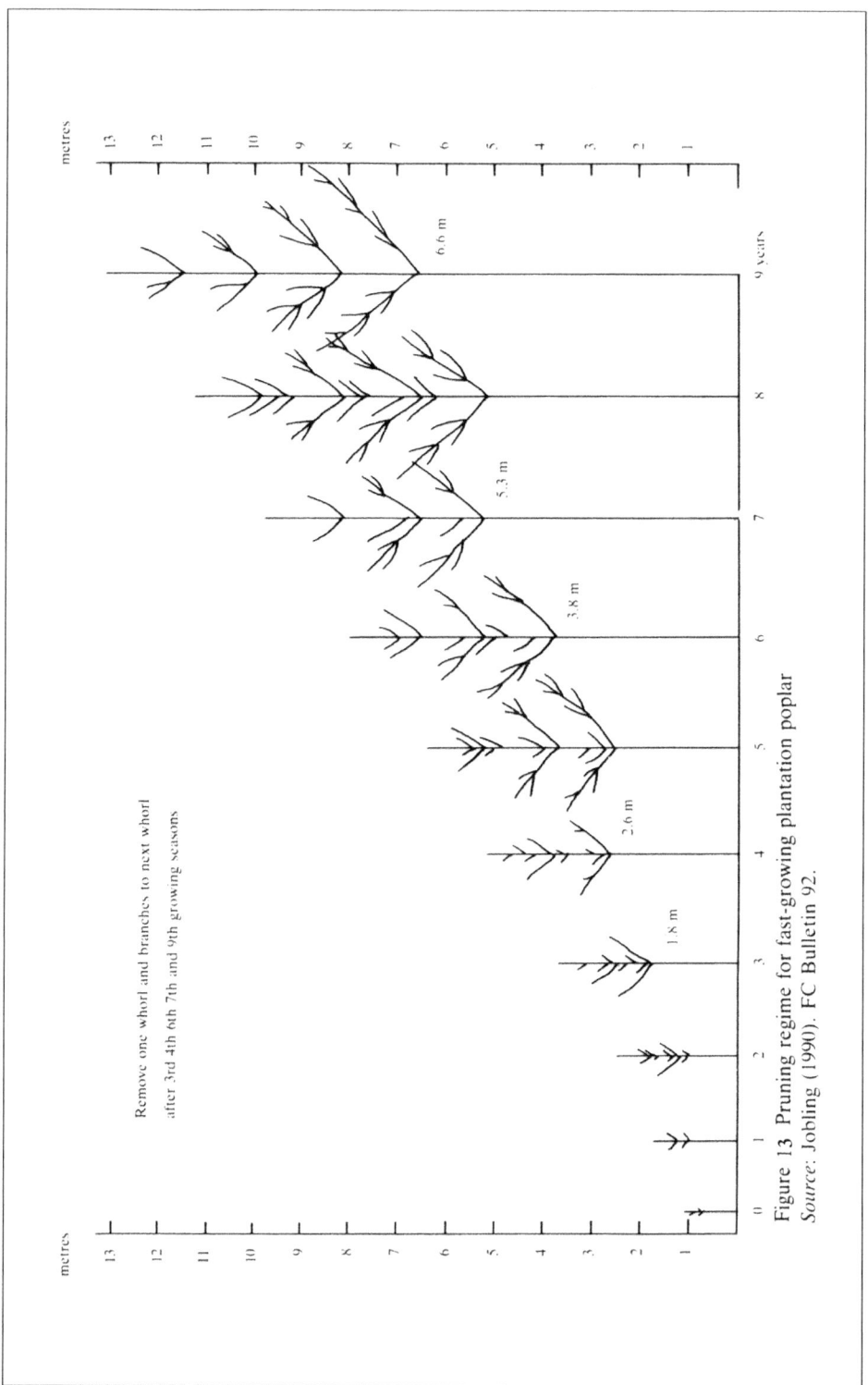

Figure 13 Pruning regime for fast-growing plantation poplar
Source: Jobling (1990). FC Bulletin 92.

minimum length 3 m). The range of yield class of *P. euramericana* hybrids is 4–14 (average 6), i.e. their maximum mean annual increment is 4–14 m^3/ha/year; and of balsam poplars 4–16 (average 6). An example of General Yield Class and of Production Class Curves (explained in chapter 12) is given in Figure 15. The early volume culmination should be noted.

Yield tables for unthinned stands of yield class 14 poplar planted at 4.6 and 7.3 m spacings are shown in Table 14. However, it should be borne in mind that volume/hectare would in practice be higher because the poplar yield tables are calculated at 7–8 m spacing. Maximum MAI at closer spacing is greatly increased with yield class.

Table 14. Yields of unthinned stands of poplar YC14 at spacings of 4.6 m and 7.3 m.

YC14 4.6 m spacing

Age yrs	Top height	Trees /ha	Mean dbh	B/A/ ha	Mean vol	Vol/ ha	MAI vol/ha	Age yrs
7	11.8	470	16	9	0.09	40	5.7	7
12	18.5	465	24	21	0.30	139	11.5	12
17	24.2	449	30	33	0.60	272	16.0	17
22	28.8	439	36	45	0.97	426	19.4	22
27	32.4	431	40	55	1.32	571	21.1	27
32	35.1	425	44	64	1.64	696	21.7	32
37	37.2	420	46	71	1.89	794	21.4	37
42	38.7	416	48	76	2.09	868	20.7	42
47	39.8	412	50	80	2.24	924	19.7	47
52	40.6	409	51	83	2.37	969	18.6	52
57	41.3	406	52	85	2.47	1003	17.6	57

YC14 7.3 m spacing

Age yrs	Top height	Trees /ha	Mean dbh	B/A/ ha	Mean vol	Vol/ ha	MAI vol/ha	Age yrs
7	11.8	185	17	4	0.11	21	3.0	7
12	18.5	185	27	11	0.43	79	6.6	12
17	24.2	185	36	19	0.89	164	9.7	17
22	28.8	185	43	27	1.42	263	11.9	22
27	32.4	185	49	35	1.95	360	13.3	27
32	35.1	185	53	41	2.41	447	14.0	32
37	37.2	185	56	46	2.80	518	14.0	37
42	38.7	185	59	50	3.11	576	13.7	42
47	39.8	185	61	53	3.36	621	13.2	47
52	40.6	185	62	56	3.54	656	12.6	52
57	41.3	185	63	58	3.68	682	12.0	57

Source: Insley (ed.) 1988. FC Bulletin 80 and Christie, J.M., 1994. FC Technical Paper 6.

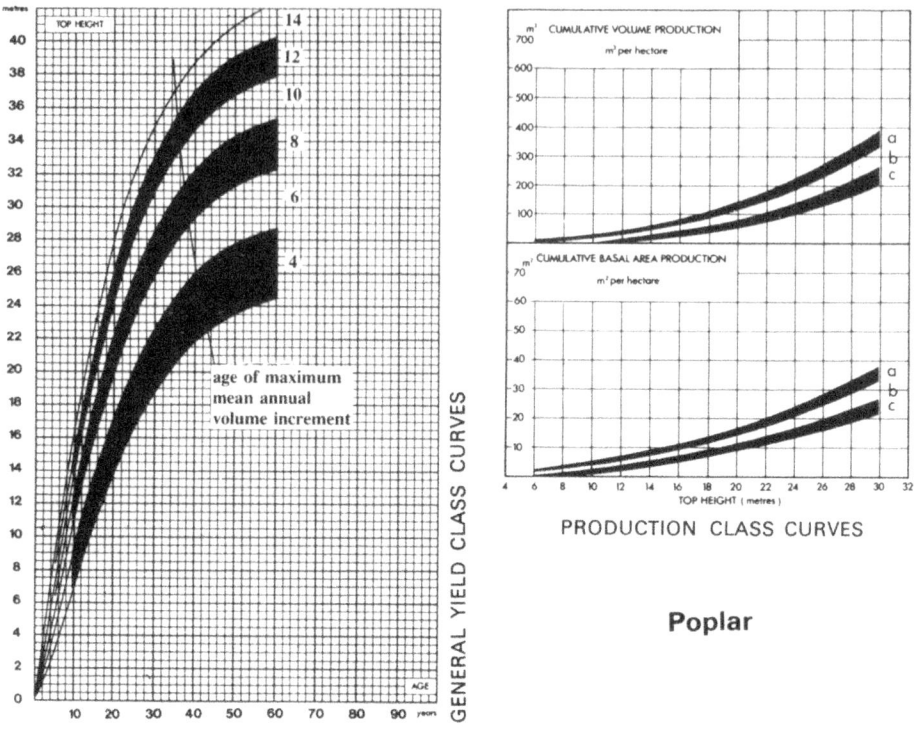

Figure 14. Poplar: Top height/age curves for estimating General Yield Class; and Production Class curves
Source: Edwards and Christie (1981). FC Booklet 48

An indication of the cash flow for poplar growing is given in Table 15. The year of planting is conventionally described as year 0. Prices for poplar timber, delivered in, are (1991) £25-£40/m³. An average return to the grower may be £10-£20/m³; and £15/m³ in sustainable markets seems a reasonable expectation. As another indication of likely income, a plantation of poplar Yield Class 14, felled at thirty-five years, would produce around 480 m³/ha of logs with a top diameter underbark of 18 cm, a total yield of about £7,200.

Wood properties of poplar are noted in chapter 9. The current and future market for high quality, pruned poplar is uncertain (FAO 1980; Campbell and Aaron 1980; Crichton 1982; Crichton and Elliott 1982; Jobling 1990); there is no substantially priced market in sight to replace the comparatively high-priced match-splint market. However, large clean poplar is needed for peeling and veneering for packages, chip baskets and vegetable crates. Much is used for pallets and some for coffin boards. A limited market is sawn mining timbers including waggon and barrow bottoms, because of poplar's high resistance to abrasion. Small-sized poplar may be readily disposed of to the hardwood wood wool, pulp and boardwood industries, and sawlogs to sawmillers and merchants.

A renewed interest is being shown in the development of farm-forestry to reduce the production of surplus food (see chapter 7). For farmers looking for alternative land uses, given even modest market outlets, poplars clearly have some distinct advantages in comparison with many traditional forest crops. High yielding pulpwood crops (200–300 m³/ha) on rotations of twelve to fifteen years, i.e. Yield

Table 15: Poplar: An indication of cash flow £/ha. Spacing 8 x 8 m triangular pattern (180/ha) guarded against deer and sheep, growing rather better than GYC 12, and producing 300 m³/ha in twenty-five years.

Year of operation or receipt of income	Costs/ha				Receipts/ha		Costs/Income/ha
	Plants Planting Guards £	Weeding £	Pruning £	Supervision £	Woodland Grant Scheme £	Timber £	Cash Flow £
0	560	112		36	156		-552
1		56					-56
2		56		36			-92
3		112	55				-167
4			110				-110
5					45		+45
6			220	36			-256
7			220				-220
8			110				-110
9			110				-110
10			55		22		-33
11			55				-55
24						4,500*	+4,500
Totals	560	336	935	108	223†	4,500	
	Costs: £1,939				Receipts: £4,723		Net Return: +£2,784

Net return over twenty-five years = £2,784 ha. IRR = 6.5% approx.

With FWS Payments ‡ = £5,634 ha. IRR = 11.5% approx.

* 180 trees pruned and cleaned to 6 m with average tree size approx. 1.65 m³ at 25 years (300 m³ @ £15 = £4,500).

† WGS: grant is calculated in the £1,375 band and rounded to £223 (£1375 x $\frac{180}{1111}$).

‡ 24 annual payments of £114 (i.e. £190 less 40% income tax) = £2,850.

Any grazing or intercropping is excluded. The costs exclude land (interest or rent), interest on capital and Management Grant.

Source: Tilhill Forestry Limited (Arnold Beaton).

Class around 30, appear quite feasible, the highest yields being obtained with trees planted 2.2 to 4 m apart. Certain poplars can be planted at even closer spacings, and on very short coppice rotations, for fuelwood and energy biomass (see chapter 7).

The use of poplars in agroforestry is noted in chapter 13, section 10. Grant-aid is available for wide-spaced poplars even when arable inter-row cropping or grazing is intended, provided the Forestry Commission is satisfied that the agricultural regime will not be detrimental to the growth or health of the trees and will not run counter to any environmental considerations. The clones used must be judged by the Commission to be suitable for the site – this implies that they can be identified and one piece of evidence which assists in this task is a certificate of origin. Applicants are advised to limit the extent of planting of the Belgian clones until more is known about their site tolerance and to restrict the area of planting of any single clone or of closely-related clones because of the potential risk from new pests and diseases which would result in the loss of entire stands (Potter *et al*. 1990; Jobling 1990). A Poplar Working Group has been set up by the Forestry Research Coordination Committee. The Secretary is P.M. Tabbush, Forestry Commission, Alice Holt.

Christmas trees

Annually just under 5 million home-grown Christmas trees are sold in Britain. In addition over 1 million are imported from Denmark, Belgium and the Netherlands. About 0.25 million are supplied by the Forestry Commission. The traditional tree is Norway spruce, but in recent years about one in five have been pines and firs.

The enterprise can be profitable, but is not without risk, particularly because of the cyclical nature of the market, and quality is becoming increasingly important. Other risks include climate, pests and cultural techniques. Methods of production range from being lucrative supplements to general forest practice to highly-specialized enterprises on their own. Christmas trees are sometimes planted in a matrix of timber species to provide early financial returns to help offset establishment costs of the latter, or planted pure on relatively small areas. Tops from thinnings cut in late November or December are sometimes saleable. The enterprise remains the only remunerative crop suitable for siting beneath transmission lines, or where views must not be obscured. Much planting is now on farmland.

Along with Norway spruce, small numbers of Douglas fir, grand fir and lodgepole pine have entered the market, but notably there have been increases in planting and marketing of Scots pine, Noble fir, Caucasian fir and a few White fir. The three firs are more expensive to produce, being costly as transplants and slow-growing during the early years, hence taking longer to mature. However, with their strong branches and attractive blue/green or grey foliage (Caucasian fir, glossy dark green) they are much sought after; and, along with Scots pine, they retain their needles longer thus commanding high prices, probably treble those of Norway spruce.

Norway spruce and firs when planted pure as Christmas trees (in which case forestry grant-aid is not available) are generally spaced at about 1 x 1 m (10,000 ha), and Scots pine at about 1.25 x 1.25 m (6,400/ha) to allow for annual shaping by shearing. Sometimes the trees are planted in various patterns of mixtures with oak, ash, western red cedar or other timber trees; or, if the crop is of one species only, say Norway spruce, the grower sometimes ensures that sufficient are left standing after each Christmas removal to conform to conventional silvicultural spacing to produce pole thinnings and a final crop of timber. In both these cases grant-aid, probably pro rata, may be available subject to specific conditions and verification.

To grow Christmas trees profitably, experience and skill are desirable. Success comes chiefly by attention to six factors: (i) a suitable site, not too fertile (although even on good agricultural land vigorous growth can be controlled by shearing); (ii) care in expenditure; (iii) constant tending, especially in total weeding; (iv) protection from insect defoliation; (v) shearing; and (vi) intensive marketing. The trees need plenty of light. They do not appreciate an overstorey but will tolerate a little shade, although in ratio to the increase of it there will be a reduction in quality and value due to thinness and softness of the foliage. The three main firs are more tolerant of shade. Poor land such as that covered with heather, or chalk downs with shallow soils, is quite unsuitable for Christmas trees. The market calls for healthy specimens that have grown steadily and not so fast as to become 'leggy' or have lengthy leaders. Trees which have become 'checked' or overshaded attract few buyers. Theft is a real risk and hence species suitable for use as Christmas trees should not be planted near public roads unless an occupied house stands nearby.

In planting Norway spruce or Scots pine, well rooted 30–45 cm 1 + 1 are ideal. In the case of the main firs, the size is likely to be smaller and the age perhaps greater. In Norway spruce, slower-growing origins/provenances (e.g. the Schwabische region of south Germany) are preferred to the faster-growing origins/provenances (e.g. Toplita in Romania) which produce wide-whorled trees. If a nursery planting system is followed, ground preparation will be by ploughing and initial weed control by simazine or glyphosate. The trees must be carefully planted, and fully weeded each year, a quite costly operation according to site conditions and method used.

Shearing (shaping) is sometimes undertaken from the fourth year, usually during the dormant period (but Scots pine at the beginning of July). Each species requires different timing of shaping and pruning, nearly always linked to the trees' different growth patterns. (Norway spruce can be sheared in August to March.) This operation is the cutting of part of the leading shoot (if too long) and ends of branches to produce a bushier tree or better shape and improve the percentage of saleable trees. Sheared trees usually command enhanced prices.

Hazards in the growing of Norway spruce include the possibility of insect defoliation. Green spruce aphid *Elatobium abietinum* may appear on the underside of needles in spring or autumn; the needles become banded yellow then turn brownish and drop off. Trees should be inspected from September onwards, and insects sprayed when seen, using Perimicarb ('Aphox'). Pineapple gall woolly aphid *Adelges abietis* may distort new shoots by gall formation (green turning to brown). As a precaution, spray in November to February before gall formation in spring, using Gamma HCH. Other hazards are drought, frost, fire, animals and man; they can reduce the eventually harvested crop to well below the number of trees planted. The firs, as yet do not have severe problems with aphid attack.

The growing cycle of Christmas trees is about four to seven years for Norway spruce and Scots pine, and six to ten years for the three main firs, depending on growth and size required. The height at first removal is around 0.75 to 1.0 m. There is an art in selection from the crop to obtain the most profitable results. Taking out one tree may allow four others to improve. Sometimes it is problematical whether trees should be removed in one particular year or allowed to remain for another and thereby realize an extra sum each. The most popular sizes appear to be 1.0–1.4 m and 1.5–1.8 m; such trees, when sold, standing, to a wholesaler, command prices of at least £2.00 (but up to £1.00 per 30 cm of height when delivered to a retail outlet).

Trees produced by specialist growers (e.g. Yattenden Estates Ltd. in Berkshire, at Leigh Sinton near Malvern and Elveden Estate near Thetford) command premium prices. Firs command prices up to £3.00 per 30 cm of height. Generally, prices are increased if freshly-lifted trees are immersed in a 0.5 per cent solution of sodium alginate to reduce needle fall. It is doubtful how effective anti-transpirant can be. Except when sold individually, trees are often bundled in quantities varying with their size and quality. Some added-value is gained by using a baling machine, inserting in sleeves of netting, or fixing in wood blocks.

An indication of the cash flow for Christmas tree production with Norway spruce at 1 x 1 m (10,000/ha), 8,000 saleable, is given in Table 16. The year of planting is conventionally described as year 0.

Table 16: Christmas trees (Norway spruce): An indication of cash flow £/ha.

Operation	Year of operation or receipt of income	Description	Costs/ incomes/ hectare* Cash Flow £
Ground preparation	0	Chemical weed	-200
Fencing			-500
Plants (Norway spruce)		10,000 @ £80/1000	-800
Planting		10,000 @ £50/1000	-500
			-2000
Beating up (10% labour)	1	1,000 @ £85/1000	-85
Plants		1,000 £85/1000	-85
Weeding		Chemical and manual	-250
			-420
Weeding	2	Chemical and manual	-275
Weeding	3	Chemical and manual	-275
Weeding	4	Chemical and manual	-200
Shearing		10,000 @ 4.5p	-450
			-650

Operation	Year of operation or receipt of income	Description	Costs/ incomes/ hectare* Cash Flow £
Shearing	5	10,000 @ 4.0p	-450
Weeding		Chemical and manual	-200
Harvesting		2,000 1.0–1.25 m @ 50p	-1000
Sales		2,000 1.0–1.25 m @ £1.75	+3500
			+ 1850
Shearing	6	6000 @ 5.5p	-330
Weeding		Chemical and manual	-175
Harvesting		4,000 1.25–1.5 m @ 55p	-2200
Sales		4,000 1.25–1.5 m @ £2.25	+9000
			+6295
Shearing	7	2,000 @ 6.0p	-120
Harvesting		2,000 1.5–1.75 m @ 60p	-1200
Sales		2,000 1.5–1.75 m @ £3.00	+6000
Reinstatement of site			-1200
			+2680

	£/hectare
Sales	18,500
Costs	11,295
Profit	£7,205

The result should be spread over 8 years, as one year's fallow is required to get the land ready for another rotation, mainly to facilitate effective root clearance.

* Excludes: Land (interest or rent); Interest on capital; Supervision/overheads; Spraying against insect attack; Theft.

IRR = 29.5% (over 7 years)

For calculation of the true financial result of the enterprise, one would need to take into account the years in which the items of expenditure and income occur: the main limitation in the Table is the exclusion of an appropriate rate of compound interest. (An example of a forestry calculation embodying expenditure and income over time, and compound interest, is given in chapter 14.)

Income tax incidence of a Christmas tree enterprise is as follows: when the trees fully occupy a substantial area, profits are taxable (as for a market garden) under Schedule D. However, the incidence is different (i) when the trees are planted in mixture with other tree species intended to produce a conventional crop of timber; or (ii) when the grower always leaves growing sufficient of a pure crop after each removal as Christmas trees, to comply with conventional silvicultural stocking and ultimately to provide pole thinnings and a final crop of timber; or (iii) when tops only of pole thinnings are sold. In all three cases the net income therefrom is tax-free.

The growing of Christmas trees, unless undertaken by specialist production, may not always be remunerative. Prices are very market and quality dependent; the current situation may change rapidly with an increase in supply. Plastic trees, and trees imported from Europe, can reduce the demand for the British grown product. However, in spite of the hazards, difficulties and uncertainties, there are some woodland owners who have made as much income from Christmas trees as from all their other minor forest products. Furthermore, the plantations have usually provided cover and nesting facilities for pheasants and partridges.

The British Christmas Tree Growers' Association, 12 Lauriston Road, Wimbledon SW19 4QT (Tel. 081–946–2695) advises members on the management of their plantations and provides marketing information and publicity.

Cricket-bat willow

Cricket-bat willow *(Salix alba* cv. *'coerulea')* is the only cultivar of willow planted for the cricket bat trade. The timber is tough, fibrous, light in weight, and white to pale yellow in colour. The cultivation complements farming rather than forestry. The attempt to grow it successfully is only worth while on fertile low-lying ground with a good supply of water in the soil, and in areas with a moderately warm and dry climate (not more than 900 mm per year rainfall). The willows grow best on deep permeable alluvial loams or fertile boulder clays of high pH. Waterlogged soil is unsuitable. The eastern counties of England, particularly Essex and Suffolk, the Thames Valley and Hampshire, are the main growing areas. The willows are relatively quick to mature (rotations of fifteen to twenty years), and hence a satisfactory profit should be realized.

Sets (long cuttings) raised in a nursery stoolbed, reach in four years a height of 3–4 m or more and a diameter of about 3–5 cm. A straight unblemished stem of 2.3–3 m is required. Sets are cut in the winter for planting in that season. The price at the nursery is about £5. They should be planted about 75 cm into the soil at wide spacings, say about 12 x 12 m, i.e. approximately 65–70 per hectare. Alternatively they may be planted at say 10 m apart, about 1 m away from one or both banks of a watercourse. In planting, the top section of soil, about 30 x 30 cm, is removed to a depth of 30 cm; a hole about 8 cm in diameter is then made with a crow-bar to a further depth of 45 cm, the set inserted, and the soil rammed firmly and the top section replaced.

The trees must be fenced against livestock and rabbits. The stem up to 3 m must be kept free of side shoots by disbudding each year in early spring, three times at least in early years, the aim being to keep the bole unblemished and free from knots. An annual top-dressing of organic manure usually promotes even more rapid growth. Detailed information as to treatment is available (Bryce 1976; Porter 1986) and significant help can be obtained from willow specialists – J.S. Wright and Son Ltd. of Great Leighs, Essex.

The serious Watermark Disease of this willow, caused by the bacterium *Erwinia salicis,* is now closely controlled by the Watermark Disease (Local Authorities) Order 1974 in Essex, Bedfordshire and Suffolk (Preece 1978; Gibbs 1987; Patrick 1991). When prevalent, the disease stains the wood and destroys the fibres making it more liable to splinter thus causing it to be absolutely unsuitable for its intended purpose. Control is chiefly by ensuring that the sets are entirely free from infection, and by burning or chemically killing the foliage, branches and stumps of diseased trees. The Watermark Disease Group is based at the University of East Anglia.

At the age of fifteen to twenty years the willow is generally ready for sale. The bole should have achieved about 120–135 cm circumference at 1.5 m from the ground (or at least greater than 35–45 cm diameter at breast height). An indication of the cash flow for cricket-bat willow growing is given in Table 17. The year of planting is conventionally described as year 0. The example assumes that the willows will be planted at 12 m spacing (say 70 per hectare), and will be felled at year 20.

Table 17: Cricket-bat willow: An indication of cash flow £/ha. The costs exclude land (interest or rent), and interest on capital.

Operation	Year(s) of operation	Labour	Machinery	Materials	Costs/hectare Cash flow £
Ground preparation	0	-50	-2		-52
Fencing and sleeving	0	-500	-50	-500	-1,050
Sets and planting (£5 + £1 each)	0	-70		-350	-420
Protection	0			-100	-100
Weeding	0 to 2	-15		-5	-20
Disbudding	1	-25			-25
Disbudding	2 and 3	-10			-10
General Maintenance (£10/ha/yr)	4 to 20	-170			-170
					-1,847

For calculation of the true financial result of the enterprise, one would need to take account of the years in which the items of expenditure and income occur: the main limitation in the Table is the exclusion of an appropriate rate of compound interest. (An example of a forestry calculation embodying expenditure and income over time, and compound interest, is given in chapter 14.)

As to income, a mature cricket-bat willow is worth about £50–£65 +. The 70 trees per hectare may yield £3500–4500. Only the basal 2–3 metres of bole are usually bought for bat-making. The worth of a tree is likely to be indicated by the number of 'rolls' (0.70 m in length) it is likely to produce. The value of a roll varies with the girth of the tree as measured at 1.4 m above ground. Usually, three, four, rarely five, rolls are cut from each tree. Carleton Wright, a widely experienced willow grower and bat-maker, indicates the following minimum values for top quality, well-grown trees: 'Circumference: 52–56 ins: £15/roll, 57–61 ins: £18/roll, 62–66 ins: £20/roll.' Thus it is often prudent to allow trees to grow on for a year or two, thereby increasing their size and value. The approximate standing price based on the above prices would amount to £110–160 m^3. Some of the remaining timber on rare occasions might be sold for chip baskets, pulp and boardwood, and firewood. Often the waste approaches 50% of the whole tree (Wright 1987).

Growing of cricket-bat willow is a specialized enterprise which can be successful only when efficiently undertaken.

Shelterbelts

Shelterbelts play important roles in providing shelter to farm crops, pastures and livestock, as well as in landscape and sporting. Using trees to reduce windspeed locally is important to both arable and livestock farming. The main beneficial effects of shelter are by way of increasing crop yields and reducing or preventing soil erosion. The benefits are difficult to quantify in financial terms. Increases in crop yields of around 5% can be expected (though the land occupied by the belt itself is unavailable for crop production). Benefits to livestock are noted *infra*. In the western lowlands excessive shelter can be harmful, slowing the drying of grain and increasing damage from disease in damp years. In eastern sandy/loamy sand/littoral soils shelterbelts prevent soil drift at sowing and germination time.

In the uplands (particularly in areas of high rainfall, acid soils and severe exposure to wind), benefits of shelterbelts include: (i) shelter for the homestead and buildings, as well as for stock (thereby reducing disease problems caused by climatic stress); (ii) increased grass production in early spring (by raising air and possibly soil temperatures); (iii) allowing the movement of stock to upland areas earlier in the spring (thereby releasing in-bye ground for silage or hay production); (iv) aiding management of upland pastures by creating sheltered sub-divisions; (v) providing poles, fencing materials, firewood and sawlogs for on-farm use or sale; (vi) providing cover for pheasants (where not too exposed); and (vii) providing wildlife habitats and landscape features.

Shelterbelts only give maximum sustained benefit if they are carefully sited, designed, established and maintained. Their subsidiary or complementary roles of enhancing landscape, improving sporting and creating new habitats for wildlife should be constantly borne in mind. ADAS Leaflets *Shelterbelts in the Uplands* (1988) and *Shelter in the Lowlands* (1988) comprehensively describe factors necessary for success. Poorly designed or wrongly sited belts can damage crops by increasing turbulence or creating overly damp conditions from too much shelter, a particular problem in the lowlands. Other guidance on management of shelterbelts is given in Ministry of Agriculture Leaflet *Shelterbelts for Farmland* (Fixed Equipment of the Farm Series, No. 15). Conservationists urge the use of native trees wherever practicable – oak, cherry, beech, alder and aspen – but site and other factors must be taken into account.

Some shelterbelts are purposefully left open (i.e. unfenced) to enable livestock to enter for shelter and perhaps to gain modest grazing. Although this object can conflict with sound silviculture, and will certainly inhibit regeneration and may lead to bark stripping, most such belts and other small woods can be managed to provide both forest produce and some shelter provided soils are coarse textured and freely-drained and the trees are large with thick bark. However, sheep and cattle are rarely compatible with sound silviculture or woodland sporting; and use of belts should be short and strictly controlled.

Many shelterbelts established as even-aged plantings become draughty and inhospitable as they mature. The key to improving their effectiveness as shelter, and their value to wildlife, is to create a graded edge incorporating shrub species. A great many shelterbelts are too narrow and uniform. When mature they do not admit of anything but clear-felling and restocking across their full breadth. Owners are consequently loath to spoil the landscape and amenity by wholesale removal; the belts become hollow and the wind sweeps unhindered between the tall trunks. With broader belts the leeward side could be felled and restocked, protected by the windward side; thereby some shelter and landscaping could be continued.

Design of shelterbelts is important, the porosity, height, length, and orientation being the main considerations (Savill and Evans 1986). The higher a shelterbelt, the greater is the distance which is protected downwind and, to some extent, upwind as well. In general, a dense barrier provides some sheltering effect to a distance of 10–15 shelterbelt heights downwind. However, by increasing porosity to about 50%, the downwind influence, though slightly reduced in degree, can be extended over a larger distance of 20–25 heights, and also the somewhat greater wind penetration reduces turbulence with important consequences for lodging of crops and incidence of snowdrifts. The longer a wind-break the more constant its influence; if it is too short, or has gaps, jetting effect may increase windspeed at the ends and near gaps.

Establishing and aftercare of shelterbelts require as much attention as that given to a somewhat similar conventional plantation. Species choice is crucial: here climate, soil and exposure will be the chief considerations. Fencing must be adequate; its cost, depending on shape and extent of the area, and what is to be fenced against, can be substantial. If shelter is sought with as little delay as possible, Japanese larch may be a sound choice, though not long effective towards the bottom; European larch or grey alder may be an alternative but require a more fertile soil and the larch are prone to canker. Sitka spruce may be ideal towards the western seaboard; further inland Norway spruce may be satisfactory. Scots and Corsican pines may also be useful. Some broadleaves, particularly beech, sycamore and grey alder, perhaps with a few rowan, whitebeam and cherry to add diversity and colour, may be a satisfactory choice either alone or in mixture with conifers. Mixtures are recommended in order to provide filtration. The establishment of a 0.5 hectare shelterbelt (30 x 165 m) is likely to cost in the region of £1,000–£1,250. Grants available (i) from the Forestry Commission, if a sufficiently large area (minimum width 30 m, and minimum extent 0.25 ha) and (ii) from the Departments of Agriculture are noted overleaf:

	Less Favoured Area Farms	Standard Farms
Shelterbelts with 50% or more broadleaves	50%	40%
Other shelterbelts	50%	15%
Enclosing grazed woods	50%	40%

The work qualifying for grant-aid can be carried out under 'Standard Costs' published by the Departments of Agriculture on the basis of the cost of farm labour and new materials. They do not include an allowance for overheads and profits: in most cases contractors would add 15–30% to cover them. Standard costs are therefore usually lower, sometimes by a larger percentage, particularly where the labour content of the item is high.

Hedgerows and hedgerow trees

Hedgerows and hedgerow trees are important landscape features and provide valuable habitats for wildlife, particularly birds and insects. Additional benefits include provision of stock, visual and noise barriers, shelter, encouragement for game birds to rise to the gun and, ultimately, timber production (Hodge 1990). They provide a distinct boundary to land and confer shelter, particularly for livestock during inclement weather. In recent years they have been much to the fore because of the removal of hedges to create bigger fields, the killing of most hedgerow elms by Dutch elm disease and the decline of ash. The impact of hedgerow removal on the appearance of the countryside has aroused widespread criticism from public attached to a traditional view of the landscape, as well as conservationists concerned for wildlife. Particular hazards to hedgerows are from spray drift, straw and stubble burning, and waste dumped into hedge bottoms.

At least 90% of hedgerow trees are broadleaved – mainly oak, ash and elm. They and the hedgerows can bring both advantages and disadvantages to an estate or farm. The chief advantages are shelter and landscape. The main disadvantages of hedges are the amount of land they occupy, the shading of crops, the harbouring of occasionally harmful mammals and insects and the need for regular management if they are to remain stock-proof and not spread unduly. Hedgerow trees are, of course, an important source of timber and fuelwood. On most farms there will often be parcels of land such as field corners which can be planted with trees. The best way to recruit trees in existing hedges is to encourage some suitable shoots or saplings to grow up. Ash is usually discouraged. Holly when berried can provide a seasonal income, and hence a few might be encouraged.

The planting of new hedges and the retention of as many as possible is becoming increasingly urged. They, and shelterbelts, are often the only permanent vegetation on farmland. This stability helps to sustain wild plants and animals. A tall thick hedge with good bottom growth is of more value to wildlife than a low thin hedge. Partridges are partly dependent on hedgerow bottoms and verges for cover, nest sites and the great variety of plants and insects which provide food for parent bird and chick. Old hedges (some serve as farm or parish boundaries) are particularly valuable because they usually have a greater variety of shrubs. (It is believed that an ancient hedge will have at least one species for every hundred years of its existence.) Almost all the main groups of Britain's wildlife are represented in hedgerows and shelterbelts.

Some farmers today, ignoring conservation values, consider that farm hedges and hedgerow trees have become obsolete; others favour their retention. Some owners dislike them, holding that they reduce the yields of crops by absorbing water and nutrients as well as by casting shade; also that the roots tend to block field drains within their reach and that the branches are rendezvous for birds intent on depredations of all sorts. Furthermore, the stretch of hedge under trees may be weak, causing gaps which have to be mended with fencing material. Owners ought to have a decided opinion on the subject. They should know how to plant a good hedge and how to maintain an established one. Treatment is of particular account because if hedges are not controlled they are best not kept at all. Bad hedges on boundaries may mean, among other things, bad neighbours for, except in some instances of common land, every man must fence in his own livestock.

On some farms and estates hedgerow trees are intentionally and well maintained; but far too often when hedgerow trees die or are removed they are not replaced. Road alterations sometimes involve a removal of wayside trees. Unmanaged hedgerow trees are generally heavy branched, of poor shape and low quality. To prevent this, trees need care, especially occasional pruning. Attention to them is rarely given and may seem not worth the cost. Often, standing trees are used merely as fencing posts; nails are driven into them and attached wire becomes embedded. When such a log is converted in a sawmill, pieces of metal may damage the saw in a costly and dangerous way. Such risks are so great that roundwood merchants will rarely pay for timber below fencing or hedge height. Trees have often been maltreated and injured and in consequence may be decayed, especially those near a road where they may have been hacked about because they overhang, interfere with overhead cables or buildings or spoil prospects. Many hedgerow trees die as a result of roads authorities dumping salt near their roots. Dying or dead trees can become a financial liability to the owner who may be compelled to remove dangerous trees near roads. Owners are responsible for the safety of their trees and liable for damage or injury caused by them. Hence safety of trees is especially important where they border roads, footpaths, wayleaves or buildings. Where there is injury to people or damage to property as a result of a tree breaking or blowing over, both the courts and the insurance companies will consider that there has been negligence if the tree had not been inspected regularly. Such an inspection can be by owners, their tenant or their agent/forester. Whoever inspects the trees should have some knowledge of signs of structural weakness and symptoms of declining vigour: these are described in Forestry Commission Leaflet *The Recognition of Hazardous Trees*. Where signs of ill health or structural weakness are found then remedial steps should be taken to make the tree safe – this may involve seeking a detailed report from a forester or arboriculturist before a decision is taken about the appropriate treatment for the tree. Yet hedgerow and other farm trees have meant some financial gain to many a farmer-owner. Much can be done to increase the amount of timber from these sources.

For planting and managing hedgerow trees guidance is given in Ministry of Agriculture Leaflet *Farm and Estate Hedges* (Fixed Equipment of the Farm Series, No. 11), Leaflet 763 *Planting Farm Hedges* (1982) and Nature Conservancy Leaflet *Hedges and Shelterbelts*. They meet both the farmer's and the conservationist's requirements. More recent detailed information on establishing and managing hedgerow trees is provided by Hodge (1989, 1990a, 1990b) including the cost of hedgerow and hedgerow tree establishment. For information on the history and ecology of hedges and hedgerow trees the reader is referred to Rackham (1976, 1980, 1986).

In planting hedges, perhaps 80% of species is likely to continue to be hawthorn with other shrubby species such as blackthorn, guelder rose, hazel, dogwood, dog rose, to add diversity and conservation value, perhaps interspersed with trees – such as oak, holly, hornbeam, rowan and crab apple to create diversity. Where planted, or recruited in other ways, such species should be established about 20 m apart. Grants for hedges and hedgerow trees from the Agriculture Departments are as follows:

Less Favoured Area Farms	Standard Farms
50%	40%

The work qualifying for grant-aid can be carried out under 'Standard Costs' as noted *supra* for shelterbelts. The two Countryside Commissions and English Nature offer discretionary grants. Also, under MAFF Set-aside (for seven southeast counties), a cash grant for five years is obtainable for establishing and managing hedges to create habitats for wildlife.

Land reclamation and forestry

Mining and other industrial processes have left much waste ground usually consisting of structureless material largely devoid of topsoil and subsoil. Such sites, epitomized by the colliery spoil-heap in industrial regions, are frequently ugly, may erode seriously and may become an environmental hazard. (A separate situation is when, following opencast mining of coal or extraction of sand and gravel, the topsoil and sometimes the subsoil is laid aside and later replaced.)

Derelict sites bring many problems. Forestry offers a significant potential land use for otherwise unused reclaimed land whether it be derelict industrial sites, spoil-heaps or sites following opencast mining (Forestry Commission 1989). Silviculture on such sites can have many benefits in addition to the production of timber. The landscape appearance is much improved and eventually wildlife is attracted to the area. Afforestation on new larger sites can include the creation of lakes and ponds, useful for fire fighting and valuable as wetland wildlife areas. Multiple land use with areas of amenity woodland is increasingly becoming the norm for reclamation sites.

The problems confronting successful afforestation or rehabilitation are outlined by Savill and Evans (1986) relating to the physical and chemical characteristics of the soil, exposure and protection. Often the physical conditions of many substrates arising are inhospitable to tree growth. Soils may have become compacted and consolidated by heavy machines and so present barriers to root growth and drainage. However, reclamation to woodland is, in general, much cheaper than to agriculture, often costing one half of that spent on reclaiming an equivalent site to grass, and maintenance costs are approximately halved. Trees are less demanding, generally save expensive levelling and elaborate drainage operations, and are a more permanent solution than farming.

The Forestry Commission over recent decades has carried out a substantial amount of research into improving tree establishment and growth on reclaimed land and much progress has been made (Jobling and Carnell 1985; Wilson 1985; Moffat 1987; Reaney 1990). Early work highlighted the most suitable trees for different site types; more recently techniques for improving soil physical conditions prior to tree planting have been developed. Trees are thus able to root deeper, allowing more nutrients and water to be taken up, and aiding stability. Following opencast mining, the ideal shape of land for tree growth has also been determined: large ridge and furrow landforms are particularly useful for sites with high water-tables in high rainfall areas because they maximize the amount of land unaffected by waterlogging (Moffat and Roberts 1988). In addition to research on alder, much has gone into examining the use of other nitrogen-fixing shrubs or plants for land reclamation (Moffat *et al.* 1989). Legumes like tree lupin and everlasting pea are successful on a range of reclaimed sites, being relatively unpalatable to rabbits, sheep and deer. They do much to cover bare spoil, are well known as soil improvers, and add colour to a site in summer. More recently the Forestry Commission has examined the use to which sewage sludge, rich in nitrogen and phosphate, can be put on reclaimed land lacking essential nutrients; a marked improvement in tree growth has been achieved (Moffat 1988).

Fertility and moisture supply of spoil-heaps will depend on type of soil. Difficulties due to exposure, instability and erosion are not usually severe. The species choice is governed, firstly, by local conditions. Secondly, there may be some further restriction to less demanding species on burned heaps and acid ones. In general, the most successful species are alder, birch, and Corsican and Scots pine. Successful reinstatement of opencast mining sites has been achieved with Corsican pine and Japanese and hybrid larches, along with alder. Planting of china clay spoils in Cornwall has received experimental attention (Moffat and Roberts 1989).

Plans for satisfactory restoration after mining are prepared as part of the whole operation. Site preparation aims to leave a safe and stable substrate, to shape the restored ground in ways which blend with the local landscape and other conditions, and to make the site hospitable for trees. The nature of reclaimed sites necessitates use of species which are tolerant of exposure, undemanding, or will fix their own nitrogen. However, occasionally there have been failures at the sapling stage. Initially, growth has been vigorous but after a few years stagnation has set in and 'distress' flowering exhibited, probably caused by drought and poor water relations generally. Many tree species will grow if the pH is at least 4, including birch, larches and several pines. Alders fix nitrogen and tolerate moderate compaction as well as acidities down to pH 3.5. However, where alder are needed for reclamation sites and for those sites lacking organic matter as a result of topsoil stripping, plants inoculated with selected bacteria – like micro-organisms of the genus *Frankia* – should be specified (McNeill *et al.* 1990), as noted in chapter 3, section 3.

The Forestry Commission provide expert advice to local mineral planning authorities under the terms of the Town and Country Planning (Minerals) Act, 1981, whereunder the Commission must be consulted whenever it is proposed to reinstate a site to forestry after mineral extraction. They give advice on whether it is practical to do this, and on the best ways to achieve a successful tree cover.

References

Anderson, M.L. (1952), 'Early Hardwood planting in Scotland and its present-day Significance'. *Scottish Forestry 6* (3), reprinted in *44* (3) 1990, pp. 195–205.
Beaton, A. (1987), 'Poplars and Agroforestry'. *Quarterly Journal of Forestry 81* (4), pp. 225–33.
Binns, W.O. (1975), 'Fertilisers in the Forest: A Guide to Materials'. FC Leaflet 63.
Binns, W.O., Mayhead, G.J. and MacKenzie, J.M., 1980), 'Nutrient Deficiencies of Conifers in British Forests'. FC Leaflet 76.
Brazier, J.D. (1977), 'The Effect of Forest Practices on Quality of the Harvested Crop'. *Forestry, 50,* pp. 49–66.
Bryce, J. (1976), 'Cultivation of the Cricket-bat Willow'. FC Bulletin 17.
Campbell, D.A. and Aaron, J.R. (1980), 'New Markets for Poplar'. *Forestry and British Timber,* June 1980.
Chadwick, D.J. (1989), 'Machinery for Ground Preparation prior to Restocking', *Forestry and British Timber,* Autumn 1989.
Christie, J.M. (1994), 'Provisional Yield Tables for Poplar in Britain'. FC Technical Paper 6.
Crichton, D. (1982), The Future of Poplars in Britain'. Proceedings 10th International Forestry Students Symposium, pp. 60–6, University of Aberdeen.
Crichton, D. and Elliott, G.K. (1982), Trends and Prospects for Marketing Poplar in Britain'. School of Agriculture and Forest Science, University College of North Wales, Bangor, Gwynedd.
Davies, R.J. (1988), 'Alginure Root Dip and Tree Establishment'. FC Arboriculture Research Note 75/88/ARB.
Dutch, J.C., Taylor, C.M.A. and Worrell, R. (1990), 'Potassium Fertiliser – Effects of different Rates, Types and Times of Application on Height Growth of Sitka Spruce on Deep Peat'. FC Research Information Note 188.
Edwards, P.N. and Christie, J.M. (1981), 'Yield Models for Forest Management'. FC Booklet 48.
Evans, J. (1984), 'Silviculture of Broadleaved Woodland'. FC Bulletin 62.
Everard, J.E. (1974), 'Fertilizers in the Establishment of Conifers in Wales and Southern England'. FC Booklet 41.
FAO (1980), 'Poplars and Willows in Wood Production and Land Use'. FAO Forestry Series, No. 10, FAO, Rome.
Forestry Commission (1989), 'Land Reclamation and Forestry'. The Forestry Commission's contribution to European Year of the Environment, 1989.
Gibbs, J.N. (1987), 'New Innings for Willow?' *Forestry and British Timber,* June 1987.
Hamilton G.J. and Christie, J.M. (1974), 'Influences of Spacing on Crop Characteristics and Yield'. FC Bulletin 52.
Hodge, S.J. (1989), 'Hedgerow Trees: Regenerating a Valuable Resource'. *Forestry and British Timber,* April 1989.
—— (1990a), 'The Establishment of Trees in Hedgerows'. FC Research Information Note 195.
—— (1990b), The Establishment of Trees in Existing Hedgrows'. FC Aboriculture Research Note 91/90/ARB.
Hodge, S.J., and Walmsley, T.J. (1990), The Use of Water-Retentive Materials in Tree Pits'. FC Arboriculture Research Note 92/90/ARB.
Hollingsworth, M.K. and Mason, W.L. (1989), 'Provisional Regimes for Growing Containerised Douglas Fir and Sitka Spruce'. FC Research Information Note 141.
Insley, H. (ed.) (1988), 'Farm Woodland Planning'. FC Bulletin 80.
Jobling, J. (1990), 'Poplars for Wood Production and Amenity'. FC Bulletin 92.
Jobling, J. and Carnell, R. (1985), Tree Planting in Colliery Spoil'. FC R & D Paper 136.
King, C.J. *et al.* (1983), The Use of Chemicals (other than herbicides) in Forest and Nursery'. FC Booklet 52.
Lines, R. and Brown, I. (1982), 'Broadleaves for the Uplands' in *Broadleaves in Britain* (eds. D.C. Malcolm, J. Evans and P.N. Edwards), pp. 141–9, Institute of Chartered Foresters, Edinburgh.

Low, A.J. (ed.) (1985), 'Guide to Upland Restocking Practice'. FC Leaflet 84.
—— (1986), 'Use of Broadleaved Species in Upland Forests'. FC Leaflet 88.
MacKenzie, J.M., Thompson, J.H. and Wallis, K.E. (1976), 'Control of Heather by 2,4-D'. FC Leaflet 64.
Mason, W.L. (1986), 'Precision Sowing and Undercutting of Conifers'. FC Research Information Note 105/86/SILN.
—— (1991), 'Improving Quality Standards for Conifer Planting Stock'. *Scottish Forestry* 45 (1), pp. 28–41.
Mason, W.L. and Biggin, P. (1988), 'Comparative Performance of Containerised and Bare-root Sitka Spruce and Lodgepole Pine Seedlings in Upland Britain'. *Forestry, 61* (2), pp. 149–63.
Mason, W.L. and Hollingworth, M.K. (1989), 'Use of Containerised Conifer Seedlings in Upland Forestry'. FC Research Information Note 142.
Mason, W.L. and Jinks, R.L. (1990), 'The Use of Containers as a Method for Raising Tree Seedlings'. FC Research Information Note 179.
Mason, W.L., Biggin, P. and McCavish, W.J. (1989), 'Early Forest Performance of Sitka Spruce Planting Stock Raised from Cuttings'. FC Research Information Note 143.
Matthews, J.D. (1989), *Silvicultural Systems*. Clarendon Press, Oxford.
Mayhead, G.J. (1976), 'Forest Fertilising in Great Britain'. Proceedings of the Fertiliser Society No. 158.
McIntosh, R. (1984), 'Fertiliser Experiments in Established Conifer Stands'. FC Forest Record 127.
McKillop, E.G., Pepper, H.W. and Wilson, C.J. (1988), 'Improved Specification for Rabbit Fencing for Tree Protection'. *Forestry, 61* (4), pp. 360–8.
McNeill, J.D. *et al.* (1990), 'Inoculation of Alder Seedlings to Improve Seedling Growth and Field Performance'. FC Arboriculture Research Note 88/90/SILN.
Miller, H.G. (1981), 'Forest Fertilization; some Guiding Concepts'. *Forestry, 54,* pp. 157–67.
Miller, K.F. and Coutts, M.P. (1986), 'A Provisional Comparison of Ploughing and Subsoiling in Relation to Growth and Stand Stability of Sitka Spruce in Upland Forests'. FC Research Information Note 104/86/SIL.N.
Moffat, A.J. (1987), 'The Geological Input to the Reclamation Process in Forestry'. Geological Society Engineering Geology Special Publication, *4,* pp. 541–8.
—— (1988), 'Sewage Sludge as a Fertiliser in Amenity and Reclamation Plantings'. FC Arboriculture Research Note 76/88/SSS. See also FC Bulletin 107.
Moffat, A.J. and Roberts, C.J. (1988), 'The Use of Large Scale Ridge and Furrow Landforms in Forest Reclamation'. FC Research Information Note 137 (see also *Forestry 62* (3), 1989.)
—— (1989), 'Experimental Tree Planting on China Clay Spoils in Cornwall'. *Quarterly Journal of Forestry, 83* (3), July 1989.
Moffat, A.J., Roberts, C.J. and McNeill, J.D. (1989), 'The Use of Nitrogen-fixing Plants in Forest Reclamation'. FC Research Information Note 158.
Nelson, D.G. and Quine, C.P. (1990), 'Site Preparation for Restocking'. FC Research Information Note 166.
Nelson, D.G. and Williamson, D.R. (1990), 'Herbicide Experimental Update'. *Forestry and British Timber,* February 1990.
Patrick, K.N. (1991), 'Watermark Disease of Cricket-bat Willows: Guidelines for Growers'. FC Arboriculture Research Note 197.
Pepper, H.W. and Tee, L.A. (1986), 'Forest Fencing'. FC Leaflet 87.
Porter, J. (1986), 'Working with Willows'. *Horticultural Week,* 17 October. 1986.
Potter, M. (1989), 'Treeshelters 10 years on'. *Timber Grower,* Spring 1989.
—— (1991), 'Treeshelters'. FC Handbook 7.
Potter, C.J., Nixon, C.J. and Gibbs, J.N. (1990), 'The Introduction of Improved Poplar Clones from Belgium'. FC Research Information Note 181.
Preece, T.F. (1978), 'Watermark Disease of the Cricket-bat Willow'. FC Leaflet 20.

Price, C. (1989), *The Theory and Application of Forest Economics*. Basil Blackwell, Oxford.
Pyatt, D.G. (1987), 'Afforestation of Blanket Peatland – Soil Effects'. *Forestry and British Timber* *16* (3), March 1987, pp. 15–17.
—— (1990), 'Long-term Prospects for Forests on Peatland'. *Scottish Forestry, 44* (1), pp. 19–25.
—— (1990), 'Forest Drainage'. FC Research Information Note 196.
Pyatt, D.G. and Low A.J. (1986), 'Forest Drainage'. FC Research Information Note 103/86/SIL.N.
Rackham, O. (1976), *Trees and Woodland in the British Landscape*. Dent, London.
—— (1980), *Ancient Woodland*. Edward Arnold, London.
—— (1986), *The History of The Countryside*. Dent, London.
Reaney, M. (1990), 'How Green is my Valley'. FC 'Forest Life', Issue No. 7 (1990), pp. 10, 11.
Savill, P.S. and Evans, J. (1986), *Plantation Silviculture in Temperate Regions*. Clarendon Press, Oxford.
Stables, S. and Nelson, D.G. (1990), 'Rhododendron Ponticum Control'. FC Research Information Note 186.
Stevens, F.R.W., Thompson, D.A. and Gosling, P.G. (1990), 'Research Experience in Direct Sowing for Lowland Plantation Establishment'. FC Research Information Note 184.
Tabbush, P.M. (1987), 'Effect of Desiccation on Water Status and Forest Performance of Bare-rooted Sitka Spruce and Douglas Fir Transplants'. *Forestry, 60* (1), pp. 31–43.
—— (1988), 'Silvicultural Principles for Upland Restocking'. FC Bulletin 76.
—— (1988a), 'Planting Stock Survival'. *Scottish Forestry, 42* (2), pp. 120–8.
Tabbush, P.M. and Williamson, D.R. (1987), *'Rhododendron Ponticum* as a forest weed'. FC Bulletin 73.
Taylor, C.M.A. (1986a), 'Nutrition Prospects in Upland Restocking'. *Timber Grower,* Autumn 1986, 23.
—— (1986b), 'Forest Fertilisation in Great Britain'. Proceedings of the Fertiliser Society No. 251, p. 23.
—— (1990), 'The Nutrition of Sitka Spruce on Upland Restock Sites'. FC Forest Information Note 164.
—— (1991), 'Forest Fertilization in Britain'. FC Bulletin 95.
Taylor, C.M.A. and Tabbush, P.M. (1990), 'Nitrogen Deficiency in Sitka Spruce Plantations'. FC Bulletin 89.
Thompson, D.A. (1979), 'Forest Drainage Schemes'. FC Leaflet 72.
—— (1984), 'Ploughing of Forest Soils'. FC Leaflet 71.
Tuley, G. (1989), personal communication.
White, J.E.J. (1990), 'Nursery Stock Root Systems and Tree Establishment'. FC Occasional Paper 20.
Williamson, D.R. (1990), 'A Brief Guide to some Aspects of The Control of Pesticides Regulations 1986'. FC Research Information Note 170.
Williamson, D.R. and Lane, P.B. (1989), 'The Use of Herbicides in the Forest'. FC Field Book 8.
Wilson, K. (1985), 'A Guide to the Reclamation of Mineral Working for Forestry'. FC R & D Paper 141.
Wright, N.J. (1987), 'Stumped for a Market – Bat-makers' Residue Dilemma'. *Forestry and British Timber,* September. 1987.

Chapter Six

AFTER-CARE (TENDING) OF PLANTATIONS

Establishment of a plantation is generally accepted as completed when weeding is no longer necessary. Thereafter management settles down to routine maintenance of fences and drains (and of roads, if installed), fire protection and payment of insurance premiums, and protection from pests, diseases, the elements and trespass. These outlays, along with any management fees and other overheads, continue annually, together with compound interest on all expenditures net of any grants. Occasional spot inspections should be undertaken, whereupon any gorse, broom and harmful climbing or trailing plants can be dealt with, and double leaders singled – often advisable in ash and oak, occasionally in beech, Sitka spruce and Lawson cypress. Singling (formative pruning) is a silvicultural operation to ensure by elimination of forks and multi-leaders that trees develop a single straight stem and leader. It can be accomplished with a suitable knife, secateurs, hacker or saw, according to size and height of tree.

Most plantations start off even-aged and equal in height, but before long changes occur: uneven height growth and horizontal branching develop, some trees become unthrifty and others die. Slowly in broadleaves, more quickly in conifers, the young trees begin to touch their neighbours. Eventually they close up into the thicketstage – reached by the eighth to fifteenth year in conifers, but much later in broadleaves. By this time weeds and other low vegetation have been suppressed.

On old broadleaved sites, coppice shoots from the previous crops may re-grow, and naturally regenerated seedlings, especially birch, often invade. The branches both of the planted and naturally regenerated trees intermingle and as they deprive one another of light some branches are killed off: this is desirable – the mutual pruning leads to the formation of clean stems, and eventually of knot-free wood.

In the thicket-stage, competition for existence among the trees becomes keen: the vigorous dominate, the weakly succumb. The *canopy closure*[1] stage is approaching or has been reached – between the tenth and twentieth year, according to species, site and management. Canopy closure takes place gradually, from first contact between neighbouring branches to a state of maximum competition between trees. Despite this competition, it is desirable, silviculturally as well as

---

1   *Canopy* can be defined as the cover of branches and foliage by tree crowns; and *closed-canopy* as a canopy in which the individual tree crowns are in general contact with one another. Tabbush and White (1988) in regard to Sitka spruce, point out that no information is available from yield tables on the timing of canopy closure, but that the timing of it is fundamental to a consideration of the effects of spacing on timber quality, since it marks the onset of competition, and hence branch suppression and limitation of the juvenile core; the attendant restriction of growth rate also tends to increase wood density. Of course, crown diameters of forest trees commonly bear a definite relationship with their bole diameters.

economically, that a crop closes canopy as soon as possible. In conifers, occasionally, effective top-dressing with fertilizer can assist early closure (discussed in section 2 *infra*).

1. Cleaning, rack-cutting, brashing and pruning

Cleaning is the removal of unwanted broadleaved trees and woody growth usually before canopy closure. It generally applies to broadleaved or mixed stands; rarely to upland conifers. Inspection following rack-cutting *(infra)* may indicate the need to liberate and favour straight undamaged trees of more valuable species. Conifers in particular are at risk of becoming swamped, or their leaders damaged by a whipping effect (especially by birch). During cleaning, honeysuckle, wild clematis (old man's beard, travellers joy) and ivy can be severed, along with invaded species such as birch, thorn and willow, and unwanted coppice shoots. The last can be quite troublesome when they splay out over planted trees, the worst being sycamore, ash, lime, wych elm, oak, sweet chestnut and hazel. Wild cherry and rowan are rarely harmful and, along with a few maiden sallow and silver birch, may be spared to induce colour and diversification in the plantation. Cleaning is more often required on richer lowland than upland sites. Nature conservation interests, where they are important, will largely dictate the cleaning treatment – see chapter 13, section 7.

Cleaning may sometimes obviate the premature suppression of side branches of desired trees, which otherwise might induce top-heavy development and instability (particularly of Douglas fir and larch on highly fertile soils). It is usually vital on difficult sites such as chalk downland and heavy Midland clay. The operation requires adequate thought, for sometimes a naturally regenerated tree or a healthy coppice shoot might be utilized as a substitute for a failed, leaning or poorly-grown planted tree, with a view to achieving a full stocking. Some moderate natural growth may be capable of affording protection and nursing to the main species. A leaning tree occasionally may be worth staking or holding upright by the branches of an unwanted live birch being wound around it. Any multi-leaders, if they can be reached, should be singled.

Cleaning is best undertaken during the bareness of winter, provided that the workers can identify the leafless broadleaves and so select those that are unwanted. Some foresters prefer to undertake cleaning in two stages, one between the weeding and thicket-stage and the other after that latter stage. The value of early or 'low' cleaning rests upon the savings of planted trees *before* they are damaged and the relative ease of entry. The second, 'high' cleaning, is then usually light.

The effect of cleaning is therefore to concentrate the growth potential of the site on to the more valuable species in the crop (an effect later pursued in each thinning). If cleaning is not undertaken, the result may be a drawn-up inadequate crop which may be difficult to treat later. However, a judgement always has to be made on the effectiveness of the operation relative to likely crop losses and the high cost of cleaning. Where the operation is undertaken, the method followed is generally either by chainsaw or clearing saw (cutting, usually to waste, or girdling to kill standing). One method is to use a clearing saw fitted with an attachment (Enso) which applies chemical to the cut surface as it is cut. However, the main method is chemical injection (eventually to kill standing), using glyphosate (Roundup) applied by knapsack sprayer or forestry spot gun, with guard. Foliar treatment by spraying is preferable whenever the unwanted trees or woody growth are in leaf and the foliage is accessible. Costs per hectare are relatively low. Note always the warnings relating to chemicals given in chapter 5 under 'ground preparation' and 'weeding'.

Rack-cutting. In or around the thicket-stage, rackways (narrow tracks) can be cut through the plantation, sufficient to discover its silvicultural state. Rack-cutting can sometimes be so arranged as to subserve access in case of fire and assistance to game and shoot management. Later the rackways will ease marking, thinning and extracting. In conifers, rackways are best created by removing the lower branches of one or two adjacent rows of trees usually in straight lines; their width and siting generally depending on the type of extraction equipment that is to be used in thinning. Dead branches of larch can be knocked off, but the more persistent branches of other conifers and broadleaves necessitate an edge or sawing tool. During the operation, any harmful climbing or trailing plants can be removed; but, where not harmful, conservation interests should be recognized.

Brashing. Complete or partial brashing (in effect 'low' pruning) comprises the removal of the dead and dying lower branches of trees to a height up to 2 m, or a man's reach. Brashing is often when the breast height diameter of the tree is at a maximum of 15 cm, and the stem is brashed up to the point of about 10 cm. It is more usually undertaken in conifers than in broadleaves, and undertaken either in the thicket-stage or delayed a few years to the pole-stage (i.e. between canopy closure and the first thinning).

Brashing, being very labour intensive and costly, nowadays is often omitted from silviculture. However, some foresters prefer to undertake it; others combine it with cleaning at the minimum intensity needed for access. In mixtures, at least some of the conifers may need to be brashed to liberate the broadleaves.

The main aim of brashing is to facilitate access for management, especially by inspection racks, enabling the crop's growth to be assessed and treatment determined. It eases marking and thinning, and sometimes lowers susceptibility to fire. An incidental effect is the development of clear timber in the first two metres of the stem, by limiting the extent of dead knots. Another benefit is the enabling of final crop trees to be selected, marked, and if need be, high-pruned. Brashing occasionally is undertaken for game management including facilitating access for beaters and sportsmen. At other times it may be done for a recreation facility or timed to coincide with an offer for sale. Money should not be wasted on brashing stems which are dead, dying or otherwise useless, nor on those likely to be removed during the first thinning. Brashing undertaken too early will mean additional compound interest; yet it should be sufficiently divorced in time from a first thinning to allow the brash to settle.

Where brashing is influenced by thinning consideration, much depends on the projected thinning method: it may be prudent in conifers intended to be selectively thinned, but is unnecessary in systematic (i.e. line) thinning. To ease a selective thinning, it has been the custom to brash, on average, a minimum of 50% of the crop to be retained, the non-brashed trees being those to be removed. Systematic thinning only benefits from sufficient brashing which allows inspection of the crop and marking of the first tree in each rack.

The persistence of branches on trees varies with species and crop spacing. In similar conditions, shade tolerant trees retain living branches for much longer than intolerant species. Among the latter, the duration of retention of dead branches also varies considerably; larches self-prune well, while other conifers and most broadleaves do not. The dead branches of larch are easily removed (they can be knocked off with a stout stick), but those of other conifers and most broadleaves need to be severed with an edge tool or sawn back to the stem. Billhooks and the like, if risked at all, need to be applied with great care, otherwise they will slice patches of bark off the stems; on European larch this may induce canker, and in any tree will induce a blemish in the timber. A short curved saw with about five teeth and a medium-short light handle, is a satisfactory tool. Undercuts to branches, to prevent snagging, are advisable for robust live branches.

The cost of brashing is based on several considerations: (i) species (larch are the easiest followed by pines; Sitka spruce is the most difficult); (ii) crop spacing (this determines the number of trees per hectare; those widely spaced are thicker and tougher branched); (iii) age (if done as soon as the canopy has closed the branches are still green); (iv) number of trees brashed per hectare; and (v) terrain (a steep gradient usually hinders the work). The yield class of the plantation, on which may depend the average number, size and coarseness of whorls per tree, may also affect the cost. If brashing of larch costs £x per hectare, pines and firs may cost a minimum of £x plus 50%; Norway spruce, western red cedar and western hemlock £x plus 75%; and Sitka spruce £x plus 100%. In western red cedar and western hemlock, part of the cost may be recouped by the sale of their undamaged foliage to florists, especially that severed from trees on the perimeter and ride-side.

Brashing should reduce the cost of marking and measuring the first selective thinning, as well as ease the felling and extraction. If combined with the marking, only those trees to remain, or a selection of them, are brashed. Felling costs differ little between non-brashed crops for systematic thinning and brashed crops for selective thinning. There is a tendency to move away from the high intensity of brashing and marking traditional selective thinning and instead to adopt a 'feller-selection' system, whereby, racks having been taken out every ten to fifteen rows depending on accessibility, skilled fellers are entrusted to select the trees to fell. The cost of brashing and marking is obviated, and a small addition for selection is made to the felling piece-work rate.

In respect of brashing (likewise of rack-cutting and cleaning), conservational and ecological aspects should be given appropriate consideration according to crop, location and objectives. Concurrently, there should be borne in mind the development of routes for future extraction. If these were not determined at planting, it may be sufficient to remove either a single or double row of trees at intervals to allow for the passage of machines or horses; the break in the canopy will rapidly close and there is little loss in production, nor is there increase in windthrow risk. Thereby the yield from the first thinning may be more profitable. In hilly terrain it would be important to choose the routes to fit in with the local undulations to ease extracting.

The economic essence of brashing (see 'Investment Appraisal' in chapter 14) is that it involves immediately costs in a labour intensive operation in order to reduce later costs and improve later revenues. The higher the discount rate, the less likely is brashing to be worthwhile. The decision as to whether or not to brash is a matter of weighing the costs against the benefits. The fact that brashing takes place comparatively early in the rotation substantially increases the total cost of compound interest.

Pruning. The objective of pruning, i.e. debranching selected standing trees to heights above the 2 m achieved by brashing, is to produce knot-free timber. If trees are intended to produce only low quality products such as pulpwood, boardwood and other small diameter roundwood, pruning is unnecessary and wasteful. However, where quality is required, pruning is a valuable aid to the early production of clear timber in the outer zone of each stem. Such timber should then have a greater potential market than that from unpruned trees because, in the case of conifers, it can compete with the higher grades of imported softwood from the viewpoint of utilization; and, in the case of broadleaves, it produces the qualities required for veneer, beams, planks and joinery, all commanding high prices. Yet it must be said that purchasers, although accustomed to pay a premium for improved quality broadleaves, are not generally accustomed to pay a premium for pruned conifers.

Pruning of conifers. Whether pruning of conifers is worthwhile, i.e. profitable, is debatable. The outlay might make their timber so costly on the market that it cannot compete in price with slow-grown, naturally pruned imports from virgin or second growth stands. A general objective of pruning conifers in comparatively recent decades whenever undertaken has been to break the long association of British softwood with knottiness. To this end it might still pay to prune Scots pine in order to eliminate the black knots, which are liable to fall out in use, and are a shortcoming in this otherwise fine timber; likewise Corsican pine to limit the prominent whorls of knots which are its main weakness as structural timber. It might also pay in fast-growing Douglas fir, an excellent timber for joinery; and in European larch for 'boatskin' quality. Yet little pruning of conifers has been undertaken.

To keep the knotty core small, pruning of conifers must be done early, probably at about twenty years of age, according to species and yield class, starting desirably when the dominant trees do not exceed 10–15 cm in diameter at about 1.5 m and, assuming that brashing has taken place, carried up from 2 m to 4 m. For economy, pruning should coincide with initial thinning. When the average diameter at about 1 m above the already pruned first length attains about 10 cm in diameter, a further length can be pruned up to 5–6 m. Although injuries to trees associated with green pruning are rare, there should be a general restriction that this length shall not include more than two green whorls. The operation must not be so drastic as to affect adversely the growth diameter or height, although the amount of live crown which can, if necessary, be removed without harmful effect may range from about 25 to 45%. Excessive green pruning may sometimes lead to stress and all its attendant dangers of increased risks of attack by insects and pathogens.

The height to which to prune conifers, if undertaken, should be governed by the length of clear timber requisite for future markets: as examples, larch 'boatskin' material might be in 7–9 m lengths, while greenhouse and garden frame material need not be longer than about 5 m. How high to prune or what diameter of pruned log to achieve, as well as what proportion of a tree crop should be pruned to develop into high quality, are matters for sound consideration. Pruning of conifers can be done at any time of the year, but in March to May aids early occlusion. The operation should be restricted to selected stems, commonly those of the potential final crop. In practice, the number of conifers pruned per hectare might range from 200 to 375; or the work may include all trees expected to form the final crop plus about 50%, the pruned trees being favoured and encouraged at each thinning.

Selected conifers, already brashed, can be pruned from the ground with a long-handled tool, usually a curved saw but sometimes a double-edged chisel. Higher pruning is done later in a like manner with longer handles, but the operation is sometimes completed more efficiently by the aid of a ladder and short-handled saw, the branches being severed while the operator is descending. The long handles may be of western red cedar, bamboo or other light sort, or may comprise socketed sections of aluminium rods. Appropriate clothing and other equipment are essential. All methods of doing the job are tiresome until the muscles of the neck, back and arms become accustomed to the labour; falling sawdust can harm the eyes. Traditional pruning, when undertaken, is normally done in up to three stages to increase the 2 m brashed stem to about 6 m of clear stem.

Usually it will not pay to prune conifers if the trees are to be harvested before a substantial amount of clear wood has developed. The top diameter of the pruned section should have grown to at least 40 cm diameter overbark at the time of felling. It will certainly be unprofitable on poor sites where growth is slow, or for the slower-growing species. (It may benefit future marketing to record the number and size of the trees at the time of pruning.) The main doubt of pruning has been uncertainty that a sufficiently higher price will be realized for timber so treated. The quantity of pruned sawlogs which have appeared on the market is insignificant; and hence no special markets have developed, predictions in price must be largely guesswork, and there remains difficulty in comparing the value of the mature timber with the immediate outlay plus compound interest.

Pruning of conifers has occasionally been practised on a few estates for some years, in many instances specifically but not always justifiably directed towards the production of transmission poles; occasionally it has been primarily to improve the crop's appearance.

Pruning of broadleaves. Broadleaves require a different pruning regime to that for conifers. Broadleaves are unlikely to have been brashed; and the branches which may require pruning are generally on the lower bole, and not in the higher lengths. Many broadleaved trees, particularly the 'decorative' ones of veneer size, command high prices when free of knots and other defects; and pruning away branches to achieve this on the lower bole may be profitable. There is greater price differential between clear and knotty broadleaved timber than there is in the case of conifers. Stem-form in broadleaves is generally poorer, and singling (early formative pruning) and pruning of boles are needed to achieve improvement. Their pruning often seeks to favour only potential final crop trees, and it is important to ensure that these are of best quality. The need for pruning depends entirely on the degree of branch development on the lower bole: little or no pruning is required for well stocked, dense crops which have encouraged early side branch suppression. A separate inducement to prune the bole via branches is (e.g. in the rot *Stereum gausapatum* on oak) a more serious problem.

For rapid wound occlusion most broadleaved species are best pruned in late winter or early spring, but there are some exceptions. To minimize sap exudation, birch, Norway maple and sycamore should not be pruned in spring; cherry should only be pruned between June and August – to minimize risk of infection from bacterial canker and silver leaf disease. Walnut should be pruned in or around August. In general, to help prevent entry of disease, pruning should not be done between mid-March and the end of May when the tree is flushing since its resistance to infection is likely to be at a minimum during that period. To restrict decay beneath pruning wounds, branches should not be severed completely flush but slightly proud of the stem to retain the barky swelling, called the 'branch bark ridge', around the bark base (Evans 1984). Cherry and walnut should be pruned to produce a single stem of 1.5–2.0 m. Everard (1985) gives recommendations for pruning sessile oak and pedunculate oak (see chapter 5).

Pruning of broadleaves, being costly, should only be considered for potential final crop trees (50 to 100/ha according to species) and only for stands expected to yield good quality saw timber or veneer logs. Costs rise rapidly with increasing branch size and with height of bole. It is rarely worth pruning branches greater than 5 cm in diameter or higher than 5 m. Pruning should start early in the life of a tree to restrict the size of the knotty core and, because branches are small, no heartwood is exposed and wounds heal more readily. Generally, pruning up to 3 m of selected stems should be done at first thinning. Pruning may be needed to control epicormic shoots – mainly a problem of oak and in growing high quality poplar and cricket-bat willow, matters discussed in chapter 5.

Conifers pruned now would not produce knot-free timber until some twenty to twenty-five years hence; in broadleaves the waiting period might be double or treble; all depending on species, yield class and markets. During these periods compound interest would be added to the cost. If at harvesting those compounded costs are inadequately exceeded by the price achieved, pruning is not worthwhile. Pruning delayed to a later stage of the rotation becomes little more than a cosmetic operation because much of the clear timber outside the pruned knots eventually will be sawn off as slab wood; there is also a risk of rot organisms entering the stem through large pruning wounds. The operation is rarely undertaken nowadays, other than of 'decorative' species of broadleaves. An alternative to pruning to reduce knots is narrow tree spacing when establishing. As to the economic advantages of pruning over narrow spacing, see comments in chapter 14. The pruning of poplars and the de-budding of cricket-bat willow are special cases, discussed in chapter 5: these species have high volume yield, reduced number of trees per hectare to be pruned, and relatively short rotations. The best guide to pruning poplars is Jobling (1990).

2. Fertilizing established plantations (top-dressing)

Fertilizing accelerates tree growth in both height and diameter. When applied at planting (see chapter 5) or at least before canopy closure, it can improve conifer crops, but thereafter their soils can usually furnish a sufficient quantity of mineral substances for the continued production of trees. There is also increased capture of nutrients from the atmosphere. However, towards the end of the rotation (when any deficiency is almost entirely nitrogen) it may be possible for a return on an investment in fertilizer application to be quickly realized by extra volume production. Nature conservation considerations are noted in chapter 13, section 7.

Top-dressing of conifers. A special type of top-dressing of conifers – the alleviating of nitrogen deficiency in a Sitka spruce stand – has been discussed in chapter 5 under '"Check" of spruce in heather'.

Many of the present closed-canopy, pole-stage coniferous crops were established without the use of fertilizers, either because the sites were relatively fertile or because the planting pre-dated later fertilizer practices. Fertilizers applied as *top-dressing* (by fixed wing plane or helicopter) might be used to increase the productivity of such crops. However, it is unusual for crops in this condition to be unhealthy through lack of nutrients – a closed-canopy crop recirculates its available nutrients efficiently. On poor sites (podzols and peats) mixtures of species are known to be efficient at increasing nitrogen availability; thus Sitka spruce mixed with suitable origins of Scots pine, lodgepole pine or Japanese larch can grow better than Sitka spruce pure.

Experiments by the Forestry Commission (McIntosh 1984) suggest distinct differences between the responses to fertilizer of pine and spruce stands. Nitrogen appears to be relatively common in established Scots pine stands, and consistent responses to the application of nitrogen fertilizer have been obtained. On the other hand no clear pattern has emerged from the spruce experiments except that, in general, stands on poor sites can be expected to respond to treatments which would have produced a response in the establishment stage. For spruces, responses to application of nitrogen or phosphate are clearly possible in stands of moderate to high productivity which are apparently free from nutrient deficiency, but no satisfactory method of predicting which stands will respond and by how much has yet been found. Taylor (1991) should be consulted.

Analysis of foliar[2] nutrient concentrations in the top whorl has provided meaningful results in Scots pine stands but nutrient concentrations in established spruce stands have all been above the levels considered as satisfactory and have shown no relationship with growth or fertilizer response. Experiments suggest that nutrient deficiencies and fertilizer responses are more likely in the planting stage than in established stands where crown development is complete, recycling of nutrients within the tree and stand begin to be more important, and capture and retention of atmospheric inputs are more efficient. Information supplied by Macintosh (1984) enables foresters to calculate the economic advantages of fertilizing Scots pine stands but he recommends that application of fertilizers to establish spruce stands should, at present, only be carried out after prudent consideration of all available knowledge.

---

2 *Foliar analysis* is a diagnostic technique in which samples of needles/leaves are analysed chemically to determine their nutrient content. Both the absolute concentration of nutrients and the ratio of one nutrient to another are compared with average values for a species. Substantial departure from the values usually associated with healthy trees may indicate a deficiency (or, rather rarely, a toxicity). In foliar sampling and analysis, current lateral shoots are collected from representative dominant trees at the start of the dormant season. These are analysed chemically and the concentrations of the nutrient elements are expressed as percentages of oven-dry weight of the needles or leaves. The relationship between height growth and foliar nutrient concentrations have been used to establish what concentration of nitrogen, phosphorus and potassium (N, P, K) are commonly associated with poor and good growth. These standards are used to interpret the results of foliar analysis of samples collected by foresters who are contemplating top-dressing. Deficiency of imbalance shown by foliar analysis can be corrected with an appropriate fertilizer. If, however, analysis reveals a serious deficiency of two or more nutrients, the cause may lie in the failure of the root system following waterlogging or fungal attack, rather than a lack of nutrients in the soil. Taylor (1991) gives up-to-date information on foliar sampling procedure and analysis.

The correct timing of fertilizing is important, for it is wasteful to invest before it is needed, while waiting until nutrient deficiency is obvious means a loss of timber and revenue (Binns et al. 1980; Taylor 1986). The early recognition of deficiency symptoms should therefore assist in planning fertilizer programmes, whether top-dressings are part of a regime or are prescribed after chemical analysis of the foliage. Binns et al. (1980) provide help to foresters to recognize those nutrient deficiencies which have been found to be important in Britain. The deficiency symptoms they describe are visible manifestations of nutrient imbalance. The addition of a fertilizer to correct an evident deficiency of one element may induce another imbalance which will then produce different deficiency symptoms. The best time of year to observe deficiency symptoms is during the autumn and early winter when height growth is complete, foliar nutrient concentrations and colours have stabilized, and cold winds have not scorched the needles. Useful observations can, however, be made at any time of year, and in spruces potassium deficiency is particularly noticeable in early summer when the new shoots are fully extended. The normal autumnal yellowing of the old needles of pines prior to these being shed should not be confused with nutrient deficiency symptoms, although early shedding of needles can be a symptom of nutrient disorder.

Ideally, stands develop at a steady rate throughout their life, without either checks or spurts of rapid growth. Not only is this important for productivity, but wood grown at a steady rate may be more valuable than wood with annual rings of widely varying width. It follows that remedial treatments should, as far as possible, be timed so as to maintain steady growth. Unfortunately a stand that shows deficiency symptoms will not have been growing at its optimum rate for two or three years previously nor is it usual for the rate to recover in the year following fertilizing; and though decreasing height growth can be a useful sign (for growth may decline before deficiency symptoms are obvious), by the time this is seen some production will already have been lost.

Thus although foliar analysis and visual symptoms can be used to identify nutrient deficiencies, it is still difficult to achieve the most efficient use of fertilizers. Here local experience and a knowledge of species performance by site types are valuable, for foresters can be on the look-out for deficiencies before growth is seriously affected. Combined with foliar analysis and site-type appreciation, deficiency symptoms form a practical basis for prescribing fertilizer treatment. However, until local experience is gained, they should not be used on their own for making decisions on top-dressing. It should also be remembered that fertilizer treatment may not cure a nutrient deficiency; and even if it does, it may not pay.

Binns et al. (1980) and McIntosh (1984) describe and illustrate symptoms of nitrogen, phosphate and potassium deficiency in forest stands up to 5 m tall for Sitka spruce, Scots pine and lodgepole pine. Less coverage is given for Norway spruce, Corsican pine, Douglas fir, western hemlock and larches, as well as for the rare deficiencies of magnesium and copper. Concentrations of nutrients in the foliage associated with deficiency are given for the different species, and the site types where deficiencies in spruce and pine are most likely are tabulated. Deficiency symptoms, used in conjunction with foliar analysis and site-types, form a practical basis for prescribing fertilizer treatments; they should not, however, be used on their own without considerable local experience.

Phosphate deficiency in Sitka spruce and possibly in lodgepole pine is common in upland peaty sites. Foliar analysis has been used for many decades for assessing P requirements but this has been found to be unreliable for stands after early pole-stage.

Rather than risk uncertain economic returns for investment in fertilizer, most foresters do not fertilize stands after early pole-stage. Experiments have been made to see whether they are losing out on better yields (Harrison and Dighton 1987). A sensitive new method, based on the responses of feeder roots, is being developed, which may improve assessments of P fertilizer responses of such spruce stands.

For Wales and Southern England, Everard (1974) recommends that, almost invariably, fertilizers should not be used in forest crops lying east of a line from Chester to Southampton; in this region nutrient deficiencies are seldom limiting to tree growth. To the west of this line, where climatic conditions are often very favourable, growth can be most severely restricted by nutrient deficiencies. Where fertilizer is required it is generally phosphate. Nitrogenous fertilizer should be used very rarely, even though nitrogen deficiency is often diagnosed. Where fertilizer is applied as a top-dressing, foliar analysis should be used to confirm the need for phosphate, which may be required on some sites between five to eight years after planting. Unground rock phosphate from North Africa may be broadcast at any season. Of urea and muriate of potash, the recommended N and K fertilizers, K can be used at any time of the year, and N at any time during the growing season. However, for P and K, application should be avoided to frozen or snow-covered ground.

To determine whether it is worth spending money on fertilizing it is necessary to carry out an appraisal of costs and benefits (see chapter 14). Top height/age curves for 0–20 years are used to convert height difference in experiments to the number of years crop development is advanced by fertilizing. This period is then used in deciding if the use of fertilizers is financially justified. In general, fertilizer is used only when the costs of applying it are less than the discounted value of the resulting increment of production.

Large areas of pure Sitka spruce in private and State forests, especially on heathlands and unflushed moorlands, are being treated by application of nitrogen (urea or ammonium nitrate) to maintain satisfactory growth rates. This programme is only carried out in plantations that have not achieved full canopy closure. Further nitrogen should not be required. Certainly there has been no consistent or predictable benefit from applying nitrogen to pole-stage Sitka spruce stands (McIntosh 1984).

Aerial fertilizing. Most fertilizing is done under contract by helicopter, allowing rapid execution of large programmes and reducing supervisory workload, although it can be prone to erratic distribution dependent on terrain, wind speed, quality of flying and equipment. Contracts should include provision for re-application to untreated areas. Monitoring distribution is difficult because of the pace of the operation, the many factors influencing it and the weather conditions for successful capture and measurement (Farmer *et al.* 1985). Poor distribution has been a widely recognized problem (Taylor 1990). More accurate systems of helicopter navigation are being sought (Potterton 1990; Kensington 1990; Taylor 1991).

Phosphate is generally applied as unground rock phosphate (mainly from Tunisia and known as Gafsa) and is characterized by its dusty nature, making it prone to drift during aerial application. Granular rock phosphates are also available. Potassium is applied as muriate of potash in the form of crystals and is admixed with unground rock phosphate. As to the notable benefits of application of potassium fertilizer to Sitka spruce on deep peat see Dutch *et al.* (1990). Nitrogen is applied as urea, in prill form. The urea is easier to spread consistently than P or PK. The following rates have been used by the Forestry Commission: 450 kg unground rock phosphate (= 60 kg P/ha); 200 kg muriate of potash (= 100 kg K/ha); and 350 kg urea (= 150 kg N/ha). Costs of fertilizer per tonne depend on analysis (N, P, K), size of order, delivery cost and season of delivery. Costs of application by helicopter depend on terrain, distance involved, extent and weight. Owner's management and 'ground control' would be additional.

Top-dressing of broadleaves. Most commercially managed broadleaved stands are on relatively fertile soils and will not normally require addition of fertilizers. Evans (1984) points out that there is some evidence of variation among broadleaved species both in their site fertility requirements and response to fertilizers when added. For example, oak is relatively insensitive to soil fertility though not to other site factors such as exposure, whereas ash and sycamore perform increasingly well with increasingly nitrogen rich sites. In soils of high pH with free calcium carbonate fragments (chalk or limestone) in the surface layers, iron and manganese become insoluble and uptake by the roots is reduced or prevented. This causes yellowing of the main part of the leaf, while the veins remain green – a condition known as lime-induced chlorosis. Tree species vary in sensitivity: Norway maple, sycamore, ash and lime are fairly tolerant; beech, oak and southern beeches are more sensitive. Before applying fertilizer to an established broadleaved crop, foliar analysis should be carried out to determine whether any nutrient appears to be in short supply. It is possible to indicate average concentrations of the most important nutrients for the main species. Evans (1984) and Taylor (1991) explain the procedure for obtaining an analysis.

In all top-dressing with fertilizers, the environmental implications must be borne in mind. Indiscriminate application can result in the enrichment of drainage water and consequent eutrophication of watercourses, lakes and reservoirs. Fertilizing of non-forest land can also harm various habitats of rare plants, for example on boglands. Useful guidelines for preventing such environmental damage have been provided by Binns (1975). Fertilizing regimes include many complex interacting variables: choice of target crop, timing, method and intensity (Taylor 1991).

3. Protection against insect pests and diseases

Forest crops need protection to ensure successful survival to their desired rotation. Damaging agents can disrupt orderly regeneration, tending, harvesting and marketing. Causes include extremes of climate (particularly excessive heat or cold, drought, wind or snow), nutrient deficiencies, air pollution (all discussed in chapter 3) and certain insects, diseases, animals, birds and fire.

The probability of damage changes over time. Control measures need to be devised and applied. Strategies to protect the crop can be classified, according to relevancy, as initial (fencing and fire-breaks), and recurrent (forest hygiene, control of rabbits and grey squirrels, fire prevention measures and treatment against insects and diseases). Nature conservation considerations are noted in chapter 13, section 7.

When the cost of protective measures is combined with loss of growing stock and growth increment, the climatic, biotic, mechanical and environmental agents place a heavy financial burden on a forest crop. Consequently, the impact of these agents on silviculture should be kept as low as possible: avoiding loss of revenue is equivalent to gaining revenues.

Pests and diseases can be carried on plants, seeds, wood and wood products. Many of these articles when being imported must meet strict landing requirements to prevent the introduction of new pests or diseases. In some instances where the nature of the risk is too high, for example unseasoned 'green' softwood with bark, the import of these goods is prohibited. (A particularly dangerous pest to be excluded is *Ips typographus* – the eight-toothed spruce bark beetle.) Guides for importers are to be found in Forestry Commission 'Plant Health Leaflet No. 5' (1990). Guides for exporters have been noted in chapter 2.

Insect pests

Many species of insects can damage a forest crop at different times and stages of its growth. There are relatively few which foresters may encounter in sufficient numbers to have to treat them as serious. Consequently it is hard to justify the need for foresters to possess a *detailed* knowledge of their life-cycle and habits. The subject is more the province of the entomologist, whose help when needed is readily available.

Forestry Commission Entomology Branch carries out basic and applied research into the ecology and control of insect pests. As a means of developing rational control programmes, studies are made of their distribution, abundance, biology and inter-relationships with the state of the host tree. At the same time, research on improvements in chemical and biological control methods are sought. Research on important defoliators is aimed at increasing understanding of the underlying factors leading to pest population outbreaks: these are controlled conventionally by insecticides but the aim is to develop control strategies encompassing several disciplines in an integrated manner. Other investigations include the use of insect viruses for pest control, the development of more efficient spray application systems, and the use of pheromones for monitoring and controlling insect populations. The Branch provides a comprehensive advisory service to foresters, both within the Commission and in private forestry. Enquiries should be addressed to the Entomology Branch as follows:

(i) South of the Humber/Mersey line. The Forestry Commission Research Station, Alice Holt Lodge, Wrecclesham, Farnham, Surrey GU10 4LH (Tel. 0420–22255);

(ii) North of the Humber/Mersey line. The Forestry Commission Northern Research Station, Roslin, Midlothian EH25 9SY (Tel. 031–445–2176).

An important basic guide to insects feeding on trees in Britain is provided by Bevan (1987); it is supplemented by other literature of the Forestry Commission. A selection of the more important insects follows.

Different insect pests are associated with different ages of the host tree crop. The early years of a conifer crop are much more critical from the point of view of insect damage than in the case of a broadleaved crop. Most serious is when a new conifer crop is to be a restocking of another one recently felled. In such cases, the Large pine weevil *Hylobius abietis* and the Black pine beetles *Hylastes* spp., which have multiplied in the stumps and roots of the previous crop emerge to feed upon the young plants and may effect heavy mortality if no protective measures are taken. Stoakley and Heritage (1990) give information on chemical methods for control of these insects. Two other weevils which damage young conifers are the claycoloured weevil *Otiorhynchus singularis,* and the small brown weevil *Strophosomus melanogrammus*. Both can be controlled by spraying with Gamma HCH.

In the first ten years or so after establishment, sap sucking adelgids may attack conifers such as Douglas fir, the larches and spruces. Although such attacks can retard growth it is seldom economically worthwhile to attempt control artificially. Pine sawflies, *Diprion pini* and *Neodiprion sertifer* are conspicuous on young pines and occasionally defoliation may be almost complete. Defoliation seldom brings about death of the tree although there may be a noticeable decrease in height increment particularly for trees attacked by *N. sertifer*. A viral insecticide, marketed as Virox, has been used successfully for control of *N. sertifer* in Scotland, applied either from the air or from the ground; it acts to prevent early damage and loss of increment. Outbreaks seldom persist for more than two or three seasons before they collapse naturally. In young broadleaved crops, defoliation by leaf beetles and by the caterpillars of moths and sawflies is sometimes encountered. Serious damage is rare and recovery is normally very good.

In the thicket- to older stages, many species of leaf feeding insects cause damage of varying degrees of severity, the most important defoliators being the caterpillars of moths and sawflies, and aphids. Sitka spruce is frequently defoliated by the Green spruce aphid *Elatobium abietinum*. Outbreaks are invariably associated with mild winters, −8°C being a threshold low temperature for winter survival. Although recovery from attack is normally good, that is unless site conditions are particularly adverse, considerable loss of increment may result from severe defoliation. Control measures are impracticable, partly due to difficulties of forecasting severe attacks and thus of taking timely action, partly for the doubtful economics of such action, and partly for ecological objections to wholesale insecticide applications (Carter and Nichols 1988; Carter 1989).

The grey aphid *Adelges laricis* can result in wholesale canopy discoloration and degrade of European larch, and may also be the prime factor in bringing about the condition known for many years as 'dieback of European larch'. Alpine origins/provenances of larch are particularly susceptible to this malaise, whilst Carpathian provenances and hybrid larches are less so, and the Japanese larch is virtually resistant.

Examples of defoliators among the moths may be found in several species. The Pine looper moth *Bupalus piniaria* has periodically resulted in defoliation and mortality, from secondary causes, of Scots pine. Epidemic populations have been the subject of control operations; the most recent in Tentsmuir in 1984 utilized the latest spray technology and the relatively specific insect growth regulator insecticide diflubenzuron (Dimilin). By utilizing ultra-low volume technology the dosage for effective kill was reduced to 25% of the recommended rate.

The Pine beauty moth *Panolis flammea* has attacked lodgepole pine since 1976 in north Scotland. Outbreaks have led to severe defoliation and deaths of trees. The organophosphorus insecticide fenitrothion, applied by aircraft, has given satisfactory control – generally about 98% mortality (Leather *et al.* 1987). As mentioned for Pine looper moth, new technology for aerial application has been developed such that both fenitrothion and diflubenzuron are used at volume rates of 1 litre per hectare thus minimizing environmental contamination and maximizing control. The same technology is being used in joint work with the NERC Institute of Virology and Environmental Microbiology at Oxford to develop a biological control using a naturally occurring virus. Indeed this approach is now the preferred option in environmentally sensitive areas where control of the Pine beauty moth is necessary.

The Pine shoot moth *Rhyacionia buoliana* is a pest causing damage to stems of Scots and lodgepole pines. The main damage results in characteristic distortion of the stem, which often becomes the point of later breakage by wind and snow. Silvicultural treatments which encourage rapid growth of young trees through the susceptible range of height will help to reduce attack and aid recovery.

Sawflies, such as *Gilpinia hercyniae* on spruces, and *Cephalcia lariciphila* on larch, can reach damaging levels in pole- and later stages.

In general the control of defoliators is a complicated operation, since usually fairly large areas are affected, and special equipment has to be used. Expert guidance should be sought from the Forestry Commission Research Division.

The Oak leaf roller moth *Tortrix viridana* periodically causes damage to oak woods, particularly older ones. The trees usually recover well, assisted by heavy lammas shoot production, but a distinct loss of timber increment results. The later flushing sessile oak is less susceptible to heavy infestation than is the pedunculate oak.

Bark and wood feeders (beetles and weevils) are, in the main, secondary pests whose numbers are dependent on the availability of suitable breeding sites in the form of debilitated or damaged trees or felled produce. Multiplication usually takes place beneath the bark. When numbers of these insects are high they can, under certain circumstances, attack and damage healthy growing crops. The most important problems are connected with bark beetles on spruce, pine and larch, and also weevils on the pines. The insects concerned include the weevils *Pissodes* spp., Pine shoot beetle *Tomicus piniperda,* and the Larch bark beetle *Ips cembrae* (confined to east and central Scotland). Control of the beetles can be achieved by maintaining a good standard of forest hygiene, and it is thus a management rather than a strictly entomological problem. As a general rule it is wise to ensure that stems which are felled in thinning and clearing operations are not left on the site long enough for a brood to be produced from them. Material, therefore, should not be left in the forest for more than six weeks from the time of felling, during the period from April through to July, in the case of *Tomicus,* and rather later in the year for *Ips.* If removal within this time-limit is not feasible the bark beetle brood should be destroyed, either by debarking the timber or by protecting cut logs with an aqueous spray of Gamma HCH. Good forest hygiene is also effective in controlling numbers of *Pissodes* weevils.

By far the most serious bark feeding pest is the Great spruce bark beetle *Dendroctonus micans,* discovered in Britain in 1982. It breeds under the bark in extensive chambers, and all stages of the beetle can often be found at any time of the year. All species of spruce are susceptible and attacks can cause not only death of large patches of bark resulting in severe damage and distortion to the trunk but also death of the tree. Attacks are generally signalled by abnormal resin bleeding with obvious tubes of resin exuding from the stem. The present population of the pest is limited to

Wales, the West Midlands and Lancashire (King and Fielding 1989). The Forestry Commission continues to pursue a vigorous campaign against the beetle. A strategy has been developed based on reducing the chances of dispersal of the beetle on infested timber combined with biological control through breeding and release of the imported specific predatory beetle *Rhizophagus grandis*. At the heart of this strategy is a legally enforced 'Scheduled Area' within which all movements of spruce are subject to licensing to approved mills equipped with facilities for debarking logs and treating the residues to prevent further infestation. At the same time surveys around the edge of the infested area provide information on the natural rates of spread of *D. micans,* which appears to move no more than about five kilometres per year.

Of stem feeders, the more important species belong either to the family of so-called scale insects or to the woolly aphids or adelgids. The Felted beech coccus *Cryptococcus fagisuga* may produce unsightly quantities of white waxy wool on the stems and branches. It is associated with the fungus *Nectria coccinea*. A joint attack of these two organisms can cause a serious canker and dieback condition in beech crops. The Ash scale *Pseudochermes fraxini* is associated with, and may be a contributory cause of, a debilitated condition of ash. Among conifers, conspicuous stem infestation may be seen on *Abies* spp., sometimes leading to a form of timber degrade. Infestation on Weymouth pine appears to have no noticeable direct effect although stems affected are often also attacked by the pathogenic rust fungus *Cronartium ribicola*.

The Control of Pesticides Legislation 1986 means that all insecticides used in forestry must be approved for the purpose. In some cases, as described for *H. abietis* control, this is in the form of 'off-label' approval since the actual methods are not covered by the manufacturer's own label. A limited number of insecticides have full label approval. There is also an 'interim arrangement' such that any insecticides used in 'ornamental' situations are covered for use in forest nurseries. However, if there is any doubt it is advisable to consult the Forestry Commission or an appropriately qualified contractor.

Insects and storm-damaged trees. Following the storm of 16 October 1987 (and equally applicable to the gales of January and February 1990) the Forestry Commission Research Division provided invaluable guidance outlining the risk of attack by insects on windthrown and damaged broadleaves (Winter 1988) and conifers (Winter and Evans 1990).

Often in forestry, quite evident insect damage has to be tolerated simply because control action would be uneconomic, or ecologically or tactically undesirable. Artificial control becomes urgent where crop survival is in jeopardy, and examples may be found both in the nursery and during the establishment phase of new crops and in older crops. The abundance of some insects and the damage they do is directly related to the general health condition of the crop. By observing the rules of good silviculture and maintaining the crop in a sound condition, the scale of damage inflicted by many pests can be considerably reduced. Good silviculture alone, embracing correct choice of species and careful subsequent tending of the crop, will not result in immunity from all insect troubles. Such measures may well help to ward off 'secondary' pests whose increase is dependent upon the appearance of some predisposing factor; but a number of insect pests, including some of the most harmful species, are capable of attacking and seriously damaging if not destroying apparently healthy and well-tended stands. The latter types of insect is commonly referred to as 'primary'. Different insect pests are associated with different ages of the host crop. The main pests which occur in nurseries have been noted in chapter 2.

The concept of biological control is an attractive one for the long term management of insect pests, and relies on sustaining a balance between the natural insect enemy and its prey. The basic tenets of biological control are the identification of a suitable natural enemy and assessment of whether it has the potential to regulate its prey. Such an approach is being adopted for control of *D. micans* by the introduced specific predator *R. grandis* (Evans, H.F. 1989). It is important to realise that biological control does not imply eradication of the pest insect but merely to keep it at levels acceptable to foresters. The ideal approach results in cycles of abundance of both natural enemy and the prey such that the pest density remains below the acceptable damage threshold. The solution to the question of which control method to use to combat tree pests is often a mix of pragmatism, detailed ecological study, and sometimes luck! (Evans, H.F. 1989).

Diseases

Several fungal and bacterial diseases can damage a forest crop. Foresters cannot be expected to possess a *detailed* knowledge of them, but should be aware of the most prominent ones; the remainder are more the province of the pathologist and mycologist, whose help when needed is readily available.

In many cases living factors (pathogens) and non-living factors (climate, site factors and air polution – all discussed in chapter 3) interact to bring about tree diseases. Some diseases develop because the trees are growing under unsuitable conditions. Others may appear following wounding, for instance after careless brashing, pruning or extracting. The least that foresters can do is to diminish losses by correct species choice, minimizing other damage, and removing diseased trees whenever noticed and always during each thinning. Sound silviculture will help to keep disease to a minimum.

An important basic guide to forest diseases in Britain is provided by Phillips and Burdekin (1982), dealing with the causes, symptoms and diagnosis of disease, and with control measures, including plant health legislation. They discuss diseases caused by living agencies with a wide range of hosts; and fungi that cause tree decay. They summarize the diseases of the main individual tree genera grown in Britain. The text is supplemented by other literature of the Forestry Commission. The Commission's Pathology Branch undertakes work on diseases and disorders of trees, including those caused by biotic factors such as bacteria, fungi and viruses (as well as abiotic factors such as winter cold, frost and drought). The Branch provides a comprehensive advisory service to foresters, both within the Commission and in private forestry. Enquiries should be addressed to the Pathology Branch, as noted under 'Insect pests' *supra*.

Diseases in woodland are best controlled by avoidance. Over the years such control has to some extent been achieved, simply by the commonsense planting of tree species which are known to thrive, in preference to those that do not (i.e. correct species choice).

If regeneration is by planting, control of disease in the forest begins with control in the nursery, because many of the pathogens of the early life of the crop may originate there and may be taken onto the forest site by infected plants. Examples are the blister rust *Cronartium ribicola* of five-needled pines, and the more pathogenic species of *Lophodermium* on Scots pine; as well as some common foliar diseases of the nursery, such as *Meria laricis* on larch and *Didymascella thujina* on western red cedar. When natural regeneration is used, the young crops are affected mainly by pathogens already present on the forest site. A frequent source of damage originating from the establishment stage lies in the residues of previous crops, particularly stumps and roots which can carry pathogenic root fungi.

Diseases of general importance. The roots of both coniferous and broadleaved trees are susceptible to attacks by Honey fungus *Armillaria* spp. There are a number of species of *Armillaria,* each with its own ecological and pathological attributes. Most infection is by means of brown to black bootlace-like strands (rhizomorphs) that grow through the soil from infested wood, often that of broadleaved stumps (Greig and Strouts 1983). In their early years conifers are particularly liable to be killed on sites which formerly carried broadleaved trees. Usually deaths are so scattered that appreciable gaps are not formed, but if they do occur it may be necessary to restock them with more tolerant trees such as most broadleaves and Douglas fir or species of *Abies.* Commonly planted conifers, as they increase in age, become more resistant to killing by the disease, but root decay may render them liable to windthrow. Honey fungus is rarely troublesome in broadleaved woodland.

Stem decays are initiated when wounds invite fungi and bacteria to enter living sapwood. There are wound-associated stem rots of both conifers and broadleaved trees. Some are initiated through extraction damage, as with *Stereum sanguinolentum;* others through brashing and pruning wounds. Currently available wound treatments are of little value in protecting against decay.

Diseases of conifers. Decay fungi in conifers have been described by Greig (1981) and Phillips and Burdekin (1982). The most serious cause of disease in British coniferous forests is a white rot fungus *Heterobasidion annosum* causing root rot, butt-rot and, on certain sites with special characteristics, death. The potential for damage is great as all conifers commonly grown in Britain are susceptible. The larches and spruces are very liable to butt-rot, western hemlock and western red cedar being the most susceptible species of all. The pines, which are more resistant to butt-rot, are particularly susceptible to root killing on high pH soils and on former arable land. Douglas fir and species of *Abies* are not particularly susceptible to butt-rot or to extensive killing of roots. However, they may be rendered liable to windthrow by partial decay of their roots. The disease is present in most areas that have previously carried coniferous crops. The development of the disease in first rotation crops and its further development in the second rotation can be greatly retarded by the prompt treatment of freshly cut stump surfaces with a solution of urea, a non-toxic, nitrogenous fertilizer. Pine stumps, but not those of other conifers, can be treated with commercially produced spore suspensions of another fungus *Peniophora gigantea (Phlebiopsis)* which prevents the entry of the pathogen and decays the stump without posing any threat to standing trees. The relatively present low level of disease in Britain's forests will only be maintained if stump treatment is continued. All conifer stumps greater than 2.5 cm in diameter should be treated immediately after felling. An infected site may be restocked with a species showing some resistance to *H. annosum,* such as Douglas fir, grand fir and Corsican pine.

Butt-rot of conifers caused by the fungus *Phaeolus schweinitzii* is of low incidence and occurs mainly on sites that have previously carried broadleaves or pines. Group dying of conifers is caused by the fungus *Rhizina undulata* the spores of which readily germinate in the soil after they have been heated, e.g. where workmen's fires have been lit; the disease spreads through the litter and upper soil damaging or killing roots as it progresses. In plantations the disease can be prevented by prohibiting the lighting of fires (Greig 1981).

Larch canker and dieback is a bark disease of European larch caused by the fungus *Lachnellula willkommii* (Phillips and Burdekin 1982). Susceptibility to canker is related to the origin of the larch, high alpine provenances being the most susceptible and Carpathian provenances such as Sudeten larch, the least. Hybrid and Japanese larch are generally resistant.

Corsican pine in the north and west of Britain is very susceptible to the disease *Brunchorstia pinea* caused by the fungus *Gremmeniella abietina* (Phillips and Burdekin 1982). Damage may vary from scattered shoot dieback to death of entire trees. The disease has sometimes appeared on Scots pine in certain upland sites in north-east England and south Scotland. (A similar disease, caused by the fungus *Ramichloridium pini* occurs in south-west England and south Wales.)

Resin-top disease of Scots pine, caused by the rust fungus *Peridermium pini*, is most prevalent in east England and north Scotland. Vigorous crops are the most susceptible, and dominant trees often the worst affected. Infected trees should be removed in the course of thinning. Needle-cast of pine can be caused by several fungi. Some of the most spectacular damage is caused by *Lophodermella sulcigena*, which attacks the current year's needles, particularly those of Corsican pine. Severe outbreaks are normally confined to western and northern parts of the country, and are usually too infrequent to cause lasting injury.

Diseases of broadleaves. The common decay fungi in broadleaved trees have been described by Burdekin (1979) and Phillips and Burdekin (1982). The most important disease of poplar is bacterial canker caused by the bacterium *Aplanobacter populi*, which gains entry through fresh leaf scars and wounds in the bark. Losses from the disease can be reduced by the use of resistant varieties (see chapter 5), and by removing all cankered trees on or next to the site that is to be planted with poplars. Ash die-back is discussed by Hull and Gibbs (1991).

Beech bark disease, noted under 'Conifers' *(supra)*, affects beech from the pole-stage onwards (Lonsdale and Wainhouse 1987). It develops where bark first infested by an insect, the felted beech coccus *Cryptococcus fagisuga*, is subsequently invaded by a fungus, *Nectria coccinea*. Prompt felling of affected trees is required if the timber is to be utilized, and recently infected trees and those markedly infested should be removed during thinning.

Dutch elm disease (Gibbs *et al.* 1977), unfortunately so prevalent during recent decades, is caused by the fungus *Ceratocystis ulmi*, spread by the Elm bark beetles *Scolytus scolytus* and *S. multistriatus*, which breed under the bark of diseased trees. Greig (1988a) gives information on the breeding of elms resistant to Dutch elm disease, as well as on English elm regeneration (1988b). In a few areas there are powers to enable the local authorities to require owners to fell infected elms. These are contained in the Dutch Elm Disease (Local Authorities) (Amendment) Order 1988. A sister order, the Dutch Elm Disease (Restriction on Movement of Elms) (Amendment) Order 1988, prohibits the movement of elm into any controlled area from any place outside it and also restricts the movement of elm within the controlled area to wood which has had the bark removed or to wood being moved in accordance with the terms of a licence issued by the Forestry Commission. Details of areas covered by these Orders are available from the Commission (to which applications for licences to move elm timber should be addressed).

4. Protection against animals and birds

Plantations require protection from semi-domestic and wild animals, also very occasionally from a few species of birds. Horses, cattle, sheep and goats can inflict substantial damage, particularly on broadleaves, by barking, browsing and trampling. The only sure remedies, apart from care always to close gates, are treeshelters or efficient fencing and, where appropriate, the construction at approaches to woodlands of grids against cattle and sheep (arranging escape routes for hedgehogs and other small mammals).

Woodland animals. Several species of wild animals damage trees and cause conflicts between management objectives, in particular those of timber production, farming and wildlife conservation Of utmost concern to silviculture are deer, rabbits, grey squirrels and, less frequently, voles. Evans (1984) provides a table detailing the main form of mammal damage that can be expected. Fencing or guards are often necessary, as is control by man. Chemical repellants[3] are rarely used. Control by man is often necessary.

Rabbits are one of the most harmful animals with which foresters have to contend – preventing natural regeneration by eating seedlings, distorting growth of leading shoots, and killing or seriously checking tree growth by stripping bark. Newly planted trees, as well as older thin-barked trees such as beech and ash, are particularly vulnerable. Rabbits will consume buds and shoots up to a height of about 50 cm, leaving a sharp angled cut on the end of the stem and branches. Damage occurs particularly during winter and early spring and especially in periods of prolonged snow cover. Control is recommended during the period from October to mid-March. Gassing is the most effective method, although snaring, springtrapping and ferreting can be followed. Shooting is best regarded not as a protection method but as a sport.

Myxomatosis, no longer always lethal, cannot be considered the answer to the rabbit problem. Their control should be pursued by every legitimate means. All concerned with silviculture should try to have all the land within their management free of

---

3 *Chemical repellants* for protecting young trees from mammals are costly, often unpractical for more than a limited number of trees, and certainly less effective than either fencing or tree guards/treeshelters (Pepper 1978). Application of them is labour intensive, the effective life is limited, and treatments need to be reapplied, at least annually. Chemical repellants such as AAprotect (winter application only against deer, hares and rabbits), and Dendrocol 17 (winter application only, against deer) can be sprayed or painted on trees to reduce winter damage from mid-November. They are phytotoxic if applied to actively growing trees and are therefore unsuitable in May and early June. Their use falls within the restrictions imposed by the Control of Pesticides Regulations 1986. Fowikal, which is only effective for about 6 weeks and needs to be reapplied, is mainly used to protect garden trees and shrubs.

rabbits and to persuade their neighbours to act likewise. Only thus can wasteful damage and the great expense of protection be alleviated. To this end rabbit clearance societies are playing a useful role in some districts. However, complete eradication of rabbits is usually impracticable. The main aim should be to reduce their numbers to levels at which losses to trees are economically acceptable. More effective results will be achieved if several adjoining properties are treated at the same time in a co-operative exercise. Control over a substantial block of land will also reduce the rate of re-infestation.

The Ground Game Act 1880 gives an occupier the right to shoot rabbits during the day and to authorize in writing one other person to do so; the person must be a member of the occupier's household or staff, or be employed for reward. An occupier can apply for authority to use a reasonable number of extra guns, if the owner of the shooting rights will neither permit the occupier to bring on extra guns, nor undertake to control the rabbits. Under the Act as amended by the Wildlife and Countryside Act 1981, the following are allowed to shoot rabbits at night: an owner-occupier with shooting rights, a landlord who has reserved his shooting rights, a shooting tenant not in occupation who has derived his shooting rights from the owner, and an occupier or one other person with shooting rights. The Firearms Act 1968 requires any person possessing, purchasing or acquiring a shotgun to obtain a shotgun certificate from the police. A fee is payable for a certificate valid for three years. Springthorpe and Myhill (eds.) (1985) explain the law relating to firearms, traps and poisons.

Gassing, when correctly used under the right conditions (i.e. in dry periods, and low wind, by two men), is highly efficient. Rabbits lying out must be driven to ground with dogs and guns before warrens are gassed. Care needs to be exercised to ensure that gassing operations are not carried out where badgers would be at risk. Gassing involves introducing either a powdered cyanide compound which gives off hydrocyanic acid gas, or aluminium or magnesium phosphide tablets which emit phosphine into the burrow systems. The most extensively tested and widely used fumigant in Britain is hydrogen cyanide, Cymag (ICI), which emits the gas on contact with moist air or moist soil. The powder can be blown into burrows by power or hand pump, or placed with a spoon just inside the entrance to the burrow.

The Pests Act 1954 permits use of approved spring-traps designed to catch and kill rabbits humanely but makes it an offence to set spring-traps other than in a burrow. The Protection of Animals Act 1911 requires all traps to be visited at reasonable intervals and at least every day between sunrise and sunset. Snaring is undertaken with snares prepared from six or eight stranded brass wires running freely through an 'eye' made in one end of the wire. A loop of 10 cm diameter is held about 10 cm from the ground with a short notched stick. The free end of the wire is securely tethered with a strong rotproof cord to a peg which is long enough to be driven firmly into the ground to prevent a snared rabbit dragging it up. Snares should not be set where livestock are present or where there is any risk to domestic pets. Care is needed in siting snares since the law requires that all reasonable precautions be taken to avoid catching protected species. Every snare must be recovered at the end of a session. The Wildlife and Countryside Act 1981 prohibits the use of self-locking snares and requires snares to be visited daily; they are best inspected at both dawn and dusk.

Fencing with wire netting is used where the rabbit harbourage makes other techniques impractical or where complete exclusion is the aim. In many situations fencing, or the use of plastic treeguards or treeshelters, is a more cost-effective prevention of damage measure when compared with control methods that have to be taken year after year. Specifications of fences and treeshelters have been given in chapter 5.

In addition to the above points of law relating to rabbits, an Order under the Pests Act 1954 declares England and Wales a rabbit clearance area in which all occupiers of land are responsible for destroying wild rabbits on the land. Where this is not reasonably practical, occupiers must take steps to prevent damage. The Minister has powers to require rabbit control to be carried out; if this is not done, he may arrange for the necessary work at the expense of the occupier, who would also be liable for a fine.

Further advice on rabbit control can be obtained from ADAS Regional Wildlife Biologists. Useful Forestry Commission publications include that of Tittensor and Lloyd (1983).

Hares travel afar, and where allowed to become too numerous will damage young plantations, biting off the shoots of young trees in a way similar to rabbits but leaving the shoots lying on the ground beside the tree. Frequently a hare will damage a group or row of trees. Fencing and treeshelters help to give protection. Brown hares are found in reducing numbers in the lowlands, and Blue hares in the uplands.

Badgers are widely distributed throughout much of the British Isles, typically in the more hilly and well-wooded districts, and are particularly common in the southern and western counties of England and Wales. Their first choice is for broadleaved woodland, nearly half the setts in Britain being in this type of habitat (Neal 1982). Badgers are beneficial in eating young rabbits in relatively large numbers; nests of young field voles and wood mice are likewise consumed. Their chief harm from the silvicultural viewpoint is the habit of pushing up, or digging under rabbit-proof fences, creating gaps whereby rabbits can enter plantations. The remedy is to fix a small 'badger gate' – free to swing both ways (Rowe 1976). None of the foregoing activities are likely significantly to affect tree survival, and the badger, fully protected, is an important component of the woodland ecosystem.

Pine martens, once widespread, survive only in the Scottish highlands and four widely separated areas in England and Wales. There are a few in the Lake District and Wales (Hurrell 1968). They may appear in places away from the areas mentioned. Pine martens are not specially attached to pines, but forests, including pine forests, can be a natural habitat. They do not harm trees, and keep in check a number of mammals potentially harmful to silviculture. This fascinating mammal should be allowed to increase in the few woods where it is found and to extend its range to other regions. Foresters are keen to protect them: they are protected by the Wildlife and Countryside Act 1981.

Polecats are generally regarded as one of the rarest of British mammals but are still common in parts of the North and mid-Wales (Poole 1970). The ferret is a polecat which has been domesticated at least since Roman times. Polecats live in burrows and are solitary animals. The practice of plantation afforestation may have influenced polecats' distribution, as young conifers surrounded by tussocky grass and abundance of small mammals provide an ideal habitat: here its prey includes voles, mice, rabbits and hares. Polecats are protected by the Wildlife and Countryside Act 1981.

Voles and Field mice are long-established natives, which may be found in and around practically every wood and plantation in the country (Rogers Brambell 1974). Woodland mice are discussed by Gurnell (1979). In times of their population increase and when coupled with scarcity of food, the Field vole in particular may do grave damage to young forest crops, stripping the bark from the roots and lower stem of trees up to the height of the surrounding vegetation. Often the tree is completely girdled at the base, and recovery is unlikely. The form of damage by the Bank vole is sometimes the barking or girdling of taller saplings higher up the stem. All tree species may be damaged but broadleaves are generally most vulnerable. Damage can occur at any time of year but is most likely when the small mammal numbers are high and their main food, green grass, is scarce as in late winter and in early spring when the first flush of grass has been delayed by cold weather. All kinds of voles and Field mice readily consume tree seed. Bank voles will climb both small and large trees, particularly pine, and eat the buds; they occasionally remove the thin bark from the main stem and branches.

Control of voles and mice in woodlands presents difficulties of an economic rather than a practical character, for when plagues occur (usually over a three to five year cycle) their numbers become so enormous that it is difficult to reduce them at an acceptable cost. At such times, Short-eared owl often appear in increased numbers and thrive on the abundant small mammals. Other predators include foxes. Methods of control are trapping, poisoning and chemical repellants. Split plastic tube tree guards, without ventilation holes and at least 20 cm tall (they must be above the height of the surrounding vegetation) can be placed around the base of young trees and pushed into the soil about 0.5 cm. They expand as the diameter of the stem increases and they do not have to be removed (Davies and Pepper 1990). The incidence and severity of damage can also be reduced by eliminating weed competition. Dormice, however, are specially protected.

Grey squirrels. Bark-stripping damage to trees by grey squirrels is a major concern to many woodland owners (Rowe 1980; Pepper 1989). High costs and many difficulties are experienced in growing some species of broadleaves in areas when populations of grey squirrels are high. Damage normally occurs during the months of May, June and July and occasionally in August. Sycamore, beech, ash, Norway maple and birch are the most frequently and severely damaged. Oak and sweet chestnut are less frequently attacked. Wild cherry notably is not often damaged. Scots and Corsican pine are the most frequently attacked conifers. Generally fewer than 5% of damaged trees are killed, but the damage often results in a degrade in timber quality. Pole-stage trees aged ten to forty years old are the most vulnerable. Trees younger than ten years are not generally large enough to attract or support the weight of squirrels, although five- to ten-year-old sycamore may be, and the bark on the main stem of trees older than forty years is too thick to be stripped. Stripping can occur anywhere on the main stem from the base to 5 cm top diameter in the crown. When damage is confined to the base it is similar in appearance to rabbit damage.

Pepper (1990) summarizes the legislation relevant to the control of grey squirrels. Tree protection can best be achieved by the annual control of numbers before and during the damage period (April to July). Destroying squirrels at any other time of year will not reduce subsequent levels of damage. Control must be concentrated in and around the ten- to forty-year-old pole-stage vulnerable crops. It is important to locate areas which hold and maintain a high squirrel population, but in which damage may not occur. These 'holding areas' are often the more mature mixed woods which may not be vulnerable, but nevertheless need effective control to protect adjacent pole-stage stands.

Shooting, with or without drey poking, is ineffective in controlling numbers sufficiently to have any impact on damage levels. Cage-trapping, spring-trapping and poisoning are the only recommended methods. Poisoning is the cheapest and most effective method. The 1973 Grey Squirrel (Warfarin) Order permits in England and Wales, with the exception of a few specified counties, the use of warfarin in hoppers of approved design. It also specifies the bait, 0.2% warfarin on whole wheat, and the design and dimensions of the hopper that must be used (Pepper 1989). The wheat-warfarin bait is available ready mixed but is expensive: it is cheaper to obtain the approved warfarin liquid concentrate available in 500 ml bottles and mix it with locally obtained wheat. Gloves and masks should be worn at all times when mixing bait and filling hoppers.

The success of the control operation depends on the ability to find sites for hoppers that squirrels will visit regularly to feed. The best sites are invariably under a large tree and where the ground is clear of vegetation. Hoppers should be spaced approximately 200 m apart and distributed throughout the control area at a density of one hopper to 3–5 ha. It is essential that a continuous supply of bait is available to the squirrels, and the hoppers must not be allowed to become empty. Every possible attempt should be made to safeguard dormice and other small mammals and birds: in order to achieve this objective, a modification of the hopper tunnel entrance has been developed. Control should stop at the end of July or second week of August at the latest. The Forestry Commission encourages and assists neighbouring woodland owners to form Grey Squirrel Control Groups to coordinate and concentrate their control effort: co-operative effort is essential. They will also give advice and practical training courses. In areas vulnerable to grey squirrels, prescriptions for their control are an obligatory part of the approved plan for which grant-aid will be paid.

Red squirrels, especially in Scotland, occasionally become numerous enough locally to do damage in pine plantations. They eat seeds, particularly of conifers, fruits, buds and shoots, and occasionally eggs and young of birds, but foresters are anxious to protect them. Their most important habitat is conifers (especially Scots pine, which produce more regular cone crops than Corsican pine, and its cones contain a greater number of seeds). Older conifers produce more cones, but trees of twenty to forty years of age provide best shelter. Whereas the grey squirrel cannot survive on coniferous seeds alone but needs also a supply of broadleaved fruits such as those of oak, beech and sweet chestnut, the red squirrel survives mainly on coniferous seeds. Hence, the management of age structure and species of conifer, as well as the proportion of broadleaved trees, will be important factors to encourage the maintenance and spread of red squirrels. They are protected by the Wildlife and Countryside Act 1981 (Tittensor 1975).

Foxes. The control of foxes may be an essential part of good neighbourly policy. Otherwise there is no economic or ecological justification to control foxes, or other predatory species (Lloyd and Hewson 1986). Foresters should be aware of the needs of sheep-farming neighbours in respect of fox control. Foxes often control rabbits and other small mammals.

Deer. Most woodlands of more than a few hectares are likely at some time to contain deer, not necessarily a resident population (Prior 1983). They rely greatly on the forest habitat, and in some instances are a welcome addition to the sylvan scene. They can be a serious problem to silviculture and entail additional costs, doing substantial damage by browsing,[4] fraying,[5] and stripping bark.[6] The most favourable habitat for roe and muntjac are young woods with much undergrowth which provides food and cover, e.g. bramble and shrubs. Thicket-stage stands of conifer provide cover but less food, reducing their carrying capacity. Red, fallow and sika prefer older or mixed-age woodland containing some pole-stage stands. Fallow, roe and muntjac cause concern to silviculture in lowland broadleaved woodland. In upland coniferous forests, roe and red deer create problems in afforestation and restocking (Ratcliffe 1987; McIntosh *et al.* 1989). However, deer are a component of the forest wildlife and they and their habitat should be managed in such a way as to minimize damage and maintain their numbers at an acceptable density.

The primary factors determining population levels of deer are the proportion of forage to cover and its arrangement in space and time. Failure to take account of their ecological needs and to design forest layouts which facilitate their management can lead to high and often ineffective expenditure on fencing, and difficult shooting (Savill and Evans 1986).

Large areas of coniferous forests have been established in upland Britain with little knowledge and experience of the potential problems arising from the creation

---

4 Browsing by deer is the removal of leaves, buds and shoots from a tree. All species will browse up to the heights they can reach (muntjac 0.5 m, roe 1.2 m, fallow and sika 1.4 m, red deer 2.0 m). Muntjac will also bend a whippy tree to the ground by walking over it and, while holding it down with their chest, will browse the top shoots. Deer will browse at any time, with peaks of activity at leaf flush and when alternative food is scarce. Browsing may result in reduced height and deformation such as bushing or multiple leaders, but rarely death.
5 Fraying by deer is the action of the males rubbing their antlers on young trees to clean them of velvet or mark territory. As a result, the bark often hangs down in tattered strips from the stem and branches. Branches are also broken and left hanging. The main fraying period for roe is from March to August and for the other species from mid-July to late October. Only relatively few trees are damaged in this way and it is most likely that they will be concentrated around the edge of the wood. Fraying is not usually of any economic significance.
6 Bark stripping is undertaken by all the large species of deer. Red, fallow and sika bite into the bark and pull it upwards leaving vertical teeth marks at the base of the wound. It is possible to distinguish the species responsible by the width and height of these teeth marks but there is some variability. Thicket- and early pole-stage trees of both conifers and broadleaves are attacked. Norway spruce, lodgepole pine, poplar, willow and ash may all be severely damaged. Brashing the trees allows deer easy access into the woods and thus increases the risk of damage. Disturbance of normal feeding patterns, for example by the replacement of cattle by sheep in surrounding fields, or increased use of the woods by the public can trigger serious bark stripping damage. Stripping occurs from January into early spring and especially during periods of snow. Wind-snap may occur, particularly in conifers. Timber degrade is inevitable.

of more suitable habitats for deer. Damage to commercial tree crops and the build up in woodlands of high densities of deer, have meant conflicts between deer and forestry and between the owner of the open hill and forestry. The *Factsheet: Deer and Forestry* (ICF 1989) has focused attention on the subject and suggested the strategy and tactics of deer control, including the influence on the forest design. If the principles of deer management are clearly understood, then the forest can be planned with deer as part of the ecosystem, but controlled so as not to incur unacceptable damage.

Good forest design can minimize damage and enable necessary shooting to be done relatively easily, and often profitably. In large areas where deer may be expected to become established, effective control through the rotation can only be possible if a percentage (5%–15%) of the area is left unplanted. The function of these open areas is to provide non-crop trees in places where the deer can be seen and controlled. For this purpose, as well as for the storage and handling of produce, landscape and nature conservation, parts of the road network should be left at least 20 metres wide. The banks of streams should be treated similarly. They often provide richer and favoured grazing. Other small plots of better soil should be planned as feeding areas: for the larger deer species these should be small and scattered (1–4 hectares), rather than large; roe prefer dappled shade with edible shrubs such as willow or rowan. Siting such areas on the edge of plantations can be beneficial. Damage can be reduced by leaving corridors of unplanted ground, or well-brashed young stands, or old stands enabling deer to move between high and low ground (Ratcliffe 1985). Deliberately planting preferred browse species of shrubs and trees, such as willows, can help if they are sited carefully and the quantities are related to animal density. Some species of commercial trees are more able to recover from damage than others: Douglas fir tends to quickly callous-over bark-stripped wounds, and the stem does not become rotten, whereas Norway spruce is more susceptible to decay. Sitka spruce recovers well from heavy browsing in the early years.

Establishing small or extensive areas of plantations where deer are present or in the vicinity, involves careful planning before planting, to avoid serious damage to the trees, coupled with an assessment of the possible benefits of control. In the conditions applying over most of the UK at present, the density of deer over the boundary is likely to be more than sufficient to replace culled animals very quickly. Protection measures in the establishment phase are expensive, but later, or in the case of larger woodlands of mixed age, part of these costs can be offset by managing the deer as a woodland asset, marketing venison and letting stalking rights. Management of deer for sporting is discussed in chapter 13.

Roe, the most widely distributed deer species, are encouraged by afforestation and by the clear cutting and restocking silvicultural system. They like clearings and forest edges for food, and thickets for cover, but dislike crops in the pole-stage and older because the understorey and vegetation are usually sparse. Populations have risen greatly in southern England, the Highlands of Scotland, Argyll, Galloway, northern Britain and mid-Wales and the damage done has often reached unacceptable levels. Increasing the number culled in the course of control measures can be balanced out by increased productivity and immigration.

Red deer are mainly found in Scotland – there are some English populations (Ratcliffe 1987) – and obtain most of their food in young plantations before the canopy closes as well as in clearings and glades. Young trees may be browsed as the deer retire to cover and during winter when food is scarce; bark stripping can occur during late summer and winter, with consequent degrade of timber. High populations can suppress or virtually ruin the broadleaved component in an otherwise coniferous forest. Simply reducing numbers is not sufficient to stop damage. Silviculture and management must be reconciled with the habitual pattern of dispersal and movement of the deer.

Fallow deer (Chapman 1982) are widely distributed in Britain, particularly in the Midlands and southern England. Some forests have held fallow for centuries, but more wild herds have become established as a result of escapes from parks. They are associated typically with broadleaved woodland but also frequent mixed woodlands. A wide range of plant species may be eaten by fallow so their diet will vary from habitat to habitat and with the season, according to availability. The acceptable density of a fallow population is dependent upon many factors which are not constant from place to place nor necessarily from year to year within one locality. A practical balance has to be found between the number of deer the habitat can support and the level of damage that can be acceptable to silviculture. Where land ownership is in smaller parcels than the home range of deer, an area management scheme involving collaboration with neighbours may be essential.

Sika deer are spreading in west and north Scotland, and there are some English populations. They prefer mixed woods, where they graze but rarely browse. Muntjac deer are spreading through southern England. They prefer woods with dense shrubs and herbs, where they browse and graze.

Woodland birds. Woodland birds have a high conservation value in both lowland and upland. They must be regarded as an amenity in woodlands and as part of a complex balancing mechanism which sound forest practice should try to preserve or recreate. Only occasionally should some of them (e.g. wood-pigeon) be treated as pests to be controlled. Bird management in woodlands is discussed in chapter 13. In this section reference is limited to those few species of birds whose activities may necessitate some control by foresters to protect trees – either in the nursery or in the woodlands.

Until a few decades ago it was sometimes believed that a fairly simple balance sheet could be drawn up to show the value or otherwise of birds to farmers, gamekeepers and foresters (Campbell 1974), some species being categorized as 'beneficial' (e.g. 'insectivorous' species) and others as 'harmful' (e.g. woodpigeon). Today there is a more enlightened approach by foresters and conservationists (Avery and Leslie 1990).

Significant damage by blackgame or capercaillie in Scotland (both in decline) is rarely encountered and then only when young plantations (principally pine) are being established in prime habitat for these species. Consequently in view of their conservation value the assumption should be that control would not be contemplated. Where this assumption is clearly contradicted by experience, the only effective measure of control is prudently to reduce the population of the offending species in and near the damaged plantation by shooting during the respective seasons. (There is a voluntary bar on shooting on Forestry Commission property.)

Pheasants cause no loss to the interests of forestry beyond consuming mast and occasionally ruining an unnetted nursery seedbed. (They do good by eating bark beetles, wireworms, and the pupae of such as Pine looper moth.) As for woodpeckers, any harm they do by pecking holes in commercial timber trees is only occasional and of little economic importance. The activities of wood-pigeon in eating acorns and beech mast are noted in chapter 13.

Starlings form communal roosts throughout the year apart from the breeding season, almost wholly in rural areas. Among preferred sites are thicket-stage spruce or mixed spruce and broadleaved woodlands. Some defoliation and breakage of branchlets and leaders of medium sized trees are due to the combined weights of many birds, but generally mechanical damage appears to be negligible. Of more concern are the chemical effects that droppings have on trees – scorching needles and twigs and probably killing the tree via the root system. Currie *et al.* (1977) explain methods of preventing damage, chiefly scaring techniques – particularly amplified distress call apparatus used during November to April. Large starling roosts can cause a rather unattractive area of damage within a woodland, but now few problems are posed.

Assessment of wildlife damage in forests. Foresters should be aware of the scale and extent of wildlife damage in their woodlands. Such information will be an aid to making decisions on policy related to wildlife management. As a quick and easy method of assessment, the 'nearest neighbour method', explained by Melville *et al.* (1983), is a handy tool for foresters to assess wildlife damage as a basis for evaluating management options. An annual assessment of vulnerable stands will provide them with information on the degree of damage, whether it is increasing, and whether or not to consider instigating preventive measures. Deciding on the stage at which damage levels become unacceptable depends on foresters' local knowledge of site and species. For example, when considering deer browsing pressure on young plantations, the tree species involved is important: Sitka spruce is able to withstand a much greater degree of leader browsing than Douglas fir, Japanese larch or broadleaves. The 'nearest neighbour method' can provide data on the severity of damage, how it is spread within a forest, and how it is changing from year to year. The impact of deer and other wildlife on silviculture can best be judged subjectively from local experience. Fencing of extensive woodland areas is usually a last resort, being costly and ineffective over long periods.

More emphasis is required on multiple objectives in woodland management, catering not only for wood production but also wildlife conservation, landscaping, recreation and shooting (deer and game birds) – instead of the present tendency to separate these activities. Foresters – as well as naturalists and conservationists – should aim to achieve high diversity by maintaining as wide a range of species (animals and birds), occupying as wide a range of niches as is possible within the limitations of the site. Animals and birds usually only become pests in particular circumstances, and species which for some reason become either scarce or numerous, may require special and urgent management aimed at artificially regulating their population size (Ratcliffe 1985, 1987).

5. Fire risk and hazard

Woodland fires are an ever-present danger during dry weather – from a combination of risk and hazard (Mayhead 1990; Hibberd (ed.) 1991). Likelihood of fire increases with the presence of people in or near the woodland, the occurrence of fires on adjoining land, and where roads pass nearby. Fire hazard refers to the quantities of vegetation and brushwood in or around the woodland and their flammability: it is high when the relative humidity of the air is low, air temperature and wind speed are high, and the moisture content of the vegetation and brushwood is low. The danger becomes extreme when high risk is combined with high hazard.

Trees react in different ways to fire, and some subclimax forests are effectively propagated by it.[7] The relevance to British silviculture is minimal.

Fire can be a spectacular and destructive cause of damage, particularly in plantations. The level of risk depends much upon the stage of development of the particular woodland concerned, its species composition, and its location. The risk is highest in young woodland before canopy closure, where in late winter and early spring the dead-dry remains of the previous year's undergrowth may provide a substantial quantity of inflammable material.

Woodland fires are costly. They can severely harm the flora (although much recovers); fauna is also harmed but, fortunately, many vertebrate wildlife can escape by flying, running or burrowing. Most fires are man-made – started accidentally by neighbours or visitors, or deliberately (arson). February to May is one of the greatest danger periods, when there may be much dry grass and bracken from the previous season before new growth begins. A second, less serious peak in fire danger occurs in high summer as a result of high temperature and low humidity making the undergrowth dry and inflammable. The composition of a woodland largely determines its susceptibility to fire – the trees themselves, undergrowth, and litter on the forest floor; in the case of plantations on peat the soil itself can burn. Broadleaved trees, except when young, are generally less inflammable than conifers; of the latter, pines are often especially susceptible, whilst larches are the least.

The risk of a fire spreading from nearby ground can be reduced by fire-breaks – semi-permanent strips of at least 10 m in width kept clear of inflammable vegetation. Maintaining them can be by way of cultivation, herbicides, grazing, mowing, or swiping. Methods aimed at prevention include: publicity, including public notices/warnings; brashing (the cut material being removed); and (becoming less frequent) fire-belts and barriers 10–20 m in width comprised of species which do not themselves readily catch fire, Japanese larch being the first choice, followed by broadleaves, especially alder. Adequate roads and rides can be useful as internal fire-breaks, and will allow access by vehicles carrying men, tools, pumps, water or foam. Good access is essential for private vehicles and for those of the Fire Service. Water is invaluable for fire-fighting and for damping down after a fire. Damming of watercourses with earth and stone, and excavating for ponds, can be undertaken subject to authorization. Chemical additives are sometimes used to increase the effectiveness of water – alginates to increase the viscosity; and wetting agents, or surfactants. Fire retardants – usually ammonium sulphate and diammonium phosphate – provide another way of attacking fires of rather low intensity.

Early detection of an outbreak of fire is vital. Small conflagrations are more easily extinguished and therefore it is important to make provision for early notification. Neighbours should be encouraged to report promptly any fires, and be advised how and to whom they should notify. Fire lookouts and patrols can only be considered for times

---

7 Savill and Evans (1986) usefully summarize Rowe, J.S.'s (1983) classification of trees according to their reactions as invaders, evaders, resisters, endurers, and avoiders of fire; pointing out that trees' adaptations to fire can, from time to time, be both a help and a hindrance when coping with fire damage in plantations.

of extreme fire danger – usually in co-operation with nearby woodlands and the Forestry Commission.

A small supply of fire-fighting equipment should be readily available, the amount and type depending on the scale and nature of the woodland. For the commonest type of fire, involving ground vegetation and small trees, beaters of various kinds are used: birch brooms, conveyor-belting on long handles, long-handled shovels, and hessian sacks, if they can be wetted. Some means of applying water is always useful: knapsack sprayers, small portable pumps, canvas or plastic buckets. Other equipment might include: wire cutters, rakes, chainsaws, axes and brushcutters. Every individual who works in woods – employees, contractors or purchasers – as well as sporting tenants, should be made aware of responsibility for taking the utmost care with all types of fire. Training of fire fighters in advance can be well worthwhile. Procedure for calling out of fire-fighter teams will be set by the owner, and be made known to woodland staff, neighbours, the Police and the Fire Service.

Research by the Forestry Commission into fire-fighting techniques has resulted in the introduction of new materials and modified equipment which greatly help the prospects for forest fire control. In particular, foam appears to be a most suitable technique for forest fire-fighting (Ingoldby and Smith 1982).

Helicopters can be an extremely useful and cost-effective aid to fire-fighting – for reconnaissance, rapid deployment of men and equipment, direct water bombing or laying barrier traces with water or foam. Throughout Scotland two commercial firms, PLM Helicopters Ltd and Gleneagles Helicopters, offer a fire-fighting service for the Forestry Commission and the private sector. The main problem with helicopters is that they cannot operate at night or for more than two hours without refuelling; nor can they operate in wind speeds over 40 knots or in conditions of high turbulence. The cost is around £500 per hour's flying time. PLM have their own trained crews and ancillary equipment (which the private sector has not). The average cost at most minor fires has been less than £1,500, normally offset by the value of the timber saved. A helicopter operates under the firemaster's direction, along with available local staff equipped with beaters. The owner or forester must decide, in conjunction with the Fire Service, whether or not to call out the helicopter as they will be responsible for the cost. A necessary provision is for a water supply of not less than 1.5 metres deep within a mile of the plantations. If a loch is not available, local streams may be adequate when coupled with a plastic dam and pump. If the Fire Service is involved they take charge of the whole operation, including the helicopter, otherwise the owner is expected to be the firemaster. Special water bombing equipment is held by some Forestry Commission Forest District Offices, in areas more prone to fires. This could be called on, if available, in the same way as a private owner calls on Commission employees for assistance, and would be charged for. In preparing a fire plan that may involve helicopters one should consult the local Fire Service, and discuss with the relevant insurance company so that any conditions as to budget or authorization is pre-agreed.

An owner or forester having assessed the fire danger on his property and in each component block of his woodland, should prepare with the advice and co-operation of the Fire Service a full fire plan, and make its contents known to all that may be concerned. Annual contact is recommended.

Woodlands can be insured against fire losses but insurance companies usually require reasonable precautions to reduce fire danger. Values for insurance are noted in chapter 16, section 6 ('Valuations') and insurance premiums in section 6 of this chapter. Insurance should not reduce the incentive to display protection.

6. Maintenance of woodlands

Once trees have been established they will normally require annually some maintenance, protection and management work. Regular expenditures may occur on inspecting the growing-stock for signs of ill-health or pest and disease attacks, assessment of growth for yield planning purposes, surveillance during the fire seasons, maintenance of the road and drainage systems and fencing. Such expenditures though small mount up. Costs do not change markedly between silvicultural options; but they are part of forest ownership, and must be included in an investment appraisal (see chapter 14) and as a debit against woodland value. As to road maintenance, major regrading is generally necessary after each thinning. The usual assumption is that roads will be maintained to a constant standard, preventive and remedial maintenance offsetting deterioration – a function of time rather than use. Average annual costs are likely to range from £15 to £50/ha.

Sometimes drains in thicket-stage crops have to be deepened or widened (see Forestry Commission Work Study Branch Report 6/90), the cost using tracked excavators being 35–50p/m, dependent on machine hire rates and expectation of earnings by owners/drivers.

The amount of fence maintenance required will largely depend on how well the fence was planned initially. Regular checks should be made to close any entrance holes that may have been made under or through the netting. The removal of any animals may then be essential to prevent further damage to the fence or plantation. Particular attention should be given to fences following heavy snowfalls or frozen rain (rime) or a hunt, and during holiday times when the general public may have climbed over them.

Insurance. It is possible to insure woodland crops against fire and storm damage. The costs vary according to age of crop, and whether it is conifer or broadleaved; locality is also pertinent. As the age increases beyond the establishment stage, total cover increases but premium as a percentage of it diminishes. Premiums tend to be cheaper for broadleaves. Average annual premiums are likely to be about £5/ha. Table 18 indicates current (1991) premium rates, based on the values noted in section 2 of chapter 16.

Table 18. Indications of insurance premium rates for woodlands.

	Premium as % of insurance value
Fire, lightning, aircraft and explosion:	
All plantations up to 5 years	0.145
Plantations 6 to 25 years	
Broadleaves	0.065
Mixed plantations (not exceeding 50% conifers)	0.120
Larch	0.120
Other conifers	0.145
Plantations over 25 years: as above, less 25% discount.	
Storm damage:	
All plantations up to 20 years	0.160
All plantations over 20 years	0.350

(Maximum age in Scotland is 45 years and in England, Wales and Northern Ireland 50 years)
Source: Willis Wrightson Limited, Dundee.

7. Costs of plantations to the thicket-stage

Cleaning and partial brashing having been undertaken, and adequate attention given to protection and maintenance, the costs of the plantation to the thicket-stage (say, the fifteenth year) may be ascertained on the pattern in Table 19 which indicates current (1991) average costs. Contractors' charges would include their profit; their costs of labour, machinery and materials usually have the benefit of scale.

Table 19. Indications of costs of **conifer** plantations to the thicket-stage.

		Cost £/hectare (Labour, machinery, materials and labour oncosts; not overheads)	
Operation	Year(s) of operation	Small lowland sites	Extensive upland sites in England, Wales and Scotland
Conifers			
Total establishment (see chapter 5, Table 11):	0–4	1,800	1,175
6th to 15th years:			
Cleaning	10	150	–
Brashing (partial)	15	125	–
Protection and Maintenance★	6–15	150	100
Total to the 15th year		2,225	1,275

★ Maintenance of fences and drains; protection against pests and fire; management; and insurance.

Grant-aid, where relevant (see chapter 15), would reduce the total costs.

The above costs exclude rack-cutting, overheads, land, roading and road maintenance, top-dressing of fertilizer and interest on capital. If roads were included the extra cost might be £270-£500/ha.

Broadleaves (other than poplar and cricket-bat willow): If all or most of the trees in Table 19 under 'Small lowland sites' were broadleaved, the total costs to the fifteenth year (assuming brashing not yet necessary), could be at least £300/ha higher than those for conifers. Compared with conifers, the cleaning and brashing of broadleaves are later in rotation time. However, the grant-aid may be higher, dependent on the scheme and band involved.

For accurate costing of Table 19, it would be necessary to take into account the years in which the items of expenditure and of grant-aid (where relevant) occur: the main limitation in the Table is the exclusion of compound interest. (An example of a calculation embodying expenditure and income over time, and compound interest, is given in chapter 14.)

Operational costs of after-care (tending) of plantations. Costs in Table 20 are generalized and will vary between scale, and individual sites and weather conditions. Contractors' charges may lie within the ranges shown but would include their profit; their costs of machinery and materials usually have the benefit of scale. More detailed indications of operational costs, as well as output guide times, are given by Insley (ed.) (1988.)

Table 20. Indications of operational costs of after-care (tending) of plantations.

Operation (Output Guides and Standard Time tables for most of the operations are available from the Forestry Commission Work Study Branch – see chapter 16)	£/hectare (Labour, machinery, materials, and labour oncosts; not overheads)
Cleaning According to terrain, species, density, age, tree size and method	
(i) Manual application	100–200
(ii) Mechanical application Swipe Clearing saw	50–75 450–550
(iii) Mechanical/chemical application	600–700
(iv) Chemical application	100–150
Brashing: According to terrain, species, age, size, density, intensity and method: Partial: 25% 50% 100%	60–150 125–300 250–450
Protection and Maintenance:	Annually: 15–20
Insurance: Storm and Fire: According to location, species, and age:	Annually: 4–6
Pruning: According to terrain, species, age, tree size, and method (300 trees/ha):	
Conifers:	150–200
Broadleaves: Generally undertaken later than the 15th year, and then in two or three stages	300–500

Operation (Output Guides and Standard Time tables for most of the operations are available from the Forestry Commission Work Study Branch – see chapter 16)	£/hectare (Labour, machinery, materials, and labour oncosts; not overheads)
Later fertilizing (Top-dressing): According to terrain, and method. Cost of fertilizer depends on analysis (N, P, K), size of order, and delivery charge. Helicopter application (according to terrain, distance involved, extent and weight):	100–150
Roads: (i) Typical cost range: £6,000-£20,000 km (ii) Cost at 12 m/ha:	£500–750 ha.

References

Avery, M. and Leslie, R. (1990), *Birds and Forestry.* Poyser, London.
Bevan, D. (1987), 'Forest Insects'. FC Handbook 1.
Binns, W.O. (1975), 'Fertilisers in the Forest: a guide to Materials'. FC Leaflet 63.
Binns, W.O., Mayhead, G.J. and McKenzie, J.M. (1980), 'Nutrient Deficiencies of Conifers in British Forests: an illustrated guide'. FC Leaflet 76.
Burdekin, D.A. (1979), 'Common Decay Fungi in Broadleaved Trees'. FC Arboriculture Leaflet 5.
Campbell, B. (1974), 'Birds and Woodlands'. FC Forest Record 91.
Carter, C.I. (1989), 'The 1989 Outbreak of the Green Spruce Aphid *Elatobium abietinum*'. FC Research Information Note 161.
Carter, C.I. and Nichols, J.F.A. (1988), 'The Green Spruce Aphid and Sitka Spruce Provenances in Britain'. FC Occasional Paper 19.
Chapman, N.G. and D.I. (1982), 'The Fallow Deer'. FC Forest Record 124.
Currie, F.A., Elgy, D. and Petty, S.J. (1977), 'Starling Roost Dispersal from Woodlands'. FC Leaflet 69.
Davies, R.J. and Pepper, H.W. (1990), 'Protecting Trees from Field Voles'. FC Arboricultural Research Note 74/90/ARB.
Dutch, J.C., Taylor, C.M.A. and Worrell, R. (1990), 'Potassium Fertiliser – Effects of Different Rates, Types and Times of Application on Height Growth of Sitka Spruce on Deep Peat'. FC Research Information Note 188.
Evans, H.F. (1989), 'Biological Balancing Acts'. FC 'Forest Life', Issue No. 6, July 1989.
Evans, J. (1984), 'Silviculture of Broadleaved Woodland'. FC Bulletin 62.
Everard, J. (1974), 'Fertilizers in the Establishment of Conifers in Wales and Southern England'. FC Booklet 41.
Farmer, R.A., Alexander, A. and Acton, M. (1985), 'Aerial Fertilising; Monitoring Spread is Vital Operation'. *Forestry and British Timber,* April 1985.
Gibbs, J.N., Burdekin, D.A. and Brasier, C.M. (1977), 'Dutch Elm Disease'. FC Forest Record 115.
Greig, B.J.W. (1981), 'Decay Fungi in Conifers'. FC Leaflet 79.
—— (1988a), 'Breeding Elms Resistant to Dutch Elm Disease'. FC Arboriculture Research Note 2/88/PATH.
—— (1988b), 'English Elm Regeneration'. FC Arboriculture Research Note 13/88/PATH.

Greig, B.J.W. and Strouts, R.G. (1983), 'Honey Fungus'. FC Arboriculture Leaflet 2.
Gurnell, J. (1979), 'Woodland Mice'. FC Record 118.
Harrison, T. and Dighton, J. (1987), 'Phosphate Fertiliser – What your Spruce and Pine Need'. *Forestry and British Timber,* October. 1987.
Hibberd, B.G. (ed.) (1991), 'Forestry Practice'. FC Handbook 6.
Hull, S.K. and Gibbs, J.N. (1991), 'Ash die-back – a survey of non-wood trees'. FC Bulletin 93.
Hurrell, M.G. (1968), 'Pine Martens'. FC Forest Record 64.
ICF (1989), 'Fact sheet: Deer and Forestry'. Institute of Chartered Foresters, Edinburgh.
Ingoldby, M.J.R. and Smith R.O. (1982), 'Forest Fire Fighting with Foam'. FC Leaflet 80.
Insley, H. (ed.) (1988), 'Farm Woodland Planning'. FC Bulletin 80.
Jobling, J. (1990), 'Poplars for Wood Production and Amenity'. FC Bulletin 92.
Kensington, N. (1990), 'Electronic Track Guidance (fertilisers)'. *Forestry and British Timber,* May 1990.
King, C.J. and Fielding, N.J. (1989), '*Dendroctonus micans* in Britain – its Biology and Control'. FC Bulletin 85.
Leather, S.R., Stoakley, J.T. and Evans, H.F. (1987), 'Population Biology and Control of the Pine Beauty Moth'. FC Bulletin 67.
Lloyd, H.G. and Hewson, R. (1986), 'The Fox'. FC Forest Record 131.
Lonsdale, D. and Wainhouse, D. (1987), 'Beech Bark Disease'. FC Bulletin 69.
Mayhead, G.J. (1990), 'Fire Protection in Great Britain'. *Commonwealth Forestry Review, 69* (1) 1990, pp. 21–7.
McIntosh, R. (1984), 'Fertiliser Experiments in Established Conifer Stands'. FC Forest Record 127.
—— (ed.) (1989), 'Deer and Forestry'. Proceedings of a Conference, Glasgow, June–July 1987. Institute of Chartered Foresters, 1989.
Melville, R.C., Tee, L.A. and Rennolls, K. (1983), 'Assessment of Wildlife Damage in Forests'. FC Leaflet 82.
Neal, E. (1982), 'Badgers in Woodlands'. FC Forest Record 103.
Pepper, H.W. (1978), 'Chemical Repellants'. FC Leaflet 73.
—— (1989), 'Hopper Modification for Grey Squirrel Control'. FC Research Information Note 153.
—— (1990), 'Grey Squirrels and the Law'. FC Research Information Note 191.
Phillips, D.H. and Burdekin, D.A. (1982), *Diseases of Forest and Ornamental Trees.* Macmillan, London.
Poole, T.B. (1970), 'Polecats'. FC Forest Record 76.
Potterton, E. (1990), 'Electronic Track Guidance (fertilisers) comes of age in British Forestry'. *Forestry and British Timber,* February. 1990.
Prior, R (1983), *Trees and Deer.* Batsford, London.
Ratcliffe, P.R. (1985), 'Glades for Deer Control in Upland Forests'. FC Leaflet 86.
—— (1987), 'The Management of Red Deer in Upland Forests'. FC Bulletin 71.
Rogers Brambell, F.W. (1974), 'Voles and Field Mice'. FC Forest Record 90.
Rowe, J.J. (1976), 'Badger Gates'. FC Leaflet 68.
—— (1980), 'Grey Squirrel Control'. FC Leaflet 56.
Rowe, J.S. (1983), 'Concepts of Fire Effects on Plant Individuals and Species'. In *The Role of Fire in Northern Circumpolar Ecosystems* (eds. R.W. Wein and D.A. MacLean), John Wiley and Sons, New York.
Savill, P.S. and Evans, J. (1986), *Plantation Silviculture in Temperate Regions with Special Reference to The British Isles.* Clarendon Press, Oxford.
Springthorpe, G.D. and Myhill N.G. (eds.) (1985), 'Wildlife Rangers Handbook'. FC. Under revision.

Stoakley, J.T. and Heritage, S.G. (1990), 'Approved Methods for Insecticidal Protection of Young Trees against *Hylobious abietis* and *Hylastes* Species'. FC Research Information Note 185.

Tabbush, P.M. and White, I.M.S. (1988), 'Canopy Closure in Sitka Spruce – Relationship between Crown Width and Stem Diameter for Open-grown Trees'. *Forestry 61* (1), pp. 23–7.

Taylor, C.M.A. (1986), 'Forest Fertilization in Britain'. Proceedings of the Fertiliser Society No. 251.

—— (1990), 'Survey of Forest Fertilizer Prescriptions in Scotland'. *Scottish Forestry 44* (1), pp. 3–9.

—— (1991). 'Forest Fertilization in Britain'. FC Bulletin 95.

Taylor, C.M.A. and Tabbush, P.M. (1990), 'Nitrogen Deficiency in Sitka Spruce Plantations'. FC Bulletin 89.

Tittensor, A.M. (1975), 'Red Squirrel'. FC Record 101.

Tittensor, A.M. and Lloyd, H.G. (1983), 'Rabbits'. FC Forest Record 125.

Winter, T.G. (1988), 'Insects and Storm-damaged Broadleaved Trees'. FC Research Information Note 133.

Winter, T.G. and Evans, H.F. (1990), 'Insects and Storm-damaged Conifers'. FC Research Information Note 173.

Chapter Seven

SILVICULTURAL SYSTEMS, THE 'NORMAL' FOREST, NATURAL REGENERATION AND ENRICHMENT

1. Silvicultural systems

Foresters in Britain for the last seventy years have been forming plantations, chiefly coniferous, to replace crops of trees felled during two world wars and to afforest bare land in the uplands. Since around 1960 there has been increasing interest in regenerating the older plantations which have become mature, and also those woodlands, mainly broadleaved, which were planted during the latter years of the nineteenth century. At the same time the public have become increasingly interested in the amenity aspects of forestry and in recreation within woodlands.

Foresters on the continent of Europe during the seventeenth and eighteenth centuries, faced with problems of regenerating woodlands, devised what are called silvicultural systems to plan the work of tending, harvesting and restocking in a systematic way. In this chapter most of these systems are described and their suitability for conditions in Britain are tentatively discussed.

A silvicultural system is a method of silvicultural procedure worked out in accordance with accepted sets of silvicultural principles, by which crops constituting a forest are tended, harvested and restocked by new crops of distinctive forms *(British Commonwealth Forest Terminology,* 1953). Matthews (1989) points out that a silvicultural system embodies three main ideas: '(1) The method of regeneration of the individual crops constituting the forest, (2) the form of crop produced, and (3) the orderly arrangement of the crops over the whole forest, with special reference to silvicultural and protective considerations and efficient harvesting of produce.'

For over half a century British foresters have adhered to the classification of silvicultural systems set out by Troup (1928) and revised by Jones (1952). Subsequent commentators on silvicultural systems drew attention to the English translations by M.L. Anderson of treatises by Knüchel (1953) and Köstler (1956). Matthews (1989) has updated the whole subject, grouping the various forms of treatment into major systems, and devised a general classification of them. The scarcity of both editions of Troup's book means that Matthews' classification and treatment of the systems are now the most quoted and followed. His classification is:

High forest systems. Crops normally of seedling origin.
    Felling and regeneration for the time being concentrated on part of the forest area only:
        Old crop cleared by a single felling; resulting crop even–aged – *Clear cutting systems*

Systems of successive regeneration fellings. Old crop cleared by two or more successive fellings; resulting crop more or less even-aged or somewhat uneven-aged:
> Regeneration fellings distributed over whole compartments or subcompartments:
>> Opening of canopy even; young crops more or less even-aged and uniform – *Uniform systems*
>> Opening of canopy by scattered gaps; young crop more or less even-aged – *Group system* Opening of canopy irregular and gradual; young crop somewhat unevenaged – *Irregular shelterwood system*
> Regeneration fellings confined to certain portions of compartments or sub-compartments at a time:
>> Fellings in strips – *Strip systems*
>> Fellings beginning in internal lines and advancing outwards in wedge formation – *Wedge system*

Felling and regeneration distributed continuously over the whole area; crop wholly uneven-aged (irregular) – *Selection systems*
> Accessory systems arising out of other systems:
>> Form of forest produced by introducing a young crop beneath an existing immature one – *Two-storied high forest*
>> Form of forest produced by retaining certain trees of the old crop after regeneration is completed – *High forest with reserves*

Coppice systems. Crops, in part at least, originating from stool shoots (coppice) or by other vegetative means:
> Crop consisting entirely of vegetative shoots:
>> Crop removed by clear felling; even-aged – *Coppice system*
>> Only a portion of the shoots cut at each felling; crop uneven-aged – *Coppice selection system*
> Crop consisting partly of vegetative shoots, partly of trees generally of seedling origin – *Coppice with standards system*

There are several kinds of 'high forest'. In the first kind the individual trees are all even-aged and the crops present a *regular* appearance with one canopy layer. Such crops may be composed of one species, and so are called 'pure', or if two or more species are present they are called 'mixed'. If a pure, regular crop of trees is thinned heavily and underplanted with a new young crop two canopies are formed and the crop becomes two-storied. The upper storey may be, for example, Japanese larch and the lower grand fir, and the result is mixed, *two-storied high forest*.

By contrast, there are crops with several stories so that the whole space from the ground to the tops of the tallest dominant trees is occupied by trees of different ages and sizes. Such a crop is termed *irregular* or *uneven-aged* and usually contains several species able to grow together in intimate mixture.

The regeneration of regular crops is tackled in different ways. If the new crop is required quickly and is of a hardy species the old crop can be removed in one felling and the bare site is restocked, usually by planting but sometimes naturally. This is the *clear cutting system* and is the one most widely used in Britain.

As an alternative, and especially if the new crop is of a tender species, part of the old crop can be retained as a shelter wood to provide overhead or side shelter for the new trees and to control the ground vegetation. There are several shelterwood systems suited to different species and site conditions and if the object is to produce a regular crop of young trees the Shelterwood *Uniform System,* Shelterwood *Group System,* or one of the Shelterwood *Strip Systems* are used. The last named have not been much applied in Britain but the first two have been used under a variety of conditions and so are described more fully later. Some foresters in Britain have been successful in obtaining natural regeneration from seed produced by the old crop but most prefer to plant so that restocking is rapid, the trees in the new crops are even-aged and the crops are regular. Sometimes the shelterwood uniform system is used for prolific seed producers, such as Scots pine: a good example was at the Crown Estate at Windsor in Berkshire.

Sometimes there are advantages in accepting the natural regeneration which accumulates in gaps caused by windthrow or earlier fellings of mature crops. During World War II when much felling was done in a short time it was not possible quickly to restock the felled areas, and the bare ground became filled with natural regeneration, sometimes of valuable species but often of less useful ones. Small mixed crops of young and older trees developed at different times and the woods became irregular in structure. Several owners and their foresters worked to maintain the structure, often because it is visually attractive. The silvicultural system is called *group selection*. Other owners and foresters also saw the group selection system as a means of converting regular crops to mixed irregular crops and examples are described later in this chapter. There is also a *single tree selection system* which is used in the mountainous parts of Europe but this is quite specialized and has not been much tried in Britain, largely because it is only suitable for extreme shade bearers of which Britain has few. Another related system is the *irregular shelterwood* which also produces mixed irregular crops, but it also has not been widely used here.

High forest systems

Regular (even-aged) silviculture. The system of clear cutting and replanting is the usual method of regenerating high forest in Britain. In this system trees are grown from seed to maturity without being coppiced. Successive areas are clear-felled and regenerated, most frequently by artificial means but sometimes naturally. Matthews (1989) gives advantages and disadvantages compared with other systems, and application in practice. The system concentrates working, maximizes out-turn of produce from a stand and ensures rapid and uniform restocking over a whole site. In carrying it out, factors to be considered include: size of coupe, damage to adjacent stands, effects on the site, and scheduling of restocking either by planting or natural regeneration.

British foresters have gained international renown for their successes in this system, particularly in 'commercial plantation forestry' – generally even-aged crops, usually of a single species, virtually all conifers – eventually to be felled in large blocks, followed by complete restocking, often with the same species. This system has been justified economically, despite some objections on ecological and visual grounds, especially in sensitive areas of high landscape importance. However, in many areas the risk of windthrow, which often coincides with the uplands and high amenity, forces a non-thin/clear-fell system to be adopted. Clear-felling creates a sudden and dramatic change in the landscape, and breaks the continuity of woodland conditions. It is increasingly recognized as being inappropriate for woodlands having a high value for landscape, recreation, and nature conservation – benefits which are discussed in chapter 13. However, it is probably the only system suitable for the light-demanding species used in plantations.

Irregular (uneven-aged) silviculture. By contrast to the system of clear cutting and replanting, irregular systems of silviculture with their wider age distribution, provide continuous woodland cover. Prompted by this, several foresters and conservationists over the past few decades have been interested in the use and development of alternatives to clear cutting systems. Matthews (1989) gives advantages and disadvantages compared with regular silviculture, and application in practice. In the ideal *irregular stand,* trees of all age classes grow in mixture. Regeneration is continuous and the canopy is also continuous throughout the stand, both vertically and horizontally. The irregular structure may be preferred where it is necessary to conserve the physical condition and nutrient status of the soil under a permanent forest canopy, or to avoid erosion on slopes or on soils that are liable to become degraded. Irregular stands also provide varied habitat for the conservation of wild plants and animals. They provide an alternative landscape, and have great recreational potential. Their management requires skill, but often proves more interesting. The system's attributes in relation to landscape, recreation, and wildlife conservation are discussed in chapter 13. The economics of the system are noted later in this chapter.

Shelterwood systems. These are high forest silvicultural systems in which the young crop is established under the overhead or side shelter of the old one; at the same time the old crop protects the site. The overstorey (the parent crop) is managed to bring about natural regeneration on the ground beneath; and consists of gradual removal of the overcrop with the object of securing good regeneration of reasonably uniform age class over a whole compartment. The term 'shelterwood systems' includes systems of successive regeneration fellings, namely:

1. *The uniform system.* The term is an abbreviation of 'shelterwood uniform system'. It implies a uniform opening of the canopy for regeneration purposes, as well as an even-aged and regular condition of the young crop produced.

2. *The irregular shelterwood system* (Femelschlag) is one of successive regeneration fellings with a long and indefinite regeneration period, producing young crops of somewhat uneven-age type.

3. *Strip systems.* These differ in detail but have one common feature, namely that the coupes take the form of quite narrow strips which have advantages, mainly of a protective nature, over large coupes. The fellings start at one end of a cutting section (unit) and proceed to the other. The systems are:

    i   The shelterwood strip system, which evolved out of the shelterwood uniform system for protective reasons.

    ii The strip and group system, a modification of the shelterwood strip system.

    iii The wedge system, a form of strip system in which the fellings begin by a strip in the centre of a cutting section and proceed outwards in both directions.

For all the shelterwood systems, Matthews (1989) gives advantages and disadvantages, and application in practice. In Britain the systems have not been widely practised though under certain conditions some can succeed with beech, sycamore and a few species of conifers. Fine examples in conifers, using irregular shelterwood, are to be seen at Longleat in Wiltshire (noted later). The system appears to be suitable for well-tended stands of potentially windfirm, prolific seeding species such as Douglas fir, larch, western red cedar, western hemlock, ash and sycamore.

Group regeneration (usually of broadleaves but equally possible for some conifers) involves working with several areas of 0.1 to 0.5 hectares in a stand which is to be regenerated, waiting until good natural regeneration has been achieved say after five to ten years, respacing as appropriate, and then further opening up existing groups and starting new ones, when seed or advance regeneration is plentiful. The cycle is repeated until the desired area is completely regenerated. The minimum size of opening is likely to have a diameter of at least twice the height of the adjacent trees. The objective of the group system is to produce new crops which are essentially regular in structure by the time the pole-stage is reached. Thus within a given area the fellings to enlarge the groups and restocking by natural regeneration or planting must follow a fairly close timetable. The long-term plan of work can be disrupted by windthrow and it is not easy to control the operation, especially in the later stages when the groups begin to coalesce. Under the conditions of private estate forestry in Britain something more flexible is needed and several owners and foresters have found *group selection* more suitable. For the Forestry Commission, however, with large areas, the shelterwood group system can be appropriate (Joslin 1982; Everard 1987). Group selection is sometimes used in lowland Britain and is suited to the complex needs of many upland woodlands in Wales.

Group and group selection require skill, patience and commitment over a long period to complete regeneration of a whole stand. The approach is probably the most appropriate to attempt when seeking natural regeneration of broadleaves since it does not depend on just one good mast year but allows a gradual regeneration process following 'nature'. It is well-suited to small-scale working, and as it avoids the sudden impact of clear-felling it is useful where there is a desire to maintain woodland conditions for reasons of silviculture, landscape, conservation, amenity or sporting. Although usually more expensive because of the small-scale harvesting, regeneration or restocking and subsequent tending, there can be considerable silvicultural gains to the protected site as well as aesthetic advantages.

The selection system. This silvicultural system differs from all others in that felling and regeneration are not confined to certain parts of a stand but are distributed all over it, the fellings removing single trees or small groups of trees selected through the area. Fellings done in this manner are termed *selective fellings;* they result in an uneven-aged or irregular type of stand in which all the age or size classes from naturally regenerated seedlings to mature trees are mixed together over every part of the area (Figure 16). At intervals ranging from five to ten years, a whole stand is worked over; the largest trees are felled and the remainder of the stand is tended, cleaned or thinned as needed. Where conditions are favourable, new seedlings appear.

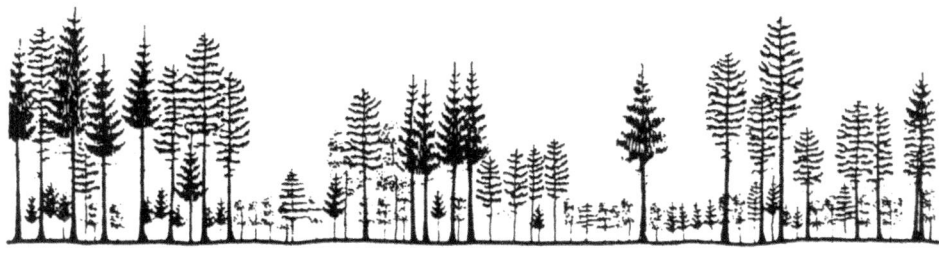

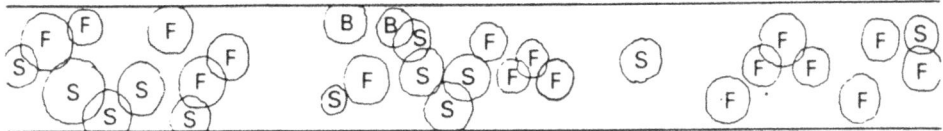

Crown Projections

Figure 15. Selection system. Side elevation and plan of forest of selection type containing three species. S, spruce; F, Silver fir; B, beech
Source: Matthews (1989).

To continually augment the crop they must be shade tolerant, conifers or broadleaves, since they are never completely free grown but always in the shade of older trees.

Under ideal conditions the process goes on year after year over the whole stand. The standing volume is fixed by rules of management. *The felling cycle, structure of the growing stock,* and *regulation of yield* are all important. In parts of Europe, particularly Switzerland, the regulation of the yield from selection forests is by the *Méthode du Contrôle*.[1] The typical form of selection fellings is more suited to shadebearing than to light-demanding species. Hence if the selection is to be applied to light-demanding species it is necessary to fell trees in groups so as to create gaps of sufficient size to enable regeneration to establish itself. Such fellings are known as *group selection fellings* and the modification of the selection system under which this type of felling is employed is termed the *group selection system.*

Matthews (1989) gives advantages and disadvantages of the selection system, and application in practice. There is no concept of rotation length or a regeneration period, as both harvesting and regeneration take place regularly and simultaneously throughout the stand. Marking and harvesting must be done with appropriate care to prevent damaging the remainder of the stand. Usually there are three canopy layers: regeneration, poles and dominants. The system has many environmental advantages. It maintains a woodland appearance and continuous ground cover. It gives maximum protection to the soil from erosion; and provides protection for frost sensitive species such as

---

1 The age-classes in irregular forestry are not distributed by area, therefore foresters on the continent of Europe rely on periodic enumerations of the growing stock. The *Méthode du Contrôle,* or Check Method, is the most widely known and its application in the forests of the Swiss Jura is described in detail by Knüchel (1953). It has been used by the British foresters noted earlier, but some present day foresters consider that a procedure more suited to conditions in Britain should be devised (Garfitt 1987). Helliwell (1985) and Pryor and Savill (1986) have proposed that a series of trial areas be established to investigate the economics and management of irregular forestry.

beech, allowing them to regenerate even in frost hollows. The system also has the advantage that seed from all seed years, however infrequent or partial, can be used. Other advantages include (Price 1989): better timber (due to more even ring width and smaller proportion of weak juvenile wood); protection of site (by reduced soil insolation); and more interesting work conditions. Disadvantages include: difficult, small-scale working, with higher unit costs; and more *dispersed* working, increasing transportation costs. There is also the disadvantage of the time taken to create an uneven-aged system: its creation by *afforestation* entails an opportunity cost, in that planting of some land is delayed for a period up to the whole rotation. Equally, conversion of an even-aged stand entails felling some trees before optimum rotation and leaving others to become financially over-mature. (It may also be costly to convert uneven-aged crops to even-aged, as it entails keeping the oldest age-classes far beyond their optimum rotations, or clearing younger trees before theirs, or both.)

There are no extensive examples of the selection system in Britain, though various forms of selective felling are often miscalled 'selection forestry'. The true system is probably unworkable here, apart from in the cases of beech and some conifers. None of our species, and certainly no broadleaves, can match the shade tolerance of European silver fir (which in Britain commonly suffers from *Adelges nusslini*), and this makes a true multi-storied selection stand difficult to achieve. Furthermore, the lower summer light intensities tend to reduce the number of photosynthetically active canopy layers. In most species, fellings would need to be heavier, or more frequent, the canopy must be sparse or tend to a group selection system with a mosaic of single-storied groups, and high pruning particularly of broadleaves is likely to be essential to obtain high quality butts.

In Britain, examples approaching the single tree selection system tend to merge with the group systems (Hart 1967). Pryor and Savill (1986) draw attention to 'selection stands' ('group selection' is probably the best term) and refer to Weasenham in Norfolk (chiefly conifers), Rossie Priory near Dundee (broadleaves), Checkendon in Oxfordshire (beech), and a few other stands in the Chiltern beechwoods. Other examples are at Dartington in Devon (conifers), at Pusey Wood in Oxfordshire (beech, western red cedar and other species) and at Wakefield in Northamptonshire (mainly ash).

Pryor and Savill (1986), discussing selection silviculture, concluded that: the proportion of small sized timber produced is probably lower; and timber quality will certainly be adequate, although in sparsely stocked stands high pruning may be necessary. Experience is necessary to harvest selective fellings without doing damage to the trees remaining. (Damage to regeneration less than about 3 m tall is minimal, provided directional felling is employed.) The felled produce has to be collected from a wide area; and its marketing may be difficult because of various sizes, and perhaps of more than one species. However, if the groups are of reasonable size and extraction racks are nearby, there need be little damage.

For landscape and amenity benefits, in terms of continuity and lack of sudden change, group selection stands are notable (see chapter 13). Approaches towards a selection system are attractive from the standpoint of silviculture, landscape, conservation, amenity and sporting; also there is the advantage of less snow damage. Sustained yields can be obtained, and the advantage to an estate of being able to draw on a regular yield of all sizes of produce from a stand may be welcome.

Accessory systems. The formation of (i) two-storied high forest and (ii) high forest with reserves, provide examples of accessory systems because they may arise from various other systems and are not dependent on any particular form of regeneration. Matthews (1989) gives advantages and disadvantages of the accessory systems, and application in practice.

Two-storied high forest. A second crop is introduced into a stand at some stage during the rotation. Often it is referred to as 'underplanting'. Such stands are then composed of an upper and lower storey of trees, growing in intimate mixture on the same site. Generally two species are involved, the upper storey usually comprising a light demander under which a shade tolerant species can grow without being suppressed when introduced at a later date. Trees of the upper storey may arise through natural regeneration or planting. They are treated as an even-aged crop in terms of stocking and thinning until they approach middle age, when a heavy thinning takes place. At this stage the crop is underplanted with the second, shadebearing species which forms the lower storey. Sometimes the lower storey is established through natural regeneration before or soon after the heavy thinning is done. Both stories are allowed to grow up and subsequent thinnings are done in the lower storey. Both stories may be felled together, or the upper storey is removed in one or a series of fellings to leave the lower as an even-aged maturing crop. Two-storied high forest is frequently limited to only one rotation because of the difficulty of forming a second understorey below the previous shade-bearing one.

Hiley (1959, 1967) experimented with the two-storied high forest system at Dartington in Devon by underplanting Japanese larch with western hemlock, western red cedar, and Douglas fir (Howell *et al.* 1983). The first two species succeeded in the understorey but the Douglas fir did not. Another possible conifer underplant species is grand fir. In broadleaves, the main examples are where beech is introduced into young pole-crop stands of birch, mature pine, or poorly-growing ash and even oak to aid its progress (notably in the Forest of Dean, using beech). Also, throughout the 1950s and 1960s it was common practice to thin oak stands heavily and underplant with conifers to obtain an early return, while at the same time encouraging rapid growth of the best oak. Sometimes the two-storey structure lasts only a few years since the overstorey is removed before its shade seriously impairs the growth of the underplants. Damage to the latter during felling is usually much less than might be expected.

High forest with reserves is an accessory silvicultural system produced by leaving selected stems of an old crop standing over a young crop established by regeneration from the old one. These stems, known as 'reserves', may be retained, scattered singly or in small groups, for the whole or part of a second rotation. The chief object is to retain selected stems to put on increment and produce large-sized timber, generally of Scots pine, European larch, oak and beech. However, the adverse effect on younger trees below or surrounding them should be remembered. Reserves may also act as seed bearers to fill blanks in crops. Another important reason is to create amenity by producing stands with a natural appearance.

The history of irregular silviculture in Britain. The history of irregular silviculture in Britain began with Schlich, whose descriptions of shelterwood and selection systems appear in his writings (1904, 1923). The first comprehensive accounts in English of the irregular systems were written by Troup (1928), one of Schlich's students who succeeded him as Head of the Department of Forestry at Oxford University. The first forester to practise group selection in Britain and to apply to it the *Méthode du Contrôle* (the calculation of increment and control of yield) was Ray Bourne, another of Schlich's students who, while lecturing at Oxford, began to convert several stands of beech on the Parmoor estate in the Chilterns to group selection (Bourne 1935, 1942, 1945, 1951). Hiley (1939, 1953, 1967) and Ackers (1939), two experienced managers of privately owned woodlands, gave thought to a limited use of some form of selection system, but little ensued of a permanent nature.

In 1950 the Forestry Commission began systematically to test methods for converting unproductive woods (mainly arising from wartime fellings) to high forest (Wood 1950). The results of this work became the basis of recommendations for State and private foresters (Wood *et al.* 1967; Evans 1984). Good examples of irregular silviculture can now be seen on more than twenty private estates and in a few Forestry Commission forests. The supplementary work of Garfitt is referred to later. Another who recognized the potential of the group selection system under British conditions was Reade (1957, 1969, 1990) who has done notable work on his Ipsden estate in Oxfordshire.

The foregoing practices related mainly to broadleaves, but at the same time attempts were made to convert regular stands of conifers to an irregular structure. A major interest was taken by M.L. Anderson (1930, 1931, 1951, 1960), whose contribution to irregular silviculture in Britain began in 1929. Between 1929 and 1932 Anderson designed trials of a method of planting small densely-stocked groups of trees at wide spacing on upland sites being afforested in Scotland and Northern England. Derivatives of his method of building up irregular forest structures by planting trees in groups arranged in systematic fashion, have been used relatively recently on the Tavistock Woodlands Estate in south Devon, work described later in this chapter.

During the 1950–1970 period, Penistan (1938, 1953) and Garfitt (reintroduced later) had a notable influence on the development of irregular silviculture in Britain. Some instances of the systems arose as a specific management policy, for example in the woodlands of the National Trust, for reasons of amenity, conservation, and visual impact (Wright 1982). Matthews (1986) and Blyth and Malcolm (1988) describe trials of the group selection system in Glentress forest in south central Scotland, at Bowhill near Selkirk, Eildon near Melrose, and on the Corrour estate in the Highlands. The group selection system has been most often applied in the lowlands of England on sites with soils capable of supporting good growth of a wide range of species. The windthrow hazard is generally low and the presence in the growing stock of a third or more of broadleaves has further lessened the risk of wind damage. This form of irregular forestry has been used to good effect in restoring unmanaged woods to a productive condition relatively quickly and cheaply. The net costs of the work are closely related to the previous state of the woods and the skills of foresters and their work force. The group selection system, used for this purpose, is very flexible. Most species can be accommodated by adjusting the size of the group. The yield of the growing stock can be improved in quantity and quality by introducing genetically superior origins/provenances and cultivars and by careful selection and tending for their improvement throughout the rotation. The possible range of produce at any one time is wide. The result of damage done during felling can readily be controlled. The major problem is the need to control damage by rabbits, grey squirrels and deer, which demands close attention to fences and keepering (Garfitt 1984).

Helliwell (1982, 1988) emphasizes that when regular forests are contrasted with comparable areas of continuous irregular forest it is evident that some benefits can be more readily obtained from the latter in Britain. These include natural appearance, permanent amenity, facilities for recreation and varied habitat for the conservation of wild plants and animals. The quality of timber produced by an irregular stand may be higher than that from a regular stand. However, this latter advantage may only begin to appear late in the conversion process if the existing growing stock is of poor or moderate quality. Some expected advantages may not be proven for many years. Managerial practices are very important in determining the level of damage by insects and fungi in managed forests and it is not sufficient to assume that mere separation of species in a mixed and irregular forest would itself reduce damage. When considering the physical and chemical condition of the soil under the continuous cover of an irregular forest, there is insufficient evidence in Britain to be certain what changes will occur.

Several foresters and conservationists, in particular Helliwell (1988) and Peterken (1981), have emphasized that irregular silviculture will become more important in areas subject to high landscape pressures largely because of its attractive appearance and its resemblance to what is commonly but sometimes erroneously regarded as the natural forest. More detailed discussion on the relationships between nature conservation and silvicultural techniques is given by Peterken (1981). Crowe (1978) and Helliwell (1982) provide further discussion on landscape and amenity benefits.

Garfitt's contribution to irregular silviculture. J.E. Garfitt, a consultant forester of wide experience, has been a vigorous advocate and successful implementer of the group selection system of silviculture in Britain since World War II. He continued and expanded much of the consultancy work of Ray Bourne, noted earlier, under whom he had been a student at Oxford.

The tendency of natural regeneration to occur in groups led him to adopt groups of comparable size when planting areas from which scrub-growth had been cleared (Garfitt 1953). He applied these methods on a number of private estates, evolving a form of the group selection system suited to stands containing a mixture of both light-demanding and shade-bearing species (Garfitt 1953, 1963, 1977, 1980, 1984). In standing woods group fellings were used to provide space for the new plantations. Variations of this were practised on estates at Hockeridge in Hertfordshire, Guiting in Gloucestershire, Thonock in Lincolnshire, Parmoor in Oxfordshire and Cirencester in Gloucestershire. At Cirencester there were considerable areas in which existing regeneration of ash and sycamore occurred in patches; here planting was carried out between the patches, using pure groups approximating in size to that of the regeneration groups. This involved the removal of existing inferior growth, including hazel coppice. European larch, Scots pine, and beech were used in these plantings and some fine pole-crops have resulted. At Thonock and Hockeridge group fellings up to 1.5 ha (in contrast to those at Cirencester which, following the size of naturally-regenerated groups, tended to be in the region of 0.1 ha). These fellings could thus be regarded as small coupes and probably represented the maximum size to which the 'groups' in this system could be expanded. At Guiting, where the crop was pure hazel coppice, fellings were made along similar lines, belts of hazel 10 m wide being left between adjacent coupes, forming a grid which provided shelter on all sides of each coupe.

Garfitt (1953) describes a form of group system which he applied at Cirencester Park in areas where the overstorey had been completely removed in wartime fellings.

Clearance of the hazel underwood was essential before replanting as described above. No planting was done under the shade of old hazel or other growth. Management has been changed under subsequent regimes but many of the stages in the original system can still be traced, although much of the planted conifer has been removed in early thinnings. This has resulted in a greater preponderance of sycamore and ash. Current management is tending towards a form of group shelterwood system leading to final crops of a 'uniform' character. Groups of regeneration are typically 0.05 to 0.1 ha in size, and the range of age within each group less than ten years. The aim of the present management is to spread the felling of the beech overstorey over twenty to thirty years, by which stage all the groups will have merged. The most abundant seedlings are ash and sycamore, but through the selection applied in three cleanings in the first fifteen years, the proportion of beech, oak and cherry is considerably increased. Ash and sycamore are, in effect, used as a matrix or nurse for the final crop species which are less abundant.

Commenting on the size of groups in general, Garfitt suggests that in a larger group one would actually find least growth in the centre and the best towards the margins of the groups. (That at the actual margin itself is depressed due to over-shading by the standing crop.) In these outer areas there would be considerable benefit from the sideshelter and other factors – leaf-fall, mycorrhiza, and others – and yet not too much competition and shading from the surrounding trees. Further investigation on the effect of shading and competition on the growth of trees and weeds is necessary before an optimum size for groups can be recommended. Garfitt (1984, 1988) further explains the management of the groups, and emphasizes the need for continuity of management. He sets out some disadvantages of the group selection system, i.e. the need to control rabbits, grey squirrels and deer; and indicates some problems of harvesting, which can largely be overcome. Earlier (1977) he recommended uneven-aged silviculture for use in the service of landscape and amenity. Some of his comments on natural regeneration are noted in section 3 *infra*.

Bradford-Hutt Plan for continuous canopy forestry. In the 1950s the sixth earl of Bradford, a distinguished forester, having pondered the potential advantages of irregular (uneven-aged) forestry decided to experiment in evolving a silvicultural system suitable for Britain or at least for its environmentally sensitive areas. Experiments carried out in conjunction with his forestry manager, Phil Hutt, on his Knockin estate in West Shropshire, led to the conclusion that European selection systems were inappropriate, and that it was necessary to evolve a new system of continuous canopy forestry which would be both commercially viable and aesthetically attractive and beneficial.

In 1959, when Lord Bradford purchased the woodlands of the Endsleigh estate near Tavistock in south Devon, an opportunity was taken to ensure their survival and utilize their potential for timber production, as well as to safeguard their conservation and landscape importance. The Tavistock Woodlands Estate lies in the Tamar and Tavy valleys straddling the borders of Devon and Cornwall. The area is one of high rainfall with soils mostly of medium to high fertility, the steep valley sides present an erosion danger, and there is risk of late spring frosts.

Almost at once Lord Bradford and Phil Hutt put into practice some of the ideas they had tested in their earlier experiments (Bradford 1981), and embarked upon the conversion of the extensive even-aged conifers to irregular forest. (Among the many plantations were fine Douglas fir, fifty to fifty-five-years-old providing high quality saw timber.) The procedure decided upon, a new concept in Britain, has become known as the Bradford-Hutt Plan for continuous canopy forestry. Its aim was to seek an alternative to traditional clear-felling and planting and create a system of continuous forest cover on much of the estate's woodlands that would be sound economically and avoid the disadvantages inherent in plantations managed under clear-fell regimes. Part of the impetus for development of the system was the desire to obviate the problem in group selection of felling and extracting an overstorey without damaging younger age classes; also to control the yield of the stands, particularly as *normality* was among the objectives. The Plan was started under some apprehension as to costs of conversion, reactions of estate staff and the national forestry 'establishment', along with the long-term problem of continuity of management.

The crown projection area of the main species of trees at maturity, particularly Douglas fir, was close to 36 m^2, and it was therefore decided to adopt a plot size of 6 x 6 m set out in a geometrical pattern, and a rotation of fifty-four years. These factors proved to be suitable also for other large conifers growing on the estate (mainly western red cedar and western hemlock).

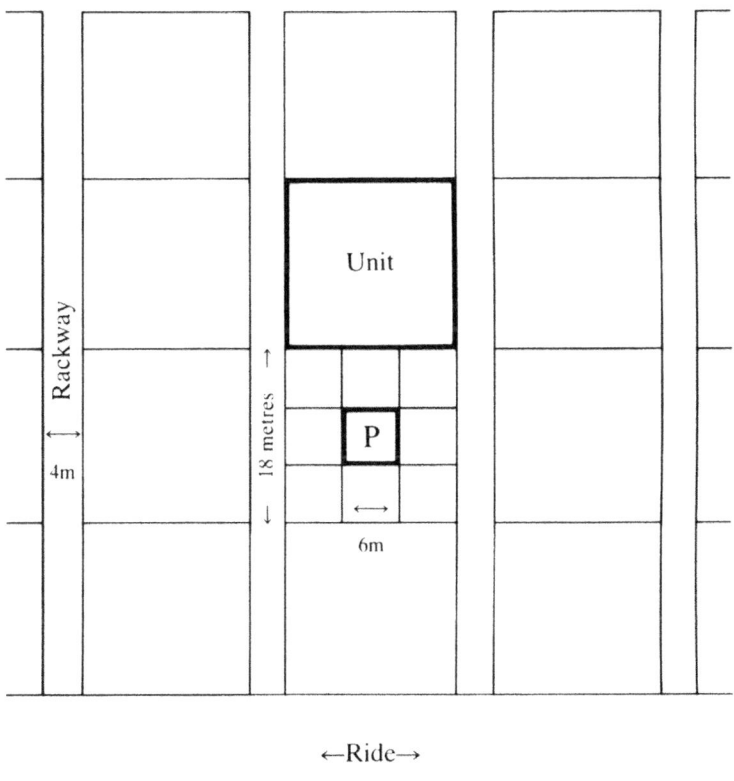

Figure 16 Bradford-Hutt Plan (Tavistock): Schematic representation with units, plots, rackways and rides. P = plot

The woodlands under the Plan (see Figures 17, 18 and 19) are divided as under conventional practice into a number of compartments and sub-compartments of appropriate area. Each area consists of a number of units, 18 x 18 m (thirty per hectare) each containing nine plots of 6 x 6 m. Each plot is intended to produce to maturity one final crop tree at age fifty-four, and is initially planted with nine plants (see Figure 17). The 4 m width extraction racks between pairs of units, i.e. every 36 m, simplify both management and extraction; and the loss of productive area is minimal as the canopy soon closes above such narrow racks.

The sequence followed is:

1 The size of the basic unit is 18 x 18 m, an area chosen because each mature conifer was estimated to have a crown area of about 36 m^2 (say 6 x 6 m) at fifty to fifty-five years (say fifty-four years). Thus an 18 x 18 m unit of nine 6 x 6 m plots could be established.

2 An important aspect of the Plan is the spiral planting scheme. A fifty-fouryear rotation within each 18 x 18 m unit is used. After felling and clearing as necessary the central 6 x 6 m plot, nine small shade-bearing trees are planted. Six years later, nine more trees are planted in an adjacent felled plot, then six years later, nine more in the next plot of the unit, and so on up to fifty-four years.

3 After a plot has been growing for eighteen to twenty years, the nine trees are thinned, so that by thirty-six years there is only one tree left – the final crop tree, to be felled at age fifty-four.

4 Once the cycle is fully established, after fifty-four years, the pattern is as shown in Figures 18 and 19, and the cycle begins again. The result is that once the system is complete, one mature tree is removed from each 18 x 18 m unit every six years.

5 The 18 x 18 m units are then repeated throughout the woodland in the manner shown in Figures 18 and 19, with 4 m wide rackways spaced 36 m apart between each pair of units for management and extraction.

The small size of the plots means that only shade-bearing species can be used at first, entirely conifers to date (1991) apart from southern beech. The Plan is essentially a 'group system' but as plots within a unit are felled and planted at six-yearly intervals, then each unit will eventually contain nine plots of all ages. If the uneven-aged units, rather than the even-aged plots are considered as the basic group, then the Plan could be considered a 'group selection system'. In summary the Plan establishes small groups on a geometric pattern in such a way that a range of nine age-classes from one to fifty-four occur in a spatially predictable way. Thereby, felling and extraction become easier, the required objective of *normality* is assured, and the adoption of a yield regulation system based closely on the Forestry Commission yield tables is possible.

From about 1960 (see Plate 6) the first 6 x 6 m plots were cut at 18 m intervals in even-aged stands of Douglas fir and larch of varying ages. (NDR calculations confirmed that the optimum time to start converting a plantation is at the time of first thinning.) The central plot in the unit was planted with nine shade-bearing plants – western red cedar, western hemlock or coast redwood. (Southern beech and Douglas fir were also planted in some of the early plots.) After six years, a 6 x 6 m plot adjacent

Silvicultural Systems   261

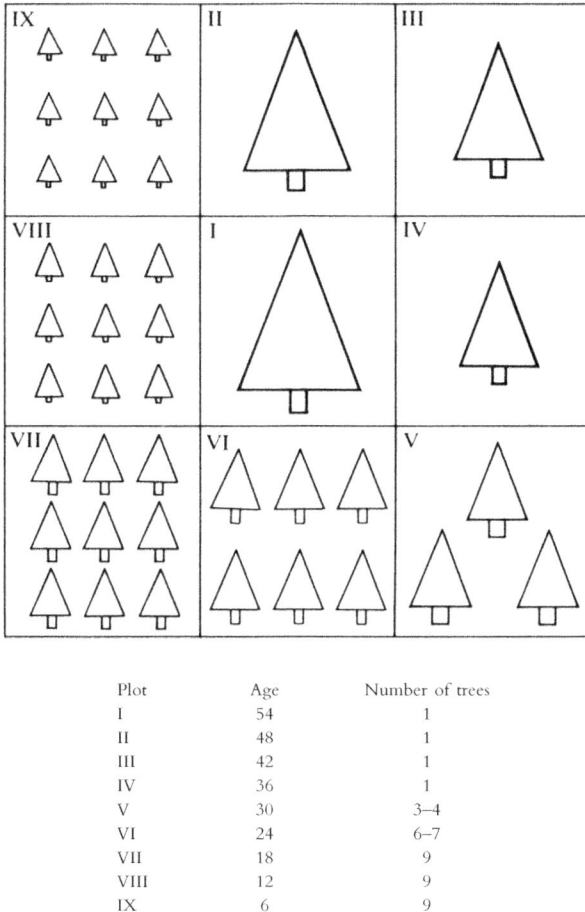

Plot	Age	Number of trees
I	54	1
II	48	1
III	42	1
IV	36	1
V	30	3–4
VI	24	6–7
VII	18	9
VIII	12	9
IX	6	9

Figure 17. Bradford-Hutt Plan (Tavistock): Schematic representation: a unit at the end of the 54 year rotation

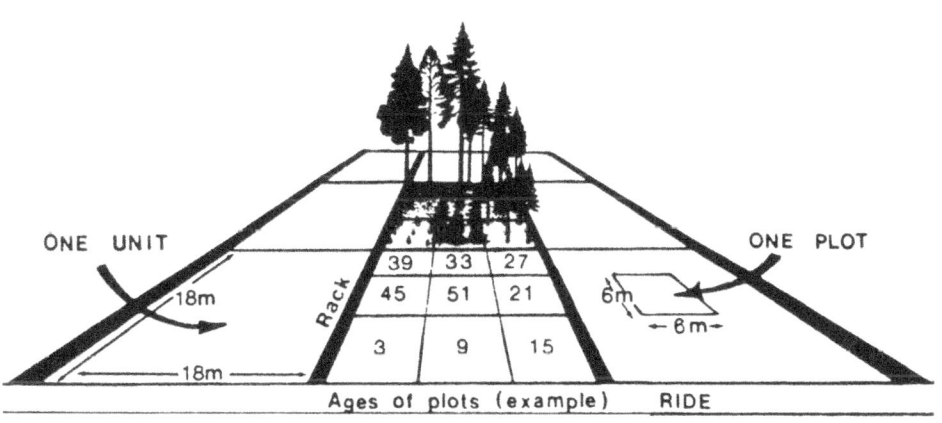

Figure 18. Bradford-Hutt Plan (Tavistock): Schematic representation

to the first central plot in the unit was likewise felled and then planted; and this pattern proceeded at six-yearly intervals in a clockwise spiral around each central plot.

Set-backs experienced during the first decade of the Plan included inadequate growth of western hemlock plots under a canopy of Sitka spruce and beech. Laying out the first plots in mature conifer stands with no clear rows proved a skilled task, and some mistakes were made. In some plots, the planted trees became crowded in by the surrounding matrix of older conifers. However, after six years, and as subsequent stages were introduced, the planted trees showed that they could grow adequately, and were quickly making up for the initial lost growth.

The overstorey matrix was thinned accordingly to conventional estate crown thinning policy to recommended stocking levels. All the new plots were measured as they approached first thinning stage; and the results showed that despite the slow start suffered by the early plots during the conversion stage, the chosen final crop trees in each plot were growing at between six and four rings to 2.5 cm, i.e. the maximum rate consistent with acceptable timber quality. (The outturn is expected to be about 80% saw timber and only 20% pole-size timber.) Douglas fir, introduced from the fourth plot onwards, are thriving. The first western hemlock plots have been thinned, and potential final crop trees are being pruned. Recently (1991) the sixth plot (of nine) of a unit is being established in some units – more than halfway to full conversion at fifty-four years. About 350 hectares have been converted.

The Plan, which Lord Bradford lived to see progressing satisfactorily, and which Phil Hutt MBE, now in retirement, frequently inspects with interest and pride, was continued by Roy Dyer for the seventh earl. The Plan is the most extensive and organized example of continuous forest cover based on uneven-aged forestry in Britain. Profitable woodland production and utilization has been continued throughout the Plan; broadleaves and conifers grow together in harmony. The whole forms an excellent example of commercial forestry complementing creative conservation. It creates a mosaic of varying habitats, thus increasing ecological diversity and enhancing botanical interest (Harris and Kent 1986, 1987). Landscape and amenity have been safeguarded: an open structure is retained with no apparent visual change.

Benefits claimed for the Bradford-Hutt Plan are described by Collin (1986). Chiefly, they are the obviation of many of the disadvantages of a clear cutting system. That system can cause severe landscape change and even after restocking has an unattractive thicket-stage; there is a narrow range of flora and fauna at most stages of rotation, and vulnerable flora is liable to extinction. The Bradford-Hutt Plan avoids the aforementioned disadvantages and claims other advantages which include producing timber of higher value with a regular flow from stands. Collin (1986) has also pointed out some of the problems to be overcome: financial, especially the investment cost of felling immature crops early and extra cost of supervision, felling and extraction, and technical (the need for skilled and committed supervisory staff and a trained workforce). Paramount is the long-term commitment needed from all concerned. The Plan represents a logical basis for developing other systems to suit the growth requirements of conifers, of mixtures, or of broadleaved species in other parts of Britain (Bradford 1981; Hutt and Watkins 1971; Hutt 1975; Dyer 1986; Collin 1986; Matthews 1989).

Economics of irregular silviculture. Compared with the management of regular silviculture, that of irregular silviculture is more time-consuming and usually more expensive. The systems require high input of expertise to work on all but a very small scale. However, few of the approaches towards irregular silviculture have been adequately costed, and their individual success has depended almost entirely on the personal interest, enthusiasm and skills of individual owners and foresters; and hence the lessons they teach may not be fully justified on a large scale. Compared with the even-aged clear cutting system, cash advantages and disadvantages claimed for irregular stands have been cited by Price (1989), as noted earlier in this chapter.

Pryor and Savill (1986) in researching irregular systems which are, or could be, used in Britain, focused on techniques and financial results rather than ecological principles. For each of the systems, they presented: (i) descriptions based on European use; (ii) an account of British examples; (iii) discussion of associated silviculture techniques; (iv) analysis of likely profitability; and (v) evaluation of benefits for landscape, amenity and conservation. Although establishment costs for broadleaved shelterwood systems were lower than after a clear cutting system, the overall profitability for oak was similar. (The net present values of some 'low-input' options, such as simple coppice, were surprisingly high, always assuming there is a market for produce.) Most profitable was an ash/sycamore shelterwood system. They recommend that the failure of adequate natural regeneration should be tackled by concentrating on prolific-seeding species such as ash and sycamore rather than oak and beech, and by more thorough preparation of overstorey density.

Lorrain-Smith (1986) in a paper on economics in relation to irregular forestry, comments on the aims or objectives of possible action, and methods of assessing options. He asserts that in certain situations forest management to favour aesthetics, wildlife habitat and species conservation will necessitate irregular systems of silviculture (i.e. the non-marketable benefits may be dominant), and hence it is important that the economic consequences are identified and evaluated. Most foresters and researchers who have written on the subject of irregular forestry agree that operations are more expensive than in a regular forest, but there is some disagreement about the scale of such differences. Shrimpton (1986, 1988) has studied the situation particularly in the Edinburgh University Experimental Area at Glentress. Crockford *et al.* (1987) have dealt with the relative economics of woodland management systems, following earlier work by Pryor and Savill (1986). Much has been updated by Spilsbury and Crockford (1989), following the changes made in 1988 of support by taxation and grant-aid.

Problems may arise in evaluating uneven-aged woodland. Available yield tables and growth models are all based on even-aged stands, making it difficult to forecast the productivity of an uneven-aged stand. It is more difficult to assess the volume of growing timber, and to decide how many trees of which sizes should be felled to obtain an appropriate steady flow of produce over the years.

A sober judgement on the economics of natural management of temperate broadleaves is that of Leslie (1989). He writes of forestry practices which may involve long rotations, low yields, slow-growing native species, and use of natural regeneration, and indicates five points of considerable significance in such management:

1 The economic prospects for natural management are greatly and, perhaps primarily, governed by the rate of interest used in calculations and projected over long periods of time;
2 The choice of that rate is almost entirely subjective;
3 There is little in economic theory or practice to guide or even constrain that choice;

4 Empirical evidence indicates that appropriate rate is more likely to be within a range of 2 to 4% than anything much higher; and

5 At discount rates within the above range, natural management of temperate broadleaved forests can be a much better financial and economic proposition than alternatives in forestry or land use generally.

Leslie (1989) asserts that his findings go a long way to dispelling the misconceptions about the economic feasibility of natural management systems in forestry. At high rates of interest, natural systems cannot compete in financial terms with alternatives in forestry such as high yield, fast-grown commercial plantations or alternative forms of land use. High interest rates are needed to demonstrate the uneconomic nature of natural management systems but at lower rates the reverse could well apply. In effect, Leslie says that the purpose of economics is to explain decisions made by practical people – owners and their foresters. Conventional economics rules out natural management but such work still goes on and the conventional economic case must therefore be wrong; and Leslie suggests what is wrong.

Subsequently Spilsbury (1990), using discount rates between 1 and 5%, has examined the profitability of a wide range of management options, ranging from intensive coniferous plantations (Sitka spruce, Japanese larch, Douglas fir) to more 'natural' systems (Douglas fir/oak, beech, oak, oak/ash/cherry, ash/sycamore shelterwood, re-coppicing of neglected oak, neglect of oak/ash group system, beech natural regeneration, and birch natural regeneration). 'The investments are investigated by the use of three different measures of long-term financial viability, benefit-cost ratio, land expectation values and forest rent. Analysis by the first method highlights the efficiency of investment for more "natural" management systems (and should be more widely applied), whilst the second method favours the "traditional" coniferous plantations. Forest rent tends to favour management options generating large revenues irrespective of their timing within a rotation.'

The need to promote visual amenity, wildlife conservation and recreation (albeit complementing timber production) has led to a renewed interest in development of a transformation from uniform plantations to irregularly structured forests (Edinburgh University 1990). Woodland owners should consider the possibility of adopting uneven-aged management. Current grants can help to pave the way for greater flexibility in woodland management, and thereby create woodlands which are valuable for amenity and nature conservation as well as producing saleable timber. In the lowlands at least, the adoption of partial irregularity may often be a reasonable objective; and Poore (1988) asserts that a practical or financial inability to achieve a full range of age-classes should not be seen as a reason for rejecting an irregular structure completely. This applies also to any system.

Coppice systems

The coppice system. This well-known silvicultural system for broadleaves, comprehensively described by Crowther and Evans (1986) and Matthews (1989), produces a crop of small or medium dimension timber raised from regrowth shoots produced by cut stumps (stools) of the previous crop. The system provides many commercial and aesthetic benefits, and is a low input: low output enterprise which needs little tending after establishment. The cycles or rotations are relatively short (fifteen to forty years), producing a regular periodic income throughout three or more cycles. The crop is even-aged; and may consist of one species (pure) or several (mixed).

Most broadleaved species will produce coppice. Besides sweet chestnut (commercially the most important), ash, oak, hornbeam, lime, sycamore and alder are the species that have been recently or are presently worked as coppice, pure or mixed. Beech, birch, cherry and some poplars, do not coppice vigorously or only have the capacity to do so when young and the stool is fairly small. (Hazel and osier, and other short rotation coppice crops, are noted later.) Few conifers can coppice, among them yew and Japanese cedar, which are not used, and coast redwood which is being experimented with (e.g. at Leighton in Powys and Longleat in Wiltshire).

The traditional coppicing system has had a profound influence on much broadleaved woodland. The fall in demand for some of its products and the decline in numbers of men who have the necessary skills of conversion, have much reduced the area of such crops (sweet chestnut is the exception), but coppice is still a significant forest type. The recent revival is due to rising firewood prices and increasing interest in conservation and landscape, for which coppice has important attributes (Fuller and Warren 1990). Nonetheless, much current coppice is a neglected resource which for several reasons has not been brought back into fully effective use.

The growth of a coppice crop is quite different from that of a plantation. Coppice stools may produce very high numbers of shoots per hectare, with rapid canopy closure and early culmination of maximum mean annual increment. The number and vigour of shoots per stool varies with species, spacing, age and size of poles. In the main species the number of stools per hectare range from 600–1200. Coppice is rarely thinned, but on occasion there may be some merit in removing stems of undesirable form or species, particularly birch, thereby improving the diameter and height growth of the remaining stems. Rotation lengths depend on species, speed of growth within species, and desired size and quality of poles. Typically they are fifteen to forty years (sweet chestnut twelve to twenty years), according to the crop's vigour and the market in view. The life of coppices (number of cycles) varies up to three or more; oak coppice is generally retained for four to five cycles (90–130 years) and ash for about three cycles (50–70 years).

Planting a new coppice necessitates well grown robust transplants spaced every 1.5–2.5 m. Following careful tending they should become established by about the fourth year, when the decision has to be made either to allow the maiden trees to continue for one rotation before cutting and coppicing, or at once to 'stump back' the trees almost to ground level to induce them to form a stool and throw up shoots, i.e. to begin coppicing. In the latter case, subsequent respacing of the shoots might improve the quality of the remaining shoots, but this operation is rarely worthwhile. The first cut of a newly established coppice should be made in March or early April so that the new shoots do not emerge until June when the risk of severe frost damage is slight.

On felling, the stools regenerate rapidly and provided they are protected in the first season from damage by farm stock, rabbits and deer their growth (oak up to 1m, and ash, sycamore and sweet chestnut as much as 2 m in the first year) exceeds that of woody weeds; and hence weeding and cleaning are rarely required. A problem encountered in recent coppice working is damage by deer (particularly muntjac and roe), especially of palatable species such as ash and lime.

Coppice management including methods of felling is explained fully by Crowther and Evans (1986). Traditionally, coppice is cut in the dormant season (October to March). Lop and top must be removed or stacked away from the stools, and all poles or converted produce extracted before the spring. Late summer felling in particular should be avoided, as new shoots may not harden off before winter.

Mixed coppice. Sale and clear cutting is usually by the hectare or by some local measure such as the 'coupe', the 'cant', the 'fall' or the 'cut'. A sustained yield may be obtained from a large area if the number of annual coupes are planned to be the same as the years of the cycle. After each felling a small amount of infilling of gaps between stools by planting or layering may be needed, along with attention to fencing where there is a serious risk of browsing damage. A well-stocked coppice at rotation age will have dominated the site so that there is very little established ground vegetation, other than the pre-vernal plants appreciated by conservationists.

Satisfactory coppicing depends on species, the fertility of the site, the age of the stools and the care taken to prevent damage to them. The vigorous growth of some coppices and the low cost of regeneration can make them profitable, especially where they are of good quality and can meet the demands of a particular market not too distant. In principle, coppice can be worked indefinitely but two factors, site exhaustion and stool mortality, could lead to declining yield with time. Cutting successive crops may eventually deplete a site of available nutrients, especially of phosphorus, though there is little evidence to show that this has led to a decline in growth. Applications of fertilizer should be considered if foliar analysis has indicated a possible nutrient deficiency.

The principal current markets for mixed broadleaved coppice are pulpwood (a large mill at Sudbrook in Gwent), boardwood, turnery, fencing, firewood, charcoal, and tanbark (oak). For smaller dimensions specialist markets are available, including pea sticks, bean rods and hedge laying material.

Yield data on mixed broadleaved coppice are not comprehensive. Evans (1984) indicates the mean annual increment over a rotation, in terms of dry wood/ha/yr as about 2.5 tonnes for sycamore, birch, ash, lime, oak and alder; and 6 tonnes for poplar and willow. Pryor and Savill (1986) suggest that yields are 3.5 to 4.5 m^3/ha/yr (on a 20–30 year cycle). A general guide to the net income per hectare is given by Insley (ed.) (1988): oak (20–35 year rotation, yield 3–7 m^3/yr) £55–£550; and mixed broadleaves (20–25, 6–10 m^3/yr) £50-£530. Some suggestions of profitability are also given by Pryor and Savill (1986) and Crockford et al. (1987). Preliminary yield tables for oak coppice are given by Crockford and Savill (1991).

The foregoing figures of yields and prices indicate a 'profit' but no account has been taken of land cost, which becomes important when comparing alternative land uses. However, coppice has high conservation and landscape values which are difficult to measure in monetary terms. But clearly the possibility of a regular cash return, which coppice working affords, improves the prospect of retaining such woodland and hence conserving the values mentioned.

Sweet chestnut coppice. This is the most successful and profitable coppice crop in Britain. Working practices in it have been described by Evans (1984) and Crowther and Evans (1986). The labour involved in conversion on the site (crosscutting, cleaving, peeling and pointing) can be intensive according to product mix. The coppice lies chiefly in Kent and Sussex (with small areas in Gloucestershire and Herefordshire) where it serves an important rural industry related to the cleft fence paling and occasionally hop-pole markets. In general the climate is relatively warm and sunny, and the annual rainfall is in the range of 500–850 mm. The coppice thrives particularly on frost-free warm sites with loamy or sandy soils; a reasonably fertile soil is needed, the ideal pH being 4.0–4.5. The best quality is usually found on greensand.

Cycles of cutting are between ten and twenty years, mainly twelve to sixteen years, well short of the age when mean volume increment reaches its maximum. They are wholly determined by the requirements of the market, for which stem diameters not below 7–10 cm are usually sought. Rollinson and Evans (1987) have reported on sweet chestnut coppice growth and relation to site and various stand characteristics, and show how volume or weight per hectare may be predicted from very simple measurements. They provide two provisional yield tables of volumes and fresh weights for a range of ages and mean diameters – see Tables 21 and 22.

Table 21. Sweet chestnut coppice: yield table based on age.

Age (years)	Volume (m^3/ha)		Fresh weight (tonnes/ha)	
	To 4 cm tdob	To 7 cm tdob	To 4 cm tdob	To 7 cm tdob
5	15	–	15	–
10	65	25	60	25
15	130	80	125	80
20	205	160	200	160
25	300	270	295	265
30	405	405	400	400

Source: Rollinson and Evans (1987). FC Bulletin 64.

Sales of sweet chestnut coppice are by private tender, private treaty or public auction. Such coppice is usually sold standing with coupes ranging from 0.5 to 2.0 hectares. The main determinants of price, other than quality, age and size, are: location (roadside frontages are preferred), availability of local markets for the lower quality produce (firewood, pulpwood and boardwood), and availability of cutters and cleavers. Prices per hectare, though chiefly dependent on the above factors, fluctuate from year to year according to the fencing market and the economic climate (Bax Standen 1990). In 1988 standing prices were in the range of £600-£4,500/ha (average about £1,300), in 1989 they were from £250-£2,000/ha (average about £1,000). During 1990 demand and prices fell dramatically. Current (1991) standing prices may lie within the following ranges: low quality cants: £500-£1,000/ha, good quality £750-£1,250/ha, and prime quality £1,000-£1,750/ha. Yield on a twelve to sixteen year rotation (with 800 to 1,000 stools/ha) may be 5–6 m^3/ha/yr. Pryor and Savill (1986) suggest yields are 3–7 m^3/ha/yr in a fifteen to twenty year cycle. The profitability of sweet chestnut coppice is usually satisfactory. Total growing costs, including auction fees when felling, amount to £75-£100/ha.

Table 22. Sweet chestnut coppice: yield table based on mean diameters.

Mean diameter (cm)	Volume (m³/ha)		Fresh weight (tonnes/ha)	
	To 4 cm tdob★	To 7 cm tdob†	To 4 cm tdob★	To 7 cm tdob†
5	20	—	20	—
7	95	—	90	—
9	160	65	150	60
11	215	135	210	135
13	260	200	255	200
15	300	255	295	250
17	330	300	325	295
19	355	335	350	330
21	370	360	370	355
23	375	375	375	370

★ Use mean diameter of all coppice with a diameter equal to or greater than 4 cm ob.
† Use mean diameter of all coppice stems with a diameter equal to or greater than 7 cm ob.

Source: Rollinson and Evans (1987). FC Bulletin 64.

Hazel underwood is a minor short rotation coppice crop, sometimes grown pure but generally an understorey to standards of some other species, usually oak. Good quality crops can prove profitable in districts in the south of England where supplemented by skilled local underwood workers, particularly makers of wattlehurdles and thatching spars, supplemented by production of pea sticks, bean rods and hedge laying materials (see chapter 9). Some is used for crate-wood and small turnery products. A modest area of it on any estate is useful and usually has good conservation value.

Ideally hazel is worked on a six to ten year cycle, yielding shoots 4–5 m long; if left longer the hazel tends to lose value, and traditional users will not purchase it. Growth rate is similar to that of young coppice oak or lime. On fertile sites and with a stocking of about 1,500 stools/ha the yields of oven-dry wood are about 25 tonnes/ha at ten years, rising to 45 tonnes if grown on to fifteen years. Hazel has no potential to develop into anything approaching high forest, and will require substantial investment if it is decided to clear and restock with another species.

Coppice with standards. This silvicultural system consists of two distinct elements: (i) a lower even-aged storey treated as coppice, and (ii) an upper storey of standards forming an uneven-aged crop and treated as high forest. The standards (according to their age termed 'stores', 'princes', 'tellers', 'standils' or 'reserves') are grown on for large size timber. The standards are usually of oak (particularly over hazel), ash, lime, sweet chestnut, cherry or larch; beech are unsuitable as they cast too much shade. The coppice is usually hazel, sweet chestnut, alder, ash, lime, sycamore or wild cherry. The crowns of the upper storey generally occupy 30–50% of the ground area; and the stocking, depending on tree size, is usually 30 to 100 per hectare (18 to 10 m spacing). They should be evenly spaced, and the crowns far apart, never touching, so as to allow plenty of light to the coppice below. Too great a number of standards overshades and depresses growth of the coppice. In a fully functioning system the standards will consist of from three to six different age classes.

At each coppicing, standards in the oldest age class are felled and a few new ones established by planting or recruited from coppice shoots or natural regeneration. The length of time a standard is retained will depend on desired log size and species growth rate; oak is generally retained for five to six coppice cycles (100–130 years) and ash for three to five cycles (60–100 years).

Standards develop large open crowns, exhibit rapid diameter growth similar to that under free growth conditions (Jobling and Pearce 1977) and reach timber size in about three to six coppice cycles. On some species, notably oak, standards often develop vigorous epicormic branches and unless these are pruned off or controlled in some other way highest quality timber cannot be produced.

The coppice with standards system is not so much favoured as hitherto, and many crops have been and are being converted either to high forest or, less frequently to simple coppice. One disadvantage of the system is that the standards tend to be short in bole and the coppice beneath them weak and partly suppressed, resulting in a deleterious effect on the value of both; and timber merchants dislike acquiring areas which have to be cut yet yield relatively little. Matthews (1989) gives advantages and disadvantages compared with other systems, and application in practice.

Stored coppice is a stand of trees derived from shoots grown beyond the normal coppice rotation, often with the intention of converting to high forest. Much coppice woodland is already past the normal age of cutting and is becoming a form of high forest. The promotion of coppice to high forest by 'storing-up' (i.e. allowing coppice to grow unmanaged) is an alternative system for most species of coppice (not hazel). An aid to attaining high forest, and the improving of the quality of the stand, can be by removing all coppice stems except the best one (straightest, most vigorous) on each stool. The operation is called 'singling' coppice and the stems that are grown to become large trees are known as 'stored coppice'. Once the operation is undertaken, subsequent treatment is as for a high forest stand for thinning and felling. Yield from stored coppice in general does not differ greatly from planted crops but the pattern of growth is different. Early growth is more vigorous than in the planted crop but this greater vigour diminishes and by seventy to ninety years the initial growth advantage is usually no longer evident.

Conversion to stored coppice has been affected predominantly through neglect. Many woodlands today resembling high forest are stored coppice owing to decline in coppice working this century; a condition which applies quite widely to oak coppice. Stored coppice may produce inferior timber due to increased curving at the base of the stem (butt sweep), and to a high percentage of decay and dark staining developing from the stools. Storing should not in general be the principal method of regeneration for stands grown for timber, but it should be borne in mind as an option. Thinning of stored woods to further the development of high forest is quite practicable, as seen in the extensive oak woods of Exmoor and Dartmoor (Penistan 1986).

Short-rotation coppice for fuelwood and energy. Short-rotation coppice is grown in a cycle, usually of less than ten years. Several broadleaves can be worked in this way either as underwood for rural crafts (hazel has already been noted) or as biomass for fuel, energy, or industrial manufacture. The results of ecological, genetic, physiological, and silvicultural research have been applied in trial coppice plantations grown on short rotations (Matthews 1989; Mitchell *et al.* 1989).

Short-rotation coppice crops are broadleaved species such as willows and poplars planted at high densities and mechanically harvested every two to eight years, for use for energy generation or industrial cellulose. Further research and development is required, but in time these crops are likely to provide a significant alternative use of agricultural land. Planting grants are available for establishing them, also management grants under certain conditions (see chapter 15). They are eligible for the Set-aside payments available for non-agricultural use of land.

The original interest in short-rotation crops was as a means of producing biomass for paper pulp, and then, following the 1970s oil crises, for energy generation. The Department of Energy took the lead in developing it. This interest has been stimulated by the potential for on-farm electricity generation within a privatized industry and the potential of these crops to contribute cellulose for use in the wood processing industries.

The concept has been tested experimentally in parts of the United Kingdom on a wide range of sites including good quality land, mainly funded by the Department of Energy (1989). Important trials have been made at Long Ashton Research Station outside Bristol and at the Horticultural Centre at Loughgall in Northern Ireland (Stott and Parfitt 1986). Of the broadleaves tested, willow and poplar have given the best results. *Salix x viminalis* Bowles Hybrid or the hybrid *S.* x *dasyclados* have advantages on better agricultural sites. Of the poplars, trials with hybrids of *Populus trichocarpa* x *deltoides* like 'RAP' are encouraging, and newer hybrids from Belgium may be even better. The coppicing of *Eucalypts* (e.g. *E. gunnii*), alder and southern beech *(Nothofagus* spp.) is under investigation (Stott *et al.* 1985). The results to date are modestly encouraging. The Energy Technology Support Unit (for the Department of Energy) have discontinued with *Nothofagus* spp. but have increased numbers of willow and poplar clones in trial, including many *S. viminalis* hybrids from Scandinavia and *Populus* clones from Belgium.

Insley (ed.) (1988) provides a planting model example for short-rotation coppice. Cannell *et al.* (1988) found that willow and poplar planted as cuttings at 0.3 x 0.3 m spacings, yielded in one year, respectively, 10 tonnes and 5 tonnes/ha, due to the ability of maiden willow cuttings to produce twice as many shoots as poplar. The general yield from coppice is expected to be about 10–12 tonnes dry wood/ha/yr (theoretical yield class 25 plus) if harvested every three or four years, equivalent in energy content to 6 tonnes of coal or 4 tonnes of oil (Stott and Parfitt 1986). Oven-dry yields of poplar biomass in five and six-year-old field experiments have ranged from 7 to 10 tonnes/ha/yr, while in old, vigorous stoolbeds, yields of 20 tonnes/ha/yr have been obtained (Jobling 1990).

Ground preparation is most effective by cultivation, usually ploughing, preceded by killing of weeds by herbicide, which aids weed control and improves yields. Optimum spacing will vary, depending on the length of cutting cycle and species chosen. For cycles of six to eight years producing larger sized products, wider spacing 1.5–3 m optimises production, while for short rotations of two to four years closer spacings 0.5–1 m are more appropriate. Willows are usually spaced closer than poplars. Cuttings 20–25 cm long are inserted into cultivated ground. Weed control at the early stages of establishment is crucial (weed competition in the early years will severely reduce final yields). Drought can lead to failures. The growth at the end of the first year is cut back to near ground level to form the coppice stool. Pests and diseases can be a problem and reduce yields, particularly rust diseases on willows.

The techniques are still very much in the experimental and development stages. Little of this material has been available up to now and as a result no market price structure has developed. Further research is needed to select the most appropriate clones so as to increase yields and the reliability of yield estimates, and to improve associated technologies. The economic sensitivity of energy forestry to changes in costs, prices and markets also needs further study and more markets need to be identified (Department of Energy 1989). The most useful comparisons which can be made are in relation to use of wood as an alternative fuel source. The demand has been helped by the development of efficient boiler systems (Mitchell 1984). Early indications are that one or two hectares of this system should be able to generate sufficient fuel to keep an average farmhouse heated on a continuing basis (Insley (ed.) 1988).

The land must be reasonably flat for mechanized working. The initial establishment is intensive and costly. Planting densities are high, adjusted according to the expected length of rotation and size to which the crop is to be grown before harvesting. At least 5,000 stools/ha are needed and more commonly between 10,000 and 20,000 for five-year willow and poplar cycles. The Department of Energy's current favoured spacing for willows is 10,000/ha at 1 x 1 m. They have an experimental coppice harvester under trial. For traditional sweet chestnut coppice on a twelve to fifteen year cycle, 800–1,100 ha is aimed for (Crowther and Evans 1986; Rollinson and Evans 1987). Irrigation to prevent water stress would improve results.

Reports on trials by Aberdeen University (Mitchell *et al.* 1989) for the Department of Energy include costings and economics. The relatively short investment period justifies limited application of fertilizers, pesticides and other inputs, even though the crop often commands a low unit price. The latter can be partly offset by mechanized harvesting. Such crops produce not only early returns, but also *more* frequent return. Price (1989) points out that to evaluate highly profitable short-rotation crops it is important to calculate net present value of a perpetual series of rotations, or to use forest rent, rather than profit on a single rotation (see chapter 14). Coppicing's economic importance improves under high discount rates, short rotations and 'costless re-growth'. Low discount rates favour more frequent replanting to maximize production. The trials by Aberdeen University noted above (Mitchell *et al.* 1989) have generated realistic costs and have demonstrated that short rotation coppice production requires intensive methods of husbandry to the extent of complete cultivation and weed control and selection of clones. Recommendations for biomass production are:

1 The site should be chosen within existing field structure so as to minimize fencing costs;

2 Preferably arable land should be chosen to reduce herbicide costs and allow spring cultivation;

3   It is advisable initially to screen a large number of clones, before choosing the most disease resistant and high yielding;

4   Weed control is vital during the establishment stage; and

5   Cutting back is advisable, both to facilitate weed control and to encourage more shoots per stool.

However, the trials highlighted the potential problems: disease has to be more fully understood before there is a biomass production system ready for adoption by farmers; and there is still the question of nutrient removals which require further investigation. It was shown that the establishment costs can be reduced to less than £1,000/ha, but there are more critical elements likely to pose a financial constraint on the adoption of energy forestry: (i) cost of harvesting, chipping and transporting; (ii) high capital cost of equipment to move and combust the wood chips; and (iii) cost of other fuels. A further constraint is the absence of adequate markets for wood chips although a market exists for industrial wood chips. Where a grower can also find a substantial on-site use for biomass without resort to external markets, then the prospects are good.

Fuel/energy crops and short-rotation biomass crops might become widespread in Britain since large tracts of fertile land could become available. On a relatively small industrial farm or estate, the potential, according to location and markets, could be attractive. It is unknown whether such crops are likely to reduce the diversity of the ecosystem, but less-intensive longer-rotation coppice systems are much favoured by conservationists.

Single stem plantations. Another short (ten to twenty year) rotation option is single stem plantations – which involves high density planting (1 x 1 m, i.e. 10,000/ha) and clear-felling of the whole crop at between twelve and twenty years' growth (Mitchell 1989). Thus, unlike coppice, single stem plantations will have to be replanted after each harvest. Almost all tree species are suitable for this use, but broadleaves such as alders, birch and southern beeches seem to offer the best potential. Productivities are expected to lie between 12 and 14 dry tonnes of fuel ha/yr.

Modified conventional forestry (MCF) alters and extends existing forestry practice so that wood is produced for fuel as well as for timber and pulpwood. The MCF yields a greater volume of wood products than conventional forestry. Coniferous plantations under MCF would have trees planted much closer than normal – say 5,000 trees per hectare – and thinnings would be taken at a much earlier stage and chipped for fuel. Harvesting of later thinnings and clear-fell would be carried out by integrated harvesting techniques. Product separation would be into sawlogs, small diameter roundwood and fuelwood. This offers opportunities for simple, low-cost production of fuelwood.

Studies have shown a high level of interest among farmers in energy forestry as an alternative land use – provided that ready markets for wood fuel can be found. Present indications are that fuelwood from MCF has the potential to compete in the industrial market, which is the most competitive market for fuels, while coppice fuelwood could become viable in the higher-priced agricultural and service sector markets, as well as in the industrial markets of the future.

Conversion of silvicultural systems

The term 'conversion' is generally taken to mean a change from one silvicultural system to another. Examples are: conversion of coppice or coppice with standards to high forest; and clear cutting to the group selection system. In effect, they are conversion of regular stands to irregular stands. Conversion also includes restoration of unmanaged (degraded) forest to a productive condition. Matthews (1989) describes some examples of conversion to show how it may be achieved. The economics of conversion are touched upon later in this chapter. A noteworthy example of conversion from clear cutting to group selection is the Bradford-Hutt Plan described earlier in this section.

2. The 'normal' forest and sustained yield

The term *normal* is applied to a forest, woodland or group of woodlands containing a regular and complete succession of age classes, up to the most mature. It embodies the idea of a fixed rotation. At its simplest the term *normality* means that a forest has an even distribution of age classes by area, ranging from newlyplanted to stands of rotation age. More precisely, *normality* implies arranging the forest so that production is sustained. The concept can be extended to cover normality of annual production by species and size classes, of annual revenues, and of labour requirements (Savill and Evans 1986).

In the management, the volume felled each year should be equal to the annual increment; the stocking then remains constant and the yield is sustained. On many private estates the woodlands are understocked for various reasons, but foresters might regard some of the principles of the *normal* forest as an ideal to be aimed at and approached as nearly as possible if their objective of woodland management is to secure *sustained yields*. But the importance here rests not on the normal forest but on the *normal growing stock*. They should always bear in mind the ideal stocking which they are trying to attain and consider how a normal growing stock can be constituted. On most estates it is possible to achieve a more normal distribution of age-classes by a number of methods used singly or in combination. These methods, involving a long lapse of time, include: (i) grouping the whole of the woodlands to form a single felling series; (ii) underplanting or replacing plantations that are unsatisfactory or have failed; (iii) taking advantage of the fact that some species grow faster than others and that the rate of growth of any one species will vary according to the site and management; (iv) felling crops before or after they would usually be considered merchantably mature; and (v) by varying the thinning treatment. The choice of method in achieving normality is according to the importance which foresters attach to each of several aspects of management. However, within a reasonable range of rotation and thinning it is possible to achieve approximately the same yield in both volume and the average size of trees felled with widely different levels of growing stock.

An early step towards securing a sustained yield may be by first planting up the better sites, to shorten rotations, leaving the poorer sites to the last. This sequence of establishment is also the more profitable one. Where the abnormality consists in an insufficient area of mature crops it will also be desirable to fell rather less than the annual increment in order to increase the growing stock. Assessment of both the growing stock volumes by size classes, and of the increment, is therefore required. After many decades of careful management the wood may approach something near to the ideal. In all this foresters, by setting such targets for planting, thinning and felling as will, eventually, tend towards normality in the growing stock, will be able to leave the woodlands in a condition nearer the ideal than that in which they found them.

Forest economists (for example Johnston *et al.* 1967) argue that *normality* has little little relevance in places where cheap and efficient internal transport is available since no one industry or community depends upon a regular supply of material from a single forest. However, this may not be true for the few private owners depending upon their woodlands for income. Perhaps the main disadvantage of too strict an adherence to (or move towards) normality is inflexibility in management, but this can be avoided by, for example, considering age classes of five years instead of annual ones (Savill and Evans 1986). It is often impracticable to spread planting of new areas over a whole or even a large part of a rotation, but a satisfactory state, approaching normality, can be reached by the end of a second rotation, even in small woodlands. This is achieved by the different species on sites of different levels of productivity, the occasional need for some premature felling, windthrow, and the possibilities of some delayed felling. Managing forests to maintain a mosaic of different age classes and, if possible, of species as well, could be an important aim of a silviculturist. Malcolm (1979) pointed out that plantations cannot be expected to function as completely stable ecological systems in the face of environmental fluctuations, biological hazards, and changing demands by man. Heterogeneity confers a measure of resilience which enable forests to persist and to continue to be productive, despite such fluctuation, and it also enables them to absorb changes in market demands, social attitudes, and levels of inputs more easily (Savill and Evans 1986).

The foregoing comments will not be relevant to all private woodlands. The private owner may aim at sustained yield of timber based on a growing stock as close to normal in age or size as practicable; because many of the markets for the produce are quite close and it may be desired to obtain an assured annual or periodic income from it; in addition the wish may be to provide steady employment for local people (Price 1989). However, for some owners a steady cash flow year by year is possibly a more important aim than the attainment of *normality* in age class distribution. On some estates the need to build up a reserve of timber to provide for inheritance tax or other expenses in the future may tend towards a want of balance in the distribution of the age classes. The concept of the normal forest is an ideal of classical management and in times of great change it may have rather limited benefit to offer. Likewise it may be a too rigid concept for private estate forestry where flexibility is desirable; but it is far better than the common practice of no concept at all. Some foresters and their woodland owners might well regard the normal growing stock as an ideal to be continually pursued, although the tribulations of expediency may have to be suffered many times on the way to reaching it, or not reaching it.

The concept of sustained yield can also be applied to the conservation of cherished landscapes, and the provision of facilities for outdoor recreation (Matthews 1989).

3. Natural regeneration

Natural regeneration, by contrast to artificial planting, implies the natural growth of young trees from seed that has fallen on the forest floor. It may be obtained from seed already on the area, or from seed disseminated by neighbouring trees. It is the main method, along with sucker shoots from tree roots, by which forests have renewed and extended themselves down the ages. The process of natural regeneration consists of a series of interrelated events, each greatly influenced by chance. Much can be done, by actions founded on observation, to assist the process and reduce the impact of factors threatening failure so that it can be brought to a successful conclusion in a reasonable time. But natural regeneration remains a precarious stage in the life of a forest stand (Matthews 1989). Deliberate combination of natural and artificial regeneration is an old and well-established practice, and foresters who are committed to artificial methods will accept suitable natural seedlings when they appear.

Natural regeneration holds a special fascination, yet in formal forest management, aside from some medieval systems, it has never featured strongly in Britain. Exceptions include oak in the Forest of Dean since about 1899 and the New Forest during the 1930s; also beech and ash in parts of the Chilterns and the Cotswolds. Brief accounts of many current examples are given by Pryor and Savill (1986). Throughout much of Europe impressive forests, both broadleaved and conifer, based on natural regeneration, have prompted British foresters to emulate such achievements. However, for various reasons, the appropriateness or relevance of such silviculture to much of our forestry is debatable. In any case, continental foresters experience significant damage by deer browsing, and necessary fencing costs are high. Much of the most successful natural regeneration appears to be on neutral pH soils and assisted by worm-casts.

Everard (1988) explains the continental natural regeneration of oak, and suggests that there are no great differences in physical conditions between parts of Europe and southern England, yet on the continent, natural regeneration of oak, beech and other broadleaves is prolific. The differences appear more related to the silvicultural experience of foresters and the continuity of forest management. Continental oakwoods usually have a good understorey of beech or hornbeam, which keeps the forest floor clean and the soil moist. Their long rotations and dense stocking typically result in the canopy being much higher than in Britain, and this has a marked effect on the light reaching the forest floor. Everard urges that in southern Britain natural regeneration of oak and beech can be an effective alternative to planting, though difficulties remain, particularly in the relation between light and the growth of trees and weeds; and cost-effective methods of mechanical weeding and cleaning need to be developed. Yet foresters could quickly overcome these difficulties in order to perfect a system that has so many potential advantages over the traditional clear cutting and planting, especially in terms of nature conservation (Peterken 1981), as well as of amenity and landscape.

Natural regeneration locally can give good results at low cost, adds interest to the woodlands, and is especially appealing to owners who do not wish to commit themselves to expensive restocking schemes. It can be planned or unplanned, accepted or discarded. There are many situations, in addition to places where it is actively sought, where it may be used or where its occurrence must be considered in the overall silvicultural treatment of a stand.

Natural regeneration is often considered to be a 'free' gift from nature, but if a complete crop is to be obtained then considerable expense is involved. Ground preparation may be necessary, weeding is often more expensive than that in a plantation, and the subsequent cleaning and early respacings of the often dense seedlings can be costly. The revenue from the felling of the final crop may be reduced due to greater care needed in its removal, and the spread of the revenue over fifteen to thirty years of regeneration fellings may also be a financial disadvantage. However, there can be several benefits, some noted above. The whole subject is complex yet interesting.

The extent to which some of the continental silvicultural systems will be pursued in Britain is a matter of current debate. Useful comments thereon have been made by Helliwell (1982, 1985), Everard (1985), Matthews (1989), McIver (1991), Yorke (1991) and Wright (1991).

Natural regeneration of broadleaves

Evans (1988) has rendered another notable service to Britain's broadleaved silviculture through his treatise on natural regeneration. His comprehensive text emphasizes that natural regeneration, when it is available, can be used for infilling gaps, adding to woodland, and stocking inaccessible pockets of ground; also for enhancing wildlife habitats, conservation and amenity in plantations as well as for regeneration operations when a heavy seed year coincides with harvesting. The process should be used essentially when conditions are right, the ideal requirements being:

1. A good quality parent crop and a suitable site, flexibility in forest management to time operations with good seed years, or to follow advance regeneration, and the necessary skill and patience;

2. A regular supply of viable seed of a suitable origin/provenance;

3. A receptive forest floor favourable to the rapid germination of seed;

4. A microclimate favourable to germination of seed and survival and growth of young seedlings; and

5. Resistance of plants to competing vegetation and to damage by animals, insects, fungi, and extremes of climate.

Much of Evans' detailed information and recommendations is included below; and thereafter is illustrated by examples of successful lowland natural regeneration together with comments of foresters who have attained them.

Preparing for natural regeneration. Where production of timber is the main object of management, natural regeneration should only be pursued if the parent crop is of good quality and is well-suited to the site. Good stem-form and a history of satisfactory growth and timber quality are important. (Perpetuation should not be risked where, for example, oak is prone to shake, beech to pink coloration, ash to blackheart, and cherry to 'vein'.) Only the best parent trees should be allowed to be a seed source. Thinnings late in the life of the stand will of course favour such trees, but thinning should also encourage development of large crowns on the most desirable mother trees; large crowns and greater exposure to sunlight increase flowering and thus seed-bearing capacity.

The quantity of ripe seed produced in any one year is amongst the most unpredictable of silvicultural phenomena. Weather conditions are probably the most important influence determining seed yields from year to year. Other important factors are tree health, defoliation by insects, and period since previous heavy seed production. Trees which seed regularly and regenerate easily under suitable site conditions and localities include ash, sycamore, birch, alder, rowan, Norway maple and some willows; sweet chestnut could be added following warm summers; also, in scattered locations, whitebeam, wild cherry and southern beeches. Oak is less frequent. Beech is unlikely to seed heavily more frequently than in cycles of 3–5–7 years. Most broadleaved species produce plentiful seed for a substantial part of their life but both in the years leading up to reproductive maturity and when trees become very old, seed yields may be small or negligible. Evans (1988) provides a table of seed production characteristics of the principal broadleaved species and draws attention to many useful relevant published texts. (As examples, the normal minimum seed-bearing age of ash is twenty to thirty years, beech fifty to sixty, oak forty to fifty, sweet chestnut thirty to forty, and sycamore twenty-five to thirty.) Matthews (1963) lists Britain's important beech mast years, and this is complemented for various broadleaves by Evans. The long interval that can elapse between mast years rules out relying only on such exceptional years, though obviously when they occur the opportunity can be taken to use the plentiful seed. Partial mast years with heavier than average seed crops occur at more frequent intervals. Natural regeneration operations both for shelterwood uniform and group selection systems, should be concentrated on these and the full mast years.

The current seed crop should provide the bulk of seed for intended regeneration, but where natural regeneration has been planned and prepared for over many years there will sometimes be some accumulation of buried seed in the soil from previous years; but this would not include oak, beech, sycamore and sweet chestnut. Advance regeneration is a sign of this but when regeneration operations begin and the stand is opened up seedlings developing from the current year's crop may be augmented from these earlier seedfalls. (This is a particular feature with ash, the seed of which is usually dormant for eighteen months and only germinates in the second year after falling.) The timing of seeding (mostly in late summer and autumn) and dispersal mechanisms (chiefly by falling, wind, birds and mammals) obviously play an important role in the success of regeneration. Some seed can be lost while still on the tree: in broadleaves this may be caused by weevils in beech nuts, knopper gall damage to acorns, and larval feeding on ash. However, most losses occur once the seed has fallen: beech, oak and ash seed is eaten by woodmice, voles, grey squirrels and many birds particularly jay, wood-pigeon and chaffinch. Such predation can be very heavy. Seed which is quickly covered by litter suffers far less predation and also is at less risk from damage by sharp frosts in the autumn.

Ideal ground conditions are for seed to fall on the surface of slightly disturbed mineral soil and become rapidly covered by litter. Physical, and to a lesser extent chemical, condition of the soil will influence natural regeneration. Species show some variation according to soil reaction, with oak, ash and sycamore regenerating profusely in base rich soils close to neutral. However, the main effect of soil conditions on regeneration is indirectly via its influence on ground flora. The optimum condition is for the soil to be bare, yet moist and not eroding or compacted, though a light scattering of vegetation, particularly under beech on acid sites, will hold litter and promote soil activity. Presence of much herbaceous or rank grass weed growth introduces many obstacles to regeneration. Exposure of loosened, mineral soil, and absence of rank vegetation are crucial to success. Rotovation, scarifying or disc ploughing in the summer or early autumn may help to achieve success. The aim is to break up the surface, incorporate litter and humus, and dislodge vegetation.

Managing the overstorey. The overstorey is one means of controlling weed growth through the degree of shade cast and root competition. Where a clear-felling regeneration system is followed the main consideration is to ensure that the preparatory thinnings in the preceding decades, undertaken to encourage crown development of the best trees, have not been so heavy that substantial weed invasion has been allowed. Where some overstorey is maintained after the main regeneration fellings to provide frost protection or possible further seed, such cover will also help suppress weeds but retention of too many trees will also slow progress of the regeneration. A balance must be struck between loss of regeneration under dense shade and the undesirable encouragement of a smothering ground vegetation by the opening of the canopy. Generally, the more light the better the growth; however, maximizing light supply has to be balanced by the need to retain a degree of overhead cover in some regeneration systems to protect seedlings from frost and sun-scorch, to suppress weed development, and perhaps to provide additional seed.

In group and clear-felling systems all parent trees are felled in the area concerned. Nearby trees of unwanted species such as birches and willows may additionally be removed to lessen seed supply. In shelterwood systems the main felling occurring in the autumn of a good seed year, should remove 50 to 70% of trees in a stand, provided it has been prepared for natural regeneration in the preceding thinnings. For oak and ash, sufficient trees should be felled to leave at least 6 m between crowns of the remainder, i.e. to leave about twenty to twenty-five trees per hectare; for beech rather more can be left with a 4 m gap between crowns acceptable. An even scattering of trees should be left over the site.

In both group and shelterwood systems further fellings will be required and the timing of these is critical; neglect, even for two or three years, can seriously harm progress of the regeneration. In the group system, enlargement of the group by felling trees around the edge is essential after four or five years. Selection of trees to be removed must depend on silvicultural judgement and amount of shade being cast. New groups are opened when the next good seed year occurs. In the shelterwood systems the rate of removal of trees depends on species. For ash regeneration, all overstorey trees should be removed within three years, for oak in five to seven years, perhaps in two operations, and for beech the work can be spread out over fifteen or even twenty years if required. Since worthwhile quantities of further seed to supplement regeneration will rarely be forthcoming, except for ash and sycamore, removal of remaining overstorey trees should be dictated by their role in frost protection and diminishing weed growth and, of course, consideration of amenity and landscape where these are important.

Tending of young natural regeneration. Sometimes a carpet of seedlings which has arisen following a good seed year diminishes within a few years. Even with much less profuse regeneration steady reduction in numbers is common. Late spring frost is a major cause of mortality in newly-emerged regeneration; oak is less damaged, and ash and beech are quite susceptible. Aspect will affect the damage. Frost heave can also cause high mortality. Another major climatic hazard is drought particularly in the late spring/early summer period. Many birds, especially wood-pigeon and pheasant, will peck tender seedlings. Browsing by small mammals, rabbits, deer and livestock particularly sheep, can cause extensive damage. Mildew and 'damping off' cause some failures of oak on damp sites. Losses from insects and other invertebrate damage include chafer and cutworm girdling of roots and the root collar, and defoliation by slugs and caterpillars.

Some weeds may smother natural regeneration; others, for example bramble, may help to protect it until borne down by snow. Weeds compete for moisture and nutrients through their root competition. The solution is not to allow weed to get too established before commencing regeneration and prevent it becoming harmful during the first few years. The main aids available are manipulation of the overhead canopy to control light and provision of favourable ground conditions for rapid growth of emerging seedlings.

Assessment of the stocking of seedlings should be undertaken in the early autumn of the first year after regeneration has begun, and gaps larger than about 7 x 7 m infilled by planting robust stock. Natural regeneration can be protected as necessary against browsing, but there is no protection against mice, voles, grey squirrels and birds. Treeshelters can be used for individual plants on a small scale, to enhance growth, give protection and permit easy use of herbicide to control weeds. They should be placed over young seedlings in September of their first year. The aim is to protect one tree every 3 m, or at least 1,000 per hectare, but ideally with no more than 5 m between any two treeshelters.

Unwanted woody growth when harmful should be removed. This is often combined with *respacing,* an operation necessary to prevent natural regeneration becoming drawn up and unstable. The aim is to achieve an even distribution of favoured young trees whose crowns are given adequate room to grow. Respacing is either by hand tools or by using a clearing-saw.

Complementary regeneration can arise as sucker shoots from tree roots, notably wild cherry, grey alder, elm and wild service tree. Although they can occasionally be used for stocking a gap adjacent to a parent tree they are otherwise of little silvicultural value owing to the species involved, and can sometimes even be a nuisance. One exception could be wild cherry where suckers and natural seedlings often come up together and could be used to advantage.

Examples of broadleaved natural regeneration. Three examples relating to lowland natural regeneration of broadleaves are noted below:

1 J.E. Garfitt (1963, 1980), benefiting from his extensive experience in natural regeneration throughout much of his silvicultural management in southern Britain, comments that small areas pose few treatment problems. Cautious respacing at an early stage can be followed by successive respacings at intervals suggested by the rate of growth until a spacing similar to that employed in planted crops is attained; and thereafter the regimes prescribed for the latter may be followed. However, with larger areas the treatment problems may become acute, because of the

extreme density of young crops particularly of beech, ash and sycamore. Garfitt advocates an initial respacing when the seedlings are about 1.5 to 2.5 m tall, to a spacing of say 0.7 to 1.0 m, followed by a second respacing, when the trees are up to 3.0 m tall, to a spacing of perhaps 1.5 to 2.0 m. Thereafter, at least two and in many cases three 'thinnings' are necessary before the stage corresponding to the first thinning of a planted crop is reached. As an alternative recommended particularly where marketing of the outturn is difficult, he gives detailed prescriptions for a method (his 'Belgium System') based on a free-growth principle which dramatically reduced the work involved at all stages of the thinning.

2   John Workman's continuing group selection in fine quality beech on the Cotswolds at Ebworth in Gloucestershire is the most impressive example of the system in Britain – the securing of advance natural regeneration, and then felling the overstorey where it is well established. Where there has been a gap in the beech canopy, sycamore and ash may have come in initially, but usually sufficient beech is awaited before undertaking fellings. Occasionally beech regeneration will appear when the overstorey is not mature (100 years), in which case thinnings are made to favour this, and eventually the overstorey may be felled before it is fully mature to ensure that the advance regeneration does not suffer. Typically it takes around thirty years to regenerate a stand, and towards the end of this period some infilling with beech transplants may be necessary. Although no weeding is done, the regeneration is cleaned once the overstorey has gone, and secondary species are only retained to nurse the beech. The groups are generally less than 0.3 ha in size.

In effect, the natural regeneration is mainly of beech of high quality parentage, with some planting, thus operating an uneven-aged system, mainly two-storey, along with groups of different ages. The whole concept is based on fostering regeneration, i.e. observing where beech seedlings are appearing and gradually felling back from these initial patches, allowing side-light to penetrate below the canopy and so provide the very critical conditions essential for successful establishment of beech seedlings. The objective is towards a sustained yield of permanent forest cover. This notable example has been achieved over many decades by skilled and patient management in the face of many problems: damage by wood-pigeon, grey squirrel, wood mice, fallow and roe deer, excessive bramble and wild clematis, and extremes of temperature (Workman 1986). Among lessons learned are: that voles respace regeneration; badgers aid by disturbing the forest floor; if canopy is too dense the floor is clean and inhospitable, yet if opened too much, dense bramble and grass smother the seedlings; the balance is precarious and depends on intimate knowledge of the overstorey and the site. The result is a woodland heritage of great national renown (see Plate 7).

3   In the Forest of Dean, excellent regeneration of oak was obtained following an abundant mast year in 1899 (Troup 1928, 1952). An area of about 200 hectares was enclosed; oak regeneration appeared in quantity, the seed-bearing parents were gradually removed, and gaps were infilled with Norway spruce, sycamore and European larch. By 1952 the oak had been underplanted with beech. The result noted by Troup (1928) and by Jones (1952) was 'an excellent uniform crop of oak and beech (153 hectares in extent), equal to anything to be seen on the continent of Europe'. Some of the finest stands remain (Hart 1966).

Everard (1987) has related the practice of broadleaved natural regeneration in the Forest of Dean from 1897 until recent times. Joslin (1982) has described the regeneration and management regime for approximately 1200 hectares of mixed broadleaves in the same forest, a system best termed a group shelterwood. Seeding fellings, removing about 60% of the volume, are made to enhance groups; these are followed five to seven years later by secondary fellings; the timing of the final felling depends partly on amenity and other considerations. After eleven years of applying this system, 12% of the area has been successfully regenerated, and 48% partially stocked with advanced groups. Many impressive crops can now be seen, established by this low-cost regime. However, in many areas the density of saplings of final crop trees is still rather sparse, with birch and coppice shoots predominating. Hopefully, within this low-grade matrix are sufficient seedlings to make a final crop. The use of the coppice regrowth as a nurse to such seedlings requires careful timing of cleanings and respacings, if the seedlings are not to be overtopped. Incidentally, light grazing by sheep is a considerable aid to the establishment by natural regeneration noted above.

Natural regeneration of broadleaves in the uplands. Broadleaved natural regeneration is highly pertinent to the uplands, and not only because birch and rowan spring up profusely on any exposed mineral soil and heath. Many semi-natural woodlands in the uplands are failing to regenerate and gradually disappearing because of grazing pressures. Natural regeneration is probably the most economical way of maintaining woodland in inaccessible locations, and on steep and rocky slopes. The dominant influence of grazing, mainly by sheep and red deer, in preventing regeneration is demonstrated by the re-growth that follows whenever it is removed or sharply declines.

The occurrence of broadleaved natural regeneration in the uplands brings some advantages. It is a safer means of increasing the proportion of broadleaves than by planting since it comes from locally adapted stock and, necessarily, where it survives it must be able to grow despite the harsh conditions. The upland environment is generally less well-suited to broadleaves than the lowlands, so that building on nature's provision appears a sensible course, especially since there is rarely any intention of growing high quality timber crops.

Natural regeneration of conifers

Much of the foregoing information on broadleaves (chiefly in the lowlands) applies equally to conifers (mainly in the uplands of Scotland, though there are some fine examples of coniferous natural regeneration in the lowlands of England, e.g. at Longleat described hereafter, at Windsor, and in the New Forest).

Most of the conifer species planted commercially can produce natural regeneration under favourable circumstances (Low (ed.) 1985). Good seed years, however, generally occur only once in three to ten years depending on species, and it is not possible to predict this occurrence with accuracy. Norway spruce and Corsican pine so rarely release adequate quantities of seed that significant natural regeneration is unlikely. Predation of the seed of a number of conifer species by seed wasps can reduce the amount of viable seed fall (occurrence of the Douglas fir seed wasp *Megastigmus spermotrophus* is so general that Douglas fir regeneration is effectively restricted to particularly good seed years). Hybrid larch does not breed true and the use of natural regeneration for this species will lead to a decline in vigour and form.

In the uplands, natural regeneration of conifers has until now rarely been used as an alternative to replanting felled areas. Insufficient is known about the combination of site conditions required to promote its successful establishment, and it would frequently impose unacceptable constraints on harvesting plans if attempts were made on any significant scale to synchronize fellings with heavy seed years. The possible use of any silvicultural system involving the retention of seed trees is highly restricted by the likelihood of the latter being windthrown. Furthermore, several years can elapse after felling for the establishment of an acceptable stocking level over the whole clear-felled area, and the achievement of this may in fact be prevented by the rapid development of weed growth on more fertile sites.

Some natural regenerated conifer seedlings can be found on most clear-felled sites, the number per hectare being related more to the previous management practice than to inherent site conditions. Premature felling of the parent stand reduces the amount of seed available. Young seedlings are vulnerable to desiccation, mammal damage and weed competition, and benefit from a certain amount of shelter and shade, although they cannot survive in dense shade. It follows that mature stands which have been heavily thinned or contain pockets of windthrow are most likely to provide favourable conditions for natural regeneration. The clear-felling of a mature, cone-bearing stand may also give rise to a dense carpet of seedlings, although this may be distributed unevenly and concentrated mainly in brash covered areas. Scarifiers have been developed in Scandinavia primarily for the encouragement of natural regeneration through redistribution of the lop and top to improve the uniformity of stocking, and provision of strips or patches of bare mineral soil. A regime designed to encourage natural regeneration would involve a combination of heavy thinning, seeding or shelterwood fellings and scarification (Low (ed.) 1985).

In some forest areas, notably in mid-Wales and south Scotland, dense young crops of natural Sitka spruce seedlings have appeared spontaneously, and it has been necessary to consider how best to manage these. Stocking is often in the region of 10,000–100,000 stems/ha, and the reduction of density through natural mortality is a slow process. Competition restricts diameter growth severely whilst having relatively little effect on height growth, so that naturally regenerated stands are likely to have a low mean diameter and commensurately low sawlog content within a normal rotation, unless some respacing is carried out. Since the stand closes canopy at a very early stage, there is an increasing penalty to pay in terms of reduced volume production the longer this respacing operation is delayed, because growth of the favoured trees will have been retarded by competition during the period between canopy closure and respacing. Respacing poses a number of practical problems and, depending on timing, may involve the use of chemicals, clearing-saws or chainsaws as follows (Low (ed.) 1985):

1. Height below 1 metre. The stand may be sprayed with herbicide either in bands or using two passes at right angles to leave small clumps at 2 m centres. The operation will cheaply provide access to carry out selective respacing at a later date.

2. Height 1–3 metres. Respace using a clearing-saw, cutting below the bottom live whorl of branches.

3. Height above 3 metres. Respace selectively using a chainsaw. Stumps greater than 2.5 cm in diameter must be treated with urea to prevent infection by *Heterobasidion annosum*.

Plate 10. Caledonian Scots pine at Glen Affric in Scotland. The native pinewoods, among the least modified woodland in Britain, are an irreplaceable ecological reservoir of adapted genetic stock of plants and mammals. They need management in order to maintain and enhance their ecosystems and aesthetic value
[G.A. Dey]

Plate 11. Springtime in Ashridge Park, Hertfordshire. A fine example of forestry supplementing flora. The shade from young beech trees still allows bluebells to flourish [W. Frances]

Plate 12. Thorpe Wood Local Nature Reserve in Cambridgeshire. Primroses and other spring flowers flourish alongside recently cut ride edges in broadleaved woodland. The attractions to other wildlife are also significant [P. Wakeley]

Plate 13. Sparrowhawk nesting in Norway spruce. Birds have a high conservation value in both upland and lowland woods. They need food sources, nest sites, territory and over-wintering areas, all of which can be provided by the diverse range of woodland habitata in Britain. Foresters have a close relationship with the RSPB and other conservation bodies: and continually exchange ideas that help maintain and improve wildlife habitats and protect endangered species [R. Wilmhurst]

Plate 14. Pheasant shooting at Wharfedale in North Yorkshire. Broadleaved woodland in the background. An example of woodland and farmland supplementing each other for shooting. Rentals for sporting can contribute significantly to the profitability of woodland management options. Many woodlands owe their past and continued existence to their sporting facilities. The capital value of a woodland is also considerably improved by its suitability for sporting

[J. Edenbrow]

Plate 15. Red deer. The interaction of deer species is of major concern to foresters: populations have a home range and many move in response to forestry operations. Control of numbers is necessary to maintain populations within the carrying capacity of the woodland

[Forestry Commission]

Plate 16. Talymaes Woodlands near Crickhowell, Powys, in Wales. Highly productive forestry started in the 1950s, harmonized into a sensitive Welsh agricultural landscape. One of the fine plots of Japanese larch [C. Hughes]

Plate 17. Broom Hill Wood (8 ha) near Huntley, in Gloucestershire. A bequest from a distinguished private forester to his agent. Some woodland owners have been prompted to establish plots of the less common exotic species – here a stand of coast redwood *Sequoia sempervirens* planted in 1955. Photographed in January 1991 following three light thinnings. Coppicing of the cut stools is evident [C. Hughes]

Plate 18. Woodlands at Huntley in Gloucestershire. Churchyard specimen trees scenically enhanced by fine plantations mainly of European and Japanese larches, with Douglas fir and grand fir projecting above. An achievement and the resting place of a distinguished private forester

[C. Hughes]

Plate 19. Blaisdon Wood (70 ha) near Longhope, in Gloucestershire. A typical ancient lowland woodland site recorded from 1282, mainly on terrain unsuitable for ploughing. A lowland investment woodland – with five changes of ownership (including two syndicates) in forty years

[J.P. Birch]

Nelson (1990) updates information on chemical control of Sitka spruce natural regeneration. Costs should be evaluated in each situation, but in general even the full cost of respacing a stand above 3 m tall using a chainsaw is likely to compare favourably with the compounded cost of planting, protecting and maintaining normal transplant stock. As a result, acceptance and subsequent treatment of spontaneous natural regeneration may be an attractive option if the species and provenance of the parent stand are acceptable. The opportunity to introduce genetically improved stock must be forgone, but against this there is an opportunity for a high level of selection during respacement.

An example of natural regeneration of conifers. Fine examples of conifer natural regeneration are to be seen at Longleat forest estate in Wiltshire. This renowned estate includes about 1,600 hectares of woodland of which about half were clear-felled during World War II, and later replanted resulting in large areas of even-aged stands, mainly conifers. From 1969 John McHardy, forest manager for Lord Bath and Viscount Weymouth, has successfully practised irregular (unevenaged) silviculture in contrast to the traditional system of clear-felling and planting, particularly with the objective of ensuring a continuous cover of commercial forest (Helliwell 1984; Poore 1986).

The aims have been to establish, respace and tend a new crop of natural regeneration every decade, though thinning/fellings of the overstorey are generally every three to five years. The strategy is to produce six to seven main age classes with a ten-year time of passage between them. The optimum size at felling is 75 cm dbh. The whole is determined by the light requirements of the understorey and the need to reduce root competition from the overcrop.

The system partly arose as a side-product of the very heavy crown thinnings applied to the twenty- to thirty-year-old larch (mainly Japanese) and Douglas fir plantations. The thinnings have allowed plentiful development of regeneration of these and other species. This is respaced with a clearing-saw at age five to eight to about 1.5 x 1.5 m when 1.3 – 2 m in height and the tops beyond the reach of roe deer. Apart from application of 'Asulox' to control bracken every few years, this is the only tending needed. (Grant-aid is obtained towards the cost of the natural regeneration.)

The stocking (i.e. number of trees per hectare) is light but the volume is normal. There is some sacrifice of volume production during the conversion process because of felling before full financial maturity. High pruning is necessary on final crop trees, but diameter increment is outstandingly high. The aim is gradually to reduce the overstorey to about forty-two trees per hectare in the 80+ age class; twenty-five to thirty dominant ones per hectare to be retained to financial maturity and twelve per hectare of the 80+ to become interesting immense 'monarchs of the forest' as well as to continue to act as seed trees. (Wind has damaged the tops of some of the large Douglas fir but they remain sound well crowned seed parents.)

About 800 ha of the managed area overlies greensand, giving a freely-drained slightly acid soil up to a metre in depth; about 600 ha of the woodlands lie on clay, and 200 ha on chalk. Most conifers seed well at least every three years and regenerate easily on the greensand at relatively low cost. The main species are Douglas fir and Japanese larch supplemented by western red cedar, western hemlock, grand fir, Lawson cypress, coast redwood, Scots and Corsican pine. In addition there has been some regeneration of oak, ash, sycamore, beech, birch and sweet chestnut. Some planting of wild cherry and oak has taken place, using treeshelters.

Control over the initiation of regeneration is partly through differential thinning intensity and partly through the selective control of bracken with 'Asulox'. Grass, which severely interferes with germination and seedling development and provides a habitat for voles, is controlled with 'Clanex'. Some bramble is retained as deer browse. Small patches of heather and other minor vegetation are retained for ecological reasons. Achievement of a balance in nature is constantly pursued.

The only pre-cultivation has been chemical spraying and bulldozing of some patches of rhododendron and laurel. Many areas only require thinning to induce natural regeneration. Respacing of the profuse regeneration is the only major costly routine operation, but this expense is no more than would have been the purchase of plants, planting and weeding. Adequate control of roe deer generates net income. Another cost arises from the considerable care taken in felling and extracting some of the overcrop, though this method of harvesting does less damage to the soil than compaction from heavy vehicles under a clear-felling system. Any soil disturbances in these operations creates a seedbed for the next regeneration. Acceptance of the costs is encouraged by the improved genetic vigour achieved by respacing the 'seed-bed' on the forest floor.

The plantations were crown-thinned from the outset to avoid excessive reduction in crown size, and all coarse or mis-shapen trees were removed at the first opportunity. A proportion of the better trees were pruned to a height of 4.5–5 m after the first thinning and then up to 6.0–6.5 m a few years later. The pruning has distinctly enhanced the value of the timber, particularly that converted in the estate's sawmill, as well as in higher prices received for such specialist markets as transmission poles, barn poles and gatemaking timber.

The amenity value of the plantations is steadily increasing, to the satisfaction of the owner, agent, manager and the public who visit the estate in large numbers. The woodlands are highly commended by amenity groups and conservationists. The extra age class serves a particularly useful and satisfying purpose in landscape terms during periods when the younger age classes predominate and in other circumstances could provide a senescent age class for nature conservation purposes. Something approaching 400 ha of uneven-aged silviculture shows distinct advantages in terms of amenity, nature conservation and recreational facilities, complementing intensive timber production and utilisation. The whole is a testament of what can be achieved from fine parent trees on suitable site conditions under sound planning and management. Truly a fine national heritage (see Plates 8 and 9).

Economics of natural regeneration

The advantages of natural regeneration particularly as it relates to safeguarding ecosystems of the site, the protection by the old crop, and the favourable microclimate at or near the ground have been noted by Matthews (1989). Furthermore, the seed bearers used as parents of the succeeding crop are well-adapted to the site; and the interruption in production associated with clear cutting is shortened or absent. The disadvantages of natural regeneration are mainly managerial and economic (Price 1989; Matthews 1989):

1 It is rarely an easy operation (except with sycamore and ash) and is always expensive in skilled man-power, time, and money;

2   The process may have to be reinforced by planting so as to avoid the risk of failure, to correct deficiencies in stocking, or to shorten the duration of individual operations;

3   Possible higher tending costs, including respacing (as natural regeneration rarely has ideal density);

4   Possible delay in establishment (because natural regeneration comes either too slowly, or too densely creating a period of intense competition); and hence delayed financial revenue;

5   Reduced final revenue, both because genetic quality is less controlled than with artificial planting, and because the desired species gives incomplete site cover;

6   Opportunity costs of retaining seed trees past optimal rotation; and

7   Operations need to be timed with good seed years not good markets.

Price (1989) points out that under low discount rates complete site utilization and high quality timber justifies the heavier investment of artificial regeneration. By contrast, high discount rates give less weight to long-term revenue, and the short-term cost-saving of natural regeneration dominates; delayed establishment is not so important under high discount rates. However, the economics have rarely been investigated and few direct comparisons made with planting.

Under favourable conditions of plentiful seed, a receptive forest floor and microclimate, natural regeneration may occasionally offer the potential of reduced establishment costs, especially in ash, beech, birch and sycamore (Pryor and Savill 1986; Crockford *et al.* 1987). Using nature's gift brings many conservation, amenity and landscape benefits, though they are difficult to value in money terms (see chapter 14). Grant-aid, which gives support to natural regeneration, is leading to this form of silviculture playing a greater role in British silviculture. Grant-aid is available for new natural regeneration and for existing regeneration for twenty years (see chapter 15).

However, foresters usually need the speed and certainty given by planting. The notable regeneration at Longleat in Wiltshire is almost totally due to a skilled forester operating on very favourable greensand soil. This exercise, as well as that with beech on the Cotswolds and with Scots pine at Windsor, are among the few outstandingly fine examples. Elsewhere, most foresters consider that the delay in establishment should be avoided, and favour resort to planting.

4. Enrichment

The term 'enrichment' embraces various measures for raising the proportion of principal species in unmanaged (degraded) stands – by planting young trees of principal species at wide spacings among existing forest growth. Enrichment is a preparatory stage in the progress towards a full silvicultural system. The term includes planting on land occupied by scrub. In Britain there are two main approaches to enrichment (Evans 1984): (i) to accept much of the existing crop and restock gaps or clearings with the same species or a more productive species; and (ii) to clear strips at intervals and restock (this is usually cheaper than using groups and is easier to manage).

Poor quality woodlands in terms of timber production usually contain gaps and bare areas. It is often possible and desirable to avoid full-scale restocking, and thereby save costs of plants and planting. Enrichment, although fairly expensive in terms of cost per tree, may be an acceptable treatment in a crop which already has a patchy though potentially useful stocking. The existing stocking is usually of trees which have coppiced or appeared by natural regeneration, though insufficient to occupy all the site, and perhaps being entirely absent from some parts. It may include standards, around which shade tolerant trees may be planted to produce useful poles. There may, too, be some trees which are too small for immediate felling, or the stocking may be of young plants which are likely to form the bulk of the next rotation. Enrichment may only be worth undertaking if the diameter of the gaps is more than about one-and-a-half times the height of the surrounding trees.

The species used for enrichment must be amenable to the site conditions, especially soil and climate; they should be able to grow sufficiently fast to keep ahead of the surrounding weeds, coppice shoots and other unwanted growth. Sometimes poplars at wide spacing, up to 8 m, are ideal, as they are fast-growing and can mature in about thirty to forty years; but their root competition sometimes makes the results unsatisfactory. Douglas fir does well on sheltered well-drained fertile soils, but it will not tolerate more than a little top shade; and because of the possibility of windthrow, exposed sites should be avoided. Other useful species are Japanese larch, western red cedar, western hemlock, grand fir, beech, sycamore, wild cherry and Norway maple.

Spacing at 2 to 4 m should be satisfactory. The height of the plants may be 0.5 to 1.0 m; and they must be robust. Where necessary to avoid damage by rabbits, hares or deer, it should be considered whether the cheaper method is to fence the whole area or only particular sections; or to surround each new tree with a plastic guard or treeshelter. The inter-planted trees should be inspected each year to ensure that the surrounding larger trees are not overly shading them. If there are inadequate openings in the canopy, some of the surrounding trees must be removed: as the planted trees grow, an ever-increasing gap size is usually necessary.

Enrichment as noted above is mainly with fast-growing species, when the aim is to produce a crop which can all be felled at the same time. An alternative is to enrich with the object of producing irregular stands to be worked ultimately by 'group selection'. Matthews (1989) gives the advantages and disadvantages of enrichment, also application in practice.

5. Basket willow coppice

The cultivation of basket willows is an agricultural or even horticultural rather than a forest industry (Stott 1956). Osier willows (*Salix* spp.) are grown as a form of low 'coppice', usually oneand occasionally two-year-old, to provide rods (termed wands or withes) for the making of baskets and garden screens and furniture. (Large willows are pollarded to produce vertical components of some of the products.)

Most of the osier crops in England are found on the mid-Somerset levels, growing on the moist, rich alluvial soils of reclaimed marshland which anciently was widely submerged; the remainder occur as isolated beds in river valleys mostly in the Midlands and East Anglia. The most important species are *Salix triandra* var. Black Maul, *S. viminalis* and *S. purpurea* (Stott and Parfitt 1986). The life of a willow bed may vary from at least twenty to fifty years, and during which most crops would be cut every year.

Rich and well-watered arable lowland is required for the beds. To establish the crop, deep ploughing and harrowing are needed. Cuttings about 30 cm long are prepared from selected one or two-year-old rods and planted in March or April at 60 cm between rows and 30 cm in the rows, necessitating about 55,000 per hectare (or at 65 × 35 cm, needing 40,000 per hectare). Chemical weed control is essential during the first two years of establishment. Thereafter the grass sward that develops is cut by hand and by small row-crop machines. The weevil *Cryptorhynchus lapathi* damages osiers by boring into the stem, causing the stem to die above the point of attack. The so-called 'Americanas' (*S. rigida*) and *S. viminalis* and its hybrids are particularly susceptible.

The first saleable crop is obtained in the third season. The annual crop of slender, flexible rods up to 3 m tall is harvested, often mechanically, between November and March. An average annual yield is about 450 bundles of rods or 10–12 tonnes green per hectare, which after processing gives about 300 bundles of 'buff' or 'white' rods. The processed rods currently cost about £14 per bundle to produce, and retail at about £17 per bundle; a bundle measures about 1 m in circumference at 5 cm from the base and weighs approximately 12 kg.

The growing of osiers is specialized. Anyone likely to be concerned with it is referred for full information to the University of Bristol Research Station at Long Ashton.

References

Ackers, C.P. (1939), 'The Selection System'. *Quarterly Journal of Forestry* 33 (3) pp. 173–6.
Anderson, M.L. (1930), 'A New System of Planting'. *Scottish Forestry* 44 (2) pp. 78–87.
—— (1931), 'Planting in Dense Groups spaced at Intervals'. *Quarterly Journal of Forestry* 25 (4), pp. 312–16.
—— (1951), 'Spaced Group Planting and Irregularity of Stand Structure'. *Empire Forestry Review* 30 (4), pp. 328–41.
—— (1960), 'Norway Spruce-Silver Fir-Beech Mixed Selection Forest'. *Scottish Forestry* 14 (2), pp. 87–93.
Bax Standen (1990), personal communication.
Blyth, J.F. and Malcolm, D.C. (1988), 'The Development of a Transformation to Irregular Forest: 35 years experience at the Glentress trial area'. Oxford Forestry Institute Occasional Papers No. 37, pp. 33–41.
Bourne, R. (1935), 'The Efficiency of Irregular Stocking'. *Empire Forestry Journal* 14 (2), pp. 215–17.
—— (1942), 'A Note on Beech Regeneration in Southern England'. *Quarterly Journal of Forestry* 36, pp. 42–9.
—— (1945), 'The Neglect of Natural Regeneration', *Forestry* 19, pp. 33–40.
—— (1951), 'A Fallacy in the Theory of Growing Stock'. *Forestry* 24 (1), pp. 6–18.
Bradford, Lord (1981), 'An Experiment in Irregular Forestry'. *Y Coedwigwr* 33, pp. 26–30.
Cannell, M.G.R., Shepherd, L.J. and Milne, R. (1988), 'Light Use Efficiency and Woody Biomass Production of Poplar and Willow'. *Forestry* 61 (2), pp. 125–36.
Collin, N. (1986), 'The Bradford-Hutt Continuous Cover Forestry System'. In *Proceedings, Discussion on uneven-aged Silviculture,* Pershore, 29 October 1986 (obtainable from D.R. Helliwell, Yokecliffe House, West End, Wirksworth, Derbyshire DE4 4EG).
Crockford, K.J., Corbyn, I.N. and Savill, P.S. (1987a), 'Management of Woodlands for Fuel and Timber'. Oxford Forestry Institute. Report for the Energy Technology Support Unit for the United Kingdom Department of Energy (ETSU B 1156).

Crockford, K.J., Spilsbury, M.J. and Savill, P.S. (1987b), 'The Relative Economics of Woodland Management Systems'. Oxford Forestry Institute Occasional Paper No. 35.

Crockford, K.J. and Savill, P.S. (1991), 'Preliminary Yield Tables for Oak Coppice'. *Forestry 64* (1), pp. 29–50.

Crowe, Sylvia (1978), 'The Landscape of Forests and Woods'. FC Booklet 44.

Crowther, R.E. and Evans, J. (1986), 'Coppice'. FC Leaflet 83.

Dyer, R. (1986), 'Practical Experience of Uneven-aged Forestry; its Benefits and Problems'. In *Proceedings, Discussion on uneven-aged Silviculture,* Pershore, 29 October 1986 (obtainable as under Collin *supra*).

Edinburgh University, Forestry Department (1990). Study sponsored by the Scottish Forestry Trust.

Energy, Department of (1989), *Quarterly Journal of Renewable Energy* 5, pp. 10–12.

Evans, J. (1984), 'Silviculture of Broadleaved Woodland'. FC Bulletin 62.

—— (1988), 'Natural Regeneration of Broadleaves'. FC Bulletin 78.

Everard, J. (1985), 'Management of Broadleaved Forests in Western Europe'. Forestry Commission internal report.

—— (1988), 'Oak and Beech Natural Regeneration'. *Forestry and British Timber,* November 1988.

—— (1987), 'Natural Regeneration of Oak'. Oxford Forestry Institute Occasional Papers No. 34, pp. 23–8.

Fuller, R.J. and Warren, M.S. (1990), 'Coppiced Woodlands: their Management for Wildlife'. NCC.

Garfitt, J.E. (1953), 'The Rehabilitation of Devastated Woodlands'. *Forestry 26,* pp. 28–32.

—— (1963), 'Treatment of Natural Regeneration'. *Forestry 36,* pp. 103–12.

—— (1977), 'Irregular Silviculture in the Service of Amenity'. *Quarterly Journal of Forestry 71* (2), pp. 82–5.

—— (1980), 'Treatment of Natural Regeneration and Young Broadleaved Crops'. *Quarterly Journal of Forestry 74* (4), pp. 236–9.

—— (1984), 'The Group Selection System'. *Quarterly Journal of Forestry 78* (3), pp. 155–8.

—— (1987), 'Yield Control of Irregular Woodlands'. *Quarterly Journal of Forestry 81* (3), pp. 181–4.

—— (1988), 'Irregular Systems of Silviculture for Broadleaves and Mixed Crops: do they work in an English context?.' Oxford Forestry Institute Occasional Papers No. 37, pp. 47–51.

Harris, M.J. and Kent, M. (1986), 'Studies on the Ground Flora under Selection Forestry in the Tavistock Woodland Estate'. In *Proceedings, Discussion on uneven-aged Silviculture,* Pershore, 29 October 1986 (obtainable as under Collin *supra*).

—— (1987), 'Ecological Benefits of the Bradford-Hutt System of Commercial Forestry'. *Quarterly Journal of Forestry 81* (3), pp. 145–57; *81* (4), pp. 213–24.

Hart, C.E. (1966), *Royal Forest.* Clarendon Press, Oxford.

—— (1967), *Practical Forestry for the Agent and Surveyor.* 2nd edn. Estates Gazette Ltd.

Helliwell, D.R. (1982), *Options in Forestry.* Packard, Chichester.

—— (1984), 'Success with 'uneven-aged' at Longleat'. *Forestry and British Timber,* April 1984.

—— (1985), 'The Need for an Experimental Study of Different Silvicultural Systems'. *Scottish Forestry 39* (1), pp. 8–12.

—— (1988), 'Uneven-aged Woodlands in Britain; Advantages, Disadvantages, and Problems'. *Arboricultural Journal 12,* pp. 273–8.

Hiley, W.E. (1939), 'Swiss Forestry'. *Quarterly Journal of Forestry 33* (3), pp. 159–63.

—— (1953), 'Irregular Forestry'. *Quarterly Journal of Forestry 47* (4), pp. 231–7.

—— (1959), 'Two-storied High Forest'. *Forestry 32,* pp. 113–6.

—— (1967), *Woodland Management.* 2nd edn. Faber and Faber, London.

Howell, B.N., Harley, R.M., White, R.D.F. and Lamb, R.G.M. (1983), 'The Dartington Story II'. *Quarterly Journal of Forestry,* 77, pp. 5–16.

Hutt, P. (1975), 'The Bradford Continuous Canopy Forestry System'. *Practical Education 72,* pp. 20–4.

Hutt, P. and Watkins, K. (1971), 'The Bradford Plan for Continuous Forest Cover'. *Journal of the Devon Trust for Nature Conservation* 3, pp. 69–74.
Insley, H. (ed.) (1988), 'Farm Woodland Planning'. FC Bulletin 80.
Jobling, J. (1990), 'Poplars for Wood Production and Amenity'. FC Bulletin 92.
Johnston, D.R., Grayson, A.J. and Bradley, R.T. (1967), *Forest Planning*. Faber and Faber, London.
Jones, E.W. (1952). See Troup, R.S. (1952).
Joslin, A. (1982), 'Management of Broadleaves in the Forest of Dean with Special Reference to Regeneration'. In *Broadleaves in Britain: future management and research,* eds. Malcolm, D.C., Evans, J. and Edwards, P.N., pp. 53–60. Institute of Chartered Foresters, Edinburgh.
Knüchel, H. (1953), *Planning and Control in the Managed Forest*. Oliver and Boyd, Edinburgh.
Köstler, J. (1956), *Silviculture*. Oliver and Boyd, Edinburgh.
Leslie, A.J. (1989), 'On the Economic Prospects for Natural Management in Temperate Hardwoods'. *Forestry, 62* (2), pp. 147–66.
Lorrain-Smith, R. (1986), 'Economic Aspects of uneven-aged Forestry'. In *Proceedings, Discussion on uneven-aged Silviculture,* Pershore, 29 October 1986 (obtainable as under Collin *supra).*
Low, A.J. (ed.) (1985), 'Guide to Upland Restocking Practice'. FC Leaflet 84.
Malcolm, D.C. (1979), 'The Future Development of even-aged Plantations: Silvicultural Implications'. In, *Ecology of even-aged forest plantations,* pp. 481–504. Institute of Terrestrial Ecology, Cambridge.
Matthews, J.D. (1963), 'Factors Affecting the Production of Seed by Forest Trees'. *Forestry Abstracts 24,* pp. 1–13.
—— (1986), 'The History and Status of uneven-aged Forestry in Europe and Britain'. In *Proceedings, Discussion on uneven-aged Silviculture,* Pershore, 29 October 1986 (obtainable as under Collin *supra).*
—— (1989), *Silvicultural Systems*. Clarendon Press, Oxford.
McIver, H. (1991), Research at University of Edinburgh.
Mitchell, C.P. (1984), 'Growing Trees for Energy'. *Timber Grower,* 1984.
—— (1989), 'Short Rotation Forestry for Energy – Single Stem Plantations'. Report for the Energy Technology Support Unit of the United Kingdom Department of Energy (ETSU B 1081).
Mitchell, C.P., Wightman, A.D., Ford-Robertson, J.B. and Ennion, R. (1989), 'Establishment and Monitoring of Large Scale Trials of Short Rotation Forestry for Energy'. Report for the Energy Technology Support Unit of the United Kingdom Department of Energy (ETSU B 1171).
Nelson, D.G. (1990), 'Chemical Control of Sitka Spruce Natural Regeneration'. FC Research Information Note 187.
Penistan, M.J. (1938), 'The Selection System – Irregular Silviculture'. *Quarterly Journal of Forestry 32* (1), pp. 51–4.
—— (1953), 'Uneven-aged Woods and the Plan of Operations'. *Quarterly Journal of Forestry 47* (1), pp. 19–27.
—— (1986), 'Oak in Wessex: an Account of Field Studies 1982–84', *Forestry 59,* pp. 243–57.
Peterken, G.F. (1981), *Woodland Conservation and Management*. Chapman and Hall, London.
Poore, A. (1986), Association of Professional Foresters' Newsletter 1985, pp. 44–8.
—— (1988), 'British uneven aged Silvicultural Systems'. Commonwealth Forestry Institute Occasional Papers No. 37, pp. 2–11.
Price, C. (1989), *The Theory and Application of Forest Economics*. Basil Blackwell, Oxford.
Pryor, S.N and Savill, P.S. (1986), 'Silvicultural Systems for Broadleaved Woodland in Britain'. Oxford Forestry Institute Occasional Papers No. 32.
Reade, M.G. (1957), 'Sustained Yield from Selection Forest' *Quarterly Journal of Forestry 51* (1). pp. 51–62.
—— (1969), 'Silviculture of Selection Forest'. *Quarterly Journal of Forestry 63* (3), pp. 197–210.
—— (1990), 'Chiltern Enumerations'. *Quarterly Journal of Forestry 84* (1), pp. 9–22.

Rollinson, T.J.D. and Evans, J. (1987), 'The Yield of Sweet Chestnut Coppice'. FC Bulletin 64.
Savill, P.S. and Evans, J. (1986), *Plantation Silviculture in Temperate Regions*. Clarendon Press, Oxford.
Schlich, W. (1904), *Manual of Forestry, 2, Silviculture*. Bradbury Agnew, London, pp. 243–7.
—— (1923), 'The Regulated Selection Forest'. *Empire Forestry Journal 2* (1), pp. 76–82.
Shrimpton, N. (1986), 'Studies of the Economics of Irregular Stands'. In *Proceedings, Discussion on uneven-aged Silvicuture,* Pershore, 29 October 1986 (obtainable as under Collin, *supra)*.
—— (1988), 'Modelling the Costs of uneven-aged Forest Management'. Oxford Forestry Institute Occasional Papers No. 7, pp. 42–6.
Spilsbury, M.J. (1990), 'The Economic Prospects for Natural Management of Woodlands in the United Kingdom'. *Forestry 63* (4), pp. 379–90.
Spilsbury, M.J. and Crockford, K.J. (1989), 'Woodland Economics and the 1988 Budget'. *Quarterly Journal of Forestry 83* (1), pp. 25–32.
Stott, K.G. (1956), 'Cultivation and Uses of Basket Willows'. *Quarterly Journal of Forestry 50,* pp. 103–12.
Stott, K.G., McElroy, G., Abernethy, W., and Hayes, D.P. (1985), 'Coppice Willow for Biomass in the UK'. Long Ashton Research Station, Bristol.
Stott, K.G. and Parfitt, R. (1986), 'Willows on Farms for Amenity and Profit'. British Association, Bristol, 1986. Forestry Section Proceedings. Forestry Commission Occasional Paper 16, pp. 36, 37.
Troup, R.S. (1928), *Silvicultural Systems*. Clarendon Press, Oxford.
—— (1952), *Silvicultural Systems*. 2nd edn. (ed. E.W. Jones) Clarendon Press, Oxford.
Wood, R.F. (1950), 'Rehabilitation of Devastated and Derelict Woodlands'. *Quarterly Journal of Forestry 44* (1), pp. 5–10.
Wood, R.F., Miller, A.D.S., and Nimmo, M. (1967), 'Experiments on the Rehabilitation of Uneconomic Broadleaved Woodlands'. FC R and D Paper 51.
Workman, J. (1986), 'Experience in the Management of Beech Woodlands'. In, *Proceedings, Discussion on uneven-aged Silviculture,* Pershore, 29 October 1986 (obtainable as under Collin *supra)*.
Wright, M. (1991), 'The Role and Implications of Irregular Forestry in Britain'. *Scottish Forestry 45* (1), pp. 13–27.
Wright, T.W. (1982), 'The National Trust's Approach to the Regeneration of Broadleaved Woodland'. In *Broadleaves in Britain: Future Management and Research,* eds. Malcolm, D.C., Evans J. and Edwards, P.N., pp. 77–91. Institute of Chartered Foresters, Edinburgh.
Yorke, M. (1991), 'A Study of Regeneration and Management of uneven-aged Coniferous Forest in Central Europe and its Relevance to Great Britain'. Forestry Commission Internal Report.

Chapter Eight

THINNING

1. Thinning classes and regimes

Thinning is the removal of a proportion of the trees in a stand, with objectives both economic and silvicultural. It has the twin purposes of providing income and giving space to improve the quality of the remaining trees, at the same time accelerating individual tree growth. Thinning does not increase the total timber yield from a stand but it does increase the proportion of the more valuable larger diameter trees. Thinning also widens the range of choice of future silvicultural operations, by making the remaining trees more able to stand alone. Nature conservation considerations in thinning are noted in chapter 13.

Trees are planted sufficiently close together that some will become suppressed and over-topped by their neighbours and ultimately die; and if the crop is not thinned, this process will continue to occur throughout its remaining life. In the course of a normal rotation such mortality would represent a very considerable loss of merchantable timber volume. By removing suppressed trees before death occurs, much loss of volume is prevented; and the remaining trees benefit. The more space a tree has, the more it will grow in volume (but not in height). Earlier marketability is thus attained. Hence there is need for the important operation of thinning, to reduce stand density, thereby reducing competition and providing more growing space for the remaining trees, to increase the total yield of usable timber over the life of the crop, and to provide an intermediate yield of timber. Even if no net revenue is attainable from the first thinning, it may be carried out in the expectation of greater returns later in the rotation.

Thinning can improve or mar a stand in both content and profitability. If done too heavily, or if too long delayed, the results can be disastrous. In a mixture it may favour any one of the components. Undesirable trees are removed, and, very important, sales are made of periodic thinnings, thus improving the financial return. The aim is the fullest productivity of the crop and the highest value of cumulative volume produced, in accordance with the grower's considered objectives. One intention may be to produce some large size timber of veneer quality; another, sawlogs for structural purposes; yet another, to supply one of the markets for small diameter material, e.g. for pulpwood or boardwood. Usually these intentions are supplementary. But whatever the objective, every thinning throughout the life of the plantation must be directed towards attaining it. The first general requirement, sufficient length of clean bole, will be partly achieved in the earlier life of the stand by encouraging competition among trees; this will reduce the size of knots, and may be supplemented later by brashing or high pruning. The second requirement, increase in girth and speed of growth, will only be attained by reducing competition; this is achieved in thinning by keeping the crowns free from interference by their neighbours, thus ensuring full vigour of development up to maturity. Thinnings may, in general, provide income equal to or more than that from the final crop, and, of course, within a shorter time; but however desirable early or intermediate cash returns from thinnings may be, they should, at least in theory, be a secondary consideration. Thus a thinning should be undertaken whenever necessary, even though no use can be made of what is removed, if delay will harm the crop and may induce windthrow or snow damage. In a mixture, delay in thinning may result in the dominance of the less desirable component over the whole or part of the stand. The most obvious effect of thinning on timber quality is on ring width; for quality sawn softwoods there is generally a limit of not less than eight rings to 2.5 cm. However, thinning can only control ring width after competition has set in, and it can have no effect on any wide-ringed core. Hence in the fastest-growing conifers the choice is between wide-ringed cores with markedly slower later wood and similar cores with a more even decline in ring width. Density and strength increase with age after the *juvenile* period of approximately ten to fifteen years; hence the traditional association of wide-ringed material with poor timber properties relates chiefly to that part of the tree which is not controlled by thinning. Rate of growth is of limited value in forecasting strength but is retained in rules for the grading of sawn timber mainly as a guide to machining properties.

Thinning is a challenging part of silviculture on account of the different rates and qualities of growth, or failure, even in a pure plantation. Thus it is necessary in thinning to be able to distinguish between the classes of tree found in a crop, based on canopy classes as indicated in Figure 20. In general, the dominants are retained for the final crop together with the best co-dominants. The sub-dominants are removed in the mid-thinnings, while the other four classes are removed at the earliest possible date.

Thinning regime. If a crop is to be thinned decisions have to be made on the age of first thinning, the thinning regime (selective, or systematic), the type of thinning ('crown' or 'low'), the intensity of thinning, the thinning cycle, the thinning yield, and the age at which thinning should cease. These choices will depend on such factors as the objectives of the owner, the markets available at the time and likely to be available in the future, the quality of the stand, its liability to wind damage and the implication of the terrain for harvesting methods and hence costs.

The ultimate objective in thinning practice is usually to obtain the largest financial return from the crop (Hamilton and Christie 1973). The financial result of one particular thinning, however, cannot be judged in isolation: the first thinning may yield very little, if any, harvesting profit, although its effect on the ultimate profitability of a crop may be considerable. As a general rule, the most profitable way to thin a crop is to remove as much volume as possible in the form of thinnings, without any appreciable reduction in total volume production. Quite a large proportion of the total production, something like 50–60% by volume or 80–90% by number of trees, can normally be removed as thinnings during the life of a crop without reducing total volume production. If at any stage excessive volume is removed in thinnings, the production capacity of the remaining crop will be reduced. The volume removed obviously has an important impact on the profitability of a particular *thinning regime,* but other factors which are also quite critical to the financial returns from the crops are thinning *cycle* and *type* of thinning. Each factor can influence profitability, but probably the most important in this respect is the *thinning intensity.*

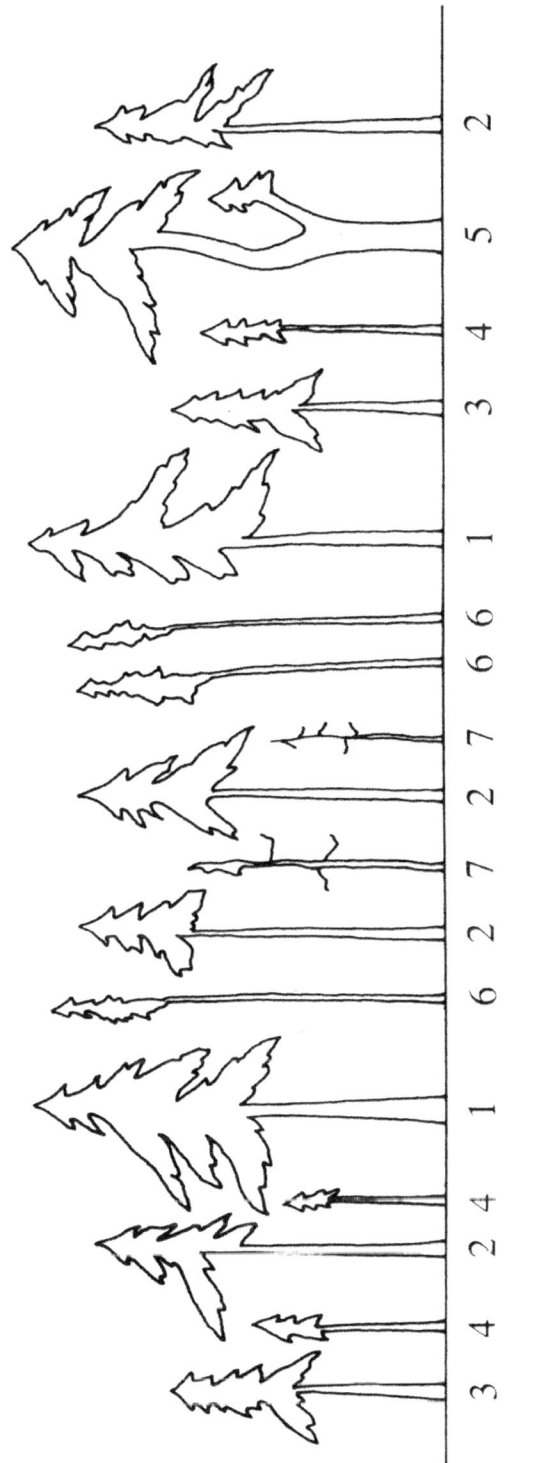

Figure 15. Thinning: Classification of the types of trees found in a crop, based on canopy classes: 1, Dominant; 2, Co-dominant; 3, Sub-dominant; 4, Suppressed trees; 5, 'Wolf trees'; 6, 'Whips'; 7, Leaning, dead or dying

Thinning intensity is the rate at which volume is removed from a stand, i.e. the annual thinning yield or the volume removed in a particular thinning divided by the intended cycle. Over a wide range of thinning intensities the total production remains unaffected. Higher thinning intensities have the effect of creating more growing space for the main crop trees which are able to respond to a greater degree, but as intensity increases the situation ultimately arises where the main crop cannot respond sufficiently to make full use of the growing space created by thinnings and the result is a loss in total volume production. The maximum intensity which can be maintained without loss of volume production is termed the *marginal thinning intensity*. The greater responses of the main crop trees induced by higher thinning intensities are reflected in greater increases in mean diameter. Generally, the value per unit volume increases as the mean diameter increases. In addition, the greater thinning yields resulting from high intensities provide greater net revenues. Taken together these features tend to make higher thinning intensities more profitable, but a maximum occurs where these gains are offset by loss of revenue resulting from losses in volume production. The intensity of maximum profit is normally slightly greater than the marginal thinning intensity. However, since the difference between the two is relatively minor and since maximum volume production is considered a desirable management objective, the marginal thinning intensity has been adopted in the Forestry Commission Forest Management Tables (Edwards and Christie 1981), see chapter 12. The marginal intensity is reasonably close to an intensity which in terms of annual rate of volume removal is 70% of the maximum mean annual increment (MAI), i.e. 70% of the yield class (see chapter 12). Thus the annual thinning yield of a stand of Yield Class 10 thinned at this intensity will be 7 m^3/ha from the time of first thinning. The Management Tables specify the period over which the normal thinning yields apply.

The *thinning cycle* is the interval in years between successive thinnings. It has an influence on profitability in that the net value of any single thinning depends, in part, on the scale of the operation. Long cycles entailing heavier single thinnings are thus usually more profitable, but may increase the risk of windthrow, and in extreme cases may result in loss of volume production. The usual range of thinning cycles is from three years for the faster-growing conifers, through four, five and six years for the majority of crops, to eight, nine or even ten years in slow-growing crops. (However, see no-thinning regimes in section 6 *infra*.)

The *thinning type* refers to the type or dominance class of trees which are removed in thinning; for example, a 'low' thinning is one in which trees are removed predominantly from the lower canopy (i.e. sub-dominants and suppressed trees), whilst in a 'crown' thinning the accent is on the removal of upper canopy trees (i.e. dominants and co-dominants). Trees of the upper canopy tend to be more efficient producers, so that, in general, predominantly low thinning is preferred. However, the removal of some upper canopy trees may be desirable, and is usually inevitable in order to release the better dominants, which will form the ultimate main crop of a stand. Foresters should always remove very coarse, spreading trees, leaning trees (which produce compression or tension wood), forked trees, trees with spiral bark (this often indicates spiral grain in the stem wood), and trees with thin, unhealthy-looking crowns. Provided they justify the cost of cutting, small suppressed trees should be felled, unless they are performing a useful function.

Thinning yield is the actual volume removed in any one thinning. If a fullystocked stand is thinned at the marginal thinning intensity, the thinning yield will be 70% of the maximum mean annual volume increment multiplied by the cycle. The thinning yield should not be so heavy that it opens up the stand to the risk of windthrow or to invasion by other (unwanted) woody species. Also, thinning should not remove all the dominant and good quality trees, leaving none to form a reasonable crop after thinning. Yield from understocked stands may be reduced although it will usually be more practicable to leave these until they have reached full stocking again. For example, it may be best to take a three-year cut even if it is proposed to thin again in four years' time. Similarly in overstocked stands, the thinning yield may be increased to reduce the stocking level so that a six-year cut may be taken even if it is proposed to thin again in five years' time. The procedure may be repeated over several thinnings so as to achieve a controlled reduction in stocking.

Stands are not normally thinned unless they are fully stocked. Although stocking may be judged by visual inspection, a more objective method is to measure the basal area of the stand (see chapter 1, section 4) and compare it with a table of threshold basal areas (see Table 23). Using the relascope method it is necessary to mark three 'sides' of the trees to be removed so that at least one mark is visible from the point of sampling, and to enable the adequacy of the thinning to be assessed. If the actual basal area is equal to, or greater than the threshold basal area, the stand is fully stocked and therefore ready for thinning.

Table 23: Threshold basal areas for fully stocked stands. Basal areas in square metres per hectare.

Species		Top height (metres)									
		10	12	14	16	18	20	22	24	26	28	30
Scots pine		26	26	27	30	32	35	38	40	43	46	—
Corsican pine		34	34	33	33	33	34	35	36	37	39	—
Lodgepole pine		33	31	31	30	30	31	31	32	33	34	—
Sitka spruce		33	34	34	35	35	36	37	38	39	40	42
Norway spruce		33	33	34	35	36	38	40	42	44	46	49
European larch		23	22	22	22	23	24	25	27	28	30	—
Japanese and Hybrid larch		22	22	23	23	24	24	25	27	28	29	—
Douglas fir		28	28	28	29	30	31	32	34	35	37	40
Western hemlock		32	34	35	36	36	36	37	38	38	39	40
Western red cedar		—	49	50	51	53	55	57	60	63	66	70
Grand fir		—	39	39	39	39	39	39	40	41	43	45
Noble fir		—	45	46	46	47	48	49	51	52	54	—
	Yield class											
Oak	4	24	24	23	23	24	24	—	—	—	—	—
	6	—	26	25	24	24	25	25	25	25	—	—
	8	—	27	25	24	24	24	25	26	26	26	—

Species		Top height (metres)										
		10	12	14	16	18	20	22	24	26	28	30
Beech and Sweet chestnut	4	—	22	23	25	27	30	31	33	36	—	—
	6	—	24	25	25	27	29	29	31	33	35	37
	8	—	—	27	27	27	28	28	29	31	33	35
	10		—	28	28	27	27					
Sycamore, ash, birch and alder	4	—	17	17	18	21	—	—	—	—	—	—
	6	—	17	18	19	22	25	—	—	—	—	—
	8	—	17	18	20	22	25	28	—	—	—	—
	10	—	18	19	20	23	26	30	33	—	—	—
	12	—	19	20	21	24	27	31	35	—	—	—

Source: Rollinson (1985). FC Booklet 54.

2. Traditional thinning chiefly in lowland stands

Much thinning practice has changed in recent years, especially of conifers, particularly where soil types restrict rooting and exposure to wind is substantial. Before describing those changes, the traditional thinning practices are noted, related chiefly to lowland sites and to *selective* ('crown' or 'low') thinning whereby trees are removed or retained on their individual merits. The commonest type of selective thinning is known as *intermediate* thinning, which involves removal of most of the suppressed and sub-dominant trees and also opening up the canopy by breaking up groups of competing dominant and co-dominant trees so as to encourage the development of the better trees and to leave an open and fairly uniform stand. (All the foregoing is in contrast to *systematic* thinning – noted in section 4 *infra*.)

To assess the need for first thinning, inspection rackways should be cut at intervals throughout the stand. With selective thinning it will usually be necessary to undertake a degree of brashing to view the trees and to mark those to be cut. It is an expensive operation, however, and even in selection thinning it is rarely justifiable to brash every tree.

Selection and marking of a thinning is an important and interesting though sometimes tedious task. The work, so far as forked or leaning trees, 'wolf trees' and 'whips' are concerned is simple, but the creation of excessive gaps in the canopy must be avoided. The marking of trees that are dead and dying is usually unnecessary where experienced fellers, trained to remove such trees as a matter of course, are employed. But the marking to provide sufficient growing space, to favour one tree at the expense of another, and at the same time to create an even distribution of those remaining, is much more difficult. If the policy is to select, mark and favour from the outset those trees likely to form the final crop, and perhaps also those to form the final thinning, the work of marking interim thinnings is somewhat eased. The canopy and the crowns of the trees must be observed; thus the selection for marking can be a task which causes aches in both the back and the neck. The best crowns are making the most wood, but a badly bent or forked tree stem is undesirable, however vigorous the crown; and long, spindly stems are unlikely to recover. Occasionally time will have to be spent in deciding which of two trees shall be favoured; sometimes a reversal of choice has to be made after viewing the possible result from another direction. Form of stem and shape of crown will be determining factors. Some inferior trees may have to be retained in order to avoid excessive gaps in the canopy, because the possibility of windthrow must be continually borne in mind. Generally, for the convenience of the fellers, marking is done on the same side of the trees throughout the stand; but it may be better to mark on two opposite sides of each tree: the marks can then be seen from almost any angle. Foresters will decide whether they can do the marking alone or whether an assistant is needed. Paint can be applied with a brush or with a paint-gun, or a timber scribe can be used to form a cross, or a slasher to slice a blaze. If more trees are to be removed than are to remain it is quicker to mark those trees which are to stand; marking should then be by painting, not by scribing or blazing which would damage the desirable trees. There is a tendency to adopt a 'feller selection' system, whereby skilled fellers are entrusted to select the trees to fell. A small addition for selection is made to the felling piece-work rate.

The first commercial thinning is usually made when the crop has reached a height of at least 8 m. In terms of age the time of first thinning varies from twelve to twenty years for fast-growing species, twenty to thirty years for the majority of species and sites, and thirty to forty years for slow-growing species. In practice, the time is often decided on the basis of a compromise between waiting until the thinnings are large enough to find a ready market and not delaying so long that the girth increment of the individual tree is greatly reduced by competition. This also provides a useful basis for deciding on the timing and frequency of later thinnings but as most of the trees will by then be of marketable size, the problem reduces to that of ensuring that the individual trees have sufficient space to maintain a reasonable rate of girth increment. Density of canopy thus becomes a useful guide to the timing of later thinnings because this automatically reflects the degree to which the crop has responded to the increased space following the last thinning, no matter how severe or how long ago it may have been.

It is usual to remove about half the total volume produced during the whole life of the crop in the form of thinnings, which may involve removing 90% of the original number of trees planted. In fast-growing crops it is possible to remove reasonably large thinning yields relatively frequently, but in slower-growing crops it is necessary either to reduce the volume per hectare removed at each thinning or to reduce the frequency of thinning if the remaining trees are not to become so far apart that they are unable to make full use of the site. The length of the thinning cycle may also be determined by the risk of windthrow (infrequent heavy thinnings increase the risk). Thinning cycles commonly used have been noted *supra*. It is important to remember, however, that thinning cycle is not independent of weight of thinning (i.e. the volume per hectare removed at each intervention). If light thinnings are carried out on a long cycle the individual trees are likely to suffer considerable competition in the intervals between thinning. At the other extreme, heavy thinnings on a short cycle are likely to result in an understocked crop which has too few trees remaining to make full use of the site, with the result that volume increment per hectare is reduced. Since, as a general rule, the most profitable way to thin is to remove as much volume as possible without reducing volume increment per hectare, the combined effect of frequency and weight of thinning is of considerable importance.

Provided that a sufficient number of the more vigorous trees in the crop are left to ensure that the vigour of the crop is not reduced, the type of thinning has little effect on total increment. It is thus practicable to suit the type of thinning to local circumstances at the time of thinning. For instance, 'crown' thinnings, in which the trees removed are not the smallest in the crop, may be more appropriate than 'low' thinnings if there is no market for small produce.

Traditional thinning of conifers. The following notes indicate the traditional thinning of the more common conifers. (Current thinning practice is further discussed in section 4.) In all thinning, stumps should be treated with urea to prevent damage by root and butt rot fungus *Heterobasidion annosum*.

Species, rotation and yield class (with average)	Thinning recommendations
Scots pine 55–100 years 4–14 (8)	Seldom fast-growing, and hence fairly long intervals may be left between successive thinning, but systematic treatment nevertheless is necessary. Removal of any 'wolf trees' is required in the earliest thinnings. Relatively intolerant of shade, and hence suppressed trees seldom survive long under taller neighbours. Very desirable to thin to produce transmission poles. Thinnings should be barked or removed from the stand before Pine shoot beetle, which breeds freely in butt lengths, can emerge to attack surrounding trees (within 6 weeks in summer months).
Corsican pine 45–80 years 6–20 (11)	In S. England is relatively fast-growing. 'Wolf trees' are relatively rare. Thinnings should be barked or removed from the stand before Pine shoot beetle, which breeds freely in butt lengths, can emerge to attack surrounding trees (within 6 weeks in summer months).
European larch 45–60 years 4–16 (8)	Of all commoner conifers is most seriously affected by insufficient thinning. Requires ample space for healthy development and, if denied, canker and checked growth usually result. Fast growing only in early life. Intolerant of shade, and hence suppressed trees quickly die and become worthless as produce. 'Whips' may also prove troublesome. Need for frequent and heavy thinnings. A thinning that does not break the canopy is of little value.
Japanese larch 40–55 years 4–16 (8)	Between the ages of 10 and 25 years may be among the fastest-growing of conifers. Unless frequently thinned a stand may get out of control. Some stands contain many' badly shaped dominants with wavy (corkscrew) stems, liable to become 'wolf trees': they should be removed at the

Species, rotation and yield class (with average)	Thinning recommendations
	first opportunity. If thinnings are delayed the whole stand may tend to develop into tall 'whips', whilst suppressed trees seldom survive long.
Hybrid larch 40–55 years 4–16 (8)	Thinning should in general follow that recommended for Japanese larch *supra*.
Douglas fir 45–70 years 8–24 (14)	On soft rich soils is very liable to windthrow, or damage by snow. Essential to start thinning early and to repeat at frequent intervals. 'Wolf trees' are common and where they cannot be removed without damage to surrounding trees it may be necessary to ring-bark them. Responds rapidly to thinning; and large gaps, such as occur where 'wolf trees' were removed, soon close over. Height growth rapid. Delay in thinning may lead to serious windthrow.
Norway spruce 55–90 years 6–22 (12)	Rate of growth is only moderate. Moderate shade bearer, requiring more light in drier areas. Sub-dominant trees that have fallen below the level of canopy have an exceptional power of recovery; if freed in time they may develop into useful trees. 'Wolf trees' are relatively uncommon.
Sitka spruce 45–70 years 6–24+ (12)	Fast-growing on favourable sites. Windthrow is likely on soft, wet soils, in exposed situations. 'Wolf trees' are rare but may develop on exposed sites through breakage of crowns. Stands are usually uniform. In normal site conditions, thinning should be moderate and frequent, but see further comments in section 4 *infra*.
Lodgepole pine, Western red cedar, Western hemlock, Lawson cypress and Grand fir. Yield classes respectively: 4–14 (7), 6–24 (12), 12–24+ (14), 8–20 (12) and 12–34+ (14)	Lodgepole pine might be thinned as for Sitka spruce. The others should be thinned frequently until from 12–15 m in height. Thereafter, if branch suppression is satisfactory, thinnings at longer intervals may be made. Where thinnings have been delayed at the outset they should be light and frequent.

Traditional thinning of broadleaves. The following notes indicate the traditional thinning of the more common broadleaves. (Current thinning practice is further discussed in section 3.) Such thinning is a more skilled undertaking than in the case of conifers. Marking must take full account of the characteristics of the species: will epicormic shoots form? will the crown respond to fill the gaps? what mixture of species is necessary? and, how to maintain the understorey? In general: commence when stands are 8–10 m tall, aim to improve stand quality, and select early those trees for the final crop (more vigorous, straight and free of defects). Rate of growth is of less importance than achieving even growth in wood quality. The ideal is to thin little and often; and to aim for quality.

Species, rotation and yield class (with average)	Thinning recommendations
Oak 　Pedunculate 　Sessile 120–160+ years 2–8 (4)	Grows slowly by comparison with conifers. Strong light demander; but if given too much room in the early stages, tends to form a bushy crown, whilst height is checked. If kept dense, usually forms a straight stem, not necessarily from the leading shoot, so singling (formative pruning) may not be required. When thinning remove epicormic ridden trees and favour ones relatively free of them. If a conventional thinning policy is followed, thin frequently and lightly. If a heavy thinning of free growth is planned, begin early in the life of the stand to build the selected tree's crown and vigorously control epicormics, preferably by annual removal, following each thinning. 'Wolf trees', especially those developed from coppice shoots, can be many and harmful. Too sudden opening up leads to formation of epicormic shoots on the stem (see *infra*), and these are a defect; underplanting with beech is helpful. Recommendations by Everard (1985) are given in the footnote.[1] See also 'free growth' *infra*.

1　Everard (1985) after describing the classical European method of thinning oak and the more 'progressive' Danish method, adds that the Forestry Commission has some 60 permanent sample plots in oak sited throughout southern England, most sites having a number of thinning treatments. On five of the sites there is a comparison between no thinning and thinning of various grades. In three experiments in young crops there is a clear benefit of thinning whereas in two experiments in older crops the effect is less marked. In three experiments, the 100% thinning intensity grade has been achieved by 'low' thinning and by 'crown' thinning, and these show an increased breast height diameter increment resulting from 'crown' thinning. Drawing on the results of the English sample plots and Danish methods, Everard includes the following recommendations for oak: (i) First thinnings should take place at about 10 m top height and should set out the extraction system, remove 'wolf trees' and break up groups of competing dominants. (ii) Three to five years later select per hectare 70 final crop trees in Sessile oak and 55 in Pedunculate oak using vigour as the main criterion for selection. Uniform spacing is highly unlikely, but no selected tree should be within 7 m of another. Carry out the second thinning, in which the selected trees should be favoured by removing competing dominants. Thin the matrix to favour any understorey of shadetolerant trees. (iii) Thin often to maintain good crowns on the selected trees. (iv) Where high pruning is to be carried out, and where an understorey is to be established, this

Species, rotation and yield class (with average)	Thinning recommendations
Beech 100–130 years 4–10 (6)	In stands up to about 80 years of age, beech is generally responsive to thinning and can be thinned even after a period of neglect. Response to thinning appears wholly related to the size and shape of the remaining crown (Evans 1984). Since beech does not suffer the problem of epicormic branching, thinning can be heavy and infrequent. Early selection of vigorous well-formed trees which are favoured by moderate to heavy crown thinning will generally ensure a vigorous and responsive stand. In very densely stocked stands it is costly to make the first thinning selective and some form of mechanical thinning, e.g. one in four line removal, may be necessary. Stem selection should then be made prior to second thinning. The procedure of identifying in advance the potential final crop trees is important because relatively few stems in beech stands are of good quality; the ones that are merit favouring and nurturing including, possibly, high pruning. However, a danger exists in heavily thinning stands with a poor selection of wellformed trees, which may result in coarse unmarketable trees; on the other hand positive neglect of thinning in poor crops may result in a higher density, eventually, of acceptable stems whose crowns will respond to later thinning. In older beech stands, especially if over 100 years old and previously neglected, thinning can have potentially very damaging effects. Once trees are drawn up and small crowned, thinning in mature or over-mature stands can lead to sun-scorch of remaining trees and sometimes to wind damage. Trees which are suppressed and underthinned may be prone to Beech bark disease and should be removed.
Ash 60–80 years 4–10 (5)	Capable of very rapid growth on favourable soils; unless frequently thinned many develop into tall 'whips'. The position of the side buds usually leads to forking of the stem whenever the terminal bud is injured; this defect can be rectified by early removal of one of the branches of the fork. Best ash is rapidly grown, so early and frequent thinning is advisable, whether in pure or mixed stands. Thinning should have been commenced by the time stand top height is 10 m; using moderate 'crown' thinning. The best stems should be afforded plenty of light and space

should be done when the oak is about 30 years old using hornbeam, beech or small-leaved lime. Britain is alone in Europe in neglecting the importance of the understorey. By careful management of mother trees the understorey can often be obtained by natural regeneration, but where this is not possible planting after the first thinning should be considered. Care must be taken that vigorous species like beech do not grow into the crowns of the main species.

Species, rotation and yield class (with average)	Thinning recommendations
	with only side shade. In this way they will be furnished with a crown down to half stem height and therefore well able to respond to subsequent thinnings. Too little thinning rapidly restricts crown size which for ash, unlike oak, is not easily increased again when stand opening does occur. Conversely, sudden opening of a stand by heavy thinning may over-expose crowns and cause partial stagnation of growth.
Sycamore 60–80 years 4–12 (5)	Responds well to heavy thinning, which should generally start by the time the top height is 10 m; on fertile soils it may be begun even sooner. Thinning cycles should be kept short and generally never longer than about 6 years. Thinning should aim to maintain a deep crown without over-exposing individual trees. If a site is sheltered, almost 'free growth' conditions may be applied to enhance stem diameter increment. Sycamore is more shade tolerant than ash and can carry more stems per hectare. Pruning may be advisable. Large straight stems, suitable for rollers and rotary-cut veneers are valuable; and it is better to produce a small number of them rather than a large number of smaller and inferior trees.
Sweet chestnut 70–90 years (coppice 12–20 years) 4–10 (6)	Thinning of high forest stands follows in general that recommended for oak *supra*. For treatment as coppice, see chapter 7.
Birch 30–60 years 4–10 (6)	Seldom have been planted. Stands naturally grown may be worth improving by thinning. However, even if thinning is commenced soon after canopy closure, birch is generally unresponsive. The species is relatively fast-growing, and is light demanding; individuals outgrown by neighbours soon become suppressed and die. Ensure that the lead is not taken by mis-shapen trees with bent, forked, fluted, or elliptical stems. Requires ample head room. After 'whips' and 'wolf trees' have been removed, the weight of thinning should be moderate to heavy. Seeded trees are normally preferred to coppice stems. As there may be a future for large veneer quality butts, preference might be given to large straight cylindrical stems free from defects. It may be worthwhile to prune side branches of these to a height of 5 m.

Species, rotation and yield class (with average)	Thinning recommendations
Alder, Wild cherry and Southern beeches (*Nothofagus* spp.) 30–70 years Yield classes respectively: 4–13, 4–10 (6) and 10–20 (12)	Each species should be judged on its merits though main thinning principles still apply. Mis-shapen trees should be removed early to secure more even stands. Light thinnings should be the rule until straight stems and reasonable height are secured. Then opening up may commence at a rate varying with species and particularly with its ability to tolerate shade.
Broadleaves grown as coppice for pulp and boardmill wood and turnery wood.	Thinning of coppice is unusual except with birch when the larger poles may be cut, leaving small stems to 'bulk out'. The rotation may be about 25 years for alder and birch, and 30–40 years for other broadleaves. General information on coppice is given in chapter 7.

Epicormic branches and knots. Development of epicormic branches is a major problem of oak silviculture (Evans 1984; Harmer 1990) and also affects sweet chestnut, poplar and cricket-bat willow. When epicormics persist for more than one year a knot is formed in the wood. For high quality timber such knots must be small, and absent altogether for ornamental veneer. Epicormic shoots arise either from adventitious buds in callous tissue, e.g. pruning scar, or, more often, from dormant buds on the stem which are stimulated to sprout by a change in the tree's environment, most commonly as a result of thinning. With adequate light, an epicormic shoot will develop into a large side branch but more usually will remain as a semi-moribund shoot neither dying nor growing vigorously. The incidence of epicormic branching is clearly under some genetic control. Sessile oak has fewer than Pedunculate oak. Control of epicormic branches has two aspects: preventing them from emerging, and preventing them becoming large branches. The most important silvicultural tool is to avoid sudden change in stand conditions. This, of course, conflicts with the desirability of regular thinning and especially the use of 'free growth' (noted *infra*). Frequent, light thinnings reduce the problem in oak. Even following heavy thinning the initial profusion of epicormics will not go on being produced year after year. As a tree adjusts to its new environment the emergence of new epicormics declines. Similarly, trees least influenced by thinning, the dominants, generally have fewer epicormics than other crown classes. Thus control of epicormics is most important immediately after a thinning or other sudden change in a stand.

Prevention of epicormic shoots becoming large branches can be achieved in several ways. Naturally, this can be done by manipulating the light regime and keeping the stem dark. The use of a beech understorey, or other shade-bearing species, is an effective though a costly means of control unless the understorey is there naturally. (This is the main tool used in French silviculture to restrict development of epicormic branches.) Epicormics can also be controlled directly by pruning; to be effective, this needs to be done regularly, preferably each year with the shoots being removed in mid-season. Experiments continue investigating chemical and other means of control (Evans 1984).

'Free growth'. Because thinning of broadleaves seeks to produce a final crop of relatively few but well-formed trees, one unconventional option is to attempt to identify such trees at an early stage and to favour just these by very heavy thinning to the exclusion of all other; the system is called 'free growth' – a technique which has been tried mainly in oak (Jobling and Pearce 1977). As noted above, an important silvicultural problem associated with open-grown oak is the stimulation of epicormic branches. The object of the free growth technique is to accelerate diameter increment on selected trees by freeing their crowns from competition. Compared with conventionally thinned oak, basal area increment of open-grown or free growth trees is doubled. Oak of almost any age will respond to free growth, but for growing timber trees the technique is best begun in plantations, pure or mixed, when twenty to forty years old. Evans (1984) gives the procedures to be adopted.

3. Thinning of broadleaves and mixtures chiefly in lowland stands

The effects of thinning on growth of broadleaved stands do not differ substantially from those of conifers (Evans 1984). Light thinning maximizes production per hectare but growth of individual trees within the stand is slower. Heavy thinning enhances individual tree growth at the expense of some loss of total production from the site. In broadleaved stands and well-managed lowland woodlands there are few problems in thinning (in contrast to problems in some upland conifers), but difficulties may occur in mixtures and overstocked stands after a long period of neglect.

Care is required in applying the general relationships noted earlier in varying thinning intensity to manipulate tree and stand growth. Species differ in their capacity to respond to thinning (light demanders such as ash and sycamore generally require more open stand conditions than oak or beech). In most stands there are usually relatively few trees of good form, which restricts the choice of ones to favour in thinning. Many broadleaved species do not have strong apical dominance and heavy thinning may depress height increment and encourage the development of spreading, heavily branched crowns; and, especially in oak, poplar and sweet chestnut, heavier thinnings stimulate undesired epicormic branching.

In broadleaved stands, the first thinning generally commences early when the trees are 8–10 m top height and basal area between 20 and 30 m^2/ha. If a market outlet exists for small sized material thinning can begin any time once top height is about 8 m or more. It is silviculturally very important to realize their potential because of generally poor stem-form, the end-use intentions for a stand, and the kinds of sites planted. Thinning should aim to produce well-balanced, even crowns on final crop trees. Poor quality trees, particularly if unduly vigorous ('wolf trees'), must be removed at an early stage. The primary objective is to improve the quality of the stand. Achieving a particular yield per hectare is generally less important than ensuring that well-formed trees are favoured in thinning. The effect of rate of growth on wood quality is generally much less important than achieving even growth. Selecting which trees to favour must be undertaken with care, and is best done at the time of the first commercial thinning; the thinning operations then favour these selected trees (the most vigorous, straight and free of defect). In conventionally (i.e. selectively) thinned broadleaves it is prudent to select two to four times as many potential final crop trees as are needed to achieve final crop stocking to allow for losses. (Final normal stocking – stems /ha – for oak is 60–90, ash 120–150, beech 100–120, and sycamore 140–170.) Later thinnings continue to be selective but increasingly some of the originally favoured trees will be removed. High pruning of 100–300 trees per hectare is discussed in chapter 6.

Whereas in conifers the aim is to maximize total production per hectare, in broadleaves the primary purpose of thinning is to favour the final crop trees. However, in selectively thinned stands, usually receiving moderate 'crown' thinnings, between 45 and 70% of the total stand volume production will be removed in thinning during the course of a rotation. In terms of stocking, only 1 to 10% of the trees originally present remain to the end of the rotation. Where thinning rates and levels of stocking follow the Forestry Commission Yield Models based on marginal intensity, during the first part of the rotation the earlier thinnings remove the equivalent of 70% of the maximum mean annual increment (yield class).

However, the rotation age of most broadleaved species is well past the age of maximum mean annual increment and thinning intensity is reduced as stands become older; this is usually effected by extending the interval (cycle) between thinnings. Typically the proportion of trees removed in each thinning declines from about 25 to 15% of the total. The proportion of standing basal area removed each time is rather more constant and averages 10 to 15% of the stand total.

The thinning cycle need not be a rigid number of years though excessive delays between thinnings should be avoided. Generally intervals between thinnings are five to seven years in young stands. The more vigorous the stand the shorter is the cycle. The need to thin can be judged subjectively in summer when the amount of crown overlap and the proportion of trees barely making the canopy can be seen. A more objective way is to determine the basal area per hectare and compare it with the recommended threshold basal area (Table 23). The longer the interval between thinnings, the greater will be the volume of produce harvested on each occasion; this has definite economic advantages, but the effects on the remaining trees are more drastic, e.g. increased exposure, greater variation in ring width leading to problems in seasoning and working, and undesirable stimulation of epicormics in some species. The silvicultural ideal for most broadleaved stands is to thin little and often; and this ideal should be followed if markets and circumstances allow.

Mixtures of broadleaves and conifers. By timely thinning, mixtures of broadleaves and conifers may be swung in favour of one component or another, provided the objective is always aimed at. If the intended final crop species grows more slowly than the other in the stand – a common feature of such mixtures – regular thinning is essential. Thinning secures the final crop species and prevents trees becoming over-topped, slender, or whippy or suppressed. Neglect or delay in thinning is more serious in such mixed stands than in pure stands since one component eventually tends to dominate and the other becomes suppressed or excessively drawn up. If this dominant species is not the desired final crop and if regular thinning cannot be assured then there will have been little point in establishing a mixture. Broadleaves in mixture with light-demanding conifers (e.g. larch or pine) generally suffer more from delayed thinning than when in mixture with Norway spruce or western red cedar. The rapid crown expansion and coarse branching of larch and pine lead to early interference with the broadleaves, particularly if planted in strips three rows wide or less. Where delay in thinning, or poor mixture design or bad species choice has led to severe suppression or failure of the desired main crop species, the chances of effecting a worthwhile recovery are low; and replanting is rarely an option because of cost. In the affected part(s) of a stand any surviving main crop trees of reasonable potential should be favoured and the secondary species accepted as a long-term constituent of the crop.

The principles of early selection of potential final crop trees, priority removal of 'wolf trees', and the timing and cycle of the thinnings, are substantially the same as for pure stands. Thinning intensity in terms of yield per hectare per year should be determined separately for each major species component of a stand (Rollinson 1985). The secondary species in a mixture is mostly removed by about the fourth thinning, i.e. by about half to two-thirds of the way through the rotation of the final crop species, though a few specimens may be retained for amenity or added value.

In strip (row) mixtures of broadleaves and conifers, the first thinning should usually remove one outer conifer row if the species are in reasonable competitive balance, or both outer rows of conifers if the broadleaved species is becoming suppressed. The broadleaves are normally selectively thinned and trees extracted using the rackways in the conifer. In the second thinning any remaining adjacent conifer row is removed, and the rest of the crop selectively thinned. Subsequent thinnings are wholly selective, with most of the remaining conifers being removed in the third thinning unless strips are very wide, such as six rows or more.

Broadleaved mixtures. Thinning is relatively straightforward in broadleaved mixtures, favouring good stems of desired main crop species and early removal of misshapen trees, fast growing species such as birch and alder (both have short rotations), and unwanted ones such as sallow, rowan and whitebeam (unless these were removed in cleaning; although somewhere in a stand there should be found a place to retain a few of these attractive species for conservation reasons). Where intermediate yields of good quality cherry, sycamore or ash are sought from mixtures with beech and/or oak, some favouring in early thinnings of both the short rotation and the long rotation species will be necessary.

Neglect of or delay in thinning, often for long periods, is not uncommon in broadleaved woodlands. In young stands such neglect or delay leads to increased mortality and a preponderance of spindly trees. No great silvicultural harm is done provided a stand can be rehabilitated by opening up gently, i.e. thinning a little and often, to reduce basal area per hectare to normal levels in (say) three thinnings over ten years. By contrast, a single heavy thinning will break up the canopy and expose crowns to damage from snow and wind; the tall thin trees will snap or whip around in the wind. An old, neglected overstocked stand but which has not reached the normal rotation age, should be considered for thinning. However, any benefit therefrom will depend on whether the remaining trees are able to respond; this ability may be reduced by both the lack of thinning in the past and increasing age for some species. Where, owing to past neglect, crowns have become small, rapid increase in crown size cannot be expected and only a poor response to thinning is likely. Some species are better able to respond than others: oak, provided crown dieback has not begun, is generally able to expand its crown following thinning, whilst ash is very much less able to do this; beech and sycamore are intermediate.

All thinning runs certain risks, but in old overstocked broadleaved stands they are exacerbated. Opening up such stands increases risk of windthrow, stagheadedness in old trees, development of epicormic branches, snow-break, sunscorch on bark of thin-bark trees such as beech, and damage to the butt and surface roots during extraction. Therefore, in the treatment of old stands the first question to decide is: will sufficient trees respond and the stand remain intact? Thinning is inadvisable if there are signs of windthrow, crown dieback or chlorosis. If thinning appears safe it should be done lightly and infrequently: no more than 5–10% of trees or 5% of stand basal area should normally be removed at intervals of ten years or more. Thinning must be selective with the aim of removing dead, dying and diseased trees, and taking care not to break the canopy more than is absolutely necessary – that is, a light 'low' thinning. If neglect has made a stand very fragile (trees drawn up with small, weak crowns) thinning may be too risky. Such stands will eventually have to be felled and regenerated; the longer they are left usually the poorer is their condition.

Supplementary to thinning, adequate drainage must be stressed. One object of thinning is to secure the stability of the crop; but a well-thinned plantation may still be subject to windthrow if drainage is neglected. Windthrow is often found to start along blocked channels. Once it has started it may spread through the entire stand. So the open drains should be attended to following each thinning to ensure that any surplus water can get away freely. Felling and extracting of thinnings must be done with appropriate care; this along with road maintenance is treated in chapter 10.

4. Current thinning of conifers chiefly in uplands

Traditional thinning as described in section 2 should be seen in contrast to current information and methods generally followed. Thinning practice has been greatly influenced in recent years by the recognition of the threat of windthrow to plantations, particularly to conifers in the uplands. Until the 1970s it was common practice to remove about 50% of the total yield of, for example, spruce crops as thinnings, but more recent practice has moved towards a smaller number of thinnings or no thinning at all. The result of this is that an increasing proportion of the annual cut is taken as clear-felling. (Within the Forestry Commission, economic appraisals have indicated that about 35% of stands should not be thinned because the potential loss of revenue from premature windthrow would render thinning economically unviable.)

The timing of first thinning varies depending on the species, yield class, initial spacing of the stand, the thinning intensity and whether thinning will yield a return above threshold surplus. The normal age of first thinning is later in more widely-spaced stands and also for a heavy first thinning, as otherwise stocking would be reduced to a level which would cause a loss of cumulative volume production. There are circumstances where the most profitable treatment may be to begin thinning later than the normal thinning age, notably where the standing value of the trees in such a thinning is low. Where the first thinning is delayed, then it will need to be heavier so that the stand returns to the normal stocking level. It may not be possible to do this in one operation as this could lead to loss of volume production or stand instability, so subsequent thinnings may need to be heavier than normal to compensate. The standard ages of first thinning are tabulated in

Table 24. The timing of first thinning is somewhere around the first age at which – in conifers – the stand can be thinned such that the volume increment will allow subsequent thinning on a five year cycle at the intensity stipulated in Forestry Commission Management Tables (see chapter 12). The economic optimum age of first thinning may be somewhat later than this under certain circumstances.

Systematic thinning. Thinning may be either *selective* (noted in section 2 *supra*) or *systematic*. In *systematic thinning* trees are removed according to a predetermined system, which does not permit consideration of the merits of individual trees, i.e. it does nothing to improve stand quality. The operation leaves parts of the crop unthinned. Two types of systematic thinning are *line thinning* and *chevron thinning*. *Line thinning* (Hamilton 1980) is a systematic thinning in which trees are removed in lines or in a series of interconnecting lines, and has three principal forms: (i) *row thinning* in which the lines of trees removed follow the planting rows; (ii) *strip (syn. corridor) thinning* in which lines of trees are removed but where the lines do not necessarily follow the planting rows; and (iii) *chevron thinning*, a series of widely spaced, approximately parallel, lines (main racks), the intervening area being thinned by the removal of regularly spaced pairs of lines (line racks) originating opposite each other in the main rack and acutely angled to the main racks. *Staggered chevron thinning* is similar to chevron thinning except that the side racks are not directly opposite but alternate.

Patterns of single row thinning considered practical include the removal of one row in every four, or three, or two. Where the original spacing is comparatively close, it may be necessary to remove two adjacent rows in order to create an adequate extraction or access lane; the practical treatments include the removal of two rows either in every four or in every five. Any thinning which leaves three or more adjacent rows unthinned denies a proportion of the crop any prospect of enhancing its diameter increment. The loss in volume production associated with line thinning enlarges with an increase in the number of adjacent rows removed.

Table 24. Standard ages of first thinning for a wide range of species, yield classes and initial planting spacing.

Species	Spacing (m)	Yield class													
		30	28	26	24	22	20	18	16	14	12	10	8	6	4
Scots pine	1.4									21	23	25	29	33	40
	2.0									22	24	27	31	35	45
	2.4									24	26	29	34	39	49
Corsican pine	1.4							18	19	20	21	23	25	28	33
	2.0							19	20	21	22	24	27	30	36
	2.4							20	22	23	25	27	30	34	41

Species	Spacing (m)	Yield class													
		30	28	26	24	22	20	18	16	14	12	10	8	6	4
Lodgepole pine	1.5									19	21	23	26	31	40
	2.0									20	22	25	28	34	44
	2.4									21	24	27	31	38	48
Sitka spruce	1.7				18	18	19	20	21	22	24	26	29	33	
	2.0				18	19	20	21	22	23	25	27	30	35	
	2.4			19	20	21	22	24	25	28	30	34	40		
Norway spruce	1.5					20	21	22	23	24	26	28	31	35	
	2.0					21	22	23	25	26	29	31	35	41	
	2.4					23	24	25	27	28	31	34	39	46	
European larch	1.7										18	20	22	26	32
Japanese larch Hybrid larch	1.7									14	15	17	19	22	26
	2.0										15	16	18	20	23	27
	2.4										16	17	19	21	25	30
Douglas fir	1.7				16	17	17	18	19	21	23	25	28		
	2.0				16	17	18	19	20	22	24	27	30		
	2.4				17	18	19	20	22	24	27	30	34		
Western hemlock	1.5					19	20	21	22	24	26	28			
Western red cedar/ Lawson cypress	1.5					21	22	23	24	26	28	30			
Grand fir	1.8	19	20	20	21	21	22	23	24	25					
Noble fir	1.5					22	23	25	27	29	31				
Oak	1.2												24	28	35
Beech	1.2											26	29	32	37
Sycamore/ Ash/Birch	1.5										14	15	17	20	24

Source: Rollinson (1985). FC Booklet 54.

Line thinning has both advantages and disadvantages when compared with selective thinning. The advantages result from easier and cheaper harvesting and from obviation of the costs of brashing and marking. The main disadvantages arise from losses of volume production and from reduced stand stability. In any particular situation, the use of line thinning will be appropriate if the advantages outweigh the disadvantages. (But note amenity considerations in chapter 10 under 'Harvesting'.) The commonest system now is racks to allow access by harvesting and extraction machinery with selective thinning between. Windthrow Hazard Classification (see chapter 3) is used to determine whether thinning is a valid option for a particular site. On some sites which are recognized as being extremely susceptible to windthrow, line thinning should not be contemplated. On windfirm sites line thinning will usually yield net benefits and will therefore be adopted (it is cheaper to mark and yields higher volume at lower harvesting costs in low value first thinnings). On sites which are not of these classes the decision will depend on more than this single factor and it is advisable therefore to assess rather more carefully both the favourable and unfavourable aspects of using line thinning in a particular crop.

Determining the interval between thinnings is usually a compromise between profitability which favours longer intervals and more timber removed at each, and crop stability which favours more regular thinnings which remove less timber.

5. Economic aspects of thinning

In deciding whether to use systematic thinning or selective thinning, it is advisable first to consider the various factors involved in each situation, to quantify their effect in economic terms so far as this is possible, and thereafter to compare the net benefits of the two alternatives. Items which are common to all the options can be ignored. The major quantifiable differences, in costs and revenues, between the two methods of thinning are: (i) in costs of brashing and marking (little marking is required for systematic thinning); (ii) in discounted revenue arising from differences in volume production and average tree size; and (iii) in harvesting costs. These costs and returns occur at different times and, in order to make comparisons, it is necessary to discount or compound these values to a common date, in this case the age of first thinning. The resultant calculations generally show a considerable net saving to be gained from line thinning (Hamilton 1980). However, some allowance should be made for the less quantifiable disadvantages of line thinning, namely landscape considerations, damage to stand and soil, and most important, the effect on stability.

Because of the variety of thinning regimes that can be applied, differences in the price of thinnings of various sizes from place to place, and differences in the costs of access and risk of windthrow, it is difficult to make general rules about the course of management which is likely to produce maximum net discounted revenue. However, given some assumptions about these factors and with the aid of some generalized tables, it is possible to calculate the most profitable course of action (see chapter 14). Appraisal of the decision on timing of thinning is often complicated by the presence in a stand of a number of different species, ages and yield classes.

The effects of thinning on growth of lowland conifers do not differ materially from those of broadleaves; but in the uplands, conifers require their own special treatment. Within the last decade, important changes have arisen in silvicultural management of upland conifers; as a result of the increasing wind damage in even-aged pole-stage plantations, much shorter rotations are now planned or expected in many forests. Under non-thin management, stands which are susceptible to wind damage can be grown on increased physical rotation lengths, and coupled to a reappraisal of the mensuration data on sawlog production under differing thinning regimes, a greatly increased area of conifers is now under non-thin management. These rotations will usually be shorter than usual rotations for stands of equal yield class in windfirm areas conventionally thinned, but longer than stands which would be thinned in unstable areas because rotation length will be determined by the physical factor of terminal height rather than tree size and growth rate. Recent developments in the mechanization of thinnings are likely to have application only in stands which are windfirm. Mechanized thinning in stands at risk from wind damage is considered particularly hazardous.

The yield of sawlogs from unthinned spruce stands on short rotation may not be markedly different compared with thinned stands. The total volume production, and the total timber production of sawlogs to 18 cm top diameter, is only a little less in unthinned than in thinned stands; for larger logs, the difference between thin and non-thin becomes greater. The value of sawlogs is also pertinent. The total timber production of smaller sawlogs, say to 14 cm top diameter, is much the same under both regimes. Even on stable sites, the small additional yield of sawlogs from thinned stands necessitates a careful appraisal of the costs and revenues associated with the thinning operation. In most Sitka spruce plantations, any thinnings which are carried out without yielding an operational surplus might be considered of doubtful value *per se,* and this is particularly the case in the lower classes with moderate windthrow risk, when returns on the thinning investment are often highly marginal. However, it is quite possible that although money is lost on early thinnings, the overall net present value (NPV) could be improved. Other factors, such as the improvement in timber quality resulting from selective thinning, and predictions of price/size differentials in the future (see chapter 11) are particularly difficult to quantify with any precision, but must also be taken into account in economic appraisals of thinning policy.

With the major part of Britain's younger coniferous plantations confined to remote sites at high elevation in the north and west, and despite high yield potential, these forests will be run on short-rotation lengths due to the high risk of windthrow. Even where the risk of wind damage is only moderate, it may be preferable to avoid thinning operations if calculations based on realistic assumptions of future timber prices do not support thinning as a worthwhile investment. However, for many areas the decision on whether or not to thin is a marginal one. With more optimistic assumptions on the future values of both small diameter roundwood and sawlogs, or with quite small reductions in harvesting and transport costs, it would be possible to increase significantly the proportion of young stands which will be thinned in the future. This has been an increasing trend since prices of small diameter roundwood have risen during the later 1980s.

It is unfortunate that the measures normally available to improve the economic surplus of thinning operations all tend to be detrimental in terms of wind stability. Delaying thinning, to increase the average tree size removed, usually results in higher wind damage levels than with thinning carried out at an earlier age. Similarly, removal of large volumes per unit area during the thinning operation also increases the risk of wind damage. The adoption of more efficient mechanized working methods, with the implied increase in systematic thinning patterns, wide access racks and high volume removal also tends to increase the susceptibility to wind damage.

Knowledge of the distribution of Windthrow Hazard Classes (see chapter 3) in a forest area can be of great value to foresters in constructing thinning programmes. In practice, areas in Classes 5 and 6 are nearly always classified as non-thin subjects. In Class 4, the timing, pattern and intensity of any thinning operations are critical, and selective thinning well before *Critical height* is recommended. The options of systematic thinning (line; chevron) or delayed thinning are virtually confined to the three lower Classes, 1 to 3. A single thinning of these may be the best option. Snow damage in unthinned stands is usually substantially worse than in thinned stands. Systematic thinning methods with high volume removal are quite inappropriate for sites with a high risk of wind damage, and such areas should be managed under non-thin regimes, to give the maximum rotation length.

6. Thinning practice

Thinning practice is governed by many different criteria; long-term profitability is the usual choice. Economic appraisals determine whether or not – and how – stands should be thinned. The incidence of wind damage has been noted earlier. As to the incidence of price, the decline during the early 1980s in *real* terms in the price of small diameter roundwood brought into question the desirability of starting thinning at traditional ages, or, indeed, of thinning at all. Economic appraisals made by the Forestry Commission in 1981 indicated that the economic case for thinning largely turned on whether or not the first thinning paid. At that time a small increase in size gave a large increase in revenue so it was desirable to delay start of thinning wherever possible. Caution about changing from traditional approaches to thinning in order to enhance diameter growth and to select for the best trees – as well as concern over the effects on levels of future production programmes – led to some reluctance by the private sector in departing from conventional thinning practice and, as a result, a larger proportion of first thinning was undertaken at a loss than can be justified on grounds of long-term profitability. Foresters have adopted two main practices to try to make the first thinning profitable. The first, to adopt a 'low' thinning or other forms of neutral thinning to raise the mean tree size for a conventional time of first thinning and to lower harvesting cost. The second, to delay the time of first thinning usually by five or ten years: this leads to the production of trees which are of a higher value when they are felled and can enhance total discounted revenue despite its delay. Unfortunately both of these actions are likely to increase the risk of wind damage.

Thinning practice in the Forestry Commission. This practice, particularly in upland conifers, is based to a large extent on assessments of Windthrow Hazard Class (see chapter 3) and determination of the marginal tree size: that is, the size of the tree at which revenue just balances harvesting cost. A further analysis of the economics of thinning by the Commission confirmed earlier ideas and introduced a concept of 'threshold surplus':

1. Where windthrow hazard is high, no thinning is the preferred option; where hazard is moderate, thinning is likely to be best done at the conventional first thinning age; on stable sites thinning can be delayed by five to ten years if this is necessary to avoid a loss making operation.

2. As a result, about 35% of State stands may not be thinned because the potential loss of revenue from premature windthrow in thinned stands would render thinning economically unviable.

3. In steep terrain, stands are unlikely to be thinned because of high costs of harvesting thinnings and of building roads.

4. On stable sites, the higher the yield class the more likely it is that thinning will give the maximum discounted revenue. In areas of moderate windthrow risk, faster-growing stands can reach the height at which they are likely to blow down more rapidly incurring a greater revenue penalty than slower growing stands which can reach the age of maximum discounted revenue.

5. Tables have been produced giving the minimum surplus or maximum loss per cubic metre which must be obtained from a first thinning operation to justify starting thinning. The surplus values are termed 'threshold surplus' values and they vary with species, growth rate, Windthrow Hazard Class and road cost. The values may be positive or negative.

Forestry Commission thinning policy has been described by Harper (1986). The Commission uses maximization of net present value (NPV) as the criterion on which to decide thinning treatments; and its thinning policy, predominantly relating to Sitka spruce, is implemented by guidelines for its foresters. Subsequently market conditions have improved and the marginal tree size has fallen so that the proportion of crops which can be thinned profitably has increased.

To thin or not to thin? The total cumulative production of utilizable volume is significantly greater from a thinned crop than from an unthinned one, except where the yield class is low or felling age is early. This difference increases the longer the rotation. Much of the extra volume from a thinned crop comes from small diameter roundwood removed during early thinnings although when crops are felled at the same age, a thinned crop will produce more log sized materials to 18 cm top.

Forestry Commission usual method for comparing options is to discount all the revenue to the year of planting, using a 6% discount rate. It is found that there is little to choose between thinning or not thinning in low yield class or for short-rotation crops. Thinning on a regular cycle maximizes revenue from fastgrowing, high yield class crops especially if they are not felled until the age of maximum discounted revenue (DR). This economic optimum felling age will be later for thinned than unthinned crops. The Commission find that crops of Yield Class 10 and above, on stable sites, will produce a significantly better return if they are thinned. When yield class is low, there is relatively little financial benefit from thinning.

In many areas windthrow dictates both the opportunities for thinning and the age of felling. On sites susceptible to windthrow, thinning increases the likelihood of damage. Where thinning is delayed, stability will be further reduced. On sites of Windthrow Hazard Classes 1 and 3 (see chapter 3) both thin and non-thin crops will reach age of maximum DR before wind damage becomes significant; these sites are stable and should be thinned so as to obtain the largest volume and DR. In higher Classes, crops which have not been thinned remain stable for longer allowing longer rotations and therefore higher volumes and DR. In general, crops on sites of Windthrow Hazard Classes 5 and 6 are so prone to damage that any thinning will seriously foreshorten the rotation and greatly reduce the revenue: it follows as a general rule that crops in those classes should not be thinned.

Economic analysis, touched upon *supra*, has been condensed by the Forestry Commission into management guidelines for use in deciding if and when to thin. The general guidelines which the Commission used for the 1987 production forecasts are:

1. If the crop is very unstable so that any form of thinning will induce windthrow, it should not be thinned. In general, crops in Windthrow Hazard Classes 5 and 6 will not be thinned.

2. If the crop is only moderately stable so that a delayed thinning will almost certainly foreshorten the rotation, thinning must start at Management Table time. If thinning cannot be done at that time or it cannot be thinned at Management Table time for less than the Commission maximum permitted loss, then the crop should not be thinned. (Tables of permitted losses have been prepared for Forestry Commission managers which are based on the comparison of DRs but also include a margin to ensure that thinning is not undertaken at a loss so large as to offset the benefits on the price and cost assumptions adopted.)

3. If the crop is stable and delayed thinning will not reduce stability, thinning can start at Management Table time (or later if this is necessary to increase the surplus from first thinning to an acceptable level). However, a non-thin regime may be best for low yield class crops, even if they are stable.

These guidelines were produced on the basis of 1987 Forestry Commission price and terminal height assumptions. The Commission's objective is to maximize discounted revenue. It may not be entirely appropriate to private foresters who may wish to use different assumptions and may be aiming for quite different objectives but the guidelines nevertheless can serve to give some basic guidance. An important effect of thinning in the private sector is on the ease of access to woods for inspection and sporting, even in low yield class crops in the lowlands and some upland areas.

Thinning prescriptions are subject to change. For example, that for Sitka spruce should now aim to select for improved stem-form and reduce the persistence of deep crowns. The whole concept of thinning continues to be a subject of profound debate. Price (1985, 1987, 1988, 1989) has explored the economic theory behind the distribution function of thinning, that is, its capability to concentrate volume production selectively onto stems of a particular yield class, and concludes that considerations other than the distribution of volume should dominate decisions on if, when and how to thin.

7. Thinning control

Before foresters can decide on an appropriate thinning regime for their crops, they require to know – beyond the area and species composition – their current and future rates of growth, a subject discussed in chapter 12.

Having determined for each of the types of stand in a forest the thinning regime calculated to maximize profitability in the long run, it is important to control the volume removed from individual stands when the thinning is being marked. The main objectives of thinning control are usually to combine maximum profitability in the long run with maintenance of a regular supply of material from thinnings – insofar as these objectives are compatible with one another. Failure to control the volume removed as thinnings can result either in over-cutting (which leads to loss in volume production) or under-cutting (i.e. overstocking, which depresses the mean diameter and hence the value increment and profitability). Such failure to control the thinning may also produce an erratic flow of timber to be harvested for the processor. It is desirable that the basal area stocking of all stands is checked before thinning to ensure that it is adequate.

Principles of thinning control are outlined by Hamilton and Christie (1973). Hamilton (1980), in his Appendix C, is specifically concerned with thinning control when various forms of line thinning are used. The practical aspects of thinning control are provided by Rollinson (1985).

Thinning can be controlled in terms of the number of trees, the basal area, or the volume. Control by the number of trees is not recommended as the result of the thinning is very dependent on the thinning type. If the smallest trees are removed, the stand will be left much denser than if the same number of larger trees are removed. Thinning should therefore be controlled by basal area or volume, and this can either be by the amount *removed,* or the amount *remaining.* Control by the amount removed is preferable for four reasons: (i) it is easier to do; (ii) it tends to produce a constant and predictable yield of timber which is useful for planning purposes; (iii) it discourages drastic reduction of the level of the growing stock which can lead to windthrow or other damage caused by a sudden opening up of the canopy; and (iv) it considerably reduces the effect of inaccurate yield class assessment, which can occur if the Local Yield Class (see chapter 12) has not been measured.

Therefore, it is strongly recommended that control of thinning should be by the basal area or the volume *removed*. It is essential that the stocking of all stands should be checked before thinning to see that it is adequate. Rollinson (1985) gives a checklist of office and field procedures to be followed in marking a thinning. Having determined the yield class of the stand and decided on the thinning treatment which will yield the maximum benefit (e.g. maximum volume production, maximum diameter increment of the main crop, maximum profitability) the procedure is:

1 Choose the thinning type (selective or systematic);
2 Choose the thinning intensity;
3 Choose the thinning cycle;
4 Calculate the thinning yield;
5 Check that the stand is ready for thinning;

6 Start by marking a representative corner of the stand;

7 Measure the volume marked (using a relascope or sample plots);

8 Repeat this procedure several times and if the volume marked is found to be consistently greater or less than the specified thinning yield, adjust the size, or number, or both, of the trees being marked accordingly; and

9 Continue marking to this revised standard, making only occasional checks.

The precision of control will depend upon the expertise and resources available, as well as the nature of the crop being thinned. Normally it should be possible to control yields to within 15% of the target thinning yield, but achieving this degree of control may be unduly expensive in mixtures or crops that are very variable in other ways. Thinning yields also need to be controlled in terms of annual programmes of production. This is relatively straightforward if the annual programme is drawn up stand by stand and if the progress of work is recorded alongside the planned cut. Control is achieved by keeping a record of the volume cut and comparing this at intervals with the planned yield for the same period.

Rollinson (1985) should be continually followed as a field guide to thinning yields. Efficient thinning practice depends upon intelligent application of the guide under a wide range of conditions. For example, it may pay to thin an understocked stand because of contractual commitments, despite the loss of increment which is likely to result.

8. Respacing of upland conifers

The effect of spacing on tree growth has been noted in chapter 5 and discussed by Edwards and Grayson (1979) and Rollinson (1986, 1988). Traditional forest practice has been to manipulate stand density by thinning; but as a result of increasing wind damage in even-aged, pole-stage plantations, a greatly increased area of upland coniferous plantations, predominantly Sitka spruce, is now under a non-thin management regime. This has important implications for planting practice: the manipulation of crop spacing and the choice of optimum spacing at the time of the rotation have to be determined at or very soon after planting. One means of manipulating stand density (Moore 1973; Rankin 1979; Edwards and Grayson 1979; Ford 1980) is to plant trees closely but to carry out a respacing (also termed 'pre-commercial thinning', 'unmerchantable thinning' or 'thinning to waste') before or about the time of canopy closure while access is still easy. The advantages of such a system have been claimed to be: earlier establishment of trees at closer spacing, and enhanced growth rates up to and possibly beyond the time of respacing, and smaller branches and a smaller juvenile core in the early life of the crop possibly leading to gains in wood quality.

Methods of respacing. In each case the unwanted trees can be chosen individually or according to a systematic pattern without consideration of the status of individual stems. If the trees are chosen individually, it is possible to remove the poor quality trees and the slower growing ones, thus improving the average quality of the remaining stand. For normally established pure crops – predominantly Sitka spruce – there are three methods of respacing (Edwards and Grayson 1979):

1   Cut the unwanted trees at ground level, with no debranching (snedding) or cross-cutting; but treatment to protect the remaining crop from *Heterobasidium annosum* may be necessary, and, if the operation is carried out early, continued weeding to protect and assist the growth of the remaining trees may also be needed.

2   Decapitate the unwanted trees at some convenient height, for example, 1m. The idea is that the decapitated stems will continue to suppress weed growth, they will help to maintain the forest microclimate, giving shelter to the remaining trees, and they will lessen the growth of the lower branches of the remaining trees. Additionally there will be less danger of fungal invasion of potential menace to neighbouring trees. Unfortunately the decapitated trees continue to grow, competing with the remaining crop for nutrients, and it is quite possible for new leaders to develop on the decapitated trees which will compete with the remaining crop. To prevent this, the remaining crop needs to be about 5 m taller than the decapitated stems. The latter subsequently create harvesting difficulties.

3   Kill the unwanted trees by girdling or by chemical injection (discussed *infra*). This method has most of the advantages of decapitating without the disadvantages, as there is no fear of competition or obstructions to harvesting.

The conclusions reached by Edwards and Grayson (1979) were:

1   For Sitka spruce, increasing initial spacing (past about 3.5 m) causes a loss of measurable volume. There is no reason to believe that this finding is changed where respacing occurs at an early stage of the crop's life.

2   Wide spacing affects revenue production in three ways: firstly through the reduction in volume, secondly by increasing the size of tree at a given age or height, and thirdly by influencing value per cubic metre through effects on the quality of the wood produced.

3   Taking account of these points and realistic costs of respacing, it appears that the spacing likely to yield maximum net discounted revenue lies in the range 1.8–2.4 m. Since this range is close to that of much recent planting [pre-1979] there appears to be no strong case for spending money on the operation.

4   Windthrow considerations, at least in relation to the incidence of windthrow on different spacings, are not thought to affect these conclusions substantially, especially when, as is usual, plantations have been established after ploughing. There may however be circumstances in which respacing is nevertheless considered worthwhile. Apart from the case of treatment designed to remove particularly coarse trees at an early stage of the plantation's life, some respacing may be justified if it is thought that, following widespread windthrow, the sale of sawlogs will be easier and prices affected less than would be the case for small diameter roundwood.

Oceanic silviculture. Claims have been made for special growing conditions arising in respaced crops of Sitka spruce where decapitation (early and intensive thinning to waste) has been adopted (Moore and Wilson 1970; Moore 1972; Rankin 1979).

The main exponent of the so-called Oceanic system of forestry (a novel approach to the problems of windthrow and a low financial return from small diameter roundwood) was the late Major General D.G Moore. It has been applied mainly to Sitka spruce, and sets out to reap the most rewarding harvest from a stable forest grown to a planned maturity on the shortest possible rotation. This is attained by maximizing the living crown of the forest and the rooted base of the individual trees. Respacing is undertaken at the time of canopy closure (about the tenth year), and there is only one crop – the final harvest. The system has been used for over thirty-five years in Northern Ireland, and followed in such places as south-west Scotland and Yorkshire. The limiting factors are ring width for quality and top height for stability. The system does not produce gains in height or basal area; nor increase the volume of timber produced from a given area of land.

The future of Oceanic forestry is problematical. Perhaps the cost of respacing may be obviated by wider spacing (doubtful) and self-thinning mixtures. The difference in timber quality produced from non-thin and respaced crops is debatable. Respacement experiments described by Rollinson (1986) indicate that it is too early to conclude whether there will be significant long term gains in yield, quality or stability by selection respacing. From an economic viewpoint, respacing is part of establishing, and pre-commercial thinning is part of harvesting; both are investments in improved long-term revenue from the crop (Price 1989).

Chemical thinning in the pole-stage. An effect of the often unremunerative price obtainable for small conifer diameter roundwood in the early 1980s, combined with increasing costs of harvesting early thinnings, was to chemically eliminate unwanted stems in the pole-stage in order to reduce stocking density, thereby combating the silvicultural penalty of 'no thin'. The Forestry Commission undertook experiments to test the effects of arboricides to kill unwanted trees, without harming adjacent crop trees through root grafts (a phenomenon known as 'flashback'). The injection of glyphosate (Roundup) showed considerable potential (Woolfenden 1986; Mann 1984; Ogilvie and Taylor 1984). The treatment was generally one 45° downward cut per tree at a convenient height, using a light hand axe or hatchet; with the cambium thus exposed, the chemical was subsequently introduced into the cut using a garden spray bottle or a special drench gun, average 2.0 ml per tree. Trees up to 10 cm diameter had one cut; trees over 10 cm diameter had two or three cuts. The time of application was from early July until the end of December to avoid the period of active sap flow.

Chemical treatment claims to demonstrate several advantages when compared with the conventional mechanical methods of control of stand density. The risk of post-thinning windthrow is minimal (no gaps are created within the crop, and healthy trees gradually overtop their treated neighbours), the presence of dead or dying trees helps to obviate canopy damage (wind energy from swaying neighbours is absorbed), a high degree of crop quality selectivity can be achieved, and no stump treatment against *Heterobasidion annosum* is required. Moreover, gradual breakdown of trees will ensure a steady nutrient input to remaining trees during subsequent years. In an intimate mixture, chemical thinning can be used to eliminate an unwanted species. However, the main benefit of chemical thinning is in situations such as exposed sites where conventional first thinning methods cannot be adopted without inviting windthrow.

Viewed against the advantages are concerns about the long-term disadvantages of chemical treatment: the effect of the presence of large numbers of dead standing trees on forest hygiene, the extent to which 'skeleton' trees may hamper final harvesting operations, and whether treatment will show a measurable growth suppression of neighbouring stems. Whether such chemical control is seen as a short-term silvicultural expedient or an accepted and tested practice, the primary objective remains clear. Chemical stand density manipulation must aim to use the minimum necessary quantity of arboricide consistent with debilitating treated stems such that untreated trees can subsequently overtop and kill their neighbours.

The most patent use for the technique is obviously on sites of high windthrow hazard. The extent to which it is accepted for other sites will depend not only on operational costs but also future prices of small diameter roundwood and demand for different forest products. Where early thinnings are likely to incur a cost penalty, chemical treatment may represent a potential alternative to conventional thinning on windfirm sites. Other specialized uses are envisaged in overcoming problems with mixture management. However, as the markets for small diameter roundwood have improved, there is less likely to be a cost penalty in conventional thinning. Furthermore, it is unknown how respacing will affect final timber quality – it certainly increases volume per tree, but that extra may not be worth much.

Current research is directed at improving the understanding of the factors affecting tree stability and tree quality so that foresters can be given clear advice on the silvicultural options to follow, but many questions remain unanswered.

Self-thinning mixtures. Possibilities are being investigated of planting self-thinning mixtures, that is, mixing the main species with a second species or provenance which will die out soon after canopy closure. The technique is to mix a slowwith a fast-growing species, or slowand fast-growing provenances of one species (Lines 1981). Such planting, when successful, gives roughly the same results as close planting followed by respacing. There are two possible objectives. One is to use a secondary species such as larch or lodgepole pine to improve the nutrition of Sitka spruce on nutrient deficient sites: some of the results are dramatic and may enable Sitka spruce to be grown on relatively poor sites without a heavy fertilizing commitment. The other objective is to achieve some degree of natural thinning by planting in mixture a species which will gradually be suppressed and killed by the major species. On some sites both functions might be served by the same trees. Such mixtures are being investigated as a potential method of producing a forest structure which is stable against wind, with the advantage of large average tree diameters at the time of clear-felling. However, the experimental planting of special mixtures (e.g. spruce/alder) in an attempt to improve wind stability on shallow rooting soils has not proved beneficial.

9. Harvesting of thinnings

Prior to harvesting being undertaken, the trees to be removed have to be marked (unless either row thinning or cutter-selection is to be used); and for some purposes the standing volume may have to be measured or estimated. Marking and/or estimating volume may cost between £50 and £100/ha, depending on age, species and crop density. In the cutting and extracting of the thinnings (see chapter 10), usually in the pole-lengths (delimbed – snedded – with minimum if any crosscutting), careful planning is necessary to select the methods best suited to the circumstances of the operations. The most vital factors are: (i) terrain (the land as a working surface); (ii) the crop (in particular, species and tree size); (iii) markets (these determine specification of produce); and (iv) machinery and labour availability. In thinning, the planning as well as the cutting and extracting are virtually similar (though usually more costly) to what is required in clear-felling.

The foregoing extensive consideration of thinning emphasizes that the operation is a vital part of silviculture. From the economics viewpoint, thinning regimes include many interacting variables – method, timing, choice of target crop and rotation.

References

Edwards, P.N. and Christie, J.M. (1981), 'Yield Models for Forest Management'. FC Booklet 48.
Edwards, P.N. and Grayson, A.J. (1979), 'Respacing of Sitka Spruce'. *Quarterly Journal of Forestry, 73* (4), pp. 205–18.
Evans, J. (1984), 'Silviculture of Broadleaved Woodland'. FC Bulletin 62.
Everard, J. (1985), 'Management of Broadleaved Forests in Western Europe'. Forestry Commission internal report.
Ford, E.D. (1980), 'Oceanic Forestry'. *Forestry and British Timber, 9* (2), pp. 26–8; (3), pp. 40–3; (4), pp. 26–8.
Hamilton, G.J. (1980), 'Line Thinning'. FC Leaflet 77.
Hamilton, G.J. and Christie, J.M. (1973), 'Influence of Spacing on Crop Characterization and Yield'. FC Bulletin 52.
Harmer, R. (1990), 'The Timing of Canopy and Epicormic Shoot Growth in *Quercus robur* L.' *Forestry 63* (3) pp. 279–93.
Harper, W.C.G. (1986), 'To Thin or not to Thin – Optimising Present and Future Returns to the Grower'. *Forestry and British Timber,* December. 1986, pp. 22–7.
Jobling, J. and Pearce, M.L. (1977), 'Free Growth of Oak'. FC Forest Record 113.
Lines, R. (1981), 'Self-Thinning Mixtures'. FC Report on Forest Research 1981.
Mann, C (1984), 'Chemical Thinning, is it the Answer?' *Forestry and British Timber,* March 1984.
Moore, D.G. (1972), 'Management of Sitka Spruce in British Forests'. *Quarterly Journal of Forestry, 65* (2), pp. 57–62.
—— (1973), 'The Oceanic Forest'. *Scottish Forestry, 27* (2), pp. 93–120.
Moore, D.G. and Wilson, B. (1970), 'Sitka for Ourselves: the 25 Year Rotation'. *Quarterly Journal of Forestry, 64* (2), pp. 104–12.
Ogilvie, J.F. and Taylor, C.S. (1984), 'Chemical Silviculture (Chemical Thinning and Respacement)'. *Scottish Forestry, 38* (2), pp. 83–5.
Price, C. (1985), 'The Distribution of Increment and the Economic Theory of Thinning'. *Quarterly Journal of Forestry, 79* (3), pp. 159–68.
—— (1987), 'Further Reflections on the Economic Theory of Thinning'. *Quarterly Journal of Forestry, 81* (2), pp. 85–102.
—— (1988), 'One more Reflection on the Economic Theory of Thinning'. *Quarterly Journal of Forestry, 82* (1), pp. 37–44.
—— (1989), *The Theory and Application of Forest Economics.* Basil Blackwell, Oxford.
Rankin, K. (1979), 'Early Final Crop Selection'. *Quarterly Journal of Forestry, 73* (1), pp. 31–5.
Rollinson, T.J.D. (1985), 'Thinning Control'. FC Booklet 54.
—— (1986), 'A Comparison of Selective and Systematic Respacing of Sitka Spruce'. *Quarterly Journal of Forestry, 40* (1), pp. 19–25.
—— (1988), 'Respacing Sitka Spruce'. *Forestry, 61* (1), pp. 1–22.
Woolfenden, D. (1986), personal communication.

Chapter Nine
BRITAIN'S FOREST RESOURCE AND TIMBER INDUSTRY
1. Britain's forest resource

The forests and woodlands of Britain have had a long and chequered history (Hart 1966; Holmes 1975; James 1981; Anderson 1967). Many native woodlands were cut down to make charcoal (for smelting ore and forging iron), to build ships, dwellings and other structures, and to use as domestic fuel; much deforestation was for grazing and arable purposes. Much of what survived was further depleted by the requirements of the Napoleonic Wars and the two World Wars of this century. During some of the time the forests of the northern hemisphere and the British Empire provided cheap supplies of wood so that there was little incentive to maintain a domestic resource.

By the time the Forestry Commission (replacing the Office of Woods) was established in 1919 less than 5% of Britain's land area was wooded. Subsequent successive governments have consistently supported the expansion of forestry. The area of woodland has subsequently been doubled, to about 2.3 million ha or 10% of Britain's land area, mostly by the Commission as the Forestry Enterprise but substantially supplemented by the private sector. About 60% (including most of the broadleaved woodland) is now privately owned.

Over the past seventy years substantial afforestation has taken place, its nature and location being influenced by Government policies, including incentives and controls, and by a range of agricultural and environmental considerations. To minimize its impact on agricultural production, almost all of the afforestation has been in the uplands (some 90% in Scotland since 1919). Only conifers grow well on these poor and exposed sites, and introduced (exotic) species have proved to be reliable and highly productive. Of the 2.3 million ha about 70% is coniferous and the remainder broadleaved. *Productive* woodlands total about 2.130 million ha (1.515 million ha coniferous high forest, 0.576 million ha broadleaves, and 0.039 million ha coppice and coppice with standards); about 1.266 million ha are privately owned (Table 25). The private sector owns almost half of the conifer plantations, and over 90% of the broadleaved woodland (see chapter 4).

In addition Britain has about 196,000 ha of *unproductive* woodland of which 82% is in the private sector. Approximately 80% of the coniferous area is under thirty-five years old, producing moderate amounts of large sawlogs but with increasing quantities in the future. The broadleaved growing stock is generally much older and the level of production is fairly stable; the majority is found in England and Wales. The most recent forest inventory data available are the Forestry Commission *Census of Woodland and Trees 1979–82,* which provides a comprehensive account up to 1980.

Government policy is that the bulk of afforestation should be undertaken by the private sector, encouraged by grant-aid and capital taxation concessions. The Foresty

Commission's comparatively modest afforestation programme is concentrated in the more remote upland areas (in 1989–90, its afforestation totalled about 4,100 ha and the private sector's 15,600 ha). In addition, substantial areas of woodland are restocked each year after harvesting (in 1989–90, 7,900 ha by the Commission and 6,300 ha by the private sector).

Employment in the private sector is approximately: estates, 14,390; forest management companies 2,180; harvesting companies (timber merchants) 5,580; and wood processing industries 10,030. The Forestry Commission total is 7,525. The total for the whole industry is 40,245 (Thompson 1990).

Table 25. Private woodlands at 31 March 1990; measurement in 000's ha.

Country	High Forest		Coppice and Coppice with standards	Total Productive Woodland	Other Woodland
	Conifers	*Broadleaves*			
England	205	393	37	635	88
Wales	52	55	2	109	9
Scotland	444	78	0	522	65
Great Britain	701	526	39	1,266	162

Source: Forestry Commission 'Forestry Facts & Figures' 1989–90.

A substantial national asset has been created which supplies wood for domestic processing, gives employment, sustains rural economies, provides a wide range of opportunities for public enjoyment, and is of increasing value for wildlife. The Government has also introduced effective measures to maintain and expand the country's broadleaved and native pine woods. As afforestation is the major cause of land use change, there has been some controversy associated with the expansion of the forest area despite its many economic, social and environmental benefits.

The maintenance and expansion of the State forests are funded by timber and other revenues, supplemented by annual grant-aid from the Government. In return, the public receive, in addition to the conservation of the country's woodland heritage, an improvement in the many benefits flowing from a multipurpose industry: an increasing supply of wood, supporting a progressive expansion of the processing industries, more jobs, many in rural areas with high unemployment, and a more diverse forest estate, better able to cater for public enjoyment, richer in wildlife and in close harmony with the landscape. The future for the whole industry is generally promising and is taking a new lease of life from an increasing recognition of the multiple benefits it supplies.

The case for national forestry. The proportion of Britain's land devoted to forestry, 10%, is among the least in the Economic Community (see chapter 17, Section 4 and Table 52). Only the Netherlands and Eire have a lower proportion; and the average for the EC is 25%. The UK's annual production (mainly conifers) is about 12% of consumption (the EC as a whole imports about 50% of its consumption). In 1990 British imports of wood and wood products cost over £7 billion.

Britain's combination of climate, soils and terrain is unsurpassed for tree growth in most of Europe and Scandinavia, tree growth being at least twice as fast as that enjoyed by many overseas competitors. With the exception of Eire, the growth rates from conifers in Britain are higher than anywhere else in Northern Europe. Assuming no increase in real prices of timber, the Forestry Commission is achieving an average real rate of return over the whole rotation of its estate in excess of 3%, with 4–5% being earned on the best sites (see chapter 14, section 4).

Most European countries have long traditions of forestry and are less concerned with rates of return than with achieving sustained yield and reasonably balanced cash flow. As a result, calculations of this nature are rare on the Continent, though those which do exist indicate rates of return of around 1% for countries such as Germany, Finland and Sweden. Even under the best conditions found in Europe, 3% is regarded as a good return. In the absence of country estimates, an indication of the rate of return is given by the mean yield class. In general terms yield class shows a decline along a line from SW to NE across Northern Europe. Sitka spruce in Northern Ireland has a mean yield class of 14. In Britain the mean is 12, but for Norway spruce in Finland and the USSR the figures fall to around 4, and 2 in the extreme NE. In Scandinavia, growth rates for conifers are in the range of 4 to 6 m^3/ha/yr whereas in Britain they are about 10 to 12 (the average for broadleaves being probably 4 to 6 m^3/ha/yr.)

The foregoing facts support the case for an expansion of national forestry in Britain (the case for complementary private forestry is given in chapter 17, section 1), and the relevance of such a policy of expansion is further illustrated by noting the following significant benefits of forestry (Francis 1989):

- The reduction on reliance upon other countries for wood;

- A renewable supply of wood to underpin the growth of the domestic wood processing industry, and established sustainable employment and economic activity, which supports rural communities and diversifies their economic structure;

- Opportunities for a wide range of sporting and recreational activities;

- Diversification of landscapes and provision of new habitats for wildlife, assuming that appropriate environmental safeguards are observed;

- Help in reduction of the burden on taxpayers and consumers created by food surpluses, when land is converted from agriculture to forestry;

- Demonstration that the UK is taking steps to reduce its reliance on the world's natural forests whose environmental importance is now seen to be so crucial; and

- Contribution towards a moderation of the process of global warming.

Discussions of the economic case for forestry are frequently beset by controversy: about discount rates; about free trade versus self-reliance; about whether downstream industrial development is beneficial if it requires subsidies to stay in operation; about the importance of multipliers, linkages, surplus labour, the terms of trade and other economic concepts (Grundy 1985). Some environmental economists assert that import-saving is not a strong economic argument for forestry. However, the contribution of forests to containing the 'greenhouse effect' is becoming increasingly important (see chapters 3 and 14). Forests have also indirect benefits, such as the watershed protection function. Their social benefits are more fully discussed in chapters 13 and 14.

The case for further afforestation is strong. Britain has significant areas of suitable and potentially surplus farmland upon which forests and woods could be developed. Contribution to self-sufficiency is steadily rising year by year but is unlikely to rise beyond 25% by the turn of the century. Meanwhile the impact of forestry on output and employment in the UK is substantial (McGregor and McNicoll 1989), and much of forestry policy continues to be the subject of intense debate.

In 1989 the Fraser of Allander Institute (University of Strathclyde) produced a report on *The Impact of Forestry on Output and Employment in the United Kingdom and its Member Countries,* showing that forestry, taken together with its processing sectors, has significant connections to many supplying industries as well as to a variety of customers (see Figure 21). It quantifies the 'knock-on' effects of the buying and selling behaviour of forests and related activities on the UK economy as a whole and on its member countries. The main results of the study, which relates to the UK in 1984, are:

- The value of output of forests was £384 million;

- The total impact of forests and dependent activities on output was £1,954 million;

- The number of full-time equivalent jobs in forests was 11,800.

- The total impact of forests and dependent activities on employment was 55,000; and

- The distribution of forestry's impact on all sectors of the economy was England 74.0%, Scotland 16.7%, Wales 6.3% and Northern Ireland 3.0%.

Subsequently, at least £750 million of investment has been attracted to the processing sector of the industry, a development which has increased the dependence on UK wood supply and has expanded the impact of forestry on the UK economy.

The Government's forestry policy as reflected through the Forestry Commission's two main aims is (i) to achieve an expansion of the area of productive forests so that the availability of the benefits conferred by forests and woods is increased; and (ii) to encourage the sustained management of both public and private sector forests and woods so that the supply of these benefits is maintained. The Government implicitly recognizes the value of the whole forest industry (producing and processing) and supports and encourages its growth.

2. Britain's timber industry

Wood has diverse uses, including paper-making, panel boards, construction, furniture, packaging, fencing, and mining timber. Among a wide range of other uses are transmission poles, wood wool, turnery and fuel. Bark and residues from sawmills are recyled to other industries.

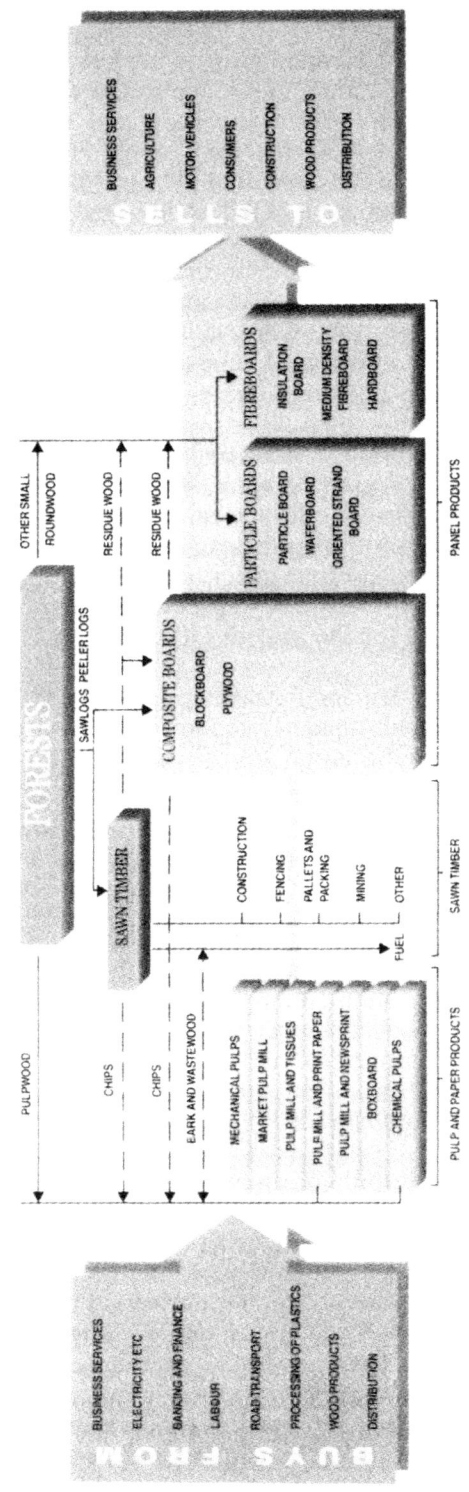

Figure 20. The structure of the forestry industry
Source: *The Impact of Forestry on Output and Employment in the U.K. and its Member Countries*. A report by the Fraser of Allander Institute (University of Strathclyde), 1989.

The UK uses annually about 50 million m³ of wood raw material equivalent (see Table 26), of which about 6.2 million m³ (about 12%) is home produced (see Table 27). Demand and supply in the wood-using industies are subject to fluctuations. The industry suffered severely in the early 1980s due to the economic recession. Three pulpmills closed (Fort William, Ellesmere Port, and St. Annes, Bristol) which together had annually consumed 600,000 m³ of small diameter roundwood. The timber market and sawmilling were depressed.

By contrast, the subsequent development of the wood-using industries in Britain is a considerable success story. During 1987–89, following a promotional study adopted by the Foresty Commission with the support of the Industry Departments and various development agencies in England, Wales and Scotland, the depressed condition was transformed by large-scale private investments in paper and woodbased panel mills and some major sawmills. A vital element in attracting the new investments was the ability of the Forestry Commission to guarantee long-term supplies of roundwood.

The increased domestic production of roundwood stimulated the establishment of a strong industrial base for wood processing, chiefly paper-making, panelmaking and sawmilling. All the main relevant products are now produced in Britain. Sawn timber, mainly softwood, is produced in some 400 sawmills, widely dispersed across the country. There are four paper mills, and ten panel-producing factories – among the most up-to-date and competitive in the world. These developments more than make up for the previous lost markets for roundwood.

Timber demand. The UK's annual demand for sawlogs (mainly coniferous) is about 3 million m³. The demand for small diameter roundwood (mainly coniferous) is likewise about 3 million m³ (confirmation of this is difficult because not all wood-processing mills publish consumption figures distinguishing between roundwood and residues.) Firewood, mainly broadleaved, amounts to about 0.2 million m³ annually.

Softwood accounts for about 90% of all the timber used in Britain: most of the wood for construction, fencing, packaging, pitprops and the manufacture of paper and panels, comes from coniferous forests. Over recent years, British-grown softwoods have been achieving a gradually increasing share of many of the market sectors, a trend which will gather momentum over this and the next decade as (i) increasing supplies of well-managed timber become available; (ii) there is an even greater awareness of the properties and uses of British-grown softwood timber; and (iii) even more widespread attention is paid to good presentation, particularly of sawn products. Whiteman (1991) has updated the trend of demand.

Timber supply. The UK's annual supply of roundwood is about 6 million m³ (see Tables 26 and 27). Of sawlogs, conifers amount to about 3 million m³ and broadleaves 0.60 million m³; of small diameter roundwood, conifers amount to about 2.3 million m³ and broadleaves 0.25 million m³. Of the 6 million m³, about 34% represents the private sector and 66% the Forestry Commission. Conifer supply is expected to rise to almost 9 million m³ in the early 2000s (about 38% private sector and 62% Forestry Commission). Together with 0.9 million m³ broadleaves the total will be about 10 million m³. Present indications are that there exists in some parts of the country greater local demand for roundwood than can be met, and a number of processing mills are buying supplies from areas well outside their normal catchment area. Whiteman (1991) has updated the trend of supply.

Table 26. Volumes of UK (a) imports, (b) production and (c) apparent consumption of wood and wood products during calendar year 1989.

			Wood raw material equivalent (millions m³ ub)
a	Imports		
	Wood: roundwood and sawnwood		18.6
	Pulp		8.9
	Panel products		6.1
	Paper		14.4
		Total	48.0
b	UK production		6.4
c	UK apparent consumption		50.3

Source: Forestry Commission 'Forestry Facts & Figures' 1989–90.

Table 27. Volumes of deliveries of British-grown roundwood to wood processing industries during calendar year 1989.

	000's m³ ob
Sawlogs: conifer	3,045
broadleaved	615
Round pitwood	35
Particleboard	955
Fibreboard	15
Paper and paperboard	1,165
(of which exports)	(160)
Posts, stakes, rails, wood wool and exports of roundwood other than pulpwood	175
Total industrial wood	6,005
Fuelwood	220
Total	6,225

Source: Forestry Commission 'Forestry Facts & Figures' 1989–90.

Future timber supply and demand. The most important influences on demand are: price, prices of substitutes, national income and investment, and levels of output in certain industries. Whiteman (1991) considers it likely that timber production will rise in line with consumption up to 2025. After that, if no afforestation is undertaken, production is set to fall, while demand will continue to rise. The sawmill capacity is adequate to accept all available sawlogs; and the current four pulpmills and ten wood-based panel mills can utilize all available production of small diameter roundwood and residues from sawlogs. The British timber industry is a boon to growers: assured and expanding markets for both roundwood and sawnwood in safe and long-term stable types of industry provide encouragement for afforestation, restocking, and the orderly harvesting of existing growing stock. Whiteman (1991) has updated the trend of demand and supply.

Conifer sawmilling. In the sphere of sawlogs (i.e. chiefly the butts of thinnings and final crop trees), the most active sector of the British processing scene is sawmilling. Domestic sawmilling, mainly softwoods, has expanded almost twenty fold in as many years (Elliot 1988; Burd 1989). There has been a massive increase in the volume of British-grown and British-sawn softwood replacing foreign timber. British sawmillers are now meeting internationally accepted standards – especially with regard to dimensional accuracy, strength grading and general presentation. Their share in the British market in construction and carcassing is increasing substantially; likewise in the production of pallets, packaging, fencing and agricultural building material. British sawmillers also meet the stringent requirements of service and delivery necessary to compete with the most efficient producers in a worldwide market. Through all these efforts British timber is gradually becoming more and more accepted in the construction industry; and the scope for further movement into this major sector is considerable since the sawnwood penetration to date only represents about 10%. The Forestry Commission helps the conifer sawnwood industry by log grading before they leave the forest (noted later in this section). This improves sawn timber recovery and the yield of timber which meets machine-graded standards necessary.

Heavier investment in kiln-drying of timber for the construction market will be essential as the single market in the European Community in 1992 moves nearer, requiring the harmonization of standards for sawnwood products. This will supplement the already followed machine stress grading.

The end uses of conifer sawnwood (about 2.7 million m^3 annually) are chiefly construction, furniture, fencing, pallets, mining and packaging. The processing of the sawlogs results in residues totalling about 1.23 million m^3 used chiefly for paper and panels.

Broadleaved sawmilling. This much smaller sector has remained stable despite the loss of elm and the advent of windthrown trees, but the grade of sawlogs is concentrated on the low value-added markets of end use such as fencing, pallets, mining and packaging; the higher value markets are joinery, furniture and construction. The whole amounts to about 0.55 million m^3 annually. The processing of the sawlogs results in residues totalling about 0.12 million m^3, used chiefly for paper, panels and fuel. The majority of the best quality broadleaved logs are exported for veneering, but many prime logs for planking and beams are processed in Britain. Supply of broadleaves is limited, and production is expected to remain fairly constant at about 0.9 million m^3 annually (both sawlogs and small diameter roundwood) mainly in the private sector. There is sufficient demand to absorb what is available, dependent on region. Broadleaves, important though they are locally, are unlikely to increase substantially except as a consequence of conservation management. A potential for increasing supplies from broadleaved woodlands are the current measures within the Forestry Commission's broadleaved policy.

Small diameter roundwood processing. The end uses of broadleaved small diameter roundwood (about 0.45 million m^3 annually) are chiefly paper, panels, fencing, turnery and firewood. Those of coniferous small diameter roundwood (about 2.3 million m^3 annually) are chiefly paper, panels and fencing. The industry is discussed in section 4 of this chapter. Here the demand has developed even more spectacularly than from sawmilling, through the major investments in newsprint manufacture, lightweight coated paper and novel board products, such as oriented strand board, medium density fibreboard and cement-bonded particleboard. Expansion and demand underline the significance of the small diameter roundwood sector (both conifer and broadleaved) to the growth of the British forest products economy. Furthermore, in recent years up to 30% of the volume of logs processed by the sawmilling industry has been recycled as chips and residues for the paper and panel mills, representing much improved utilization of the harvested resource. The contribution to the profitability of sawmilling has been and continues to be notable.

Future timber production. Britain's output of timber from existing forests will increase substantially during the present and following decades as felling of mature timber approaches the maximum for the area planted. However, production is unlikely to rise beyond 20–25% of consumption by the turn of the century. Forecasts of the roundwood production are indicated in Table 28.

Table 28. Forecasts of roundwood production in Britain; volumes in 000's m^3 ob.★

Period[†]	1987–91[†]	1992–96	1997–2001	2002–2006
Conifers:				
Forestry Commission	3,400	3,700	4,600	5,400
Private Sector	1,700	2,300	2,800	3,400
Total	5,100	6,000	7,400	8,800
Broadleaves:				
Mainly Private Sector (approximately)	900	900	900	900
Total Conifers and Broadleaves	6,000	6,900	8,300	9,700

★ One m^3 ob equals approximately 0.875 m^3 ub and weighs 0.9 tonnes when freshly felled.
[†] An average annual figure for each of the 5 year periods quoted.
Source: Forestry Commission 'Forestry Facts & Figures' 1989–90.

Coniferous wood production, as indicated in Table 28, in both state and private forests is forecast to rise from an average 5.1 million m^3 ob per annum in the quinquennium 1987–1991 to 8.8 million m^3 ob per annum in the quinquennium 2002–2006; broadleaved production is expected to remain constant at about 0.9 million m^3 ob per annum, mainly in the private sector. Hence, production from existing forests will increase to almost 10 million m^3 annually in the next century. Production from the current conifer planting will begin to supply small diameter roundwood from thinnings in about twenty years. The forecasts are standing volumes in cubic metres over bark. They have been computed in three parts: first, the volume of wood of 18 cm diameter and above; second, the wood between 14 cm and 18 cm diameter; and third, the wood volume between 7 cm and 14 cm diameter. A convention has been used whereby the volume in the 18 cm category is classed as sawlogs, while the remainder of the 14 cm – 18 cm is regarded as being composed of small diameter roundwood. Of necessity, this is a rather arbitrary division of the standing volume between sawlog and small diameter roundwood which in practice will vary according to the technical requirements of individual processors. In order to estimate the extracted volume, allowance must be made for the gross difference between standing and extracted volumes which includes not only waste but covers the effects of different measurement methods and conventions. Where sawlogs are sold in underbark terms, a further reduction must be made to allow for the volume of bark.

The Commission's 1987 forecast was based on its current management practices and takes account of crop stability, recreation and landscape considerations. Such factors generally but not invariably have the effect of deferring yield but they may be further modified by market and contractural obligations. Further, the growing stock was adjusted to take account of the disposal of plantations under the Commission's rationalization programme.

The wide variation in the management objectives of private owners makes the private sector forecast necessarily less precise. The forecast was prepared by TGUK in consultation with the Forestry Commission and was based on the 1972–82 Census of Woodland updated to take account of the estimated felling and planting which has taken place since then. Commission disposals to the private sector have been added to the database. The 1982 forecast was of *potential cut* and took no account of retentions by estate owners for amenity, conservation or fiscal reasons. Experience showed that this caused some confusion when interpreting the data. Therefore a forecast of *potential removals* was produced by making a deduction of 10% from the potential cut to allow for timber unlikely to be removed from the forest.

Overall, assessments of coniferous and broadleaved roundwood production indicate that, until the mid-1990s and perhaps slightly beyond, there is an approximate match between the installed and planned industrial capacity and the growth of supplies. The Forestry Commission therefore found it necessary in 1988 to urge caution in the consideration of future plans for any additional processing capacity. There is, however, a lack of markets in the south-east of England.

Though many species are grown, commercial planting of conifers has been largely of nine species providing four timber types – two spruces, three pines, Douglas fir and three larches. (The most important of the so-called 'minor species' are western hemlock, western red cedar, grand fir and noble fir.) Annual production of the nine major conifers is expected to increase from the current (1991) 4 million m^3 to more than 7 million m^3 by the year 2000. Well over half of the total production will be of Sitka spruce and Norway spruce. European larch is expected to provide only a modest volume and decreasing proportion of the British timber supply; and further supplies of larch will be mainly Japanese. Because of limited seed supply, planting of hybrid larch has been at a low level and timber supply for some decades will be small in comparison with that of Japanese larch. Lodgepole pine is of comparatively recent use in commercial British forestry and as yet is not a significant contribution to the market.

Forecasts of demand for small diameter roundwood and wood residues are indicated in Table 29. Compared with the forecasts of availability based on the quinquennial review made by the industry through the Home Grown Timber Advisory Committee, the potential supply variance is expected to be as indicated, but various factors may create changes in the situation (see Whiteman 1991).

Table 29. Forecasts of demand for small diameter roundwood and wood residues, excluding pitprops, fencing, export, firewood and miscellaneous.

	Million green tonnes					
	Actual	Forecast of demand				
	1989	1990	1991	1992	1993	1994
Available supply						
Small diameter Roundwood	2.067	2.156	2.530	2.733	2.897	3.034
Wood residues	1.297	1.414	1.488	1.535	1.565	1.627
	3.364	3.570	4.018	4.268	4.462	4.661
Potential supply variance small diameter roundwood (+ = surplus; − = shortfall)	+0.092	+0.276	−0.012	−0.135	−0.201	−0.262

Source: UK Wood Processors Association (January 1991).

Forecasting and achieving production of roundwood. The production (output) of timber involves forecasting how much will be produced (grown), and harvesting the crop. This has been noted earlier. It is important to be able to predict future timber yields, both for the development and usage of timber-using industries and to plan for the optimum labour and machinery requirements within forests.

Sawmillers and other processors of roundwood have the right to expect service as do their customers. Foresters being the first link in the long chain to the eventual wood product user, must take a responsible attitude and must be market orientated. A modern sawmill, and even more significant a paper or board mill, adequate to meet customer demands for quality and efficiency of production, represent a huge financial investment. Hence the timing of such a development is very critical and absolutely dependent upon the accuracy of forecasting future sawlog and small diameter roundwood production; also upon the careful control of harvesting to achieve the forecast not in the limited sense of the overall volume in Britain but to volume within a given location, species, size and quality (Burd 1989).

All four factors are vital to the sawmillers and processors. Given this information and surety of production the grower can rest assured that the industries have the technical knowledge and skill to convert Britain's forest resource into first class products.

It is possible to predict future production from the growing stock with a reasonable degree of confidence given certain assumptions about growth rates (yield classes), management regimes and, of course, price levels. But all these can and do change, so forecasting future production is not an exact science; at best it gives a considered view of the future of supplies as seen at a particular time.

The realization of the full potential cut depends on many factors which influence harvesting and marketing decisions, especially the existence of markets and the opportunities to develop new ones. Historically, the Forestry Commission has been able to harvest annually between 90% and 105% of its five-year average forecast. On the other hand in the private sector, composed of a large number of individual owners who manage their woods for a variety of purposes, the level of actual cut compared with potential is less certain (Kupiek and Philip 1989).

Since production forecasts are dependent on assumptions about future treatment of crops, particularly rotation lengths and estimates of yield class, there is a need to monitor current practice so that if practice begins to change and differ significantly from the assumptions used in the forecasts then the forecast can be quickly revised. However, it would seem prudent to base such assumptions on recent and current behaviour rather than on statements of intentions which may never be fulfilled (Dewar 1988; Malyon 1989). The next production forecast is due to be published in 1991 to take effect from the year 1991–92.

3. Properties of British-grown timbers

The properties of any timber show considerable variation depending on its seed origin, condition of growth, age and other factors. The uses obviously depend on the properties. Data for commercially important British softwoods have been made available by Harding (1988). Explanations of the terms used in classifying the basic properties are given below, relying mainly on information from the Timber Research and Development Association (1987) and the Building Research Establishment (Harding 1988).

Density is a measure of dry wood substance per unit volume in the wood. It also varies with the amount of water contained in a piece of wood and for this reason it is important that when the density of a timber is stated, the moisture content of the timber should be cited. For most purposes the nominal specific gravity is used, which is defined as the green volume over the oven-dry weight; however, in this chapter the densities of the timbers when dry relate to a moisture content of 12%.

Surface texture is classified into fine, medium or coarse – an arbitrary classification involving opinions. *Shrinkage* in the tangential and radial directions is obtained when plain-sawn and quarter-sawn boards are dried from a green condition. Conventionally, shrinkage is expressed as the percentage reduction in the green dimension on drying to a moisture content of 12%.

Movement refers to the dimensional change which takes place when seasoned timber is subjected to change in atmospheric conditions; and is classified as small, medium, or large. Movement values give some indication of how the dried timber will tend to behave when subjected to atmospheric changes in service. A so-called 'stable' timber is one that exhibits comparatively small dimensional change. Most British softwoods are in this category. *Working quality* refers to the ease of machine working and is generally classified as good, medium, or difficult. A difficult classification does not mean that a timber is unworkable but indicates that particular care should be taken in machining. Additional comments on physical characteristics of wood are noted below (Johnston 1979, 1982, 1983).

Knots can degrade the value of timber. Tight live ones are better than loose dead ones, but all are defects. Knot size, largely a function of species, and, to a lesser extent, the number and shape, can be influenced by foresters. The shape of knots is determined by branch angle so that a horizontal branch of a given diameter produces a smaller area of knot in a longitude plane than a similar branch growing in a more nearly vertical plane. This aspect of tree form is highly heritable and can therefore be influenced in a tree-breeding programme. To a small degree the number of knot whorls is under the control of foresters because branching can be affected by fertilizing. But the most important aspect of knots is their diameter and this is highly correlated with initial spacing and with thinning intensity. The more competition a tree encounters, the smaller will be the knots in the timber. From crops having a wide initial spacing and an intensive thinning regime, clear timber can only be produced if the trees are manually pruned. Knots reduce strength by occupying part of the cross-section of a timber member, and further reduce strength, bending and tensile by causing the surrounding grain to become distorted. Of perhaps even greater importance is the effect of knots and their associated grain on the seasoning properties of timber and on its stability in service. Another serious disadvantage of knots is their effect upon timber working properties: they quickly blunt tools, exude resin and are an indirect result of the tearing which results when their associated contorted grain is machined or planed.

Grain angle. There are two major causes of irregular grain – genetical and silvicultural. Many species or individual trees within species produce various forms of irregular grain. Some – such as 'fiddle back figure' in sycamore – have a commercial value but the more commonly occurring spiral grain has no advantage. It is, in part at least, genetical and can be partly eliminated in tree-breeding programmes and in thinning operations. It is also particularly associated with juvenile wood *(infra)* so that if there is a large proportion of juvenile wood in a log there is a greater chance of the wood twisting after it has been sawn and is in service. Another form of irregular grain is the sloping grain associated with a high degree of taper in the stem of the tree. Trees are almost invariably sawn parallel with the long axis of the stem so that a tapered tree produces timber having a sloping grain. This is a disadvantage both in sawing and machining and also in service where the longitudinal movement will be greater and less regular than with straight grain. The strength of a structural member having sloping grain is also less than one having straight grain. Faulty sawmilling can cause an undesirable steep slope of grain.

Juvenile wood, that is, wood from near the centre of the tree, tends to be less dense, less strong and more knotty than the mature wood, and in conifers tends to have a more spiral grain. The boundary between the two is gradual but the juvenile characteristics are only pronounced in the ten to twenty growth rings in the centre of the stem.

Reaction wood, i.e. *tension wood* in broadleaves and *compression* wood in conifers, is not usually considered to be of much significance although in craft furniture making it can be troublesome. Irregular spacing and irregular branching lead to stresses in the tree, and the same effect may be caused by wind sway. Compression wood is most likely to be serious, however, in leaning trees or those subject to butt sweep.

Rot of many kinds affects wood, the most important in conifers being *Heterobasidion annosum*. There has been a considerable volume of research on this disease and effective measures have been developed to limit its effect upon British forests (see chapter 6). Most other rots are of little importance under *forest conditions*. Decay invariably involves a loss of quality through the breakdown of cellulose and lignin by various wood rotting fungi. In advanced decay, the breakdown is obvious. In its early stages it may show principally as a discoloration, often reddish-brown, of the wood. *Sapstain,* often affecting windthrown trees and logs delayed in conversion, is the term given to a discoloration of wood that does not involve a reduction in strength. It is often blue in colour and is typically due to the colonization of the wood by species of fast-growing non-decay fungi. Rarely it may be of mineral origin. For structural timber, stain is not considered to be a defect, but the appearance of the wood could lead to a reduction in value. *Insects* such as longhorn beetles can be the cause of direct degradation through the wood-boring activities of their larvae. However, bark beetles are chiefly important as a means whereby stain fungi are transmitted to fallen or felled trees. Natural durability and preservative treatment are discussed in chapter 10. Other defects in round timber which generally cause degrade in value include *resin pockets, fluting* and *shake*. Oak and sweet chestnut in particular are prone to shake. It causes an important reduction in timber value. In oak, the soil type in which it is grown appears to be the factor most strongly linked with the occurrence of shake. Highest shake incidence has been found on stony soils derived from a variety of parent materials (Henman and Denne 1988). The effect may be heritable, and external indicators of shake-prone trees exist (Savill and Mather 1989, 1990).

Chemistry of wood. *Cellulose* (about 50% of the wood) is a glucose polymer with 1 to 4 glucosidic linkages. It is the major component of the cell wall, formed as long unbranched chain molecules which build into the ultra structural architecture of microfibrils. Its configuration is that of a tough crystalline form not so readily attacked by chemical or biochemical agents as the carbohydrate fraction of starches or sugars. *Hemicellulose* (about 25% of the wood) is a polymer of mannose (principally in conifers) and xylose (principally in broadleaves). Although simpler in chemical structure it has a complex physical shape of a short well-balanced chain molecule arranged in close proximity to the long chain molecules of cellulose. *Lignin* (17–20% of the wood) is a polymer of substituted phenyl propane units of considerable variety, serving as a cement between adjacent wood cells, a stiffening agent within the cell walls, and a barrier to enzymatic degradation of the cell wall. It has an extremely complex physical structure intimately linked with the cellulose of the cell wall. *Ash* (2–5% of the wood) consists of a wide variety of elements, Ca, K, Mg, P, Si, S and trace elements. *Water and solvent extractives* include some of the simpler lignin compounds and consist of a considerable mixture of organic chemicals, including acetic acid, alcohols and phenols. *Tannin* is included in the mixture and is principally a phenol-based compound. (In some conifers overseas, extractives are tapped for the naval stores industries as oleo-resins which are important industrial chemical bases, e.g. for oil and turpentine.)

Chemical pulping attempts to delignify the cell wall leaving the physical structure and cellulose content of the cell intact. Spent cooking liquors therefore contain large quantities of chemically treated lignin and are a rich store of chemical by-products from wood. Recovery of specific chemicals from spent liquor is a specialized science.

Properties of the main British-grown timbers together with a brief summary of their uses are listed below. It should be remembered that the properties of the sapwood are different from those of the heartwood of the same timber. For additional information on conifers (softwoods) see Building Research Establishment (1983), and on broadleaves (hardwoods), Farmer (1986). Other useful sources are Aaron and Richards (1990) and Brazier (1990). For general treatises on commercial timbers, see Desch (1986), Patterson (1988) and Wilson and White (1980). The uses briefly indicated (and supplemented in section 4 of this chapter) do not constitute a market, and while a particular type of timber may be very suited for a specific purpose a demand may not exist. Even where a market exists for a particular type of timber, it does not follow that it is open to all potential suppliers.

CONIFERS (SOFTWOODS)

Properties of timber	Uses (in addition to firewood)	
D= Density (average kg/m^3) at 12% moisture content. T= Texture. MM= Moisture movement. WQ= Working quality. ND= Natural durability of the heartwood. PP= Permeability of heartwood to preservative treatment. Colour and other features as aids to identification of timbers are given in chapter 1.	I= Impregnate with preservative★ P= Peel (debark)	
	Small diameter roundwood and poles (including tops of sawlogs)	Sawnwood; or Veneer (rarely conifers, and for export only)
Scots pine D: 510; T: coarse; MM: medium; WQ: medium; ND: non-durable; PP: moderately resistant (but sapwood permeable). Versatile, suitable for a wide range of uses. Used for many structural purposes, including some of the more demanding applications. Suitable for joinery and furniture. Although not durable, easily treated with preservative and ideally suited to applications where there is a high decay hazard. Preferred species for telephone and transmission poles. Very suitable for the production of wood wool cement slabs because it shreds well and does not inhibit cement setting. Works easily and cleanly in most hand and machine operations, but sharp cutting edges required for the faster grown timber with its wider bands of soft springwood. Knots can be troublesome when dry	Posts, stakes – (P and I); rails; wood wool (P); pitprops (P); transmission; poles (P and I); chipboard; pulpwood; orientedstrandboard; fibreboard; bark products.	Furniture. Construction – (trussed rafters, general carcassing and framing); flooring; joinery, sleepers (I); fencing (I); gates, posts (I); pallets; mining timber – (chocks, coverboards, sleepers, lids, etc.); boxes; general purposes.

Properties of timber	Uses (in addition to firewood)	
D= Density (average kg/m³) at 12% moisture content. T= Texture. MM= Moisture movement. WQ= Working quality. ND= Natural durability of the heartwood. PP= Permeability of heartwood to preservative treatment. Colour and other features as aids to identification of timbers are given in chapter 1.	I= Impregnate with preservative* P= Peel (debark)	
	Small diameter roundwood and poles (including tops of sawlogs)	Sawnwood; or Veneer (rarely conifers, and for export only)
as they are liable to fall out during sawing and planing. Finishing and gluing, and nailing generally satisfactory. Dries very rapidly and well, with little degrade, but, because of its susceptibility to blue stain, logs should be converted and the sawn timber loaded into the drying kiln without delay. In demand as sawlogs; and the small diameter roundwood for pulpwood and boardwood.		(The 'redwood' of the imported timber trade.)
Corsican pine D: 510; T: coarse; MM: small; WQ: medium; ND: non-durable; PP: moderately resistant (but sapwood permeable). Similar in appearance to Scots pine, but larger proportion of sapwood. For most structural and other purposes can be used instead of Scots pine, but is less commonly used for transmission poles. Working properties, including nailing and finishing, similar to Scots pine, but the knots are slightly less hard and usually hold better during cutting operations. Shrinkage and movement values are slightly less than in Scots pine. Dries rapidly and well, though it is subject to stain if drying is delayed. In demand as sawlogs; and the small diameter roundwood for pulpwood and board wood. The preferred species for wood wool.	Posts, stakes – (P and I); rails; pitprops (P); transmission poles (P and I); chipboard; pulpwood; orientedstrandboard; fibreboard; bark products.	Construction – (trussed rafters, general carcassing and framing); joinery; wood wool; boxes; pallets; fencing (I); gates, posts (I); mining timber; general purposes.
Lodgepole pine D: 460; T: medium; MM: small; WQ: good; ND: non-durable; PP: moderately resistant. Knots are generally small and tight. Dries rapidly and well, with no checking or splitting. Normally free from distortion in drying. Resin exudation is fairly common. Strength properties between those of Scots pine and spruce. The heartwood is more resistant to treatment than that of Scots pine. Works easily with hand and machine tools. Finishes cleanly and takes nails satisfactorily. Sawnwood is used in	Posts, stakes – (P and I); rails; pitprops (P); chipboard; pulpwood; Orientedstrandboard; fibreboard; bark products.	Construction – (general carcassing and framing) joinery, flooring; wood wool; fencing (I); pallets; packaging; gates, posts (I); mining timber; general purposes.

Properties of timber	Uses (in addition to firewood)	
D= Density (average kg/m³) at 12% moisture content. T= Texture. MM= Moisture movement. WQ= Working quality. ND= Natural durability of the heartwood. PP= Permeability of heartwood to preservative treatment. Colour and other features as aids to identification of timbers are given in chapter 1.	I= Impregnate with preservative* P= Peel (debark)	
	Small diameter roundwood and poles (including tops of sawlogs)	Sawnwood; or Veneer (rarely conifers, and for export only)
building construction work and some types of joinery. In limited supply as sawlogs; and the small diameter roundwood for pulpwood and boardwood.		
Douglas fir D: 530; T: coarse; MM: small; WQ: good; ND: moderate; PP: resistant to extremely resistant. Straight. Fair loadbearing. Fair strength-to-weight ratio in compression and bending. Tendency to wavy or spiral grain. More difficult to treat than pine. Incising before treatment gives satisfactory results with items of larger cross-section, such as railway sleepers. Given an effective preservative treatment, will give excellent service in demanding exterior conditions. Dries rapidly and well, with little checking or distortion. Strong and, while not quite the equal of pine in bending strength, is comparable in stiffness. Works readily with hand and machine tools, but not as easy as pine. Hard, loose knots are liable to damage cutting edges. A good finish can generally be obtained, although there is a tendency for fast-grown material to splinter and break away at the tool exit where the cut is across the grain. Takes nails and screws satisfactorily, but care needed to avoid splitting. Available in large sections and is primarily a construction wood. Selected timber is used for joinery purposes, where its stability in service and appearance are particularly advantageous. In great demand as sawlogs; and the small diameter roundwood for pulpwood and boardwood. Cleaves satisfactorily when young.	Posts, stakes – (P and I) rails; telegraph poles (P and I); pitprops (P); flag poles (P); chipboard; pulpwood; fibreboard; bark products.	Construction – (general carcassing and framing) joinery, flooring; sleepers; fencing (I); packaging; plywood (mainly imported); vats and tanks; gates, posts (I); mining timber; general purposes. (The 'Oregon pine' of the imported timber trade.)
Sitka spruce D: 390; T: coarse; MM: small; WQ: good; ND: non-durable; PP: resistant. Most Sitka spruce is fastgrown. It is non-durable and both the sapwood and heartwood are difficult to treat	Posts, stakes; rails; pitprops (P); chipboard;	Construction – (general carcassing and framing); joinery (interior; not

Properties of timber	Uses (in addition to firewood)	
D= Density (average kg/m³) at 12% moisture content. T= Texture. MM= Moisture movement. WQ= Working quality. ND= Natural durability of the heartwood. PP= Permeability of heartwood to preservative treatment. Colour and other features as aids to identification of timbers are given in chapter 1.	I= Impregnate with preservative★ P= Peel (debark)	
	Small diameter roundwood and poles (including tops of sawlogs)	Sawnwood; or Veneer (rarely conifers, and for export only)
with preservative solutions by pressure impregnation. Roundwood can be treated by sap displacement methods, but complete sapwood penetration is not always achieved. Used for a wide variety of applications. Its light weight, strength, nailability and non-tainting characteristics make it ideal for packaging, particularly for fruit, foodstuffs and similar commodities. Because of its pale colour, good fibre characteristics and low resin and other extractive content, spruce is the preferred wood for making high quality mechanical pulp used in newsprint and other paper products. Dries rapidly, but care is needed in drying if distortion (twist and cup) and degrade (splitting and loosening knots) are to be minimised. In demand as sawlogs; and the small roundwood for pulpwood and boardwood. Residues command a premium.	pulpwood (first class); fibreboard; bark products.	high grade); building; sheds; wood wool; mining timber; boxes; crates; cable-drums; packaging; pallets; gates, posts (I); general purposes.
Norway spruce D: 390; T: medium; MM: small; WQ: good; ND: non-durable; PP: resistant. Properties similar in many respects to those of Sitka spruce. Non-durable and resistant to preservative treatment by pressure methods. An effective penetration of the sapwood can be obtained when roundwood is treated by sap replacement methods. Dries rapidly and well, with little tendency to split or check but, like Sitka spruce, with some risk of distortion. For most application, the spruces are used interchangeably, but Norway spruce is the more widely used for power transmission poles since the sapwood is more easily treated than that of Sitka spruce. In demand as sawlogs; and the small diameter roundwood for pulpwood and boardwood. Residues command a premium.	Posts, stakes – (P and I); rails; pitprops (P); ladder poles (P); chipboard; pulpwood; fibreboard; bark products.	Construction – (general carcassing and framing) building sheds; joinery (interior); mining timber; wood wool; fencing (I); boxes; crates; packaging; pallets; flooring; gates, posts (I); general purposes. (The 'Whitewood' of the imported timber trade.)
Larch: European; Japanese; Hybrid D: 540; T: medium; MM: small; WQ: medium; ND: moderate; PP: resistant (sapwood	Posts, stakes, rails;	Construction – (general carcassing and

Properties of timber	Uses (in addition to firewood)	
D= Density (average kg/m³) at 12% moisture content. T= Texture. MM= Moisture movement. WQ= Working quality. ND= Natural durability of the heartwood. PP= Permeability of heartwood to preservative treatment. Colour and other features as aids to identification of timbers are given in chapter 1.	I= Impregnate with preservative★ P= Peel (debark)	
	Small diameter roundwood and poles (including tops of sawlogs)	Sawnwood; or Veneer (rarely conifers, and for export only)
moderate). Resinous. There is little difference, technically, between the wood of the three larches when made between material of comparable age and grade. However, hybrid larch can grow vigorously under particularly favourable conditions, with a consequent reduction in wood density and in the characteristic strength and decay resistance of other larches. Timber of old growth larch is amongst the heaviest of British softwoods, but that from younger trees is somewhat lighter. Larch is amongst the hardest and toughest of British softwoods. In other strength properties it is similar to Scots pine and has the same working stresses. Larch is one of the most durable of the British softwoods, but is resistant to preservative treatment; even the sapwood is moderately resistant. Has a tendency to spring off the saw on conversion, incurring some wastage. The timber dries fairly rapidly but with some tendency to distort, split and check, and for the knots to split and loosen. The dried timber saws and machines fairly easily and finishes cleanly, but the hard knots have a blunting effect on tools. Pre-boring is necessary to minimize splitting on nailing. Japanese larch is milder to work than European larch. Larch has traditionally been regarded primarily as an estate timber, used for a variety of purposes. Currently it is extensively used for panel, interlaced and lap fencing. It is the preferred softwood for boat-building. Much is used for construction purposes, although its hardness and tendency to distort and to split on nailing can present problems. Greater use may come in applications requiring larger sizes, where the inherent strength and durability of larch give it a distinct advantage.	transmission poles (P and I); pitprops (P); rustic work; chipboard; pulpwood; fibreboard; bark products.	framing); garden furniture; boat-building; mining timber; fencing (I); pallets; panel, interlaced and lap fencing (I); gates, posts; general purposes.

★ This relates particularly to (a) where much sapwood is present; and (b) small roundwood posts and stakes with relatively permeable sapwood.

In addition to the nine main British conifers (softwoods), noted above, information is given below on the properties and uses of several other conifers which are usually planted for timber, as well as some species (e.g. yew, true cedars, coast redwood and wellingtonia) which, although planted for amenity, landscaping and other reasons, produce useful timber particularly in their older ages and larger sizes.

CONIFERS (SOFTWOODS)

Properties of timber	Uses (in addition to firewood)	
D= Density (average kg/m^3) at 12% moisture content. T= Texture. MM= Moisture movement. WQ= Working quality. ND = Natural durability of the heartwood. PP= Permeability of heartwood to preservative treatment. Colour and other features as aids to identification of timbers are given in chapter 1.	I= Impregnate with preservative★ P= Peel (debark)	
	Small diameter roundwood and poles (including tops of sawlogs)	Sawnwood; or Veneer (rarely conifers, and for export only)
Western red cedar Low density (average 390 kg/m^3), comparatively light in weight, coarse texture, small moisture movement, good working qualities, durable and heartwood resistant to preservative treatment.	Posts, stakes; rails; pitprops (P); ladder poles (P); chipboard; pulpwood; fibreboard; bark products.	Exterior cladding; greenhouses; joinery; roofing shingles (usually imported); weatherboarding; mining timber; sheds; fencing (P and I); seed-boxes; general purposes.
Western hemlock Medium density (average 500 kg/m^3), medium in texture, small moisture movement, good working qualities, non-durable and heartwood resistant to preservative treatment. Fairly strong. Subject to fluting in the round. End uses similar to the spruces.	Posts, stakes (P and I); rails; pitprops (P); chipboard; pulpwood; fibreboard; bark products.	Construction; joinery; fencing (P and I); boxes; mining timber; general purposes.
Lawson cypress Strong. Comparatively light in weight. Fine texture. Heartwood moderately durable.	Posts, stakes (P and I); rails; pitprops (P); chipboard; pulpwood; fibreboard; bark products.	Fencing (P and I) mining timber; joinery; general purposes (known in the imported trade as 'Port Orford cedar').

Properties of timber	Uses (in addition to firewood)	
D= Density (average kg/m³) at 12% moisture content. T= Texture. MM= Moisture movement. WQ= Working quality. ND = Natural durability of the heartwood. PP= Permeability of heartwood to preservative treatment. Colour and other features as aids to identification of timbers are given in chapter 1.	I= Impregnate with preservative* P= Peel (debark)	
	Small diameter roundwood and poles (including tops of sawlogs)	Sawnwood; or Veneer (rarely conifers, and for export only)
Silver firs: Grand; Noble; European Timber similar to the spruces but less strong and of lower value. Not naturally durable out of doors, but can be treated with preservatives.	Posts, stakes (P and I); rails; pitprops (P); chipboard; pulpwood; fibreboard; bark products.	Construction – (limited); joinery (interior); boxes; packaging; mining timber; general purposes.
Coast redwood; Wellingtonia (Giant sequoia) Rather soft. Light in weight. Considered strong for its weight. Dries rapidly. Heartwood naturally durable. Wide sapwood. Texture moderately coarse to coarse. Coast redwood sometimes available as thinnings. Readily treated with preservative.	Posts, stakes (P and I); rails; chipboard; pulpwood; fibreboard; bark products.	Joinery (limited); weatherboarding; garden furniture; gates; general purposes.
True cedars: Atlas; Deodar; Lebanon High density (average 580 kg/m³), medium texture, small/medium moisture movement, good working qualities, durable and resistant to preservative treatment. Fragrant cedar odour. Rarely available as thinnings.	n/a	Joinery (limited); garden furniture; gates; general purposes.
Yew High density (average 670 kg/m³), fine texture, small/medium moisture movement, difficult working qualities, durable and resistant to preservative treatment. Tough, elastic and heavy.	Excellent long-life posts, stakes and rails, but rarely available.	Joinery (interior); furniture; tables; cabinets; turnery; veneer.

* This relates particularly to (a) where much sapwood is present; and (b) small roundwood posts and stakes with relatively permeable sapwood.

For broadleaves (hardwoods) grown in Britain, comprehensive technical information as to the properties of their timber are less commonly available as that for conifers (softwoods), but the main properties and uses of many individual timbers are listed overleaf:

BROADLEAVES (HARDWOODS)

Properties of timber	Uses (in addition to firewood)	
D= Density (average kg/m^3) at 12% moisture content. T= Texture MM= Moisture movement. WQ= Working quality. ND= Natural durability of the heartwood. PP= Permeability of heartwood to preservative treatment. (Colour and other features as aids to identification of timbers are given in chapter 1.)	I= Impregnate with preservative★ P= Peel (debark)	
	Small diameter roundwood and poles (including tops of sawlogs)	Sawnwood: or Veneer (for export only)
Oak: Pedunculate; Sessile D: 670/720; T: medium/coarse; MM: medium; WQ: medium/difficult. ND: durable; PP: extremely resistant. May stain from iron, or corrode metals in damp conditions. Relatively heavy in weight. Cleaves satisfactorily. Evans (1984) gives the features distinguishing the wood structure of the two native oaks.	Excellent long-life posts, stakes (especially when cleft); rails (especially when cleft); pulpwood; boardwood; 'tanbark'.	Furniture; joinery (external and interior); flooring; cooperage; fencing; gates; veneer.
Beech D: 720; T: fine; MM: large; WQ: good; ND: perishable; PP: permeable. Bends well. Good strength. Relatively heavy in weight.	Turnery; pulpwood; boardwood.	Furniture; joinery (interior); flooring; pallets; plywood (mainly imported); veneer.
Ash D: 710; T: medium/coarse; MM: medium; WQ: good; ND: perishable; PP: moderately resistant. Flexible; bends very well. Cleaves satisfactorily. Tough. High resistance to shock loading. Relatively heavy in weight. Tendency to split when felled. White ash is much prized.	Rails (especially when cleft); turnery; handles of striking tools; bar hurdles; pulpwood; boardwood; excellent firewood.	Joinery (interior); vehicle framework; pallets; sports goods (if white or pale pink) for skis, oars, cricket stumps; veneer.
Sycamore D: 630; T: fine; MM: medium; WQ: good; ND: perishable; PP: permeable. Straightgrained. Odourless and, with its pure white colour, makes it the preferred species for use in contact with food. Tendency to split in nailing. A superior turnery wood.	Turnery; pulpwood; boardwood.	Joinery (interior); kitchen utensils; textile requirements; butchers' blocks; turnery, veneer (especially if figured/ rippled).

Properties of timber	Uses (in addition to firewood)	
D= Density (average kg/m^3) at 12% moisture content. T= Texture MM= Moisture movement. WQ= Working quality. ND= Natural durability of the heartwood. PP= Permeability of heartwood to preservative treatment. (Colour and other features as aids to identification of timbers are given in chapter 1.)	I= Impregnate with preservative★ P= Peel (debark)	
	Small diameter roundwood and poles (including tops of sawlogs)	Sawnwood: or Veneer (for export only)
Sweet chestnut D: 560; T: medium; MM: large; WQ: good; ND: very durable; PP: extremely resistant. May stain from iron in damp conditions. Works better than most hardwoods, but with a tendency to split in nailing. Occasional logs have an ornamental wavy grain. Often a substitute for native oaks but is less strong and lacks the ornamental silver grain figure.	Excellent long-life posts, stakes, poles (especially when cleft); rails; fencing; hop poles; pulpwood; boardwood. Potential energy crop grown on very short rotations.	Joinery (external and interior); furniture; coffin boards; veneer.
Alder: Common; Grey; Italian; Red D: medium; T: coarse; MM: medium; WQ: good; ND: non-durable; PP: permeable.	Posts, stakes (P and I); turnery (not Grey); pulpwood; boardwood.	Fencing (P and I); pallets; general purposes.
Birch: Silver; Downy D: 670; T: fine; MM: large; WQ: good; ND: perishable; PP: permeable. A rather dull, featureless surface. Good strength properties. Relatively heavy in weight.	Turnery; pulpwood; boardwood; branches for besom heads and steeplechase jumps.	Furniture (within veneers); turnery; pallets; plywood (mainly imported).
Wild Cherry (Gean) D: 630; T: coarse; MM: medium; WQ: good; ND: moderately durable; PP: resistant. Inclined to warp. Increasingly in demand for quality furniture and veneers.	Turnery; pulpwood; boardwood.	Furniture; decorative joinery; cabinet making; turnery; veneer; instrument cases.

Properties of timber	Uses (in addition to firewood)	
D= Density (average kg/m^3) at 12% moisture content. T= Texture MM= Moisture movement. WQ= Working quality. ND= Natural durability of the heartwood. PP= Permeability of heartwood to preservative treatment. (Colour and other features as aids to identification of timbers are given in chapter 1.)	I= Impregnate with preservative★ P= Peel (debark)	
	Small diameter roundwood and poles (including tops of sawlogs)	Sawnwood: or Veneer (for export only)
Horse chestnut D: 510; T: fine; MM: small; WQ: medium; ND: perishable; PP: permeable. Brittle. Comparatively light in weight.	Pulpwood; Boardwood.	Turnery; boxes; fruit trays; coffin boards.
Elm: Common D: 560; T: coarse; MM: medium; WQ: medium; ND: non-durable; PP: moderately resistant. Cross-grained – difficult to split. The preferred species for pallet blocks. Wych elm (less common) is also a splendid wood but with different properties.	Posts, stakes; dock piles and fenders; pulpwood.	Furniture; weatherboarding; boat-building; packaging; pallets; coffin boards; turnery; garden furniture; general purposes.
Plane D: 640; T: fine; MM: medium; WQ: good; ND: perishable; PP: medium. Relatively heavy in weight.	n/a	High class furniture; decorative joinery; inlay/work; veneer. (Termed 'lacewood' when quartered.)
Walnut D: 660/670; T: coarse; MM: medium; WQ: good; ND: moderate to very durable; PP: resistant. Fine burr veneers can come from the main root area. Relatively heavy in weight.	n/a	High class furniture; decorative joinery; cabinets; turnery; gun/rifle stocks; table ware; veneer.
Lime D: 560; T: fine; MM: medium; WQ: good; ND: perishable; PP: permeable. Uniform grain.	Turnery; pulpwood; boardwood; besom handles.	Carving; beehives; turnery; veneer; hat blocks; piano keys.

Properties of timber	Uses (in addition to firewood)	
D= Density (average kg/m^3) at 12% moisture content. T= Texture MM= Moisture movement. WQ= Working quality. ND= Natural durability of the heartwood. PP= Permeability of heartwood to preservative treatment. (Colour and other features as aids to identification of timbers are given in chapter 1.)	I= Impregnate with preservative★ P= Peel (debark)	
	Small diameter roundwood and poles (including tops of sawlogs)	Sawnwood: or Veneer (for export only)
Poplar D: 450; T: fine/medium; MM: large; WQ: medium; ND: perishable/non-durable; PP: extremely resistant. High resistance to abrasion. Low flammability. Free from taints and odours. Tendency to wooliness. Straight grain. Relatively light in weight. Low bending strength. Medium crushing strength. Very low shock resistance. Increasingly used for non-returnable pallets. For additional information see Jobling (1990).	Posts, stakes; rails; pulpwood; boardwood. wood wool. Potential energy crop grown on very short rotations.	Pallets; boxes; turnery; wagon floors; coffin boards; veneer peeling for packaging, chip baskets and vegetable crates; mining timbers (selective).
Willow (other then cricket-bat willow) D: 450; T: fine; MM: small; WQ: good; ND: perishable; PP: resistant. Straight grain. Relatively light in weight.	Posts, stakes; pulpwood; boardwood. Potential energy crop grown on very short rotations.	Boxes; crates; artificial limbs.
Southern beeches *(Nothofagus* spp.) Moderately dense and fairly strong. Not of the same quality as beech. *Nothofagus procera* is likely to display more attractive timber properties than *N. obliqua*.	Posts, stakes (P and I); pulpwood; boardwood.	Furniture, cabinets and flooring when large; joinery; general purposes.
Hornbeam Very hard and heavy. Dense and strong. The strongest, hardest and heaviest British timber.	Small turnery; pulpwood; boardwood.	Turnery; chopping blocks; carving; cogs; wooden screws; industrial floors.
Norway maple White-grey wood with characteristics similar but slightly inferior to sycamore.	Similar purposes as sycamore.	Similar purposes as sycamore.

Properties of timber	Uses (in addition to firewood)	
D= Density (average kg/m^3) at 12% moisture content. T= Texture MM= Moisture movement. WQ= Working quality. ND= Natural durability of the heartwood. PP= Permeability of heartwood to preservative treatment. (Colour and other features as aids to identification of timbers are given in chapter 1.)	I= Impregnate with preservative★ P= Peel (debark)	
	Small diameter roundwood and poles (including tops of sawlogs)	Sawnwood: or Veneer (for export only)
Turkey oak Relatively heavy in weight. Prone to warping and shrinkage. Lacks the strength, durability and seasoning of native oaks. Virtually unsaleable for sawing. Not a substitute for native oaks.	Posts, stakes (P and I); pulpwood; boardwood.	General purposes.
Hazel Cleaves satisfactorily. Pliable when young. Small dimensions.	Small turnery; fencing materials; wattle hurdles; thatching spars; pea & bean sticks; pulpwood; boardwood.	n/a
Holly Tough, heavy and a fine texture. Prone to distortion. Takes stain evenly.	Posts, stakes; pulpwood; boardwood.	Turnery; carving: inlay-work.
Holm oak Tough, heavy and hard. Highly figured. Inferior quality to native oaks.	Posts, stakes (P and I); pulpwood; boardwood.	Carving.
Red oak Open texture with large pores. Lacks the durability of native oaks. Not a substitute for native oaks.	Posts, stakes (P and I); pulpwood; boardwood.	Cheap furniture and flooring.
Box Hard with a fine even texture. Small dimensions.	Wood engraving; small turnery; rarely available.	Turnery; craft work; drawing instruments; inlay-work.

★ This relates particularly to (a) where much sapwood is present: and (b) small roundwood posts and stakes with relatively permeable sapwood.

Determinants of wood quality. The general properties of British-grown timber have been noted in this Section. Most important to a forester is a knowledge of quality in general of the major British timbers; and how to grade them in the round and perhaps when initially sawn. Features which affect the use of timber in the round include:

1. Straightness. Often this is the single most important factor.
2. Weight. Some woods are heavy, others light. Green wood weight is largely due to the moisture content.
3. Hardness. Oak, hornbeam, holly, sweet chestnut and yew are harder than poplar, willow, lime, alder and most conifers.
4. Elasticity. Ash and yew are tough and resilient.
5. Ultimate strength in bending. That of ash, beech, yew and larches is much higher than that of grand fir.
6. Cleavability. This useful feature applies to sweet chestnut, ash, oak, hazel, Douglas fir, western red cedar, Lawson cypress and coast redwood.
7. Freedom from: (i) shake (particularly in oak and sweet chestnut); (ii) resin pockets (often in larch); (iii) drought cracks (particularly in grand fir); (iv) fluting (often in western hemlock); and (v) rot and other defects.
8. Natural durability and permeability, and usefulness after preservation (discussed in chapter 10).
9. Decorative 'elements'. This may relate to colour, lustre, grain and figure (all provide quality for veneers), but can rarely be ascertained when in the round.

All the foregoing features affect value, as will be noted in chapter 11.

Improving quality of British conifers. Britain's forests today provide some 16% of the nation's softwood timber requirements, a proportion that is rising as the forests planted in the post-war decades mature. By the year 2000, although still heavily dependent on imports of softwood, the most important single source of timber to Britain's industry could be her forests. The construction industry is the most profitable market for that timber and it is essential to grow the quality which meets that need.

Spacing in the forest plays an important role. Fast growth of individual trees at wider spacing causes an increase in the size of the core of juvenile wood in many conifers; branches and hence knots are bigger and may be associated with more compression wood and spiral grain. All these characteristics affect the quality of the wood. A high wood density in conifers is usually associated with many desirable properties and a suitability for end uses requiring strength. The single most important factor affecting density is the genetic variation which exists from tree to tree and, for Sitka spruce, some 60–70% of the total variability in density is associated with tree-to-tree variation. However, in that species, which has strength properties close to the lower limits for some industrial purposes, wide spacings may make the timber unacceptable for uses which command high prices. Rapid growth of widely spaced trees results in relatively wide bands of lower density springwood being formed, which alternate with narrow bands of higher density summerwood. There is a lowering of strength for structural uses and a combination of coarse texture and high incidence of knots makes the wood unsuitable for purposes such as joinery (Savill and Evans 1986).

The Forest Products Research Laboratory (formerly at Princes Risborough)[1] has liaised with the Forestry Commission to ensure that British forests produce wood which is acceptable for structural purposes. Past research has been mainly on Sitka spruce, to which Brazier has contributed several decades of notable work – see in particular his early concern for growing quality conifers (Brazier 1979) and broadleaves (Brazier 1985). Much work has examined the effect of planting distance on structural wood yields from Sitka spruce stands which are left unthinned. There appears to be a clear link between structural grade output and planting distance (Brazier *et al.* 1985). Closer spacings yield for a crop a higher proportion of sawnwood suitable for structural use, but they also yield fewer logs of larger diameter per hectare. It is now considered prudent, if the production of structural timber is the aim, to ensure by planting and appropriate beating up that Sitka spruce spacing at establishment does not exceed 2 m. Thinning prescriptions should aim to select for the best stem form and reduce the persistence of deep crowns.

Research seeks to examine the effects of thinning forest stands on the yield and quality of wood. The work has attracted support of the European Commission and is one of four international EC projects in programmes to assess the effect of modern forestry practices on the wood quality of fast-grown plantation spruce. A second project supported by the Forestry Commission has made considerable progress towards assessing the effects of growth features on timber strength so that these can be minimized by careful forest management, to produce higher yields of structural timber in the future. The project is establishing the effects of knots, grain alignment, wood density and cross-sectional growth characteristics, including the presence of juvenile wood, on wood stiffness.

Sawn softwood: stress grading and strength classes. Timbers derived from different tree species (e.g. Douglas fir and Sitka spruce) are different in character and strength. Mechanical properties vary within a single species. This can be attributed to variations in the strength of the clear wood substance, and the presence of knots, slope of grain and other features which reduce strength.

Systems of stress grading have been developed which enable a parcel of sawn softwood to be divided into stronger and weaker members, based on visual or machine sorting (Harding 1988; publications of the Building Research Establishment 1984, 1986, and TRADA 1989). The considerable number of alternative combinations of species and grades can be confusing for the designer, and present problems for suppliers; and the use of British-grown timber, some of which is not directly equivalent to grades of the major imported species, can be restricted. Some of these problems have been overcome by the introduction of a system of strength classes, contained in Code of Practice BS 5268: Part 2:1988. The system groups together grades and species of similar strengths into nine classes, each with a set of strength properties. The strength classes for BS 4978 visual and machine grades of British-grown softwoods are given in Table 30.

One of the advantages of the strength class system is that grading machines are able to grade directly to any strength class boundary stress level and mark the timber

---

1 The Timber Division of the Building Research Establishment at Garston near Watford, now carries forward more than 60 years of invaluable research at the former Forest Products Research Laboratory at Princes Risborough.

with a strength class number. Machine stress grading has proved advantageous to British timber uses by providing greater yields of higher strength timber than that provided from visual grading systems. This is certainly true for some important British-grown species for which the grade strength properties are well above the minimum properties of the class to which they are assigned. Machine grading to strength classes will enable these species to compete more effectively with imported timber. But machine grading assesses the performance of timber from crops managed under past silvicultural regimes. Because of the production times in forestry, a judgement must be made on the likely outcome in a knowledge of the relationships between growth on the one hand and those features which influence the technical behaviour of wood on the other.

Table 30: Sawn softwood visual grades.

Species	SC1	SC2	SC3	SC4	SC5
Douglas fir		GS	SS/M50		M75
Larch			GS	SS	
Scots pine			GS/M50	SS	M75
Corsican pine		GS	M50	SS	M75
Norway spruce	GS	M50/SS	M75		
Sitka spruce	GS	M50/SS	M75		

Key: SS – Special Structural; GS – General Structural.
Machine grades: M50 – Machine 50; M75 – Machine 75.

Source: Fewell 1983; TRADA 1989.

Softwood sawlog grading. Forestry Commission log sales of conifers comprise only two categories – 'Green' and 'Red', set out in Table 31. The aim is to improve the quality standards of sawmill outputs. The 'Green' category, drawn up with straightness and limited knot size in mind, gives a better recovery of sawnwood, and secures a higher pass rate for stress graded material. Sawlogs meeting this specification will not require further cross-cutting. The 'Red' category provides a less exacting standard of straightness, knot size and frequency; and accommodates logs which fail to meet the 'Green' specification. It embraces longer length sawlogs produced in pole-length harvesting where subsequent cross-cutting at the sawmill is a normal requirement. It provides a set of minimum standards which are applicable regardless of the nature or origins of the sawlog and which is of mutual benefit to buyer and seller.

Table 31: Forestry Commission conifer sawlog categories

Log category	'Green'	'Red'
Species	Any conifer – species to be stated	
Minimum top diameter	To be stated. Normally 16 cm but not less than 14 cm (12 cm in certain localities).	
Length	Minimum – 1.8 m. Maximum – 8.3 m.	Minimum – 1.8 m. Maximum – to be stated.
Cross-cut steps	0.3 m normal practice to 8.1 m maximum (0.1 m by request, to 8.3 m maximum)	0.3 m normal practice to 8.1 m maximum (0.1 m by request, to 8.3 m maximum) Truly random for longer lengths.
Straightness	Bow not to exceed 1.0 cm for every 1.0 m length and this in one plane and one direction only. Up to 5% of individual logs in any one load can be outside the specification to the extent that bow may be up to 1.5 cm for every 1.0 m length. Bow is measured as the maximum deviation at any point of a straight line joining the centres at each end of the log from the actual centre line of the log.	Capable of being cross-cut into straight lengths of at least 1.2 m without significant waste.
Knots	On any individual log 80% of knots will not exceed 5 cm in diameter. However, up to 5% of the logs will be allowed outwith this specification but those of an excessively coarse appearance will be excluded.	No restriction on knot size and frequency.
Trim	For manual felling, root spurs well-dressed, felling cuts as square as practicable, snedding flush to stem. With mechanized harvesting, exactly the same standards of snedding may not prove practicable, but only a modest relaxation will be acceptable. Splits, tear-outs, and double tops are not permitted.	
Scars/decay	Significant visible decay and significant scars will not be permitted.	

Log category	'Green'	'Red'
Insect damage and staining	Visible insect damage or staining indicating incipient decay will not be present when made available for loading.	
Blue stain	To minimize the infection of pine logs with blue stain, logs will be brought to the loading point within four weeks of felling. In view of this undertaking, the Commission will not accept blue stain as a defect or entertain claims in respect of it.	
Metal	Logs suspected of containing metal will not be included.	
Mean dbh	The mean dbh of *all* the standing trees which are to be removed by felling or thinning and from which the parcel of logs are to be taken, should be stated.	

Source: FC Field Book 9: Classification and presentation of softwood sawlogs (1990).

4. Utilization of British-grown roundwood

The profitability of forestry lies largely in the intelligent utilization of woodland produce, whether underwood, coppice, thinnings or mature trees. The wide choice in both utilization and markets is indicated in the following schedule, which supplements the list of uses given in section 3 *supra*. It excludes bark and residue products, chips for riding surfaces and horticultural use, and foliage. Details of some uses are given by Aaron and Richards (1990) and Brazier (1990).

Utilization	British-grown roundwood commonly used to fulfil the stated utilization	
	Conifers	Broadleaves
Building construction BS 5268: 1985, 1989. BS 4978: 1988.	Scots, Corsican and lodgepole pines, Norway and Sitka spruces and Douglas fir.	Oak for beams, cills, flooring blocks and shingles in high class houses and public buildings.
Joinery BS 1168: 1986, 1988.	Scots, Corsican and lodgepole pines, Douglas fir, western hemlock and *Abies* spp.	Oak, walnut and cherry for high class work in houses, banks, churches and public buildings.

Utilization	British-grown roundwood commonly used to fulfil the stated utilization	
	Conifers	Broadleaves
Furniture	Scots and lodgepole pines, and Norway spruce for 'carcassing'. Coast redwood and Wellingtonia (Giant sequoia) for garden furniture.	Oak, ash, beech, birch, cherry, sweet chestnut, sycamore, walnut and plane. Elm for house and garden furniture.
Veneers, 'decorative' and 'face' (currently export only)	Yew.	Walnut, oak, sycamore, sweet chestnut, ash, cherry, beech, plane and elm.
Plywood (currently export only)	Douglas fir and Scots pine could be used.	Birch, poplar and beech could be used if there was a demand.
Chip-baskets	Occasionally Scots pine.	Most well-grown poplars and willows. Occasionally hazel.
Telegraph poles (British Telecom pic) BS 1990: 1984. Amendment 5043: 1986.	Scots pine.	None.
Power poles (Electricity Companies) BS 1990: 1984. Amendment 5043: 1986.	Scots pine. Very occasionally Corsican pine, the larches and Douglas fir	None.
Flag poles, scaffold poles, and rugby goal posts	Norway spruce, Douglas fir and western red cedar.	None.
Miningtimber (British Coal Corporation) — Pit props, round or split	Most species.	None.
Miningtimber — For sawn timber BS 5750	Most species.	Most species (not willow and poplar, except poplar for waggon and barrow bottoms).
Paper-pulp — Most types of pulp currently produced	Sitka and Norway spruces preferred. Scots, Corsican and lodgepole pines, limited use only.	None.
Paper-pulp — Semi-chemical	Almost all species.	Most species (Sudbrook).

Utilization	British-grown roundwood commonly used to fulfil the stated utilization	
	Conifers	Broadleaves
Particleboard, cement-bonded panels and moulded-wood	In general, spruces preferred, then pines, and then all other species.	Most species in some mills; nil in others.
Fibreboard	Spruces and pines.	Willow, birch and poplar.
Cricket-bats	None.	Cricket-bat willow. Ash (for handles).
Sports goods	None.	Chiefly ash; some beech and walnut.
Wood wool: (a) for packaging: (b) for cement slabs: BS 2548: 1986.	Scots and Corsican pines, Sitka and Norway spruces. Corsican pine is the preferred species.	None.
Boats	European larch (for planking) and Sitka spruce (for oars).	Oak (for frames). Ash, sycamore and wych elm (for frames of small boats). Ash (for oars).
Railway sleepers (a much reduced demand)	Douglas fir and Scots pine.	None.
Railway rolling stock (a much reduced demand)	Scots and Corsican pines, spruces, larches and Douglas fir.	Ash, oak, beech, elm and sycamore.
Cable-drums	Pines, spruces, Douglas fir, and larches.	Elm (for heavy duty).
Pallets BS 2629: 1967. Amendment 2721: 1978.	All species.	All species, but usually oak, elm and poplar.
Boxes and Packaging	Norway and Sitka spruces, Scots and lodgepole pines, and *Abies* spp.	Elm (for heavy duty), poplar and birch.
Crates	Various.	Various. Hazel in particular.
Fencing BS 1722: 1986.	Larches (for posts and rails). Scots, Corsican and lodgepole pines if treated with preservative. Mixed species (for interlaced and lap fencing, mostly larch).	Oak (heartwood) and sweet chestnut. Hazel (for wattle hurdles, and hedging and garden materials). Oak sapwood and most other broadleaves suitable if treated with preservative.

Utilization		British-grown roundwood commonly used to fulfil the stated utilization	
		Conifers	Broadleaves
Gates BS 4092: 1966. BS 3470: 1978.		Larches and Douglas fir.	Oak and sweet chestnut.
Coffins		Scots pine and spruces (for cheapest quality).	Elm, oak and sweet chestnut. Horse chestnut and poplar (for cheapest quality).
Motor vehicles		Limited quantities of Douglas fir, larches and Scots pine.	Limited quantities of ash, beech, elm, oak, walnut and birch.
Carving woods		Yew.	Lime, apple, pear; also many species on a small scale.
Piling for river and sea defences, jetties and wharves		Occasionally pines, spruces and Dougals fir.	Elm, oak and (occasionally) other broadleaves.
Tight cooperage	Casks	None.	Oak.
	Vats	European larch and Douglas fir.	Oak and elm. Some sweet chestnut.
Musical instruments		Species variable.	Beech, sycamore, lime and laburnum.
Charcoal		Very limited quantities of most species.	All species; best is beech.
Turnery	General (Toys, brush-backs)	None.	Alder (not grey alder), ash, birch, cherry, beech, sycamore and Norway maple.
	Textile rollers.	None.	Sycamore and beech.
Firewood		Limited quantities of most species.	Most species, especially ash and beech.
Ladders	Poles (most are imported)	Norway spruce and western red cedar.	Ash
	Rungs	None.	Oak, ash, sweet chestnut and sallow (having "non-slip' property).

Utilization	British-grown roundwood commonly used to fulfil the stated utilization	
	Conifers	Broadleaves
Handles, Striking	None.	Ash and hornbeam.
Tent-pegs	None.	Ash and beech.

Western red cedar, western hemlock, Lawson cypress, and grand and noble firs – all relatively new to the market – are generally accepted for the same purposes as spruces, except for paper making.

Practical utilization. Uses of timbers and other wood products were introduced in section 3 *supra* and have been supplemented by the foregoing schedule indicating the British-grown roundwood used to fulfil the needs of the markets. Hereafter is provided further information relating to the utilization of, first, broadleaves and, second, conifers; in each case differentiating *Roundwood* (small and large) and *Sawnwood*. The term 'small diameter roundwood' is usual for those stems, or parts of the stem, which are below the minimum diameter limit acceptable for sawlogs (25–38 cm for broadleaves and 14–18 cm for conifers) but usually greater than 6 cm in diameter. It is the produce of small coppice and thinnings, and the tops of larger trees. Although the character of the wood differs between these sources, they are not always separated commercially. The market for small diameter roundwood has undergone significant improvements in recent years. A major restructuring of the relevant wood processing industry has seen the closure of obsolete plants and their replacement by large-scale reinvestment in new processing technology and products. The locations of the major users of small diameter roundwood, i.e. the paper and board mills, are shown in Figure 22, on page 369.

Broadleaved utilization categories

Roundwood (Broadleaves). *Small broadleaved thinnings,* particularly of beech and oak, are not always easy to dispose of at an adequate profit; sometimes they have to be sold as fuelwood. If sufficiently straight and effectively treated with preservative some species can be used as small fencing material *(infra)*. Utilization is easier when the broadleaves reach the size required for turnery, pulpwood and boardwood. But some of the mills are not within adequate profitable reach of all growers.

Underwood, mainly hazel – see chapter 7 – is sometimes difficult to market or utilize. It can be worked to produce pea sticks, bean rods, hedging material, wattle hurdles, flower stakes and similar products. For many decades underwood working has been neglected because of the decline in traditional markets and relevant skills. In Buckinghamshire, Dorset, Hampshire and Wiltshire occasionally about £60-£70 ha can be obtained for standing hazel, depending on age, condition and whether run on a rotation. Elsewhere underwood is virtually unsaleable, and costly to clear for restocking. The most likely purchasers are the few remaining makers of wattle hurdles, and other local craftsmen.

Small hedging material generally comprises: hedge stakes, in bundles of twenty, about 1.5 m long by about 5 cm top diameter; heathering (binders) of up to 4 m in length by 4 cm butt diameter; and rough-filling (tynut) for gaps in hedges. Pea sticks and bean rods are usually in bundles of twenty-five; the sticks being about 1.5 m and the rods 2 m or more by about 3 cm top diameter. Thatching speakes and crate rods may be produced; also fascines and faggots – used to prevent erosion by sea and river, and occasionally to support temporary tracks over soft ground.

Coppice – see chapter 7 – poses utilization problems, depending on its species, age, condition and location. In young crops, outlets for its produce are usually local and limited – the hardwood pulp and board mills being the main market apart from fuelwood and charcoal. In medium age crops, utilization is easier – turneries, tanneries (oak bark), fencing and charcoal being possible outlets, but still with a heavy reliance on hardwood processing mills. Older crops have the benefit of demand from sawmills, especially for fencing and mining-timber. The following suggested outlets are infrequent and local. Birch branches can be made into besoms or bundles for use in steeplechase jumps, or supplied for use in the manufacture of vinegar and steel-plate. Lime is difficult to utilize, but it can be used for horticultural stakes and the handles of besoms (its bark was formerly stripped for tying the heads). Ash is needed for bar-hurdles, tent-pegs, tool-handles and fencing rails; and, along with alder and sycamore, it can be sold to turneries. Mixed broadleaved poles of small size are accepted by metal refineries, but the demand is modest. Peeled oak branches are sometimes saleable for making garden furniture and rose arches.

Sweet chestnut coppice – see chapter 7 – is in fluctuating demand in Kent and Sussex (with some in Essex), and occasionally in Herefordshire and Gloucestershire, mainly for cleft fencing-pales, posts and stakes. Its durability and cleavability make it a sought-after species. However, prices were depressed during 1990. Names and addresses of buyers can be obtained from the Chestnut Fencing Manufacturers Society, 37 Lebanon Road, Twickenham, Middlesex.

Fencing and gate material. Here the most likely outlet is among amateur and market gardeners, nurserymen, garden centres and farmers. Some prefer broadleaves, particularly if cleft, others prefer conifers. The products may be peeled or unpeeled, pointed or not pointed, round or half-round, quartered, cleft, and perhaps treated with preservative. Prices usually depend on the species, length and top or butt diameter; they fluctuate according to demand. Fencing and gate materials most commonly in demand, where no treatment is required, are of oak and sweet chestnut. They contain a high proportion of durable heartwood. Cleft oak or sweet chestnut posts and rails and cleft ash rails are required in some rural districts. (There is a considerable variation in specifications for post-and-rail and post-and-wire fences – see particularly under **Sawnwood (Softwood),** later in this section.)

Hop-poles, usually of sweet chestnut but sometimes of European larch, are about 5–6 m in length, with a top diameter of 7–15 cm. Their butts are usually treated with preservative. The demand has greatly decreased in recent years.

Bark. Oak tanbark is used by only two tanneries (at Colyton, Devon and Grampound, Cornwall). The bark should be stripped in late April or early May just as the cambium becomes active, and then stacked to dry in the open, rough surface upwards. Sometimes the price paid for bark from coppice is higher than that off old trees. The peeled logs should be removed and converted quickly before they check and crack. To avoid the possibility of timber deterioration it may be advisable, when possible, to strip the bark from standing trees; the trees themselves can be felled after the rush of the bark-stripping season is over. About 8 tonnes of oak roundwood produce one tonne of bark. Much processed bark is used for riding surfaces, horticultural purposes, litter for cattle and fuel (Aaron 1970, 1982).

Mining-timber for sawing. British Coal Corporation purchases all its miningtimber from British growers and producers. Broadleaves are accepted as sawnwood (but not as pitprops). The availability of lower quality broadleaved mining timber now exceeds what the Corporation can accept and hence the surplus has to be disposed of as pulp and boardmill wood. Sawnwood mining-timber being mostly of conifers, is discusssed later under **Sawnwood (Softwood).**

Turnery poles. A turnery when situated within economical (i.e. profitable) reach of broadleaved stands is a great help in the marketing of thinnings and coppice. The required species are alder (other than grey alder), ash, beech, birch, sycamore, cherry and hazel; sometimes Norway maple, holly and hornbeam are accepted. (Small to medium sized sawlogs, particularly ash, are also required to produce sawn 'squares' for further processing.) The poles are sold in lengths of about 2 m and up, and butt diameter of at least 10 cm; small poles are usually topped at about 7 cm diameter, and large at 8–10 cm. Each turnery has its own specification and species requirements. The main turneries using British timber are in Kent, Gloucestershire, Worcestershire and Cardiff. Their names and addresses can be obtained from the British Wood Turners' Association.

Firewood. The value of broadleaves for firewood (likewise for charcoal, noted below) depends mainly on species, quality, presentation and locality (Keighley 1987; Pearce 1987). The best commercial species is ash, followed by oak and beech; some other broadleaves are satisfactory. Alder, lime, sweet chestnut, elm, poplar and conifers are not favoured. The market can be met in the form of blocks, short billets, poles or cordwood. (A cord traditionally measures about 2.4 m x 1.2 m x 1.2 m, and may represent 1–2 tonnes dependent on species and moisture content.)

In many parts of the country, the firewood market, mainly broadleaved, pays more than sale of industrial small diameter roundwood. The preparation is easily undertaken with simple equipment. Up to half of the total volume of large open grown broadleaves may be in the form of branchwood for which firewood forms a useful market. However, the market is somewhat of a 'start stop' nature, varying with winters and cost of alternative fuels.

Charcoal. Charcoal has a wide range of uses in many present day industrial processes as well as for domestic fuel connected with barbecues and charcoal grills. It is formed by heating wood under conditions where there is insufficient air for complete combustion (Aaron 1980; ILO 1988). Four methods are used: earth kilns (charcoal 'pits'), portable batch kilns, fixed batch kilns, and retort or continuous kilns. (The extraction of wood tar and pyroligneous acid during the process of carbonization ceased several decades ago; so too did the separation of pyroligneous acid into acetic acid, naphtha and oils.)

Any wood can be carbonized, but the properties of the charcoal differ with the species. The broadleaves used, in order of preference are: beech, birch, hornbeam, oak, ash and elm. (Charcoal made from conifers is too friable for many purposes.) Roundwood raw material comes mainly in the form of lop-and-top, and from the debris of clear-felled areas from which otherwise saleable wood has been removed. The main source is in the form of cordwood *(supra)* and sawmilling offcuts. The size specification varies with the individual producer and the method used: lengths generally preferred are of about 0.5 m–2m, and diameters 3–20 cm. Between 3 and 6 tonnes of wood are required to produce 0.5 tonne of charcoal, with production rates varying between 0.5 and 1.0 tonne per man-day in the portable kiln system (holding about 3 tonnes of wood).

Charcoal would appear to be an ideal adjunct to firewood production, the latter operating in winter, the other in summer. Foresters may be well advised to resort to these markets when possible, because roundwood merchants and other buyers pay a relatively higher price for broadleaved timber if they are relieved of the task and expense of disposing of lop-and-top, especially when they purchase parcels far from their headquarters.

Consumption of charcoal in Britain is currently about 60,000 tonnes annually, almost all imported. It is a potentially good market for low grade broadleaves. A charcoal appraisal group formed by the Forestry Commission and other interested bodies is currently exploring the possibility of establishing a continuous production in East Anglia. This may provide an invaluable market for broadleaves from the south-east of England, where markets for small diameter roundwood are scarce.

Paper and panels from broadleave small diameter roundwood are referred to later in this section. The same applies to *wood residues and chips*.

Broadleave sawlogs. Relatively long lengths are generally preferred, not least to reduce the costs of loading, stacking in the yard, and positioning during conversion. Sometimes certain multiple lengths are preferred. The demand, other than for sawing into mining-timbers and pallets, is for good quality and medium to large sizes. Inferior qualities if accepted, often somewhat reluctantly, command much reduced prices. The small capacity hardwood sawmills concentrate on the production of sawnwood for fencing, gates, pallets, cable-drums, carcassing, boxwood and mining-timber. Those of medium capacity may produce some of the foregoing as well as a limited amount of sawnwood for building and construction, joinery, furniture, cooperage and, occasionally, for shipbuilding and vehicle-body building. There is a need for more markets for sawn lower grade boards. Residues find a ready outlet, mainly in the pulp and boardwood mills.

Of broadleaved sawlog production about three-quarters is in England and Wales reflecting the distribution of the broadleaved woodland resource. The grower can obtain particularly attractive prices for superior quality logs (prime, planking, beam) particularly of oak, ash, beech, sycamore, cherry and sweet chestnut; also for veneer logs – for export. Prices, indicated in chapter 11, reflect a scarcity value.

Veneer logs. There is a large unsatisfied demand for broadleaved logs suitable for veneer wood sliced or rotary-peeled into sheet form. Most veneer logs are for decorative purposes. The thin veneer is glued to plywood, chipboard, fibreboard or sawnboard as facing. There is, however, a difference between 'decorative' veneers and 'face' and 'core' veneers; the prices paid for the logs will also differ according to species, specification and quality. Walnut, walnut burr, elm burr, oak, brown oak, ash and ripple-grained sycamore command a much higher price than 'constructional' veneer logs of beech and birch.

The suitability of a log for veneering is determined by a number of characteristics. Size has a considerable bearing on value (though quite small pieces of burr walnut are valuable). In general, a log should be straight and clean, free from external and internal defects. Walnut is usually felled very low and often taken away complete with main roots, because in the base is found the 'head' which has additional value. Oak logs, if free from brown or mineral stain, are always in demand. The price paid for sycamore depends on whether it is 'plain' grained or 'ripple' grained; the latter quality commands high prices. Sycamore is used much younger than walnut or oak but its many peculiarities, including the green sap stains that show in early spring, as well as 'pips', can make it quite a gamble for the veneer merchant. Ash, when white at both ends, and free of brown heart and mineral spot, commands a good price.

A tree bought for its veneer content is usually taken as it stands. Its quality, and therefore, its true value is not known until it is felled and converted. Buyers although fair are wary of purchasing a tree not up to its face-value. Reputable veneer merchants prefer to pay a satisfactory price based on the value of the log ascertained after felling. A special type of veneer industry in Britain is that for poplar and willow, rotary-peeled for making chip-baskets and fruit and vegetable crates (see chapter 5). Prices and specifications of veneer quality broadleaves are indicated in chapter 11. The market is relatively excellent, both in demand and price. Competition for supplies is very keen. All veneer logs are currently exported, mainly to Europe. Much re-enters Britain as veneers.

The long-term needs of the veneer industry should influence the growing of certain broadleaves, particularly oak, walnut, ash, sycamore, beech and cherry. Pruning is an important factor of their management.

Sawnwood (Hardwood). *Sawnwood* is the product of the sawmill – planks, boards and beams. Usually it is not further manufactured than by sawing and the product is of various sizes: thickness 2–10 cm, width 8 cm and above, and length 3–6 m. The product is parallel-sided and square edged and its manufacture from roughly cylindrical, tapered logs involves considerable wastage – in the order of 30–50%. It can be shaped, planed and moulded to specific requirements. Broadleaved sawnwood dominates the decorative, solid timber markets in furniture panelling and fitments. Wood residues and chips from sawmilling can be either burned to produce industrial energy or chipped for pulp and boardwood manufacture. Once produced, sawnwood, according to end use, is usually dried, either by open stacking with minimum protection from the elements, or by artificial means in a kiln in which air temperature, humidity and circulation are controlled. Sawnwood which is to be used in contact with the ground is treated with wood preservatives, usually under pressure, to provide protection against biodegradation. (Building and construction sawnwood, joinery, mining-timber, and pallets, being mainly of conifers, are discussed later in this section under **Sawnwood (Softwood).**)

Current research on utilization of broadleaves includes a study of the hardwood market chain from grower to consumer in order to identify the most important studies required to improve hardwood utilization, especially for intermediate and lower grades. The study comprises a survey of the volume of roundwood offered for sale by members of TGUK, a look at volumes of hardwood trade in sample areas, a study by TRADA to examine the hardwood trade, with particular emphasis on sawmilling and downstream utilization, and a further study by TRADA of those utilization processes of final uses where British hardwoods might be used in increased volumes or substituted for imported timber or other material.

Conifer utilization categories

Roundwood (Conifer) *Small conifer thinnings,* by contrast to broadleaves, are relatively easy to market in most districts. They are straighter and more uniform in shape. Effective preservation treatment of most of them is advisable to ensure an extended life if they are to be used in ground contact or in other conditions favouring decay; even some of the 'durable' species hold a high proportion of non-durable sapwood and only their heartwood is durable. Until conifers reach the sizes required for wood wool and pitprops their main use is as wood for paper, board, and fencing material. The agricultural, horticultural and general estates' demand for posts, stakes, rails and poles of many kinds, is a substantial market to which utilization should be oriented.

Rustic poles. This relatively modest market usually requires larches. In suburban districts they sell well and fetch high prices. The poles range in length up to about 7m, usually topped at about 3 cm in diameter and have butts of 5–15 cm diameter. The small sizes may be used for 'filling in'. Autumn or winter felled poles are preferred as they then have a firmer bark. Pergola-poles, particularly of Douglas fir or grand fir, with a 15–25 cm length of their branches adhering, find a limited sale.

Bark. The use of conifer bark has shown a steady rise over recent decades. Much processed bark as sawmill residues goes to pulp and boardwood mills, but large quantities are used for riding surfaces, landscaping, mulching, growing-mediums for horticultural purposes, cattle litter, and as fuel (Aaron 1982). There is a brisk demand for bark where it accumulates in quantities exceeding 1000 tonnes per annum. Some pulp and boardwood mills burn their bark for steam-raising.

Foliage of western red cedar, western hemlock, grand fir and Lawson cypress usually finds a ready but generally local market in the florist trade. (Likewise do sprays of sallow, spindle tree, rowan, box, Eucalypts and ivy – see chapter 5.) Sales of conifer foliage may be on the basis that a reliable purchaser undertakes the cutting and bundling; they are sometimes a means of recouping part of the costs of brashing and pruning. (Holly sprays, particularly if berried, are in demand towards Christmas-tide.) The market for Christmas trees has been noted in chapter 5.

Transmission poles. Telephone and electricity transmission poles are an important use of coniferous roundwood of small sawlog size (Aaron and Oakley 1985). The annual UK requirement is for about 150,000 poles of which 30,000–40,000 are currently supplied from British forests, mainly of Scots pine. Though criteria of length and stem-form are demanding, poles offer an adequately profitable return to the grower. BS 1990: Part 1: 1984 gives dimensions and quality requirements for overhead power and telecommunication poles, and gives the relative strength of species. However, the two main consumer industries, Telecommunication and Electricity supply, have each stated their own specifications and have limited the number of species and size categories. The methods of procurement vary between the two industries, but long life is a prerequisite for the poles, and it is essential that seasoning/preservative treatment is carried out 'by specialist companies using pressure and vacuum cycles and employing specialist techniques.

Scots pine, the preferred timber for poles, is the yardstick for assessing the performance of other species. It is now the only species, purchased from British forests, acceptable to British Telecom plc. Other species, mainly Corsican pine, representing only a small fraction of the total requirements, are sometimes purchased by pole producers for treatment by sap displacement techniques for electricity transmission poles.

Both British Telecom plc and the Electricity Companies buy their poles only after they have been seasoned, fabricated and treated. This is performed by a limited number of specialist companies: the names and addresses of these pole purchasers can be obtained from the Secretary, Sleeper and Pole Section, The Timber Trade Federation, Clareville House, 26/27 Oxendon Street, London SW1Y 4EL. Members of that Section can also give advice about the standards required and explain their inspection procedures. In any particular stand only a comparatively few stems will be acceptable: under-thinned and drawn-up poles may be ideal. Price per pole depends on the market, size and transport distance. It is invariably more attractive than sawlog prices, even after selection costs. The most likely requirements are Light and Medium classes noted in Table 32.

Table 32: Transmission pole classes: measurements in mm unless otherwise stated.

Nominal Length m	Light poles			Medium poles		
	Diam. at top		Min. diam. 1.5 m from butt	Diam. at top		Min. diam. 1.5 m from butt
	Min.	Max.		Min.	Max.	
6	135	160	160	—	—	—
7	135	160	170	150	180	—
8	135	160	180	155	190	—
9	135	160	190	160	190	230
10	135	170	195	—	—	—
11	135	170	205	160	200	250
12	—	—	—	160	200	260
13	140	180	220	170	210	270
15	—	—	—	175	215	300

The foregoing are ub dimensions. Butt diameters are accepted above the minimum, but top diameters cannot exceed the diameters stated. The main requirements are in 7 m, 8 m and 9 m Lights, and 9 m Mediums. The poles must be straight, free from defects such as rot, big knots, rings of big knots, inbark and physical damage. Slight single sweeps and reverse sweeps are allowed within strictly defined limits; sweeps in two planes (double sweeps) are not permitted. Most poles are still imported and the number of poles from British forests has declined primarily because of the restriction of species acceptable to the main users. It is, however, a prestigious market offering good returns from quality stands and it is pleasing to know that one's trees are assured of another forty to fifty years 'life' (or even twice that length of time) serving a critical structural function in our major utilities.

Other poles. These include poles for scaffold, ladder, rugby and flag; also for river and sea defences, jetties and wharves.

Pitprops. British Coal Corporation[1] purchases all its pitprops from British growers. Only conifers are accepted. Demand is declining steadily as uneconomic mines are closed and because of changes in deep mining technology and increase in opencast mining; but the market is still much appreciated. The pitprops are chiefly peeled, but unpeeled are still used in small quantities in South Wales. Almost any conifer is acceptable to the following specification:

	Lengths (m)	Top diameter (cm)
Peeled props	0.6–3.6	6–23
splits	0.9–3.0	10–20
Unpeeled props	0.9–2.7	8–18
splits	1.35–3.6	11–20

British Coal Corporation Specification 695:1985 sets out the full requirements for the props and splits, relating to dimensions, straightness, knots, peeling, butt flare and seasoning. Suppliers are required to obtain BS 5750 registration (noted later under 'Sawn mining-timber'). The Corporation has introduced a series of package standards for timber which have assisted greatly in terms of the use of mechanical handling equipment, transportation and storage. Suppliers have to conform to the appropriate package standard which in the case of props and splits necessitates the use of self off-loading vehicles.

British Coal Corporation award contracts for the pitprops and splits based on competitive tender. Prices (based on £/m^3 but converted in accordance with agreed packaging requirements into 'price per pack' for ordering and delivery) vary and are not disclosed, but they are believed to represent a reasonable financial yield to the suppliers. In such circumstances it is difficult to ascertain the standing value (£/m^3) to the grower for the conifers from which the props are produced.

Foresters who consider tendering to supply props are advised to seek guidance from experienced suppliers. Specifications are rigid and demands somewhat erratic.

---

[1] British Coal Corporation, Headquarters Supply and Contracts Department, Coal House, Doncaster DN1 3HJ.

Usually it is a market best left to 'middlemen' or forestry companies, but remains an invaluable outlet for appropriate conifer thinnings.

Conifer sawlogs. Consumption of conifer sawlogs (termed 'millwood' in Scotland) has increased substantially. The Forestry Commission's two categories have been noted in Table 31 *supra*. The demand is for relatively long lengths and medium to large diameters (smallest in Scotland), but some sawmills prefer certain multiple lengths, with small to medium diameters. In broadleaves there are two categories, 14–18 cm and 18+ cm, with an increasing proportion of logs cut to lengths in the forest: for 14–18 cm, in 1.8 m, 2.4 m, 2.7 m and 3.1 m; and in 18+ cm, in 3.6 m, 4.9 m and 5.4 m. The main markets for sawlogs are noted later in this section. The grower is fortunate in that the market for sawnwood mining-timber, pallets and fencing materials can absorb a considerable volume of relatively lower grade sawlogs. Medium sized thinnings are required for some of the foregoing as well as for boxwood and packaging. Larch and Douglas fir are in demand for interlaced and other fencing panels, the 'fencing bars' being required in 1.8 m lengths with a minimum top diameter of 14 cm. There is a need for more markets for lower grade sawn boards. The grower can obtain particularly high prices for superior quality sawlogs, especially Douglas fir, larches, spruces and pines. Indications of prices of conifer sawlogs are given in chapter 11. They fluctuate according to the supply/demand situation, the species and the location, and the prevailing price of sawnwood.

Paper and panels from coniferous small diameter roundwood are referred to later in this section. The same applies to wood *residues and chips.*

Sawnwood (Softwood). *Sawmilling.* The structure of the sawmilling industry has undergone major improvements in recent years. While the number of mills has fallen substantially, large units have risen in number, output and efficiency. There are some 400 sawmills in Britain (only about 100 principally engaged in sawing broadleaves) ranging from small estate mills sawing less than 1,000 m^3 a year, to large modern mills with an input in excess of 50,000 m^3 a year. However, some 65% of the total output of sawn softwood is produced by about 18% of the sawmills. Most of the larger sawmills are designed for continuous flow-line production and several are integrated with paper and boardwood mills, enabling wood residues and chips to be recycled. Thereby the pulping and reconstituted wood-based panel industries play an important role in the overall viability of sawmills.

Sawnwood timber output in the UK has increased dramatically. The main end use is construction (particularly builders' joinery, carcassing and civil engineering), supplemented by pallets, packaging, fencing and mining-sawnwood as indicated in Table 33.

In 1985, British sawn softwood accounted for an estimated 18% of the total consumption. The construction industry is by far the largest market sector for sawn softwood, taking about 70% of the total supply, but only 5% of this is British-grown. Conifer sawlog production is expected to increase from its 1990 level of about 2.4 million m^3 to about 4 million m^3 per annum by the end of the century. Expansion of the capacity of the larger sawmills is expected to cope with this increased volume.

Table 33. Forecast for market share development of sawn softwood by end uses 1993–97, compared with 1983–87.

	Units x 1000 m³ consumption					
	1983–87			1993–97		
Market sector	Imports	British	Total	Imports	British	Total
Construction						
Carcassing	1,880	205		1,620	550	
Builder's joinery	2,010	40		2,185	45	
Civil engineering	825	50		880	90	
Others	845	40		890	40	
	5,560	335	5,895	5,575	725	6,300
Pallets/Packaging	580	435	1,015	680	465	1,145
Fencing	575	300	875	610	320	930
Mining-sawnwood	–	200	200	–	60	60
Miscellaneous	245	40	285	185	30	215
	6,960	1,310	8,270	7,050	1,600	8,650

Source. Dr G.K. Elliott (1982).

The introduction of the strength class system of grading and the increasing attention being paid to good presentation of the sawn product, have assisted in the wider use for construction of British softwoods, particularly Sitka spruce. There is still scope for a significant increase in market penetration by British-grown softwoods into the very large construction sector (this is in addition to fencing, sheds and agricultural use; pallets and packaging, mining, rolling stock manufacture and repair, boat building, furniture manufacture and joinery).

Activity in the fencing, sheds and agricultural buildings sector is dependent upon levels of agricultural investment; in a smaller market, and one in which British timber already has more than a 40% share, there may be less scope for further penetration than in other construction uses. The demand for crates, cases and cable-drums is not expected to increase substantially, but the pallet sub-sector could absorb a higher proportion of British timber.

The profitability of sawmilling depends largely on the recovery rate of sawnwood achieved from the roundwood. The sawnwood recovery depends on the saws used, the species, sawlog cutting size, straightness, taper, and quality; in turn, the recovery determines the percentage of saleable residues. The Forestry Commission's two categories of sawlogs noted in Table 31 *supra* have been devised with the foregoing factors in mind. With a sawn recovery from a conifer log of 50%-70%, and a delivered-to-mill cost of £40 m³ the 'wood cost' at mill would be £60-£52 m³. If sawing cost is £50 m³, the sawnwood must sell at £110-£102 m³ to break even. The greater the recovery, the lower is the amount of saleable residues. From each m³ of sawnwood, residues may fetch £6-£20.

Building and construction. Here, several timber properties are important, including the ease with which it can be worked, its dimensional stability, and in certain cases its natural durability or treatability with preservatives. Its structure, strength and stiffness are of primary concern, and timber for this use is marketed in terms of its strength characteristics. Systems of stress grading and strength classification have evolved and are defined in British Standards (BS 5268: 1984; and BS 4471; 4978). Some timbers are intrinsically stronger than others and this is reflected in the strength class table. Thus the pines are stronger than the spruces. Virtually all British softwoods can be used for general framing and carcassing purposes, provided they are appropriately sawn and graded. Where enhanced durability is required Douglas fir and the larches are rated as moderately durable, and other species can be upgraded by preservative treatment. With the forecast increase in production, Sitka spruce can be expected to displace a substantial proportion of imported 'whitewood' currently used for carcassing and framing. However, cut-price imports and 'dumping' present problems.

Joinery. Timbers for joinery must finish well, be stable in use and, if used out of doors, be durable or accept preservative treatment. The requirements are set out in BS 1186: Part 1: 1986 which specifies limits for knot size and frequency, slope of grain and rates of growth. (See also BRE Digest 321: Timber for joinery.) Requirements are difficult to meet with short rotation, plantation-grown timber: the small proportion of British softwood suitable for joinery is obtained from older and larger trees, mainly Scots pine and Douglas fir.

Fencing and gates. Requirements for the three main types of wooden fencing are covered by BS 1722: Part 6 (palisade), Part 7 (post-and-rail) and Part 11 (woven wood and lap boarded panel). Motorway fencing is covered by the Department of Transport's *Specification for highway works,* 1986. The requirements for field gates and posts are set out in BS 3470: 1975. Because of its strength and durability, larch is the popular softwood timber for almost all types of fencing, gates, and similar farm and estate uses. Larch and Scots pine, suitably treated, are commonly used for posts and all of the softwoods are used for fencing rails, again with appropriate preservative treatment. Spruce is not accepted for motorway fencing posts. Though the pines do not have a high natural durability, they are easy to treat with preservatives and the treated timber is very suitable for situations where there is a high decay hazard; larch, on the other hand, has a higher natural durability than pine, but is much more resistant to preservative treatment. BS 3470: 1975 lists European larch, Scots pine and Douglas fir as being suitable for field gates and posts. The fencing trade fluctuates; currently (1991) it is depressed.

Farm buildings. The requirements for the various types and classes of farm buildings are set out in BS 5502: 1980–86. The pines and spruces, suitably treated, are used extensively for vertical space boarding and for purlins. The pines are easily treated with preservatives and can be used, suitably treated, in high decay risk situations. The heartwood of larch and Douglas fir is suitable for constructional purposes, without preservative treatment, for a twenty-year design life, in situations exposed to the weather, but not in ground contact or where the wood is liable to remain damp for long periods. If sapwood is present, appropriate preservative treatment is required. Large poles are often used for barns.

Pallets. The requirements are covered in BS 2629. Most main conifers can be used provided that due regard is given to the occurrence of the strength-reducing characteristics, e.g. knots, slope of grain, amount of wane, decay or insect attack. Pines are used extensively because of their strength, nailing and nail-holding properties. The spruces, with their good strength-to-weight ratio are suitable for light-weight pallets. Larch is the strongest for pallets, but it does not take nails and other fastenings as easily as pines and spruces.

Wood residues and chips. Consignments of wood residues and chips to pulp and boardwood mills have increased in quantity and play an important role in the profitability of sawmilling and of forest management. Currently, in sawmilling, some 1.23 million m^3 residues are produced annually from conifers and 0.12 m^3 from broadleaves. Wood residues consist principally of industrial residues, e.g. sawmill rejects, slabs, edgings and trimmings, veneer log cores and rejects, sawdust, bark, and residues from carpentry and joinery production. They have not been reduced to small pieces. Chips and particles are wood deliberately reduced to small pieces from wood in the rough or from industrial residues, suitable for pulping, for particleboard and fibreboard production, for fuelwood or for other purposes. At least one pulpmill and one particleboard mill cannot accept wood residues. Spruce chips command a premium of £7-£8 per tonne over those of larch. Some residues are used for fuel; others (e.g. bark and chips) for riding surfaces, horticultural, and other purposes.

Sawn mining-timber. British Coal Corporation[2] purchases the majority of its sawn-wood mining-timber from British-grown sources. Demand is declining steadily as uneconomic mines are closed. Yet in 1988 mining sawnwood consumption represented over 17% of the total domestic sawnwood output; and hence remains an invaluable market for the grower. The quality of the roundwood required to produce the sawnwood has needed to be improved as a result of the 'Quality Assurance Scheme' under BS 5750. Most of the sawnwood is softwood, but a proportion of some categories can be hardwood (poplar and willow not accepted). The main categories are: chocks, boards for self-advancing supports (SAS boards), sleepers, and miscellaneous uses including crowns/baulks, pillarwood, coverboards and lids/blocks. Most conifers are acceptable, especially pines, spruces, larches and Douglas fir. The higher quality of roundwood needed to meet the quality requirements of the sawnwood has been reflected in somewhat higher relevant prices to the grower.

The 'Quality Assurance Scheme' identifies basic disciplines and specifications. Applications for Registration have to be made to British Standards Institution, P.O. Box 375, Milton Keynes MK14 6LO. Fees have to be paid. The sawmiller must keep records of production and source of supply, and be able to identify decay and fungal growth; the machinery must be capable of producing to specification; and a named person must be responsible for quality. BSI Inspectors make periodic surveillance visits. Intending suppliers, when registered under the terms of BS 5750, Part 2, can approach British Coal Corporation with a view to being included in future Tender invitations; and of course must obtain a contract before starting to prepare produce.

---

2 British Coal Corporation, Headquarters Supply and Conracts Department, Coal House, Doncaster DN1 3HJ.

Registration in terms of BS 5750 does not guarantee that business will be awarded – contracts are given following annual competitive tendering.

Foresters are unlikely to become involved in producing mining sawnwood but many will desire to supply sawmillers with appropriate roundwood either conifer or broadleaved.

Wood processing industries using small diameter roundwood

This industrial sector mainly comprises paper mills, board mills and manufacturers of cement-bonded panels and moulded-board. Their interests are promoted by the United Kingdom Wood Processors Association whose members have invested over £1 billion in the industry; and most support the FICGB by a levy on supplies of roundwood. Their forecast of demand for 1990 to 1994 has been shown in Table 29.

The total capacity may eventually exceed the available supply of British roundwood in the short to medium term. A shortfall of some 200,000 tonnes annually has been predicted, in the context of a total production of 5 million m^3. Much depends on whether the mills operate to their maximum capacity, which may not necessarily happen. Some of the additional demand, for a limited period, may have to be met by modest imports of roundwood or wood chips (the cost might be double that of home-grown supplies). The industry, understandably, strongly urges an increase in afforestation.

These industries have been gearing up to secure requirements of small roundwood and residues throughout the 1990s and beyond. Competition is active within the sector as well as with those sawmills which use small diameter sawlogs. Some processors have immense harvesting organizations. A few have invested in their own or associated woodlands.[3] They are serious purchasers of productive or immediately pre-productive plantations.

Periods of over-supply of small diameter roundwood sometimes occur, especially after long spells of dry weather, the yards becoming fully stocked, creating temporary curtailment of intake. Other causes may be high inflation and cost of borrowing; and excessive overseas competition. A few particleboard mills can accept stored small diameter roundwood to assist harvesters. There is always recognition by the mills of the need to safeguard the services of harvesting and marketing

---

3 Shotton Paper Company plc in North Wales, endeavouring to secure future supplies of small diameter conifer roundwood, in 1988 began entering into a 10-year Forest Management Scheme with private owners including institutions. Thereunder they give a free management and contract supervision package in return for the sole right to harvest all timber from participant woodland owners. The latter's return is enhanced with lower unit costs and higher timber prices. Shotton are in close association with BSW Timber plc which owns eight sawmills throughout Scotland and Wales and which are always seeking sawlogs (40% returns to Shotton in the form of chips for pulping). Shotton has over 20,000 hectares under its scheme; and also owns outright several thousand hectares of plantations in Wales and Scotland. (Shotton Forest Management, Carlisle Sawmill, Rockliffe, Carlisle CA6 4BA. Tel. 0228-748934.)

contractors. A sudden glut of supplies can (not inevitably) restrain prices paid to growers; but will disrupt harvesting and marketing.

Paper and paperboard. The production of pulp for paper and paperboard currently consumes between 40% and 50% of Britain's total available small diameter roundwood. The purpose of pulping is to separate the wood fibres so that they can be re-formed into newsprint, magazine paper, cardboard, paperboard, stationery and many other domestic paper products. The preferred species are those with pale-coloured wood; long, strong fibres; and low resin and other extractive content. The processes used to achieve the fibre separations are (Elliott 1988; Harding 1988):

> *Mechanical:* a fibre pulp is obtained using a grindstone (groundwood pulp) or by the defibration of wood chips by means of rotating grooved discs (thermomechanical pulp). Yields are high, at around 95%. Mechanical pulp is usually blended with a proportion of chemical pulp when a better quality finish is required.

> *Chemical:* a fibre pulp is obtained by cooking the wood chips with solutions of sulphate or sulphite to remove most of the lignin and hemicellulose. Pulp yields are low, at around 50%, but the paper produced has a high quality finish.

> *Semi-chemical* (for hardwoods): softening is achieved by a mild chemical treatment followed by mechanical separation of the fibres in a disc refiner. The pulp is used for corrugated board and cardboard.

The fibre characteristics of all the main British-grown conifers make them suitable for pulp production. Because of their chemical composition, some are more amenable to pulping than others, but by judicious choice of pulping conditions and after beating and bleaching all species except yew may be processed successfully. Spruce, because of its pale colour, good fibre characteristics and low resin and other extractive content, is the preferred wood for high quality mechanical pulp used in newsprint and other paper products. A wider range of species is used for processes involving chemical treatment, including semi-mechanical pulping (an important outlet for broadleaved small diameter roundwood), but the presence of extractives can cause problems in processing.

Preliminary steps in production of pulp are debarking and sometimes chipping; these are followed by a choice of pulping techniques as already noted. Of the species used for pulp about seven-eighths are conifers and one eighth broadleaves. The favoured conifers are Sitka and Norway spruce. Occasionally pines are acceptable but not Douglas fir and the larches. The wood is required fresh, and the grower should deliver as soon after harvesting as possible, so that weight loss to his disadvantage is reduced.

Information on the current (1991) four pulpmills using British-grown small diameter roundwood – three using conifers and one using broadleaves – is given in chapter 11 (see Figure 22). Their products are well-matched to the market opportunities. The mills themselves – one in Scotland, two in Wales and one in England – are appropriately located (except for the south-east of England) to take advantage of current and increasing supplies of conifer pulpwood; the broadleaved supply is about static.

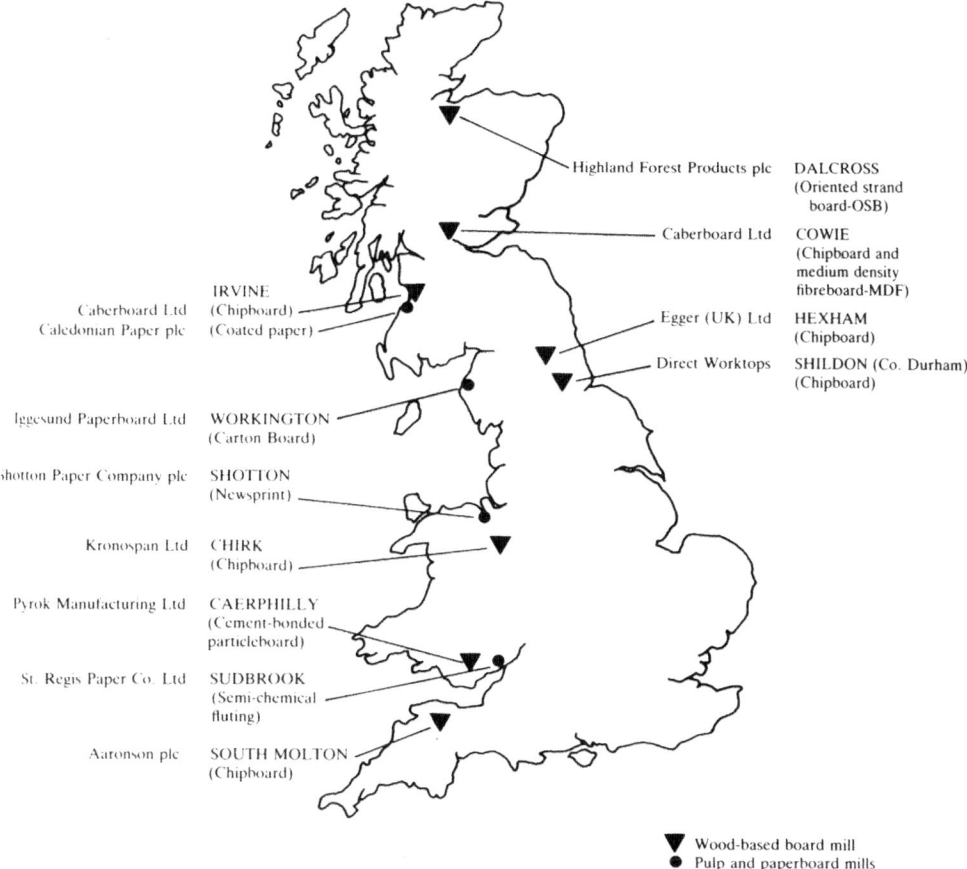

Figure 21. Locations of the major wood processing mills using small diameter roundwood (1991). Note the lack of relevant markets in the south-east of England

The export pulpwood market. The Scandinavian market, particularly Sweden and Finland, helped to maintain British-based harvesting resources following the contraction of the domestic conifer pulp industry at the beginning of the 1980s, at one time taking around 20% of British-grown conifer production, but is now reduced to a much smaller per cent of Britain's entire conifer production. (In 1989, 168,000 tonnes were exported.)

Wood-based panel board. These materials fall into three main groups: particleboard, fibreboard and oriented strand-board; in addition are cement-bonded panels and moulded-board. (Plywood is imported, and is not discussed herein.) Panel products are used by the construction industry for sheathing and shuttering. Reconstructed panels (particleboard, fibreboard, waferboard, oriented-strandboard and medium-density fibreboard) are used in furniture, for flooring, and DIY markets. Relevant information on the wood-based panel products is provided by the Building Research Establishment in 'Wood-based panel products – a specifier's guide' (BRE Folder AP31, 1987) and 'Selecting wood-based panels' (BRE Digest 323).

Particleboard (chipboard) is a sheet material which uses a substantial quantity of small diameter roundwood and sawmill residues (chips, flakes, edgings, offcuts and sawdust) and much otherwise 'waste' wood. Properties of the wood are of importance for their effect on board qualities, which are also determined by aspects of manufacture, such as particle size and distribution and adhesive type. Wood density is of some relevance for its effect on cutters during the chipping process, and wood colour perhaps more so since, in many applications, a pale-coloured board is preferred and the amount of dark-coloured heartwood in the finish must be restricted; bark, too, detracts from the appearance of boards but usually this is not removed from roundwood by the manufacturer. The presence of extractives in the wood can adversely affect the setting of adhesives, leading to a reduction in internal bond strength. Most main conifers are used to some extent in the production of chipboard, but manufacturers prefer the spruces and pines. Smaller quantities of larch and Douglas fir are used because of their darker colour and the presence of extractives; for this reason they are mixed with other species. Price and availability of regular supplies are usually the main factors influencing the choice of wood raw material.

Oriented strand-board (OSB) is similar in some respects to waferboard, which is manufactured in the UK from large flakes of mainly Scots pine wood bonded together with a phenolic resin. The flakes used in OSB are narrower than those used in waferboard allowing some measure of alignment somewhat like that of adjacent veneers in plywood.

Fibreboard is a panel produced by the adhesion of fibres and fibre-bundles to form sheets, achieved by one of two processes depending upon the board type required: (i) a wet process, in which a pulp slurry, after draining, is pressed using heat (the adhesive properties of the lignin bind the fibres to form a board); and (ii) a dry process, in which a dry mix has resin added to bind the fibres under heat and pressure. This type of board is known as medium-density fibreboard (MDF). The need to reduce the wood to fibre-bundles requires a high power consumption, favouring the use of the lighter-weight woods. The presence of resin in the wood can be troublesome for its effects on the cutting and beating operations. Although the presence of bark is tolerated in some measure, it affects the manufacturing process and can also detract from the appearance and quality of the board. Several types of fibre building board are produced, differing primarily in the density achieved in pressing. The quality of the wood raw material is normally agreed between supplier and manufacturer. All the conifer species are used to some extent in the manufacture but the spruces are preferred to the more resinous species.

Cement-bonded board, moulded-wood, and wood wool. Wood wool is aggregated strands of wood which are used to protect fragile goods from damage by shock loads; it is also used in the manufacture of wood wool-cement slabs (Aaron and Richards 1990). The most important properties for processing wood wool are light weight, straight grain and freedom from knots. Corsican pine is the preferred species. Tough but soft and resilient fibre is the main consideration for packing items. For foodstuffs, fruit and upholstery, freedom from odour, gums and resin is essential to avoid staining the packaged items. When used for wood wool-cement slabs, high extractive contents can retard or impair the setting of the cement. Decayed wood must not be used but the presence of blue stain does not have a marked effect on the cement-setting process. The spruces and pines are the preferred species for wood wool products, the spruces because of the pale colour and low extractive content, the pines because they shred easily. Colour and extractive content make the use of Douglas fir and larch less acceptable. The timber has to be straight and peeled, but not basted, be free from large or protruding knots, and the ends must be sawn at right angles to the lengths. The logs have to be cross-piled in the open for seasoning, for a period of four months if felled from October to February and of two months if felled from March to September. Some cracks are permissible.

Currently (1991) there are seven wood-based panel board mills using Britishgrown small roundwood, some of conifers, some of broadleaves, and some of both. Additionally there are three manufacturers of wood wool, cement-bonded board and mouldedboard. In terms of volume of small diameter roundwood used, the industry is second only to sawmilling and, including wood residues utilized, some 2 million tonnes annually are processed into varying forms of board for housebuilding, factory reconstruction, shuttering, furniture and flooring. The mills are appropriately located (except in the south-east of England) to take advantage of current and increasing supplies of boardwood raw material (though broadleaved supply is virtually static). Information on the current (1991) boardmills using British-grown small diameter roundwood (see Figure 22) is given in chapter 11.

References

Aaron, J.R. (1970), 'The Utilisation of Bark'. FC Research and Development Paper 32.
—— (1980), 'The Production of Wood Charcoal in Great Britain'. FC Forest Record 121.
—— (1982), 'Conifer Bark: its Properties and Uses'. FC Forest Record 110.
Aaron, J.R. and Oakley, J.S. (1985), 'The Production of Poles for Electricity Supply and Telecommunications'. FC Forest Record 128.
Aaron, J.R. and Richards, E.G. (1990), *British Woodland Produce*. Stobart Davies, London.
Anderson, M.L. (1967), *A History of Scottish Forestry*. Thomas Nelson & Sons, London.
Brazier, J.D. (1979), 'Never mind the Trees, What about the Wood?' Building Research Establishment Information Paper IP 12/70.
—— (1985), 'Growing Hardwoods: the Timber Quality Viewpoint'. *Quarterly Journal of Forestry 79* (4).
—— (1990), 'The Timbers of Farm Woodland Trees'. FC Bulletin 91.
Brazier, J.D., Hands, R. and Seal, D.T. (1985), 'Structural Wood Yields from Sitka Spruce: the Effect of Planting Spacing'. *Forestry and British Timber,* September 1985.
Building Research Establishment (1983), *A Handbook of Softwoods;* (1984), *Specifying Structural Timber;* (1986), *Timber Drying Manual* (2nd edn.); (1988): See Harding (1988).
Burd, C. (1989), 'British Softwood: Going through the Mill'. FC 'Forest Life' (5).
Desch, H.E. (1986), *Timber – Its Structure, Properties and Utilisation,* 6th edn. revised by J.M. Dinwoodie. Macmillan, London.
Dewar, J. (1988), 'Private Sector Production Forecasting in Scotland'. *Timber Grower,* Winter 1988.
Elliott, G.K. (1982), 'Wood: an Essential Renewable Raw Material'. *Span,* 25 March 1982, pp. 128–30.
—— (1988), 'A Perspective of the British Timber Scene'. FC 'Forest Life' (4), August 1988.
Evans, J. (1984), 'Silviculture of Broadleaved Woodland'. FC Bulletin 62.
Farmer, R.H. (1986), *Handbook of Hardwoods,* 2nd edn. BRE.

Fewell, A.R. (1983), 'Strength Classes for Structural Timber'. Building Research Establishment (1983) BRE News 59.
Francis, G.J. (1989), 'The Case for Investment in Forestry'. In *Proceedings of a Discussion Meeting on UK Forest Policy into the 1990s,* Bath University 31 March to 2 April 1989. Institute of Chartered Foresters, pp. 87–108.
Grundy, D.S. (1985), 'Developing the Economic Arguments for Investment in Forestry'. FC Research and Development Paper 145.
Harding, T. (1988), 'British Softwoods: Properties and Uses'. FC Bulletin 77.
Hart, C.E. (1966), *Royal Forest: A History of Dean's Woods as Producers of Timber.* Clarendon Press, Oxford.
Henman, G.S. and Denne, M.P. (1988), 'Control of Shake in Oak'. FC Report on Forest Research 1988.
Holmes, G.D. (1975), 'History of Forestry and Forest Management'. *Philosophical Transactions* of the Royal Society of London B271 pp. 69–80.
International Labour Office, Geneva (1988), 'Fuelwood and Charcoal Preparation'.
James, N.D.G. (1981) *A History of English Forestry.* Basil Blackwell, Oxford.
Jobling, J. (1990), 'Poplars for Wood Production and Amenity'. FC Bulletin 92.
Johnston, D.R. (1979), *The Craft of Furniture Making.* Batsford, London.
—— (1982), 'Getting the Best from British Forests'. *Forestry and British Timber,* March 1982.
—— (1983), *Wood Handbook for Craftsmen.* Batsford, London.
Keighley, G.D. (1987), 'Wood as Fuel: a Guide to Burning Wood Efficiently'. FC Occasional Leaflet.
Kupiek, J. and Philip, M. (1989), 'Production Forecasting for Private Sector Scottish Forestry'. A study for the Scottish Forestry Trust. *Scottish Forestry, 43* (2), pp. 99–104.
Malyon, H. (1989), 'The Private Sector Forecast – a Practical Approach'. *Scottish Forestry 43* (4), October 1989.
McGregor, P.G. and McNicholl, I.H. (1989), 'The Impact of Forestry on Output and Employment in the UK and its Member Countries'. Fraser of Allander Institute and Department of Economics, University of Strathclyde.
Patterson, D. (1988), *Commercial Timbers of The World.* (5th edn.). Gower Technical Press, Aldershot.
Pearce, M.L. (1987), 'Fuelwood Demonstration Plots at Stoneleigh'. FC Occasional Leaflet.
Savill, P.S. and Evans, J. (1986), *Plantation Silviculture in Temperate Regions.* Clarendon Press, Oxford.
Savill, P.S. and Mather, R. (1989), 'Oak Shake'. *Forestry and British Timber,* October 1989.
—— (1990), 'A Possible Indicator of Shake in Oak: Relationship between Flushing Dates and Vessel Sizes'. *Forestry 63* (4), pp. 355–62.
Thompson, J. (1990), 'Employment in Forestry and Primary Wood Processing in Britain'. FC Occasional Paper 27.
Timber Research and Development Association (1987), TRADA wood information Section 2/2, Sheet 10. (A companion TRADA Sheet is: 'Wood, decorative and practical' 1985.)
—— (1989), 'Guide to Stress Graded Softwood'.
Whiteman, A. (1991), 'The Supply and Demand for Wood in the UK'. FC Occasional Paper 29.
Wilson, I. and White, D.J.B. (1980), *The Anatomy of Wood – Its Diversity and Variability.* Stobart, London.

Chapter Ten

HARVESTING

The forest being a renewable resource, harvesting is part of a natural and necessary sequence of events in forest management. Trees can be harvested either part-way through a rotation as thinnings, or at the end of a rotation by clear-felling. The method of harvesting depends to some extent on the silvicultural system being followed. This chapter relates particularly to the clear-felling system; but part could be applied to the group and selection systems, though for them extra care and higher cost may be applicable. Thinning, a large component of harvesting, has been discussed in chapter 8, and part of the present chapter supplements it. Utilization of the harvested produce and consideration of the markets to be supplied for processing are dealt with in chapter 11. The interaction of clear-felling and restocking, and environmental considerations in harvesting, are discussed, respectively, in sections 4 and 5 *infra*.

1. The rotation length

First to be considered is the age of felling, i.e. the rotation. The economics relevant to it are noted in chapter 14. Forestry being a long-term investment, the length of the rotation is a paramount consideration. Two important external factors which exert an influence on the rotation are growth rate (which depends on species and site factors) and the market conditions (i.e. the demand for either small diameter roundwood or sawlogs). An equally important influence may be the owner's objectives in relation to non-wood benefits such as amenity, landscape, wildlife conservation and sporting, which may reflect either personal wishes or perhaps the pressure of public opinion if the woodland lies in a sensitive location.

Subject to any constraints, there are six main ways of determining the length of the rotation, each of which has been developed under varying conditions to serve different objectives. First is the *physical rotation*: the trees are grown until they die of old age (the conditions in which this rotation may usefully be adopted is in 'protection forest' or 'scientific forest' where timber removal is undesirable or perhaps impossible). Second, the *silvicultural rotation*: the trees are grown until their natural vigour abates or until conditions are reached when restocking is most easily accomplished (this rotation may be of value for silvicultural or ecological research but, like physical rotation, it has little application in commercial forestry practice). Third, the *technical rotation*: the trees are grown to supply a specific market, the crop being felled when it reaches a marketable specified size (this is probably the most common type of rotation in private forestry). Fourth, the *rotation of highest net annual income*: the forest is managed to yield ultimately the highest net annual income per hectare regardless of the time taken to achieve this state or the capital involved (for this reason it cannot be accepted as economically sound). Fifth, the *rotation of maximum mean annual volume increment*: the forest is managed to yield the highest volume of timber per hectare (this rotation may be unsatisfactory for two reasons – no allowance is made for difference in the unit value of timbers of varying sizes, and, as in the rotation of highest net annual income, input is ignored). Finally, the *financial rotation:* the rotation which is the most profitable, i.e. giving the highest return on the capital invested.

The first three types of rotation do not aim directly at maximization and of them only the technical rotation is commonly used in private forestry. The three remaining types aim at maximization but the rotations of highest net annual income and of maximum mean annual volume increment may be unsatisfactory for the reasons given. Thus only two of the six types are usually applicable to private commercial forestry, namely, the technical and financial rotations. The advantages of the technical rotation are that it is straightforward to manage as it involves no detailed financial calculations; but the disadvantages are that it allows no check on the efficiency of the use of capital and land. There are two ways that the financial calculations may be of value to foresters: (i) in determining, for planning purposes, the return on the capital invested and the rotation at which this is maximized; and (ii) in determining, at any stage during the plantation's life, the relative advantages to be gained from felling the crop at once or retaining it for a further period. Methods of appropriate appraisal are discussed in chapter 14.

Rotation length is the main silvicultural 'tool' available for foresters to influence final tree size. A range of rotations for various species has been indicated in chapter 8 under 'Thinning recommendations'. Windthrow Hazard Classification (see chapter 3) is widely used by foresters to predict probable rotation lengths for forest stands. Table 34 illustrates how the economic rotation age can be determined. The rotations shown for conifers are for thinned and non-thinned models of crops based on a discount rate of 5% and using 1988–89 price-size curves (see chapter 11).

Table 34. Felling ages for conifers. Average age of maximum DRo at 5%.

Rotations for thinned conifers

Yield Class	SP	CP	LP	SS	NS	JL	DF	EL	GF	NF	WH	WRC
4	77		64			57	58					
6	74	64	63	64	75	56	56					
8	69	63	62	63	71	54	62	53				
10	63	60	59	59	66	50	57	48		65		
12	58	56	54	56	61	46	53	44	56	63	63	65
14	53	53	49	52	56	43	48		53	61	59	62
16		49		49	53		46		50	58	54	58
18		46		46	51		45		48	54	51	56
20		45		44	49		44		46	52	48	52
22				43	49		44		45	49	46	50
24				42			44		44		44	48
26							44					
28							44					
30							44					

Rotations for non-thinned conifers

Yield Class	SP	CP	LP	SS	NS	JL	DF	Other Conifers
4	71		64			51		No information available. Estimate as 5 years before 'thinned' *supra*.
6	65	58	53	58	66	46		
8	61	53	50	56	62	44	56	
10	57	51	47	54	59	43	54	
12	55	48	45	52	57	41	52	
14	53	47	44	50	55	41	50	
16		46		49	53		47	
18		45		47	50		45	
20		44		45	48		42	
22				43	46		40	
24				40			39	

Source: Forestry Commission Investment Appraisal Handbook, unpublished but reproduced by permission.

Rotations for broadleaved high forest are usually longer than for conifers. The normal rotation age for oak may be 120–160+ years, beech 100–130, and sycamore/ash 60–80. However, broadleaves even more than conifers, have relatively rigid lower diameter limits for particular uses (noted in chapter 9). The large minimum diameter limits to attain broadleaved prime timber and veneer qualities, combined with relatively slow diameter increments (rarely more than 55 mm/year for oak and beech, or 88 mm/year for sycamore, ash, cherry and walnut) give rise to the long rotations associated with growing high quality broadleaves, according to site quality, as noted in Table 35. However, broadleaved rotation length is principally influenced by technical considerations, usually diameter of log (coppice being an exception). In oak and beech, rotation age is well past the age of maximum mean annual increment. For poplar, see Jobling (1990).

When arranging felling, it may be prudent to retain some stable scattered timber trees, particularly broadleaves, past their normal rotation age for reasons of landscape, amenity and nature conservation. However, there are attendant risks in such a policy. Overmature trees are prone to disease, dieback, windthrow and snowbreak; they can become unsafe. They have very little height growth, and basal area increment is less than 1% – a loss because site growth potential is not being realized. Furthermore, they may adversely affect restocked plantings. Evans (1984) provides a table of ages generally indicating significant over-maturity in broadleaves

There may be an increasing trend to regulate rotations to provide a measure of continuity of supply to wood-processing industries. Steady buying of woodlands by wood processors and sawmillers is likely to continue, particularly to ensure short or medium term supplies of small diameter roundwood and sawlogs. This is encouraged by the knowledge that premature felling of stands is an economic decision for the forest owner, and cannot be dictated by the Forestry Authority.

Where a mill is using small diameter roundwood (and particularly when associated with a sawmill using the lower sizes of sawlogs) the rotation may be shortened from

Table 35. Broadleaves: minimum rotation lengths (years).

Species	Prime timber			Veneer*	
	Site quality			Site quality	
	fertile	average	poor	fertile	average
Oak	80	100	130	120	160
Beech	75	95	120	110	140
Sycamore	45	60	75	55	70
Ash[†]	55	70	90	70	90
Sweet chestnut	50	65	85	65	85
Cherry[‡]	50	65		60	80
Walnut*	50	65		60	
Poplar[§]				25	30

* On poor sites (Yield Class 4 or less), slow-growth and often defective stems rule out most possibilities of obtaining veneer quality material.
[†] Sports quality ash may be obtained on shorter rotations than shown.
[‡] Cherry and walnut are rarely felled only for prime timber: though both are used for high-class joinery, material is mostly obtained from trees felled for veneer.
[§] Shows rotation for peeled veneer for vegetable crates.

Source: Evans (1984). FC Bulletin 62.

the conventional. The occurrence of a heavy mast year (perhaps complemented by advance regeneration) should not be overlooked (see chapter 7). Another dictate of felling age may be the expected extent of shake in oak or sweet chestnut. Cutting to take advantage of temporary high prices (marketresponsive or price-responsive cutting) may be better than waiting till the planned end of a rotation. But an expectation of improving markets may justify prolonging a planned rotation. (The faster the predicted increase in real price, the more prolonged the planned rotation would be.) Thus the age of felling of stands, under various circumstances, can be wide in range. Broadleaves in particular are rarely determined by the long-term economics of the investment. However, foresters (when thinking commercially) will be aware as to when increment is declining or arrested; also when the size and condition of trees required for the most profitable market are reached.

2. Techniques of harvesting

A harvesting system comprises the combination of methods used to fell trees, extract the produce to roadside, and transport it to market. The components of the system are the tree crop, the road network, the machines, vehicles, and skilled workers. Supplementary are the markets for the produce, and the specifications they require.

The high level of salaries and cost of social security, strict ergonomic standards, specific harvesting conditions, and strong influence of environmental and conservation aspects, have led to need for improvement of methods, especially mechanization. Techniques are changing rapidly to increase output and reduce costs. Methods are tending to become cost-effective. The greatest hope for future productivity increases is likely to lie in the felling, delimbing and conversion phases. Relevant research is undertaken by the Work Study Branch of the Forestry Commission for traditional forest practices. Their relevant Output Guides and Standard Time Tables are available (see chapter 16). The Department of Energy is assisting in relation to the harvesting of biomass material. The growth of mechanized harvesting technology calls for an increasing need for specialists trained in forest engineering techniques. The Forestry Engineering Group of the Institution of Agricultural Engineers, chaired by the Forestry Commission's engineering director, acts as a forum for liaison between manufacturers, distributors and users as well as those who build the roads and bridges and haul the timber. The Group is helpful in defining the industry's real needs into the next century.

The main constraints encountered when choosing a harvesting system are connected with climate, soil, terrain, and the need to protect the site (especially for restocking) and any trees to remain on it. The produce moves from the forest to market in several stages. That from stump to road is usually relatively short, seldom exceeding 0.5 km, but it is the most difficult and, on a weight basis, the most costly component of the harvesting system. Each silvicultural system produces a distinctive form of crop which generates particular problems in felling and extraction. The harvesting system chosen by foresters is the one that gives the minimum cost of timber delivered to market while causing least possible damage to the site, and to the interests of restocking.

Harvesting is usually by three methods: (i) *tree length system* (using a skidder) which retains the delimbed tree as the unit of load until the roadside stage; (ii) *shortwood system* whereby the trees are cut into product specifications in the wood and thereafter extracted as large bundles (using a forwarder); and (iii) *whole tree harvesting* (discussed later) which has not been used to any extent in this country. The main factors which influence the choice of systems (i) and (ii) are: access, terrain, product mix and specifications, species, availability of labour, machinery planning, previous thinning systems, positioning of roads and tracks, size of area, and landscape considerations. Choice of tree length versus shortwood could be affected by the following (Taylor 1981):

- Shortwood extraction by forwarder is generally cheaper in unit cost, usually in the order of 10%, than most tree length systems under similar conditions;

- Tree length extraction by skidder requires higher density road systems, while forwarders can operate with lower road densities and to some extent with lower road standards;

- The ability to carry timber economically over fairly long distances, entirely on wheels, makes forwarders particularly attractive for work in estate woodlands. They can travel over intervening fields with minimum damage to grass, in contrast to the ploughing effect of skidders;

- The tree length system, a three-phase operation, is at its best with tree sizes greater than 0.1 m^3 average and where there is close supervision to ensure good integration between work at stump and conversion at roadside; it is more or less essential where there is to be a complex product mix or where one of the products is of small unit size;

- The shortwood system, a two-phase operation, is efficient irrespective of tree size and is easier to organize, being at its best with a simple product mix, e.g. sawlogs and pulpwood;
- The tree length system requires ample working space at roadside for conversion and stacking of produce, while the shortwood system requires stacking space only and products can be spread out over longer distances at little extra cost; and
- The shortwood system implies some manhandling of timber pieces at stump to facilitate forwarder loading; however, the tree length system also implies manual piling at roadside or the cost addition of a mechanical loading device.

On balance, the choice between the two systems may rest on other considerations involving the whole harvesting package rather than the extraction phase only. A single set of rules for selecting a harvesting system is impossible to suggest because different factors affect particular situations. For example, a stand where the principal product is pitprops is almost bound to need a tree length system, even if other factors favour shortwood. Similarly, foresters in charge of a large area might decide that the savings on road investment associated with a shortwood forwarder system outweigh all other factors. The two main components of harvesting – felling and extracting – are now discussed in greater detail.

Felling. Clear-felling having been decided upon, and the interaction with extraction, environmental and restocking interests duly considered (see sections 4 and 5 *infra*), concern should be given to ensuring an efficient and tidy operation. Organization and control are necessary. Felling will be affected by such local factors as: crop type (species, spacing, size of tree), terrain, and proportion of partially thrown, hung-up or snapped trees. If felling is to be followed by conversion, the specifications required by the market will be important. In general the larger and fewer the specifications, the higher the output and the lower the cost. The output is higher and the cost lower in clear-felling than in thinning. Harvesting following windthrow *(infra)* is the highest cost.

Felling and extracting can damage trees that are to remain, leading to decay and other degrade. In pole-stage and older crops, much mechanical damage can occur during thinnings because the space available for harvesting is limited. The extent of injury to the butts and superficial roots of standing trees depends on the resistance of the bark to abrasion, the diameter of the stems, the nature of the terrain, and the method of harvesting. The risk of damage to stems and roots rises with increasing slope and is also higher where an extraction path or rackway changes direction or where the ground becomes soft. Thus rackways should be as straight as possible, aligned up and down slopes, and kept to drier, firmer ground; they should also be widened at curves and at exits to main extraction routes.

Silviculturally, the choice of *season for felling* (other than of coppice) normally matters little; the operation is generally possible throughout the year. However, there are notable exceptions: broadleaved poles for turneries need to be felled before the sap starts to rise; oak, when its bark is to be stripped on the ground, will have to be felled during a very short period in spring, usually early May; felling beech in summer is bad practice for many purposes; oak and ash for planking and veneer should be winter-felled; and sometimes in order to facilitate the manual peeling of medium and large conifers it is necessary that the work be done when the sap has risen. Except in those cited instances the time of felling is usually governed by the consideration of the least costly and least damaging extraction. Sporting considerations sometimes affect felling times. Most felled broadleaves do not deteriorate to any great extent by a short delay in extraction and conversion, but ash, beech and sycamore, as well as most conifers, may suffer considerably from delay; this is because, though superficial fungal attacks cause staining rather than real decay, the deterioration is sufficient to lessen the attractiveness of the parcel to the buyer and so cause loss in value. Some foresters prefer to fell their broadleaves when the leaves have fallen. Clear-felling during summer should usually not be earlier than August if hardy weed growth is expected to be troublesome later and hinder restocking.

Felling (by chainsaw) is easier and less costly per unit when trees are large and there is plenty of space to move (as in late stage thinning and clear-felling). By contrast, in early thinnings, which combine small trees and confined working space, the cost of cutting and extracting will substantially reduce the standing value of the produce removed. The underlying cutting principle is that the operation should provide the best possible presentation for extraction. Felling should follow established techniques, the cutters exercising prudent caution and minimizing movement of the timber. Maximum use should be made of aid tools, of which there is a wide range (Howard and Hayes 1989). The Work Study Branch of the Forestry Commission have grouped various tools according to function, indicating within each group the advantages and disadvantages of the various types available. Many of the aid tools when used correctly can be of great assistance to the chainsaw operator. The techniques involved in the use of some tools may be self-evident, but the more sophisticated, such as winches and the felling cushion, require correct operator training and practice if they are to be used successfully. The main current aids are: breaking bars, pulp hooks, felling wedges, the Nordfeller felling cushion, turning hooks, cant hooks, tongs, and hand winches. (Frequent critical appraisal of the contents of the various trade catalogues is recommended to all interested in the subject of aid tools.)

Felling and delimbing (snedding), also conversion, demand great care. Lightweight chainsaws are most commonly used for small trees. Chainsaws are fitted with anti-vibration and noise systems to protect the health of the operator: and with safety measures to obviate accidents. Before starting any work it is essential that all operators concerned must have been trained adequately in the use, fuelling and maintenance of their respective machines and aid tools, in the correct working techniques, and be equipped fully with appropriate protective clothing. Details of these aspects are given in Forestry Commission FSC guides: numbers 10–15 cover chainsaws.

Skilled cutters fell close to the ground, keep breakage of the butt length to a minimum by felling on to smooth ground and not down-hill, and avoid damage to other trees. They fell in a direction that will best suit extraction; this is particularly important in thinning and selective felling and where there is an understorey. The quality of their work, particularly in large broadleaves, is shown in panelling the butts, in rounding out limb-scars, in topping, and in leaving the stump tidy. Special care is taken with stools intended to coppice: bad workmanship destroys the hope of good re-growth. Cording of branches of broadleaves and disposal of their lop and top will be effectively carried out.

After felling, cutters calculate the volume of the trees, and in the case of broadleaves record the number of cords. Experienced cutters are accurate in measuring felled trees, and on large trunks will usually scribe broad-arrows at stops or set-offs where they have begun and ended each measurement of length; sometimes they indicate where they have taken a mid-measurement. A number is usually allotted to each large tree and recorded on the butt either by nailing a metal disc or by scribing the number. Occasionally marking chalk is used but it is not always satisfactory through being indecipherable when the log reaches the sawmill. Disputes between sawmiller and grower can arise through faulty or inadequate marking or numbering. The cutter's measure will usually be accepted. The number of cords of hardwoods will be 'taken up'. Each cord will contain about 3.6 m^3 stacked volume; a common dimension is 2.5 m x 1.2 m x 1.2m. If burning of lop and top has been carried out, the extent will be paid for by the hectare.

Indications of ranges of rates paid for felling and delimbing (snedding) in thinnings, including treatment of stumps with urea (supplied) against *Heterobasidion annosum* are: small sized trees, £6-£10 m^3 or tonne, medium sized trees £4-£8 m^3 or tonne. Rates for conifers and broadleaves are generally about equal. They should be related to workmen's travelling expense, to the terrain, and to ownership and running expenses of the chainsaw. The lower end of the ranges might reflect favourable conditions, or the advantages of some previous brashing. The rates would increase slightly for minimal cross-cutting, and increase by £2-£5 m^3 or tonne for converting to small diameter roundwood products of specific lengths and diameters (e.g. bars, posts, stakes, rails, pulpwood and boardwood). The conversion component cost would be according to average tree size, or to product mix and specifications. The more complicated the latter the higher is the rate. In clear-felling, the comparable felling rates for large dimension trees are £2-£5 m^3 ob. Cording broadleaves costs about £5-£10 a cord. Burning a substantial amount of broadleaved lop and top might cost £100-£300/ha. Clearing by bulldozer is much cheaper. A firewood merchant might be willing to remove without charge, or even pay something for the produce.

Finally, it is worth repeating that an important factor in felling is presentation to the extraction method and equipment – largely governed by terrain classification (Rowan 1977) noted in section 3 *infra*.

Harvesting windthrown trees. Experience of harvesting windthrown timber has been built up over recent decades. Compared with normal clear-felling, harvesting of windthrow is inherently more difficult and slower in various ways, not least because of restrictive movement and increased danger to operators. Windthrow – either catastrophic or endemic (as discussed in chapter 3) – is notoriously variable and the harvesting problems must be considered in relation to each individual site. Scattered windthrow or snapped trees occur in many parcels of standing timber, and adequately trained and experienced operators will take these in their stride. More complex harvesting problems arise when the trees are windthrown in groups, sometimes extending over whole sub-compartments: Jones and Smith (1980) give detailed guidance (see also chapter 3, section 1). Harvesting windthrown timber is different from thinning or clear-felling, in that a greater degree of organization and control is required. The choice of method, and manpower and machinery requirements are determined according to local factors such as crop types, terrain types, distribution and type of windthrow (proportion of partially thrown, hung-up or snapped trees, degree of tangle), and marketing possibilities (i.e. what can be recovered, and what is subject to degrade if not harvested quickly). Another factor is protection: to be effective, stumps should be treated with urea against *Heterobasidium annosum* within twenty minutes of felling or, in the case of windthrow, of cutting off the tree from its 'plate'. If extraction does not follow almost immediately then access to the stump surfaces is obstructed and treatment delayed. In pine cr ops, where the usual stump treatment is by suspension of oidia of the competing fungus *Peniophora gigantea,* care should be taken not to contaminate the butt ends of the trees as this may lead to degrade of the timber. The likely incidence of beetle attack should be borne in mind (see chapter 6).

Following the great storm of 16 October 1987 in south-east England, a 'Forest Windblow Action Committee' was formed – comprising representatives of the Forestry Commission, Timber Growers United Kingdom, British Timber Merchants' Association of England and Wales and the UK Wood Processors Association. In the field was established the Committee's action team – the 'Windblow Task Force' – and together they quickly issued sound guidance on the felling, measuring and marketing of the windblown trees, safety measures, and restocking (Grayson (ed.) 1989). The storms of January 1990 led to renewed efforts on similar lines (see chapter 3, section 1).

With few exceptions (notably the pines) there is a time-scale of two to three years, exceptionally up to five years, in which trees which are not dangerous can be harvested and marketed in a systematic and economical manner. The Forestry Commission give the following general guidance on the sequence in which tree species should be cleared:

First: The pines (pine species are likely to harbour insects harmful to conifers during the summer and suffer from early staining of timber)

Sycamore, poplar, beech, ash, lime, plane and birch (some degrade may occur in the summer)

The spruces and silver firs

Douglas fir and western hemlock

The larches

Last: Oaks, sweet chestnut and yew: if blown over and intact they may be left for up to five years.

Trees of species which degrade quickly but have a limited value need not be given priority if they are only to be used for firewood or pulpwood. They should be removed as part of an orderly clearance programme. A sprinkler method for the wet storage of sawlogs can work successfully (Webber 1990; Thompson *et al.* 1990) providing a method of stockpiling timber surpluses that can occur after serious storms, without accompanying fungal degradation. Costs are likely to be £4-£6 m^3 over one or two years.

The felling of windthrown and broken trees is particularly hazardous and, whether the operation is undertaken by the owner's staff, or by contractors or merchants' labour, it is essential that Forestry Commission's FSC Guides are observed. Output Guides prepared by the Commission's Work Study staff provide guidance for foresters dealing with windthrown situations. The Commission's Education and Training Branches have joined in preparing the guidance available.

Harvesting, as already outlined, can be carried out by workers employed direct by the woodland owner, or by roundwood merchants to whom the standing trees have been sold, or by specialist harvesting teams who undertake all or some of the operations – felling, extracting, and transporting, perhaps along with some conversion. Many contractors operate on a sufficiently large scale to make worthwhile the full training of their operatives, to hold adequate machinery and other equipment, and to keep production teams employed throughout the year. Only owners of extensive woodlands can be expected to follow suit, but many medium sized woodland enterprises employ direct labour on small-scale harvesting, especially at times when other forestry work is scarce. Adequate training in safe working techniques is always essential.

Whole tree harvesting: forest biomass. In Britain the forest industry may in due course face a position where the demands of the forest products industry outstrip the raw material availability: this situation may particularly apply to small diameter roundwood. One of the ways in which potential shortfall on the supply side may be overcome is through the greater utilization of the existing forest resource (Hummel et al. 1988).

Whole tree harvesting and recovery of forest residues are significant developments in the concept of greater utilization of forest biomass. Essentially, these techniques involve the recovery of the majority of the above ground components of the tree in a harvesting system. Hakkila (1989) traces the developments in harvesting systems over the last twenty years or so. He covers the whole field of utilization of forest residues, detailing the determination and inventory of forest residues, technical properties of residual tree components, recovery and comminution of forest residues, transport, examples of harvesting systems through to utilization for pulp and paper, fibreboard, particleboard and structural flakeboard, energy, fodder, protein and vitamins as well as chip upgrading. Finally the problem of the ecological consequences of residue removal on the forest is considered. In traditional harvesting only the merchantable stemwood is recovered; the residues, consisting of branches and foliage, are left on the forest floor – either to disintegrate or, for restocking, to be crushed and removed as part of site preparation. But there is increasing interest in the prospect of raising the fibre yield by making use of branches, tops and possibly stumps and roots which are left behind in harvesting residues. There is an estimated potential gain in dry fibre yield of at least 20% by using branches and tops, and a further 10% if foliage is used. A further 40% of fibre is available if stumps and roots could be harvested (Holmes 1977).

Whole tree harvesting can about double a crop's yield in terms of weight, while leaving the ground free of obstruction other than stumps and a few branches. Foresters would welcome cleared ground for restocking – by contrast to a site strewn with brash and debris of the previous crop, removal of which is often slow and costly. The Forestry Commission has been researching some aspects of whole tree harvesting (Anderson 1985) in collaboration with other authorities. The potential benefits of whole tree systems are set out by Gardner (1990). A concern in the systems is the ecological consequences of residual removal.

The additional biomass obtainable, in the form of tops, branches and roots, from whole tree harvesting systems is very much a factor of tree size, species and harvesting system adopted. Additional recoverable biomass can range from 200% (expressed as a percentage of stem weight) in early conifer thinnings and broadleaves to 20% in clear-fell of mature crops at the end of rotation. Trials carried out by the Wood

Supply Research Group at Aberdeen University (Mitchell *et al.* 1989), investigating the potential of wood as a fuel on behalf of the United Kingdom Department of Energy, have established relationships between stem weight and additional biomass for a range of species and ages. Cannell (1982) assembled data from numerous papers relating to world forest biomass. (Apart from this, a considerable quantity of wood residues is formed in the process of preparing and processing timber – cross-cutting of trunks, sawing production of boards and veneer – amounting, for example in sawing to 30–40% of the raw material.) Utilization of the whole wood biomass involves the following measures (ECE/FAO, Geneva, 1988):

- Application of progressive kinds of equipment and technology in felling, transporting, primary conversion and woodworking;
- Rational use of the felled timber in processing;
- Utilization of all species of wood, including low-value broadleaved trees and shrubs;
- Utilization of green matter, stumps, roots and bark; and
- Collection and processing of the wood residues produced during felling and primary conversion (twigs, branches, broken ends, bits left over from sawing operations and sawdust).

Analysing the above measures it can be concluded that in the present stage in the development of the different branches of the forest industry, the greatest results can be achieved through the utilization of wood residues.

Three main harvesting systems can be used to maximize biomass production: (i) whole tree comminution, where the whole tree is harvested and chipped or chunked to provide a single product; (ii) integrated harvesting, where the whole tree is extracted in one pass to landing for product separation, the stem into conventional roundwood products and the branches and tops comminuted; and (iii) residue harvesting, where after conventional shortwood harvesting of the roundwood in clear-fell, the tops and branches are chipped either at the stump (terrain chipping) or at the landing (landing chipping). Markets for the chips produced are dependent on chip quality; a major market is mulch for horticultural purposes, and ground surfacing for riding, paths and play-areas. For higher quality white wood chips there is potential for supply to particleboard and pulp manufacturers. There is a limited use in the UK of chips for energy purposes; existing installations are generally domestic heating but there is increasing interest for utilization of biomass for electricity generation.

Portable or semi-portable chipping machines, available in a wide variety, assist in clearing cordwood, lop and top, branchwood, scrub or small unwanted coppice – generally wood of low grade which cannot otherwise be marketed economically. The chipped produce can be sold either locally for fuel or along with pulverized bark, or delivered to pulp or boardwood mills. However, the costs are substantial, and can be difficult to justify. Contractors charge £100-£320 per hectare, according to site and crop conditions. Cost effectiveness can only be seen in the light of comparative costs for preparing ground for planting or restocking. Penrose (1990) discusses in-forest wood chipping, Keighley (1989) wood chips as fuel, and Hudson *et al.* (1989) wood chipping harvesting systems.

Extracting. In extracting, the felled and delimbed trees, whether or not simply cross-cut or fully converted, are moved to a permanent transport network. The operation much depends on terrain classification (Rowan 1977), access, distance to roadside, method used and the magnitude of the operation. If the felling has been a thinning, or made under a selection silvicultural system of management, special care is required to avoid damage to the remaining trees and, as far as possible, to any natural or planted understorey.

Methods of extraction differ as noted earlier in this section. The traditional method depended upon horses and the skill of the horsemen; this partnership has now virtually disappeared, yet is still viable where the use of vehicles is impossible or environmentally unacceptable. The horse was replaced by the agricultural tractor, and for several decades general tree length loads have been chained behind tractors and their winches and ground-skidded (tushed) from woodlands. This system has gradually improved along with tractor reliability and as terrain capability has been advanced. Use has also been made of custom-built framesteered forest tractors instead of simple agricultural tractor conversions. All are still used. Locally use is made of grapples and tongs closely coupled behind the tractor.

Whatever the method of extraction it will have a bearing on stacking and loading for transport. If the trees to be extracted are to be transported in their full length a log-deck will be useful, but a mobile crane or some other method of loading will be necessary. Should the produce consist of small or medium sized poles, or of pitprops, pulpwood and the like, loading may be by lorry-mounted loaders, e.g. the 'Hiab' hoist, or by crane, or by timber-loading grabs. The produce will be stacked on the side of a ride or internal road so as to facilitate loading according to the method being used.

More up-to-date methods of extraction are used (some may be inappropriate for small woods). There has been the development of shortwood forwarders, skidders and cable-cranes with terrain capability which is superior to any agricultural tractor. These machines are described below (Hughes 1984; Hibberd (ed.) 1991). The all important factor is presentation to the extraction equipment:

> *Forwarders* are tractors with a trailer and loading crane which extract converted products lifted entirely clear of the ground. The relevant Forestry Commission FSC guide is number 23. Because of their larger load carrying capacity, the cost of the actual movement of timber is less per cubic metre for forwarders than for skidders. This affects requirements for forest roads, as forwarders extract economically over long distances and so roads can be spaced further apart. The amount of stacking space required at roadside is little because the crane can make high stacks. Timber extraction by forwarder is generally clean.
>
> *Skidders* are usually purpose built tractors which extract by lifting one end of the load of full length delimbed trees clear of the ground and pulling (tushing) it to the roadside with the other end dragging on the ground. The relevant Forestry Commission FSC guide is number 22.
>
> *Cable-cranes* are ropeway systems where timber is extracted by means of moving cables, powered by a stationery tractor-powered winch on the roadside. The timber load is usually carried wholly or partially clear of the ground. In almost

every case it is preferable to use forwarders or skidders if at all possible owing to the high cost of extraction using cable-cranes. The latter need a team of at least two men and their outputs are generally well below efficient tractor operations; however, they can extract timber on the most difficult and awkward sites where all other methods fail. The relevant Forestry Commission FSC guide is number 25.

Loaders. Timber handling, stacking and loading by hand, is extremely hard work and should be replaced by mechanical handling wherever possible. The maximum size of billet for manual handling should not exceed 30 kg.

Extraction from single or blocks of small woods, because of the lower volumes of timber cut, is comparatively expensive through a higher cost of fetching-in equipment per m^3 handled; and the cost of any road at all may be prohibitively high. The use of forwarders allows extraction over agricultural land at comparatively low cost, with minimum damage to fields, and this has eliminated the necessity of building roads to outlying blocks of woodlands.

For large-scale harvesting, systems generally available (Evans, G. 1989) are:

1. Motor manual tree length felling, with extraction by skidder or skyline;
2. Motor manual shortwood felling, with extraction by forwarder or skyline;
3. Motor manual tree length felling and tree length delimber, with skidder extraction;
4. Motor manual tree length felling, with whole tree extraction by skidder, or clam bunk skidder, or skyline; followed by processor at roadside to delimb and convert; and
5. Processor or harvester, with forwarder extraction.

Processors and harvesters are discussed later in this section. Various factors influence the choice of harvesting systems, particularly where large volumes are involved (Evans, G. 1989). They relate particularly to conifers:

Terrain. This is very variable – from flat dry areas to wet steep slopes. The position of roads, rides and tracks is important. So too is whether the harvesting is of a thinning or a clear fall. In the case of the former, previous thinning regimes may be pertinent – stability of the crop can be a factor.

Species. Keeping species separate (where essential) is very difficult, especially with forwarders.

Product mix. Where the market requires a product mix not exceeding three specifications, the most widely used and cost-effective system is shortwood/forwarder. The important exception in product mix is where fencing material graded to length and diameter is involved. Cutting and grading of this is a roadside operation, which means that either random length small diameter roundwood is cut, using a semi-shortwood system and forwarding, or tree length working, using a skidder. Often involved is the cutting of up to nine specifications –

two of sawlogs, two of bars, four of fencing material, and pulpwood lengths. Forwarding such a mix can be very difficult, and hence tree length skidding is used extensively: working costs are higher but this is usually more than balanced by the increased sale prices obtained.

Labour. Harvesting is usually done by contractors, whose employees and supervisors must be well-trained to carry out the work at a competitive price, with an output that is consistent and can be forecasted. They must be capable of producing timber to correct specifications and maintaining good standards of workmanship with due regard to health and safety.

Forest machinery. The size of the working area obviously influences the machinery used. Machinery, because of its specialist nature, is very expensive. Most is owned by contractors. If they are under-capitalised, their machinery is purchased on loan capital. Contractors can vary in size from two/three man gang operators using skidders or forwarders to a larger harvester and forwarder combination costing a huge sum. Such a combination needs a volume of around 200 tonnes each week on small first thinnings to around 500 tonnes on clear falls; also continuous work on suitable terrain with an appropriate tree size for the duration of their capital loan, usually four or five years.

Mechanical harvesting machines (noted *supra*), with 'sheers', 'snippers' or 'saws', now being tested are common in some parts of the world. Mechanical systems of delimbing are also being tried. A machine which delimbs and cross-cuts felled trees is called a *processor*; if it also cuts the trees down or severs the trees from their roots at this stage, it is called a *harvester*. The felling, delimbing, cross-cutting and extracting (also transport) can be undertaken by machines either as a complete operation or undertaking each of the tasks as a single or multiple operation. Developments in specialized harvesting machines have been rapid during the past two decades. Many of the more advanced machines are highly efficient and reliable, but their capital costs are so high that economical operation requires large harvesting programmes and intensive use. This precludes private woodland owners but some of the larger timber merchants and harvesting companies and several medium sized contractors are operating in this field. Under this scale of working, reductions in cost of harvesting are possible. Machine costs depend on the capital expenditure and interest thereon, the expected life, the number of hours gainfully employed, and the running costs including depreciation, fuel and maintenance. Relevant information is available from the Forestry Commission Work Study Branch – see chapter 16.

Costs of extracting. Costs of extraction vary widely according to many factors, the most important being whether the produce is in tree length or in converted lengths. Other factors are access for machines, terrain, species, average tree size, or product mix and specification, and distance from roadside (usually 100–300 m). Costs in a thinning are higher than in a clear-felling. Generally, costs of extraction range between £1 and £2/m^3 or tonne for every 100 m involved. Further indications of costs £/m^3 or tonne are: skidder (for tree lengths) £3.50-£7, forwarder (for shortwood) £3-£6, and cable-crane £5-£10. Costs in a thinning may lie between £3 and £10 m^3 or tonne; and in a clear-felling between £3 and £8.

Combined costs of harvesting. The three components of harvesting (felling, converting and extracting) have been discussed separately, but are usually undertaken by a contractor as one combined phase, i.e. completed to roadside, ready for transporting. Combined costs are indicated in Table 36. Costs are much dependent on the factors noted earlier under 'Felling' and 'Extracting', in particular: system, terrain, average tree size, or product mix and specification, and distance from roadside; also, whether a thinning or a clear-felling.

Table 36. Indications of combined costs of harvesting.

Harvesting costs	£m^3 or tonne		
	Thinnings		Clear-fellings and large sized thinnings
	First	Subsequent	
Felling	6–10	4–8	3–6
Extracting	3–10	3–9	3–8
Converting	3–5	2–4	2–3
Combined costs	12–25	9–21	8–17

Costs much depend on whether the conversion is simple or complicated, or whether undertaken in the wood or at the roadside. If in tree lengths the extracting costs would be at the lower end of the ranges. The converting component cost would be according to average tree size, product mix and specification; this would have an interaction with extraction costs. The combined costs of harvesting are particularly dependent on the product mix. A first thinning of a mixed plantation might cost say £12/m^3 for the pulpwood component, £14/m^3 for sawlogs and £18/m^3 for fencing material.

Transporting. Road haulage accounts for the movement of most of the timber produced in Britain. A little is transported by rail and by sea. For commercial reasons the vast majority of road transport is by 38 tonne articulated vehicles. Rigids, because of their lower load carrying capacity and higher running costs, are usually for short hauls, or where access is restricted, or is too difficult for articulated vehicles. The majority of lorries are self-loading, purpose-built vehicles which are used continuously for the haulage of roundwood, residues and wood products. In many parts of the country hauliers carry such produce long distances to markets on general purpose vehicles, and back-load with steel, fertilizer and other materials.

Methods of transporting timber on highways have not changed significantly in recent years except in size of vehicle and improved loading and unloading. Road haulage has usually been competitive and adequately placed to meet most of the demands of the timber industry. There is a sufficiently wide variety of load carrying vehicles to provide a standard one to meet almost any problem of transport once the road is reached. (The concern regarding damage to highways and bridges is noted in section 3 *infra*.) Shipping of moderate quantities of pulpwood takes place to Scandinavian and occasionally to some UK mills, particularly from small ports on the west coast of Scotland. The measure of shipping is quite variable. Movement of timber by rail has decreased in recent years.

Costs of transporting, including loading and unloading, obviously reduces the return to the grower, and can be quite high if long distances are involved without return loads (they figure prominently in long distance haulage rates). Indications of costs of road haulage (dependent upon distance to markets) are given in Table 37.

Table 37. Indications of costs of road haulage.

Kilometres	Pitprops, pulpwood, boardwood and poles £/tonne	Sawlogs and Veneer logs £/m^3
First 50	2.00–4.00	2–3
51–100	4.50–6.00	4–5
101–150	6.50–8.00	6–7
151 +	8.50+	8+

Rates are lower if there is a return load.

Combined costs of harvesting and transporting. The fragmented nature of many private woodlands makes it impossible to ascertain realistic average combined costs of harvesting and transporting. However, they probably lie within the range £15-£25/m^3 or tonne, dependent on access, terrain, specification, quantity, distance to road and market, and methods used in the operations. The level of costs largely determines the return to the grower. The standing value can be determined by deducting the costs at any point of sale (felled, extracted, or delivered) from the price obtained at that point of sale. Table 38 is an indication how, from a delivered price, a standing value of produce can be determined. The example relates to conifer first thinnings. Calculations relating to later thinnings and to clear-fellings are more complicated because the butt length(s) produce sawlogs whereas the upper length(s) produce small diameter roundwood and possibly bars.

Table 38. An example of costs of harvesting and transporting whereby the standing value of the produce can be determined.

Conifer first thinning	£/tonne	
Harvesting costs		
Felling	6–13	
Converting	3–5	
Extracting	3–10	
Cost at roadside		12–25
Transporting	3–5	
Cost delivered		15–30
Sale price delivered		20–42
Standing value		5–12

Private growers when considering harvesting have to decide on the point of sale (see chapter 11) and whether to undertake the work by direct labour or to use a contractor. Insley (ed.) (1988) explains the relevant advantages and disadvantages. Growers need to have appropriate labour and equipment; they retain full control over operations in their woods; but the work can be dangerous and costly for inexperienced and untrained people. Health and Safety at Work regulations must be adhered to. By contrast, a competent contractor has the skilled labour and adequate equipment, as well as knowledge of specifications, prices and markets. Where harvesting is entrusted to a contractor, an agreement should incorporate such points as:

- Location and area of woodland involved, and how the trees which are to be cut are identified;

- Access to, in and from the wood, haulage routes, stacking places and working areas;

- Steps to be taken to protect remaining trees from both physical damage and spread of rootand butt-rot;

- Price agreed per m^3 or tonne of wood harvested, or for each tree;

- Quantity of wood to be harvested, or an agreed method of calculating the amount as harvesting proceeds;

- Degree of conversion to be undertaken;

- Dates when the work is to start and finish; and

- Condition in which the site must be left after harvesting, e.g. burning of debris, or tidying of haulage routes.

Some of the above points are also relevant to Conditions of Sale (discussed in chapter 11).

Contractors who undertake harvesting include the several major forestry management companies, specialist harvesting companies, the few co-operatives, and a substantial number of firms and individuals within the Association of Professional Foresters. To assist small owners, the Forestry Commission has produced a series of county directories covering England and Wales, which lists sawmills, timber merchants and contractors who are involved in harvesting, marketing and processing of trees. They highlight the information that contractors or buyers will want to know on the species, dimensions and quantity of trees to be harvested. The lists, 'Marketing for Small Woodlands', are available from the Commission's Publications Department, price 40p for each county.

3. Forestry roads

The network of forest roads, along with made-up tracks, rides, rackways and paths in a woodland provide access for people, vehicles and machines engaged in regeneration, tending and harvesting operations, as well as for fire-fighting and other protective measures. Where forestry and agriculture are integrated, stretches of roads (likewise of fences) may serve both enterprises, and the costs can be allocated appropriately.

Roads are essential in all but the smallest woods. Those which serve the needs of harvesting will almost certainly satisfy most if not all other requirements. A good road system will reduce the amount of cross-country movement of timber in extraction operations, and may allow greater use of large, more economical vehicles. Roads are costly to construct, and hence should be planned, designed and constructed with care. The purpose of planning roads for removal of timber is to try to achieve the combination of road and extraction costs (and sometimes road transport cost as well) which gives the lowest overall cost. The costs incurred are the capital costs of construction and the recurrent ones of maintaining the road itself, the bridges, culverts, fords, passing and turning places, and the stacking sites. The value of the investment depends on such factors as the volume of timber produced by the woodland, the kind of harvesting system envisaged, and the standard of road required.

Terrain. The planning of forest roading requires a knowledge of terrain (i.e. the land as a working surface). Foresters have to be able to assess the extent of different types of ground to be worked each year, so that budgeting and cost control can be accurate, and to deploy machinery on the sites best suited to its particular capabilities. The Forestry Commission's *terrain classification* system (Rowan 1977) has three key features with five sub-divisions to each:

Ground conditions	*Ground roughness*	*Slope*
Bearing capacity of the soil, which is determined by soil type and moisture regime.	The presence of obstacles to movement across the land surface.	Gradient and topographic form of the slope.
1 Very good	1 Very even	1 Level
2 Good	2 Slightly uneven	2 Gentle
3 Average	3 Uneven	3 Moderate
4 Poor	4 Rough	4 Steep
5 Very poor	5 Very rough	5 Very steep

The terrain class of any site can be expressed as three index numbers corresponding to the three criteria above. For example, a site described as 2.1.3 means: ground conditions class 2 (good), ground roughness class 1 (very even) and slope class 3 (moderate). Each of the three key features, and application of the system, are described fully by Rowan (1977). Terrain class is an important factor in the choice of forest machinery.

When to construct roads. Roads can be provided either before establishment or delayed until harvesting. Roads for access prior to planting, or for other purposes in advance of harvesting, should be evaluated in economic terms (see chapter 14), justified chiefly on the basis of savings in costs of cultivation, fencing, planting and tending. They are required to give access for workers, fencing materials, fertilizers, cultivators and ditching machines. Relevant factors which have to be taken into account are availability of capital, current interest rates, and the requirements for landscape, recreation and sporting, and any restrictions on access which they demand.

The difficulty of assessing the economics of roads is greater at harvesting (Hay 1986). The discounted value of timber from the final crop over fifty or more years is a very small quantity and might not justify roading at year 0 in the crop's life. It is therefore necessary to assess the roading requirements in relation to the expense of carrying out the establishment operations, by quantifying the work to be done and the distances to be travelled on foot or by cross-country vehicle to carry out that work; and to compare the cost with that of the assessment with roads in place, to determine the economic benefits of the various systems.

Roads, like land itself, are part of the capital asset which may be available for several rotations of crop. Therefore, it can be argued that the cost of constructing them should not be charged wholly against a single rotation, in which case the timing of construction is not so critical from the point of view of compound interest/discount. However, roading is generally thought of as serving a single rotation, where costs are all chargeable against it, which course can make a great decrease in the NPV. The later in the rotation the roading costs occur, the less they are discounted in the financial calculations; and from the financial point of view, construction of roads should be delayed as long as possible to ensure large volumes of timber will be available to pay for the road costs as soon as possible after construction. If the first thinning is to be done in year 20, volumes from future thinnings and final felling will be discounted to the time of construction.

In the early years of an extensive afforestation, the relevant operations can be fairly continuous and intensive, and use of the access provided may be regular and fairly heavy. Once the whole area has been cultivated, fenced, planted, weeded and established, there follows a period during which access is required only for inspection, protection and maintenance. Hence, as comparatively few roads are required at the afforestation stage (or at least only of a low standard relative to that required for harvesting) foresters should consider how much of the total road system envisaged can be justified prior to establishment. The main principle is that savings in operating costs must exceed construction costs to the greatest possible degree. Because there is a long period between the establishment and harvesting phases, the total cost of early provisional roads is sometimes charged to the establishment operations.

Road considerations in regard to harvesting system. Thinning may not begin for at least twenty years in conifers and much later in broadleaves, and since relevant technology of harvesting is rapidly advancing, foresters are unable to predict with certainty the precise system that will be used so far ahead. However, thought should be given to the expected system even before planting, because it determines the optimum final spacing of the roads; and it will be prudent to place access roads on alignments that are likely to fit in with the ultimate harvesting.

The standard of road construction is directly influenced by the method of extraction and transport. If the timber is to be carried direct to the customer on large lorries from any part of the woodland, high specification roads giving access in all weather will be required. If the timber is to be removed by tractor and trailer to a collecting point alongside a public road, an intermediate specification may meet the case. The roads built to intermediate or high specifications generally form the permanent element of the road network. Temporary roads or made-up tracks laid out for tractors and the like can be built to low specifications; this part of the network tends to be more flexible, and can be modified as new harvesting methods and machines are developed.

Decisions on the method of extraction and standards of road construction together with an assessment of the volume of timber likely to be produced now and in the future leads to selection of road spacing (Rowan 1976). Roads are seldom straight. The actual length that has to be constructed exceeds the theoretical straight length, by amounts ranging from 25–35% in lowland woodlands and favourable upland terrain to 45% on difficult terrain (Matthews 1989). The need to conform to topographical and terrain features, the use of bends to gain height at acceptable gradients, and the limits on watercourse crossings all increase the length of road required to get from one point to another. Junctions and stacking sites, connections with public roads, rights of access, and the size and shape of the woodland also affect the length of road that must be built and maintained.

Road specifications. The detailed design of road alignments and cross-sections, and of bridges, passing and turning places, and other works is a civil engineering task, as also are the precautions taken to prevent erosion and pollution of watercourses. Designs which reduce the impact of harvesting systems and road networks on the landscape require greatest attention in areas of high scenic value. The earlier practice in extensive afforestations of leaving wide swathes or rides unplanted to form fire-breaks has been superseded by good access road systems. (The old geometric ride systems and many of the fire-breaks are often grossly artificial and visually intrusive. They need to be planted or otherwise modified to blend in with the other parts of the woodlands.) The layout of both roads and rides should be designed to reflect landform. Width, shape and direction should be varied, and long straight stretches avoided. The Forestry Commission *Forest Landscape Design Guidelines,* 1989 (see chapter 13), should be followed.

The width and specification of forestry roads, and hence cost, will depend on the required use, topography, terrain, underlying formation, and availability of local road-making material. Financial justification of the expenditure will depend on the volume and value of the crop, and the type of harvesting system envisaged.

Low specification roads are normally about 3 m wide, involving excavation to hard formation level followed by laying 15–40 cm of crushed stone which is then blinded and rolled. Suitable sized drainage ditches to outfalls and parallel to verges are required to improve the long-term performance. Costs range from £10 to £20 per linear metre. Such roads would be suitable particularly for four-wheel drive vehicles in dry weather conditions. Guideline costs for roads suitable for farmland woods, are provided by Insley (ed.) (1988); when based on 12 m of road per hectare, the cost may be in the region of £300-£400 ha.

High specification roads capable of carrying heavy vehicles and machinery are normally 3.2 m wide. Adequate side drainage is essential. The ground should be excavated to hard formation level, and stone bottoming laid from 30 to 50 cm thick. The surface is then blinded and normally rolled to a camber of 7–9 cm and this changes to a crossfall of 15–20 cm where the road is on a steep slope. The road width can increase to 6 m at sharp curves with a minimum radius of 15 m. Turning and passing places should be provided as necessary. Costs are in the region of £20-£35 per linear metre. Such roads would be suitable for vehicles with a gross weight of up to 38 tonnes in all weathers. In deep peat involving the use of geotextile fabric, or excessive depths of bottoming, rock cutting/blasting, drainage works, and large culverts or bridges, the cost will be much higher.

Road planning and construction. The 'density' of roading is usually expressed as so many metres of road per hectare of woodland. The average spacing between roads is the reciprocal of density. The higher the density of roading the higher the overall cost of construction per unit area. At the same time, the higher the density, the lower the cost of removing timber from stump to the roadside. The optimum density is thus a balance between the two. The variables in determining the optimum roading density are cost of construction, cost of removal to the road, and volume production per hectare. The most economical density is where the sum of costs for roads and removal is at a minimum. The big advantage of no-thin is the postponement of the high cost of construction and elimination of maintenance.

A detailed calculation of the best spacing of roads can be made for any given woodland. In small woods, density of 20 or more metres per hectare may well be appropriate. Rowan (1976) and Hay (1986) discuss the economics of road planning, and the timing of the investment. The internal forest road system should be planned with the public highway layout in mind; the public roads to which connection is to be made must be of a standard sufficient to carry the timber generated by the woodland. The maximum-weight vehicles allowed on forest roads are those which meet Construction and Use Regulations. A significant difference between conventional road design and road designated for forestry purposes is that a great deal of traffic by timber extraction vehicles from adjoining woodlands enters the road at 90°.

Road location is greatly affected by topography and terrain. The conventional procedure on cross-sloping ground is to locate the road alignment in such a way that excavation is minimized, but the terrain may well introduce the problems of both horizontal and vertical curvature. In areas of high rainfall, the existence of watercourses poses a special problem of road location, and it is not common for a bridge or culvert crossing to dictate the position of the future road.

Once the location of the road in terms of a broad corridor has been planned, the road survey can take place, involving the preparation of a longitudinal section, with cross-sections, and a plan. These provide the basis for detailed designs of the road, taking into account such aspects as the specification data, water crossings, soil conditions, construction methods, the type of plant to be used on construction, and the availability of suitable road-making material (Hibberd (ed.) 1991). Efficient drainage systems for the road in the form of side drains, lateral water crossings and road camber or crossfall, are essential. The road formations resulting from the various types of construction are compacted, so far as possible, using a vibratory roller. The next consideration is that of pavement construction, chiefly by motor grading and compaction by roller.

Maintenance will be essential, especially after heavy rain or severe frost, or following extensive vehicular use. Annual maintenance is chargeable to the single rotation. In determining the NDR of a crop, the annual maintenance is more easily used as a capitalized sum which can be added to the construction cost (giving the capital that would have to be invested to produce the annual return).

Timber traffic on highways. Regulations and constraints on the use of highways and bridges in timber haulage are an increasing problem. (They can reduce the market value of isolated or otherwise adversely affected woodlands.) The infrastructure of many minor roads in rural areas with maturing woodlands is quite unsuitable for modern timber loads. Lorries weighing up to about 38 tonnes gross may cause havoc, with road surfaces breaking up and ditches and verges crumbling. The bridges and culverts were never built to take such loads.

A vexed problem which has arisen recently concerns the upkeep of public roads used to transport from woodlands to mills. Many woods as well as planned extensive afforestations, particularly in the uplands, lie in isolated areas of low population, where most local authority roads were constructed for relatively light volume traffic. This situation did not pose a great problem when timber-carrying vehicles were small, but the advent of large articulated self-loading vehicles has brought problems to many areas, and the cost of upgrading roads and bridges to a standard capable of taking heavy timber traffic is causing concern. Some local authorities have designated specific roads as 'preferred for forestry' use and placed restrictions on other minor roads, due mainly to the restricted weight-carrying capacity of the relevant bridges and culverts.

In many cases the forest roads are built to a higher load-bearing specification than the country roads serving the area. Elsewhere, in the older afforestations, roads constructed to carry a gross vehicle weight of about 12 tonnes are now having to be upgraded to carry a gross vehicle weight in the region of 38 tonnes.

4. Interaction of harvesting and restocking

Harvesting can have a profound interaction with the subsequent restocking. The repeated passage of logs and mechanical equipment over a clear-fell site can compact the surface and upper horizons of the soil, reducing permeability and microscopic pore space which in turn increases bulk density and impairs infiltration (Matthews 1989). These effects are greatest on soils containing much silt or clay, especially in wet weather, and least on sandy soils. Harm to the restocking site can be reduced substantially by orderly felling by trained cutters, extraction along prepared routes, and speedy restoration of the physical condition of the soil and drainage system.

During felling of extensive coupes of conifers on soils of low bearing capacity, such as peaty gleys, the logs should be separated from the brashwood/slash in such a way that the produce is correctly presented for extraction, and the skidder or forwarder moves on a cushion of the debris. Machines are available that exert low ground pressure and can be manoeuvred in restricted space.

The cost of restocking can be affected by the planning and organizing of the harvesting. The interaction between the two phases should aim to minimize their combined costs (Low (ed.) 1985). Extraction should be timed to coincide with periods when the bearing capacity of the soil is firmer than usual. Important factors are the design and use of machines and equipment relative to their effect on the site, operator training to minimize adverse effects, sympathetic forest road design to minimize adverse effects on soil erosion, water run-off and stream flow, and choosing the appropriate harvesting system for the local conditions.

Careful consideration should be given to the job specification in harvesting operations. In felling, the stem cut-off diameter should be as small as is practical, stumps should be cut low, tops should be cut into short lengths, and intentions should be specified for dead or otherwise unmerchantable material. Unextracted logs, needlessly high stumps, and unprocessed tops cut to an extensive diameter can be obstacles to man and machine in both extraction and in subsequent scarifying operations.

In the various 'organized felling methods' for harvesting, in particular extensive coupes of conifers, there is a formal separation of timber and brashwood/slash into distinct zones. Their use can provide various important benefits during harvesting, including greater safety for workers, improved ergonomics, higher outputs during felling and extracting, less waste as produce is clearly visible to the operator of the extraction machine, and, in difficult soil conditions, improved access for wheeled machines on mats of thick debris. The implications for restocking will vary. Work in the brash-free zones will be easier but may be more difficult within the brash zones. If brash is to be treated (e.g. by burning, chopping or raking into heaps), the zoning is likely to assist the brash treatment, but is otherwise likely to hinder site preparation by plough or scarifier. Traditional felling methods tend in practice to produce an uneven brash cover, resulting in local concentrations which often conceal unextracted produce and in consequence may be as much of a hindrance as the brash zones in organized felling.

In extensive harvesting operations, the extent of manual felling and conversion seems likely to diminish as large processors and harvesters are introduced. It is generally possible to concentrate the brashwood/slash ahead of such machines in a belt 3–4 m wide, leaving much of the site debris-free. As the processor or harvester, and then the extraction machine, pass over the debris, it tends to be crushed and mixed with soil, leading to its rapid breakdown. Most harvesters are now equipped to treat stumps with chemicals against rootand butt-rot. In the wetter areas, and particularly on soft soil types, the original drainage system is likely to be seriously disrupted by the passage of harvesting machines, and by brashwood/slash accumulations. The provision of temporary culverts, using lightweight piping, will reduce damage to main drains. Site damage due to rutting and soil compaction will be minimized if harvesting operations on wet sites can be carried out during the drier seasons.

Foresters should recognize the seriousness of the problems for restocking created by harvesting, and wherever possible reduce the impact.

5. Environmental considerations in harvesting

Harvesting when undertaken imprudently or carelessly can adversely affect some of the environmental benefits claimed for woodlands, such as their role as landscape and recreational features, wildlife sanctuaries, and as water purifiers and regulators. Felling involves significant landscape changes. It should take into account all the consequences so as to ensure that any potential conflict with environmental objectives is identified and reconciled to the best possible extent. Particular attention is required as to the effects on the landscape, the appearance of the harvesting site and routes, the wildlife habitat and the forest ecosystem. The concern should be for both the woodland and the general public locally and within viewable distance (see chapter 13).

When issuing a felling licence the Forestry Commission consult the Local Planning Authorities on the visual and amenity aspects involved. Every opportunity should be taken by foresters to demonstrate how well-planned clear-felling and subsequent restocking can harmonize with the environment. Furthermore, clear-felling can provide an opportunity to improve design in forest layout and to introduce variety to the sylvan scene and habitat. Clear-felled shapes should reflect those of the landform, and the size of coupes should be in scale with the landscape. Shortening some rotations combined with lengthening others can be used to create a significant appearance of mature forest, not least in the vicinity of recreational facilities.

Sometimes the felling of small copses and spinneys and even small clumps of trees may be of greater concern to the general public than clear-felling of 'commercial woodlands', especially on hillsides and in sensitive lowland landscapes. Felling can usually be acceptable in the landscape provided positive steps are taken to maintain the overall wooded effect by regulating the progress of felling and restocking, and the afforesting of other sites (see chapter 13).

On nature conservation grounds, objection to thinnings or clear-fellings seems to be largely confined to operations in broadleaved woodlands and a few old established coniferous woodlands, and mainly because an established ecosystem is being upset or even destroyed. Informed conservationists in some cases may welcome harvesting as a means of improving the diversity of wildlife, provided that the operations take place regularly so as to provide a continuing variety of habitats. Regard should be paid to maintaining a measure of habitat stability in the vicinity to reduce the impact on local wildlife; and operations should be timed, as far as possible, to minimize disturbance.

By contrast to clear-felling, selective thinning provides many aesthetic advantages: it usually has no visible effect on the landscape. However, line thinning, in most forms, creates a visible pattern in plantations prominent on hillsides. The visual effects are most apparent when entire rows are removed, and become more evident as the number of adjacent rows removed increases. Lines can be aligned obliquely to contours to reduce their impact but opportunities for doing so may be limited by considerations of operator safety and damage to remaining trees. Chevron thinning, particularly staggered chevrons, tends to provide an irregular edge and thus reduce the visual impact. The trees edging the rows in line thinning close up within two or three years, especially in faster-growing crops, and if followed by selective thinning for second and subsequent thinnings, the visual effect will be acceptable; but careful consideration should be given to selective thinning in sensitive landscapes.

Most roads and other extraction routes in low-lying woodland are hidden within, or at least partially masked by, trees; on hillsides they are often prominent. In the early stages of upland afforestation, roads are particularly significant; later in the rotation their impact on the landscape is much reduced. The most unsightly scars caused by forest roads, giving rise to adverse criticism, frequently are those on hillsides which run to or through afforestations. However, roads deployed mainly to serve agricultural or shooting purposes, may appear equally unsightly. Internal roads when constructed mainly with harvesting in mind should be designed to harmonize with the landscape. Their visual adverse effect is usually short-lived because of the growth of trees and vegetation. Artificial seeding of grass can accelerate the process.

The tidiness of the harvesting site and extraction routes is important – not only to contribute to efficient working but also to give satisfaction to the public.

Operations which adjoin public roads require special care to avoid damage to drains, verges and road surfaces; and debris and mud should be removed. The safety and convenience of the users of public roads and footpaths is a paramount consideration.

Noise from harvesting machinery generally gives no rise to complaint. Where the intermittent noise of chainsaws or mobile machines for conversion of products is intrusive, even near small centres of population or recreation, some restrictions on their use may be necessary at certain times of the day or week. Due attention must be given to the effect of harvesting on watercourses. Management for environmental objectives is further discussed in chapter 13.

6. Conversion; and estate sawmilling

Aware of the properties and uses of timber, and having given adequate thought to utilization, foresters in harvesting have to decide whether to sell their trees standing, felled or converted. Adding value is an important factor in determining the viability of forestry. Besides selling medium to large sized timber to roundwood merchants or direct to sawmillers, foresters may have to market small diameter roundwood from thinnings or top lengths of larger trees, underwood and coppice, as well as lop and top, semi-decayed trees and a range of minor produce. Occasionally they may have to remove a windthrown or other tree that no merchant will take away; unless a full load can be made up, such items usually show a low return.

Conversion of small diameter roundwood is basically into products of specific lengths and diameters – sometimes requiring ripping, cleaving, debarking and pointing – perhaps to be followed by preservative treatment. From large roundwood, the chief products will be sawlogs; however, there will always be second or third lengths too small for sawlogs. From small diameter roundwood, the products will include (in addition to lengths for pulp, fibreboard and particleboard) a wide range of fencing materials, garden products, pitprops (conifers only), turnery (broadleaves only) and other poles. The products will need to be converted to purchasers' specifications as to length, minimum and maximum butt and top diameters, and the tolerance of these. Species and straightness are likely to be specified; and adverse features, e.g. rot, fluting and splits, ruled out. Tops and inferior grades will meet less discriminating markets, including firewood, depending on the expected respective costs and sale prices. The more diverse the products which are converted, the more is the skill needed from the workers; outputs fall and costs rise. The extra management input and costs of entering the highest specification markets, e.g. transmission poles, must be weighed against the attraction of higher sale prices. Prices of converted products are indicated in chapter 11.

Conversion in woodlands. Timber in its round unconverted state is costly to extract, load, haul and unload; therefore the more that can appropriately be left in the woodland of unavoidable uneconomical material especially when dealing with produce of low unit value, the greater will be the savings. It is imprudent to spend money on any item whch is unprofitable or may even incur a loss (however, the welcome demand for fuelwood and for wood chips must be borne in mind).

Conversion in woodlands, therefore, sometimes means the difference between a profit and a loss, or at least the covering of expenses on produce which might otherwise be unsaleable. It may, too, be a boon by providing useful work during inclement weather. Some conversion can, in reasonably good weather, be done *in situ* anywhere within the woods, but, for static work, shelter in some form is generally required. An old hut, or a tarpaulin tied to poles to give top and side cover from wind and rain, is sometimes sufficient for temporary needs.

Conversion with hand tools such as chainsaw, axe, billhook, peeling-spade and draw-knife, can be applied to many varieties of produce. Simple wooden 'breaks' for cleaving, and wooden 'horses' for cross-cutting, may be necessary. Such conversion (with certain reservations to be explained later) given the necessary skills, appropriate materials and markets, may include:

1. Making hazel wattle hurdles.
2. Tying birch bundles for steeplechase jumps, vinegar works or steelworks.
3. Tying birch or heather besoms, including stripping of lime poles for the handles.
4. Preparing, and tying where necessary, hazel pea sticks and bean rods, thatching speakes, hedge stakes, heathering, tynut and crate rods; also coniferous pergola and rustic poles.
5. Making simple (not turned) handles from ash.
6. Preparing transmission, scaffold, ladder and other poles.
7. Tying fascines and faggots.
8. Cutting and stacking cordwood.
9. Chainsawing or axing-out gate and other posts.
10. Cross-cutting stakes, posts and rails; and ripping rails.
11. Pointing stakes.
12. Peeling stakes and posts.
13. Stripping and stacking oak bark (April/May only).
14. Cross-cutting and grading turnery poles.
15. Converting and peeling pitprops.
16. Converting timber for pulp, particleboard and fibreboard.
17. Converting and peeling wood wool billets.
18. Converting and peeling hop-poles.
19. Cleaving oak, sweet chestnut, ash, willow, western red cedar, Lawson cypress and Douglas fir into stakes and posts. This can be done with the use of a 'froe' (or 'dull-axe') and striking it with a wooden 'beetle' (or mallet). Large material is cleft with a sledge-hammer and wedges.
20. Charcoal-burning in stacks ('pits') or in portable steel kilns.
21. Cutting and bundling foliage for the florist trade.
22. Lifting, grading and tying Christmas trees (November/early December).

It must be noted, however, that although hand-peeling of long poles in the woodland may pay in the case of such as (6) transmission and other poles and (18) hop-poles, the smaller sizes of (15) pitprops and (12) stakes and posts, may be peeled much more quickly if a de-barking machine is used. With the addition of a mobile saw-bench (driven by petrol, diesel or tractor) operations (6), (10), (14), (15), (16) and (18) can be speeded up; along with the ripping of rails and pointing of stakes. In all the foregoing conversion, it is necessary to realize that workers with suitable skills are difficult to find.

Due attention must be given to the best place for the work of conversion, and here three matters demand consideration. The first is the amount or nature of the conversion. If it is intensive, e.g. with much peeling and to short lengths, it is better done centrally in a depot. If it is simple, e.g. to pulpwood or boardwood, it may be best done at stump or beside a ride or road. The second consideration is the method of extraction (discussed in section 3 *infra*). If extraction is possible by machine in one shift from the stump to the main transport or depot, handling in the largest pieces will pay. However, if a machine cannot be taken to the stump but can be moved through closely spaced rackways, conversion at the stump, followed by man-handling to a rackway, may be most economical. In contrast, if a horse is used to pull produce 100 m or more from the stump to a ride or road, conversion, except perhaps for severing a sawlog off a large thinning, should be after extraction. The third consideration concerns the amount of waste. In general, the poorer the crop, the greater the potential waste, and the more advantageous is conversion at stump; but against this must be set the fact that the more pieces handled at each stage the higher are the costs. In every operation foresters must ponder and decide the cheapest combination. Work study principles can here play an important role.

Whatever operations or conversion are to be carried out in woodlands, tools and equipment must be assembled and operators made available. Considerable experience is needed to choose the most appropriate types of tool and equipment. The wrong types or inferior ones, or any in untrained hands, can lead to financial loss (and possibly accidents). The work in many instances will call for a degree of skill on the part of the operator. There must be appropriate supervision; a competent forester can make a success of converting woodland produce where an inexperienced one will fail and incur loss. It is important that operators are instructed regarding the specification of produce to be prepared; unless they are experienced, skilled supervision is required to ensure the best use of each tree or pole and to reduce waste to a minimum. Too many specifications can be troublesome.

It is to be regretted that workers experienced in the conversion of underwood, small coppice and broadleaved thinnings are now so few. In consequence it is not always possible for some woodland crops to be adequately and profitably utilized. The art of making a hazel wattle hurdle, an ash bar-hurdle, of tying besoms, fascines and faggots, of twisting withes for tying pea sticks and bean rods has dwindled. So too, with a few notable localized exceptions, has the art of cleaving pales, posts, stakes and tent-pegs. Felling and cross-cutting with an axe and bow-saw are wholly displaced by the use of sophisticated chainsaws. Gone are the days when a feller gave pride to the accurate and neat panelling with his axe of the butt of a large tree and the careful rounding-out of the branch scars. Foresters must adjust themselves to these changed conditions. They should encourage woodland workers of the right type and recognize the skills of those few who still find conversion in woodlands a worthwhile and rewarding occupation.

Peeling and pointing by machine. Machines for the cross-cutting, peeling and pointing of posts and stakes, and for ripping rails, provide many opportunities for adding value to small diameter roundwood. (The same intention applies equally to equipment for splitting broadleaved roundwood into firewood.) Such conversion can be undertaken at rideside, roadside or in a wood-yard or other depot. Sometimes its quantity can justify a form of operating where the work can be spread out in linear fashion. Adequate width of rides makes possible the accommodation of the full range of conversion, but it is normal to encounter considerable interference between operations. Any delay caused can often be accepted as there are no strict rules as to when such conversion (giving shorter extraction distances but greater conversion difficulties) should be replaced by conversion at a depot (with longer extraction distances but more efficient conversion). Whichever is chosen, the principles of efficient working are the same. Part of the objective is to leave unwanted waste in the wood.

Simple conversion is often achieved by chainsaw; but most conversion is undertaken on mobile self-powered sawbenches or 'peeler-pointer' machines, all of which enable accurate cross-cutting of small diameter roundwood into fencing material, pitprops, wood wool billets and the like. Output Guides and Standard Time Tables are available from the Forestry Commission Work Study Branch – see chapter 16. In contrast, cleaving roundwood into posts, stakes and pales is usually undertaken on site.

The principal outlets for so-called 'stake material' are largely found within the agricultural and horticultural industries. Low cost, tractor-mounted equipment offers potential for the small grower and on-farm user, while the larger grower and sawmiller can invest in specialist static machinery. The range of capital investment required will vary from the small farmer producing his own stakes for on-farm use, to the sawmiller producing substantial quantities of machine-peeled roundwood products. Usually preservation treatment is required (see section 7 *infra*).

Foresters should consider how an investment in equipment can fit into their woodland enterprise. Mobile pointers used either free-standing or powered by the tractor's hydraulics can have an output of a hundred or more stakes an hour; but to achieve this the machine must have a ready supply of material. Therefore, effective organization and supervision of the operation is essential. An additional advantage of machines powered by the tractor's diesel engine is that the tractor can simultaneously operate the peeler and power the full stake producing operation.

Nowadays, the 'stake market' relies mainly on peeled, pointed and preservative treated products. (A well-prepared stake, pressure-treated, may have an average life span of at least twenty years – some three times that of untreated material – and command a sale price of at least 25% above that of unpeeled stakes.) The decision as to whether to invest in a 'peeler-pointer' rests largely on the current and expected output of thinnings from a woodland enterprise. Much will depend on the associated treatment facilities. A 'peeler-pointer' may cost anywhere between £2,000 and £6,000. A skilled operator, with a steady supply of stake blanks, should be expected to peel up to 1,000 1.7 m by 8–10 cm top diameter stakes each day. The quality of end product generally aimed for is a 90% de-bark. There is also a market for stakes processed by Bezner-type rounders – an 'urban' type stake. Sale of bark and peelings is an important factor contributing to the viability of the operations – for use in equestrian and horticultural establishments and playgrounds. In large sawmills, residues may be sold to boardwood mills. Log splitting equipment, of which there is a wide variety, costing £600 or more, can greatly facilitate the presentation of material for firewood and is best suited to the conversion of broadleaved branchwood.

For operators using chainsaws in conversion, the Forestry Safety Council gives the following guidance for cross-cutting and stacking (FSC13): (1) In *preparing to work*, clear any debris which will interfere with the cross-cutting process, wear a safety helmet, visor, ear defenders, gloves, leg protection, safety boots and carry a first aid kit, and plan the work so that the lightest produce moves furthest. (2) In *cross-cutting,* maintain a balanced stance, do not stand on the stack, reduce any excessive tension by first making a cut on the compression side of the log, and when it is necessary to use a boring cut do not start with the top of the guide bar and ensure that it does not strike other stacked material as this can cause kick back. Switch off the chainsaw if it jams, be ready to step back quickly if the log being cut starts to roll; stack cut material frequently so that this does not create a hazard underfoot; and switch off the chainsaw or engage the chain brake when walking on difficult ground. (3) In *stacking,* use appropriate aid tools for lifting and moving timber (if one has to struggle then the material is too heavy: use another method). Use other logs as a pivot for moving timber; keep the back straight when lifting, and when stacking on sloping ground the stack should be built up from the bottom and supported to prevent rolling. Timber stacks should always be left in a safe condition particularly on roadsides frequented by the public. When a mobile saw-bench or mobile peeler-pointer machine is being operated, FSC30 and FSC31 provide useful guidance.

Conversion in wood-yard or depot. Many of the items and operations mentioned above under 'Conversion in woodlands' can alternatively be accomplished in a wood-yard, carpenters' shop, estate sawmill-yard or other depot. Having to haul some eventual waste from the woods is then inevitable. Products which need accommodation include gates, ladders and ash bar-hurdles, for which mortising is required, fencing, sheep-cribs, tree-guards, scaffold and other poles, rustic seats and arches, flower-boxes, and tent-pegs. One advantage of working in such depots is that overhead cover is usually available, and operators can always be gainfully employed who otherwise would be entitled to wet-time wages while standing by. The depots will probably cme within the provisions relating to the guarding of machinery. Safety precautions are paramount.

Efficiency in marketing, likewise in the management of woodlands, requires the depot being placed in the most suitable situation and its being appropriately designed and equipped. On all wooded estates except those with only a small woodland area the depot should be distinct from though near, if justifiable, the estate-yard which is sometimes controlled by a clerk of works. It may contain a garage and repair shop for vehicles, plant and other equipment, also store-sheds for wire netting, treeshelters, tools and fuel. A separate store should safely house chemicals. There may be a saw-bench for cross-cutting, ripping and pointing, storage space for stacking minor forest produce and round, sawn or cleft products. There may be an office for a forester, and nearby if safely sited, a modest preservative tub, drum or tank. If the depot is alongside a road it is useful for the display and sale of woodland products; particularly good displays should be set out on market days where a route is traversed by farmers and contractors. Unless someone is always in attendance a notice should state when a member of the staff is available, or when and where to apply. Price lists should be available, and the displayed products should be clearly priced; advertising may be helpful.

The agricultural goods or products for which timber prepared in the depot can be used include boxes, tool handles, materials for use in the construction of barns, cattle-sheds, horse-boxes, feeding-troughs, tool-sheds, bridges, silos, pig-sties, duckboards, sides and bottoms of farm vehicles, and wooden parts of farm implements. If the workmanship is good and the material is not skimped in quality the products should attract buyers, particularly those anxious to use British-grown timber and wood products.

Estate sawmilling. A few large, well-wooded estates run modest sized sawmills. Usually the intentions are to provide sawn products for use on the estate and to add value to the estate's production by outside sales. Occasionally they may buy-in roundwood to supplement the estate's supply; and may even contract to saw the timber of an adjoining estate.

Some estate sawmills are more of a convenience than to save money or add value by production and sale – that is unless they are kept reasonably fully employed and efficiently run. Nevertheless, they are a means of utilizing low grade timber which some of the trade mills do not want; and they may provide wet-weather work where only a limited amount of sawnwood is required by the estate. In other circumstances the outlay on machinery and running costs together with the skilled supervision required rarely are justified. Old, decrepit sawmills, usually of obsolete types, can be a menace to life and limb although they may appear to be convenient and useful to the estate. A few private estates still use circular rack saws, and the owners may not realize how wasteful and costly they are to run. Tradition and the convenience of sawing one's own timber on the estate are sometimes the justifications, but in many instances foresters would do well to dispose of the mill, sell the roundwood to a sawmiller or merchant, and purchase such sawnwood as meets the estate's requirements. Some owners, however, feel that it is still justifiable to continue with bouts of sawing on old machines where all the costs of running them, together with necessary labour and roundwood, are allowed as a maintenance expense against taxable estate rents.

In the provision of a sawmill to convert roundwood from non-elected (for tax) woodlands, no tax-reliefs are available on the capital or running costs, or on any trading loss, but the profit on conversion is tax-free. In the case of roundwood from elected Schedule D woodlands (this concession applies until 5 April 1993), or of roundwood purchased outside the estate, profits on conversion are taxable, while capital costs and losses are tax-relieved. A few estates have run their sawmills partly on their own non-elected roundwood (profits being tax-free) and partly on purchased roundwood (profits being taxable). In such cases separate accounts must relate to each source of roundwood. The same principle applies when an estate sawmill uses roundwood both from its non-elected and Schedule D woodlands.

Advantages of a well-run estate sawmill may include the enabling of a woodland owner to sell timber in a far greater range of markets than if sold unconverted, and also to dispose of, in the form of posts, stakes and rails, inferior timber which may be difficult or impossible to sell standing or in the round. In addition, difficulties, delays or disputes in connection with felling, extracting and hauling, which may occur when the timber is sold standing, are obviated. When the price of home-grown roundwood is low, estate conversion may be advantageous. Waste is reduced to a minimum, as sawmill conversion can be down to small sized products, and residues can be sold (necessitating loading facilities for large vehicles). Sometimes, too, an estate sawmill provides a freedom of action in woodland management which can scarcely be secured without it.

However, estate sawmills call for skilled labour and supervision; both are very costly and are rarely available. Again, to be efficient, economical plant is required, and this too is costly; additional haulage equipment is also usually required. To keep the mill running it may be necessary on occasion to over-cut the timber on an estate, or to buy-in timber, which may be costly in competition with established sawmills and merchants. The difficulties are patent.

Much consideration therefore is necessary before deciding to set up an estate sawmill (Wardle 1990). One important criterion is whether there is sufficient timber due to be cut annually on the estate to make the mill worthwhile, and, if not, from what other sources additional timber can be purchased at economical prices. An estate mill of even moderate size may convert 20 m³ to 50 m³ of roundwood each week; a larger mill may convert 100 m³ or more. Another consideration is: how much converted timber is required annually on the estate itself? The incidence of taxation, too, is important, for a mill supplied substantially from Schedule D woodlands or bought-in timber may not be an attractive proposition. Furthermore are the considerations of the availability of a site convenient to the woodlands, the necessary power, water, labour force, supervision, accommodation, access and transport? The difficulties are again patent.

If an estate mill is projected, the question of power-supply must be settled. Although steam-power enables wood residues to be used as fuel, and may appear cheap, there will be delays in starting; and it will take a man much of his time to attend to it. Diesel power is economical and quick starting; this may provide direct drive or generate electricity. Electricity where available is the best driving power; it provides instant starting, is clean and safe, allows a motor to be provided for each saw, and occasional power-cuts are usually not serious. Sawmills come within the very stringent safety provisions of several government regulations. Precautions in respect of belting, shafting, guards, fire-precautions, sawdust, chemicals, working conditions and the like are essential.

The choice of type of plant and equipment must be made. Usually band-saws will be chosen. Though more expensive to buy than circular saws, they cut faster, require less power and labour, are less noisy and cause less waste in sawdust. Band-saw blades are costly; they might turn out some erratic sawing if not adequately maintained and well-run, and they need skilful maintenance, especially in sharpening and tensioning. A skilled sawyer can sometimes be trained for the latter tasks, but usually the saw blades are attended to by reasonably quick van-service operated by a local firm.

The economics of sawmilling much depend on the size of the log and the dimensions being sawn from it. With good quality large logs the cost of 'throughand-through' sawing is low, but cutting small dimensions from small and inferior logs is quite high. Waste in conversion is substantial. This depends on several factors, such as straightness, taper and quality of the log, the presence of rot, shake and other defects, the type and sharpness of the saw, and the skill of the sawyer. Mills buy, or are debited with, roundwood on conventional Hoppus measure, which is approximately 21% short of true measure. The sale of fuelwood, slabs, offcuts, board ends, and sawdust helps towards the profitability of sawmilling. In some cases the residues represent the profit of the whole enterprise. A separate matter is the use of mobile saws, costing £220-£250 per day, whereby two men can convert on site or in yard 6–8 m³ per day. Knight (1989) and Andrews (1990) provide useful guidance.

Unless an estate by sawing can obtain a good margin above what it could receive for its roundwood (related to a price equivalent to that delivered to the estate mill) the mill would not be worthwhile. New portable mills often improve the economic picture. Much depends on the capital and running costs of the equipment, the amount of roundwood put through it and the sawnwood sale price.

The siting and layout of an estate sawmill require special planning. The site should be carefully selected, particular care being taken that it should be easily accessible from a hard road, and the ground selected should be adequately drained so as to avoid any possibility of flooding (some sawmill machinery requires excavation to

1.5 m below ground level). A metalled road should be laid completely round the sawmill itself, and hard-standing extended to serve the piles of logs and poles as well as the stacks of sawnwood. Easy access for logs to the machines should be provided, as much labour can be wasted in bringing logs from the pile. Sizes and types of roundwood should be segregated in the piles to avoid the necessity of moving several logs in order to reach a particular one. Cross-cutting should be undertaken as necessary, but the machines should not be kept idle awaiting the next log to be cross-cut. Conveyors are of great value at times, especially for small roundwood, and can provide a rapid feed.

The specification of roundwood to be generally handled will have a direct bearing on the access roads, the building and the machinery. If the roundwood is expected to be lengthy the roads must be sufficiently wide for loads to "swing' where necessary. The tables or carriages of the log-cutting machines must be sufficiently long for their task. If large diameter logs are to be handled, substantial and consequently much more costly machinery will be required.

The mill should be open as much as is prudent to afford uninterrupted movement from the log piles to the log-saw. As a protection against inclement weather long sliding-doors can be hung along the mill sides, but any restriction of free access to the main machine can only result in increased costs.

Capital expenditure on suitable mechanical lifting facilities should be carefully considered. Overhead electric gantries running the whole length of the yard and mill, and capable of unloading from timber trailer to piles, and conversely from piles to the log-saw, are ideal, but the capital cost renders them uneconomic to most estates. A fork-lift truck with suitable attachments would be more useful. All can be expensive, and often can be difficult to justify.

The labour requirements will depend upon the input and output, the number and type of machines, and the extent of their use. The minimum required, when all the equipment is in full production, might be four men. All would help to grade, stack, wheel and load and also remove sawdust and other residues: it is essential to keep the sawmilling floor clear of sawnwood and waste. A clerk might have to be employed part time on measuring, despatching and invoicing. Overall supervision, probably by a full-time manager, would be needed. He might have to spend part of his time as a local salesman, marketing the finished products.

Labour requirements are much reduced by new types of equipment now being installed in some commercial sawmills particularly those using conifers. Advances have been made in applying new engineering techniques, in electrics, hydraulics and pneumatics, to the handling of sawlogs. These advances and labour-saving devices are necessary where profit margins are low. Conifer logs are led to the mill by powered log chains. Other powered conveyors move the logs through the entire conversion process. One man only is required to control the round log, the actual sawing and re-sawing, the turning and the discharge of the sawnwood output on to various conveyors. (In modern sawmills, surveillance and operation of practically all the main sawing is by one very skilled and alert man – such is the rate of progress.)

In summary, sawmilling is a highly specialized, skilled and competitive business and an estate sawmill should not be provided unless a strong case is made for it. A decision has to be made as to whether the mill is intended (i) primarily to provide sawnwood from estate roundwood for the estate's use coupled with some local sales; (ii) to enter into very limited competition by using some bought-in as well as estate roundwood; or (iii) to enter into full competition with established local mills. If the intention is (i), there may be some hope of economic success; if (ii), those who have the decision must tread warily; and if (iii), the proposal, under the present competitive state of the sawmilling trade, should either be discarded or treated as a commercial undertaking separate from the estate. The current demand is mainly for sawn products of good quality, accurately sawn, often adequately seasoned, sometimes treated with preservative, and delivered promptly at competitive prices. It must be emphasized that expert advice on the set up or improvement of a sawmill is essential.[1] So many factors enter into the project that careful forethought will be amply rewarded. Sawmilling is a business best left to the trade. A healthy and fully efficient sawmilling industry, capable of paying adequate prices for roundwood, in the large majority of cases is far more satisfactory to the grower than his entering this hazardous enterprise.

7. Preservation of wood

Preservative treatment not only considerably extends the serviceable life of most timbers, but costs of maintenance, repairs and replacements are generally much lower than if untreated timber is used. However, preservation is not always necessary, since many environments in which wood is used do not constitute a hazard. Some timbers – especially if sapwood can be excluded – already possess sufficient resistance to decay to give an adequate service life without treatment. In some cases when only a short service life is required for a product, preservation is again unnecessary. However, where conditions are such that treatment is needed it is important to know what methods are available and the degree of protection afforded by each one. The following information relates particularly to British-grown timber species (Morgan 1975).

Decay in service. The rate at which wood decays depends on a number of environmental factors (such as temperature and moisture condition), but decay can be completely arrested if the organism is starved of either water or air. Thus if wood is kept dry (i.e. at a moisture content below 20%) there is no danger of fungal attack; such conditions prevail for most interior timbers in well-designed buildings. Equally, if timber is kept saturated with water, as it may be if submerged and deeply buried, *fungal* attack will be inhibited through lack of air, though slower *bacterial* attack may become significant under these conditions. There are, however, many service situations between these two extremes where the moisture content of timber remains above 20% for long periods of time and conditions are ideal for decay. All timber used in contact with the ground comes into this category; and in such circumstances it is necessary to protect susceptible timber by preservative treatment if anything other than a short service life is contemplated. This is recognized by timber users who require a long service life and it is customary to specify the treatment of transmission poles, sleepers and roadside fencing.

---

1 Advice and estimates can be obtained from: (1) Stenners of Tiverton Ltd, Lowman Green, Tiverton, Devon EX16 4JX (0884 253691); (2) 'Forestor', Anton Mill, Andover, Hants SP10 2NW (0264 334142).

The classification of natural durability used in the Building Research Establishment at Garston is based primarily on the average life of *heartwood* stakes (50 x 50 mm in cross section) buried to half their length in the ground. Timber of larger cross-section will have a longer life; the life of buried stakes is roughly proportional to their narrowest dimension. Thus timber with a life of ten years in 50 x 50 mm dimension stakes would be expected to last fifteen years in 75 x 150 mm size. The classification in Table 39 relates directly to British-grown timber in ground contact, but the same order of resistance to decay can be taken as applying to other service situations, though the time scales for average life will of course be different:

Table 39. Natural durability of the *heartwood* of some British-grown timber species.

	Conifers	Broadleaves
Perishable Less than 5 years		Alder Ash Beech Birch Lime Poplar Sycamore Willow
Non-durable 5–10 years	Firs: Silver Douglas Pines: Scots Corsican Lodgepole Spruces: Norway Sitka	Elm
Moderately durable 10–15 years	Larches Lawson cypress Western red cedar Coast redwood	Turkey oak
Durable 15–25 years	Yew	Oak Sweet chestnut

Source: BRE Digest 296. Timbers: their natural durability and resistance to preservative treatment.

It can be seen that most of the British-grown species fall into the perishable or non-durable range, and consequently for many purposes the added protection of preservative treatment will be required. In addition it must be borne in mind that this classification refers only to the *heartwood*, and that the sapwood of all species (even the durable ones) has no resistance to decay. This is particularly important in British-grown timbers since the ratio of sapwood to heartwood will be greater in young trees. Its effect can be seen in the average life of roundwood oak and sweet chestnut fence posts: whilst the mature heartwood of these timbers is classed as durable, untreated fence posts (7–10 cm) have a life of about ten years. The same considerations apply to sawnwood fence posts, where sapwood cannot be excluded, and this is one of the reasons that it is necessary to treat oak posts used in motorway fencing.

Preservative treatment. The following terms are used to describe the extent which a timber can be impregnated with preservatives: *permeable* (can be penetrated completely under pressure without difficulty, and can usually be heavily impregnated by the open tank process); *moderately resistant* (fairly easy to treat, and it is usually possible to obtain lateral penetration of the order of 6–19 mm in about 2–3 hours under pressure); *resistant* (difficult to impregnate under pressure and require a long period of treatment – incising is often used to obtain a better treatment); and *extremely resistant* (can absorb only a small amount of preservative even under long pressure treatments). Guidance on the need for preservative treatment in particular situations is given in a number of British Standards including BS 5268 Part 5 (1989) and BS 5589 (1989).

The preservative treatments which can be applied to timber fall into five main categories: (i) hot-and-cold open tank using creosote; (ii) vacuum-pressure processes using creosote or copper-chrome-arsenic (CCA) salts; (iii) diffusion treatment using water-soluble boron salts; (iv) double vacuum treatment using organic solvent preservatives (the main commercial process is based on tributyltin oxide); and (v) immersion brushing and spraying treatments using organic solvent preservatives of various types, e.g. pentachlorophenol, tributyltin oxide, copper naphthenate, and creosote. All these treatments, except diffusion, must be carried out after the timber has been dried below the fibre saturation point so that no free moisture remains in the capillary structure of the wood to hinder the entry of preservative fluid. The diffusion treatment operates by a different mechanism and requires that the moisture content of the timber be as high as possible at the time of treatment. In addition, for all treatments it is important that the surface through which the preservative is applied is clean and free from bark.

Hot-and-cold open tank treatment. This process is an alternative method for obtaining deep penetration of creosote in the more permeable timbers where a pressure plant is not available. It consists essentially of heating the timber in a 'bath' of creosote so that the air inside is forced out of the wood. On cooling, the residual air contracts and draws creosote in to replace the air that has been lost. The process is used mainly for treating fence posts, stakes and rails, and is an effective alternative for this purpose.

Pressure treatment. Timber which is to be used under severe hazard conditions (e.g. ground contact) requires the deeper penetration and loading which can be achieved using pressure processes. Pressure creosoting, which has been used traditionally for the protection of transmission poles and railway sleepers, is of proven effectiveness. Pressure treatment may be carried out using coal tar creosote or water-borne preservatives. Almost all used in Britain are of the copper-chromearsenic type. British Standards covering pressure processes (BS 913 – Creosote; BS 4072 – CCA) specify how the treatment should be carried out, and treatment requirements for commodities are to be found in BS 5589. In addition, where particularly severe environment is involved or where a long service life demands a high standard of treatment, some major users have laid down specific processing requirements. Pressure treatment does, of course, require special equipment – a pressure cylinder, storage tank and associated pumps and control equipment; information on this can be obtained from the British Tar Industry Association for creosote, and from the main operators of the commercial processes of CCA. General information can also be obtained from the British Wood Preserving Association.

Although pressure treatment of timber is one requirement for severe hazard conditions another is that the species of timber chosen should be amenable to treatment. For transmission poles and round fence posts it is necessary that the sapwood is permeable in order that the heartwood zone is surrounded by a protective band of treated sapwood. For sawn timbers, however, the treatment of the heartwood must also be considered, and in general the heartwood is more resistant to treatment than the sapwood. (Tables are available – see BRE Digest 296 – where the treatability of the common British-grown species are listed in accordance with a classification which compares the degree of treatment achieved in wood samples of fixed dimensions treated by a standardized treatment schedule.)

Whilst *permeable* timber can be deeply impregnated under pressure, timber in the *extremely resistant* group is only slightly penetrated by preservative. In between these two extremes lie the *moderately resistant* and *resistant* groups. The effectiveness of the treatments depends upon the natural durability of the wood concerned and the severity of the environment in which it is to be used. With resistant timber of larger dimension, improved penetration can sometimes be obtained by incising the wood prior to treatment. This consists essentially of making a series of regular cuts in the lateral surfaces of the timber through which the preservative can enter. In this way better treatment results by taking advantage of the greater longitudinal, as opposed to the lateral, permeability of timber. Improvement of sapwood permeability can also be achieved by water storage, which has the added effect of reducing 'bleeding' in creosoted pine poles.

The importance of drying timber which is to be preservative treated (excluding treatments designed for green timber) cannot be over-stressed. Preservative fluid cannot be forced into capillaries already blocked with water, and the most common cause of inadequate treatment is that the wood has been treated at too high a moisture content.

Selecting a preservative treatment. As far as the purchaser or specifier is concerned, the choice should be made against the following check list: (i) do building or other regulations exist governing treatments allowed?; (ii) what service life is required for the treated products?; (iii) what adequate treatments are available; and (iv) what does it cost? In general, all round, half-round and cleft wood, and all sawn wood except that from the heartwood of oak, sweet chestnut, yew, larch and western red cedar, should be treated if it is to be (i) in contact with the ground; (ii) enclosed in brickwork, masonry or concrete; (iii) exposed to the weather and unpainted; (iv) liable to remain damp for long periods; or (v) inadequately ventilated. In some localities there is an increasing demand for non-durable species, especially pines, treated with an effective preservative. The other species noted earlier, along with Lawson cypress and coast redwood, usually contain a high proportion of naturally durable heartwood and do not always call for treatment; some are difficult to impregnate. Oak, however, contains a considerable amount of sapwood which is perishable when used in contact with the ground; it is very permeable, can be readily impregnated, and so given a useful life. Spruces and Douglas fir take preservatives well only under pressure, whereas other species, not naturally durable, such as pines, ash, beech, birch and sycamore, can, with effective treatment, be given a life at least as long as the more durable species. The Manual of the British Wood Preserving Association is of great value to specifiers of treated timber. Much work is being undertaken by the Association on harmonizing standards to meet EC regulations. Other important sources of information are TRAD A (1990) and Aaron and Richards (1990).

References

Aaron, J.R. and Richards, E.G. (1990), *British Woodland Produce*. Stobart Davies, London.
Anderson, M. (1985), 'Whole Tree Harvesting'. *Timber Grower,* Autumn 1985.
Andrews, R.J. (1990), 'Mobile Sawmills and Rural Employment'. *Quarterly Journal of Forestry, 84* (2), pp. 103–6.
Cannell, M.G.R. (1982), *World Forest Biomass and Primary Production Data*. Academic Press.
ECE/FAO, Geneva (1988), 'Economic Aspects of the Fuller Use of the Forest Biomass'. Geneva, GE.87–42663/2055G.
Evans, G. (1989), 'Organizing the Harvesting Operation in Wales and the Marches'. ICF Report of Regional Group Meeting 30 November – 1 December. 1989, pp. 17–20.
Evans, J. (1984), 'Silviculture of Broadleaved Woodland'. FC Bulletin 62.
Gardner, D.N.A. (1990), 'The Potential for Whole Tree Harvesting in the UK'. Aberdeen University.
Grayson, A.J. (ed.) (1989), 'The 1987 Storm: Impacts and Responses'. FC Bulletin 97.
Hakkila, P. (1989), *Utilization of Residual Forest Biomass*. Springer-Verlag, Berlin.
Hay, R.M. (1986), 'The Role of the Civil Engineer in the Forestry Commission'. Proceedings of the Institute of Civil Engineers, Part 1, 1986, 80, pp. 707–29.
Hibberd, B.G. (ed) (1991), 'Forestry Practice'. FC Handbook 6.
Holmes, G.D. (1977), 'Whole-tree Marketing'. *Forestry and British Timber,* January 1977.
Howard, D.J. and Hayes, F.W. (1989), 'Aid Tools for Timber Harvesting'. FC Bulletin 81.
Hudson, J.B., Mitchell, C.P. and Gardner, D.N.A. (1989), 'Wood Chips as Fuel: II. Harvesting Systems'. *Quarterly Journal of Forestry 83* (3), pp. 161–9.
Hughes, A.J.G. (1984), 'Nomenclature of Harvesting Machines'. FC Research and Information Note 82/84/WS.
Hummel, F.C., Paiz, W. and Grassi, G. (eds.) (1988), *Biomass Forestry in Europe: A Strategy for The Future*. Elsevier Applied Science, London
Insley, H. (ed.) (1988), 'Farm Woodland Planning'. FC Bulletin 80.
Jobling, J. (1990), 'Poplars for Wood Production and Amenity'. FC Bulletin 92.
Jones, A.T. and Smith, R.O. (1980), 'Harvesting Windthrown Trees'. FC Leaflet 75.
Keighley, G.D. (1989), 'Wood Chips as Fuel: I. Technical Consideration'. *Quarterly Journal of Forestry 83* (3), pp. 157–60.
Knight, B. (1989), *Mobile Bandsaws: Productivity and Costs*. Hatfield Polytechnic, Herts.
Low, A.J. (ed.) (1985), 'Guide to Upland Restocking Practice'. FC Leaflet 84.
Matthews, J.D. (1989), *Silvicultural Systems*. Clarendon Press, Oxford.
Mitchell, C.P., Hudson, J.B., Gardner, D.N.A., Storry, P.G.S. and Gray, I.M. (1989), 'Wood Fuel Supply Strategies'. Report for the Energy Technology Support Unit of the United Kingdom Department of Energy. Volume 1: The Report (ETSU B 1176–Pl) Volume 2: Harvesting Trials Summaries (ETSU B 1176-P2).
Morgan, J.W.W. (1975), 'Preservation of Timber: General Considerations'. *Timber Grower,* February 1975.
Penrose, A. (1990), 'The Case for In-forest Wood Chipping'. *Country Landowner,* February 1990, 31.

Rowan, A.A. (1976), 'Forest Road Planning'. FC Booklet 43.
—— (1977), Terrain Classification'. FC Forest Record 114.
Taylor, G.G.M. (1981), 'Modern Extraction'. *Timber Grower,* June 1981, pp. 29–31.
Thompson, D.A., Gibbs, J.N., Nisbet, T.R. and Webber, J.F. (1990), 'Water-restored Timber after 12 Months'. FC Research Information Note 178.
TRADA (1990), 'Preservative Treatment for Timber – a Guide to Specification'. Wood Information sheet 16, section 2/3.
Wardle, J. St. A. (1990), 'The Estate Mill'. *Timber Grower,* Summer 1990, 15.
Webber, J.F. (1990), 'Guidelines for Water Storage of Timber'. FC Research Information Note 174.

Chapter Eleven

MARKETING AND PRICES

1. Timber and other woodland produce

Marketing affords one of the principal opportunities of maintaining woodlands on a profitable basis. Efficient marketing ensures the allocation of the available timber to its most suitable overall set of end uses, and may be the most cost-effective means of increasing revenue.

There are great differences between sales from small private woods and those from the major stands of the State, large estates, the forestry management or harvesting companies and the financial institutions. The private sector benefits from the marketing possibilities achieved by the Forestry Commission, but only large owners/operators possess economies of scale and bargaining strength; yet, the private sector's oncosts and overheads may be lower.

Marketing is a perennial problem in small private woods. (Similar difficulties apply to harvesting.) Most of their owners have not the necessary experience, parcels placed on offer may be too small to attract buyers, and co-operatives have proved hard to organize in most regions. However, as noted in chapter 10, there are many large and small scale contractors willing to undertake both harvesting and marketing. For counties in England and Wales, the Forestry Commission's 'Marketing from Small Woodlands', lists sawmills, timber merchants and contractors who are involved in harvesting, marketing and processing of trees.

Owners of small woods may benefit by serving the local and perhaps specialist markets, adding value wherever possible, since it is these markets which the larger growers and roundwood merchants seldom supply. In various parts of the country, pea sticks, bean rods, hedging material, fencing products, rustic, scout and other poles, and the like, can be marketed – frequently at unit prices which are far above those for equally sized industrial small diameter roundwood and sawlogs.

Gainful marketing requires a sound knowledge of what is to be sold, what the customer wants, and what it is worth, and marketing in a planned and enterprising manner. The first requirement means that the grower should have estimates of production from the woodland, knowing for the current and following year the approximate planned cut, and whether it is to be generated by thinning, clearfelling or selective cutting. For each of the following three years, estimates likewise would be helpful; market conditions may change, but plans should be sufficiently flexible to allow for this. Longer term estimates, in broad terms, will indicate the size of the marketing task expected in subsequent years. All may help to decide the intensity of forestry and harvesting to be practised, which in turn determines the requirements of labour and machinery, and affects the cash flow.

Determinants of sale value. The properties and uses of various species of timber have been noted in chapter 9; also timber quality which affects value to a varying degree, according to the end use. The value of a log depends on end use and processing cost. If well-marketed, each log is allocated to the end use offering the best price. The main factors which affect value are: species, tree size, quality, size of parcel, ease of harvesting, access by road haulage vehicles and proximity to markets; also whether a thinning or a clear-fall. These factors, together with local and regional demand, will have to be taken into account when determining the market value of a standing or felled parcel of timber.

The main categories of sale are sawlogs and small diameter roundwood (conifer and broadleaved), veneer logs (broadleaved) scattered infrequently in a stand, conifer pitprops and transmission poles, broadleaved turnery poles, fencing timber, rustic poles, and minor products such as bark, chips and foliage. For converted products, the market may vary considerably in different parts of the country, and from time to time. Foresters should therefore ascertain what markets are currently available before felling any trees and certainly before preparing specific products.

The volume offered is important: sizeable lots often attract more and substantial buyers. Lots of 250–500 m^3 are probably ideal; quantities of 1,000–2,000 m^3 restrict the competition to larger buyers who have the necessary finance. The vendors of the larger parcels, other than the Forestry Commission, are well-wooded estates, forest management companies, or groups of owners, the sales being effected by negotiation, tender or public auction.

The species of timber has a direct bearing on demand. Good ash, oak, beech, and sycamore are always wanted. Large conifers, and veneer or prime quality broadleaves, attract buyers from long distances. Medium sized conifers draw merchants mainly interested in sawlogs. Smaller conifers attract particularly the pulpwood, boardwood and fencing trades. Rustic, post, stake and rail size conifers are locally in demand; less so are small and medium sized broadleaves – the market for them is sometimes saturated with low grade parcels, and only a turnery or another type of processor using broadleaves will alleviate some situations.

High quality timber can certainly bear heavier transport costs, thereby interesting a wider market. External factors of quality in sawlogs include: large size, straightness, concentricity of bole, and adequate length; absence of fluting, twisting, ingrown bark, coarse branch scars, and large knots; without iron, wire, woodpeckers' holes, and cracks caused by sun, drought, wind or lightning. Internal factors of quality have been noted in chapter 9.

The location of the timber is an important factor. Alongside a hard road is ideal. Extraction difficulties impose a considerable drag on sale-value; these include: steep slopes, wet soils, the need to remove and replace walls, to make or renovate a road, and inadequate routes. Selection-felling, or felling among regeneration, where damage has to be kept to a minimum, also detract from the standing value of the timber.

The sale of timber, particularly to local enterprises, carries with it the rather indefinable, but none the less extremely important, element of regular supply. It may be more profitable in the long run to offer continuity of supplies at a reasonable figure, than to attempt to secure the highest possible price on an *ad hoc* basis. Especially, but not exclusively, is this true of sales of converted products where customers will look for regularity and punctuality of supply and an adherence to the specification.

Seeking the markets. Foresters should aim for the maximum net financial return from each thinning and final crop. Their first concern may be to seek local markets for converted products among farmers, market gardeners, nurserymen, garden centres, DIY centres, firewood merchants, charcoal-burners, householders, local authorities and sundriesmen. The minor products may include processed bark or chips for riding surfaces, horticultural and other purposes; also foliage and Christmas trees. However, for substantial quantities of unprepared produce (poles/trees in the length) the grower will need to explore some of the markets set out in the schedules hereafter. Any given pole/tree may satisfy more than one market.

Foresters should seize and make use of all opportunities, seeking out markets and anticipating their requirements, even if they are only on a modest scale, studying the trends of utilization as far into the future as they can; and know the kinds of material and the quality that are required by the processing industries, along with the prices they pay for small or large diameter roundwood. They should practise prudent salesmanship in disposing of their produce, studying each week the 'wanted' advertisements in the *Timber Trades Journal,* and each month in *Forestry and British Timber.* If they have to dispose of the lop and top of broadleaves they can seek out firewood merchants and charcoal-burners. For the best quality broadleaves and for most of the medium and large conifers they will usually find a ready sale. For small and second and third grades of broadleaves and quite small diameter conifers, and for products of underwood and small coppice, they must seek markets if their selling policy is to be wholly successful and fully profitable. They should explore the possibilities of cleaving for sale those species which can be cleft easily; and take advantage of the outlets for branches of birch and the foliage of some of the conifers. Preservative treatment of non-durable broadleaves and conifers can make these timbers profitable. Thinnings of most conifers provide substantial returns and their intermittent total monetary value may be greater than that of the final crop.

Foresters should consider opportunist thinning when this is sound, desirable and remunerative. Opportunist marketing is particularly important. Many a stand can be reserved to await sale when prices are good in general and when a local consumer/merchant has a particular need. To this end sales, particularly of large timber, can sometimes be brought forward or postponed a year or two to suit markets. Sometimes consideration should be given to having some of broadleaves planked, seasoned and stored; the enterprise is often highly remunerative.

In today's market foresters' best course may be to sell small and medium sized thinnings through management or harvesting companies and the like who specialize in such produce, and their large thinnings and mature trees direct to sawmillers or to roundwood merchants. Of the last there are many who are constantly in touch with several markets and know the requirements of the main sawmillers and processors; they are in effect distributors knowing where various species and grades can be most profitably directed. In Scotland foresters can as an alternative sell through Scottish Woodlands Ltd., if the owner of the estate is a member. In England and Wales they can likewise receive valuable help from the TGUK who will put them in touch with established marketing organizations willing to undertake marketing for them at an agreed charge; they aim at ensuring that there is an efficient forestry and marketing service at the disposal of every joining woodland owner in England and Wales. Some regions have the long experience accumulated in marketing by established 'woodland owners' co-operatives'; a few have been functioning for over twenty-five years and have built up wide contacts in the timber trade and considerable goodwill, all of which are available to their members. If growers prefer to sell by public auction they will usually choose a firm of auctioneers which has a forestry department specializing in such sales and which is very well-informed on the needs of the timber trade and can offer sound advice on lotting. Some foresters like to do all their marketing without direct outside assistance; in this case, after supplying the needs of the estate, through either the estate sawmill or wood-yard, they will have to do some intense seeking of markets. How near suitable markets are will make a difference to both ease of marketing and profitability. The three schedules which follow are intended to be aids to marketing.

I: MARKETS FOR CONIFERS

Species (Scientific names are given in the Index)	Small thinnings[1]	Medium thinnings[2]	Large thinnings[3]	Final crop[4]
Scots pine Corsican pine Lodgepole pine	Fencing manufacturers. Merchants for, or manufacturers of, wood for wood wool, pulp, particleboard and fibreboard.	Fencing manufacturers. Merchants for, or manufacturers of, wood for wood wool, pulp, particleboard and fibreboard. Transmission poles (not lodgepole pine).	Sawmillers. Box-makers. Pallet-makers. Roundwood merchants.	Sawmillers. Box-makers. Pallet-makers. Roundwood merchants.
European larch Japanese larch Hybrid larch	Manufacturers of fencing, rustic-work and garden furniture. Merchants for, or manufacturers of, wood for pulp, particleboard and fibreboard.	Fencing and gate manufacturers. Merchants for, or manufacturers of, wood for mining, pulp, particleboard and fibreboard. Transmission poles. Hop-growers. Clay mines.	Sawmillers. Fencing and gate manufacturers. Box-makers. Pallet-makers. Boat-builders (European larch). Roundwood merchants.	Sawmillers. Box-makers. Pallet-makers. Roundwood merchants.
Douglas fir Norway spruce Sitka spruce Noble fir Western hemlock Western red cedar	Fencing manufacturers. Merchants for, or manufacturers of, wood for wood wool, pulp, particleboard and	Fencing manufacturers. Merchants for, or manufacturers of, wood for mining, pulp, particleboard and	Sawmillers. Fencing manufacturers. Box-makers. Pallet-makers. Roundwood merchants.	Sawmillers. Box-makers. Pallet-makers. Roundwood merchants.

Species (Scientific names are given in the Index)	Small thinnings[1]	Medium thinnings[2]	Large thinnings[3]	Final crop[4]
Lawson cypress Coast redwood	fibreboard.	fibreboard. Ladder poles (spruces).		

1. Lower lengths of small thinnings may sometimes be used in the same markets as medium thinnings.
2. Top lengths of medium thinnings may sometimes be used in the same markets as small thinnings.
3. Top lengths of large thinnings may sometimes be used in the same markets as medium thinnings.
4. Top lengths of final crop trees may sometimes be used in the same markets as large thinnings.

II: MARKETS FOR BROADLEAVES

Species (Scientific names are given in the Index)	Small thinnings[1]	Medium thinnings[2]	Large thinnings[3]	Final crop[4]
Oak Pedunculate Sessile	Manufacturers of peeled oak garden furniture. Merchants for, or manufacturers of, hardwood pulp, particleboard and fibreboard. Metal refiners.	Fencing and gate manufacturers. Merchants for, or manufacturers of, hardwood pulp, particleboard and fibreboard. Metal refiners.	Sawmillers. Roundwood merchants. Fencing and gate manufacturers. Barrel-stave cleavers.	As for large thinnings. Also veneer merchants and manufacturers of furniture and coffins. (Consider the possibility of bark stripping for tanning.)
Beech	Merchants for, or manufacturers of, hardwood pulp, particleboard and fibreboard. Metal refiners.	As for small thinnings. Also, wood turneries. Metal refiners.	Sawmillers. Wood turneries. Roundwood merchants. Furniture manufacturers.	As for large thinnings. Also, veneer merchants.
Ash Sycamore	Bar-hurdle makers (Ash). Merchants for, or manufacturers of,	Sports goods manufacturers (Ash). Wood turneries. Tent-peg	Sawmillers. Sports goods manufacturers (Ash). Wood turneries.	Sawmillers. Veneer merchants. Roundwood merchants.

Species (Scientific names are given in the Index)	Small thinnings[1]	Medium thinnings[2]	Large thinnings[3]	Final crop[4]
	hardwood pulp, particleboard and fibreboard. Metal refiners.	makers (Ash). Merchants for, or manfacturers of, hardwood pulp, particleboard and fibreboard. Metal refiners.	Roundwood merchants.	Furniture manufacturers.
Sweet chestnut	Merchants for, or manufacturers of, hardwood pulp, particleboard and fibreboard. Also see under 'Coppice' *infra*.	As for small thinnings.	Sawmillers. Roundwood merchants.	Sawmillers. Roundwood merchants. Veneer merchants. Manufacturers of furniture and coffins.

1 Lower lengths of small thinnings may sometimes be used in the same markets as medium thinnings.

2 Top lengths of medium thinnings may sometimes be used in the same markets as small thinnings.

3 Top lengths of large thinnings may sometimes be used in the same markets as medium thinnings.

4 Top lengths of final crop trees may sometimes be used in the same markets as large thinnings.

The following three broadleaved species are rarely available as thinnings, but as individual medium to large trees or final crop trees, the markets which may be explored, in addition to sawmillers and roundwood merchants, are:

Poplar: Basket and packaging manufacturers (see Jobling 1990).

Elm: Manufacturers of furniture, coffins and garden furniture.

Cherry (Gean): Veneer merchants.

III: MARKETS FOR COPPICE

Species (Scientific names are given in the Index)	Small poles[1]	Medium poles[2]
Hazel	Local workers of underwood. Wattle hurdle and barrel-hoop makers.	Wood turneries. Crate-wood users.
Sweet chestnut	Local workers of underwood. Fencing manufacturers. Hop-growers. Merchants for, or manufacturers of, hardwood pulp, particleboard and fibreboard. See also under 'Broadleaves' *supra*.	Fencing and gate manufacturers. Hop growers. See also under 'Broadleaves' *supra*.
Birch Silver Downy	Wood turneries. Besom-makers (branches). Toy-makers. Local steeplechase jumps (branches). Merchants for, or manufacturers of, hardwood pulp, particleboard and fibreboard.	As for small poles.
Alder Hornbeam Lime	Wood turneries (not grey alder). Toy-makers. Merchants for, or manufacturers of, hardwood pulp, particleboard and fibreboard.	As for small poles. Also, clog sole makers (Alder), woodcarvers and beehive manufacturers (Lime) if sufficiently large.
Mixed broad-leaves	Wood turneries (not oak). Fencing manufacturers. Charcoal-burners. Firewood merchants. Merchants for, or manufacturers of, hardwood pulp, particleboard and fibreboard. Metal refiners.	As for small poles.

1 Lower lengths of small poles may sometimes be used in the same markets as medium poles.
2 Top lengths of medium poles may sometimes be used in the same markets as small poles.

2. Forestry Commission marketing policy

The private sector will benefit by understanding the Forestry Commission's objective and response relating to its marketing (Hughes 1989). As Forestry Authority the Commission is charged to develop and ensure the best use of the country's forest resources, and promote the development of the wood-using industry and its efficiency. From this, it follows that the Commission has a duty to provide production forecasts on which industry will base investment decisions, and as Forestry Enterprise will market, within reason, in accordance with its production plan through good times and bad. This might conflict with another objective, of obtaining the best financial return, in that holding back on sales when prices are down is a basic marketing ploy; but looking at it in a broader sense, the health of the British industry (which is in a world-wide competition position) is essential to the health both of the State and private forestry. In any case, the Commission is constrained by the requirement for it to operate on a cash limit basis; and it is necessary to market timber to generate the income to fund other activities of the Forestry Enterprise, i.e. afforestation, restocking and maintenance of the forest estate. Some changes in organization are imminent (see Figure 1).

Disruptions to the market occur; e.g. world and home economy, and severe windthrow. The latter brings to the market unprogrammed quantities of all categories of timber which have to be marketed, the resultant excess of supply over demand bringing a substantial fall in value. (For example, the windthrow in south-east England in 1987 induced price falls in Wales and the West of England of conifer sawlogs from about £46 to £39 m^3, bars from £34 to 28 m^3. There was a similar effect after the windthrows of January and February 1990 in Wales and south-west England.) Such occurrences may make necessary the laying off of some cutters to reduce the build-up of stocks, limiting new offers of timber, and holding over auctions and contracts. The problem in these situations is to perceive the change in market circumstances, and to be in the position to respond when the market improves.

Forestry Commission planning of marketing is based on its Production Plan. Given a good data base, and in particular correct assessment of Yield Class and Windthrow Hazard Class, the allowable cut can be determined. Standard thinning regimes and financial rotations form the basis, but a meaningful forecast is only achieved by fitting in the exceptions to normal treatment, e.g. retention for landscape/conservation, low/high stocking levels, and where further thinning may lead to windthrow. The Production Plan will detail stands for either thinning or clear-felling; but subsequent ground inspection is vital to enable allocation to appropriate thinning/felling areas to ensure well-planned marketable parcels as, for example, roadside sawlogs or standing sales.

Forestry Commission policy is to sell on the open market unless there are sound reasons for negotiation normally to encourage wood-using industries. In practice competitive sale of sawlogs has virtually been the normal procedure with few long-term negotiated contractors. In contrast, Commission production of small diameter roundwood is dominated by long-term negotiated contracts because the high investment pulp and board mills required the stimulus of a guaranteed supply of a significant part of their wood resource to encourage their establishment. With standing sales, competitive sale via auction or tender has formed the core but negotiated contracts on a continuity basis have been used extensively to encourage the development of harvesting companies in those areas where timber output was not matched by a harvesting resource. This system has now served its purpose and is currently being phased out. With a large programme and a large number of customers, regularity of the market is necessary, the major part of the Commission's harvesting programme being offered at set intervals based on selling ahead over a period of up to six months.

3. Points of sale

There are several choices of point of sale, in sequence: (i) standing; (ii) at stump (i.e. felled, topped and delimbed); (iii) at rideside; (iv) at roadside; (v) loaded onto transport (vehicle, rail or ship); and (vi) delivered. A delivered price is inevitably the highest. Calculations can indicate the comparative price at each of the other points of sale, and can determine the standing value to the grower. Thus the main options in marketing are: (i) to sell standing; (ii) to sell felled (at stump, or extracted, or delivered); and (iii) after some conversion – even if it is minimal cross-cutting to appropriate lengths – to sell the products at a chosen point of sale.

Whether to sell standing can be determined by comparing with the estimated results of selling felled or converted. If the produce is economically harvested a more rewarding price is achieved. Thereby, too, merchants are kept out of the wood, and damage may be reduced. On the purchaser's part, there is no waste or clearing up to deter him nor stringent conditions with which to comply. However, the vendor may not have the labour and equipment necessary for felling and extracting, or for converting and transporting.

A most difficult decision is whether to undertake some amount of conversion – chiefly cross-cutting, and perhaps debarking, splitting or pointing. Relevant factors to be heeded are:

- Adequate planning is necessary, based on forecasts of production, by species, volume and specifications of converted products;
- Secure markets are needed for all the different specifications necessary for the profitable conversion of all trees;
- Sufficient conversion work must be available gainfully to keep employed skilled men to justify the cost of equipment;
- Careful organization of the work is necessary so that workers are on one kind of job sufficiently long to become adequately productive;
- Team-work based on piece-work rates for final products at the point of sale is desirable; and
- Adequate costings and efficient salesmanship are needed.

Standing sales. Many foresters prefer (or have no other option than) to sell their timber standing. The felling is done by the purchaser having skilled labour and specialized equipment, and who accepts the risk of defects and breakages during felling (but will deduct an undisclosed allowance for these factors when bidding). The volume to be paid for may be estimated before the felling or ascertained afterwards. In the case of large standing trees it is sometimes advisable to place a painted number on those to be felled. It should be made clear whether a purchaser must fell and remove scrub, underwood, yew and other normally unwanted trees, or to clear lop and top or brashwood/slash. Disposal of the debris is a difficult problem where the vendor is unable to cope with the quantities involved. To insist that the purchaser of the timber has to remove it decreases his bid. One solution adopted in broadleaves is to make the cording and the burning of the lop and top the purchaser's responsibility, the vendor paying a sum for each cord on completion and retaining the cordwood.

Sales standing involve the grower in the least outlay, work and commercial risk. The financial return is known. The trees to be sold are either individually marked or the boundaries of the area to be worked are delineated in an adequate way. If the trees comprise timber of widely different types, e.g. conifer thinnings and mature broadleaves, they should be offered separately. Desirably, each parcel should be described separately, stating actual or estimation information on number of trees, total volume and average volume per tree, for each species and category. Other conditions under which the trees are to be sold must be clearly defined, so that the vendor's interests are safeguarded and contingencies catered for. The tariff table system is a well-proven method for conifers (see chapter 1).

Sales of felled trees. Some growers, for various reasons, may wish to sell their trees felled – to do the felling at a particular season, or to provide work for their direct labour. The felled trees can be sold at stump or at rideside or roadside, according to whether the vendor has the necessary labour and equipment. The sale agreement may stipulate that the volume shall be the felled measure, with appropriate allowance for defects. The method of measurement must be agreed, e.g. to what top diameter, the allowance if any for bark, together with the price per m^3 or tonne, and whether the price varies for different species and categories.

This method of sale is usually the most accurate, as the full volume is ascertained, but it will disclose otherwise unknown defects which together with breakages will have to be borne by the vendor. The cutter's measure is usually acceptable to both vendor and purchaser. An important point in selling timber felled is to effect a sale *before* felling, otherwise the vendor may have it on his hands for some time, which could be serious with species which sap-stain readily, e.g. Scots and Corsican pines, larches, ash, beech and sycamore.

Sales of converted products. As the method of sale of trees shifts from standing or felled to converted product, so increasingly sophisticated skills, equipment and organization are required. Where the trees are to be felled and converted into such products as sawlogs, bars, pulpwood, boardwood and mining timber, it is essential to secure markets *before* felling and converting. Trained workers and adequate equipment must be available. The converted products can be sold in the wood, or at rideside or roadside; or loaded, or transported. Sale of products can be made direct to the markets or through a merchant or company who will arrange collection and transporting.

If timber or products are sold by weight instead of by volume (see chapter 1), all should be removed quickly after conversion to ensure the greatest weight to the grower. But if the grower is to bear the transport cost based on weight, or on volume given their equivalence to weight, it must be determined whether it will pay to allow the products to dry out before being hauled. The vendor will gain by the reduced transport charges but may lose by the reduced weight of the delivered load. If the charge for transport is based solely on volume it does not matter whether the produce is light or heavy.

4. Grading of roundwood

More has been written on the grading of sawnwood than on the grading of roundwood (except in the case of Forestry Commission conifer sawlogs). Without knowing something of the former and much about the latter it is difficult for foresters always to obtain the best price for either thinnings or final crops. It is necessary to have an appreciation of each tree's or each parcel's particular grade or grades and the value of them – a knowledge which every competent roundwood merchant and sawmiller possesses. Grading and offering in separate lots are no easy matters. Often the wisest course, particularly the simplest, is to sell the whole stand or parcel to a reliable roundwood merchant who will undertake the grading and distribution to appropriate markets.

Grading and lotting of felled timber make or mar the financial result of a sale. The size of lots has been referred to in section 1 *supra*. Lot composition needs care; it is best not to include broadleaves with conifers, and the trees in each lot should usually be one or allied species. Separate lots should be made of veneer logs (they are usually found scattered irregularly in a parcel), of sawlogs of various qualities, and of roundwood suitable for fencing material, mining timber, pulpwood and boardwood; also of poles suitable for rustic work, posts, stakes and rails. Inferior produce should be separated from the remainder. Where possible, lots should be separated on the ground so that only one buyer is operating in a particular section of a wood or stacking area; thereby the responsibility for any damage is easier to ascertain.

The foregoing applies to felled trees: if the trees are standing, and intimately mixed, such physical lotting is impracticable, but in general the principle of selling like with like applies. Clear-falls should not be mixed with small thinnings, nor spruce with larch, because of different markets.

Where offers for the same produce are received in different unit values they should be compared: for example, an offer of a lump sum with an offer per m^3 or tonne; or an overall offer per m^3 or tonne with separate offers for butt lengths and other lengths. Again, if one potential buyer offers £x/tonne, and another £y/m^3, it must be calculated what the first offer amounts to before the second is accepted or dismissed. The comparison changes as the produce dries out (seasons), so that the time at which the produce is to be weighed is important. These situations are best avoided by ensuring that offers are on one specified basis.

In any stand or parcel, reserving the best trees, or even the best lengths, will have an adverse effect on the value of the remainder. Conversely, including a poor quality parcel (in order to sell it) along with a good quality parcel will devalue the latter.

Grading of broadleaves. The grading of broadleaves, standing or felled, is not easy. They may fall into two or more distinct qualities or grades (see Table 46 in section 8 *infra*). A mature crop of oak may contain a mixture of various qualities: mining timber, fencing, planking and veneer. Size, presence of shake or other major defects will influence the grading: they may remove a butt of planking size into a grade only suitable for mining or fencing. (The loss to the grower can be substantial if the sale is based on felled sound measure.) When grading after felling, first to be classified are the butt lengths, usually 2 m and up in length, of planking, beam or veneer quality; they must have no visible defects such as flutes, burrs, knots, shake, rot or excessive spiral grain, and must be reasonably straight and cylindrical. Above the butt length there is usually a second quality length of 1.5 m or more between large branch scars; here small burrs and knots under 5 cm may be acceptable for constructional and fencing timbers. Within the crown may be a third grade length with branches closely spaced; here the quality may be acceptable only for mining and larger fence posts.

Ash is probably the most difficult to select for quality. An attempt should be made to grade into two or more of the following (excluding small diameter poles suitable for rails, bar-hurdles, tent-pegs or turnery): (1) sports quality, prime tool-handle quality, and prime planking quality; (2) second quality; and (3) third quality. Grade (1) is the most valuable but is most difficult to find: even a first class sports parcel may have only 25% of its value in this grade. Few trees contain more than one 2 m length of sports quality. The butt must show even growth-rings, central heart, and no discoloration. (Obviously the features cannot be verified before felling.) There must be no visible scars on the bark indicating occluded pin-knots. Smoothness of the bark often indicates the appropriate grade. In the grade for prime tool-handle production, it is permissible to have occasional pin-knots, slight discoloration and eccentric heart. Planking ash should have a first clean minimum length of 2.5 metres and a mid-diameter of at least 40 cm. This length should be free from all visible defects. Slight discolouration is not generally regarded as a defect. Lengths from the crown are worth much less and may only be suitable for mining, pulpwood, boardwood, or firewood.

Beech is generally classified into: (1) prime, white, with smooth clean bark, 40 cm and up mid-diameter, free from excessive flutes and ingrown bark, and reasonably straight; (2) prime, some discoloration, with smooth clean bark, 40 cm and up mid-diameter, suitable for merchantable grade for framing under upholstery, free from knots to allow bending; and (3) veneer quality for rotary peeling, 50 cm and up mid-diameter, free from all defects. Then there are second and third quality lengths suitable for cheap chair manufacture. Finally there remains low qualities suitable only for mining, pulpwood, boardwood, or firewood.

Other broadleaves are subject to some of the general principles of grading indicated above. The classifications are usually: veneers (see chapter 9), prime butts, and second and third qualities suitable mainly for fencing, mining, pulpwood, boardwood or turnery.

Grading of conifers. The Forestry Commission current grading classification of conifer sawlogs has been included in chapter 9.

5. Methods of marketing timber

The three main methods of marketing timber are: (i) private treaty (negotiation); (ii) tender; and (iii) public auction. In all three, the vendor should first obtain a realistic valuation and set a 'reserve' below which a sale will not proceed.

In a sale by *private treaty (negotiation)* there is no competition between prospective purchasers, but it is the least expensive method of sale, eliminating advertising and commission. It is particularly appropriate for small parcels; also for co-operating with a local industry.

Sales by *tender* are competitive, and may either be public (particulars and invitations to tender are published) or private (particulars are circulated privately to selected possible purchasers). Offers are requested by a stated date. This method is inexpensive. Buyers do not always welcome it because the reserve is unknown, and the highest or any tender may not necessarily be accepted. Some buyers prefer the method: prospective purchasers may tender a relatively high bid for produce they are anxious to secure in the knowledge that others will be unaware of what they are prepared to pay. (At a public auction they might hesitate to disclose their willingness to pay what appears a high price.) Sales by tender stand or fall by the vendor's reputation with the trade: the date for receipt of tenders must be rigidly observed; if the vendor can notify the result to all tenderers the same day or at latest the following day, the sale is seen to have been properly conducted.

Auction sales are usually only justified if a substantial amount of produce is to be sold. This competitive method is popular and usually ensures satisfactory sales. It is occasionally used by individual or groups of owners, or at an auction where Forestry Commission produce is on offer. Appropriate classification and adequate advertising are necessary. Expenses include those of arriving at a reserve figure, advertising, sale particulars and conditions, and commission. A wider circle of potential purchasers is reached, and prices are likely to be near or above those prevailing at the time.

Particulars and Conditions of Sale. Whatever method of sale is adopted there must first be conditions under which the produce is offered: these Conditions of Sale are usually set out in the Particulars of Sale giving the location and general description of the trees, the method of identification, and the situation relating to a felling licence. (If the trees for sale are felled, and perhaps converted, classification and identification must be adequate.) It is for the vendor to decide what Conditions of Sale to impose. The most satisfactory course is to treat each sale on its merits and to draw up conditions to cover its special aspects. It is not possible to list every detail which might be covered by Conditions of Sale, but some of the following suggested headings will be appropriate:

1. Schedules
   Location
   Description, identification and number of trees (or quantities of products), including species, categories, volumes (and how arrived at)
   Access and extraction routes
   Overhead lines, and concealed objects (pipes, lines, culverts)
   Gates, fences, walls to be removed and reinstated

2. Plan indicating locations of trees/products, access, routes, loading or stacking spaces, overhead lines, concealed pipes, lines, culverts

3. Purchase money
   Whether lump sum, or m^3, or tonne

When to be paid: (a) deposit, (b) remainder
Default of payment

4 Felling, extracting
   Date of entry and completion
   Chemical treatment of stumps after felling
   Loading and stacking spaces
   Cordwood, lop and top, or brashwood/slash

5 Damage
   Purchaser's responsibilities
   Responsibilities under the Health and Safety at Work regulations Liability for late completion
   Indemnity against claims by third parties
   Reinstatement of damage to drains, ditches, fences, gates, hedges, walls, watercourses, culverts and extraction routes

6 Miscellaneous
   Suspension of work in connection with a shoot
   Suspension of removal during a thaw
   Re-selling and sub-contracting
   Defective timber
   Felling of trees not in the sale
   Dogs, guns, traps, employees
   Fires, fire precautions
   Insurances

7 Arbitration

The use of the above headings, as appropriate, will much depend on whether the trees are standing, or felled/converted, and at what point of sale. If the Conditions of Sale are too stringent, prospective purchasers will reduce their offer; if too lenient the vendor may find that a good bid is eroded by unrecompensed damage.

Contracts. An agreement to sell/purchase must be confirmed by the signing of a legally binding contract based on the Conditions of Sale. Insley (ed.) (1988) provides a Specimen Contract for the Sale of Felled or Standing Timber. Hibberd (ed.) (1991) provides information on items normally included in an individual contract for standing sales of trees. The Forestry Commission has its own Standard Conditions of Sale of Standing Trees (Short Term Contract).

Charges for marketing. Fees and commissions charged for marketing vary throughout the country. As a general guide, the charges for the sale of timber and forest products are likely to be: 7.5% on the first £5,000, 5% on the next £10,000, and 2.5% thereafter. This would include the application for a Felling Licence where required, preparation of Sale Particulars, exposure for sale by private treaty or tender, and preparation of a Sale Agreement. The charges would not cover the marking and measuring of timber, nor the subsequent control of felling operations. The sale of freehold woodlands throughout the country is usually undertaken by specialist firms or individual consultant foresters (see chapters 16 and 17). The charges for seeking and negotiating the sale or purchase of a forestry property are likely to be: 2.5% up to £50,000, 2% on the next £150,000, and 1% thereafter.

6. World prices of timber

Wood is unusual among raw materials in that its price in *real* terms has increased worldwide over the very long term (Johnston *et al* 1967). However, since the 1950s there has been no discernible trend: wood prices have moved in sympathy with those of other raw materials and fuels, and have responded to changes in economic activity, but they have not tended to rise or fall relative to the general level of prices in the economy.

In most developed nations, wood product prices have kept pace with inflation or outrun it, so in real terms, prices have risen. For example, the price of sawn softwood has increased in real terms by between 1 and 2% per annum for periods up to 120 years long, and for pulp and paper there have also been rises although much less marked (Forestry Commission 1986). In the USA, starting from a pioneering condition with wood freely available, prices have risen substantially in real terms this century. This is also true in Scandinavia, where the average annual increases have varied between 1 and 3%. Analysts considering the prospects for timber in the future have universally indicated the likelihood of continued price increases. The following list summarizes fairly recent statements about price prospects in real terms to the years 2000 and 2025:

Year	Source	Product	To 2000	To 2025 +
1979	World Bank	Sawn softwood	+2½% per annum	
1980	Centre for Agricultural Strategy, Reading	Roundwood in general	+30%	+ 100% to 150%
1988	Binkley and Vincent (1988)	Southern USA stumpage prices	+ 10% to 75%	+20% to 200%
1989	US Forest Service (1989)	Sawlogs	1.03% to 1.14% per annum to 2030 (Less for broadleaves)	
1990	Resources for the Future (Sedjo and Lyon 1990)	Roundwood	+0.2% to 1.2% per annum over the next 50 years	

In the United Kingdom, during the period 1958–89 real prices for broadleaved logs have been remarkably stable and have only deviated from their average figure by more than 10% in six years. In the 1950s, prices of conifer roundwood were fairly buoyant, and fell only slightly in the early 1960s into a period of mild depression. In the 1970s, however, prices were much more unstable, reaching an all-time high in the first five years, to fall dramatically in the last half of the decade to their lowest ever in 1981–82. They subsequently recovered.

Thus over the past forty years there has been no real increase in the price of roundwood in Britain although prices have been subject to short-term fluctuations. At present, prices are about the same as the long-term (thirty-year) average. There are variations within the market so whereas conifer sawlog prices have hardly moved in real terms, prices for small diameter roundwood have risen dramatically in real terms over the last few years, from a low level in 1981–82.

Movements in timber prices over time occur for many reasons. In Britain prices are strongly influenced by the price at which timber is imported. These, in turn, reflect changes in demand and supply across the world. Factors affecting world demand for timber include: economic growth rates, population growth rates, the availability of substitutes for wood and the scope for wood to substitute for other products, the development of new wood products, and the development of new technology which can be used in producing traditional timber products. Prices are also influenced by movements in exchange rates between sterling and the currencies of our major trading partners. These factors change over time, and timber prices adjust accordingly.

Outlook for timber prices. Prices of wood vary with economic activity (the low price level of home-grown wood witnessed in the early 1980s reflected the general difficulties faced by British manufacturing industry and its suppliers at that time, as well as the direct effect of sterling's relative strength on prices of imported wood products). Wood products for which Britain has to compete in the international market are likely to be increasingly in demand, and their supply is becoming less certain. However, there is much debate about whether real prices for wood will continue to rise in the mediumand long-term future. The effects of product substitution may be greater than expected; and, in the UK, with a gradually maturing forest, prices may be held down. There is therefore no sound economic reason for the real price of a renewable resource such as timber to rise continually over the next century.

The United Nations Economic Commission for Europe (ECE) and the Food and Agriculture Organization (FAO) publish monthly the *Timber Bulletin: Monthly prices for forest products*. This and complementary publications are noted in chapter 17, section 5.

Imports meet about 90% of Britain's requirements of wood and wood products. The price of imports of wood products, whether in semi-processed or manufactured form, thus sets the level of prices for British output of wood goods and hence for the roundwood consumed by the domestic industry. Therefore, in considering the outlook for price, it is essential to consider the world scene. Consumption of wood for industrial purposes has increased throughout this century. Growth in consumption of up to 3.5% compound per annum may occur over the next two decades. The world supply of wood currently about matches demand. For this to be expanded to meet increased demand, higher prices may be required to offset the costs of extracting timber from previously unmanaged reserves. It is unlikely that demand will outstrip potential supply in the near future, as the large boreal forests have plenty of scope for increased output. However, it cannot be ruled out that higher prices will be required to ration demand should environmental constraints slow down or change the pattern of felling in some regions. In conclusion therefore, it is possible that some increase in prices could be expected in the future, if supply is constrained, or extraction costs rise.

In British forestry, the estimate of financial return generally assumes that there will be no increase in timber prices in real terms. There is little evidence to support either a rise or fall in real softwood timber prices in the future (Whiteman 1990). This, incidentally, is in line with recommendations by Fraser *et al.* (1985) and Levack (1989).

7. Timber price–size relationship

The commercial value of a tree depends on its end use and processing cost. The price ruling at any one time is not necessarily a sound indication of the final value of a stand which may not be thinned or clear-felled for several years. However, information on past price movements can be used in helping to form a view about the likely level of future prices. Once a long run price can be derived, it must be remembered that price will also depend on such factors as location, species, dimensions, quality, size of parcel, point of sale, and the state of the market.

A large sound tree at any point of sale generally commands a higher price per unit than a small one, because it is less costly to harvest, handle and process, and the larger and straighter the tree, the higher is the potential sawnwood recovery. These factors can be combined to produce a relationship between price per unit and tree size – the price–size relationship, having great significance in silvicultural and harvesting decisions. Since the ease of harvesting, standard of roads, and haulage distance to processing mills vary, the price–size relationship differs between forests *(spatial price variation),* even when the same markets are supplied. In remote stands with difficult terrain, surplus is much lower than in stands near processing industries and with easy harvesting conditions. Some mills offer a premium on delivered price to offset transport costs from remote timber resources. Generally, mills prefer a price structure attracting sufficient roundwood to fulfil processing capacity, while paying each grower only what is needed to compete successfully with other potential purchasers, or to ensure forward suppliers.

Price–size curves for roundwood. For conifers, the Forestry Commission has made a detailed analysis of its standing sales, resulting in its price–size curves which show the relationship between standing timber value and average tree size. They are used to estimate the future value of timber crops. The Commission's current price–size curve analysis for conifers is very similar to earlier work by Mitlin (1987) who explained the construction of such curves. Whiteman (1990) has improved the original curves in two main ways: (i) they are now based on data over thirty years; and (ii) they take into account the revenue obtained from both standing sales and direct working, and thus provide a more appropriate guide to future revenues from Commission plantations.

The price–size curves are built on the assumption that there will be no long-term tendency for prices to rise or fall in the future (Whiteman 1990). Because of this, they should be used only for long-term planning. In the short term the state of the timber market can result in prices that are quite different from those of the price-size curves. Market forces will play a large part in setting timber prices in any one year. They should be taken into account when valuing timber to be sold in the near future.

The assumption of long-term constant prices is not universally accepted and reports made by some other organizations predict real price increases (see section 6 *supra*). However, in some cases these predictions were based on the rapid rise in the price of timber products which occurred in the early 1970s but which was followed by a dramatic fall in the early 1980s.

Where there is a case for valuing a stand using a higher or lower price-size assumption, this should be done to enable the right management decisions to be made. Foresters may be able to obtain better prices than those shown in the price-size curves, by adding value such as converting into fencing, hedging materials and other small dimension products.

The larger the mean tree size the greater the price m^3 within certain limits. At the bottom end, the price for small diameter roundwood is governed by demand. The price for larger roundwood improves in relation to its suitability for sawing, its sawnwood recovery potential being an important factor. The price-size curve for conifers flattens off because most British sawmills are geared to convert logs up to 1 m^3 average size. However, high quality sawlogs of Douglas fir and European larch continue to rise in value.

For any tree size, the harvesting cost of thinning may be greater than that of an equivalent clear-felling. The thinning price differential is a flat rate per m^3 reducing the value of thinnings compared with clear-felling; a likely range being £0.50-£1 m^3. Overheads, probably £1-£3 m^3, should also be deducted from the price-size curve, which will reduce discounted revenue.

Price-size curves for conifer roundwood. Forestry Commission price-size curves for conifers, given in Tables 40 and 41, have been designed for general long-term valuations and are not ideal for location specific or short-term valuations. They should not be used for valuing timber if the sale is to take place in the near future (three to five years is probably an appropriate cut-off point between the short and long-term price assumptions). Short-term price assumptions should be influenced by current prices but, when valuing immature conifer plantations which are being considered for sale or purchase, the long-term price-size assumptions should be used. Stumbles (1985) gives reasons for considering that valuations made on the basis of past prices will underestimate the true value of forestry as an investment.

Table 40. Conifers: England and Wales: price-size curve at 1990-91 prices.

Average volume per tree (m^3)	£/m^3	Average volume per tree (m^3)	£/m^3	Average volume per tree (m^3)	£/m^3
0.01	-7.80	0.16	14.37	1.30	37.44
0.02	-6.54	0.17	15.01	1.40	37.44
0.03	-4.70	0.18	15.65	1.50	37.44
0.04	-2.47	0.19	16.27	1.60	37.44
0.05	0.00	0.20	16.89	1.70	37.44

Average volume per tree (m³)	£/m³	Average volume per tree (m³)	£/m³	Average volume per tree (m³)	£/m³
0.06	2.53	0.30	22.47	1.80	37.44
0.07	4.98	0.40	27.35	1.90	37.44
0.08	7.16	0.50	31.32	2.00	37.44
0.09	8.90	0.60	34.09	3.00	37.44
0.10	10.05	0.70	35.88	4.00	37.44
0.11	10.83	0.80	36.89	5.00	37.44
0.12	11.58	0.90	37.33	6.00	37.44
0.13	12.30	1.00	37.44	7.00	37.44
0.14	13.01	1.10	37.44	8.00	37.44
0.15	13.70	1.20	37.44	9.00	37.44

Source: Whiteman (1990).

Table 41. Conifers: Scotland: price–size curve at 1990–91 prices.

Average volume per tree (m³)	£/m³	Average volume per tree (m³)	£/m³	Average volume per tree (m³)	£/m³
0.01	-7.80	0.16	11.41	1.30	33.47
0.02	-6.54	0.17	11.91	1.40	33.47
0.03	-1.70	0.18	12.40	1.50	33.47
0.04	-2.47	0.19	12.87	1.60	33.47
0.05	0.00	0.20	13.33	1.70	33.47
0.06	2.46	0.30	17.57	1.80	33.47
0.07	4.49	0.40	21.29	1.90	33.47
0.08	6.12	0.50	24.64	2.00	33.47
0.09	7.34	0.60	27.58	3.00	33.47
0.10	8.14	0.70	30.02	4.00	33.47
0.11	8.73	0.80	31.88	5.00	33.47
0.12	9.30	0.90	33.07	6.00	33.47
0.13	9.85	1.00	33.47	7.00	33.47
0.14	10.39	1.10	33.47	8.00	33.47
0.15	10.91	1.20	33.47	9.00	33.47

Source: Whiteman (1990).

Price-size curves for broadleaved roundwood. Forestry Commission price-size curve for broadleaves, given in Table 42, has been designed for general long-term valuations and is not ideal for location specific or short term valuations. The curve is provisional and should be used with caution. Current work will refine relationships to take into account variations between species and ease of harvesting (Whiteman *et al.* 1991). These factors are much more important for broadleaved prices than they are for conifer prices.

Table 42. Broadleaves: price-size curve at 1990–91 prices.

Average volume per tree (m³)	£/m³	Average volume per tree (m³)	£/m³	Average volume per tree (m³)	£/m³
0.01	3.36	0.16	11.06	1.30	27.21
0.02	4.53	0.17	11.35	1.40	28.09
0.03	5.38	0.18	11.63	1.50	28.93
0.04	6.09	0.19	11.91	1.60	29.74
0.05	6.71	0.20	12.17	1.70	30.53
0.06	7.26	0.30	14.48	1.80	31.29
0.07	7.75	0.40	16.40	1.90	32.03
0.08	8.21	0.50	18.04	2.00	32.74
0.09	8.64	0.60	19.51	3.00	38.96
0.10	9.03	0.70	20.85	4.00	44.10
0.11	9.42	0.80	22.09	5.00	48.53
0.12	9.77	0.90	23.23	6.00	52.49
0.13	10.11	1.00	24.31	7.00	56.07
0.14	10.44	1.10	25.32	8.00	59.38
0.15	10.75	1.20	26.29	9.00	62.47

Source: Forestry Commission: Whiteman *et al.* (1991).

The price-size curves provide a reasonable expectation of long-term prices, but in the short term foresters in their investment appraisals should use whatever indications they can obtain from recent sales by referring to the forestry press. Summaries of all Forestry Commission conifer sales are published periodically in *Forestry and British Timber* and *Timber Grower* (TGUK). Current (1991) prices of roundwood (conifer and broadleaves) are indicated in section 8 *infra*.

8. Prices of roundwood

Home-produced roundwood accounts for only a small part of Britain's total needs, and hence its price is strongly influenced by the prices of imported sawnwood and wood-based products.

The standing *(stumpage)* prices received for roundwood significantly influence the profitability of forestry. They vary from one region of Britain to another. Widespread storms, as experienced in 1987 and 1990, suppress prices. High interest rates lead to reductions in demand and have a moderating effect on roundwood prices generally. Conifer prices are relatively well known: their level is largely determined by that of domestic processed wood and wood products, which in turn is dominated by that of imports.

The value of roundwood used in most economics calculations is the standing value; harvesting and marketing costs are not included. In theory, the value is derived from the price of final or processed products by deducting costs (including profit margin or mark-up) for each stage of converting the tree to a consumable product: this includes harvesting and processing, as well as wholesaling and retailing. In sawmilling, much of the tree may become a waste or low value product: thus typically one m³ of roundwood log becomes only about 0.5 m³ of sawnwood, the remnant being offcuts and sawdust, which some forms of further processing convert to useful products. The residual value to the grower is usually only a small fraction of the ultimate sale value of consumer products. In a delivered price (before processing) harvesting and extracting may take a third, and transport a further third. The small fraction of consumers' (or even of mills') price accruing to growers makes the forest economy susceptible to fluctuations in final product price. Occasionally growers have the benefit of a disproportionate price rise. Sometimes a reduction in product price may be briefly absorbed by the harvesting and processing industries operating at reduced profit rather than allowing machinery and workers to be idle. If the price offered to growers is too low, they at least have the option of leaving trees to grow on until times and prices improve (yet, if this critically delays thinning, the result may harm the crop and reduce the return). Higher prices enable extension of extracting and transporting limits, and exploitation of smaller tree sizes.

Major processors of small diameter roundwood, particularly the pulp and boardwood mills, negotiate prices with their main suppliers at prices reflecting the imported finished product, as well as the species, specification, and haulage distance to mill. Pitprops, and sawn mining timber, are sold to British Coal Corporation by competitive, undisclosed tender; this likewise applies to transmission poles to British Telecom plc and the Electricity Supply Companies.

Standing trees, as well as sawlogs, are usually sold direct to sawmillers or roundwood merchants. The prices may be affected by their particular requirements: if a prospective purchaser has a full order book and his roundwood stocks are low, he may be prepared to pay higher than usual prices; conversely, if trade is slack and stocks are high, he is unlikely to offer enhanced prices for further purchases.

There are no comprehensive statistics of prices of standing timber in Britain covering the private sector. However, the Forestry Commission regularly publish the average prices it receives for standing sales of conifers in England, Wales and Scotland. Tree sizes are separated, but not species, quality or region. The Commission also publishes results of its sale of sawlogs by auction, as well as indices showing changes in log and standing sale prices. These schedules appear in forestry journals and the timber trade press. Knowledge of roundwood value requires a study of local, regional and national markets, both of price levels and current demand. Such knowledge is important, even where trees are sold standing, as the price the buyer can afford to pay is determined by his markets.

As noted in chapter 1, for hardwood sawlogs and veneer logs (and less frequently for softwood sawlogs) some roundwood merchants, sawmillers and contractors continue to use imperial units in which lengths are expressed in feet, quarter girth (QG) or diameter in inches, and volumes in Hoppus feet (h ft). One h ft = 1.27 true cubic feet; 27.74 h ft = 1 cubic metre (m³); and 1 h ft = 0.036 m³. A h ft is about 21% short of a true cubic foot – the reduction helping to compensate the processor for his loss in converting roundwood to sawnwood. A useful conversion factor is: 1 cm diameter = 0.31 inches QG.

When timber is sold by weight (e.g. pulpwood, boardwood, turnery poles) the felled or converted produce should be despatched to market quickly in order to achieve the best return. The weight of any stack or load of timber will decrease as the timber dries out, and the rate of drying depends on the weather and the way the timber is handled. In particular, if the bark is removed the timber will dry out much more quickly. Coniferous timber with the bark on is likely to lose about 2% of its weight each week during the summer and about 0.5% during the winter. Table 2 in chapter 1 gives an estimate of the density of *fresh felled* timber in tonnes per cubic metre. Sales by weight operate in favour of the vendor if despatched promptly and weighed in a green state; conversely, against the transporter and processor if the produce loses weight by drying out. The situation is different when timber is sold by measurement of the stack or load converted by an appropriate factor to weight. The measured stacked volume of the produce should be multiplied by the *worked out* solid/stacked conversion factor to give the solid volume of timber as noted in chapter 1. Conversion factors might be expected to be about 55%–65% for broadleaves, 65%–75% for average quality conifers, and 75%–85% for short, straight and good quality conifers.

Prices of conifer roundwood. Prices of conifer roundwood vary greatly according particularly to species, quality and dimension. Other relevant factors are quantity, point of sale (standing, felled, extracted or delivered), distance from main markets, availability of local markets, and the state of the demand and supply balance. Current (1991) price ranges are indicated in Table 43. Prices assume no conversion other than minimal cross-cutting. Any tree may satisfy more than one market; for example, in small diameter roundwood, bars, fencing material, pulpwood and boardwood; and in medium to large roundwood, not only one or more qualities of sawlogs but also bars, fencing material, pulpwood and boardwood.

Table 43. Indications of prices of conifer roundwood.

CONIFER ROUNDWOOD		Standing £	At roadside £	Delivered £
Small diameter roundwood★ (a) Pulpwood, boardwood (Delivered prices are noted in detail later in this chapter):	m^3 or tonne:	4–10	10–20	20–32
(b) Fencing:	m^3 or tonne:	6–18	15–30+	25–40
(c) Bars: 1.8 m x min. 14 cm top diameter ub (Larch and Douglas fir have the highest value):	m^3 or tonne:	17–22	25–32	30–40
Sawlogs† (the two grades of the Forestry Commission are noted in chapter 9) (Douglas fir is the highest value):	m^{3}ob:	20–33	25–45 +	30–55 +

'Boatskin' larch (European). Straight. Min. 5 m x min. 40 cm mid-diameter x min. 30 cm top diameter: £65–£70 m^3 felled.

* Small diameter roundwood is usually sold by overbark sound measure (m³) or by weight (tonne), or by tonnes calculated by solid/stacked conversion factors. Where sold by weight, the felled produce should be removed to market quickly in order to achieve the best return.

† Conifer sawlogs are usually sold by underbark sound measure. Bark accounts for 10–15% of volume according to species and age (see chapter 1). An important factor in the value of sawlogs is the sawnwood recovery potential.

Prices of broadleaved roundwood. Prices of broadleaved roundwood vary greatly according particularly to species, quality and dimension. Other relevant factors are quantity, point of sale (standing, felled, extracted or delivered), distance from main markets, availability of local markets, and the state of the demand and supply balance. In planking quality, another important factor is colour; and in veneer quality, both colour and figure. Current (1991) price ranges are indicated in Table 44; those for veneer quality logs are given in Table 45. Prices assume no conversion other than minimal cross-cutting. Any tree may satisfy more than one market; for example, in small diameter roundwood, fencing material, pulpwood and boardwood; and in medium to large roundwood, not only one or more qualities of sawlogs but also fencing material, pulpwood and boardwood.

Table 44. Indications of prices of broadleaved roundwood.

BROADLEAVED ROUNDWOOD		Standing £	At roadside £	Delivered £
Small diameter roundwood* (a) Fencing, pulpwood, boardwood (Delivered prices of the two latter are noted in detail later in this chapter):	m³ or tonne:	2–8	8–25	20–35
(b) Turnery:	m³ or tonne:	8–20	15–30	25–45
(c) Firewood:	m³ or tonne:	1–8	8–15	15–30
Sawlogs: (a) Medium quality:† Mining, pallet, fencing, packaging (Fencing is the highest value):	m³ ob:	12–25	20–35	30–45

	Approximate minimum mid-diameter ob cm	£/m³ ob (felled sound measure)
(b) Planking quality: Ash: 'White'. Some brown heart accepted. Min. 2 m 'Tool quality'	35	60–90+ 35–40

	Approximate minimum mid-diameter ob cm	£/m³ ob (felled sound measure)
Beech: Prime 'White'. Min. 2 m	40	40–80+
Coloured	40	30–35+
Cherry: Min. 2 m	40	90–100+
Elm: Min. 2 m	45	50–90+
Oak: clean, free from knots; min. 1.8 m. 'Shake' is unacceptable.	45	70–150+
Sweet chestnut: 'Shake' or brown heart is unacceptable.	40	60–100+
Sycamore	40	50–120+
(c) Special category sawlogs: Veneer quality: see Table 45 Oak for boatbuilding: straight, 15 m & up	55	55+
Poplar for pallets and general sawlogs:	30	15–25

* Small diameter roundwood is usually sold by overbark sound measure (m³) or by weight (tonne), or by tonnes calculated by solid/stacked conversion factors. Where sold by weight the felled produce should be removed to market quickly in order to achieve the best return.

† Broadleaved sawlogs are usually sold by overbark sound measure. Bark accounts for 10 – 15% volume according to species and age (see chapter 1). An important factor in the value of sawlogs is the sawnwood recovery potential.

Table 45. Indications of prices of broadleaved veneer quality logs.

BROADLEAVED VENEER QUALITY LOGS

(For additional information see chapter 9, section 4)	Approximate Minimum mid-diameter ob cm	£/m³ ob (felled sound measure) Quality is paramount
Ash: 'White'.	40	100–225+
Beech: 'White'.	40	55–150+
Elm: Scarce. Burr elm is particularly in demand	50	90–100+

(For additional information see chapter 9, section 4)	Approximate Minimum mid-diameter ob cm	£/m³ ob (felled sound measure) Quality is paramount
Cherry: 'green tinge' is not accepted	40	90–150+
Oak: 'Shake' is unacceptable. Burr oak is particularly in demand. 'Pippy' (small burrs/epicormics) is less valuable	60	150–300+
Brown oak: Scarce Min. 2 m		250–400+
Sweet chestnut: 'Shake' is unacceptable, also brown stain.	40	80–180+
Sycamore:	40	90–125+
'Rippled' ('Fiddle')	40	150–250+
Walnut: Scarce. Preferred 'dug up'.	40	100–275+
Poplar: For rotary peeling.	35	20–30
CONIFERS: Yew only: generally 2 m & up. 'Pippy' (small burrs/epicormics) often increase the value. Although knots and fluting detract from values, merchants sometimes accept low grades and small sizes. Becoming scarce.	30	60–300+

Standing parcels of broadleaves. An average stand or parcel of broadleaves usually contains a variety of quality categories and possibly of species. Table 46 gives a comparison between quality, quantity and value for average parcels of oak, ash and beech ('M/P' represents the value of mining and pallet quality):

Table 46. A guide to the comparative values of the quality categories of standing parcels of oak, ash and beech.

An average parcel by species and quality category	Value Scale	% Volume	% Value
Oak			
Veneer butts	M/P x 10	20%	53%
1st quality butts	M/P x 6	30%	20%
Beam quality	M/P x 2½	35%	23%
Fencing quality	M/P x 2	15%	4%
Mining/pallet quality	M/P		
Ash			
Veneer butts	M/P x 5½	5%	10%
1st quality: White	M/P x 4	55%	74%
Coloured	M/P x 3½	15%	7%
2nd quality	M/P x 1½	25%	9%
Mining/pallet quality	M/P		
Beech			
1st quality: White	M/P x 2½	50%	66%
Coloured	M/P x 2	50%	66%
2nd quality	M/P x 1½	30%	23%
Mining/pallet quality	M/P	20%	11%

Source: Henry Venables Hardwoods Limited, Stafford.

Wood processing industries using small diameter roundwood. These processors have been referred to in chapter 9, and their locations indicated in Figure 22. Current (1991) specifications and prices are provided below, but because of their variation, foresters should, before undertaking conversion, ascertain the up-to-date requirements of individual mills and seek an order for supplies.

Pulpmills using conifer small diameter roundwood.

Shotton Paper Company plc at Shotton, Deeside in North Wales. (Owners: United Paper Mills of Finland.) Integrated pulp and paper mill.
Products: newsprint.
Species used: spruces (90%), pines and wood residues (also waste paper). Annual input: *c.* 700,000 tonnes small roundwood; *c.* 80,000 tonnes wood residues.
Specification: lengths 2.3 m and 3.0 m.
Price indication: £22-£25/tonne at roadside according to location.
Timber purchasing executive: Jim Sutton (0244–830333)

Caledonian Paper plc at Irvine in Ayrshire. (Owners: The Kymmene Corporation of Finland.) Integrated pulp and paper mill.

Products: lightweight coated paper ('Clydecote') from both mechanical and chemical pulps, the latter being imported.
Species used: spruces, also small quantities of western hemlock, grand fir and noble fir; not wood residues.
Annual input: c. 150,000 tonnes.
Specification: fresh, unpeeled; lengths 2.7–3.0 m or 4.5 m x min. top diameter 6 cm.
Price indication: £27.50–£30/tonne.
Forestry director: Toby Beadle (0294–312020).

Iggesund Paperboard Ltd at Workington, Cumbria. (Owners: The Modo Group of Sweden.) Paper and paperboard mills; also Iggesund Sawmills.
Products: mechanical pulp for cartonboard.
Species used: spruce; also wood residues.
Annual input: c. 180,000 tonnes.
Specification: lengths 1.8 m – 2.3 m x min. top diameter 7 cm, and max. diameter 35 cm.
Price indication: £25–£30/tonne.
Harvesting executive: Gordon B. Little (0900–601000).

Pulpmills using broadleaved small diameter roundwood.

St Regis Paper Co. Ltd. at Sudbrook, Gwent. (Owners: David S. Smith Holdings plc.) Semi-chemical fluting paper mill.
Products: fluting medium
Species used: all broadleaves.
Annual input: c. 225,000 tonnes.
Specification: lengths 2 m – 2.4 m x min. top diameter 6 cm, and max. diameter 40 cm.
Price indication: £21–£23.50 tonne.
Harvesting executive: M.R. Henderson (0291–425500).

Boardmills using small diameter roundwood (some also use wood residues).

Highland Forest Products pic at Dalcross near Inverness. (Owners: Noranda Forest Inc. of Canada.)
Products: Oriented Strand Board (OSB); 'Sterling Board'; and added value products.
Species used: Scots pine 83%, the remainder being other conifers; not wood residues.
Annual input: c. 200,000 tonnes.
Specification: length 3 m, plus 10 cm or minus 20 cm. Min. top diameter 5 cm ob. No maximum diameter limit.
Price indication: £21–£23/tonne.
Timber buyer: Douglas Lamont (0463–792424).

Caberboard Ltd. at Cowie near Stirling and Irvine in Ayrshire. (Owners: The Glunz Group of Germany.)
Products: Medium Density Fibreboard (MDF), chipboard and particleboard. (Irvine, chipboard only.)

Species used: mixed conifers and broadleaves. Can accept ageing stock.
Annual input: *c.* 900,000 tonnes.
Specification: length 1.9 – 2.1 m. Min. top diameter 35 cm.
Price indication: £19–£26/tonne.
Purchasing manager: Alan Bloomfield (0786–812921).
Timber buyer: George Webb (0786–812921).

Kronospan Ltd. at Chirk, Clwyd, North Wales. (Owners: Kaindl Group of Austria.)
Products: particleboard, melamine faced chipboard, and cut and edged shelves.
Species used: conifers and broadleaves. Can accept ageing stock, also residues.
Annual input: *c.* 600,000 tonnes.
Specification: length 3.1 m preferred, from 1 m – 6 m by arrangement. Min. top diameter 6 cm. Max diameter 60 cm.
Price indication: conifers £18–£23/tonne, broadleaves £16–21/tonne.
Purchasing director: Dr David C. Wood (0691–773361).

Egger (UK) Ltd. at Hexham, Northumberland. (Owners: Egger of Austria.)
Products: particleboard, melamine faced chipboard.
Species used: most conifers and broadleaves (except oak and beech).
Annual input: *c.* 600,000 tonnes.
Specification: length 1.7 – 2.0 m (up to 3.0 m by special arrangement). Min. top diameter 5 cm. Max. diameter 40 cm.
Price indication: £18–£24/tonne.
Roundwood buyer and harvesting manager: Alan Massey (0434–602191).

Aaronson Bros, plc at South Molton, Devon. (Owners: The Glunz Group of Germany.)
Product: chipboard, Contiboard, and Contiplas.
Species used: mixed conifers and broadleaves.
Annual input: *c.* 125,000 tonnes, 64% being residues.
Specification: length 2.25 m. Min. top diameter 8 cm. Max. diameter 35 cm.
Price indication: £16–£20.50/tonne.
Harvesting executive: Keith Thornley (07695–2991).

Pyrok Manufacturing Ltd. near Caerphilly, South Wales.
Product: cement-bonded particleboard.
Species used: Sitka spruce.
Annual input: *c.* 15,000 tonnes.
Specification: 2.3 m x min. 6 cm diam. and max. 36 cm diam.
Price indication: £25–£32/tonne.
Harvesting executive: Julian Waszczak (0443–816008).

Direct Worktops at Shildon, County Durham. (Owner: George Reynolds.)
Product: chipboard for worktops.
Species used: Pine, spruce, larch; some broadleaves.
Annual input: *c.* 200,000 tonnes.
Specification: Most small roundwood sizes.
Price indication: £15–£20/tonne.

Wood wool and Moulded-wood mills using small diameter roundwood.

Formwood Ltd., Coleford, Gloucestershire GL16 8PR (0594–33305).
   Chipboard mouldings for ceilings, screens, trays and table tops.
   European larch 2 m x 5–40 cm top diameter, peeled of bark and cambium: £30–£35 m^3 d/d.
   Beech 2 m x 5–40 cm top diameter, unpeeled: £25–£30 m^3 d/d.

Torvale Building Products Ltd., Pembridge near Leominster HR6 9LA (05447-262). For wood wool and wood wool cement slabs.
   Pines and spruces: peeled and seasoned: £40–£45 m^3 d/d.
              unpeeled: £35–£40 m^3 d/d.

Before embarking on conversion, whether of conifers or broadleaves, foresters must first ensure that markets are in fact open at the time. Thus a distinction must be made between the possible uses to which British-grown timber and other woodland produce may be utilized on the estate and the actual existence away from it of a market for any particular assortment of woodland produce at any one time in any one place. The pulpwood and boardwood mills do not accept casual deliveries: most purchase only through contracted suppliers. Furthermore, the specifications and prices must be verified and an order obtained before starting to prepare produce. This also applies to coniferous pitprops and transmission poles.

9. Prices of converted roundwood products

Conversion of roundwood to products of particular dimensions has been discussed in chapter 10. The main types of conversion are to lengths for pulp, fibreboard, particleboard and moulded board, for which specifications and prices are indicated in section 8 *supra*. The specifications for conifer pitprops and transmission poles are indicated in chapter 9, but prices of them are by (undisclosed) tender. Broadleaved turnery poles, usually 2 m and up, and of varying diameter, command prices of £25–£45/tonne delivered. Firewood fetches around £10/tonne collected; cut into short lengths and delivered £25–£30/tonne; or, by the bag, for about £2, which may represent close to £50/tonne. Charcoal prices vary greatly dependent on size, quality and purpose. Oak bark, air dried for tanning, fetches around £260/tonne collected, buyers being J. Croggon & Son Ltd., Manor Tannery, Grampound, Cornwall TR2 4QW (0726–882413), and J. & F. Baker & Co. Ltd., The Tannery, Colyton, Devon EX13 6PD (0297–52282).

Minor forest products from broadleaves include (all prices are 'collected'):

Ash for walking sticks (Fossel 1986):
   bends with swells: 1.4 cm and up x *c*. 3 cm: bundle of 36: £8.
   thumbsticks and bends: 1.5–2.1 m x *c*. 3 cm: bundle of 24: £8.
   sticks with knob: 1.4 m and up x *c*. 3 cm: bundle of 12: £9.

Buyers: Cooper and Sons Limited, Wormley, Godalming, Surrey GU8 5SY (042879–2251).

Ash scout poles: 1.5–2.1 m x *c.* 3 cm: bundle of 24: £10.

Ash cleft bar hurdles: *c.* 1.8 x 1.2 m (6 bars): £10–£15 each.

Birch branches for horse-jumps: 1.8–2.2 m: bundle of 24: £1.50–£2.

Sweet chestnut for walking sticks: *c.* 1.4 m x *c.* 3 cm: bundle of 30: £8.

Hazel for walking sticks with knob: *c.* 0.9 and up x *c.* 3 cm: bundle of 12: £5.

Hedging material:

	Length (m)	Unit	£
Stakes	1.5 m	20	3–4
Heathering	4.5 m	25	4–5
Tynut	variable	25	2–3
Barbed wire stakes	1.7 m	each	1–1.25

Bean rods: *c.* 2.2 m: bundle of 25: £2–2.50.

Pea sticks: variable: bundle of 25: £1–£2.

Hazel wattle hurdles:

	£/each
c. 1.80 x 0.9 m	17
x 1.2 m	20
x 1.5 m	25
x 1.8 m	30

Support stakes extra.

Hop-poles (little demand): sweet chestnut:

Length	Min. TD	£/each
m	cm	
5.0	8–9	7
5.3	8–12	8
5.5	8–12	9
6.0	10–13	12

Oak, peeled, for rustic work: 2–6 m x 4–10 cm butt diameter and 6 cm top diameter: £0.50–£1 m.

Minor forest products from conifers include (all prices are 'collected'):

Larch for rustic work: 2–6 m x 4–10 cm butt diameter and 6 cm top diameter: £0.50–£1 m.

Christmas trees: see chapter 5 for species, sizes and prices.

Foliage ('Greenery'): see chapter 5 for species used. The market is largely fragmented and unco-ordinated. In some localities purchasers will trim the trees and pay £100–£250/ha depending on species, quality, access and undergrowth. Noble fir commands the highest price and is often exported to Europe.

Bark: processed bark is used for riding surfaces, horticultural purposes, litter for cattle and fuel. Prices are extremely variable.

Fencing materials (wood) can comprise either broadleaved or coniferous species. There are considerable local and regional variations in demand, specifications and prices. Some indications of the two latter are given below (all prices are 'collected').

Stakes, posts and rails. A wide range of specifications and qualities are in common use especially in farming districts. Sweet chestnut, cleft oak and larch command the highest prices. Indications of prices are given below (all prices are 'collected'):

(A) Chestnut stakes and posts:

Stakes: mixed cleft and round, 7.5–10 cm butt diameter, pointed:

Length (m)	£/each
1.2	0.65
1.4	0.70
1.5	0.80
1.7	0.85
1.8	0.95

Straining and corner posts: *c.* 15 cm butt diameter, round and peeled:

2.3	6.00

Struts: minimum top diameter 6 cm, all round and peeled:

1.7	0.85

Anchor stakes: half round, minimum 7.5 cm face, peeled and pointed:

0.75–0.9	0.45

(B) Other fencing stakes and posts: Conifer:

Stakes: mixed cleft and round, 7.5–10 cm butt diameter peeled, pointed and pressure treated:

Length (m)	£/each
1.2	0.70
1.4	0.80

Length (m)	£/each
1.5	0.90
1.7	1.00
1.8	1.10

All round:

1.5	1.20
1.7	1.35
1.8	1.50
2.4	2.00

Straining and corner posts: c. 15 cm butt diameter, round, peeled and pressure treated:

2.3	5.80

Struts: minimum TD 6 cm, peeled and pressure treated:

1.7	1.30
1.8	1.50

Anchor stakes: half round, minimum 7.5 cm face, peeled, pointed and pressure treated:

0.75–0.9	0.45

(C) Stakes and posts to MAFF standard cost: Conifer:

Stakes: mixed cleft and round, 6 cm top diameter if round, or 10 cm across the face if half round, or 6.5 cm across each face if quarter-round, peeled, pointed and pressure treated:

Length (m)	£/each
1.5	1.20
1.7	1.35
1.8	1.80

All round:

1.5	1.60
1.7	1.85
1.8	2.00

Straining and corner posts: minimum 12.5 cm top diameter, round, peeled and pressure treated:

2.3	7.00

Struts: minimum 10 cm top diameter, peeled and pressure treated:

1.8	3.30

Anchor stakes: half round, minimum 7.5 cm face, peeled, pointed and pressure treated:

               0.75–0.9   0.45

Tree and netting stakes: Conifer: mixed cleft and round, 5–7.5 cm butt diameter, peeled, pointed and pressure treated:

Length (m)	£/each
0.9	0.45
1.1	0.50
1.2	0.60
1.4	0.70
1.5	0.80
1.7	0.90
1.8	1.00

All round:

2.1	1.40
2.4	1.55
2.7	1.80

Heavy duty, round, 7.5–10 cm butt diameter peeled, pointed and pressure treated:

2.1	1.85
2.4	2.00
2.7	2.45
3.0	2.80

Rails: Conifer: 3–4 m x 5 cm top diameter:

        Round: 2.50–3.00
        Half round: 2.00–2.50

References

Binkley, C.S. and Vincent, J.R. (1988), Timber Prices in the US South: Past Trends and Outlook for the Future'. *Southern Journal of Applied Forestry,* Vol. 12 No. 1, pp. 12–18.
Forestry Commission (1986), 'Background to Investment in Forestry Commission Assets', 1980; partly up dated to 1986.
Fossel, T. (1986), *Walking and Working Sticks.* Apostle Press, Beaconsfield.
Fraser, T., Horgan, G.P. and Watt, G.R. (1985), 'Valuing Forests and Forest Land in New Zealand: Practice and Principles'. Forest Research Institute Bulletin 99, New Zealand Forest Service.
Hibberd, B.G. (ed.) (1991), 'Forestry Practice'. FC Handbook 6.
Hughes, D.A. (1989), 'Marketing: the Forestry Commission View'. ICF Report of Regional Group Meeting 30 November–1 December 1989, pp. 39–41.
Insley, H. (ed.) (1988), 'Farm Woodland Planning'. FC Bulletin 80.
Jobling, J. (1990), 'Poplars for Wood Production and Amenity'. FC Bulletin 92.

Johnston, D.R., Grayson, A.J. and Bradley, R.T. (1967), *Forest Planning*. Faber and Faber, London.
Levack, H. (1989), 'Development of Valuation Procedures for Commercial Planting in New Zealand'. *Commonwealth Forestry Review 68* (3), 1989, pp. 147–53.
Mitlin, D.C. (1987), Price-size Curves for Conifers'. FC Bulletin 68.
Sedjo, R.A. and Lyon, K.S. (1990), 'The Long-term Adequacy of World Timber Supply'. *Resources for The Future,* Washington.
Sinclair, J. and Whiteman, A. (1992), 'Price-size Curve for Conifers'. FC Research Information Note 226.
Stumbles, R.E. (1985), 'Forestry as an Investment'. *Quarterly Journal of Forestry,* 94 (4). Oct. 1985.
United States Forest Service (1989), 'RPA Assessment of the Forest and Rangeland Situation in the US in 1989, General Technical Report WO-56. US, Forest Service, Washington'.
Whiteman, A. (1990), 'Price-size Curves for Conifers'. FC Research Information Note 192.
Whiteman, A., Insley, H. and Watt, G. (1991), 'Price-size Curves for Broadleaves'. FC Bulletin (in prep.).

Chapter Twelve

FOREST PLANNING

Forest planning is a discipline not well defined; yet it embraces many disciplines, and concerns almost all practising foresters. The relevant standard treatise is *Forest Planning* (Johnston, Grayson and Bradley 1967). Part of planning is decisionmaking, or investment appraisal, the most helpful texts being *Investment Appraisal in Forestry* (Busby and Grayson 1981), *Decision-making in Forest Management* (Williams 1988), and *The Theory and Application of Forest Economics* (Price 1989). All four publications relate particularly to even-aged (high forest) conifer plantations.

Management is about making and implementing decisions. It comprises three main functions: setting the objectives of the enterprise, planning and organizing operations, and controlling operations. Usually it involves choosing between two or more possible courses of action: at the initial stage of planting trees, and in their later management (see chapter 14).

The policy objectives are the criteria used by foresters when choosing between alternative courses of action and by which the success of management is judged. Planning is the task of organizing the forest operations to achieve the policy objectives, and comprises three main phases: collection of data, analysis of the various possible courses of action, and the formulation of plans. Such plans will be long-term, or medium-term (usually five-year forecasts of prescription of work), or short-term (usually a single year's operation). The latter will become the annual programme of work and thus form the basis of the annual budget. Control of all operations ensures that short-term plans are implemented by checking the progress of work throughout the budget period, and monitors the performance of mediumand long-term plans at more general levels.

This chapter does not supplant the above recommended texts, nor the notable earlier management classics (Jerram 1945; Hiley 1967; Osmaston 1968; Philip 1983). First are described the physical control of the growing stock and the regulation of yield.

1. The yield class system

The growth rate of any even-aged crop varies throughout the life of the stand in a characteristic and predictable pattern. Rates of growth are important because they affect the way in which the stand may be treated; and they enable predictions for planning purposes. Timber yields are conventionally measured in terms of volume (cubic metres). Rates of growth are conventionally defined through yield classes as described below.

Yield Class is an estimate of the maximum *Mean Annual Increment* (MAI) of stem volume per hectare per year. It is a specific growth rate category to which a crop can be assigned relatively easily. Identification of yield class is by age and top height, read directly from a height/age curve (see examples for Sitka spruce and oak in Figures 24 and 25). The growth of trees may be quantified in terms of increases in height, diameter, weight, volume or dry matter. Only height and diameter are relatively easily measured. Volume is the most meaningful for purposes of management. Measurable volume is conventionally defined as stemwood of at least 7 cm diameter overbark, or in the case of broadleaves, to the point in the crown where the trunk is no longer visible. Knowing the effective stocked area, species, age and yield class foresters are able to decide on the cutting regime which will produce maximum benefits according to the particular management objectives for the stand. They are also able to implement and control the regime and to forecast the long-term implications for timber output, and then to adopt appropriate harvesting and marketing strategies.

Volume increment. The pattern of volume increment in an even-aged stand is shown in Figure 23. After planting, the annual volume increment of a stand increases, reaches a peak after some years and then falls off as shown in the curve labelled CAI (*Current Annual Increment*). This curve represents the annual volume increment at any point in time. The average annual volume increment from planting to any point in time is shown by the second curve labelled MAI (*Mean Annual Increment*). For example, at n years, the current annual increment is x, while the mean or average annual volume increment from time of planting to n years is y.

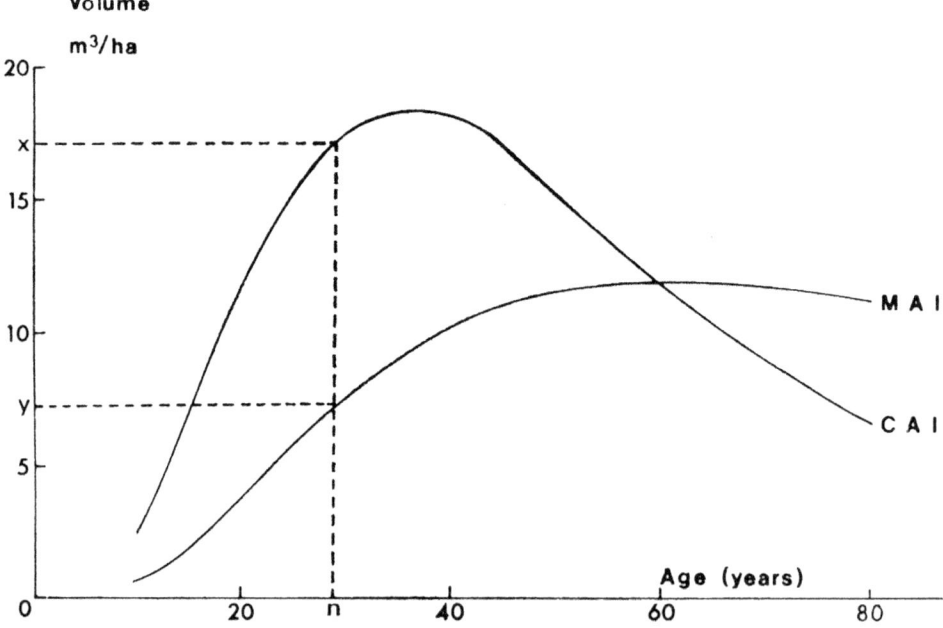

Figure 22. Patterns of volume increment in an even-aged stand
Source: Edwards and Christie (1981). FC Booklet 48.

Maximum Mean Annual Increment. The MAI curve reaches a maximum where it crosses the CAI curve. This point defines the maximum average rate of volume increment which a particular stand can achieve, and this indicates the *yield class*. For example, a stand with a maximum MAI of 14 m^3/ha has a yield class of 14. In theory, if the trees on an area were repeatedly felled at this stage, replanted, and managed in the same way, and there was no loss in site productivity, then this maximum average rate of volume production would be maintained in perpetuity. This general pattern of growth is typical of all even-aged sites. For any one species, differences usually follow a pattern wherein the faster-growing stands have higher maximum MAIs, and these maxima occur earlier. Again, although the same general pattern of growth is true of all species, there may be important differences between species. For example, maximum MAIs of different species may be of the same magnitude, but may occur at totally different times in the rotation. The important point here is that the maximum MAI is the maximum average rate of volume attained by a stand, irrespective of the time at which this maximum is achieved, and it is this feature which is the basis of the Yield Class System.

Classification by yield classes. The range of maximum MAIs commonly encountered varies with individual species, and may be 4 m^3/ha or even lower for many broadleaves, and 30 m^3/ha or more in the case of some conifers. Yield classes are created simply by splitting this range into steps of 2 m^3/ha, and numbering the steps with even numbers accordingly. Such classification is of limited use if it can only be used to categorize stands which have already reached their maximum MAI, since part of its purpose is to predict the *future* rate of growth of younger crops. Ideally, stands which have not yet reached the age of maximum MAI would be classified by reference to the MAI curves for the species. This, however, would necessitate establishing the mean annual increment of the stand, information which is seldom available because previous thinning yields have not been recorded. Even where thinning records are available, the measurement of the main crop volume can prove a relatively expensive procedure if it is required only for yield class assessment.

The yield class does not indicate when timber yields will be realized: this depends mainly on the thinning regime employed and the rotation length, which will vary, usually depending on the owner's objectives. Yields will vary depending on such factors as soil type, exposure, elevation and management treatment. Typical yields for slow growing broadleaves such as ash, oak and beech are in the range Yield Class 4–8, faster-growing broadleaves such as sycamore and cherry Yield Class 6–10, and poplar higher. Among the conifers, pines and larches are usually lower, Yield Class 6–16, while spruces and North-West American conifers are Yield Class 10–24.

General Yield Class. Since a good relationship exists between the top height and cumulative volume production of a stand, top height/age curves have been constructed from which General Yield Class (GYC) can be read directly. This obviates the need to actually measure or record cumulative volume production. These curves, produced for all major species, are available on index cards as part of Forestry Commission Booklet 48 (Edwards and Christie 1981, *Yield Models for Forest Management*) noted in section 2 *infra*. Two examples of GYC curves are shown in Figures 24 and 25.

Assessment of General Yield Class. A series of at least five sample plots should be randomly located throughout the stand, and the height of the tree of largest breast height diameter in each plot (radius 5.6 m) measured; and the average height of these trees calculated to give the top height of the stand. The number of sample plots to be measured will depend on the extent of the stand and its uniformity. In even-aged stands, top height is defined as the mean height of the 100 trees per hectare of largest diameter at breast height. The age of the stand is defined as the number of growing seasons since planting. Once top height and age are known, GYC can be established from the top height/age curves on the index cards. For example, if the top height of a stand of Sitka spruce is 19 m at an age of forty years, then using the top height/age curve (shown in Figure 24) the GYC is found to be 14.

Where there is more than one species in the stand, the GYC of each species should be assessed separately using larger plots of 0.02 or 0.05 ha (radius 8 m or 12.6 m) and measuring the height of the tree of largest dbh of each species in each plot. The average yield class of the stand can be obtained by averaging the component yield classes weighted according to the proportion of the canopy each occupies. Uneven-aged stands are treated in a similar way in that the yield class of each age category is assessed separately, and the average yield class again obtained, weighted according to the proportion of the canopy occupied by each category. Where, for any reason, the rate of height growth has changed appreciably in the life of the stand, for example because it has been 'in check', or because it has been fertilized, an adjusted age should be used instead of the actual age.

Estimating yield class of unplanted land. Estimating yield class of bare plantable land, i.e. potential productivity of the site, is important in land acquisitions in combination with price, in investment decisions, and in land-use evaluation. It may be difficult to estimate the yield class, particularly on sites which have carried an agricultural crop for some years. Generally the estimation is obtained by noting the yield class of crops of like species and site conditions in the neighbourhood. Worrell (1987) has provided a means of estimating the potential General Yield Class of QCI provenance Sitka spruce on upland sites in northern Britain from five easily accessible site factors: location, elevation, aspect, soil type and topex (the sum of the angles of elevation to the skyline at the eight cardinal points of the compass).

The minimum acceptable yield class, and consequently *upper planting limits,* will vary according to the economic criteria applied by foresters. Gale and Anderson (1984) recommended a minimum yield class of 10 for conifers based on a comparison of likely discounted costs and revenues in the Galloway region, but values as low as Yield Class 7 have been suggested (Hummel and Grayson 1962; Malcolm and Studholme 1972). Guidance is also given by Pyatt *et al.* (1969) and Busby (1974) for upland sites.

Production Class. As there are local variations in the relationship between top height and cumulative production for any one species, these variations have been largely accommodated by employing three top height/cumulative volume production functions rather than one. These three levels of cumulative volume production for a given height are termed Production Classes 'a', 'b' and 'c'. Production Class 'b' is the normal top height/volume production relationship embodied in the curves. The effect of using Production Class 'a' is to raise the yield class by one class over that indicated by the GYC curve, i.e. to raise the maximum MAI two cubic metres per hectare. The effect of using 'c' is to lower the GYC estimate by one class. Stands in exposed sites tend towards Production Class 'a', as their height growth is depressed relatively more than their volume growth. Conversely, Production Class 'c' may occur on sites where there is a moisture deficit in the latter part of the growing season, but not in the earlier part. The growth patterns described above assume that height growth remains vigorous throughout the life of the stand. Production Class 'c' may also be found on sheltered valley sites where vigorous shoot extension occurs. Top height/age related to yield class can indicate a yield class in excess of volume production. Production classes are devices which may be used to provide an improved estimate of yield class. An example of the Production Class curves for Sitka spruce is shown in Figure 24.

Assessment of production class. General Yield Class is usually adequate for most management purposes, but a better estimate of yield class can be obtained by assessing production class (Rollinson 1986). This is generally an expensive and time-consuming operation which is normally restricted to the major species in a forest. The factors which influence production class tend to be macroclimatic rather than specific to individual stands. For these reasons, it is best to apply production class for a given species to defined site types and elevation zones rather than individual stands.

Production class can only be assessed before a stand has been thinned, unless the total volume or basal area removed in thinning is accurately known. Production class is assessed by measuring either cumulative volume production per hectare, or cumulative basal area production per hectare. The second method is really a substitute for the first, but as cumulative volume production is seldom known, and generally too expensive to obtain for this purpose, the first method is seldom used. On the other hand, it is the preferred method should information on cumulative volume production be already available.

Cumulative basal area production is relatively easily obtained, and for this reason it is the method commonly used. In practice the assessments are carried out in fully-stocked unthinned stands, as records are seldom available of basal area previously removed in thinned stands. In sampling an unthinned compartment for total basal area production it is advisable to lay out at least three plots of 0.01 hectares, in which all live trees, excluding those of less than 7 cm diameter, are measured for diameter at breast height, and the average basal area per hectare is calculated. Alternatively, at least six relascope sweeps should be taken.

Production class is derived from the cumulative volume/top height curves or the cumulative basal area/top height curves printed on the index cards. For example, given that the top height of a stand of Sitka spruce is 16 m, and the cumulative volume production is 250 m^3, then by referring to Figure 24, the production class is found to be 'c'. Similarly, in a Sitka spruce stand with a top height of 12 m, and a cumulative basal area production of 50 m^3, the production class is 'a'.

The first stage in establishing a production class for a forest is to sample for production class, as described above, in about ten compartments for each major species in the forest. Taking each species separately, the production class assessments should be plotted on a small scale map of the forest, to see if there are any trends or patterns in the distribution of production class. For production class for Douglas fir see Christie (1988).

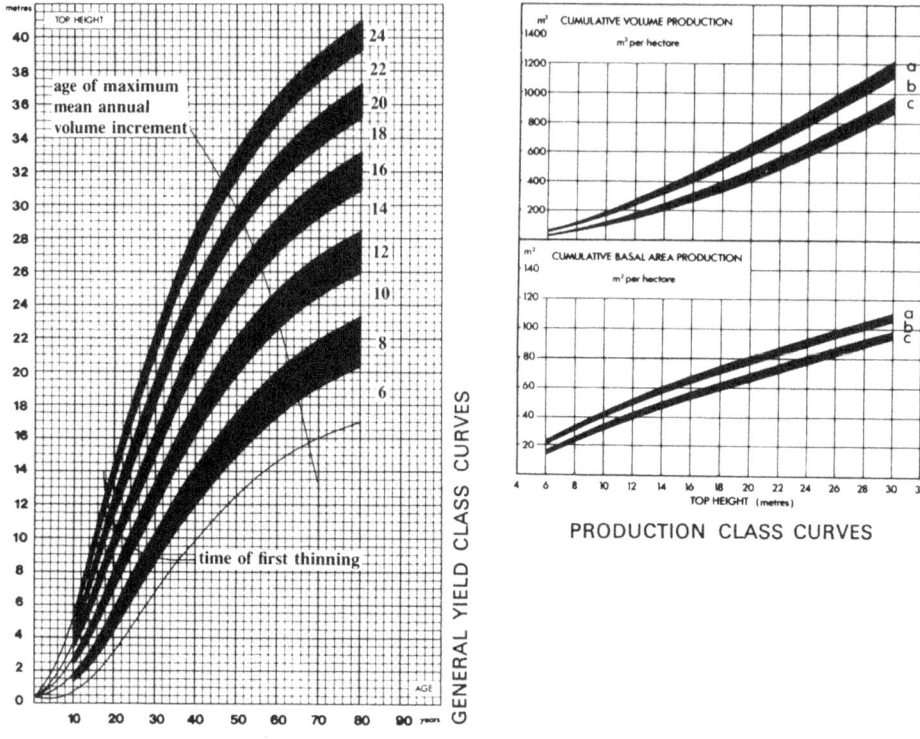

Figure 23. Sitka spruce Top height/age curves giving General Yield Class; and Production Class Curves
Source: Edwards and Christie (1981). FC Booklet 48.

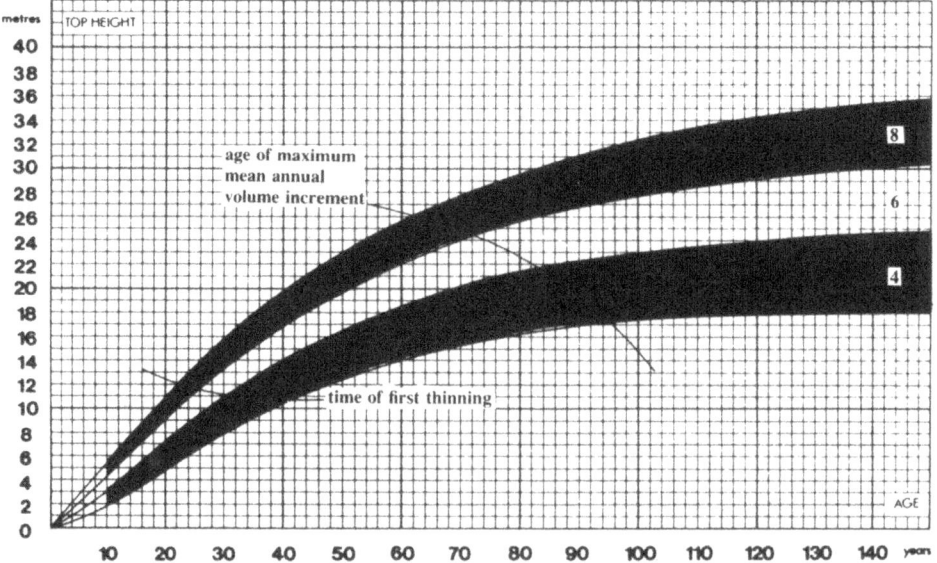

Figure 24. Oak: Top height/age curves giving General Yield Class Curves
Source: Edwards and Christie (1981). FC Booklet 48.

Local Yield Class. Where production class has been taken into account the yield class is termed a Local Yield Class (LYC). For example GYC 14, Production Class 'a' = LYC 16; 'b' = LYC 14; and 'c' = LYC 12.

The effect of variations in growth rate. An individual stand will not always follow the growth rates in any particular yield class: for part of its life, a stand may grow faster than its yield class suggests, and at other times it may grow slower. For example, a Sitka spruce stand may suffer from 'check' for the first few years, and it will therefore grow very slowly indeed. Once the stand has grown out of the 'check' phase, its growth rate, and hence yield class, will increase markedly. A second example is the effect of fertilizing the stand. This will often increase the growth rate of the stand, although sometimes only for a few years if the treatment is not repeated. The correct way to allow for these changes in growth rate is to combine the current growth rate with an 'adjusted age'. As yield classes are used for forecasting, it is very important that the recorded yield class gives the best possible estimate of future growth.

The effect of different treatments. The General Yield Class curves and the Production Class curves are based on the assumption that the stands have been planted at spacings of 1.2 m (Oak, Be), 1.4 m (SP, CP), 1.5 m (LP, NS, WH, WRC, NF, Sycamore, Ash, Birch), 1.7 m (SS, EL, JL, DF, NF), 1.8 m (GF) and 7.3 m (Poplar), and thinned (except for poplar) at the marginal thinning intensity.[1] The foregoing spacings were the ones most commonly used before the Management Tables were published but since then there has been a tendency to use wider spacing for most species especially broadleaves. This will cause a reduction in the cumulative volume production of a stand, while narrower spacing will marginally increase it. For example, if a wider spacing has been used that will reduce volume production, then by delaying the time of first thinning, the average level of growing stock implied by marginal thinning intensity can still be maintained and the normal annual thinning yield implied by the yield class removed. No further loss of volume will occur. The effect is similar to a small change in production class. Assessment of production class in the normal way may be misleading, and it should be assessed by referring to the cumulative production given in the appropriate yield model (see *infra*). Wherever possible, the yield model closest to the actual spacing and thinning treatment should be used for estimating the potential production. This model may show that the maximum MAI is different from that suggested by the yield class.

2. Yield models for forest management

The yield models available from the Forestry Commission with Booklet 48, *Yield Models for Forest Management* (Edwards and Christie 1981) – see two examples in

---

[1] Marginal thinning intensity (see chapter 8) is the maximum annual volume/ha that can be removed without incurring any loss of cumulative volume production, and implicit in this is the maintenance of an average level of growing stock. It is defined as 70% of the yield class. For example, for a stand of YC 16, thinning to marginal intensity would remove 11.2 m^3/ha/yr or 56 m^3/ha on a 5 year thinning cycle.

Figures 26 and 27 – are tabular presentations of models of stand growth and yield. The booklet replaces Forestry Commission Booklet 34, *Forest Management Tables (Metric)* (Hamilton and Christie 1971). The models have been prepared for all the major commercial species in Britain, and for a wide variety of treatments including a range of initial spacings, thinning at marginal intensity, line thinning, delayed thinning and no thinning. For each model, one particular treatment regime has been assumed. Any deviation from this regime, or any deviation from the average growth pattern will produce a different set of stand characteristics. It is inevitable that an individual stand will vary in one respect or another from the model, and so direct comparisons are not very meaningful. However, the trends of growth which are given in a model can be used to estimate the probable development of any particular stand. Only live trees have been included in the models, and all information relating to trees that have died has been excluded. Volumes derived from yield models do not include branchwood, which for broadleaved trees can form a significant quantity of the total saleable volume, especially where there is a strong demand for firewood. (For large mature broadleaves, branchwood volumes may be 50% of stemwood volumes.) Unless stated otherwise, all the models, irrespective of spacing and treatment, are based on Production Class 'b'. If a model for a different production class is required, the models for the appropriate Local Yield Class should be used. If this is done, the figures for top height will be misleading. Booklet 48 gives further information about the models, indicating the ages shown, thinning treatment assumed, accuracy, spacings, thinning intensity, thinning type, time of first and subsequent thinning, effect of thinning weight, and choice of thinning time.

The yield models assume full stocking, and the figures when applied to whole forests must be adjusted to allow for gaps resulting from crop failures of one form or another and for reduction in area resulting from roads, rides, streams and buildings. This reduction factor varies with conditions, but is usually assumed to be 15%. In practice, any stand is unlikely to grow in exactly the way represented in the models or to have had exactly the same treatment included in them. So any estimate of yield for an individual stand may be subject to an error up to 20%. The Forestry Commission do not recommend the use of yield models for individual stands but only for a number of stands in a forest where errors can be expected to cancel out.

Using yield models. 'Yield table information' has become a vital tool in forest management. Yield models, being models of stand growth and yield, are the basis for forest planning, usually by means of economic appraisal (see chapter 14). Foresters use the yield models to compare the results of alternative treatments, before deciding how to manage a particular stand. Their choice of regime will be influenced by several external factors, the availability of markets and labour, and possible methods of extraction. Foresters also need to forecast the timber production from the forest, so that they can arrange suitable markets and plan the harvesting. They should choose the most appropriate yield model, and then use it to forecast the production from the stand, using stand assortment tables, given in Booklet 48, as a guide to the likely produce assortment. Yield models do not always reflect the precise growth of individual stands, but they do accurately describe the differences between different treatments, and so they are very suitable for comparisons. They can be used to decide which initial plant spacing will be better in a particular situation, whether to thin, and if so when, how frequently, how heavily and in what way, and, finally, when to fell the stand. The choice of treatment usually depends on the economics of the alternatives – the most profitable one is that usually selected.

To calculate this, the first step is to construct a price–size curve giving the value per cubic metre of standing timber of a stated mean diameter or mean volume (see chapter 11). The yield model shows the mean size of the trees, and so by using the price–size curve the standing value of each thinning and the final felling can be calculated. These values can then be discounted back to a common date, such as the time of first thinning, and then the total discounted values for each treatment can be compared. The calculations must also take account of possible changes in the price-size curve with treatment (e.g. using wider spacing may produce knotty, wide-ringed timber which is of lower value), and difference in costs (e.g. an unthinned stand may not need any roads before it is felled).

		MAINCROP after Thinning					Yield from THINNINGS					CUMULATIVE PRODUCTION		MAI	
Age yrs	Top Ht	Trees /ha	Mean dbh	B A /ha	Mean vol	Vol /ha	Trees /ha	Mean dbh	B A /ha	Mean vol	Vol /ha	B A /ha	Vol /ha	Vol /ha	Age yrs
18	7.3	2311	11	24	0.03	66	0	0	0	0.00	0	24	66	3.7	18
23	10.2	1351	15	24	0.07	90	895	12	11	0.05	49	35	139	6.0	23
28	13.0	951	19	28	0.14	133	400	15	7	0.12	49	46	231	8.3	28
33	15.7	732	23	31	0.26	188	220	19	6	0.22	49	56	335	10.2	33
38	18.2	595	27	35	0.41	246	137	22	5	0.36	49	64	442	11.6	38
43	20.4	496	31	37	0.60	300	99	25	5	0.50	49	72	545	12.7	43
48	22.4	422	34	38	0.82	345	74	28	5	0.66	49	78	639	13.3	48
53	24.1	374	37	40	1.04	388	48	31	4	0.85	41	83	723	13.6	53
58	25.5	341	39	41	1.25	426	34	33	3	1.05	35	87	796	13.7	58
63	26.7	316	41	42	1.45	457	25	36	2	1.25	31	91	858	13.6	63
68	27.7	297	43	43	1.63	484	20	37	2	1.40	27	94	912	13.4	68
73	28.6	281	45	44	1.81	508	16	39	2	1.51	25	96	961	13.2	73
78	29.4	267	46	44	1.98	529	13	41	2	1.66	22	98	1003	12.9	78

Figure 25: Yield Model: Sitka Spruce, YC14: Intermediate thinning; and 2 m spacing

Source: Edwards and Christie (1981). FC Booklet 48.

		MAINCROP after Thinning					Yield from THINNINGS					CUMULATIVE PRODUCTION		MAI	
Age yrs	Top Ht	Trees /ha	Mean dbh	B A /ha	Mean vol	Vol /ha	Trees /ha	Mean dbh	B A /ha	Mean vol	Vol /ha	B A /ha	Vol /ha	Vol /ha	Age yrs
20	8.4	5115	7	19	0.01	42	0	0	0	0.00	0	19	42	2.1	20
25	10.4	3529	9	20	0.02	62	1195	6	4	0.01	8	24	71	2.8	25
30	12.2	2144	11	19	0.04	76	1385	8	7	0.02	21	30	106	3.5	30
35	13.9	1459	13	19	0.06	95	685	10	5	0.03	21	35	145	4.1	35
40	15.4	1073	15	20	0.11	116	386	11	4	0.05	21	40	187	4.7	40
45	16.8	822	18	21	0.17	137	252	13	4	0.08	21	44	229	5.1	45
50	18.1	645	20	21	0.24	157	176	16	3	0.12	21	48	271	5.4	50
55	19.2	524	23	22	0.34	176	121	18	3	0.17	21	51	311	5.6	55
60	20.2	436	25	22	0.44	193	89	20	3	0.24	21	54	348	5.8	60
65	21.1	370	28	23	0.56	208	66	22	2	0.32	21	57	384	5.9	65
70	21.9	319	30	23	0.69	221	51	24	2	0.41	21	60	418	6.0	70
75	22.7	279	33	23	0.83	231	40	26	2	0.53	21	62	450	6.0	75
80	23.3	246	35	24	0.98	240	33	28	2	0.63	21	65	480	6.0	80
85	23.9	218	37	24	1.14	248	28	30	2	0.75	21	67	508	6.0	85
90	24.4	194	39	24	1.31	254	24	32	2	0.89	21	69	535	5.9	90
95	24.8	173	42	24	1.49	258	20	35	2	1.03	21	70	560	5.9	95
100	25.2	156	44	23	1.67	261	17	37	2	1.21	21	72	584	5.8	100
105	25.6	142	46	23	1.85	264	14	39	2	1.37	20	74	606	5.8	105
110	25.9	130	47	23	2.04	266	12	41	2	1.53	19	75	627	5.7	110
115	26.2	120	49	23	2.22	267	10	43	1	1.76	18	76	646	5.6	115
120	26.4	112	51	23	2.40	268	9	45	1	1.94	17	77	664	5.5	120
125	26.6	105	52	22	2.57	268	8	46	1	2.03	16	79	680	5.4	125
130	26.8	98	54	22	2.75	269	7	48	1	2.22	15	80	696	5.4	130
135	27.0	92	55	22	2.92	269	6	50	1	2.41	14	81	710	5.3	135
140	27.2	87	57	22	3.10	269	5	52	1	2.58	13	82	724	5.2	140
145	27.3	82	58	22	3.27	269	5	54	1	2.78	13	82	736	5.1	145
150	27.4	78	59	22	3.43	268	4	55	1	2.97	12	83	747	5.0	150

Figure 26: Yield Model: Oak, YC6: Intermediate thinning; and 1.2 m spacing

Source: Edwards and Christie (1981). FC Booklet 48.

GLOSSARY OF TERMS USED IN THE MODELS:
Figures 26 and 27

Age: The number of growing seasons that have elapsed since the stand was planted.

Top Ht: Top height; the average height of a number of 'top height trees' in a stand, where a 'top height tree' is the tree of largest breast height diameter in a 0.01 ha sample plot.

Maincrop after Thinning: All the live trees left in the stand, at a given age, after any thinnings have been removed.

Yield from Thinnings: All the live trees removed in the thinning.

Trees/ha: The number of live trees in the stand, per hectare.

Mean dbh: The quadratic mean diameter (the diameter of the tree of mean basal area) in centimetres, of all live trees measured at 1.3 m above ground-level.

BA/ha: Basal area. The sum of the overbark cross-sectional area of the stems of all live trees, measured at 1.3 m above ground-level, and given in square metres per hectare.

Mean vol: The average volume, in cubic metres, of all live trees, including any with a breast height diameter of less than 7 cm.

Vol/ha: The overbark volume, in cubic metres per hectare, of the live trees. In conifers, all timber on the main stem which has an overbark diameter of at least 7 cm is included. In broadleaves, the measurement limit is either to 7cm, or to the point at which no main stem is distinguishable, whichever comes first.

Cumulative Production: This is the main crop basal area or volume, plus the basal area or volume of the present and all previous thinnings.

MAI: The mean annual volume increment; i.e. the cumulative volume production to date divided by the age.

Note: All trees which die through natural mortality are excluded.

Choosing the right yield model. Almost 1,000 yield models are available, but there are many cases where there is no suitable model for the crop. Forestry Commission Booklet 48 suggests the most appropriate models to use for less common species not covered by the yield models. No models exist for mixtures; it is usual to consider each component separately. Where mixtures are planted with a view to producing a pure final crop, for example Sitka spruce/lodgepole pine mixtures on peat, it is likely that the volume production per hectare will be less than a pure crop of the more productive species but the mean tree size will be greater.

Little information is currently available for the development of descriptive models which are able to predict how forests will respond to completely nonstandard treatments. Foresters would like to be able to make predictions under a wide range of hypothetical regimes including close and wide spacing, respacing, no thinning, thinning once and then not again, different patterns of systematic thinnings, and all these options combined with an application of fertilizer at various stages of the thinning cycle or with pruning at intervals.

Volume predictions for broadleaves have to be based on very approximate yield models. For instance, ash, birch and sycamore are all covered by one model; all broadleaved models now give wrong predictions about tree sizes because they are based on much closer spacing than is currently used; there is only one set of assumptions about silvicultural treatment; and there are no models for the common practice of growing broadleaves in mixtures with a conifer nurse.

Management tables can be off-putting to some foresters who have to deal with them only occasionally. However, for many foresters they are an essential tool of management, and often indispendable in such subjects as thinning, investment appraisal and valuation. Foresters intent on using Yield Models, Yield Class or Production Class curves will of course need to acquire the appropriate Forestry Commission publications wherein can be found full explanations of the concepts and usages.

Using yield models to forecast production. Forecasts of production from a forest should be calculated by totalling the forecasts of production from each individual stand within the forest. For each stand, the following information is needed: species, age, yield class, area, past treatment including plant spacing, and proposed future treatment. The first three are relatively easy to discover. Accurate maps are required to determine the area, and it is most important that this is the net area of fully stocked forest, excluding roads, rides, and any other unproductive areas. Finally, details of past treatments must be recorded, and the proposed future treatment must be decided. The expected volume and other stand characteristics at each thinning can be read directly from the yield model, and the figures for the felling can easily be calculated by combining the figures for the thinning at that age with the main crop after thinning at the same age. The volume estimates are for one hectare, so they must be multiplied by the net area to give the forecast for the whole stand. Alternatively the gross area of the forest may be used (i.e. the area inclusive of roads, rides and unproductive land) and multiplied by the Yield Table values less 15%, as suggested earlier.

3. Data collection and use

For optimum planning it is necessary to possess quantities and qualitative data on the woodland resource. Surveys are necessary to know the legal boundaries of the woodland, public rights of way, distance from a public highway, and the load carrying capacity of access and internal roads and tracks. Basic planning and record maps should be the latest 1:10,000 (metric) or 1:10,560 scale Ordnance Survey sheets. The 1:25,000 scale OS maps are a convenient and supplementary base for production planning in extensive areas, and in particular, when drawing up a sales contract to show access routes in relation to the public road system, loading bays, and stacking and conversion sites. Aerial photographs can be of assistance; they provide good quality vertical cover, and this should be combined with ground checks (Betts 1985). They are generally available at a scale of 1:7,000 to 1:10,000; as well as black-and-white prints at a scale between 1:15,000 and 1:5,000, usually obtainable from St. Andrew's House, Edinburgh, or the Ordnance Survey, Southampton. Soil and geological maps are also available. For production planning it is essential to have stock and road maps for the identification of felling coupes, for the preparation of sales plans, and for the planning of harvesting and road systems. The stock map, normally at a scale of 1:2,500 or 1:10,000 will need to show external boundaries, access routes, internal tracks, location of wayleaves, species boundaries at the time of planting, and the extent and number of each subcompartment.

Consideration should be given to the collection and analysis of data, without which it is impossible to judge which course of action is most likely to fulfil the objectives of the enterprise. Although all activities on a single wooded estate interact with one another to some extent, the critical element concerns the management of the growing stock of trees. From this flows the requirement to find markets, to programme restocking, and to design a system of access routes for general management and harvesting. Surveys and records derived from them will concentrate on the distribution and composition of the growing stock, particularly age class and yield class. Other kinds of survey may include the assessment of disease occurrence or damage; and a soil survey to assist in choosing new planting or restocking options and operations. Surveys designed to serve other objectives than timber production may be required – for example to provide the best layout for pheasant shoots or to record interesting wildlife habitats, and picnic places.

As an aid to collection of the necessary data, recourse can be made to any earlier surveys, records of planting, treatment, thinning and felling, sub-compartment schedules and forest histories. Foresters rarely take over the management of woods without discovering some records left by a predecessor. There may be additional records – of volume removed in thinning, markets supplied and income received. A preliminary reconnaissance enables foresters to make an assessment of the character of the terrain and the broad crop classification, assisting them to decide how much other work will be needed to carry out the inventory survey.

Sub-compartment notes. Compartments are permanent management units which as far as possible are based upon the road system and on well-defined features such as watercourses, paths and other natural features; their size will depend on the terrain and the area of the woodlands or plantations. The basic management unit in forestry is either the sub-compartment or the stand – an area comprising a more or less homogeneous crop in terms of age, species, composition and condition. This unit will not necessarily be permanent since it will probably change as the woods develop through harvesting and restocking. The following list comprises suggested headings and contents necessary for the production of meaningful subcompartment notes. In appraising each stand foresters should consider all of these points as they provide an aid to both day-to-day and long-term management:

1 Area, map reference, topography, terrain class, aspect, Windthrow Hazard Class, previous crop, natural vegetation, soil type, drainage, fertilizing, thinning regime.

2 Species, origin, pure or mixed, proportion of mixture (area occupied by the canopy, not number of stems), and silvicultural system (seed source will be essential if seed is ever to be collected commercially from the mature crop).

3 Age (by means of records, or whorls of branches or height) and original spacing (important in assessing the yield of the species at a later date, using Management Tables).

4 Stocking (basal area and top height by species), areas 'in check' or unproductive.

5 Understorey, or overstorey. Advance natural regeneration, if any.

6   Yield class of each species (found by age and top height, and the appropriate yield class curve) and intended rotation length.
7   Damage (insect, fungal, animal, fire).
8   State of fences, drains, roads.
9   Crop protection: (a) fire (risk, hazard, fire-plan, access, fire-beaters, water supplies), and (b) animals: wild or domestic.
10 Information on cleaning, brashing, pruning and thinning.
11 Grants received (and conditions which must be honoured).

To assist in a general appraisal, survey sheets can be based on the following suggested headings:

General:
    Name of wood. Compartment No. Sub-compartment No. Stand.
    Extent: (a) fully productive, (b) partly productive, (c) unproductive, (d) total.
    Locality: Altitude, aspect, slope, exposure, frost, geology and soil.
    Boundaries.
    Extraction routes.
    Stability.
    Fire risks and hazards.
    Previous crop.

Present crop:
    Silvicultural system and forest type: high forest (broadleaved, mixed, coniferous), coppice, coppice with standards, selection forest, scrub and felled areas, bare land and unplantable areas.
    Undergrowth and ground flora.
    Species: (a) main crop, (b) understorey (including natural vegetation if any).
    Year of planting.
    Condition.
    Diseases and pests.
    Past treatment (brashing, pruning, and thinning).
    Amenity, sporting and shelter.
    Volumes and values: (a) per hectare, (b) of stand or sub-compartment:
        Stems per hectare.
        Volume (by size and quality).
        Value (by size and quality).
        Volume and value to be removed in the next thinning.

Silvicultural prescriptions for the next 5 years (in brief, for the 6th to 10th years) urgent work being emphasized:
    Preparation of ground.
    Drainage.
    Fencing.
    Planting (including enrichment and possibilities of natural regeneration).
    Weeding.

Beating up.
Rack-cutting.
Clearing.
Brashing.
Pruning.
Thinning.
Selection-felling.
Clear-felling.
Maintenance.
Protection.

Use will be made of Ordnance Survey maps whereon the crops will be plotted by sub-compartments or stands. On completion of the survey it is a comparatively easy task to transfer information to a Compartment Register. Separate records should be made of the objectives of planting (timber production, non-wood benefits), expected rotation length, and dates of grant receipts, and of felling licences, where applicable.

There are no generally accepted standard forms for data recording. The design of forms or schedules depends upon the information to be collected and on the way it is to be stored and presented. For large woodlands or well-wooded estates, automatic data processing on computers allows management data to be sorted and summarized in a wide variety of ways, but it may still be desirable to retain some form of manual data retrieval system, unless access to the computer is fast enough for local operational use.

Prescriptions of work. Based on the collected data, prescriptions of annual work can be drawn up for each stand, and these can form part of the Plan of Operations under any grant-aid scheme, or other woodland plan. Labour and machinery requirements can be assessed – much dependent on the proposed forestry techniques, the chosen 'point of sale', the intensity of the forestry being practised, constraints (e.g. amenity, sporting), and cash flow (budget) incidence.

Planning, organizing and controlling wood production. The principal objectives of any forest enterprise affect the composition, growth and yield of the growing stock. Control of the latter is achieved by applying specified cutting regimes to the woodlands, based on the area, age and species composition of the stands, together with their standing volume and rate growth (yield class). With all this information available foresters are able to plan, organize and control the cutting regimes and consider how to control the thinning and felling yields.

Forecasting production. This has been discussed in section 2, but needs reemphasizing because of its importance. Forecasts of future volume production are essential for market planning, for the planning of labour and machinery resources, and to provide the basis for the valuation of the growing stock. To forecast production it is necessary to know the present content and condition of the woodlands (inventory), the pattern of present and future growth (yield class) and the thinning and felling proposals (cutting regimes). Forecasts of production from the woodlands are calculated by summing the forecasts of production from each stand therein. Details of past treatments and the proposed future treatments are needed to select the most appropriate yield model. The expected volume and other stand characteristics at each thinning can be read directly from the yield model and the figures for the felling can be calculated by combining the figures for the thinning at that age with the main crop after thinning at the same age. However, where assessments show a marked deviation from the yield model figures, then these should be modified. The volume estimates are for one hectare so they must be appropriately multiplied by the area to give the figure for the whole stand. Mixtures or two-storied stands are most conveniently dealt with by separating the component species or storeys and deriving an effective net area of each based on the proportion of the canopy it occupies. The forecast thinning and felling volumes can be separated into volumes of large timber to stated top diameters and volumes of smaller timber, using Stand Assortment Tables (Forestry Commission Booklet 39, Hamilton 1975, *Forest Mensuration Handbook*). Further tables based on wide spacing and for unthinned stands are available in Forestry Commission Booklet 48.

Crop assessment summaries should give a realistic picture of the growth potential of the woodlands. Areas which are markedly under or over-stocked or which, owing to delayed early growth are unlikely to be ready for a first thinning at the standard (marginal thinning) age, should be identified individually and their predicted yield modified accordingly. The reliability of forecasts of future yield depends upon (i) the accuracy of the growing stock data and of the growth predictions, and (ii) the thinning and felling policy being carried out as planned.

References

Betts, A.J.A. (1985), 'Forest Use of Aerial Photography'. FC Research Information Note 99/85/FS.
Busby, R.J.N. (1974), 'Forest Site Yield Guide to Upland Britain'. FC Forest Record 97.
Busby, R.J.N. and Grayson, A.J. (1981), 'Investment Appraisal in Forestry'. FC Booklet 47.
Christie, J.M. (1988), 'Levels of Production Class of Douglas Fir'. *Scottish Forestry 42* (1), pp. 21–32.
Edwards, P.N. and Christie, J.M. (1981), 'Yield Models for Forest Management'. FC Booklet 48.
Gale, M.F. and Anderson, A.B. (1984), 'High Elevation Planting in Galloway'. *Scottish Forestry 38* (1), pp. 3–15.
Hamilton G.J. and Christie, J.M. (1971), 'Forest Management Tables (Metric)'. FC Booklet 34.
Hiley, W.E. (1967), *Woodland Management*. Revised edn. Faber and Faber, London.
Hummel, F.C. and Grayson, A.J. (1962), 'Production Goals in Forestry with Special Reference to Great Britain'. Proceedings of the 8th British Commonwealth Forestry Conference, East Africa.
Jerram, M.R.K. (1945), *A Textbook on Forest Management*. Chapman & Hall, London.
Johnston, D.R., Grayson, A.J. and Bradley, R.T. (1967). *Forest Planning*, Faber and Faber, London.
Malcolm, D.C. and Studholme, W.P. (1972), 'Yield and Form in High Elevation Stands of Sitka Spruce and European Larch in Scotland'. *Scottish Forestry 25* (4), pp. 296–308.
Osmaston, F.C. (1968), *The Management of Forests*. George Allen and Unwin, London.
Philip, M.S. (1983), *Measuring Trees and Forests*. Aberdeen University Press.
Price, C. (1989), *The Theory and Application of Forest Economics*. Basil Blackwell, Oxford.
Pyatt, D.G., Harrison, D. and Ford, A.S. (1969), 'Guide to Site Types in Forests of North and mid-Wales'. FC Forest Record 19.

Roilinson, T.J.D. (1986), 'Don't Forget Production Class'. *Scottish Forestry* 40 (4), pp. 250–8.
Williams, M.R.W. (1988), *Decision-making in Forest Management*. Research Studies Press Ltd., Letchworth.
Worrell, R. (1987), 'Predicting the Productivity of Sitka Spruce on upland Sites in Northern Britain'. FC Bulletin 72 (supported by Fountain Forestry Limited).

Chapter Thirteen

FOREST MANAGEMENT

1. Choice of forestry intensity

For many private woodlands, particularly those on a well-wooded estate, the intensity of the forestry practised has a significant impact on management. *Intensive forestry* is meant to comprise a wide range of co-ordinated activities on a woodland property. Some foresters prefer the term *integrated forestry,* others prefer *intensive management.* It serves to increase the production and 'added value' from a woodland property. Included may be the cultivation of trees right from seed through to maturity, together with many supplementary operations, among these drainage and road construction, and the full utilization of underwood, coppice, thinnings and final crops. Furthermore, it may include the production and sale of minor forest products, such as Christmas trees, foliage, tanbark and pulverized bark; and perhaps wood preservation, sawmilling and the manufacture of gates, garden arches, seats, interlaced panels and the like. Some of these woodland enterprises have hitherto been separate undertakings; for example, an estate forest tree nursery, a sawmill, or a carpentry department, occasionally sufficiently large to justify independent management. This section concerns the co-ordination or what may be termed the mutual support of all these enterprises.

Some woodland crafts are capable of being organized separately, for example the conversion of hazel underwood and sweet chestnut coppice. Other crafts, notably those of the beech-bodger, and of the makers of ash hurdles and tent-pegs, alder clog-soles and birch besoms are fast dying out. The modern tendency is to specialize, as in the conversion of thinnings into pitprops, pulpwood and boardwood. All these crafts and enterprises can be dealt with connectedly by direct labour on an estate sufficiently large to practice intensive forestry. More often the crafts relating to underwood and coppice have hitherto been undertaken by extraneous woodmen; and the conversion of thinnings undertaken by contract or carried on by separate trades organized by roundwood merchants or contracting firms. However, a great difference lies between contractors undertaking to convert an estate's roundwood and those who are simply purchasers of that material. Contractors are paid to undertake specific tasks, which may include thinning, extracting and cross-cutting to required specifications; purchasers on the other hand are those who buy standing woodland crops in bulk, and convert them to their own required products and benefit.

Conversion and utilization constitute only a part of intensive forestry. A few large well-wooded estates undertake some or all of their own plant raising and execute their own drainage, fencing, planting, weeding, cleaning, brashing, thinning, pruning, felling, extracting and transporting; they also undertake their own maintenance and protection. All this is in addition to achieving utilization including the preparation of minor produce. Some such estates have, too, a modest sawmill, preservative plant, wood-yard and carpenter's shop. Some or all of these enterprises are organized under one management.

Forestry is, however, rarely so intensive as that outlined. There are many competently managed wooded estates which buy their planting stock from trade nurseries and sell their timber as it stands. Underwood and coppice are sold to extraneous woodmen and the marked thinnings to roundwood merchants. Such estates, therefore, confine their forestry mainly to pure silviculture. Even this on a large area requires a considerable staff, although much less than that of an estate practising fully intensive forestry. Some estates lie between the two extremes, their forestry being only partly intensive. The reasons advanced for estates being induced to adopt a fully intensive forestry (Hiley 1954) include:

- The belief that if extraneous woodmen and merchants can make a living out of woodland enterprises, the estate should be able to make a profit out of such undertakings (the scale of operations, the 'getting in' on markets, the amount of roundwood coming to hand and the availability of the necessary skills and 'know-how' being important considerations);
- The desire to avoid selling unconverted roundwood and having to buy back for the estate's use finished products, at prices which include transport both ways, as well as the overheads and profit of the converter;
- The need to start local industries to utilize material for which there is no available market within convenient reach or where the estate is dissatisfied with the price it obtains for its roundwood when sold unconverted;
- The desire to retain full and effective control of silviculture and harvesting in the woodlands and of planning operations within them with the benefit of the estate solely in mind instead of being influenced by the requirements of extraneous buyers;
- The avoidance of damage, and of delay in restocking and other operations (sometimes an estate is not satisfied with the quality or speed of the work undertaken by contractors);
- The advantage of an estate raising its own planting stock;
- The advantage of an estate's having its own sawmill, with or without a preservative plant;
- The consideration that the larger turnover obtained from intensive and higher profit forestry enables more skilful workers to be attracted and enhanced salaries and wages paid (thus an estate may be justified in employing a forester of higher calibre); and
- Diversity of enterprises makes possible the achievement of full employment in most weathers and throughout all seasons.

By contrast, reasons advanced against attempting intensive forestry, especially on small estates, include:

- For intensive forestry there is need for a great variety of skill and 'know-how' as well as expert departmental supervision and sound marketing (particularly is this so in nursery work and sawmilling; some foresters, even if they are skilful silviculturists, may not have the ability and interest, or indeed the time, to run a nursery, a sawmill, or a manufacturing department);

- Many considerations related to whether conversion should be undertaken;
- The need for adequate supervision may involve the employing and accommodating of an additional forester or woodman (this might be against the profitable working of the woodlands as a whole);
- General office work, accountancy, costing and other overheads are substantially increased; and
- The estate has to endure several additional statutory regulations and accept many risks (the latter include the financial and other hazards accepted by merchants, processors and nurserymen).

Intensive forestry where practicable and justifiable allows a desirable flexibility in woodland management and provides a wider scope for enterprise. Futhermore, the challenge becomes interesting and quickens the keenness of foresters and their staff. However, it is a matter for each estate prudently to decide how far it should (i) undertake with direct labour its many operations, and (ii) attempt to combine enterprises and processing supplementary to forestry. The combinations instanced proffer many hazards though some rewards. The degree of intensity may not be altogether a matter of free choice: in some districts local skilled craftsmen no longer exist; foresters might then need, in order to dispose of some of their produce, to enter various trades and train men in the appropriate skills. Budgeting of estimated expenditure and income on any given enterprise is a prerequisite to prudent decisions, but only later, by careful accountancy and costing, can it be ascertained whether different sections of the intensive forestry are making profits or holding their own, or are losing money. Paramount will be the energy, enterprise and skill of foresters supplemented with like qualities in the staff under them. Finally, intensive forestry should not be confused with imprudently enthusiastic and financially wasteful forestry.

2. The management scheme

Most woodlands need some plan of management, particularly depending on their size and the objectives set for them. The detail of the plan generally increases with size and complexity of woodland. For small woods, or for relatively small areas of planting, simple plans suffice (that compulsory for the Woodland Management Grant Scheme is noted in chapter 15). For large complex woodlands, particularly for well-wooded estates, comprehensive plans of management are advisable (equivalent to the Working Plans of earlier decades) and to them this section is chiefly addressed. Each forester is likely to have preconceived ideas of layout and content of a mangement plan. The following suggestions are intended to be of assistance.

Woodland management can be assisted by what can be called a Management Scheme comprising Plan of Operations, Organization Plan and Financial Plan showing the forestry enterprise as a unit of business management and drawing attention to steps that can be taken to commence and continue it. Preparation involves a forecast of work to be undertaken, how it is to be funded, and the calculation of data in terms of timber volumes, man-days, machinery and materials. The whole is in fact a Working Plan.

In proposing a Management Scheme, an outline Plan of Operations will of necessity be the first of the three parts to be prepared, but the third, the Financial Plan, will probably be the first to be considered by the owner – to decide whether he should and could fund the management. If turned down, a *full* Plan of Operations will not be required. The Organization Plan will show the arrangements to commence and continue management putting into effect the Plan of Operations, and will include such items as: supervision and control, labour (whether employed directly or by contract), equipment, and possibly the creation of a system of roads and rides. Until these have been studied, together with possible alternatives, a *full* Plan of Operations is premature. The Financial Plan is more difficult to prepare and involves a wide knowledge of (i) the complexities of forest management; (ii) the incidence of capital taxation and grant-aid; (iii) estate management where, for example, co-ordination with agriculture is possible and desirable; and lastly (iv) the overall financial circumstances of the owner. In (iv), a competent forester, with an accountant, sometimes a lawyer, can often combine to provide the overall picture. Briefly, the Financial Plan should show clearly whether sufficient money can be derived from the sale of timber and other products to pay for the progressive management of the woods, or whether an injection of new capital will be needed to make the woods productive, and, in the latter case, how much money will be required each year if the woods are managed with reasonable competence. The cost of a gradual development of the woods might be covered by receipts from sales ('woodland improvement out of woodland revenue') while a more rapid development might require injection of capital. So a number of alternative schemes may be outlined in the Financial Plan, and when the owner has decided to adopt one of these the Plan of Operations can be completed in detail.

The owner by these means, if he goes ahead with forest improvement, will know whether it is likely to prove remunerative, whether he will have to advance money in order to make his woods productive, and, if so, whether it will be a good investment, and whether sporting and amenity features will be maintained or enhanced, if these are among his objectives.

The Plan of Operations. Here the first task is to complete a general survey of the woodlands with the object of ascertaining their nature, extent and contents – in fact, to make an inventory – a procedure noted in chapter 12, section 3. Then must be determined the prescriptions of silvicultural work to be accomplished – establishment, tending, harvesting and restocking – for at least the first five years.

The Organization Plan. Here the first necessity is an estimate of the labour required. This may be based upon the maximum requirement of direct labour, or alternatively on the minimum requirement of such labour supplemented by work to be done by contractors. The amount of labour will depend chiefly on the intensity of the forestry to be practised and on the work prescribed as set out in the Plan of Operations; also on (i) the measure of co-ordination with agriculture; (ii) the structure and age of the woods; and (iii) the rotations.

If the policy at the one extreme is to be reliance on contract for cultivation, fencing, planting, bought-in plants (no nursery required), thinnings and timber sold standing (no cutters, converters, machinery, sawmill, preservative or haulage equipment needed), the labour required will be minimal – sufficient only to cover beating up, weeding, cleaning, rack-cutting, brashing, pruning, protection and maintenance. At the other extreme, if all operations are to be accomplished with direct labour, and a nursery, sawmill, wood-yard and preservative plant are to be included, and timber sold converted, then the machinery and labour requirements, including office staff and supervision, will be substantial. Should the intention be financial and physical co-ordination of forestry and agriculture there is the possibility of a reduction in woodland labour and machinery. All these considerations, perhaps involving such aspects as housing and transport, must be given their due place.

Labour. Once the principle of the labour-structure has been decided the number of workers required can be assessed from the operations to be undertaken. The extent to which operations can, or should be carried out by direct labour or contractors or purchasers of timber is of immense import and will bear heavily on the Organization Plan and particularly on the requirements of labour and machinery. Usually, efficient direct labour, wisely chosen, planned and supervised, may be cheaper than contracting out, as contractors rightly add their overheads and profit. Against this, however, must be posed the question of estate overheads which are much higher on direct labour than on contract work. The charges of co-operative associations are sometimes less than those of contractors of the conventional kind.

Forestry undertaken using direct labour is much different (and much more capital intensive) to forestry undertaken through contractors, for example when all produce is sold standing. Labour and machinery requirements vary with the woodland's stage of development. Following a high requirement during the establishment phase, needs throughout the tending phase are low; the level of requirements rises again when the harvesting phases are reached. The labour requirements will further depend on variables such as age class distribution, geographical location, site factors, and accessibility. A near *normal* age-class distribution will have predictable labour and machinery requirements; but where far from *normal*, will need relatively higher requirements either because of the above-normal establishment or of the above-*normal* harvesting, or both.

The labour requirements are increased where work is needed relating to amenity, wildlife conservation or recreation.

Contract work. Contract work is the undertaking of a particular operation. It is the result of an owner having a fairly precise idea of what is to be undertaken; and may even produce a specification of the work required. The contractor will work either on an all-in sum which includes his overheads and profit, or on a cost-plus-percentage basis; some prime costs may be involved. When undertaking major afforestations, operations are usually by one of the forestry management companies, or contractors working for them. In smaller schemes the work is either by direct labour or through local contractors. (Contractors are drawn from local centres of population: some have invested in sophisticated machinery – particularly for cultivation and harvesting – to cater for the demands for their services. There are, too, many small teams of contractors.)

Machinery. The equipment required will be determined by the degree of the intensity of the forestry to be practised, the extent of the co-ordination with agriculture, and the out-turn of produce. For most forestry, a tractor with trailer is the minimum transport. A lorry in addition is sometimes a necessity; the basis of deciding on the need for it being the estimated kilometres it would be expected to run, loaded, each year. Lorries, except for short hauls, are cheaper than tractors because of their speed on the road and capability of larger loads; they can also transport workers and equipment more quickly, but cannot always negotiate the woodland stands.

The Financial Plan. The Plan of Operations and the Organization Plan having been outlined, the Financial Plan can be prepared comprising, firstly, expected expenditure, namely: (i) labour, supervision and direct overheads; (ii) capital costs and depreciation of machinery and equipment; (iii) cost of fencing, materials and plants; (iv) running costs of machinery; and (v) indirect overheads including top-level supervision/management. Secondly, the Plan will include the expected income. The volume of timber and produce calculated during the survey for the Plan of Operations will have a value placed upon it. Expected increment and the capital appreciation or depreciation of the woodland property would be calculated on the principles stated in chapters 14 and 16.

Outlines of alternative methods of development should be considered and an indication given in the Financial Plan of the cost which each would involve. Note should be taken of grant-aid from the Forestry Commission. Capital taxation incidence should be noted. An attempt should be made to prepare a five-year budget (cash flow) and, if possible, for a longer period. The Plan will bring into relief the implications of the alternative schemes proposed. The owner must be consulted as to whether the total funding, if any, can be found within the estate, or may have to be borrowed. Only when one of these schemes has been agreed and overall directions given is it possible to put into final form the Plan of Operations and the Organization Plan, to co-ordinate them with the chosen Financial Plan, and then to put the whole Management Scheme into operation.

Finally, it must be remembered that a Management Scheme, however well prepared, cannot itself ensure sound economic execution of operations. Given an environment in which success is reasonably possible of attainment, the most important single factor will be the constant measure of enterprise and skill put into the management at all levels. It is, therefore, important that a method of control of woodlands should be devised which will ensure the maximum of co-operation, eagerness and enterprise in the management, combined with prudence and foresight in expenditure and in the efficient marketing of all produce. The workers' enthusiasm must be nourished, particularly by securing happy relationships. Periodic accountancy and progress reports should be a great help in applying sound business methods. There must be a constant endeavour to make the woodlands more productive as well as to improve the amenity and sporting of the estate, so far as these are among the objectives of management.

Charges for management. Each woodland or forest property differs from all others, hence the charges for surveying and managing it should be related to the work involved. Fees and commissions vary widely. Fees for valuation of timber and forestry properties are indicated in chapter 16, and for seeking and negotiating the purchase or sale of a forestry property in chapter 11 as well as for the sale of standing timber and thinnings. Preparation of Plans of Operation might cost from £40/ha for the first 5 hectares to £10/ha for over 200 hectares. This would include field work, preparation of compartment, sub-compartment and stand notes, updating management map records, consultation on objectives of management, preparation of prescriptions, financial estimates for five years, and completion of relevant forms. The annual management of woodlands might cost from £15/ha for the first 40 hectares, to £5/ha for extents over 80 hectares. However, if a woodland property requires intensive management it would be appropriate to make an estimate and agree a figure in advance. Management would usually include: ensuring proper implementation of the Plan of Operations, maintenance of woodland records and accounts, financial estimates and budgets, periodic progress reports, grant collection and statutory returns. The management of woodlands being so exacting an occupation, since they differ so widely in degree of their integration, and disposal of the unconverted and converted products, the fairest charges from the point of view of both owner and forester are those assessed on a time basis. Where the management is not undertaken by a full-time forester, the time charges for consultant foresters are likely to be about £60/hour for principals, £40 for senior consultants, and £30 for qualified assistants.

3. Management primarily for wood production

Forest management is about making and implementing decisions. Rational decision-making means choosing courses of action that tend to achieve objectives. The commonest specific objective is to maximize timber production; however, often that objective may be supplemented, replaced or restrained by multiplepurpose objectives – landscape, recreation, nature conservation or sporting – subjects discussed in sections 4–7 *infra*. This may induce conflicts in one or more forms: compromise and flexibility may be necessary.

Successful management to fulfil set policy objectives is achieved by careful planning, organization and controlling all forest operations. Even for small woods management will need a simple Plan of Operations; and for large woods and well-wooded estates a detailed Management Scheme, as suggested in section 2 *supra*. Important elements of it include organizing mensurational and other surveys (see chapter 12) and the incorporation of an adequate system of records and accounts (see chapter 16) integrated with work as it is accomplished. Planning therefore depends on a full assessment of the site and growing stock, supplemented by consideration of owner's wishes, prejudices and, in particular, his objectives.

Foresters should possess adequate knowledge of the composition of the woodlands, in particular (i) volumes and how they are distributed by species, age and size; (ii) the extent that is in need of thinning, mature, understocked, or bare ground; and (iii) the silvicultural system being used. They can then judge what form the growing stock should have to satisfy the objectives; and thereupon appropriately to plan, control and mould the woodlands accordingly. The prescriptions may include silvicultural operations such as fertilizer application, cleaning, brashing, thinning, and clear-felling, with subsequent restocking.

Where management has been neglected, there will be a priority of rehabilitation. Efficient use should be made of available funds, but where these are limited, or constraints (e.g. on labour) may not allow faster conversion, there will need to be a choice among treatments, and several years may elapse before an ideal condition is approached. There is obviously a cost of delay when operations cannot be undertaken at the ideal time, e.g. when mature trees are allowed to decline, when stands lose increment for want of thinning, and when plantable land is left to grow unwanted vegetation. In both neglected woodland and tree-colonized areas, assistance with improvement operations is provided for under Grant Schemes, where the crop contains adequate stocking of suitable species and is under twenty years of age (see chapter 15).

Chapter 12 was devoted primarily to principles relevant to conifers, with particular attention to maximizing economic return from wood production. Additional relevant detail for their management was provided in chapters 5 to 11, particularly as to establishing, tending, thinning, harvesting and restocking. Wind damage in upland coniferous forests, particularly of extensive windthrow well before normal economic rotation, is a serious problem for the forest. Windthrow Hazard Classification (see chapter 3) is widely used to predict probable rotation lengths for forest stands and thus assist production forecasting and economic appraisal, and to determine whether thinning is a valid option for a particular site. The economic return largely relates to species, yield class, thinning regime, rotation length, stability, location and markets.

For broadleaves, the general principles of forest planning and management apply equally as for conifers. However, broadleaves require a higher level of technical skill and commitment. Their management for wood production invariably embraces several supplementary objectives. Their growing entails more risks and more uncertainty of ending with a valuable crop. Also, because of the slower growth rates of most broadleaves and longer cycles of production, profitability is usually much lower than with conifers. The essential feature of broadleaves for useful commercial production is their timber quality, requiring consistent management throughout the rotation. The many features and attributes of broadleaved trees and woodland have been noted in chapter 4, and their establishment in chapter 5. Natural regeneration and coppicing have been discussed in chapter 7, and thinning and harvesting in chapters 8 and 9. This section 3 relates to their management for wood production. (Non-wood values are discussed in sections 4–7 *infra*.)

There are several legitimate objectives for managing broadleaved woodland (Evans 1984). Though widely occurring, many species are treated differently in different parts of Britain; this arises from the large variation in sites and soils, the importance of local markets, the policy objectives, and even tradition itself. Categorical assertion that one way is right and another wrong can rarely be justified. Almost any stand, of whatever age or species, can be treated in more than one way but still with the primary objective of wood production. (Only in the management of ancient semi-natural woods and other special purpose woodlands are the constraints particularly vital.) Good quality broadleaves can be produced either in even-aged or uneven-aged crops, and by pure or mixed stands, also as hedgerow and park trees. Being aware of the options available allows flexibility in management, and enables other management aims to be embraced as well.

The capacity to regenerate by coppicing enables broadleaved woodland in general, (beech and Italian alder excepted), to be managed either as coppice on a short (low input) rotation, i.e. less than ten years, or as coppice on a mediumlength rotation, i.e. ten to thirty years, for pole and fencing products. Even in neglected woods, coppicing can be resumed if the neglect has been less than fifty years, and sometimes after much longer periods. Storing coppice is a feasible method of conversion to high forest (see chapter 7).

The standard guide for the silviculture of broadleaved woodlands, is Evans' comprehensive text (1984). His recommendations for broadleaved high forest silviculture in brief are: (i) decide objectives of planting and management; (ii) for timber production direct the silviculture at enhancing tree and stand quality rather than just yield; (iii) generally provide beneficial side shelter and nursing when young; and (iv) seek rapid establishment, which depends on getting right the species choice, type of planting stock, handling, planting, weed control and protection. Necessary thereafter are appropriate cleaning, pruning, thinning and final harvesting.

The profitability of growing broadleaved high forest has traditionally been considered low (0–2%) compared with conifers (2–9%) owing particularly to prolonged and hence relatively costly establishment, slower growth, generally poor first thinnings and longer rotations. The average growth rate is close to 5 m^3/ha/yr by contrast to about 11 m^3/ha/yr for coniferous forest. However, making direct comparison and concluding that broadleaves grow very much slower is somewhat misleading. Unlike conifers, most broadleaves, and especially oak, are managed on a rotation much longer than that of maximum mean annual increment in order to grow large dimensional logs; and volumes derived from yield class do not include branchwood, which in broadleaved trees can form a significant quantity of the total marketable volume, especially where there is a demand for fuelwood (branchwood volume for large oak and beech averages about 50% of stemwood volume). Although tending and thinning of broadleaves are more complex tasks than those for conifers, the value of conifer logs may lie between £20 and £50/m^3 whereas with broadleaves the range might be £10–£200 or more/m^3. Foresters' skill will mainly determine the difference. By contrast to conifers, fast diameter growth of broadleaves leads to strong timber, so rapid diameter growth is a worthwhile aim. There is a ready demand for high quality broadleaved sawlogs and veneer butts. For small diameter roundwood and medium sized sawlogs the markets fluctuate, depending substantially on the level of demand by paper and board mills and by the fencing and mining timber trades; but the present reasonably satisfactory relevant markets appear set to remain acceptably stable in demand and price.

Silvicultural problems in growing broadleaved high forest include their restriction to more fertile sites, the prevalence of mammal damage, especially grey squirrels, and difficulties in resolving conflicting objectives. These factors are recognized in the present preferential grant-aid schemes. In establishment, as broadleaves generally benefit from side shelter, spacing on bare land should not be wider than 3 m (1,111/ha); nurse species, either conifers or pioneer broadleaves, should be used, planting being carried out during autumn or early spring, using sturdy stock (root collar diameter more than 5 mm, height 20–40 cm), and competition from weeds must be rigorously controlled. Fertilizers are rarely necessary. Treeshelter techniques can be extremely useful. Cleaning may be needed to release main crop species, and formation trimming of multiple leaders can be considered at the same time.

Thinning should be undertaken when the stands are 8–10 m high, aiming to improve stand quality, and early selection for the final crop made of vigorous, straight trees free of defect. The ideal is to thin a little and often. Rate of growth is of less importance than achieving even growth in wood quality. Pruning, to no more than 5 m in height, following the first thinning should be confined to potential final crop trees (probably about 75/ha for oak and 100/ha for beech) cutting branches just proud of the branch bark ridge. Timing should be late winter/early spring except for birch, sycamore, cherry, Norway maple and walnut (they are liable to suffer from sap exudation). Any nurse species can be removed as and when appropriate. The rotation length of the final crop will be determined mainly by log size required, timing it with good markets where possible. Damage to the butt log needs to be minimized, especially by considering the felling direction of large crowned trees.

In restocking, planting is usual, but natural regeneration (see chapter 7) is feasible for all major broadleaves, for which Evans (1988) provides concise guidelines: briefly, await a prolific seed year and ensure that the forest floor is in a receptive state, protect regeneration that comes, together with any advance regeneration, open up the overstorey within two to three years by heavy thinning or felling, and in due course appropriately respace the regeneration.

Currently there is a substantial under-production in many broadleaved woodlands, and hence there is potential for much increase. Most represent neglected assets – particularly true of numerous small farm woods and remote copses. Among factors which cause below-optimum yield may be the constraints of landscape, conservation and sporting and, in small woods, the difficulties of working them. Much of the neglect is due to inferior woodlands rather than inherent poor site fertility; and has often arisen from not knowing how to manage, and being unaware of the potential value of the crop or the site. Very few areas looking like woodland have no utilizable value: many neglected woodlands offer opportunities for considerable improvement in production. (Unduly satisfying other objectives may of course consign some woodland into a relatively nonproductive category.)

Many of the hardly productive broadleaved woodlands in the south of England are neglected coppice with standards, commonly oak standards over either hazel underwood or a mixed coppice of ash, sycamore and other species. Often there are too many standards whereby the understorey is suppressed and weak in structure. Possible treatments include clear-cutting and restocking, or underplanting with shade tolerant species (beech or hornbeam under the broadleaved policy; otherwise western red cedar or western hemlock). Neither treatments are favoured by conservationists. Substantial areas of Wales and southern England carry mostly unthrifty oak woods of coppice origin, where the recommended treatments are: fell and re-coppice, or fell and restock, or convert to a group felling system.

Evans (1989) has drawn attention to the neglected asset instanced by many of Britain's smaller broadleaved woods, which have suffered neglect in recent decades as markets for underwood, coppice and other small diameter roundwood produce have declined. At least half have received no formal management since the 1950s. Yet, this neglect does not mean that such woods are valueless. Most have some standing value, and the labels 'useless scrub' or 'derelict woodland' are often misnomers; even what really is 'useless' for wood production is of great value for conservation and game management. However, it is the *managed* woodland which maintains or builds up an asset and best ensures its long-term survival and well-being. Post-war rehabilitation experiments have shown that a period of neglect need not be ruinous, that 'scrub' is more a perception than a scientific fact and that development into a reasonable woodland can occur almost unaided, especially where there is protection from grazing and browsing. Thinning for firewood might secure this 'scrub' into a worthwhile crop. Furthermore, a resumption of coppicing power in many broadleaves and the response to thinning of stored oak, ash, sycamore, cherry and lime (but not beech, birch or hazel) makes rehabilitation worthwhile. So too does enrichment and underplanting (see chapter 7).

New woodland occasionally arrives in natural ways. Wherever bare or disused land is protected from grazing and browsing, woody growth will soon develop. Colonization can be rapid (it is sometimes termed secondary woodland). Much has the potential to develop into high forest or as a matrix of shelter into which trees can be planted. The main silvicultural needs will be respacing and cleaning to sort out the stand in the thicket-stage; also enrichment as appropriate.

The management of broadleaves to include non-wood benefits, particularly under the Forestry Commission's 'Guidelines', is discussed in section 4 *infra*. Much therein is relevant to supplementary wood production.

4. Management to include non-wood benefits

Woodlands supplement their wood production by several non-wood benefits. According to their location, accessibility, age, content and management, they often attract intense public pressures – usually the seeking of benefits of access, recreation, amenity and wildlife conservation, irrespective of who bears the cost. Society's increasing demands for aesthetic, environmental and recreational requirements and their evaluation are noted in chapter 14: from them has come development of the economics of, particularly, landscape, wildlife conservation and recreation. Some environmental economists suggest that the carbon-fixing role of forests may have values approaching those for timber production (see chapters 3 and 14). The need to promote visual amenity, wildlife conservation and recreation (albeit supplementing timber production) has led to a renewed interest in development of a transformation from uniform plantations to irregularly structured woodlands (see chapter 7).

All aspects of environmental quality are much influenced by forest management decisions (Campbell, D. 1990). The non-wood benefits also affect the investor: often they are seen as of greater importance than financial return. These could include amenity, opportunities for sporting and other satisfactions: only the owner can say what the benefits are worth.

Environmental restraints on private woodlands can be substantial. Many are overcome: TGUK's *Forestry and Woodland Code* (1985) is adopted in practice by many of its members. Lorrain-Smith (1989) emphasizes that not all woodlands are capable of providing the same blend of benefits; and that private owners differ both in what they themselves expect of their woods (and how, therefore, they manage them) and in how they may respond to public demands and national forest policy instruments. Commercially-minded owners may react quite differently from, for instance, owners of small woods used mainly for shooting, or conservation charities with woodlands, or woodlandowning local authorities. The woodlands of any of these categories may be small (though locally important), but together they may form a substantial total.

Aims of forestry, and of forest policy, have changed over recent decades, so that woodlands will have been established and managed with varying intentions. Many woodland management decisions have permanent effects and it may not always be possible to induce an existing woodland (especially if old) to respond to new demands. Species cannot easily be changed once planted, and scale may be difficult to alter, but other management decisions also affect the character of a wood and its potential for providing benefits. Woodlands vary in extent, age, species, composition and condition. Their geographical location is important, especially their prominence and proximity to people. All have very different patterns of economic, social and environmental benefits demanded of them. In landscape terms, there is a great difference between upland plantations remote from view and those prominently visible from tourist routes. There are differences relating to traditional lowland woods, farm woods, 'forests for the community' in green belts or urban fringes, and tiny woods, copses and spinneys in urban areas themselves. Additionally different will be lowland multi-purpose new plantings, e.g. the Central Scotland Forest and the New Midland Forest.

Faced with such demands for benefits, woodland owners may have difficulties in responding adequately. For timber production, the trees have to be thinned or clear-felled whereas for other benefits they may have to be kept standing. Managing woodland in a way to reduce or overcome conflicts (e.g. by small scale felling, or uneven-aged forestry) may be technically possible, but probably not without some loss of financial profit. Most benefits from woodland are outputs from timber, but land, labour and materials are inputs with a cost. Conventional management practice is to reduce inputs while increasing outputs to maximize net benefits. Many of the benefits provided by woods cannot be sold for money: often the public can freely enjoy landscape, climatic and environmental benefits, much wildlife conservation and even recreation. Benefits cannot all be maximized simultaneously on the same site (Lorrain-Smith 1989). Thus although there is a wide economic and social demand for the benefits that forestry can provide, some difficulties may have to be faced in managing woodlands for both timber and non-wood benefits.

Most woodlands, whatever their objective, will yield some commercial timber, and even where they are managed primarily for non-wood purposes it is clearly in the owner's interests, within the constraints, to achieve as good a financial return as possible, to offset at least some of the costs of management. Obtaining some yield of forest produce is important as the only long-term source of revenue from woodland, other than possibly from sporting rental and, occasionally, from charged for recreational facilities. The non-wood aspects of management may not always conflict seriously with the production of timber, though compromise often will be necessary. There may be the need to retain flexibility to allow opportunity to change objectives later.

Mixed woods. Management as discussed in the remainder of this section generally relates to broadleaved woodland, but much could apply equally to mixed stands, particularly where the conifer component is removed relatively early, leaving a broadleaved crop. While the ecological value of most woods is usually lower than that of ancient semi-natural woodlands (noted *infra*), their management likewise may be important for nature conservation and landscape values. For grant-aid (see chapter 15) any broadleaved species may be planted or encouraged, although native or traditional species are preferred, and for restocking up to half of the trees used may be conifers, and afforestation on land previously agriculture may be carried out with any species with no restriction on proportions of mixtures provided that the landscape aspects are acceptable.

Guidelines for broadleaved management. The Forestry Commission, following its Broadleaved Policy announced in 1985, set out the principles for managing different types of broadleaved woodland, and for different objectives. Its guidelines do not seek to provide detailed prescriptions for every type of woodland or for every type of activity; the wide range of local conditions makes this impracticable. Rather, they indicate the way to approach certain key elements of management, and are based on the following management premises and principles: (i) woodland which is now broadleaved is expected to remain so; (ii) there is a presumption against clearance of broadleaved woodland for agricultural purposes; (iii) the present area of broadleaved woodland is expected to increase; (iv) special attention will be given to ancient semi-natural broadleaved woodlands to ensure continuance of their special features; and (v) managed woodland is more likely to survive than that unmanaged. For the purpose of (v), adequate income from woodlands is essential; this can be obtained with little detriment to the landscape, wildlife or recreational interests, all of which can be better met by healthy, valuable trees than by neglected and moribund growing stock. (However, a few old trees per hectare could remain standing to benefit woodpeckers as well as many invertebrates and fungi.) This is obviously in the grower's and the national interest, in view of the importance of continuity of supply of the major broadleaved species to the wood-processing industry.

Guidelines for management of ancient semi-natural woodlands. These woodlands (discussed in chapter 4) need special management; the Forestry Commission's guidelines for them are noted below.

In *high forest on ancient woodland sites* the aim should be for diversity of stand structure, with old, mid-aged and young trees present in the wood either as individuals, groups, clumps or as roughly even-aged stands. To achieve this, a small proportion of trees may have to be left to grow on beyond the economically optimum rotation on parts of the area. Best are distinct clumps or groups, not as scattered stems left as fringes or over the felling area: groups provide a certain amount of mature woodland conditions which scattered trees do not. Where clear-felling is necessary, felling coupes should be kept small, but bearing in mind that they have to be large enough for efficient harvesting. When felling is done in 'groups', each should generally be within the range of 0.25–0.5 ha, but care is required to relate the size of the felling area to the scale of the landscape. Natural regeneration should be encouraged, where: (i) species and quality of the parent source are satisfactory; (ii) advance growth natural regeneration is present; and (iii) there is no urgency to fell and restock in one year. These provisos are necessary to avoid perpetuating poor quality stands and to avoid disappointment in waiting for natural regeneration which fails to appear through lack of seed or other reasons. Where it is decided to restock by planting, this should be done with broadleaved species present on the site and native to the area. Where a 'nurse' species is required to shelter the young broadleaved crop, this should itself be broadleaved and could take the form of coppice. There will be circumstances where a conifer might be used in mixture with broadleaves for silvicultural or land management reasons, e.g. frost sites, exposed ground, and in the uplands. If so, the conifer component should not exceed 25% of planted stock – to ensure that if the wood becomes neglected at some time in the future, its broadleaved character is not imperilled. Any planted conifers should be removed before they adversely affect the growth of the broadleaves. Thinning should be done to favour trees with good stem-form and freedom from defects in the lower bole, absence of forking in the crown, good vigour, freedom from upper stem and crown defects, low incidence of epicormic shoots, and spacing in the stand, in that order. Thinning should also aim to retain the range of species present, not only those of high timber value, to the exclusion of other species.

In *coppice* and *coppice with standards on ancient woodland sites* where much of the crop is currently worked as coppice, in many circumstances the best treatment is to continue this system. In some parts of the country the revival of coppice is desirable, particularly if the area has been coppiced within the past forty years. This method of working broadleaves should be encouraged where it is practicable. The growth of timber trees, established as scattered standards above coppice, is encouraged. They may arise from natural regeneration, promotion of stool shoots, or planting. If planted, only native species growing on the site or in the locality should be used. If coppice working is not a practical proposition, the old coppice should be 'stored' by being allowed to grow on stems singled to one per stool, and so converted to high forest. (Hazel underwood or hornbeam coppice will not develop into high forest.) The only option, if underwood or coppice working is not practicable, is to fell and replant; the restocking guidelines noted above for high forest apply.

Plantations on ancient woodland sites may contain introduced species, but having been woodland since early times, such sites may retain features of wildlife value. Management should aim to identify these features and to preserve and enhance them where practicable on the following lines. Where nature conservation value is particularly high, broadleaved woodland should be managed on the same lines as ancient semi-natural woods. Any planting in broadleaved woodland should normally be with native or traditional broadleaved species. It may be possible for up to 50% of compatible conifers to be accepted in any restocking provided they are managed in such a way that the continued long-term broadleaved character of the wood is assured. Thinning in these broadleaved woods should favour good trees, with any conifers being removed progressively. A good representation of the broadleaves present should be maintained in the upper canopy. Some plantations on ancient woodland sites are mixed or wholly coniferous. Management should seek to conserve the broadleaves present, along with their associated flora and fauna, by appropriate treatment of glades, rides along with open spaces. Conifer crops on such sites may be replaced by conifers. However, foresters may choose to replace conifers by broadleaves, wholly or partly, when the time comes for restocking. Considerations of stand diversity in such conifer plantations are important.

General management considerations. Newly planted or restocked woodland requires tending, particularly in the early years. Herbicides give good weed control at low cost but, particularly where the ground flora is itself of conservation value, the herbicide should be applied only to the immediate vicinity of the young trees, leaving the bulk of the site untouched. Dense bracken, bramble or rhododendron may require more intensive herbicide treatment. In general, ancient woodlands and sites on which broadleaves might be established do not require the addition of fertilizers to obtain satisfactory growth; but it may be necessary on restored man-made sites, impoverished heath, moorland and chalk downland soils, or when nutrient depletion has taken place after prolonged coppice working. In these cases the need for fertilizers can be confirmed by foliar analysis before treatment. Otherwise the use of fertilizers should be avoided because of adverse effects on ground vegetation. Rides and other grassy areas, glades, and internal roadside verges give diversity of habitat, attractive to a wide range of woodland and meadowland flora and fauna. Since the widespread destruction of old pasture, these areas have become some of the last reserves of unimproved grassland. Some contain quite rare, poor colonizing species. Similarly, rides and glades within conifer woods may hold areas where earlier woodland plants and tree species are still present when protected from grazing. Foresters should not unwittingly allow important fragments of habitat to be destroyed by encroaching trees or by insensitive woodland operations. Grass rides are often best maintained by mowing portions, e.g. strips, in alternate years, preferably after ground-nesting birds have finished raising their broods. Any shrubby strip between grass and woodland should be cut at intervals of three to five years, aiming to cut, in any extensive system of rides, a proportion of the shrub strip each year. In addition to the use of ponds for wildfowl and fishing, the range of woodland habitat can be diversified by retaining wet places, pools and marshy or bog areas. A simple measure is to thin out trees and shrubs on the margins of such areas, to admit more light and improve vegetation.

Progress of the Broadleaved Policy. The progress of the Broadleaved Policy since 1985 (see chapter 4) is continually monitored both by the Forestry Commission and the private forestry sector. In 1989 the Commission reviewed its policy. *Broadleaved policy: Progress 1985–88* showed the very considerable success the policy had in extending the area of broadleaves throughout the country (in 1984–85, some 9% of grant-aided planting by private owners was with broadleaves; by 1988–89 this had increased to around 17%). The policy had also been successful in arresting the loss of broadleaved woodland and in focusing attention on the irreplaceable contribution that the ancient semi-natural woodlands made to the environment. However, Lorrain-Smith (1988) had measured the economic impact of the broadleaved 'Guidelines' by analysing changes in profitability of five types of broadleaved woodland based on sampled estates throughout Britain, indicating that many owners following the 'Guidelines' might be suffering substantial reductions in the viability of their broadleaved planting.

Generally acknowledged was the high cost of managing broadleaved woodlands over the very long period between establishment and the first returns from the sale of thinnings, and of bringing neglected woodlands back into production, also the lack of sufficient incentives to encourage owners to adopt practices that would have environmental benefits over and above general maintenance work. Particular concern was expressed by some growers' organizations that owners were not receiving the necessary help required to enable them to meet the more exacting multi-purpose objectives and the associated higher costs of managing ancient semi-natural woodlands. Also, although Forestry Commission 'Guidelines' had generally served their purpose well, some commentators asserted that the 'Guidelines' were not being interpreted by all parties concerned with sufficient sensitivity and flexibility to regional or local variations.

The Government, following the Commission's review of the policy, confirmed that the 'Guidelines' should be interpreted flexibly, and that although the then planting grants were adequate, there was a need for additional grant-aid to management and other conditional support. The aim was still to be encouragement of good multi-purpose management of, particularly, the various types of broadleaved woodland – as well as to make the broadleaved policy even more purposeful as well as appropriately flexible. In consequence, a Woodland Management Grant Scheme was announced, effective from 1 April 1992. This grant is discussed in chapter 15, together with other conditional support and conditions.

Having in general terms introduced the management of woodlands (particularly broadleaved) to include non-wood benefits – albeit usually supplemented by timber production – attention is given hereunder to management as it relates more specifically to landscape, recreation, wildlife conservation, water quality and sporting.

5. Management for landscape

Trees and woodlands are among the most attractive features of a landscape (Miles 1967; Crowe 1978; Lucas 1991). Their significance is related to many factors, notably topography and outline. A large wood is a more prominent feature in the landscape, but there are regions where the pattern of small woods is vital to its character. Small woods often have an importance in the landscape far greater than their size would suggest, especially in small scale landscapes or very open ones (e.g. clumps on chalk skylines). Wooded landscapes tend to be seen either from viewpoints where they form part of the general vista, or from within. This has implications for design (see *infra*). Woods close to, or visible from main roads or towns have especial importance as valued landscape features. In designated areas, such as an Area of Outstanding Natural Beauty or a national park, woods and their management are also of particular importance. Some small woodlands or clumps of trees regarded as a major amenity are protected by a Tree Preservation Order. The Forestry Commission through the grant schemes and felling control procedures, is able to encourage the adoption of good standards of landscaping at the planting, thinning and clear-felling stages.

Coniferous woodlands in the landscape. Such areas have their special attributes, particularly when sympathetically designed and managed. To some of the public they generate concern when first planted (displacing open landscapes and ecosystems), and likewise when they are felled. There will always be individuals and organizations that approve or dislike them. Amenity groups have frequently used aesthetic and economic arguments against afforestation of hillsides and upland sites. Some have shown concern in upland afforestation with conifers (particularly Sitka spruce) – instancing the gloomy, unvarying coloration of most species with an inability to take light and shade; and of lines of conifers monotonously regimented, with aggresive striations of line-thinnings and grids of fire-breaks which were characteristic features of much early afforestation. Beauty and acceptance are often in the eye of the beholder, and relate particularly to the location and terrain in which the conifers are growing. Likes and dislikes often change with time. Mitchell

(1980) usefully assesses the role of conifers in the landscape: in Highland scenery of a rugged nature they really come into their own; and in the lowlands their attributes include the individual spire-topped conical crowns which often provide character and a focal point to the distant scene. (Larches, being deciduous, are valued in the sylvan scene through the changing relief of colour they give throughout the year.)

Broadleaved woodlands in the landscape. Broadleaved woodlands usually have even greater attributes than conifers. A useful summary has been given by Workman (1982). Most woodland in the lowlands is predominantly broadleaved in character and much of it is as small scattered woods and copses which may often coalesce in the view to create a much more wooded feel to the landscape than would be guessed at from their extent. Such woods are usually diverse in habit and appearance; and nearly all are deciduous, introducing variety and change through the seasons to woodland appearance, internal views and lighting from the canopy. Evans (1984) provides a classification of broadleaved woodland where landscape and recreation are of particular importance; and indicates general guidelines for maintaining existing woodland as well as creating new landscapes and amenities.

Forest landscape design. Forestry, like agriculture, is not subject to planning control. Scenic beauty is an aspect of environmental quality much influenced by forest management decisions. The uniform conifer plantations established in the uplands during this century have sometimes received criticism. Some efforts have been made in the past both in design of forests and occasionally in the selection of species; what may now be looked upon as poor design was really an effective answer at the time to a particular forest policy. Opportunities are now arising, as these forests come to be felled, and others established, to create a much more varied woodland both in age structure and species mix. In the 1970s, Dame Sylvia Crowe (1978), as landscape consultant to the Forestry Commission, gave invaluable guidance for woodland design.

In 1989, the Forestry Commission published *Forest Landscape Design Guidelines* prepared by their landscape architects. In brief, the guidelines are intended to provide an outline of the principles and practical application of key design principles: scale, diversity, visual force, unity, and the 'spirit of the place'; and to indicate what an intending planter should look out for, and situations which need special attention. As regards forest operations, intrusive fence lines should be reduced; cultivation by rippers, scarifiers and mounders is normally much less obtrusive than ploughing. Attention should be given to the creation of naturalistic shapes and to species choice. In an appraisal of the landscape, factors which should be carefully assessed before design work starts include sensitivity, character, heritage, diversity and any special features. In applying design principles due attention should be given to woodland shapes (naturalistic); this may include improving poor designs of existing woodland, and attention to upper, side and lower margins, edges, skylines, streamsides and open spaces. Ride and road systems need careful designing. Species choice and pattern are important; and the correct use of treeshelters is essential. Felling involves significant landscape changes and should be adequately considered when harvesting is planned (see chapter 10). Additionally, it is important to conserve and renew traditional landscapes of fields, hedgerows, clumps of trees and small woods. The foregoing factors and considerations are enlarged upon later in this section.

The role of woodlands, both coniferous and broadleaved, in the landscape influences silvicultural practices. If proper attention is given to landscape considerations a woodland's potential for recreation, and for conservation value, will usually be satisfactory as well. Where woodland is the principal land use, forest operations can often enhance rather than detract from the landscape. Thus ensuring both woodlands and forestry activities look right in the landscape should be an integral part of forest management. This is all part of good integrated environmental design for multiple objectives.

Good woodland design in the landscape is specially important in hilly or mountain terrain, where the shape and composition of the woodland as a whole is more easily seen and the natural landforms are more obvious. In flatter areas the foreground view and internal views, e.g. from roads, are important since the woods are mostly seen edge-on. Woodland design and silvicultural operations should seek to: (i) create irregular forest shapes that reflect the form of the land, e.g. rising uphill in hollows and falling downhill on convex slopes; (ii) be of a scale in proportion to the size of the landscape; (iii) enhance visually important natural features such as watercourses, gullies or crags; (iv) encourage diversity in species, ages and stand composition; (v) integrate visually with adjacent farmland and semi-natural vegetation patterns at the woodland margins and become increasingly irregular near to water; (vi) avoid unnatural straight lines, regularity of geometric pattern and shape; and (vii) minimize intrusive effects of forest operations, e.g. roads.

Forestry is part of country life and tradition, but on the whole people do not like to see sudden change. Clear-felling is the greatest change in the woodland, and this needs careful planning to improve rather than detract from its appearance. The aim should be to: (i) ensure a harmonious relationship of shape and scale between each successive felling where a hillside or particular landscape composition is to be felled over a number of years (phasing the felling helps to minimize the adverse visual effects of a sudden loss of mature woodland); (ii) design felled areas, and any retained woodland within them, to reflect landform shape and scale, and to create an appropriate shape to the woodland edge; (iii) avoid leaving long belts of trees but rather to retain groups irregularly sized, shaped and sited; and (iv) use felling to re-open views, expose significant landscape features, and to break up intrusive straight lines and other geometric shapes. In woods much used for recreation, maintaining large trees is desirable, with any felling being done selectively or in small groups. Judicious felling of some trees may open up a new vista and such opportunities should be considered as a way of increasing the interest of the woodland area. Uniform woodland, even if mature, can be enhanced in character by introducing gaps, glades and other open spaces.

The introduction of some broadleaves into conifer forests, particularly in the uplands, can relieve uniformity and improve the landscape, both externally and internally. Allocating a small proportion, say 5%, of the woodland area to broadleaves involves minimal loss of timber production, and has important benefits for wildlife conservation. In doing this the following principles should be observed: (i) ensure that species are suited to the site so that the trees will grow satisfactorily; (ii) use broadleaves to emphasize the natural landform by planting up gullies or areas beside watercourses; (iii) plant in clumps and irregular shapes related to landform and avoid straight edges, aiming for simplicity and appropriate scale (care is needed to avoid small clumps 'floating', so they should also relate to designed lines, e.g. coupe edges); and (iv) make use of or augment existing broadleaved trees and woodland, e.g. on exposed crags, along a lower boundary to link with hedgerows, or at higher elevations to soften the upper margins.

Mixtures of broadleaves and conifers are often attractive but may present problems in sensitive landscapes, particularly in the early years of establishment. These effects can be minimized by: (i) in row planting, avoiding 'pyjama stripe' patterns of alternating species; (ii) in group planting, varying the number of trees, shapes and separate groups, or using drifts; and (iii) avoiding pattern thinnings which emphasize and perpetuate the mixture pattern.

In woodlands which are prominent in the landscape, especially on sloping or hilly terrain, forms of line thinning are undesirably visible, owing to the creation of regular patterns of thinned lines (Hamilton 1980). In sensitive landscapes selective thinning should be used instead.

Silvicultural systems (see chapter 7) can have a great impact on landscape. *Even-aged high forest,* under the clear cut and replant system is widely regarded as the least desirable for landscape value. The lack of continuity of forest cover, and the sudden change from natural forest to bare-ground are major disadvantages. Little can be done about the impact from felling; however, when a mature crop is removed from an understorey of established natural regeneration or from a previous underplant, any adverse effect on the landscape is somewhat reduced. The *coppice system* is generally beneficial to landscape interests: although rotation length is short, re-growth is relatively rapid; and many such crops are small and scattered. *Coppice with standards* is likewise beneficial to landscape, because the overall appearance of such stands does not appear to change significantly even after cutting the coppice, which soon rises again.

In the *selection system* the appearance of the woodland as a whole remains unchanged. Its visual advantages externally and internally are retained; and to the casual observer the woodland appearance remains constant. In terms of continuity and lack of sudden change selection stands, as well as some group systems, are welcome. The canopy is textured and yet uniform, making it very desirable. From within, the abundance of mature or maturing trees, with isolated well-developed crowns, standing above young trees of many ages, is very attractive. The multi-storied structure is highly desirable for breeding-birds, and within any stand there will be small scale spatial variations in shading and hence of flora and shrub layer communities. The intimate mixing of old and young trees, and the continuous woodland conditions, are ideal for sustaining a diverse and epiphytic flora. However, although appearing 'natural' and perhaps foresters' fancied ideal (Evans 1988) the system is sometimes viewed as highly artificial and unnatural (Peterken 1981): Europe's ancient wildwood mostly consisted of 'patchy evenagedness'.

Afforestation: landscape principles and guidelines. Afforestation, on any but the smallest scale, will be a major change in the landscape. Extensive afforestation on hillsides in particular has a significant and sometimes dramatic effect upon the landscape, first during the cultivation and establishment stages and later as the plantation develops in height. Insensitivity to these effects in the past has led to criticism; some has been justified. With the hindsight of modern attitudes to the countryside, it is easy to see that many woodlands planted on undulating terrain or hillsides during the first half of this century paid what seems now to be scant regard to ecological, aesthetic and environmental requirements. However, most of this has been due to the Forestry Commission having been under pressure for many decades, especially following massive depletion of timber resources during two world wars, to afforest or restock as much land as feasible as quickly as possible. This led to priority being the maximum return on timber from the land available. Planting, almost wholly coniferous and in the uplands, was right up to the boundaries of land which usually followed the geometric enclosure patterns of the eighteenth and nineteenth centuries. The chief criticisms have been of harsh straight line boundaries, large blocks of dark uniform conifers, 'pyjama stripe' row mixtures and 'chequer-board' groups. Plantations were neither always in harmony with the landform nor in scale with landscape.

Modern forestry has a quite different approach to forest design. Afforestation, when designed thoughtfully against straightforward rules (noted *infra*) should create a pleasure to the eye for a generation or more. Large open upland landscapes need relatively large woodlands tailored to the landform. Blocks, whether square, rectangular or circular, on a bare hillside are to be avoided. The legal boundary need not be a rigid constraint; it can be softened by leaving parts unplanted. The alignment of roads and cultivations (particularly where ploughing is used instead of ripping or scarifying), the delineation of boundaries between species, and the rate at which large blocks are planted up, can often add variety to the landscape, particularly in the short term. Careful attention to plantation margins, and leaving of some areas in a natural state particularly around watercourses are other measures that can be considered.

Woodland edges should be made to look as natural as possible and be varied in scale with the landscape by: detailed shaping of the edge, using species with different growth rates to vary tree height on the edge, widening tree spacing at appropriate points towards the edge of the wood (anything less than three to five times normal spacing is usually insignificant), establishing irregular outlying groups and single trees, and thinning to create a softer edge to felling areas. Woodlands alongside roads and footpaths should give an interesting sequence of views with a succession of varied spaces. A motorist cannot take in a wealth of detail so design has to be on a broad scale related to the average speed of the traffic. Species, spacing and edge distance should be varied. Open space should be used to allow views into or over the woodland. Openings should be acute angles; narrow openings at right angles will be missed. Views are important landscape features requiring that open space be retained. Viewpoints should be identified and incorporated in the woodland design. Trees in the foreground and middle ground should provide a setting for the view rather than compete with it. A framing of trees is appropriate for a view of a particular feature, but with panoramic views there should be little restriction in the foreground or the middle ground.

The Forestry Commission has published in addition to its already noted *Forest Landscape Design Guidelines* (1989) the undernoted guidelines for woodlands in the landscape (Environmental Leaflet, 1989) comprising minimum standards required of all applicants for grant-aid. They should be read in conjunction with Figure 28 on page 482.

1   Walls, hedgerows or belts of trees may be a stronger pattern, especially for smaller woodlands in areas of weaker landform. New woods should conform to this pattern and regular shapes may be easier to fit in.

2   The upper margin is usually a prominent part of the design. It should rise in hollows and fall on convex ground or spurs and reflect the landform quality, with jagged shapes in rugged country and smooth shapes in rolling

topography. The upper margin should be positioned so that any open ground above the woodland is of sufficient size to reflect the scale of the hill cap or ridge.

3   If fields at the lower margin have a strong hedgerow pattern this can be reflected by extending broadleaves into conifer woodland, especially up hollows and gullies. Where there is little or no hedgerow pattern, adopt a lower margin of gently curving lines, rising in hollows, falling on spurs.

4   Unless there is a strong pattern of hedgerows, walls or dykes, side margins should be gently curving and preferably on the diagonal. Avoid straight vertical edges. Woodland should have a natural place to stop, such as a stream, crag or depression.

5   Forest roads should be located to minimize visual impact: follow landform as far as practicable, cross skylines at the lowest practicable point, and minimize the size of cuttings and embankments.

6   Unplanted streamsides should be irregularly shaped with the edges designed to link across the stream at key points. The aim should be to have an irregular distribution of broadleaves with about 50% of the stream in full sunlight, the rest receiving dappled shade.

7   Intrusive fence line effects can be reduced by positioning fences where there is least impact, for example in hollows or away from skylines, running fences diagonal to the contours, making fences follow landform, following the woodland edge closely or else leaving an intervening space of appropriate scale and shape.

8   Ploughing should be planned so that upper margins and unploughed areas are irregularly shaped to follow landform, unploughed access routes follow natural breaks in the slope and are kept to a minimum, and furrows follow natural changes of slope in gentle curves. Cultivation by rippers, scarifiers and mounders is less obtrusive than ploughing.

9   The species pattern should reflect the broad pattern of the landscape, the pattern of ground vegetation, or the local landform, whichever is the more dominant in the landscape. Too much variety of species can appear confusing, an interlocking pattern of a few contrasting species looks best. Avoid straight lines where species meet, especially on slopes.

10 When planning felling, coupes should be asymmetric and irregular, shaped to follow landform with edges diagonal to the contour, rising in hollows and descending on spurs, and smaller on lower slopes and larger on higher ground. Coupes should be comprehensively planned for the whole landscape composition. Timing of felling of adjacent coupes is vital. Wait for the new crop to close canopy when differences of colour and texture become apparent. Eliminate intrusive straight lines and other geometric shapes in the previous crop.

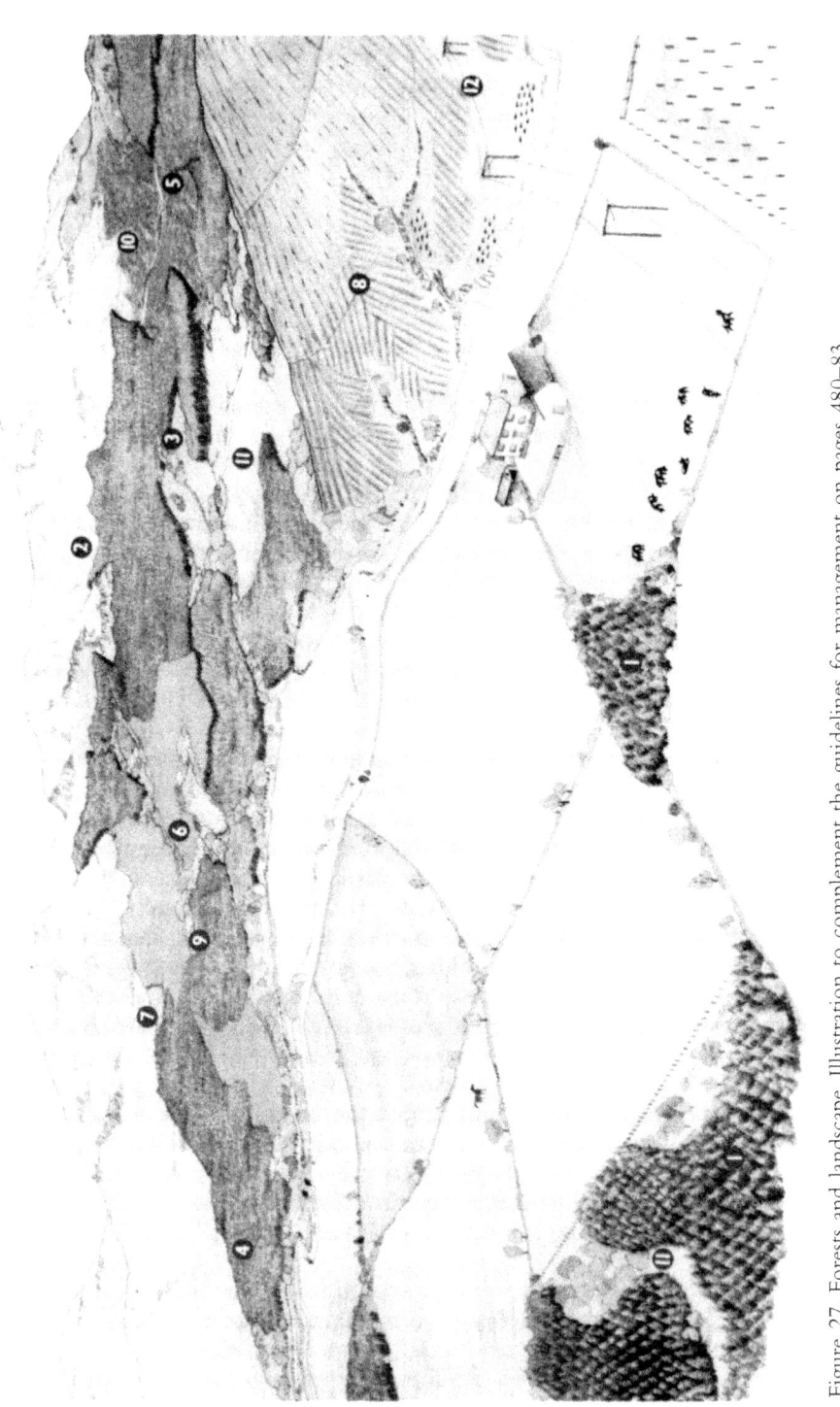

Figure 27. Forests and landscape. Illustration to complement the guidelines for management on pages 480–83
Source: FC Environmental Leaflet, 1989: *Forests and Landscape*.

11   Open space is of great value for visual and ecological diversity. This is especially important in extensive upland forests where diversity of species is difficult to achieve. Farmland within a forest may echo patterns of fields and enclosures outside the forest. In all woodlands, sufficient open space should be left around prominent rocks and crags so that they are not obscured when trees are fully grown.

12   Power lines should seem to pass through a series of irregular spaces and the woodland should appear to meet across the open space in some places. Trees can be planted closer to the line opposite pylons than in mid-span while small trees and shrubs can be grown closer still.

For some Areas of Outstanding Natural Beauty (for example, the Cotswolds, which comprise three main landscapes – steep scarp slope, deep slope valleys and plateau) guidelines for new planting have been devised by interested authorities. For new planting on farm land in general, guidance is given by Insley (ed.) (1988) and Hibberd (ed.) (1988).

It is important to find a balance between the economic demands of forestry and the requirements of the landscape. Afforestation need no longer be an unacceptable blot on the landscape: sensitively designed it can improve the landscape and add to the attractiveness of the environment. Furthermore, much can be done in regard to existing plantations. One of the more immediate visual demonstrations of good forest design is in rectifying the poor shape and species mix. Such restructuring involves the staged felling and restocking of trees planted half a century or more ago, to introduce a wider range of age classes and species. This process is particularly welcomed by today's foresters because it enables them to create a forest which is in harmony with the landscape and being constantly renewed and improved, increasing diversity of species and conditions which will also benefit wildlife, amenity and recreation.

However, attention to landscape design can be costly in revenue forgone if not properly planned at the outset. Correcting past mistakes is more expensive than careful integrated design in the first place. It is difficult to put a value on the amount of goodwill that is generated by attention to the landscape, but it is necessary for foresters to give consideration to the extent to which the owner is prepared to forgo revenue in order to carry out more than the minimum relevant measures. Woodland management which caters for landscape may increase costs and/or decrease income from wood production. It appears equitable that financial recompense should be forthcoming for any social benefits that sensitive management creates (Lorrain-Smith 1991). This is one of the factors recognized by current woodland grant schemes, though there is much debate as to whether the recompense is adequate, or sufficiently site-specific. The attention given to landscape planning is becoming evident, both in small and large scale planting. Visually attractive woodlands often command a better sale price than less attractive but more economic ones.

Landscape design being a complex subject professional help from qualified landscape architects may be needed in certain situations. The Landscape Institute is a helpful source of information. Comprehensive landscape plans are necessary when substantial scale afforestation is undertaken or when extensive felling and restocking is planned. The patterns established at these times may persist for half a century or more. 'Countryside Stewardship', a new national countryside initiative announced by the Countryside Commission in December 1990, will be a major force for landscape and wildlife habitat conservation throughout England, though not yet in Wales. The economics of landscape are briefly touched upon in chapter 14.

6. Management for recreation

Forests and woodlands are a major recreational asset. The Forestry Commission has a long-established policy of encouraging recreational use of its forests wherever this is feasible. People are very welcome on foot and on cycles in all of them, provided this does not conflict with forestry operations and that there are no legal or other constraints on public access. It lets sporting rights commercially over a substantial proportion of its forest estate. Some private woodlands are also used for commercial and informal recreation as well as for hunting and shooting (however, public access is not always so openly welcomed).

The extent to which opportunities are provided for different forms of recreation depends on the location of the woodland relative to towns, cities and tourist routes and on the objects of management. Woodland management is patently a desirable land use to meet the strong public demand for recreational facilities in the countryside (Irving 1985). People enjoy visiting woodlands as one of their leisure activities, particularly where there are picnic sites and way-marked paths. They like to see and walk through woods particularly those with variety of age-class and species, chiefly because of their appearance, changing with the seasons. The flora and fauna found therein, even when management for wildlife is not being actively promoted, are a major attraction. Birdsong, small creatures rustling in the undergrowth, the beauty of spring or summer flowers, the autumn colours, all contribute to people's enjoyment.

The Forestry Commission continues to develop recreation opportunities in the forest, encouraging more use of the extensive network of tracks, trails, open spaces and water areas – principally for the enjoyment of quiet pursuits. In addition where appropriate the Commission provides camp sites and forest visitor centres. All the facilities are designed to be in harmony with the forest environment and not in conflict with conservation. Much woodland in private ownership supplements the notable achievements of the Commission. Areas of great landscape and recreational importance are frequently identified in national and local authority plans; and they attract numerous visitors.

The recreational importance of a woodland depends on its age, structure, visual quality, accessibility, nearness to towns, and the importance of tourism in the locality. Woodlands in accessible countryside may receive some visitors, and possibly substantial numbers where there is a public right of way. Access is important. Most countryside visitors like to know where they are free to go, and dislike trespassing. Unintended trespass can be reduced by way-marking paths and providing stiles. This, together with careful siting of car parks and picnic places, however modest, encourages people into less sensitive areas and thereby reduces risk of fire and disturbance of wildlife. Conflict can be largely avoided by such zoning of the woodland for recreation. All public rights of way should be kept clear, with stiles or gates provided in fences, and undergrowth and low branches cut back along each pathway. It is sometimes important to clear access through woodland, particularly newly-planted woods, to allow walkers to reach hill or other open country lying beyond.

In managing for recreation, the first stage is to consider what people want. What facts are available about general trends? For instance national surveys show that the majority of people arrive at the forest by car, travelling distances of up to 50 km from home, some, particularly groups of children, arrive by coach or public transport, and others simply walk through the forest. The next stage is to find out who are the people who want something from any particular woodland and what it is that they want. Then to appraise the recreational potential. From this may emerge a tentative decision to provide recreational facilities of a kind which the woodland can support and the owner is prepared to provide. Not all types of recreation can be accommodated in one area, and it may be necessary to zone recreational activities that are incompatible (e.g. walking and horse riding). Where facilities are provided, whether on a simple scale or an elaborate one, they should be tidy and attractive, yet designed to blend into the landscape without being obtrusive. Care should be taken to ensure that these facilities are not over-used, as excessive use might destroy the very facility which people come to enjoy.

Recreation provides an opportunity for foresters to encourage owners to appreciate and care for the woodlands, but particularly in areas where population density is high there is a considerable cost to bear, not only in terms of litter and of wear and tear on the forest floor but also the danger of damage to the wood in other ways such as fire. However, public interest in recreation in woodlands has built up over the last few decades and will go on increasing. Foresters have to live with this and can seize the opportunity taking recreation into account alongside conservation, silviculture and harvesting. In some regions (e.g. south-east England) it is sometimes difficult to follow an adequate form of woodland management due to unacceptable trespass and damage. However, there are many ways to minimize damage, and even benefits to be gained, when the public are invited in and welcomed subject to adequate provision. A good example is the Boughton Estate, and the Living Landscape Trust founded by the Duke of Buccleuch and the Earl of Dalkeith in 1985 promoting a greater awareness among the public of the vital place of the countryside in the national heritage (the whole resulting in a contribution of timber and food production, conservation, recreation and sporting).

Advice on recreational planning in private woodlands can be obtained from the Countryside Commissions, Regional Tourist Boards, Sports Councils, some local authorities, the Ramblers' Association and the Foresty Commission. *The Forestry and Woodland Code* published in 1985 by TGUK provides notable guidelines. A report by the Royal Institution of Chartered Surveyors, *Managing the Countryside* (1989), is likewise helpful. Useful information on recreation is also provided by Hibberd (ed.) (1991), and the Forestry Commission intend to publish forest recreation guidelines. The economics of recreation are briefly touched upon in chapter 14.

7. Management for wildlife conservation

Wildlife conservation is diverse and complex, and few foresters in managing it can expect to be an expert in more than a few of its fields. Much helpful information and experience is available and should always be sought. The relevant literature is immense. The outcome of information and recommendations has notably resulted in Forestry Commission 'Forest Nature Conservation Guidelines' (1990), referred to throughout much of this section.

Woodlands provide valuable wildlife habitats for a great variety of plants, lichens, fungi, mosses, mammals, birds and insects. Wildlife conservation is the sustainable management of habitat to maintain or enhance their value (Steele 1972; Harris 1982; TGUK 1986; Springthorpe and Myhill (eds.) 1985; Gilbert and Dodds 1987; Marren 1990; Watkins 1990; Forestry Commission 1990). Woodlands are collections of woody plants of interest in themselves and provide a great number and variety of habitats which are vital for a substantial proportion of wildlife. Broadleaved woodland in particular is valued (Rackham 1980; Peterken 1981; Evans 1984). Most native trees are broadleaved and they make up the great majority of ancient semi-natural woodland types recognized botanically. Many have developed a stable woodland flora and often support plants, as well as mammals and birds, which can only survive in such situations. Most such woodlands occur on rather fertile sites and hence support richer floras than plantations generally established in less fertile conditions.

An ebb and flow of mammals and birds occurs as woodlands are created and change in structure with time. The interactions between such wildlife and their woodland environments are important. Likewise are the prediction and prevention of damage to the trees on the one hand and the management of the wildlife for conservation on the other. The interaction of deer species with woodlands is of major concern: populations have a home range and may move in response to forestry operations; and different stages of woodland development, as well as climate and site factors, dramatically affect reproductive performance, body size and numbers. Controlling grey squirrels to prevent serious bark-stripping has necessitated detailed studies of their biology; so too has the conservation of the red squirrel and its habitat. There are distinct relationships between small bird communities and plantations of conifers. Also between numbers of raptors and the availability of their prey species.

Organizations and individuals deeply concerned with wildlife conservation have expressed alarm at the depletion of woodland habitats. Criticism has related to felling of mature broadleaved stands without due concern for wildlife, replacing ancient semi-natural broadleaved stands with plantations, partially or wholly of conifers, reducing the abundance of native species (especially oak), failing to manage semi-natural woods leading to their impoverishment, clearance for agriculture or other non-forestry purposes, and permitting inappropriate grazing. Enlightened critics, however, acknowledge the need for management for timber production and the legitimacy of harvesting merchantable crops. Equally enlightened foresters seek to achieve a reasonable balance between wood production and nature conservation; and they may have a statutory duty to do so (Wildlife and Countryside (Amendment) Act 1985). Incentives to private owners are available by way of grant-aid (see chapter 15). Besides the Forestry Commission, the successors to the Nature Conservancy Council, and the National Trust, organizations and agencies paying particular attention to conservation include the Woodland Trust (national), Project Sylvanus (Devon, Cornwall and Somerset), National Small Woods Association (England), ESUS Woodland Management (East Sussex and Kent), Cumbria Broadleaves Group, Coed Cymru (Wales) and the Forestry Trust for Conservation and Education (England). Others are noted in Forestry Commission 'Forest Nature Conservation Guidelines' (1990).

Woodland animals. Many small and large wild animals, as noted in chapter 6, depend on woodlands for food, shelter and cover to help them escape from predators (Steele 1972; Springthorpe and Myhill (eds.) 1985). Regeneration, tending, and harvesting are disturbances which favour some animals but may create problems for others. Silviculture in traditional woodlands, and more particularly in plantation afforestation, change and create a variety of habitats for animals. Some mammals and especially deer, rabbits and grey squirrels can cause serious damage to trees and have to be controlled.

Afforestations, dissected by rides and other open spaces and associated edges, are becoming increasingly more varied in age and species composition as they mature. Increased diversification of habitat, compared with the extensive grazings which many upland forests partially replaced, provides a significant contribution to wildlife conservation, particularly when the habitat replaced was of low conservation value. The value for wildlife habitats will be enhanced by the planting of smaller woods in lowland by farmers and others, and by the opportunities for new planting 'lower down the hill'.

Effect of silvicultural systems. Silvicultural systems (see chapter 7) and management decisions (see chapter 14) can have a great impact on nature conservation. *Uneven-aged (irregular) stands* offer the permanence required by many animals and plants. Primrose, wood anemone, wood sanicle and other 'ancient woodland indicator' species are able to survive within an ancient woodland (Helliwell 1982). By contrast, *even-aged (regular) high forest* under the clear-cut and replant system can be less conducive to wildlife. In the closed canopy before each thinning the plant life may be virtually eliminated, particularly so with plantations of shade bearing or shade casting species. (The flora can in some measure be the same under conifer canopy as in deciduous woods: it is the density which varies rather than the content.) At each thinning more light is increasingly admitted to the forest floor (facilitating the breakdown of litter) and the plants which return are those with long-lived seeds (in the 'seed bank'), or which are easily dispersed – such as foxglove, willowherb and rushes. Much of the other original ground vegetation will not return. The same applies to some beetles, snails and other invertebrates which have difficulty in migrating any substantial distance to find suitable conditions if the woodland in which they are living is suddenly changed.

The practice of artificial restocking after clear-felling precludes any ecological or historical continuity with the previous crop. Weeding is usually intensive, often involving the application of herbicides; alternatively, uncontrolled weed growth may become rank and monospecific (particularly bramble and bracken), which again will impede the development of a rich woodland flora. (Sparing ivy, wild clematis, honeysuckle and sallow will benefit wildlife.) However, all these disadvantages can be reduced by the use of small coupe or group-fellings. This will lead to a mosaic distribution of age-classes in the woodland giving continuity and diversity which are desirable for wildlife; and can lead to an increase in the species diversity of subsequent crops if the increased amount of natural regeneration is accepted. The smaller the coupes or groups, generally greater is the benefit to conservation. Stands managed under a *uniform shelterwood system* can have many features in common with the foregoing.

Coppice and *coppice with standards* – though they are not 'natural' systems – are often regarded to be among the most desirable silvicultural systems from the point of view of nature conservation. In coppice, the alternation of light and dark phases on a short cycle (10–30 years, rather than 50–150 for even-aged high forest broadleaves) encourages a rich woodland flora. The dense thicket-stage (at seven to ten years) is a good habitat for a wide variety of woodland birds, as well as dormice in south Britain. The fact that only small parts of a coppice may be cut annually (in 'cants') results in a mosaic of group shelter which gives good habitat diversity within a single wood. Many traditional woodland spring flowers are at their best within a year or two of coppicing; they decline after about five to seven years. Coppice with standards has somewhat similar relevant attributes to those of pure coppice; moreover it provides a second layer of canopy as well as continuity of woodland conditions, particularly if it is uneven-aged. Conservationists make strong appeals for the retention of coppice and coppice with standards sytems in the knowledge that they are the traditional management of much broadleaved woodland (Rackham 1980; Peterken 1981; Fuller and Warren 1990; Warren and Fuller 1990).

The *selection system* is recommended for its continuity of habitat conditions and lack of sudden change. However, such stands are not necessarily 'natural' being the product of sustained, skilled and highly regulated forestry. From within, the abundance of maturing and mature trees, with isolated well-developed crowns, standing above younger trees makes them very attractive; however, to some people it appears too uniform and monotonous. The lack of diversity from one part of the wood to another and of cyclic changes of light and dark phases are also possible disadvantages for wildlife conservation particularly for many woodland butterflies. However, the multi-storied structure is highly desirable for some breeding-birds, and within any stand there will be small-scale spatial variations of shading and hence of flora and shrub layer communities. The intimate mixing of old and young trees, and the continuous woodland conditions, are ideal for sustaining a diverse epiphytic flora.

In *group systems,* wildlife conservation is favoured by the two attributes of continuity of forest cover and diversity of structure and species. Of all the silvicultural systems, the group shelterwood and group selection are therefore increasingly being recommended by conservationists for ancient semi-natural woodlands. The systems have never been traditional management systems in this country. They may, therefore, be regarded as less desirable than a reinstatement of coppice with standards. However, a group-felling system would give a similar mosaic structure as that produced by the 'cants' of a coppice system, which is so valued for nature conservation, and might produce timber of greater value.

Using *natural regeneration* brings many wildlife conservation benefits (Peterken 1981; Evans 1988), favouring species already on the site, tending to generate mixed stands, with a more irregular structure than plantations, the natural genetic variety can be better maintained, and the natural distribution of tree species in relation to soil types is favoured with the whole habitat better preserved. A disadvantage is where aggressive species are a threat, of which rhododendron is the worst. Where the principal aim is wildlife conservation, natural regeneration of woody species should be encouraged; and when among planted trees it should only be removed where it is a real interference, not because of a desire for tidiness (Evans 1988).

Large old trees often support an abundance of flora and fauna; they are particularly important for some epiphytes and represent irreplaceable habitats for some highly specialized invertebrates. They may also have holes for nesting-birds and bats. Retaining part of a stand for many years past its conventional rotation will lead to some loss of revenue; and the leaving of individual large trees can have some – but only insignificant – adverse effect on the growth of the restocked crop. However, this is recognized as good wildlife conservation practice.

Conservation of habitat. In lowland (chiefly broadleaves) and much upland (mainly native Scots pine and birch) where conservation of the existing ecosystem is considered of great ecological importance, many woodlands are designated as National Nature Reserves (NNR), Sites of Special Scientific Interest (SSSI), Forest Nature Reserves (FNR), or Local Nature Reserves administered by the Forestry Commission or local authorities, or are under the management of organizations such as woodland trusts and local naturalists' trusts). Many of these sites are multispecific, of predominantly native species, and have supported woodland for many centuries. During the past few years the Nature Conservancy Council compiled an Inventory of Ancient Woodland Sites, distinguishing those which continue to carry semi-natural woodlands (see chapter 4). Ancient semi-natural woodland always merits conservation as a community of plants and animals.

Important elements of wildlife conservation value are: (i) age range of trees and shrubs, and the structural layers present, including a stock of old trees and some dead wood; (ii) range and size of habitats within or near the wood, such as open glades, rides and forest edges, streams and ponds, wetlands and bogs, and rocks and crags; (iii) range of species and communities of plants and animals present, their extent and distribution, including the semi-natural communities of trees and shrubs; (iv) unusual species or features, e.g. rare plants, or a woodland type at the limits of its natural range; and (v) continuity of management and woodland structure. In addition, geological, physiographical and archaeological interests must be borne in mind.

Maximizing the wildlife conservation value of a woodland will only occasionally mean non-interventions to 'allow nature to take its course'. Four distinct aspects to management are: (i) on a small scale, improvement of habitats and ecological diversity by adopting often simple changes in conventional silviculture (e.g. one-side cutting of rides alternating each year, not planting right up to the edge of a stream, retaining a small percentage of trees past rotation length, and providing open space and edges); (ii) on a large scale, maintaining certain woodland types, either because they are an important ecotype or are ancient in the sense that the land has been woodland for many centuries; (iii) diversifying age-class structure by adjustments to planting or restocking programmes; and (iv) managing a habitat specifically to conserve rare plants and/or animals at risk from unsympathetic treatment as may result from normal forestry operations.

Conservation principles and guidelines. Well-designed, well-managed woodland contains a diversity of habitats for wildlife. Some will have special conservation value and will require particular attention, but all have wildlife value which needs to be understood and managed. Planning of afforestation should be preceded by an inventory of the conservation value of the area. Opportunities should then be sought to create diversity by leaving open spaces, by accepting natural regeneration, by planting broadleaves and understorey species and by other means.

(For the implications of and guidance on wildlife conservation in new planting of farmland, see Insley (ed.) 1988 and Hibberd (ed.) 1988.) Similar considerations apply in existing woodland and where there will be new opportunities to create structural diversity and to allow some old growth. The Forestry Commission has published the following guidelines for conservation in 'Forests and Conservation Leaflet' (1989) comprising minimum standards required of all applicants to the Woodland Grant Scheme and the Woodland Management Grant Scheme. They should be read in conjunction with Figures 29 (a) and (b) on pages 492 and 493.

1. All well-designed woodlands should have a proportion of open space within them.

2. Special areas such as streamsides, wetlands, flushes and grasslands rich in plant species, should be left undisturbed. They should be identified and marked out early in the planning stage.

3. An understorey of woody species such as hazel, hawthorn, holly or juniper is valuable for wildlife and game, particularly at the edge of the rides or of a wood.

4. In the lowlands, many plants, animals and insects benefit from the presence of sunny rides. Ideally they should have a width equal to eventual tree height and preferably be oriented east-west. The rides may need to be maintained by mowing at intervals.

5. All upland conifer plantations should contain some broadleaves. A minimum of 5% is required but more may often be established to advantage. Even individual trees and small groups have conservation value but in large upland forests, large groups may be appropriate. Spacing between trees should vary and space should be left for future natural regeneration. Wherever practical, native trees and shrubs typical of the locality should be used. Any existing broadleaves should be retained, and space left around them to allow for adequate development.

6. Streamsides always offer opportunities for diversity of habitat. About half should be left open, and the remainder should be planted sparsely with light-foliaged species. Open glades along streamsides are particularly valuable for wildlife. Where an existing woodland is too close to streams consideration should be given to removing some of the trees at an early date while trees are still small, or by thinning if windthrow risk permits.

7. Forest operations should be scheduled as far as possible to avoid breeding sites in season.

8. Special attention is required to ancient and semi-natural woodlands and to ancient woodland sites. (See the guidelines for the management of broadleaved woodland, stated earlier in this chapter.)

9. An over-grazed, draughty woodland can be much improved by excluding grazing animals from at least part of the wood and by planting or encouraging the natural regeneration of trees and shrubs.

10 Felling coupes should be planned to increase diversity of age class and structure. Use as small a coupe as practical in terms of landscape and management. Restocked areas should be well established before adjacent areas are felled. Where practicable, a permanent network of broadleaves should be retained throughout the woodland. Old, and very old trees, are particularly valuable for wildlife and a reasonable number of such trees should be retained and allowed to disintegrate. Dead wood left to rot in shady locations is valuable for insects and fungi.

11 Planting should be kept well back from the shores of lochs and lakes. The distance should always vary. A minimum of 20 metres will usually be appropriate with up to 50 metres or more in some places, especially on southern shores.

12 Where open space is lacking in existing woods, opportunities should be sought to create spaces at the earliest practical opportunity, especially near crags, around pools or lochs, as deer glades and as rides. The vegetation in open spaces, including natural regeneration, may need management techniques such as mowing or grazing, or exceptionally, by burning.

Detailed information is provided in the Forestry Commission's already mentioned 'Forest Nature Conservation Guidelines' (1990), which give practical guidance on the maintenance and enhancement of the wildlife conservation value of existing woodlands and on taking opportunities to develop wildlife value in new planting. What is being managed is the totality of the woodland area, and not merely the trees within it. Preparing a conservation plan is a useful way of helping to ensure that sites where conservation is the primary objective of management are clearly identified, mapped and given due consideration. The main purposes of such a plan are: (i) to identify, describe and evaluate the present and potential conservation value of the woodlands; (ii) to define conservation aims for sites of special interest and prescribe management regimes which will achieve the aims; (iii) to define how operations should proceed in relation to conservation aims in the wider woodland or forest; and (iv) to develop a programme of conservation action.

Obtaining a yield of forest produce and managing the woodland for conservation need rarely be in direct conflict. Even from woodlands where much attention is given to wildlife, there can be a substantial yield of timber, though rarely is it the silvicultural maximum the site might theoretically produce (Lorrain-Smith 1991). This difference is the 'price' of conservation, but comparing on economic grounds the timber yield forgone with rarity, abundance or variety of wildlife, is to use only one criterion (Evans 1984). Woodland management which caters for nature conservation may increase costs and/or decrease income from wood production and/or sporting. It appears equitable that financial recompense should be forthcoming for any social benefits that sensitive management creates – one of the factors recognized by the Woodland Grant Scheme and the Woodland Management Grant Scheme.

Ferris-Kaan (1991a) summarizes the methods of woodland management for conservation used by non-government organizations, particularly the National Trust, the Woodland Trust, the Royal Society for the Protection of Birds, and County Trusts. The economics of wildlife conservation are briefly touched upon in chapter 14, by Jeffery *et al.* (1986), and by Lorrain-Smith (1991).

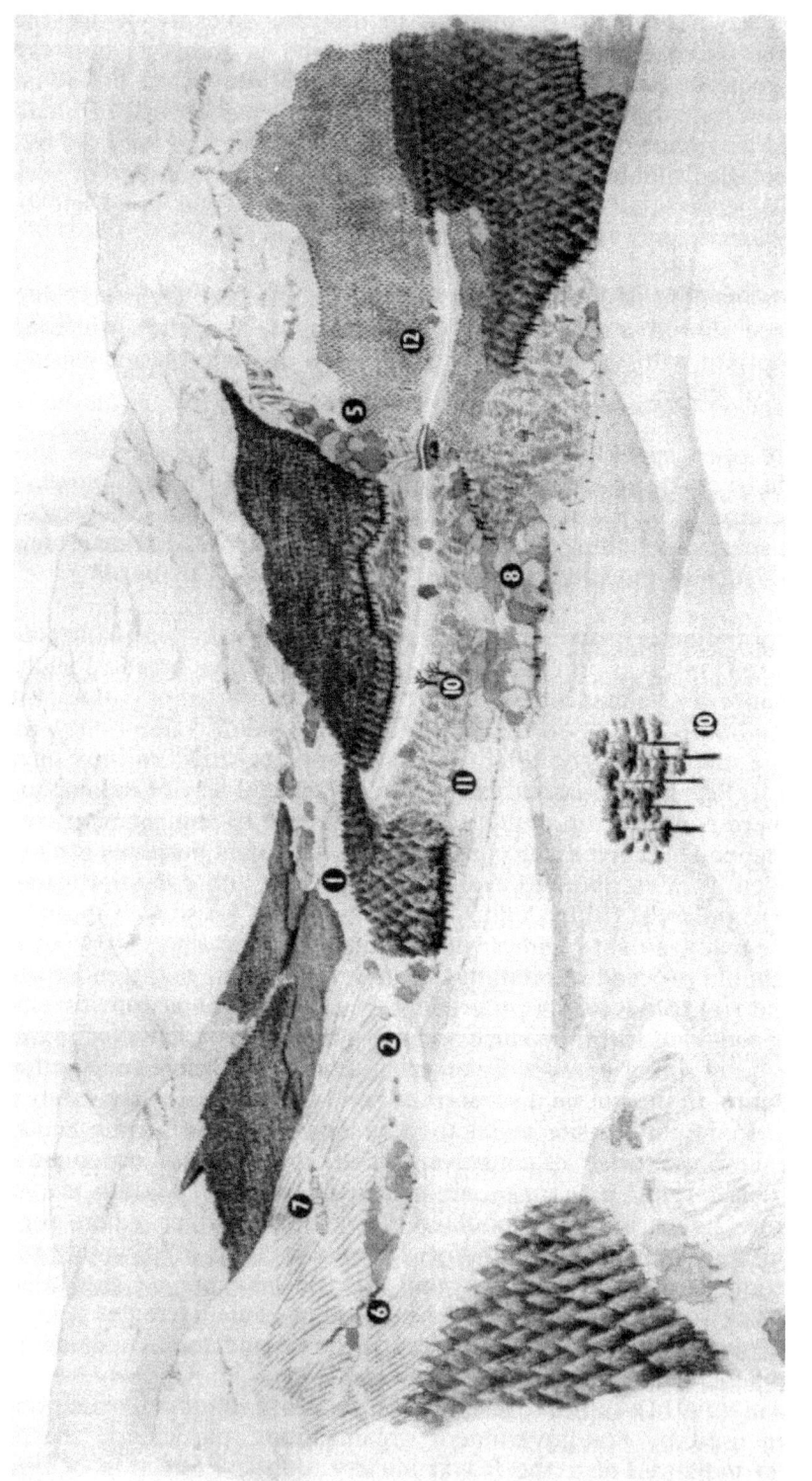

Figure 28(a). Forests and conservation I. Illustration to complement the guidelines for management on pages 490–91
Source: FC Environmental Leaflet, 1989: Forests and Conservation.

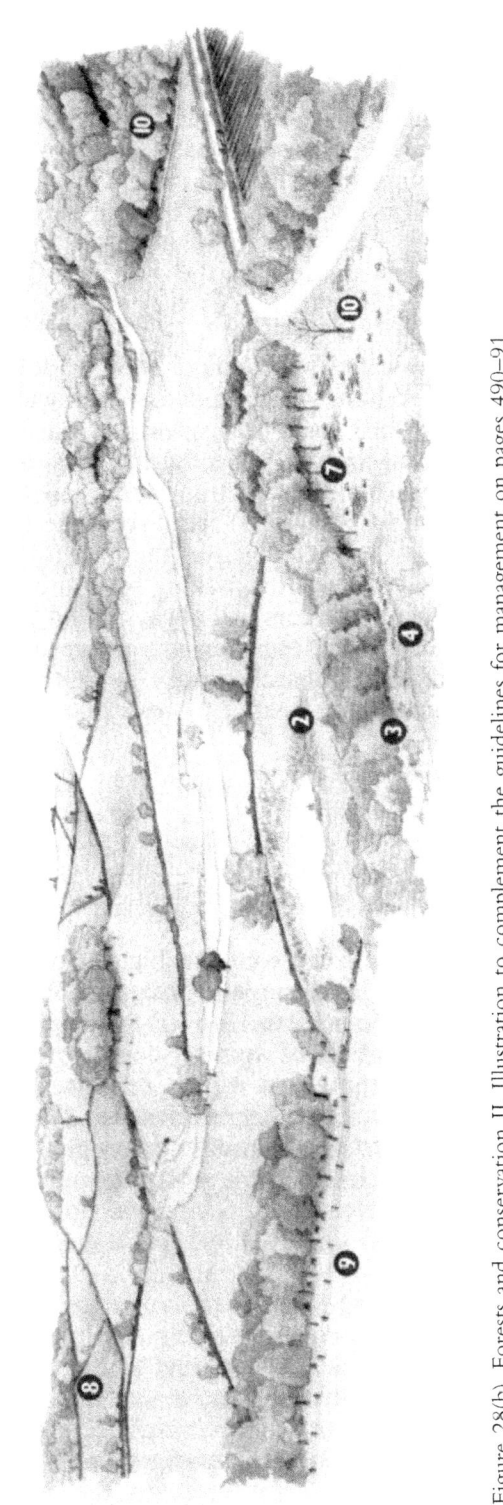

Figure 28(b). Forests and conservation II. Illustration to complement the guidelines for management on pages 490–91
Source: FC Environmental Leaflet, 1989: *Forests and Conservation*.

The concept of positive wildlife conservation within productive woodland is gaining ground as more foresters begin to understand the requirements of wildlife. There is an awareness of the importance of positive management to diversify age structure and tree species in medium aged woods. Foresters have a close working relationship with the Countryside Commissions, the successors to the Nature Conservancy Council, the Farming and Wildlife Trust (managing body for sixty-five FFWAGs), the Royal Society for the Protection of Birds and other conservation bodies, continually exchanging ideas that help maintain and improve wildlife habitats and protect endangered species. In the Forestry Commission this cooperation is effected by conservation panels in each Forest District. Well-presented information is available relating to the different forestry stages, and their associated wildlife. Several 'Duke of Cornwall Awards' have been made to foresters for the management of commercial woodlands in conjunction with wildlife and landscape interests. Similarly 'Scottish Woods and Forests Awards' have been made by the Royal Highland and Agricultural Society of Scotland each year (since 1986); also the 'Dulverton Flagon Award' for forestry and woodland management under TGUK's *Forestry and Woodland Code* (1985). However, the balance in emphasis may still be changing. Attitudes have changed in the past two decades, and may continue to do so. McPhillimy (1989), after making seven notable case studies in Scotland, avers that even the best foresters will always admit that there is more to learn and more that they can do to further secondary objectives – wildlife conservation, landscape, recreation and archaeology. As to the last, see Proudfoot (ed.) (1990) and Forestry Commission Leaflet 'Forests and Archaeology' (1990).

Managing wildlife in woodlands. Assessment of wildlife damage in woodlands as a basis for evaluation management options is discussed in chapter 6. The main objectives of managing wildlife are: (i) preventing damage to forest and farm crops by animals resident in woods; (ii) utilizing forest wildlife for sport, meat, recreation or amenity; and (iii) conserving species and communities which are becoming rarer locally, regionally and nationally.

The wildlife management objectives given the highest priority in commercial timber production should usually be damage prevention. The methods by which any or all of these objectives can be achieved are: (i) by keeping damaging wildlife away; (ii) by controlling the numbers of animals doing, or expected to do, the damage; and (iii) by altering components of the ecosystem to reduce risks of damage or enhance survival of crops or other species. Foresters need to be able to predict possible problems arising from the traditional sequence of forest practices. Wildlife management can only work effectively when priorities among the management objectives have been clearly stated. They are not necessarily mutually compatible.

Woodland bats. These fascinating small mammals depend to some extent on woodland habitats for foraging and roosting. Their habitat deserves appropriate and sympathetic management: Mayle (1990a, 1990b) discusses how this can be achieved. Bat habitats can be improved by increasing the structural and species diversity of woodlands, planting native broadleaves, retaining a few old trees beyond rotation age, and by the sensitive management of waterside habitats, scrub and open grassland. All bats are protected by the Wildlife and Countryside Act 1981. Recommended boxes are discussed by Springthorpe and Myhill (eds.) (1985).

The approach to nature conservation. All woodlands, including their associated open ground, are of value for nature conservation. The key to good conservation practice is to maintain and add to the range of habitats. Forestry Commission 'Forest Nature Conservation Guidelines' (1990) supplements its earlier noted 'Forests and Conservation Leaflet' guidelines (1989) by stressing:

- Give priority to maintaining and, where practicable, to extending existing habitats of conservation value;

- Aim to conserve and improve conditions for a whole community of plants and animals rather than for particular members and groups, unless urgent action is required to conserve rare, sensitive or endangered species;

- Aim for a balance of wooded and open habitats. Open ground is often more valuable than a change of tree species, both for nature conservation and for landscape purposes. In the uplands it may often be preferable to retain or create open ground habitat than to plant broadleaves;

- Develop a greater diversity of structure where wooded areas adjoin open ground, be it within or on the edge of the wood, so as to effect a gradual transition between them;

- Aim to create a mixture of stands of different ages throughout the woodland. This automatically increases the quantity of the valuable edge habitat;

- Diversity is, however, not always a desirable goal. Woodland diversity is naturally greater where soils are more fertile, the climate more favourable, towards the centre of the natural range of the woodland type, and in the later states of ecological succession. Woodlands comprising pioneer species, on infertile soils, and/or towards the altitudinal or latitudinal limits of the range are of conservation value because of the structure of species composition and, where extensive, because of the structure at a relatively large scale.

Across Britain as a whole the highest nature conservation value is assigned to woodlands which appear to be nearest their natural undisturbed state. Since all have been subjected to management to some extent over many hundreds of years there are no absolute models, but it is clear that high value can be attached to woodland consisting of native species, especially when the assemblage of species appears close to that which probably occurred naturally. Such species are host to a wealth of indigenous fauna and flora. Semi-natural woodlands are more valuable if they are large enough to ensure the long-term viability of the populations of the plants and animals within them. It is these considerations which give value to such semi-natural systems as the Caledonian pinewoods *(infra)* or the sessile oakwoods of the west coast of Britain.

The first priority in conservation planning is therefore to identify semi-natural woodland and to prescribe a regime which will ensure its healthy continuance. It will often be the case that the exclusion or restriction of grazing animals, both wild and domestic, will be a prerequisite to the desirable renewal of the woodland flora. The removal of trees or shrubs that are not native to the area may also be an early requirement, especially if they are threatening to suppress the native species. Opportunities should be sought for extending such woodland by fencing-in adjacent open ground or sparsely stocked areas, and undertaking such silvicultural operations as are appropriate for securing regeneration. Natural regeneration is preferred but the practical difficulties and cost may rule this out.

A second priority is to consider the future of plantations on ancient woodland sites. Having been woodlands since early times, such sites may still retain ground flora and other species of particular conservation value. The aim of forest management should be to identify these features and to preserve and enhance them. Where nature conservation value is particularly high, these woodlands should be managed on the same lines as ancient semi-natural woodlands.

Where planting is carried out with nature conservation as a primary objective, the ideal is to use trees and shrubs which are native in the area. A detailed guide to the appropriate species, together with a specification for use compiled by the Nature Conservancy Council, is given in the Appendix to Forestry Commission 'Forest Nature Conservation Guidelines' (1990).

Native pinewoods. Detailed guidelines for the management of native pinewoods are set out in Forestry Commission 'Native Pinewoods: Grants and Guidelines' (1989). The native pinewoods, among the least modified woodland areas in Britain, are an irreplaceable reservoir of adapted genetic stock of plants and animals. They should be managed so as to maintain and enhance their ecosystems and aesthetic value, enlarge their area, particularly of the smallest pinewoods, maintain the genetic integrity of the native population and, so far as practicable, maintain identifiable sub-populations, and produce utilizable timber.

Small lowland woods and *shelterbelts.* Small woods, when considered in terms of broad ecological principles, may be generally less valuable to wildlife than large woods because they are likely to contain fewer species, smaller populations of species, fewer woodland types and scope for less range of structure and habitat. However, normally set in intensively managed countryside, these small woods are often among the few areas of relatively undisturbed land, providing a sanctuary for many species of wildlife and retaining protected natural landforms or important geological sites. Often created within the last 100–200 years, they are predominantly broadleaved or mixed woods, though in some upland areas there is a long tradition of conifer shelterbelts and blocks. Many of these small woods were created, and are managed, for game shooting. Woods designed for pheasants, and which have well-structured edges, shrubby cover within the wood and open feeding areas, are often very good habitats for wildlife. Conversely pheasants stocked at too high densities can have very damaging effects. Practical advice is available (see section 9 *infra*). Many shelterbelts established as even-aged plantings become draughty and inhospitable as they mature. The key to improving their effectiveness as shelterbelts, and their value to wildlife, is to create a graded edge incorporating shrub species.

Conifer high forest. Most productive high forest in Britain, particularly in upland, is composed predominantly of conifers, of uniform age over fairly large areas, and with little diversity of tree species. Extensive research is typified by that of Ratcliffe and Petty (1986) and Staines *et al*. (1987) who describe the mammals and birds associated with the sequential growth stages of conifer forests in Scotland and discuss management practices and options which might maximize this wildlife while maintaining the profitability of the timber crop. The aim, from the wildlife conservation viewpoint, is to effect a transition from even-aged plantation to well-balanced woodland. The desired result is a woodland which has a sustained yield of the fullest range of habitats. The transition will be effected by a number of strategies, including:

- Varying the thinning regime;
- Increasing the spread of age classes by some early felling, and especially by delaying felling and by retaining about 1% of the area to be grown on to full physical maturity;
- Taking advantage of windthrow when it occurs by phasing the restocking and leaving more open space;
- Improving the structure of the woodland at the time of felling and restocking;
- Introducing or increasing broadleaves on appropriate sites, either by planting or preferably by the natural regeneration of local native species (scattered or clumped broadleaves have a significant conservation value but beware of introducing large-seeded broadleaves if red squirrels are present and grey squirrels are a threat);
- Continuously reviewing the amount and distribution of open ground, and how it relates to other open ground, both within and outside the woodland (the opportunity is greatest at restocking);
- Giving special attention to the transition zones between wooded and open areas;
- Widening the range of conifer species, particularly to include some with light canopies, and some with dependable cone crops often in mixture with the principal species; and
- In appropriate places, encouraging the development of a shrub layer.

Operational guidelines. The following notes, set out in Forestry Commission 'Forest Nature Conservation Guidelines' (1990) summarize their earlier guidelines and introduce further points in relation to specific silvicultural and harvesting operations.

Planning new planting. When land is to be afforested, opportunities must be taken to make provision for nature conservation from the start; survey the area carefully to establish the presence of communities or habitats of special conservation value and retain these as open ground within the woodland. Ensure that such areas have a sufficient buffer zone so as to protect them from modifications to adjoining sites; take particular account of semi-natural woodland for its own sake as a potential seed source. Lay out the open ground required for roads, water protection zones, service corridors, recreation, deer glades, and other purposes.

Fencing. Where appropriate, install badger gates and deer leaps; and in capercaillie areas consider the use of high visibility fence demarcators to reduce the hazard to flying birds.

Cultivation. Use the minimum cultivation necessary to achieve the desired ends, for example do not plough if ripping or scarifying will suffice. Ensure that cultivation practice complies with Forestry Commission 'Forests and Water Guidelines' (1988) see section 8 *infra*.

Drainage. Provide sufficient cross drains on ploughed ground to prevent erosive volumes of water accumulating in plough furrows. Angle drains at no more than 3° off the contour to prevent erosion, ensure that cross drains are sufficiently deeper than plough furrows to fulfil their function, do not allow drains to enter riparian buffer zones, angle drain ends to reduce water velocity and taper their depth, and ensure that drainage practice complies with Forestry Commission 'Forests and Water Guidelines' (1988) see section 8 *infra*.

Afforesting or restocking. Decide which areas should be left to regenerate naturally, whether immediately or later, for example with birch, and leave these unplanted. Plant only light-foliaged broadleaves in riparian zones. Where broadleaves are being planted primarily for conservation, use species found naturally in the area. Ensure that planting stock is suited to the site conditions, and that plants are properly handled and well planted so that there is a high initial survival of vigorous first-year growth.

Fertilizing. Because of the risk of rapid run-off into watercourses, never apply fertilizer to frozen or snow-covered ground. Plan operations to minimize drift to riparian zones.

Herbicide weeding and cleaning. Herbicides can be a valuable tool for vegetation management. Choose a method of application which gives the desired level of control with the minimum quantity of active ingredient. Overall application is seldom necessary: band or spot treatment is usually sufficient. Where there is a choice of suitable herbicide, choose one with low toxicity and which breaks down quickly in the soil. Ensure that the use of herbicides is minimized by correctly timing other establishment operations.

Thinning. Adjust the proportions of species present in a mixture. Remove trees from around watercourses and water-bodies in a manner consistent with good modern practice. Begin the process of feathering woodland into adjacent open areas. Encourage the development of a shrub layer of appropriate species throughout the crop. Encourage the development of an appropriate woodland flora where this is absent and particularly where desirable species will spread as a consequence.

Felling. Keep brash away from streamsides, open spaces, rides and roadsides by felling trees into the stand near these locations. Keep urea used for stump treatment well away from streams. Schedule operations to avoid raptor breeding sites in season.

Extraction. Plan extraction routes to minimize stream crossings. Use temporary culverts on all soft-bottomed or soft-banked stream crossings. Never extract along stream beds. Use brash mats to reduce soil damage. Plan extraction routes to avoid valuable habitat. Avoid long downhill extraction routes – use frequent changes of direction. Where there is no alternative, provide cross drains to prevent water flowing down wheel ruts. Keep site refuelling and maintenance areas well away from watercourses. Select firm, dry ground as stacking sites. Repair wheel ruts where these have unavoidably formed on erodible slopes.

Road-building. Check conservation plans before commencing operations to identify sensitive areas. Keep valley bottom roads as far back from streams as possible. Protect riparian vegetation in all operations. Avoid metalliferous or sulphide-rich material for road construction near watercourses. Design roadside drains to avoid direct discharge into natural watercourses. Adhere to the practices set out in Forestry Commission 'Forests and Water Guidelines' (1988) see section 8 *infra*.

Forest protection. If creating fire traces, use tractor-mounted swipes in preference to bulldozers. Ensure that measures adopted to control grey squirrels and rabbits affect only the intended target species. Control deer numbers to maintain populations within the carrying capacity of the site.

Local offices of the Forestry Commission or of the successors to the Nature Conservancy Council will always be helpful. Rangers, keepers and others engaged in the management of wildlife locally, usually have a fund of local knowledge not available elsewhere. The Commission's Wildlife and Conservation Research Branch specializes in woodland conservation management and makes general advice available through publications on many aspects. They are always willing to discuss major projects or problems. The British Trust for Conservation Volunteers, and Scottish Conservation Projects, carry out practical conservation work: they are able to undertake work with a high labour content, unsuitable for mechanization. Addresses of all relevant conservation bodies are set out in Forestry Commission 'Forest Nature Conservation Guidelines' (1990).

Woodland birds. Birds have a high conservation value in both lowland and upland (Petty and Avery 1990). Many are legally protected – particularly raptors (birds of prey) – red kite, hen harrier, goshawk, buzzard, osprey, kestrel, merlin, long-eared owl and short-eared owl. A few are not protected, e.g. capercaillie (except on Forestry Commission property). An extensive number of other birds are legally protected. There is a right under licence to shoot magpie, crow, jackdaw, jay, wood-pigeon, rook, starling, house-sparrow, collared dove and three named gulls. Steele (1972) lists the family and English and scientific names of woodland birds, their breeding season, nesting sites in woodlands, preferred woodland type, and main food. Comprehensive information is also provided by Avery and Leslie (1990). Petty and Avery (1990) discuss forest bird communities and provide invaluable information on management. Any conflict relating to birds between forestry, agriculture and wildlife conservation is generally in the uplands. Afforestation inevitably alters the fauna, but later there are many compensating benefits (TGUK 1986).

Petty (1987) in discussing the management of raptors in upland forests emphasizes the importance for foresters to safeguard, and if possible increase, the potential of the woodland, suggesting management techniques to achieve good results – in particular (i) conserving natural nest sites, providing artificial nest sites, nest-boxes, platforms and artificial nests, and (ii) improving food supplies – artificial feeding, and improving natural food resources through habitat improvement. Special management is recommended for goshawks (Petty 1990). Foresters should also minimize disturbance of raptors by care in forest operations, particularly when harvesting.

As part of multi-purpose forest management the Forestry Commission along with some woodland owners in the private sector has achieved much in the interests of birds. They need food sources, nest sites, territory and over-wintering areas, all of which can be provided by the diverse range of forest habitats. Part of foresters' jobs is to help provide these benefits by the way the woodlands are managed and to keep to a minimum disturbance of birds from silvicultural and harvesting operations.

Different species of birds are associated with different types or ages of woodland: newly-planted conifers to old mature broadleaves all have their contribution to make as bird habitats. Whitethroat and nightjar prefer young newly-planted woodlands or recently clear-felled areas. Nightingale, blackcap, and robin prefer the thicket-stage as the tree canopy of the young, dense woodland begins to close. Pied flycatcher, hawfinch, woodpecker and goshawk are found in mature woodlands. Some important species of birds thrive in conifer forests – siskin, crested tit, Scottish crossbill, common crossbill and firecrest. Others tend to be found mainly in broadleaved woodland – pied flycatcher, wood warbler and nuthatch. Most birds, however, prefer the diversity offered by a mix of conifer and broadleaved trees, and an uneven-aged woodland structure along with areas of open ground and open watercourses. Structural variety in a wood appears to be more important than the tree species. Where possible, it is part of the policy of foresters to achieve this species mix and to manage and harvest timber in a way that creates such a structure. Retaining old and dying trees as bird habitats when harvesting is also vital for some birds – to provide permanent nesting sites for large birds such as raptors and for small hole nesters.

However, Ratcliffe (1988) points out that there is a range of bird species which are uncommon or absent from conifer woods – nuthatch, marsh tit, nightingale, lesser spotted woodpecker, hawfinch, spotted flycatcher, pied flycatcher, redstart, garden warbler, wood warbler and blackcap (although the latter six are regularly found in conifer forest in Wales, in the early growth stages, or in crops older than normal commercial rotations, or where broadleaves or nest boxes are present – Currie and Bamford 1982). Goldcrest and coal tit are usually more abundant in coniferous rather than broadleaved woods.

Woodpeckers are interesting to have within the woods; and their insectivorous habits may be of help to silviculture. Dead wood is invaluable to them. Barn owls can be encouraged into woods by provision of artificial nesting sites in young forests (Shaw and Dowell 1990). Tawny owls approve of old decrepit trees. Woodlark and nightjar thrive on restocked land in forests, notably in Thetford Forest Park. Larch plantations are particularly valuable for conifer seed-loving birds such as siskin and crossbill; goldcrest and coal tits benefit from the large numbers of insects found on the foliage of larch, and black grouse feed on larch buds. Sparrowhawk, buzzard and goshawk favour larch for nest building often sited near the trunk on a branch about 8 m above the ground. Prime habitat for capercaillie is mature pinewood, whereas blackgame will increase in an area of extensive young plantations. Blackgame will often use plantations (particularly birch trees) as perching places adjacent to lekking grounds. In view of conservation value of both species, the assumption should be that control is unnecessary (see chapter 6).

Broadleaved woods often contain more birds (species and number) than conifers; broadleaf/conifer mixtures are even richer. Within the broadleaves, oak appears to hold higher bird densities than other species. It is generally believed that native broadleaves provide better bird habitats than most exotic trees, as the insects on which many songbirds feed are more abundant on native trees. Within the conifers, spruce crops appear to have higher bird densities than the pines.

Woodland edges adjacent to farmland are a valuable habitat for birds. Many which feed primarily on farmland, such as crow, rook, kestrel and mistle thrush, tend to nest near the edge of woods. Well-developed ground vegetation alongside such edges appears to be beneficial to barn owl, little owl and kestrel, which feed on the abundant small mammals and large invertebrates. Pheasant and partridge may also take advantage of these areas for feeding, nesting and cover from predators. Woodland edges next to moorland are valuable for merlin, Britain's smallest breeding falcon.

The woodland itself may be divided into four main zones: field layer, shrub layer, the trunks and main branches of the forest trees, and finally their canopy and crowns. Broadleaves hold a wide range of birdlife, according to their size, age and composition. For example, birds of mature oakwood benefit from a field layer of bramble, bracken, bluebell, foxglove and grass, and a shrub layer of holly and various small deciduous trees. (In such woods, the jay is an important disseminator, by burying, of acorns.) By contrast, younger oakwoods are poor in fauna, whereas coppice with standards has a rich birdlife. Beechwoods, because they cast significant suppressing shade, generally have a very weak community of breedingbirds. In good mast years they are an important food source in winter. They have poor field and shrub layers, except for a few specialist plants and an occasional yew.

The conifer forests scattered throughout Britain, ranging from the pinewoods of Sutherland to the yews of the southern chalklands, have wide ranges of composition and age. The ground cover is diverse; for example, in the north it may be a field layer of heather and blaeberry and a shrub layer which may include juniper, willow and bog myrtle. Harris (1983) has assembled and analysed much of the relevant scientific data published on the status of birds in coniferous plantations in both upland and lowland, and on the comparison with broadleaved woodland on similar sites. Where plantations are created on previously bare land, there has been an increase in the diversity of bird species, yet some species are lost. Buzzard use forests for refuge, or to nest in, but need large areas of open land for hunting. Others like skylark depend upon open habits for both food and nest sites; they disappear as a tree cover develops. Comparing bird populations found in broadleaved woodland with those of coniferous woodland shows that in the lowlands, broadleaved woodlands, especially when they include mature and overmature trees, support a wider range of species (Evans 1984). Densities, but not always varieties of species, are highest in mixed woods followed, in descending order, by woods of oak, birch, beech and pine.

Forestry, particularly when practised in mixtures, rarely harms birds and often actually increases both their number and their species. Mixed tree crops can show a wide range of birds. Large blocks of pure conifers may affect them adversely, though not as much as is commonly supposed; birds are to be found in such plantations, though probably in diminished numbers, but as the crops enter and then pass out of the thicket-stage and become pole crops a judicious use of nesting boxes to attract hole-nesting insectivores will prove successful and diversifying. Special types of nest boxes are available: details are published by the British Trust for Ornithology (see also Bolund 1987; Springthorpe and Myhill (eds.) 1985).

Woodland, old and new, provides an interesting situation for the ornithologist and for the increasing numbers of foresters who are also naturalists and conservationists. Foresters have a close working relationship with the Royal Society for the Protection of Birds (RSPB) and other conservation bodies; and continually exchange ideas that help maintain and improve wildlife habitats and protect endangered species (Smart and Andrews 1985). There is much development in bird and habitat conservation projects within forests (Cadbury and Everett (eds.) 1989; Forestry Commission 1990; Avery and Leslie 1990).

8. Management for water

Water quality and water catchments, and their relationship with forests have been noted in chapter 3. Management of forests for water is now discussed. The Forestry Commission Environment Leaflet 'Forests and Water Guidelines' (1988) is designed to assist foresters who manage woodlands in the catchments of streams and water systems. (It also serves to inform the water industry about those forest operations which may affect their responsibilities.) Measures are set out for protecting and improving the aquatic environment, and advice is given as to how forest operations can be carried out in harmony with the maintenance of good water quality. There is particular emphasis on the uplands, where most afforestation is currently taking place and where rivers, lochs, lakes and water supply reservoirs are most likely to be affected by land changes (these could also be important in the lowlands). The guidelines, which should be read in conjunction with Figure 30 on page 503, are:

1. Do not plough unnecessarily. Consider scarifying or mounding.

2. Provide cut-off drains on ploughed ground. Do not drain deeper than is necessary. Keep drain gradients below 3% and do not divert natural watercourses into new drains.

3. Stop furrow and drain ends well short of watercourses, providing and maintaining sumps or pools where necessary to trap debris and sediment.

4. Establish protective strips along all watercourses. Do not plant conifers or heavily foliaged broadleaved trees within about 5 metres of headwater streams. Larger streams and rivers need each bank to have a strip two or three times as wide as the stream bed. Avoid having a regular width of strip.

5. Aim to have at least half of the stream open to sunlight with the remainder under intermittent shade from light foliaged, broadleaved trees and shrubs such as birch, willow, rowan, ash, hazel and aspen, according to site and locality. Periodic cutting may be necessary to maintain open space in the face of invasive, naturally established trees of any species.

6. Do not locate roads close to streams.

7. Prevent roadside drains from discharging directly into watercourses.

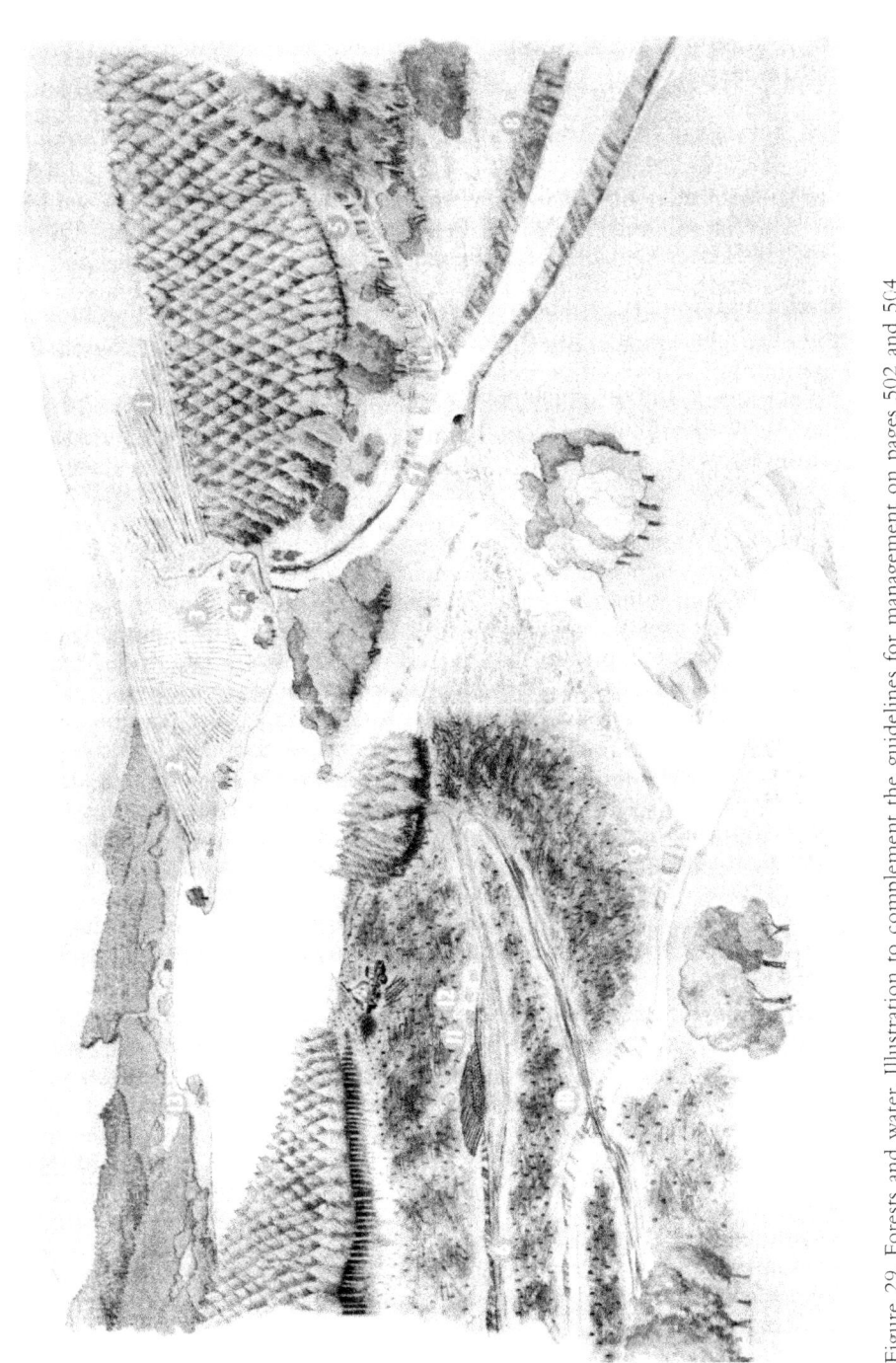

Figure 29. Forests and water. Illustration to complement the guidelines for management on pages 502 and 504
Source: FC Environmental Leaflet, 1989: *Forests and Water*.

8   Make roadside banks resistant to erosion and maintain vegetation on embankments.

9   Fell trees away from streams and keep branches and tops out of streams.

10 Plan extraction to minimize stream crossings and use pipes where extraction routes cross streams. Never let machines work in streams and avoid long extraction routes on steep ground.

11 Stack timber on dry ground away from watercourses. Ensure that fuels, oils and all chemicals are stored safely away from watercourses.

12 Refuel and maintain machinery well away from watercourses and guard against on-site spillage.

13 Do not apply chemicals within 10 metres of streams or 20 metres of reservoirs.

It is essential that foresters at all times meet their legal obligations under the *Control of Pollution Act 1974,* the *Food and Environment Protection Act 1985,* the *Water Act 1974* and other relevant legislation. Supplementing the subject is information by Mills (1980) on the management of forest streams, emphasizing that any water features are important assets deserving foresters' attention. Detailed information is given on relevant forest management and stream conservation.

Advice on matters relating to the water catchment in their area can be obtained from water companies in England and Wales, and in Scotland from the water departments of regional and Island Councils, the Central Scotland Water Development Board, River Purification Authorities, and Scottish electricity companies.

The association between forests and increased surface water acidification in sensitive areas of acid soils and geology has been accepted by many as a proof of cause and effect, and used to restrict future afforestation in such areas. There is a large degree of uncertainty surrounding the significance of a forest's acidification effect over and above that which is occurring anyway as a result of acid deposition (see chapter 3).

9. Management for sporting

Woodlands can have great sporting value – an important consideration to woodland owners and to estate and farmland management in general. In those parts of our islands where gamebirds thrive, few wooded estates pay no attention to sporting interests, although relatively few owners keep their woods solely for sport. Some estates in hunting districts allow the fox limited sylvan sanctuary. In the north, capercaillie, blackgame, red grouse and deer culling and stalking are important. (The capital value of one red stag annually is currently about £12,000.) Wood production may be the first financial objective and commitment, but sound forestry and good sporting can go hand in hand: woodlands can afford scope for sporting as well as produce satisfactory timber yields.

Management of sporting in woodlands differs between lowland and upland locations. It is a large subject and foresters concerned with the provision and management of habitat can refer to specialized literature of, for example, the Game Conservancy (1981, 1989); Prior (1983, 1987a, 1987b); McCall (1986, 1988a, 1988b);

Petty and Avery (1990); the British Association for Shooting and Conservation; and the Forestry Commission. There are abundant publications concerning shooting, stalking and hunting.

Many woodland owners, particularly in lowland, have established or retained coverts, copses, spinneys or belts of trees largely because of their interest in sporting, particularly in relation to pheasants, and some have taken that interest into account when choosing species of trees to plant. The sporting value of such wooded areas can add several hundred £/hectare to the capital value of a property and therefore can be a greater or equally significant factor favouring their retention as that of amenity or wildlife conservation. For many owners the sporting value of their woods can equal or exceed in importance that of forest produce. Management for sporting may be important as an additional source of income especially in the early years when no timber revenue accrues; but its effect on wood production and on wildlife conservation should be borne in mind. Even where sporting is the principal objective, it need not preclude all yield of forest produce. Equally, in woodlands managed for timber, sporting value can be enhanced with only little serious conflict with silviculture.

Almost all woodland has potential to provide some sporting, but mixed and broadleaved woods are generally more valuable than coniferous crops for a number of reasons (Evans 1984). The greater variety of species, both in the tree crop and ground and shrub layers, provide better cover, shelter, and sources of food for most kinds of sporting animals and gamebirds. Also in broadleaved woodland more sunlight tends to reach the forest floor and the woods themselves are usually in more sheltered areas favoured by wildlife for warmth and protection from exposure. The small size of many broadleaved woods and presence of glades and openings increase woodland edge which is so important in the life of many animals and birds: very few 'woodland species' spend all their time actually in dense woodland. In addition, all types of managed coppice, especially when with standards, provide particularly favourable habitats for many game animals and other wildlife.

Gamebirds. The pheasant is the most important woodland gamebird, particularly in the lowlands (McCall 1988; Robertson 1990). Woodcock and snipe are important in some localities. Partridge is partly dependent on hedgerow bottoms and verges for cover, nest sites and the great variety of plants and insects which provide for parent bird and chick. In the uplands pheasant, blackgame and capercaillie are locally important. Red grouse are moorland birds. Large scale afforestation radically alters and effectively reduces the habitat of agricultural and moorland birds. Wood-pigeons are traditionally regarded more as agricultural pests, but their shooting can command a significant sporting rental in woodland, especially so in small roost woods in arable areas.

For pheasants to thrive, whether wild or reared, the woodland must not be draughty; the presence of ground cover, and a perimeter hedge, improves the necessary shelter and warmth. The canopy must not be so dense as to suppress all ground vegetation and create a cold forest floor. Food, to supplement grain or corn, can be provided by seed and berry-producing trees and shrubs. Oak, beech and sweet chestnut eventually provide food for pheasants and other wildlife. Cherry, ash, birch and rowan will allow a shrub layer to survive below them, but sycamore might be avoided as it seeds profusely and may be too oppressive to other species.

For the understorey and ground cover there are many natural species that may seed in – for example, blackberry, elder, holly, hawthorn, blackthorn, dogwood and raspberry; others can be planted, among them: snowberry, berberis, cotoneaster and laurel, Rose of Sharon, shrub honeysuckle, guelder rose and wayfaring tree. *Rhododendron ponticum* and privet provide good cover but the former tends to be too invasive.

The rent from the letting of shooting in woodlands is the most conspicuous non-wood revenue; alternatively, for the owner who retains the shooting himself, the worth to him is often substantial. The value can be of a disproportionate value compared with the area of woodland because a series of small woods of suitable design strategically placed may enhance not just the rental value of the land occupied by woodland but that of a whole estate or farm.

Small woods, that is 0.5–5 hectares of whatever form, along with hedgerows, are generally the most valuable for pheasant shooting. Part of the reason is because the pheasant, like so many other wildlife species, is primarily a bird of the woodland fringe. The small areas, particularly belts, are in many ways the ideal shape because there is a high ratio of edge to area – all provided that there is adequate width to offer warmth for the birds (Ferris-Kaan (ed.) 1991b). Care needs to be taken when establishing such areas, particularly belts, to site them so that they can show high birds by making the best use of any contours. They should also supplement the natural landscape from a visual point of view; and some of them may serve a second role as shelterbelts.

The potential of woods as shooting coverts is largely determined by their suitability for holding pheasants in late autumn and throughout the winter months, together with the opportunity they offer to show high birds. Flushing cover is also required – this need only be a very small proportion of the total woodland area and can comprise low shrubs, small Christmas trees, young natural regeneration or low coppice. Rides are needed for access and feeding, but they should not extend in a straight line to the wood edge, to avoid cold winds and the inquisitive gaze of poachers.

The valuable attributes of predominantly broadleaved woodland have been noted earlier. However, such woodland, if with no shrub understorey and no provision of marginal shelter, once the leaf has fallen in autumn can be even more inhospitable to game than old dark conifers. Mixed woodland, once past the thicket-stage and regularly managed and thinned, usually provides the best pheasant holding habitat. A probable order of preference is: (i) managed traditional coppice; (ii) young mixed broadleaved/conifer woodland; (iii) light canopy broadleaved woodland with a shrub layer; (iv) mature heavy canopy broadleaves with a bare forest floor; and (v) mature heavy canopy conifers with a bare floor.

Sophisticated designs are available for provision of cover in woodlands to enhance pheasant sporting and game management; they will not maximize the site's timber growing potential, but a worthwhile yield of forest produce will generally be obtainable. For predominantly broadleaved woods in the lowlands the main requirement is to limit the extent of pure stands particularly of beech, hornbeam and sycamore, and to provide some peripheral shelter so that the woodland is not draughty. Ideal tree species include sweet chestnut as coppice, oak, ash, cherry and birch. These admit sunlight; also, after the first thinning, they will admit enough light for bramble and shrubs to grow. Where conifers are included, pines and larches are preferred to firs and spruces (though a few are ideal for future roosting). An outer hedge, along with a few rows of Christmas trees or Lawson cypress near the edge of the main belt of trees may be ideal. Wide rides, never completely shaded, are desirable. Reasonable access is necessary for management and beaters, therefore some cleaning and brashing may be required. Care should be taken where pheasant shooting is important, that trees are planted in the expected direction of each drive.

The letting (to responsible persons) of informal *rough shooting* in broadleaved or mixed woodland can aid woodland management if it helps in the control of rabbits, hares and grey squirrels. It should not be permitted concurrently in woodlands where sporting is the principal or secondary management objective (except that it might be allowed during the close season for other sports, in order to control predators); neither where the general public have access, nor during forestry operations or nesting periods. Letting of shooting needs to be approached with caution: control over silviculture and harvesting should not be relinquished or hindered; and possible effects on other wildlife, amenity and recreation should be duly considered. Forestry plans should always be discussed with the shoot tenant, to avoid conflict and to encourage co-operation.

In the uplands, with coniferous species usually the main choice of tree, the sporting value may be enhanced by wide rides with a fringe of broadleaves such as birch and rowan, supplemented by sallow or thorn in sheltered places and around rocky knolls. In uniform plantations of spruce, introductions of pine and larch will be beneficial where site conditions permit. The most common requirements are to diversify structure and to introduce broadleaved species in what will usually be areas of pure spruces; or to create 'game strips' of broadleaves, or, if impractical, of pines and larches. Encouraging re-growth of broadleaved vegetation along the edges of fire-breaks is also beneficial, as well as retaining and encouraging scattered broadleaves up gullies and beside watercourses. Most importantly such areas which by definition are low lying should be left largely wide-open as shooting rides.

Wildfowl, fishing and foxhunting. Although not woodland birds, duck and geese will often use ponds and lakes in woodlands, so giving them some sporting potential. For wildfowl to be attracted to woodland waters, they must be clearly visible from overhead and not heavily shaded. Where such waters are managed for wildfowl they can double up as fire ponds – the only requirement is good road access to one point on the pond edge. *Fishing* is not, of course, peculiar to watercourses in woodland but the presence of broadleaved trees and shrubs along the water's edge can enhance the potential provided fishermen can cast without obstruction. Establishment of broadleaves and introducing openings next to waters flowing through coniferous forest can beneficially reduce the amount of heavy shading. *Fox hunting,* a widespread country sport, depends on areas of woodland for its quarry. Woodland management to facilitate the sport sometimes comprises actually preserving a fox population, and making provision for the movement of the hunt through the wood. Encouraging foxes may in part conflict with other sporting requirements of a wood, e.g. releasing pheasants. Where hunting is the dominant consideration in woodland management, there must be some loss in woodland income from timber: a few large hunts own or rent woods and spinneys in their area specifically to manage in this way. A dense impenetrable wood is unsatisfactory for hunting (and for most wildlife).

Supplementary to the Woodland Grant Scheme and Woodland Management Grant Scheme of the Forestry Commission, the MAFF Farm Woodland Scheme offers 'compensation' for taking farmlands out of conventional production for (instead) tree-growing. These are incentives not only to forestry, landscape, and wildlife conservation, but also to the establishment of game habitats. There is an increasing awareness of the beneficial effect that habitat improvements can confer on the annual rental income from shooting rights over landed properties; also of increase in capital values of estates and farms through the establishment of woodlands. The planting of a series of small woods, selectively sited on undulating land, could enhance sporting values while at the same time provide for timber production, shelter and landscape improvements. In combination this can dramatically increase the capital value of a property especially in favoured residential rural areas.

An important consideration in correlating forestry and shooting is the relationship between foresters and gamekeepers. Harsh words have been written and spoken of this relationship in the past, but in these enlightened times when both timber-growing and shooting-rents have their correct place and importance the situation has much improved. Gamekeepers are aware of the importance of sound forestry and of the limits to which it can go to meet the interests of shooting. They should do nothing to hinder successful establishment; for instance, rabbit netting must not be raised even if it is the intention to avoid young gamebirds being trapped; a sloping plank run up from inside the plantation should suffice. Foresters should understand the requirements of sport, and plan the woodlands and the operations within them not only to guard against harming the shoot but to enhance it wherever possible. They should also know the effect of forestry operations upon the habits of gamebirds. Gamekeepers can assist forestry by controlling vermin, particularly rabbits and grey squirrels. As some predators are beneficial to forestry and harmful to sporting, they may occasionally lead to disagreement between the two interests. Some bird species (particularly crow, magpie and jay) are regarded by gamekeepers as harmful to sporting interests because they eat eggs. Their attitude towards birds of prey is sometimes hostile and illegal. A prudent balance has to be struck, though nothing usually deters their destroying of stoats and weasels. Gamekeepers can assist by notifying foresters of windthrows and damage which they may observe in trees and fences; also in the prevention of theft of Christmas trees and of trespass during periods of fire risk. Foresters on their part can schedule planting, weeding, cleaning, brashing and harvesting in such a way as not to harm nesting. They can report gamebird nests found, provide beaters when they are needed, loan machines when required, and avoid workers being in areas just before a shoot. Pheasants quickly acclimatize themselves to slight disturbances; they can sometimes be seen scratching about quite happily in proximity to workers, and even the noise of a powersaw does not always drive them far away. Foresters can co-operate with gamekeepers and farmers against wood-pigeon which take acorns, beech-mast and other tree-seed: the best results against them are obtained by the systematic destruction of their nests, over a wide area, by poking, and by organized shoots from the end of January to mid-March and from July to September.

To ensure the greatest mutual success for their departments foresters and gamekeepers should co-ordinate their plans from the beginning and keep in regular touch afterwards. A wood is always changing: nesting cover growing out, roost trees being felled, flushing points being altered, and so on. Foresters can usually arrange harvesting and restocking so that they will maintain environment for game. Belts of standing timber can be left round newly-planted areas; they will give protection to both young crops and pheasants, and can be felled at a later date.

Brashing can be stopped short of the woodland edge to obviate chilling draughts in the covert. Weeding sometimes presents a problem because delay is not always possible during the nesting season; but nesting pheasants and partridges run far greater hazards at the hands of mechanical farm harvesters than ever they do from woodland weeding. In fact the great majority of nest sites are on field edges or in the perimeter of woods within a few metres of open ground, and not in the centre of woods. If reasonable consideration and forethought are given to all that is to happen in the woods there is nothing to prevent foresters and gamekeepers co-operating. With intelligent planning and the forestry operations scheduled by consultation and carefully supervised, the interests of forestry and shooting need rarely conflict.

Deer management. Information on deer has been provided in chapter 6, in particular the effect of deer on silviculture and the control measures necessary. Management of deer for sporting can add to income. (Deer on or complementing farmland is discussed by Insley (ed.) 1988.) Standard texts on management are those of Prior (1983, 1987a, 1987b). Relevant information is also provided by CAS (1986) and ICF (1987, 1989). The British Deer Society provides helpful facilities and an advisory service.

Ratcliffe (1985) suggests ways of creating deer glades in order to increase the opportunities of culling deer. Selective shooting can result in a spectacular reduction in damage whereas indiscriminate shooting can make it worse. Where deer in the uplands are in close proximity to existing plantations, the main problem is caused by roe or red deer which can quickly colonize and severely harm afforestation and restocking. Rigorous culling is essential if crops are to be established without costly deer fencing. A professional stalker can be employed, although enthusiastic and trustworthy amateurs are also employed in some areas. Co-operation in deer management is often achieved between the traditional wooded estates, the new private afforestations, and the Forestry Commission. Letting of stalking rights benefits the local economy through demand for accommodation and other needs.

There are specific close seasons for deer (Springthorpe and Myhill (eds.) 1985). Shooting is the only permissible method of culling deer and this must be done humanely and within the terms of the relevant Acts (Deer Acts 1963 and 1980, England and Wales as amended by the Wildlife and Countryside Act 1981; and 1959 and 1982, Scotland). Only rifles of specified calibre and muzzle energy can be used although under certain circumstances a shotgun may be permitted. The owner or the tenant of property, or person authorized by them (preferably a fully trained deer control expert), may shoot deer and then only in the open season. The control of numbers and breeding potential can only be achieved by killing sufficient females of breeding age during the open season to balance the annual increment from reproduction. Night shooting is controlled in Scotland but is illegal in England and Wales. The aim for control should be to maintain deer populations within the carrying capacity of the woodlands.

Economics of game management. Many woodlands owe their past and continued existence to their sporting value. Rentals for sporting can contribute significantly to the profitability of a management option (McCall 1986, 1988; Insley (ed.) 1988), commonly ranging from a few £/ha/yr for modest vermin shooting to about £25/ha/yr for a woodland particularly suited to pheasant shooting. (For each

£5/ha/yr sporting rental the increase in Land Expectation Value at 3% would probably be about £170/ha.) Rentals for broadleaved or mixed woodlands, and for uniform conifer plantations are unlikely to differ more than £5/ha/yr: rentals are usually affected as much by location and suitability of structure, extent, terrain, age and design for sporting purposes as by the species composition. The accessibility to shooting parties is important, as also are the surroundings, whether additional woodland or agricultural land. The capital value of a woodland is also considerably improved by its suitability for sporting (see *supra* and chapter 16).

Where the sporting can be evaluated, the net portion applicable to the woodlands should be credited to them. The position as to the taxation of sporting value is noted in chapter 15. The cost of keepering applicable to vermin control is eligible for income tax relief either through a claim against rent received or in the computation of a woodland Schedule D assessment where in force until 5 April 1993.

10. Urban woods, community forests, and new national forests

Woodlands in or near towns and cities are a vital component in the environmental upgrading of a district or region and can be an essential part of the Government's plans for regeneration of inner cities and urban conurbations. In consequence, urban and urban fringe woods, community forests and new major forest initiatives are now much to the fore in forest thought and planning. The upsurge of interest in the environment in recent years has meant that there are probably more people today interested in trees and keen to grow and conserve them than ever before. This fact coupled with the urgent need to rehabilitate the remnants of Britain's industrial past in and around built up areas mean that urban forestry has found a new impetus.

Urban and urban fringe woods. Today there is a special need to attract new development into once derelict urban areas and their neighbourhood. The provision of green and pleasant working and living environments in and near towns and cities is an important contribution in that effort. The use of trees in the creation of urban green space can enhance its value and eventually provide the basis for a financial benefit through the advent of new industrial investment, a gain which is quite as commercial as growing timber for industry. Viewed in the light of Britain's total sylvan area and the amount of wood produced therein, urban forestry may seem quite insignificant; however, in terms of public use the potential for urban forestry appears limitless. The need for community involvement in tree planting schemes in and around urban areas must be seen as essential; also as a major way of avoiding or at least reducing intentional damage such as vandalism as well as unintentional damage due to ignorance about trees and their needs.

Such woods tend to have lower nature conservation value than other woodland types because: they tend to be well used by man and dogs so that the wildlife interest is dominated by the most resilient species of birds and mammals; fragile ground flora tend to be very restricted in occurrence – in some areas air pollution has affected the ability of lichens to colonize these woods; and due to safety requirements and high public usage, dead wood is very uncommon in them. For these reasons very particular skills are required to maintain and increase conservation value in woodland where the social benefit is high. Expert advice should be sought, not only for the management of existing woodland but when contemplating the development of new urban woodland.

Urban forestry embraces trees grown in and close to urban areas for their value in the landscape, for recreation and wildlife habitat, and includes trees in streets, avenues, urban parks, on land reclaimed from previous industrial use, as well as those in urban woodlands and gardens (Hibberd (ed.) 1989). The products often include the possibility of some commerical timber. Urban forestry very often relies upon the close co-operation of the local authority, private owners, voluntary groups and the local community. Forestry officers in the local authority occupy a key role in ensuring the success of urban forestry schemes. Keen and knowledgeable individuals operating within the various voluntary bodies also contribute. But because of the wide variety of interests and aims involved, there is more danger of projects failing in urban forestry than in less specialized forms of forestry. The danger is increased if there is a lack of understanding of the basic technical rules for successful tree establishment or later maintenance. Urban forestry has a reputation for being much more expensive than forestry in rural areas, especially at the critical stages of initial planting and early aftercare. Some high levels of expenditure may be unavoidable because of a small scale of operations and high costs of site preparation. Healthy growth can only be achieved if the biological needs of the trees are catered for: the appropriate species choice, site treatment and subsequent management on each location are carefully related to the trees' needs and the nature of the site.

Numerous woods in and around towns and cities are becoming increasingly important and appreciated by the public, particularly the local community. Most of these urban woods are predominantly broadleaved, usually small in size but locally important being the most natural amenity people can enjoy near to their homes. As well as having some limited timber potential they confer the benefits of visual relief, recreation, wildlife conservation, education, reduction in pollution, and enhancement of the microclimate. Whether urban woods are in public ownership or belong to private individuals, people are interested in their preservation, and thus any work carried out in them may arouse much local interest, along with some concern. Many are subject to Tree Preservation Orders.

Some urban woods generate little income because there have been few opportunities to many of them for timber production or sporting, and quite often their enjoyment by the public has traditionally been free. Consequently, expenditure on them tends to be the minimum needed to maintain their present condition, or at least to satisfy safety requirements. Their maintenance and perpetuation are essential. Caring and prudent authorities, organizations and individual owners should seek and follow sound silvicultural guidelines for positive management of their urban woods. Much depends on the size, age, extent of neglect or exploitation, public pressures, safety, vandalism and other factors. A *laissez-faire* approach to management will contribute to the demise of woods.

Urban forestry practice. Hibberd (ed.) (1989) deals comprehensively with the many and varied disciplines and requirements involved in urban forestry. In seeking to manage an existing urban woodland the first step is to determine the amount of public use. Most silvicultural problems relate directly to usage, and woods should be classified according to this: little visited, moderately used, well used, and heavily used. The best way of deciding how to care for an existing wood is to prepare a management plan including four ingredients: (i) assessing and recording woodland condition (size, age, species, public use, landscape and conservation values); (ii) deciding and setting down objectives (priorities between for example, visual importance, recreational value and timber production); (iii) identifying and drawing up work prescriptions needed to satisfy the main objective, and to ensure the general health and well-being of the woodland, with schedules and budgets for the programmes; and (iv) making provision for regular review of progress and how well objectives are being met or specific projects implemented. The overriding silvicultural need is to maintain and perpetuate tree cover. The general skill and experience of foresters will enable this to be achieved. They will find that most people are happy to approve operations which will help perpetuate the trees they enjoy, and that will maintain them so that public safety is not endangered.

Hibberd (ed.) (1989) provides information on sites including improving and ameliorating their conditions in respect to compaction and the correction of nutrient deficiencies by fertilizer application. Likewise on tree requirements – adequate supply of water, oxygen and nutrients; and anchorage – as well as choice of species, types of planting stock, establishment, early maintenance and protection. This information, supplemented by literature on arboriculture provided by the Forestry Commission (e.g. Patch (ed.) 1987; and Davies 1987) and the Department of the Environment Arboricultural Advisory and Information Service (in particular, Arboriculture Research Notes), located at the Alice Holt Research Station, is invaluable for all connected with urban trees. Enrichment, underplanting, thinning, coppicing, and the acceptance of natural regeneration, may all play a part. Many local and national authorities and organizations employ foresters, tree surgeons and arboriculturists adequately qualified and skilled to advise and undertake appropriate management.

The traditional skills and knowledge of foresters and arboriculturists need to be applied to urban tree planting schemes. Whatever the objectives in planting and managing trees, healthy growth can only be achieved if the biological needs of the tree are catered for. Many urban environments and reclaimed sites are hostile to tree growth, sometimes because of pollution but more often because of difficulties in ensuring that tree roots have an adequate supply of water and oxygen. Often soils are compacted, ill-drained or of quite unsuitable substrate for trees. (Consideration of them has been given in chapter 5 under 'Land reclamation and forestry'.) Work carried out by the Forestry Commission Research Division on behalf of the Department of the Environment has demonstrated that site preparation ensuring adequate aeration and drainage is essential. The growth of well-balanced trees, say, 90–120 cm, will in a short time usually outstrip halfstandard or standard trees. Essential too are correct planting and staking, as well as adequate control of weeds.

Community forests. The Forestry Commission and the Countryside Commission for England and Wales have launched a programme eventually to develop twelve community forests, on the outskirts of major cities, starting with Thames Chase, east of London, the Great North Forest in southern Tyne and Wear and north-east Durham, and the Forest of Mercia in south Staffordshire. Cleveland, Merseyside, west Manchester, South Yorkshire, Nottingham, Bedford, Bristol, Swindon and south Hertfordshire will fill the remaining places in the programme. Each is expected to cover between 10–15,000 hectares, with substantial areas of the forest being coincident with Green Belt areas. The aim is to develop a rich variety of landscapes, including woodlands, farm land, open spaces, nature reserves and water areas. Over the development period required to create a community forest, that is, twenty-five to thirty years, the intention is to increase broadleaved and coniferous tree cover to between 30% and 50%. The proposals are a bold and imaginative vision in which the whole community can share and help to turn into reality. They will help to provide a wide range of leisure apportunities and jobs and to restore derelict areas; and will complement the urban woods discussed earlier. A 'Valleys Forest' is projected for Wales.

Commercial forestry is likely to be a main purpose throughout considerable areas of each community forest. This will mean choosing species that grow well and have a good timber production potential. The forests will comprise a carefully designed mixture of woodland, farmland and open spaces, along with areas set aside for coppicing, craft demonstrations, nature reserves and water features. Bodies including MAFF, CLA, NFU and local authorities are showing a positive response, and the projects will also bring response from landowners and farmers. In parts of the forest, particularly close to the fringe of the town or city, there will be opportunities for people to enjoy a wide range of sport and, particularly, leisure activities. Here it is more likely that broadleaved species will dominate, probably with some conifer nurses. The challenge is to design and develop a forest of varied landscapes and utilities, and in doing so maximize the attractiveness of the area and its wildlife potential. Community forests will not be a success unless people regard them as beautiful areas with an interesting flora and fauna, and willingly use them as accessible open space for recreation and relaxation (Kirby 1990). Community groups and amenity societies are invited to take part in designing, establishing and managing the forests, and there are ways in which schools and local people can help too.

The three lead project areas are characterized by a mix of land-ownership patterns – land in public ownership (local authority, statutory undertakers, and nationalized industries), as well as land in private ownership, a good deal of which is being farmed. Large-scale changes in land ownership are not envisaged; the main approach will be to discuss with farmers and landowners the business opportunities that might be available to them for diversifying, in whole or in part, into leisure and forestry. At the same time it is important for landscape, land use, wildlife and recreation that a significant proportion of the area within each forest remains in productive and profitable agriculture. Sound business partnerships between the private and public sectors will be a key approach for those responsible for planning, developing and managing the community forest.

A structured system for proper environmental planning must be developed for the big cities. Recognizing this fact, the Countryside Commission for England and the Forestry Commission decided that it was essential for the local authorities, with their planning responsibilities, to take the key role in steering the community forests concept. The next step has been to establish project teams in each of the three lead areas, charged with the responsibility for developing a community forest plan within the first two years and to initiate projects. While this is a non-statutory plan, there will be full consultation with all interested parties before the draft plan is prepared; later to be considered and endorsed by the local authority.

The finances for the development of community forests are expected from a number of different sources, including grants from the Forestry Commission and the Countryside Commission, the MAFF and the Department of the Environment. Some funds will result from local government and major industries within the public sector such as British Coal Corporation. Private investment will be attracted from companies who see a leisure market potential with the community forests. Landowners and farmers may well wish to diversify, wholly or in part, out of agriculture into land use enterprises which are sympathetic with the objectives.

The Forestry Commission's contribution is by way of planting and management grants and advice combined with the participation of its local staff. Wooded areas such as Cannock Chase, Epping Forest and Chopwell Woods serve as good examples of areas providing much-cherished recreational opportunities on the doorsteps of large conurbations, even though they were not created for this purpose. With its unrivalled experience of managing the woods at Cannock and Chopwell, along with the Forest of Dean and the New Forest, the Commission is bringing essential knowledge to the initiative. A wide range of literature spearheaded by the Commission's *Urban Forestry Practice* (Hibberd (ed.) 1989) and supplemented by the arboricultural publications referred to under 'Urban woods' *supra,* is available to help those involved in the practical side of the project. New incentives may well need to be designed and introduced during the long development timescale, in order to reflect changes in the rural economy and society's leisure needs (Kirby 1990).

New national forests. A recent important initiative of the Forestry Commission and the two Countryside Commissions, is to symbolize and demonstrate the national approach to a new, multi-purpose mixed forestry, by establishing two major new forests, the New Midlands Forest and the other in Central Scotland between Glasgow and Edinburgh.

The aim is to achieve over twenty years a substantial increase in woodlands, by creating sylvan areas chiefly in order to improve the environment and provide facilities for recreation, together with considerable scope for timber production. Each immense forest will be a blend of trees, fields, open spaces, water areas, wetlands, villages and towns. The whole nation will share in the economic and environmental benefits – in particular landscape, amenity, recreation, wildlife conservation and education. Success will largely depend on the co-operation of local authorities, farmers, woodland owners, and not least the general public. The two Commissions will assist in appropriate ways, including finance. The European Community through its Forestry Action Programme might also assist.

For the Scottish Forest Project, a new company, Central Scottish Woodlands Ltd, has been set up, with a number of directors drawn from both the public and private sectors, including one from the Forestry Commission.

11. Farm forestry

About 300,000 hectares of woodlands in Britain exist on farms – at an average of four hectares per farm. Farm forestry has generally suffered from its small scale and much of it has been so neglected that it can scarcely be classed as productive. Yet, farm forestry in Scandinavia and parts of Europe is sometimes the best of domestic wood production. In Britain's tenant farming that was general a few decades ago, farmers managed fields for food, while their landlords held any woodlands, frequently for sporting and amenity; and where the farms and woods were sold they were sometimes depleted of their best timber, leaving the farmer with neither silvicultural experience nor worthwhile asset. Yet, even where estates have been fragmented, particularly in the lowlands, the farms often carry small woods, mainly broadleaved – coverts, spinneys, copses, clumps or belts. These, along with the few well-managed traditional farm woodlands, play an important role not only in the economy of the whole farm but also in provision of shelter, amenity and sporting. Their contribution to landscape is often immense. However, a substantial number remain a neglected asset, generally used only for firewood and low grade fencing material. Neglect has arisen more from being unaware how to manage and of the potential value rather than inherent poor quality of the growing stock: most areas looking like woodland have utilizable timber value.

Despite the traditional separation of agriculture and silviculture, farmers' knowledge of crop and land management should stand them in good stead for learning the new skills required for forestry. Many have the capacity to master new techniques; and silviculture, particularly establishing and tending trees, is no more complicated than modern farm crop husbandry. Positive management is required to prevent further natural deterioration of farm woods: and new ones should be created. If farmers decide to manage existing woods or to establish others they will do so efficiently. Encouragement available to them is discussed later in this section.

Integrating forestry and farming. In recent years there has been increasing interest in promoting the idea of integrated forestry and farming. Instead of regarding forestry as an activity which competes with farming, it is argued that under favourable circumstances the two enterprises can be combined to mutual benefit. (The beneficial impacts of the Farm Woodland Scheme and the Set-aside Scheme are discussed later.) Such integration is particularly appropriate on marginal land in upland areas of Britain where more than one hectare of land is required to support each ewe (Nix *et al.* 1987). Whether the introduction of forestry is feasible will depend on the size of the holding and whether sufficient land remains to form a viable farming unit after withdrawing part of the land for afforestation. The principal benefits of integration result from the provision of fences and roads when plantations are established, which give improved access and facilitate more extensive management of the remaining land for livestock production. Provided owners or occupiers possess the necessary skills and versatility, it is also possible to envisage labour and machinery sharing between farming and forestry operations. Since most forestry can be carried out in the winter months, work in the woods can employ farm labour and equipment when other work is less available.

A mixture of the two enterprises, particularly relative in the uplands, has other economic advantages. Their different cash flow characteristics give a flexibility not obtainable by one in isolation. For foresters, in the early years of a plantation an agricultural element offers income; for farmers, when agriculture is depressed a woodland is a source of income and a place to work, yet it can be left untouched for long periods when agriculture is thriving. The different financial characteristics also allow farmers to raise money for agricultural improvement by selling land laid to forestry, or by using woodland as loan security. Since poor quality land acceptable to forestry may be surplus to the agricultural enterprise, farmers can thus increase winter and spring grazing potential of valley bottom (inbye) land, which may be the limiting factor on stock numbers, without reducing usable summer grazings. The selling-off strategy concentrates land management, with individuals and agencies already possessing the requisite skills and equipment. Contractual obligations of permanent fencing, pest control, and maintenance of simultaneously-used roads, are important. Effective integration is not necessarily simple: it depends on positive management.

Agroforestry is a term used with a wide range of precise meanings, but generally refers to land use systems in which forestry and agriculture interact beneficially. It is the establishment of wide-spread trees (generally broadleaved) with agricultural crops or grazing between – the intimate mixture of trees and farm crops and/or livestock on the same land management unit (CAS 1986). There is a gradual change from agriculture to forest production. As the trees grow, the timber value increases; the grazing or agricultural cropping values decline as the grass is progressively shaded.

The aim of the forestry component is to grow trees with a saleable butt log of high quality timber with a minimum diameter of 45 cm at breast height in thirty-five to forty-five years, with a minimum reduction in agricultural production during the first ten to fifteen years. Pruning is essential to produce quality timber, but should never remove more than a third of the live crown at one time; an ultimate branch-free height of 4–6 m should be aimed for. Careful choice of species and spacings should allow trees to reach maturity and be harvested over a long period of time (thirty-five to eighty years). Spacings are usually 5–15 m. This wide spacing compared with conventional woodlands results in reduced overall yields of timber, although individual trees are larger. Yields will vary greatly depending on a range of factors including site, tree and crop species, and spacings.

The main suitable broadleaved species are ash, cherry, sycamore, walnut and poplar (although the 'soil-robbing' nature of ash and poplar should be remembered), while the best conifers may be Douglas fir, Corsican pine and hybrid larch. Most rotations are likely to be thirty-five to forty-five years, but for poplar it will be around twenty years and for walnut sixty years. As to poplar see chapter 5 (Jobling 1990). The best quality planting stock should be used, and pit-planted, with the intention of a 100% 'take' as there is no chance of selection through thinning. Initial protection against grazing animals will be necessary, using either strip fencing or individual tree protection. A weed-free spot of about 1 m diameter around each tree should help to achieve successful establishment.

Forestry Commission research has concentrated on systems with grazing between the trees (silvopasture) mainly using sheep, although some work has also been carried out on tree protection against cattle. Desk studies in the lowlands (Doyle *et al.* 1986) and in the uplands (Sibbald *et al.* 1987) indicated that for certain planting densities and pasture quality, combining wood and sheepmeat production could be financially attractive.

Economic models have shown that the yield from agroforestry can be expected to be slightly higher than that from pure sheep grazing or dairy cattle in the lowlands, and higher than that for a woodland crop; but these models are based upon assumptions which have not been field tested (Hibberd (ed.) 1988). Insley (ed.) (1988) provides a planting model example for agroforestry, assuming a broadleaved crop at 10 m spacing (100 trees/ha). It excludes the inputs necessary for the agricultural part of the system. However, field testing is necessary. Short-term and long-term research by the Forestry Commission are being continued (Potter and

Taylor 1989). Many 'unknowns' associated with the system require further research – not least the timber quality affected by wide spacing and associated pruning. In the private sector, notable practical experiments have been undertaken by Newman and Wainwright (1990).

Agroforestry is eligible for support under the Woodland Grant, Management Grant and Farm Woodland schemes. (As to poplar see chapter 5.) The system is a potential solution to some of the problems associated with integration of farming and forestry. In the uplands it has the additional benefits of stock shelter and extension of the grazing season; and may be a useful system for farmers to gain silvicultural experience and diversity into timber production.

Short rotation coppice crops. The Forestry Commission, under contract to the Department of Energy, is conducting research on short rotation crop systems to produce biomass for energy (see chapter 7). Species under trial include poplar, alder, willow and *Eucalyptus* at 1 and 2 m spacing and on two- and four-year rotations. These are potentially very productive systems, particularly on fertile land, but require further development on clonal testing, harvesting and utilization.

Longer rotation coppice (eight to twelve years) or single stem plantations on a similar rotation length are also being tested. These will produce larger average stem size and enable a wider range of end uses, e.g. pulp, boardwood, fencing and firewood. Establishment systems are being investigated and the range of species widened (Potter and Taylor 1989).

A number of factors will influence a farmer's decision on whether to consider such coppicing systems as a source of income. Thought will also be given to the ecological factors of energy forestry. The skills and availability of the workforce are pertinent; some employees may already have worked in conventional forestry and have experience to offer. A suitable local market outlet for the fuelwood and other biomass uses is another factor, and land tenure is a third. Then there is the question of predicting with reasonable accuracy a rate of return on a specific crop within a given period of time. If this seems attractive, the farmer must look into the finances of purchasing any specialized equipment or services needed. Only then is it possible to judge whether or not the project has the promise of being viable.

Energy forestry is probably easiest to integrate into a large owner-occupied farm having excess labour during the winter months and some areas of existing woodland or low opportunity-cost land suitable for planting. The farm may have a need for fuelwood itself or be located close to a ready urban or industrial market. There are possibilities of 'feeding into' an electricity grid system. The practice appears to be competitive with agriculture now, and future technical improvements should lead to reductions in costs. Revenues will depend on energy prices and may be enhanced if energy forestry attracts financial support or benefits by changes in taxation policy from Government. Energy coppice is eligible for the woodland option of 'Set-aside' and for grants under the Woodland Grant Scheme and Management Grant Scheme. It may, however, be planted on 'Set-aside' land under the non-agricultural use option, which attracts annual payments: £130/ha/yr in Severely Disadvantaged and Disadvantaged Areas; and £150/ha/yr elsewhere (see chapter 15).

Farm woodland. Farm forestry is the establishment of any plantation on agricultural land, i.e. to include closer or conventionally spaced trees. Due to surpluses of agricultural production within Europe, present Government policy aims to remove some better quality land from production by 'Set-aside' or by developing alternative enterprises such as farm woodlands. (EC reimbursement of Government expenditure on the Woodland Grant Scheme and Farm Woodland Scheme could eventually amount to £6 million a year.) The Farm Woodland Scheme was introduced in 1988 for an initial period of three years; grant-aid was made available to plant a target area of 36,000 hectares during the period 1989–91 on land previously under arable crops or improved pasture. Besides timber production, the schemes aim to enhance landscape and wildlife potential, create recreational benefits and contribute to supporting farm incomes and rural employment.

The Farm Woodland Scheme offers farmers annual payments (taxable as income) over a number of years to help bridge the gap between the planting of trees and the likely first income from thinnings. These payments are additional to, and depend on, planting grants under the Woodland Grant Scheme operated by the Forestry Commission (or the Department of Agriculture for Northern Ireland, Forestry Service). Other options for woodland planting on farms are available but are not eligible for the Farm Woodland Scheme.

The high fertility and broad weed spectrum of such farm sites present establishment problems for the new tree crops, which are exacerbated by many farmers' inexperience in forestry. The Government is funding research on relevant problems and systems of establishing. Forestry Commission demonstration and education in sound silviculture is playing a primary role in the initial development of farm forestry. A series of twenty-two demonstration plots for farm woodland establishment have been set up throughout Britain at agricultural colleges, showgrounds and ADAS establishments, with the purpose to demonstrate the effects of good and bad establishment techniques on survival and growth of trees (Potter 1989).

Many farmers in recent years have shown interest in receiving silvicultural advice – as to how they can improve their existing small woods or create new ones. To indicate the many benefits of woods, and how to carry out the necessary work to achieve them is the subject of a series of eight leaflets issued in 1986 by the MAFF Agricultural Development and Advisory Service (ADAS) and the Forestry Commission. They cover the practical aspects of work in woodlands and assume that the farmer has little forestry knowledge. Several welcome texts relating to 'trees as a farm crop' published in 1985 (Dartington Institute 1985), in 1987 (Blyth *et al.*) and in 1988 (Richards *et al.;* Insley (ed.); and Hibberd (ed.)) pointed out that established farm woodlands represent a largely neglected resource, and some drew attention to the available grant-aid schemes. *Farm Woodland Planning* (Insley (ed.) 1988) provides farmers who intend to undertake tree planting with the information required to plan and manage, tailored to suit their individual circumstances and relating to the wide range of sites likely to be found on farms throughout Britain. Each brings its own special requirements and costs for successful establishment, and management techniques which will influence the value of thinnings and final crop. Practical guidance on silvicultural practice is given in the supplementary *Farm Woodland Practice* (Hibberd (ed.) 1988). Information on the timbers of the farm woodland trees is supplied by Brazier (1990).

Most farms have ground where tree planting can be carried out to good effect, typically field corners, beside roads, tracks and watercourses, on banks and small wet areas not meriting drainage; also on bare areas within their existing woods. Management would bring notable benefits. All woods satisfy more than one purpose but deciding which is the most important is the first requisite. Possible objectives include: timber production, provision of stock shelter (reduction in exposure, and provision of winter grazing) and shelterbelts (for improved crop growing conditions and reduction in soil erosion), increase in sporting value, enhancement of the landscape, and provision of wildlife habitats. The approach to management will vary according to the objectives and involve different kinds of work to make the most of the woodland. Achieving success is in itself a rewarding activity; and woods can become an asset contributing to the farm enterprise.

Farmers aiming for timber production might also be prepared to grow and market a succession of other products, e.g. Christmas trees, berried holly, foliage, posts, stakes and poles, eventually to attain, already in cash surplus after grants, a valuable and attractive stand of maturing or mature trees, desirably for high quality saw or veneer logs. In the interim there is likely to have been marketed pulpwood and boardwood, firewood and perhaps pitwood. Some species of trees are of no value for timber: if an existing wood consists of thorns, elderberry, field maple, hazel, dogwood, holly, sallow, whitebeam, rowan and other largely unsaleable species, it may have to be consigned wholly to landscape or wildlife conservation or sporting objective; otherwise it might be cleared and restocked, preferably in phases, or enriched with productive species. Location and access are of prime importance where wood production is the main objective. The quality of access determines the kind of vehicles which can reach a wood and hence the ease of undertaking forestry and other operations. In woods over about 2 hectares and more than 100 m wide, internal tracks and rides improve access and reduce damage to trees left standing by encouraging vehicles to re-use the same routes.

Many of the new farm woods are being established on highly fertile lowland soils compared with those occupied by traditional forestry, and the opportunity is being taken to grow valuable high quality timber supplementing other benefits. They are also giving farmers the chance to diversify their business and improve the appearance as well as the capital value of their farm.

Farmers before committing themselves to a woodland investment or to making decisions on how the existing woods should be managed must consider carefully the possible end uses of the timber that must be grown. Insley (ed.) (1988) provides useful relevant advice, recommending the aiming for high value products from high quality material. Looking further ahead, he suggests consideration being given to forming a co-operative group of neighbouring farmers for marketing, able to offer larger volumes or continuity of supply and so attract lower costs for harvesting and higher prices from buyers. He then provides information on market identification, specifications, current prices and trends, and indicates the advantages and disadvantages of farmers undertaking the harvesting in contrast to hiring a contractor.

The Forestry Commission is involved in several research initiatives on farm forestry (Potter and Taylor 1989). Much of this work is in collaboration with The Farm Woodlands Unit at the National Agricultural Centre, Stoneleigh; also with ADAS, agricultural colleges and university departments. As well as experimentation using new production systems on fertile agricultural sites, many continuing projects on large-scale broadleaved and coniferous forestry are being re-evaluated for the application to farm forestry.

Reversion to agriculture. Trees being a long-term land use, the economic circumstances at the end of a rotation are difficult to predict. The consequences of the tree crop on future agricultural use, and any effects on surrounding fields, are important and could be a consideration before deciding on the initial site to be planted (Hibberd (ed.) 1988). Physical changes below a tree crop are likely to lead to the benefit of an increase in soil air space, due to incorporation of leaf litter and root channels. There is often a reduction in water flow through streams in afforested areas, compared with grassland areas, but this is unlikely to be important if only small blocks of woods are planted. In general, soil changes under a woodland rotation are unlikely to affect future agricultural use and there may be benefits in terms of improved soil structure. Stump removal will obviously be a problem in reversion to agriculture, particularly with broadleaves, and a drainage system might have to be restored before agricultural cropping. There is, of course, a long history of conversion of woodland to agriculture with perfectly acceptable results, so only the cost of conversion and any relevant legislation regarding change of land use need infuence the decision.

Advice on farm forestry. Besides advice obtainable from the Forestry Commission and professional consultants, the Agricultural Development Advisory Service (ADAS) provides a comprehensive, independent woodland consultancy service through its woodland advisers located throughout England and Wales. These advisers are backed by national and regional woodland specialists.

12. Amenity woods

'Amenity' and 'policy' woods are terms used to imply those woods which (in fiscal jargon) are ' not managed on a commercial basis with a view to profit' – although some timber is produced, even if it will not be fully marketed. Such woods, usually relatively small, particularly including those complementing mansions and other large dwellings, are managed or newly planted for landscape, beauty, privacy, screening and shelter. They may also include woods in close sensitive view from the dwelling, particularly when covering or draping rising ground. Along the sensitive perimeter the trees are usually kept clothed with foliage down to ground level. Such woods have traditionally been exempt from income tax but not of capital taxes.

A prime objective is to maintain a continuous canopy cover. Hence clear-felling is avoided as far as possible. To ensure permanency the wood must be either two-storied or comprise a range of age-classes in groups or intimate mixture. Permanency can be achieved by enrichment, natural regeneration, or selection or group schemes. The selection system, having trees of many ages and sizes growing intimately, provides an aspect which hardly ever changes.

Within an amenity wood the owner looks for aesthetic delights, and so the beauty even of individual trees or the tracery formed by adjacent ones is appreciated. The eye will be engaged not only by the general beauty of carefully sited species but also by that of any naturally regenerated silver birch, sallow, rowan, dogwood, spindle tree, guelder rose, wayfaring tree and whitebeam. Other ornamental trees, among them wild cherry and Norway maple, can be introduced along the margins of such woods. Occasionally the owner will accept attractive areas of mixed broadleaved 'scrub' or coppice, particularly if they serve as habitats for bluebell, foxglove, wild anemone, primrose or daffodil. The general choice of tree species will express the owner's likes and dislikes, and perhaps follow local tradition. Some owners may welcome the larches for their attractive green foliage changing in spring and summer, their yellow autumnal hues and, in Japanese larch, their winter russetcoloured stems and branches. Some may avoid evergreens, whether broadleaved or coniferous.

Most such woods require some form of active management to keep them in acceptable condition; such management can be supplemented by steps to enhance habitats for animals, birds, bats and plant life. Many lovely amenity woods, along with trees in parks and avenues were established with striking imaginativness in the eighteenth and nineteenth centuries. More recently, owners of estates – the National Trust in particular – have devoted much effort to the conservation and enhancement of the sylvan components of their parks and pleasure grounds.

Helliwell (1989) has written specifically of the establishment and management of amenity woodlands. Grants available for amenity planting are noted in chapter 15, and valuations of amenity woodlands in chapter 16.

References

Avery, M. and Leslie, R. (1990), *Birds and Forestry*. Poyser, London.
Blyth, J., Evans, J., Mutch, W.E.S. and Sidwell, C. (1987), *Farm Woodland Management*. Farming Press, Ipswich.
Bolund, L. and Insley, H. (eds.) (1987), 'Nest Boxes for the Birds of Britain and Europe'. Sainsbury, Nottinghamshire.
Brazier, J.D. (1990). 'The Timbers of Farm Woodland Trees'. FC Bulletin 91.
British Association for Shooting and Conservation, *Wildlife Conservation on the Farm*.
Cadbury, C.J. and Everett, M. (eds.) (1989), 'RSPB Conservation Review'. RSPB, Sandy, Bedfordshire.
Campbell, D. (1990), 'Forestry and the Environment – a Balanced Approach for the 1990s'. *Scottish Forestry 44* (3).
Centre for Agricultural Strategy (CAS) (1986), 'Alternative Enterprises for Agriculture in the UK'. CAS Report 11, Centre for Agricultural Strategy, University of Reading.
Crowe, Sylvia (1978), 'The Landscape of Forests and Woods'. FC Booklet 44.
Currie, F.A. and Bamford, R. (1982), 'Songbird Nestbox Studies in Forests in North Wales'. *Quarterly Journal of Forestry 76* (4), pp. 250–5.
Dartington Institute (1985), 'Potential Income to Farmers from a Woodland Enterprise'. Dartington Institute, Totnes.
Davies, R.J. (1987), 'Trees and Weeds'. FC Handbook 2.
Doyle, C.J., Evans, J. and Rossiter, J. (1987), 'Agroforestry; an Economic Appraisal of the Benefits of Intercropping Trees with Grassland in lowland Britain'. Agriculture Systems *21*, pp. 1–32.
Evans, J. (1984), 'Silviculture of Broadleaved Woodland'. FC Bulletin 62.
—— (1988), 'Natural Regeneration of Broadleaves'. FC Bulletin 78.
—— (1989), 'Small Woods: Neglected Asset'. *Forestry and British Timber,* March 1989.
Ferris-Kaan, R. (1991a), 'Conservation Management of Woodlands by Non-government Organizations'. FC Research Information Note 199.
—— (ed.) (1991b), 'Edge Management in Woodlands'. FC Occasional Paper 28.
Forestry Commission (1985), 'Guides for Broadleaved Management'. Edinburgh.
—— (1988), 'Forests and Water Guidelines'. Edinburgh.
—— (1989a), 'Forest Landscape Design Guidelines'. Edinburgh.
—— (1989b), 'Broadleaved Policy: Progress 1985–88'. Edinburgh.
—— (1990), 'Forest Nature Conservation Guidelines'. Edinburgh.
Fuller, R.J. and Warren, M.S. (1990), *Coppiced Woodlands: their Management for Wildlife*. NCC.
Game Conservancy (1981, 1989). Reviews. Fordingbridge.
Gilbert, F.F. and Dodds, D.G. (1987), *The Philosophy and Practice of Wildlife Management*. Chartwell-Bratt. Bromley, Kent.
Hamilton, G.J. (1980), 'Line Thinning'. FC Leaflet 77.

Harris, E.H.M. (ed.) (1982), Royal Forestry Society of England, Wales and Northern Ireland. Centenary Conference on Forestry and Conservation, Tring.
Harris, J.A. (1983), 'Birds and Coniferous Plantations'. Royal Forestry Society of England, Wales and Northern Ireland, Tring.
Helliwell, D.R. (1982), 'Options in Forestry'. Packard, Chichester.
—— (1989). 'Planting and Managing Amenity Woodlands'. Arboricultural Association.
Hibberd, B.G. (ed.) (1988), 'Farm Woodland Practice'. FC Handbook 3.
—— (1989), 'Urban Forestry Practice'. FC Handbook 5.
—— (1991), 'Forest Practice'. FC Handbook 6.
Hiley, W.E. (1954), *Woodland Management*, Faber and Faber, London.
ICF (1987), 'Deer and Forestry'. Proceedings of Institute of Chartered Foresters Conference 1987, Edinburgh.
—— (1989), 'Deer and Forestry', factsheet. Institute of Chartered Foresters, Edinburgh.
Insley, H. (ed.) (1988), 'Farm Woodland Planning'. FC Bulletin 80.
Irving, J.A. (1985), *The Public in Your Woods.* Land Decade Educational Council.
Jeffrey, E.H.W., Ashmole, M.A. and Macrae, F.M. (1986), 'The Economics of Wildlife in Private Forestry' in *Trees and Wildlife in the Scottish Uplands,* Jenkins D. (ed.). Institute of Terrestrial Ecology, Huntingdon, pp. 188–90.
Jobling, J. (1990), 'Poplars for Wood Production and Amenity'. FC Bulletin 92.
Kirby, M. (1990), 'Greening the Green Belts'. *Country Landowner,* February 1990, pp. 24/5.
Lorrain-Smith, R. (1988), 'The Economic Effects of the Guidelines for the Management of Broadleaved Woodland'. TGUK, London.
—— (1989), Memorandum 1.6.89 to the Agriculture Committee on Land Use and Forestry HC16–1: Vol. II: Minutes of Evidence and Appendices, 10.1.90, pp. 444–51. HMSO.
—— (1991), 'The Cost of Conservation in Woodland Management for Timber Production'. *Quarterly Journal of Forestry 85* (1), pp. 43–9.
Lucas, O.W.R. (1991), *The Design of Forest Landscapes.* O.U.P.
Marren, P. (1990), *Woodland Heritage.* David & Charles, Newton Abbot.
Mayle, B.A. (1990a), 'Habitat Management for Woodland Bats'. FC Research Information Note 165.
—— (1990b), 'Bats and Trees'. FC Arboricultural Research Note 89/90/WILD.
McCall, I. (1986), 'The Economics of Managing Farm Woodland for Game'. *Timber Grower,* Autumn 1986.
—— (1988a), 'Woodlands for Pheasants'. The Game Conservancy, Fordingbridge.
—— (1988b), 'Farm Woodlands for Sport'. *Forestry and British Timber,* June 1988.
McPhillimy, D. (1989), 'Conservation in Forests: Case Studies of Good Practice in Scotland'. Countryside Commission for Scotland.
Miles, R. (1967), *Forestry in the English Landscape.* Faber and Faber, London.
Mills, D. H. (1980), 'Water'. University of Edinburgh.
Mitchell, A. (1980), 'Conifers in the Landscape'. *Country Life,* November 6, 1980, pp. 1646/7.
Newman, S. and Wainwright, J. (1990), 'Trees Can Be Farmer's Best Friend'. *Forestry and British Timber,* March 1990.
Nix, J., Hill, P., and Williams, N. (1987), *Land and Estate Management.* Packard Publishing, Chichester.
Patch, D. (ed.) (1987), 'Advances in Practical Arboriculture'. FC Bulletin 65.
Peterken, G.F. (1981), *Woodland Conservation and Management.* Chapman and Hall, London.
Petty, S.J. (1987), 'The Management of Raptors in upland Forests' in *Wildlife management of forests,* Jardine D.C. (ed.). Proceedings of a discussion meeting, University of Lancaster 3–5 April 1987. Institute of Chartered Foresters, Edinburgh.
—— (1990), 'Goshawks, their Status, Requirements and Management'. FC Bulletin 81.
Petty, S.J. and Avery, M.I. (1990), 'Forest Bird Communities'. FC Occasional Paper 26.
Potter, C.J. (1989), 'Demonstration Plots for Farm Woodlands and Amenity Tree Establishment'. FC Research Information Note 156.

Potter, C.J. and Taylor, C.M.A. (1989), 'Farm Forestry Research'. FC Research Information Note 155.
Prior, R. (1983), *Trees and Deer.* Batsford, London.
—— (1987a), 'Roe Stalking'. The Game Conservancy, Fordingbridge.
—— (1987b), 'Deer Management in Small Woodlands'. The Game Conservancy, Fordingbridge.
Proudfoot, E.V.W. (ed.) (1990), *Our Vanishing Heritage; Forestry and Archaeology.* Countryside Commission for Scotland.
Rackham, O. (1980), *Ancient Woodland.* Edward Arnold, London.
Ratcliffe, P.R. (1985), 'Glades for Deer Control in upland Forests'. FC Leaflet 86.
—— (1988), Chapter 6 of 'Farm Woodland Practice', (ed.) Hibberd B.G. FC Handbook 3.
Ratcliffe, P.R. and Petty, S.J. (1986), 'The Management of Commercial Forests for Wildlife', in *Trees and Wildlife in the Scottish Uplands,* (ed.) D. Jenkins. Institute of Terrestrial Ecology, Huntingdon, pp. 177/8.
Richards, E.G., Aaron, J.R., Savage, J.F. D'A, and Williams, M.R.W. (1988), *Trees as a Farm Crop.* BSP Professional Books, Oxford.
Robertson, P.A. (1990), 'Woodland Management for Pheasants'. FC Research Information Note 194.
Royal Institution of Chartered Surveyors (1989), *Managing the countryside.*
Shaw, G. and Dowell, A. (1990), 'Barn Owl Conservation in Forests'. FC Bulletin 90.
Sibbald, A.R., Maxwell, T.H., Griffiths, J.H., Hutchings, N.J., Taylor, C.M.A., Tabbush, P.M. and White, I.M.S. (1987), 'Agroforestry Research in the Hills and Uplands' in *Agriculture and conservation in the hills and uplands,* (eds.) Bell, M. and Bunce, R.H.G. Institute of Terrestrial Ecology Symposium No. 23.
Smart, N. and Andrews, J. (1985), *Birds and Broadleaves Handbook,* RSPB, Sandy, Bedfordshire.
Springthorpe, G.D. and Myhill, N.G. (eds.) (1985), 'Wildlife Rangers Handbook'. FC. Under revision.
Staines, B.W., Petty, S.J. and Ratcliffe, P.R. (1987), 'Sitka Spruce Forests as a Habitat for Birds and Mammals'. Proceedings of the Royal Society of Edinburgh, 93B, pp. 169–81.
Steele, R.C. (1972), 'Wildlife Conservation in Woodlands'. FC Booklet 29.
Timber Growers United Kingdom (1985), 'The Forestry and Woodland Code'. TGUK, London.
—— (1986), 'Afforestation and Nature Conservation: Interactions'. TGUK, London.
Warren, M.S. and Fuller, R.J. (1990), 'Woodland Rides and Glades: Their Management for Wildlife'. NCC.
Watkins, C. (1990). *Woodland Management and Conservation.* David & Charles, Newton Abbot.
Workman, J. (1982), 'Management Objectives for Broadleaved Woodland – Landscape, Recreation and Sport'. In *Broadleaves in Britain: Future Management and Research,* (eds.) Malcolm, D.C., Evans, J. and Edwards, P.N., pp. 104–10. Institute of Chartered Foresters, Edinburgh.

Chapter Fourteen

MANAGERIAL ECONOMICS OF PRIVATE FORESTRY

1. Introduction

Forest economics is a specialized subject, but some of its components, notably its application to forest management, can be of very great help to foresters. This is particularly the case in making and implementing decisions and choosing between alternative courses of action which have financial implications. However, economics is also a behavioural science, providing an explanation of why people act as they do.

Foresters can benefit from the recent text by Price (1989) on *The Theory and Application of Forest Economics,* which augments his many other writings on the subject; as well as earlier books by Hiley (1930, 1964), Johnston, Grayson and Bradley (1967), Lorrain-Smith (1969) and Busby and Grayson (1981). Another contemporary publication, on forest investment appraisal, is an internal one of the Forestry Commission (Insley, Harper and Whiteman 1987). Helliwell (1988) approached forest economics in a less orthodox but quite helpful manner.

Price (1989) covers much of the traditional ground of forest economics, as well as the managerial economics of silvicultural operations, harvesting and marketing; also the bases on which forest decisions can be made in pursuit of the wider objective of social well-being. The quantity of the available data and the complexity of models, through the widening availability of computers, tend to bring economics within the reach of foresters; they should be aware of the economic and social significance of their management.

Forestry is a long-term investment, its economics being dominated by the cost of time and the associated uncertainties. The time intervals between planting the trees, thinning and (eventually) clear-felling are long. The production cycle or rotation will frequently take 50 years for a conifer plantation and can exceed 100 years for broadleaves. Foresters must establish a crop without knowing what demand there will be for the species, size and quality of the trees they can grow. They can but gaze into a crystal ball and calculate what their expected future costs and revenues are worth today. They face risks – wind, fire, biotic damage – and uncertainties. Residual prices to the grower are subject to many influences (noted in chapter 11), as well as to the relative bargaining strengths of grower, processor and consumer.

By contrast, there is a brighter aspect. Private forestry based on sound silviculture and economic judgement can be an acceptably attractive investment. Flexibility of silviculture, harvesting and marketing techniques is desirable, and accordingly aiming for too specialized a type of forest management may not always be prudent. Despite some substitution a continued demand for wood and wood products is likely; and risks can be reduced by prudent planning and management. Grants, and concessions relating to capital taxation of woodlands, further brighten the picture. Likewise does recognition of the importance of the many non-wood benefits of forests, some of which can be evaluated in cash terms (sporting rentals, [1] fees for recreation) but all of which can be added to cash benefits in considering the worth of forestry. Private forestry can often be at least as enterprising, productive and remunerative as large-scale State forestry, because of (often) better sites, grant-aid, lower overheads, and when competent and enthusiastic foresters make use of every local market for their products.

The economics of the forestry investment decision are chiefly determined by the relationship between the initial net costs of establishment, balanced against income from thinnings and final felling as and when they occur often far in the future. Cash flow is also a very important consideration in some cases. Therefore, economics enters forestry in all operations and decisions. Much depends on the location, the intensity of the forestry practised and the chosen 'point of sale' of produce, all of which will largely determine requirements of labour, materials, machinery and contracting. Forestry practised using direct labour is much different from, and may require more capital than, forestry undertaken through contractors and especially where all produce is sold standing. Different, too, is traditional lowland forestry by contrast to extensive upland afforestation.

Decision-making. Forest economics has an important part to play in assessments of the expected economic performance of options in policy and management. Rational decision-making means choosing courses of action that should best achieve stated objectives. Forest management (see chapter 13) is also about implementing decisions. From the economic viewpoint, silviculture involves making and implementing a set of interrelated investment decisions. Foresters, as decision-makers, are concerned with changes of action, or with undertaking one course of action rather than another — some of whose benefits are not realized for many years.

One of the commonest specific objectives in forestry is to achieve the highest possible financial return; another is to maximize timber production. However, objectives may often be supplemented or replaced by multiple objectives — e.g. including landscape, recreation, wildlife conservation or sporting. This may induce conflicts in one or more forms. Where profit-maximization is the objective, one way to determine the best course of action is to evaluate costs and revenues for every possible combination of activities, and to adopt the combination which maximizes net revenue. Where factor constraints operate, the cost of the constrained factors is an *opportunity cost*.[2]

---

1   Some owners may not receive any sporting rental since they may consume all the benefits themselves. In these cases the value remains the same even though because owners consume it themselves the benefit is never expressed directly in cash. At present on many private estates the non-wood benefits are more important than those which are marketable, especially timber.
2   *Opportunity cost* is 'the revenue (or benefit) foregone when a factor of production [classically: land, labour, capital, enterprise, raw materials] is withdrawn or withheld from an alternative course of action. The most obvious example is the opportunity cost of land for plantations, the forgone revenue in this case usually being its value to agriculture'. (Price 1989, p. 23.) The true opportunity loss is the *net* revenue or *profit* forgone of the best alternative.

Most elements of decision-making interact, examples being: (i) the cost of thinning may depend on whether a plantation has been brashed; and (ii) whether pruning is worthwhile depends on the likelihood of a 'premium' for knot-free timber, which in turn depends on whether it is destined for quality markets (pruning adding nothing to the value of pulpwood or boardwood). Therefore no decision can be taken in isolation, but each is made given certain assumptions about other elements in the decision-making process. Decisions on rotations are made for a given species on a given site managed under a given thinning regime, while species choice is made with some idea of rotation and end use in mind. Silviculture analysis for decision-making is often best undertaken in reverse chronological order. Thus when some hypothesis about end uses has been formulated, the desirable rotation length can be considered for each species suitable for that use. When rotation length is known, thinning regimes can be decided. Only when the most profitable regime for each species has been determined can species choice be made, and the courses of action be selected that will best meet the objectives.

Profitability in forestry is heavily dependent on assumptions about future costs, revenues, non-wood benefits, growth rates, prices, rotations and interest rates; nonetheless, calculations of *relative* profitability are useful. Over the same period of time, usually the full rotation, the choice between two species requiring similar expenditures and expected to sell at similar unit prices will depend more on their relative rates of volume production than on fluctuations in future prices.

The effect of time-scales on financial calculations. Decisions in forestry normally involve a time dimension; and hence it is frequently necessary to compare costs and revenues at different times. The period between planting and production is the heart of forestry economics and dominates forestry under any but the shortest rotations.

In both afforestation and restocking the bulk of the expenditure is at the beginning of the rotation whereas income from timber comes much later. Additional to the first year establishment costs, there is subsequent expenditure: beating up and weeding; later, cleaning and, occasionally, top-dressing with fertilizer, brashing and (occasionally) pruning; also, at some time, roading. Concurrently, there are annual maintenance costs of fences, drains and roads, as well as protection (see chapter 6). There may be intermediate income – grants, Christmas trees, foliage, sporting rental and sales of thinnings – but income from the final crop is usually forthcoming only after several decades. Consequently, early revenues, and prudent economizing in early costs, tend to have very beneficial effects on profitability.

Discounting. All costs and revenues, whenever they occur during the rotation, can be made directly comparable between species and options by compounding, or more conventionally discounting them to a common point in time – usually with discounting to the present which may be the beginning of the rotation or any point within it when a decision is being made. Discounting is the opposite of compounding. The concept of discounting for time is based on the premise that, in general, individuals prefer to receive benefit now rather than later. Just how much, depends on individuals' attitude to time – their *time preference*. *Time preference rate* is the discount rate reflecting consumers' preference for consumption now rather than at some future date; and hence the private investor's time preference will be reflected in the rates of return of forestry projects which are considered acceptable.

The *discount rate* is that rate by which future sums of money need to be discounted to give them present (or equivalent current) value. The longer the period of time and the higher the discount rate used, the greater the impact on future sums, and generally the less likely are long-term investments to be able to show a profit. The choice of discount rate is of paramount importance in forestry. It permeates every aspect of silviculture and influences the weight given to long-term costs/revenues and non-wood benefits. (The Forestry Commission currently uses a discount rate of 6% in decision-making and investment appraisal.) In the private sector, the choice of discount rate is a matter for investors, i.e. the rate which they consider acceptable.

In forestry economic appraisal (section 3 *infra*), the general range of rates used is 3% to 7%. Tables for a range of discount rates and *discount factors* are to be found in *Parry's Valuation Tables and Conversion Tables* (Davidson 1978). A few representative factors are given in Table 47. For costs and benefits occurring every year or for a period of several successive years (i.e. an annuity), tables are available which show the present value of £1 per annum spent or received each year over a period of years.

Table 47. Some representative discounting and compounding factors at 3%, 5% and 7%.

Years	COMPOUND INTEREST ('Amount of £1')			COMPOUND DISCOUNT ('Present Value of £T)		
	3%	5%	7%	3%	5%	7%
1	1.030	1.050	1.070	0.971	0.952	0.935
5	1.160	1.276	1.403	0.863	0.784	0.713
10	1.344	1.629	1.967	0.744	0.614	0.508
15	1.560	2.079	2.759	0.642	0.481	0.362
20	1.806	2.653	3.870	0.554	0.377	0.258
25	2.094	3.390	5.427	0.476	0.295	0.184
30	2.427	4.322	7.612	0.412	0.231	0.131
35	2.814	5.516	10.676	0.355	0.181	0.094
40	3.262	7.040	14.974	0.307	0.142	0.068
45	3.782	9.000	21.002	0.264	0.111	0.048
50	4.384	11.467	29.457	0.228	0.087	0.034
55	5.082	14.635	41.315	0.197	0.068	0.024
60	5.892	18.679	57.947	0.170	0.054	0.017
65	6.830	23.840	81.273	0.146	0.042	0.012
70	7.918	30.426	133.990	0.126	0.033	0.009
75	9.179	38.833	159.876	0.109	0.026	0.006
80	10.641	49.561	224.234	0.094	0.020	0.004
100	19.219	131.501	867.716	0.052	0.007	0.001

The discounted cash flow of a project is constructed by multiplying each figure of expenditure and revenue by the discount factor shown in the tables under the appropriate discount rate and at the appropriate point in time. When summed the products will give the total discounted expenditure (DE) and total discounted revenue (DR). When put together the resulting figure will give the net present value (NPV). Calculators with power functions and electronic spreadsheets (see chapter 16) have made the calculations easy.

Non-discounting criteria. Price (1989, pp. 100–1) explains the *non-discounting criteria* (i.e. where the time dimension is not taken into account): (i) *maximum forest rent*[3] (the *mean annual net revenue* over the rotation, making no distinction between revenues and costs occurring at different times); and (ii) the *payback period* criterion which 'requires that projects *eventually* repay investment monies. The most desirable course of action is that which pays back in the shortest time'. He further points out (p. 103) that 'one way to include both magnitude and timing of all revenues and costs is to calculate profit after compound interest has been paid on loans. Profit after interest produces results which are intuitively accepted in silvicultural decision-making: once rates of interest are adjusted for inflation, some silvicultural practices seem profitable, others appear unprofitable. It is adequate as a selection criterion of whether a forest investment is acceptable but is deficient as a selection criterion, because the point in time at which profit after interest accrues depends on the duration of the investment'.

The effects of compounding and discounting on the economics of forestry are dramatic. Compound interest or discount seems to dictate the adoption of plantation systems of fast growing species on short rotations and to denounce those who persist with natural management systems (see chapter 7) for their disregard of economic realities (Leslie 1989). But these conclusions are not all inevitable. Everything actually depends on the discount rate used. Relatively high rates favour short rotations and hence fast-growing species in plantations, but at low rates the reverse could just as easily apply (Leslie 1989). Relevant comments by Spilsbury (1990) are noted in section 4 *infra*.

2. Measures of profitability

Four measures of profitability used in private forestry, embodying rate of return concepts, are discussed below.

1   *Net annual income* as a measure of profitability may be related to a hectare, a stand or to a woodland. It may be calculated as the net income annually derived from a woodland (or group of woodlands) having a reasonably *normal* distribution of age-classes (see chapter 7). In order to assess the current rate of return on capital rather than the return to the land, an attempt may be made to measure capital value and to relate the net annual income to this. It is difficult (though not insuperable) to assess the capital value, particularly of immature stands (see chapter 16). However, as net annual income ignores interest, the calculation of capital value

---

3   *Forest rent* 'provides a measure of the average annual net revenue which is often more useful than the net revenue per rotation (NPV at 0% discount rate) as the bias of rotation length is removed. Its use as a criterion for the selection of investments is traditionally confined to forests of *normal* age-structure, with annual costs for planting an area being met by revenues from felling another equal area in the same year'. (Spilsbury 1990).

becomes an arbitrary proceeding, and the measure is useless for management purposes. Firstly, it pays no attention to the times at which expenditures are made and revenues received, and therefore ignores the cost of capital. Secondly, although in a *normal* forest the net annual income in any one year may be compared with that in another to give a measure of relative success in management, such a comparison is useless in a developing forest where the relative levels of expenditure and revenue depend largely on the age of the forest rather than on the efficiency of management.

2 *The indicating percent* (or current annual value increment percent) shows the rate at which the value of a stand or whole woodland increases in a year in comparison with the value of the growing stock. Dividing the current value of the crop by the rate of value increase, gives the indicating percent. No interest rates are involved. Its advantage is that it does not depend on historic costs or future expectations. It is of value when dealing with a mature stand for which no financial records are available: if the internal rate of return (IRR) *infra* of a stand is known it is simply a matter of seeing if the indicating percent is higher or lower than this figure; if it is lower the stand should probably be felled; if it is higher it can probably be prudently left for a period before felling, after a further measurement. Perhaps even more relevant is its use for comparison with investment opportunities outside the forest.

Shade-bearing species grow slowly at first and only gradually build up to peak annual increment. Hence *increment percent* remains above the required overall productivity. Increase in price/m^3 raises the rate of return represented by current crop increment: if *volume increment percent* is 3% and the price/m^3 for the crop is increasing by 1.5% for each percentage point increase in tree volume, the indicating percent is just over 7.5%. Adding volume increment percent (3%) to price increment percent (4.5%) gives indicating percent approximately (Price 1989, p. 133). An initial rapid increase in price/m^3 up to a price plateau gives fast fall-off in price increment percent and short rotations, while prolonged, steady increase favours long rotations. The lower the discount rate, the further the indicating percent can fall before extending the rotation ceases to be a worthwhile investment. Low discount rates usually lengthen rotations. (In general, low discount rates, short rotations and early returns from thinnings increase the attractiveness of an investment, while high discount rates, high early expenditure and long rotations do the reverse.)

3 *Internal rate of return (IRR)* is the rate of discount which makes discounted expenditure (DE) equal to discounted revenue (DR). This measure of profitability takes full account of the time interval which elapses between investment and return. In calculating IRR, no account is taken of market rates of interest or time preference rates: the IRR refers exclusively to the investment's *internal* ability to generate a rate of return. IRR is the maximum interest rate charged on investment funds at which an investment could break even. Usually it is found by trial and error calculations. Among compatible investments, the IRR criterion selects investments in descending order of IRR. Among incompatible investments, the investment with the highest IRR is selected.

IRR is the best measure to use when comparing the profitability of forestry with that of quite dissimilar investments; for example, where forestry can produce a real rate of return of 5% compared with hill sheep farming at 3% or a grouse moor at 10%. Such figures can also be compared with industrial or commercial investments and show immediate advantages or disadvantages. Foresters should at least know how to calculate IRR for an investment which they wish to implement. Price (1989) indicates two important objections in the forestry context to selecting projects by IRR, and advises against judging a forest investment by it. However, financial yield (IRR) is widely used in the private sector since many investors do not have a specific discount rate to use.

4 *Net present value (NPV)* at a chosen rate of discount is the most widely used and most satisfactory measure of profitability in forestry, or for expressing returns from an investment project. The discounted sum of all revenues (DR), minus the discounted sum of all expenditures (DE), is variously known as NPV, net discounted revenue (NDR), net present worth, and net discounted cash flow. The result is a present equivalent of future values. If DE is greater than DR the NPV will be negative, indicating that the project has an IRR of less than the discount rate. Conversely the NPV is positive when IRR is greater than the discount rate. A negative NPV does not mean that a project or investment is necessarily undesirable, simply that the rate of return is lower than the discount rate. Projects from a group of compatible investments would normally be selected in order of highest NPV. Sometimes maximizing NPV is secondary to other objectives, for example, increasing the conservation value of a particular area. Discounted cash flow provides a method whereby the opportunity cost of such decisions can be considered.

For ease of calculation, all expenditures or income in one twelve month period can usually be assumed to take place at the end of the year. Year 0 is usually the first year of the project, but may be later such as in planting models, where year 0 is the year of planting and year –1 is the year of ground preparation. Thus there may be negative years of operation, the values from which have to be compounded up to year 0. However, normally the time of decision is called year 0, i.e. the beginning.

NPV provides a simplified method in assessing the relative profitabilities of different species, afforestation or restocking methods and other silvicultural or management techniques which involve different patterns of expenditure and revenue over time.

However, there is a difficulty with NPV in comparing two investments of different durations, for example, the use of a given site by oak for 100 years against Norway spruce for sixty. The normal procedure is for a whole succession of oak rotations to be compared with the consequences of a whole succession of spruce rotations i.e. to take account of all future costs and revenues to infinity. The convention of viewing expenditures and revenues in perpetuity is a realistic one because it ensures that like is compared with like. Different levels of yield, cost or price may be assumed for future rotations, but the effect of different assumptions for the second and subsequent rotations has very little influence on the results obtained.

The figures calculated are only as good as the assumptions which have to be made, but their value as indicators of relative profitability remains unimpaired, and they may have to be accepted as absolute measures in some circumstances, e.g. in fixing a reserve price when selling, or a maximum price when purchasing. Initial costs per hectare are usually not difficult to assess, and the greatest difficulty concerns the evaluation of price for timber of different tree sizes and species. The Forestry Commission provide their forest managers with guidance on these subjects; private foresters will provide their own prices, costs and rate of interest.

The figure of NPV per hectare calculated is one that has practical significance. Since discounting in most cases is to the present day, the magnitude of the final figure is something which is easily appreciated. NPV as described above can help the intending investor decide whether a forest investment is beneficial or not. Either the rate of return can be calculated to compare with alternative rates of return from other possible investments, or the NPV at a given or desired rate of return (i.e. the chosen discount rate) can be derived. Maximization of NPV is the criterion used by the Forestry Commission and it is important to appreciate that this may give very different answers to maximization of IRR. For example Oak YC6 would under the NPV criterion have a rotation of around 120 years but, to maximize IRR, nearer 200 years – because IRR would be much lower.

NPV and benefit-cost ratio with limited funds. Both profit after interest and NPV assume that investment funds which can earn the going rate are available without restriction. But if funds are limited, even when they can earn the going rate, a more efficient allocation of funds is achieved by using *NPV per £ invested,* or *benefit-cost ratio,* that is: the PV of revenues or benefits divided by the PV of costs. Investments are acceptable if benefit-cost ratio exceeds unity. This always gives the same result as accepting investments whose NPV is positive (Price 1989, p. 106). However, NPV alone does not measure efficiency of fund allocation.

Annuitizing. An alternative approach to the problem of comparing unequal rotation lengths is to measure NPV per year. NPV cannot merely be divided by the number of years in the project since figures farther into the future have a lower present value. Hence, the annual figure derived has to be larger to take into account the (average) effect of discounting these annual 'profits' over time. A table of 'Annual charge to write off £1 over a number of years at various rates of interest' gives suitable factors for performing this task. The principle used is an *annuity,* i.e. the annual amount which has the same value (when summed and discounted to the present) as the cash flow from which it is calculated. To annualize or annuitize is to express the NPV of a project as equal payments over a certain number of years at a given discount rate (e.g. with a discount rate of 5%, £1 is equivalent to five annual payments of 23.1p). The payment is called the *annual equivalent cost.* The *annuity factor* is the figure which, when multiplied by the NPV, gives the annual equivalent cost.

Inflation. Cash flows deal with sums of money at different times, and so inflation has to be borne in mind. In investment appraisal it is essential to calculate using a common price base (technically referred to as 'at constant prices'), so that the effect of inflation does not artificially reduce or increase the net return from a project or option. To do this, a price base is chosen (usually the current year), and all the costs and returns from a project are expressed at the prices one would expect in that year. So for example, a rotation could be reported to have a NPV of £1,000 at 1991 prices. More simply, predicted cash flows do not have to be corrected for inflation, but instead a real rate of discount is used, i.e. the % is corrected for inflation. However, in gathering data from the past, to base predictions on, correction for inflation does have to be made.

The Gross Domestic Product (GDP) deflator or price index is one way to adjust data gathered in different years from nominal to real values. GDP is a measure of how the value of money has changed over time, with one value (the base) equal to 100, and the other values being the equivalent number of £s in each year that would be required to buy the same amount of goods as £100 in the base year.

(The index used by the Forestry Commission is based on the prices of goods sold in the UK including raw material and industrial inputs, i.e. a GDP deflator. This is considered to be more applicable to the Commission than the Retail Price Index – the consumers' price index – which is calculated from the cost of a basket of selected retail goods rather than industrial outputs – although RPI is widely used and understood in the private sector.)

Price (1989, p. 64) points out that 'in practice it is impossible to predict changes in inflation and monetary interest rates over a forest rotation. Since the two tend to fluctuate in tandem, especially in the long term, the real interest rate (representing approximately the difference between them) is much more stable. Thus it is easier to relate future revenues to the current purchasing power of the £ – that is, to project costs and revenues on the basis of current prices – and to discount at the real discount rate. Increases in price are included only if they are real: that is if they are increasing *in relation to the price of goods generally*'. Hence, in investment appraisal only real values are dealt with, so it is assumed that all costs and revenues will be relatively the same; and it is only necessary to adjust for inflation when there are grounds to suspect that something will experience a real or relative price change.

Once all values of income and expenditure have been adjusted to constant prices the rate of return is a real rate of return received after adjusting the effects of inflation. A nominal rate of return is the rate unadjusted for inflation. Only if inflation is zero will real and nominal rates of return be the same. In some appraisals it is expected that there will be a relative price change between factors (e.g. general inflation might be 10% and wage increases 8%). As long as it is expected that this difference will continue throughout the life of the project at the same *rate,* then an adjustment can be made to the discount rate used for that specific item (or the factor value can be changed and the discount rate retained).

Risk and uncertainty. *Uncertainty* describes situations where several outcomes may result, but the various probabilities cannot be calculated. *Risk* describes situations where several outcomes may result, but the various probabilities can be calculated. Uncertainty obviously has a bearing when considering future events. Often in investment appraisal it is assumed that cash flows can be predicted with certainty. Where there is a number of possible outcomes resulting from an investment it may be possible to measure the risk involved by considering what has happened in the past. It may then be possible to allocate probabilities to each possible outcome. Often in forestry such probabilities of risk (e.g. windthrow) will have to be estimated subjectively.

Forestry is a relatively safe investment: timber in some form or another will always be needed. If sound silviculture is practised, trees will continue to grow satisfactorily whatever the monetary or political situation. The investment tends towards being inflation-proof. However, throughout a forest rotation there is a range of uncontrollable and unpredictable factors operating, exacerbated by timber's long production cycle. Uncertainties include: wind, rainfall, snow, global warming, pests, diseases, world timber prices, grant levels and capital taxation incidence. Price (1989, pp. 113–4) points out that in practice, most decision-making in forestry occurs under risk or uncertainty, the five main sources being: (i) the natural environment (hazards of climate and biotic attack); (ii) technological advances; (iii) human factors (arson, or accidental fires); (iv) markets (labour and timber); and (v) changes of government, political theories, and taxation and grant regimes. So it is surprising that forestry has continued as steadily as it has. The factors 'underline the robustness to which evolution has conditioned forests, and the versatility of their products in use' and underline the realization of the need for a stable political background for forestry (Price 1989, p. 114).

In private sector investment appraisal, the choice of discount rate fundamentally depends on the risk inherent in the project to be undertaken. In its simplest form the rate is made up of two elements: the risk-free rate of interest, the rate of return required on a project assuming that its future net cash flows are certain, plus a risk premium positively related to the level of project risk (McKillop and Hutchinson 1990). The risk free rate of interest can be measured by index linked gilts; and most financial institutions look for 2% over what that figure is (in 1991 4.2% plus 2% = 6.2% approximately). An alternative is to vary the size of the risky cash flow items according to their riskiness.

Foresters may ignore the risks and uncertainties in their calculations of NPV, yet most take some precautions against them. Techniques which allow a rational appraisal to forestry decisions whose results are unsure are set out by Price (1989, pp. 118–9). The first need is to assemble known facts in an informative way (situations where absolutely no assessment of probabilities can be made are rare). A common practice is adding a 'risk premium' to the discount rate used in NPV calculations for risky projects. But not all risks increase over time: fire risks are often greater earlier in a plantation's life; wind instability may not be a problem for thirty to forty years, but may increase dramatically thereafter. This does not distort relative discounted cash flows at different points in time. An obvious and familiar strategy is to adopt risky investments but to insure against risk. Fire insurance is often sought; insurance against wind damage is less available and less common, and insurance against biotic risk is rarely available.

3. Economic appraisal for management decisions

Economic appraisal is a term used to describe the evaluation of different options, in order to meet a specified objective. In forestry appraisals the costs of different projects or operations are considered together with the likely volume production from the trees. Cash flows are discounted in order to allow for differences in timing. Investment appraisal supplements the measures of profitability discussed in section 2 *supra,* and the profitability of private forestry in section 4 *infra.* The purpose of appraisal may not be simply to discover the course of action likely to yield the highest profit but rather to determine the cost of departing from it in order to satisfy some non-commercial objective.

Busby and Grayson's *Investment Appraisal in Forestry* (1981) is an important relevant text with particular reference to conifers in pure even-aged stands, and uses NPV. Eight appendices provide: a set of compound interest tables at 3%, 5% and 7%; a price-size schedule for standing conifers (now superseded by Forestry Commission price-size curves – see chapter 11); a general index of prices (factors to adjust for inflation); discounted revenues calculated to year of planting for seven main conifers at 3%, 5% and 7%; discounted volumes for those seven conifers at 3%, 5% and 7%;

discounted revenues for different thinning regimes of Sitka spruce; optimum felling ages for the seven conifers at 3%, 5% and 7%, thinned and unthinned; and age of first thinning as specified in the Forestry Commission Management Tables. The text is in effect a do-it-yourself kit for investment appraisal and outlines the principles, specifies the assumptions, provides tables of relevant use, and worked examples of their application to a variety of cases. If a particular decision area is not illustrated in the examples, sufficient guidance is given to enable the undertaking of the appraisal from first principles.

Williams' *Decision-making in Forest Management* (1988) explains the basic concepts, methods, applications and limitations of financial appraisal and includes appropriate practical examples of uses to which the 'tools' of forest economics can be applied. A third publication, the Forestry Commission's internal *Investment Appraisal Handbook* (Insley *et al.* 1987) provides comprehensive techniques and their applications. In addition to relevant chapters by Price (1989), several economists abroad have written extensively on investment appraisal in forestry. These publications relate almost wholly to conifers in pure even-aged stands, but management of broadleaves merits the same treatment of appraising options (indeed, the range of options may be wider), and there is now less difficulty of nominating plausible and generally acceptable assumptions on the value of broadleaved trees of different sizes and qualities.

4. Principles of economic appraisal

Investment appraisal requires consideration of the following steps. First, set the *objective(s)*. Second, state the *options*. Third, for each option consider: (i) *costs* (planned expenditure and when it occurs); (ii) *benefits* (the revenue benefits or financial savings and when they occur); (iii) *non-wood costs and benefits* which can be identified but not quantified (noted later in this section); and (iv) *relative price* (if all costs and benefits are likely to increase by the same amount with inflation, use today's prices throughout; if not, adjust for relative price changes). Thereafter, (v) by *discounting*, obtain the present value of all costs and benefits for comparison. If any of the options have different time lengths, *annualize* the NPVs. Should any of the values of costs and benefits be subject to *uncertainty*, revise the assumptions by *sensitivity analysis* using optimistic and pessimistic estimates. If there are budgetary constraints, appraise by *cost–benefit ratios*. In conclusion, clearly indicate the preferred option and the extent to which its performance exceeds that of its rivals.

The time lag between incurring costs and receiving returns means that it is not always easy to evaluate the costs and returns. In particular, difficulties arise in predicting future inputs and outputs in physical terms (e.g. man-hours, cubic metres), assessing the future value of these inputs and outputs and comparing the value of the costs and returns arising at different times. These difficulties mean that it is often not immediately apparent that a particular investment (it may be a whole rotation, or merely a single operation) is worth doing or which is the best investment to choose from among several alternatives. However, this is no excuse for not attempting to make any appraisal of the investment but emphasizes the need for such appraisals rather than relying on intuition.

The remainder of this section outlines the basic concepts of investment appraisal. Space does not allow the inclusion of Tables of use in appraisal; nor of worked examples of the application to representative cases in forestry; for these, foresters are referred to the specialist texts noted earlier.

The process of investment appraisal examines all the prospective costs and benefits of a new project, discounting them at the appropriate discount rate, choosing between options on the strength of these results, then forming conclusions as to which option will best satisfy the project's objectives; also evaluating the desirability of committing resources to them. An *objective* is the aim of the investment project, usually either to maximize benefit from a project or operation or to minimize its cost. (Other less easily defined objectives may exist, such as to provide rural employment.) *Benefits* are the positive or good results (or aspects) of a project; the opposite being *cost*. Benefit is a wider concept than revenue, as it encompasses other non-financial results of a project. *Options* are the alternative ways of fulfilling one or more objectives. Investment appraisal examines as many options as possible to find out which is most satisfactory in fulfilling the appraisal's objective(s). Keeping the status quo, or 'do nothing' must always be considered as an option, as it may be the best way of satisfying an objective.

Because of the long-term periods involved in woodland management, investment appraisal calculations are done in real terms, i.e. it is assumed that all future costs and revenues increase due to inflation by the same amount, so that inflation can be ignored. (The figures in an appraisal cannot be used for budgeting, unless inflation is taken into account when the budget is drawn up. Percentage figures used in an appraisal, such as the rate of return or interest rate, must also be in real terms, and therefore be net of inflation.)

Money that is invested in a project could have been used for something else. In the alternative uses, it would have earned some rate of return or rate of interest. By investing it into a project, that interest is lost or forgone in order to reap the returns that the project has to offer. One way of deciding whether the returns from this project are higher than those in the alternatives is to calculate NPV. This enables alternative options to be ranked and compared. If NPV is greater than 0, this suggests that the project is earning more than the assumed alternative interest rate. If NPV is less then 0, this suggests the opposite, that the actual rate of interest is less than the assumed alternative interest rate; and that taking on the project would incur a loss, as higher returns could be obtained by investing in an alternative.

It is necessary to have a clear definition of the *objective* which the project being appraised is intended to achieve. In some cases there will be a single objective which can be clearly defined and quantified; in others the objective may not be so easily measured. In many cases there will be multiple objectives where it is helpful to list the objectives in order of priority, and then to optimize the main objective within the constraint that secondary objectives achieve certain defined levels. There may be many ways in which the objective can be achieved and it is necessary to reduce the list selecting only those options which are most likely to be suitable for further investigation.

All the costs and benefits associated with each option should be identified (unless a factor is common to all the options, in which case it can be ignored). The value of all items should be taken as their *opportunity cost*. All previous costs (termed *sunk costs*) should be ignored. Only opportunity cost should be used and, as sunk costs are unreclaimable, their opportunity cost is zero. The only costs that should be taken into account are ones proposed for the future. Some might include non-wood costs.

Non-wood benefits and costs. Woodlands, besides their timber values, have many non-wood benefits (see chapter 13). Those that can be evaluated include: rental of grazing, fees for shooting/hunting, charged-for recreation, benefits to water catchments, and to agriculture increases in crop and livestock yields. Others commonly relate to landscape, wildlife conservation and recreation; there is no 'market' for any of these 'goods' and 'services' and so it is not easy to establish the prices at which they could be 'sold'. This is a classic case of market failure. There is an established social good, but the owner receives no payment or reward. However, an attempt to evaluate them in money terms gives them an appropriate emphasis in comparing courses of action. One way of approaching the measuring of the value of non-wood benefits is to calculate the opportunity cost. Price (1989) discusses evaluation of environmental quality and reviews the capabilities and limitations of evaluation techniques for non-wood benefits, noting specifically the methods advocated by Sinden and Worrel (1979) and outlining possible bases of value along with some of their problems. In determining the values, costs of facilities must be deducted – all the values have the same residual nature as wood revenues. Most revenues generated by the facilities provided depend on the woodland's location, accessibility, age and structure: residual values are less when remote from consumers. In addition to all the foregoing benefits is the contribution of forests to containing the 'greenhouse effect' (see chapter 3 and section 5 of this chapter). For more recent comments see Bateman (1991).

Non-wood benefits divide into three groups: (i) items whose monetary value can be estimated; (ii) items which can be quantified but where money values cannot be estimated; and (iii) items which cannot be quantified. In an appraisal it is important to include all relevant information. If there are items for which monetary values can be estimated then these should be included in the net present value (NPV) calculation with a statement indicating the degree of uncertainty associated with them. If items cannot be valued then an attempt should be made to quantify them by some other scale. If neither of the above is possible then the appraisal should discuss the impact of the different options, trying to rank the options according to their effect on the item in question. *Sensitivity analysis, infra,* allows the implications of various possible values of non-wood costs and benefits to be explored.

The values of responding helpfully to public concern or protest over major changes during the rotational cycle of a woodland are inaccessible to ordinary techniques of evaluation. However, the economics of landscape have been discussed comprehensively by Price (1976, 1978, 1989). So too have the economics of recreation (Price 1989) whereunder are discussed such factors as travel cost analysis and the public's willingness to pay. He points out that recreational facilities offered free in a woodland (perhaps because usage does not justify cost of collection of fees) may be sold commercially where use is heavier, and the commercial fee may be taken as a measure of benefit; but that it is questionable whether the product in the woodland is the same as that sold commercially, and the type of consumer (and therefore willingness to pay) may also differ greatly.[4] There is the need for further research, on both State and private forestry, to examine issues concerning the non-wood benefits.

---

4 The values of recreational use at six Forestry Commission forests have been calculated by Willis and Benson (1989). Using the travel cost method, values varied from £1.26 per visitor at Grizedale to £2.51 at Thetford with a weighted mean of £1.90. The wildlife attributes of each forest were estimated to contribute about 38% of this value.

5. Data and formation of assumptions for economic appraisal

It is necessary to identify which costs and revenues are to be included in an investment appraisal.

Costs. Of *costs*, one classification is into traditional factor-of-production categories: land, labour, capital, enterprise, technology and overheads; and three concepts are important: *total cost* (the cost of carrying out the complete operation on a hectare or stand), *marginal cost* and *incremental cost* (more useful in comparing two or more courses of action). Foresters can decide what concepts and classifications to use; and what changes result from using a factor of production. In many instances there are costs associated with an enterprise which are unlikely to change whether a particular investment is undertaken or not. These are *fixed costs* which do not vary with the investment and thus can be omitted from the appraisal: into this category fall most *overhead costs,* but not *labour oncosts* (see comments on overheads in chapter 16 section 3). *Variable costs* need to be considered because they will vary depending on whether the investment is undertaken or not: the further into the future a cost is incurred, the more likely it is to be variable. Cash flows which have already occurred *(sunk costs),* and so will not change as a result of the appraisal, should be ignored.

Overhead costs (see chapter 16) should be included in the appraisal if a full assessment of the financial implications of a course of action is required. They should also be included if they are expected to differ between alternative projects. It is desirable to consider cost levels in recent years as well as the present in order to establish whether current costs are appropriate (they may be abnormal as a result of unusual weather) and also whether there is any significant trend, either up or down. To do this it is necessary to inflate the unit costs of previous years into current £s.

Operational costs vary according to site conditions, access, silvicultural system, scale and region; also between conifers, broadleaves, or a mixture, and between afforestation and restocking. There are interactions in costs between some of the operations; for example, between cultivation, drainage and weeding, between plant species/type and planting method, and between roading and most other operations. In the private sector insufficient information is known of true levels of oncosts and overheads for any of the major cost components, i.e. labour, machinery and materials. Forest management decisions, because of these significant gaps in data, have sometimes to rest on an undue degree of guesswork (Lorrain-Smith 1989). The economic surveys of private forestry carried out annually by Oxford and Aberdeen Universities, and from 1991 to be joined by Bangor, provide reliable costings data for the first five years of a plantation. Current (1991) ranges of operational costs of establishing plantations have been indicated in chapter 5, of tending plantations in chapter 6, and of harvesting (comprising felling, extracting and transporting) in chapter 10. They can be supplemented by costs derived from forestry operations currently being undertaken.

Revenues. Revenue is the positive side of profit, the most important being derived from thinnings and final crop. In decision-making, so far as *revenues* are concerned, three concepts are important: *total revenue* (relevant to whether a defined course of action is worthwhile or not), *marginal revenue* and *incremental revenue* (more useful in comparing two or more courses of action). Foresters can enhance the values of revenues by increasing quantity and quality of output by silviculture, improving marketing of given output, and changing timing of output (Price 1989, pp. 17–18). They must calculate what revenue is obtained by adopting a certain course of action, and what changes in revenue are obtained in adopting another course.

If there are additional sources of money income which can be foreseen (Christmas trees, foliage, recreation fees, sporting rentals) these should be included. The standing tree price *(stumpage)* always is used (harvesting and marketing costs are not included). Resort is usually made to Forestry Commission price-size curves (see chapter 11). In the event of the need to inflate past revenue to current £s, recourse is necessary to a table of factors that adjust for inflation (see section 2 *supra*). To overcome the difficulties posed by short-term price fluctuations, which can be considerable, figures for a number of successive years can be averaged after correcting for changes in the value of money over the period. These results will vary according to the period of years which is considered. Current (1991) prices of timber are indicated in chapter 11.

Yields. Yield models (see chapter 12) are used to forecast the growth information necessary when calculating discounted revenue from a crop. For any harvesting operation, information must be supplied on the volume produced per hectare and the average size of tree, either as mean tree volume or diameter at breast height (dbh). The tree size is matched with the revenue per cubic metre from an appropriate price-size curve (see chapter 11). This revenue is then multiplied by the volume production per hectare, usually reduced to 85% to allow for unproductive areas, to give the total revenue from that operation. For a given rotation length, the discounted revenue in year 0 (DR_0) is calculated by discounting the revenue from each thinning and from clear-felling back to year of planting, and summing these values.

Grants (see chapter 15), affect costs and revenues to the private sector grower and thus influence commercial decisions. A private individual or company is most interested in the cash flows of costs and revenues, and their present value after grants have to be taken into account. Hence, it is necessary to evaluate costs adjusted for grants as and when they are received.

Taxation. Taxation of woodlands is discussed in chapter 15. Since 15 March 1988 the remaining taxation incidence (subject to transitional arrangements for income tax to 5 April 1993) affecting private forestry is that on capital, i.e. capital gains tax on land and inheritance tax on both land and crop; and Value Added Tax (VAT). The main problem in estimating the incidence and effect of capital taxation concerns the identification both of the investor and the beneficiary of the investment under appraisal. (In many cases of afforestation or restocking, the same individual will not receive the revenues flowing from the original investment.) A judgement on capital taxation should be made on the incidence and effective rates arising which, applied to revenues from the class of woodland investment under consideration, is equivalent to the expected burden arising from a transfer during lifetime or a disposal at death.

6. Applications of economic appraisal

Appraisals of particular interest to the private investor may relate, for example, to whether or not to invest in forestry, or to buy a specific woodland investment, or to choose between several woodland investments that are available. Again, to making a financial choice between forestry and some other use: for instance, sheep grazing, or whether to undertake afforestation and, if so, what can be paid for plantable land, or to invest in an estate sawmill, preservative plant or forest tree nursery. For most of these major projects, resort can be made to calculating its NPV at a chosen discount rate.

An appraisal for individual silvicultural and harvesting operations is hardly different in principle from that for the major projects; as examples: whether to improve the site factors, and what species to plant. (There are interactions between the costs of the foregoing and the type of plant, and the techniques of planting, weeding and beating up.) Again, there are options as to how to reduce wind damage, and whether or not to fence. Similar principles, along with relevant interactions, apply to the operations of cleaning, brashing, pruning and topdressing with fertilizers. Appraisals are also relevant to thinning or non-thin, respacing, and the type of thinning; also to clear-felling, choice of rotation, method of harvesting, restocking, and when to replace an uneconomic crop. In all the foregoing, the expenditure can only be justified if the financial benefit is greater than the cost.

Below is given information on Land Expectation Value, and on options in establishment, after-care (tending), thinning, clear-felling and restocking, roading and extraction, and machinery.

Land Expectation Value (LEV). Traditionally, foresters have tended to think of NPV to infinity as representing the value of the planting site, as indicated by the 'soil expectation value' and Faustmann's formula (Gane and Linnard 1968). Thereby, if the cost of the land were included in the discounted expenditure and the project broke even in financial terms, it would mean that the project had just earned the rate of interest used in the discounting calculations. In these circumstances, up to the amount of the LEV of the crop (without the land cost) could be paid for the land in the first place and still break even; if less than this were paid for the land, there would be a surplus at the end of the rotation. If more had to be paid, there would be a deficit. However, this obscures an important fact. Unlike money spent on establishment and maintenance, which is really an operational cost, the value of the land (in real terms) at the end of the rotation is frequently assumed to be the same. In some circumstances, the value could be higher in real terms (the land may have been improved over the rotation through various forms of cultivation, through semi-permanent fencing/walling, or roads may have been built). So it could be said that the land value is the true capital involved, in which case the LEV represents the gain on the land value. From each viewpoint, it is evident that more can be spent on land with high productivity, than on land with low productivity. (The inferior land might be bought cheaply and improved to achieve higher productivity by cultivation or fertilizing.)

The value of a parcel of land for forestry to a prospective purchaser depends on the expected costs of establishing and maintaining the projected tree crop, and the value of the revenues and benefits expected. The sum of these costs and revenues discounted back to the time of purchasing the land, which may be a year or two before the time of planting, represents the expectation value of the crop at whatever discount rate is deemed appropriate by the purchaser. Assuming this expectation value or NPV is positive, then this is the sum which the purchaser can afford to pay for the land and make a rate of return in real terms equal to the rate of discount used. If the expected NPV is lower than the market price of the land, or worse still is negative, then the land is only worth buying if the prospective purchaser expects either the value of the land to increase enough in real terms to offset the loss (although this can be included explicitly in the appraisal), or non-wood benefits are considered to offset the shortfall.

The appraisal desirably should take into account the value of: (i) a successor crop, if the area is to continue in forestry; or (ii) the land at the end of the rotation for other purposes; and (iii) any fixed assets such as roads remaining after final felling. Assumptions about the value of successor crops or fixed assets many years ahead can be no more than conjecture. However, such matters do affect the calculation of the land value and there are two ways of dealing with them: (i) assume that the area will be restocked with the same species and hence produce a similar NPV; or (ii) attribute *residual value* to the land and other fixed assets at the end of the first rotation. Whichever way is chosen to do the calculation, the answer tends to be the same.

If a project were to be repeated again and again, it would produce a stream of cash flows, each of which could be condensed into a new NPV recurring at regular intervals. These NPVs will all be the same, since it is assumed that the operation is repeated exactly, but they occur in different years. It is possible to discount all these NPVs back to the beginning of the first operation, because, as they get farther into the future, the discount factors approach zero. This is not a very useful technique when appraising capital projects because technology is changing so rapidly. It is, however, used occasionally when looking at forest crop rotations.

Options in establishment. The usual objective of establishment is to achieve a crop on the site by sound silviculture which meets the needs of the owner. In most cases, maximizing NPV will be appropriate. In other cases the main objective may be to provide or enhance landscape, nature conservation or sporting values. A balance between several objectives is often required, and it may be useful to consider options which would maximize each objective in turn as a guide to the opportunity cost of making the compromise, in terms of both revenue forgone and reduced environmental benefits.

In planning establishment, the cost and benefits relevant to each operational option should be considered. For each option, the likely cost per hectare is estimated. The later an expenditure prudently can be delayed the less is the discounted expenditure (DE). The volume growths are matched with appropriate prices to calculate discounted revenue (DR).

Species choice. The selection of which species of tree to plant will vary with the site and the objective of management (see chapter 4). Yet different species on the same site may produce different yield classes. All other things being equal, the choice should be made for the crop that gives the highest NPV. Where two or more species are likely to grow equally well, each of the same yield class, the financial return expressed by the NPV can still be quite different, because of the pattern of growth and the age of culmination of maximum mean annual increment. The appraisal is simply a question of comparing the alternative levels of expenditure and consequent returns.

Spacing. Spacing has a notable effect on tree growth. Most yield models cater for different spacings. Beyond a certain point which varies slightly with species, as spacing increases so does the main tree size, and the volume production per hectare falls. Spacing may also result in a deviation from the price–size curve, due to reduced unit value and increased harvesting costs. The effects of wider spacing all influence the DR from the crop. Changes may also occur in the DE required to establish the crop. DR (at age of maximum DR_o, or at a given top height) increases with wider spacing up to about 2.0–2.4 m and thereafter levels off or falls slightly at 3 m spacing. The spacing at establishment may vary either from a deliberate choice at planting or through scattered losses early in the crop life. In later life, respacing by one of several methods may be carried out (see chapter 8).

Cultivation. Cultivation can be achieved by several different techniques, as examples, scarifying, ripping, mounding, or ploughing, in which case there may be different costs of planting and weeding. If ploughing is decided upon, it is generally to prepare an easier surface on which to plant, and to suppress weeds; it may also improve initial drainage. Calculations can be made to relate cost of cultivation to savings in plants, planting, weeding and beating up, in order to aid the decision. Thompson (1984) considers the economics of ploughing.

Drainage. Crops on some sites are unable to reach their full potential yield because of expected windthrow, often before maximum DR_o is reached. If by appropriate initial drainage it is possible to ensure the crop standing to a greater age, then the cost can be compared with the benefit. The same applies to planting a less-productive but more windfirm species. Sometimes drainage is essential. Thompson (1979) discusses the economics of the operation.

Initial fertilizing. Wherever fertilizer application at planting is essential for acceptable growth, the decision must be to undertake it. Taylor (1991) has updated economic appraisal.

Fencing. To fence or not to fence will depend on the alternative courses of controlling or tolerating damage (see chapters 6 and 13). Costs of fencing will depend on what is to be fenced against, the appropriate type of fence (or treeshelter), and the size, shape and terrain of the site. The DEs of each option can be appropriately calculated and used for comparison.

Plants and planting. Various techniques can be compared by relating costs to the expected benefits. As examples, instead of conventional bare-rooted plants, using either whips in sleeves or small plants in treeshelters: both cost more to buy and to plant, but usually reduce the costs of weeding and protection. Another technique is to use container grown plants which may cost more to buy and to plant, but may have benefits in some circumstances. The DEs in each case can be compared to show which technique gives the greater benefit.

Weeding and beating up. Weeding cost has an interaction with costs of cultivation, plants and planting. As to the need for beating up, the economics of spacing, as well as of the timing of first thinning, suggest that there should not be undue concern over scattered losses in a young crop providing the stocking at first thinning age is of the order of 1,500 stems per planted hectare; but much depends on species. Where the early failures are grouped such that there will be an unacceptable gap in the canopy that will endure for a large part of the life of the crop, then an appraisal of whether or not to beat up is desirable.

Options in after-care (tending). *Annual maintenance,* discussed in chapter 6, includes maintenance of fences or treeshelters, drains and roads. Usually the costs are summed and accounted for per hectare, probably annually.

Protection. The need for protection from animals, insect pests, diseases and man, and the measure of damage by a particular 'agent', are liable to be peculiar to the locality and crop and may be sporadic (see chapters 6 and 13). Estimation of the risk of damage for a given crop, degree of protection and external circumstances, is a major difficulty in appraisal of options. Much depends, therefore, on the forester's local experience. One of the simplest ways of arriving at an approximate optimum level of expenditure is to consider a number of options (as examples, to fence, or to employ a gamekeeper, or to arrange regular shoots) and to estimate the probable reduction of damage associated with each option. These reductions of damage must be converted into gains in DR which can then be set against the relevant costs (DEs) of each option. The same concept of minimizing cost plus losses applies to fire protection. (Fire insurance can be arranged, but it often reduces the incentive to deploy protection.)

Expenditure on protection is pursued in principle up to the point where the extra costs are not more than the extra value of the crop saved, although it is often difficult in practice to judge with any precision when this point has been reached. The optimum level of expenditure depends on the value of the crops to be protected, the likelihood of damaging conditions occurring, and the susceptibility of the forest to damage. Protection can be regarded as an insurance, and the optimum level of premium could be calculated *if* the relationship between expenditure on protection and degree of damage was known. That is the principle but finding the balance is not easy in practice, yet is often greatly assisted by foresters' intuition and experience. Whatever costs are expected, they should be summed and accounted for per hectare, probably annually.

Top-dressing with fertilizer. The recommended method is to appraise the fertilizer investment in terms of the number of years saved, as defined by Everard (1974). A crop may show a growth response for seven to ten years after fertilizer application. In an appraisal an estimate should be made about the extra increase in top height which might be expected to result from the treatment. The extra top height increase divided by the average annual increase in top height over the rotation gives the number of years by which the fertilizer will shorten the rotation, thus drawing closer all subsequent thinnings and clear-fellings, and increasing DRs.

In pre-thicket or pole-stages, applying a top-dressing of fertilizer to increase volume growths is (as for all investments) only justified if the financial benefit is greater than the cost. Different nutritional regimes may enhance the yield class of the crop, producing a consistently different height-age relationship throughout the remainder of the rotation. The benefit can result either from shortening the rotation or increasing the yield. The response to fertilizing in terms of volume can be found by reference to increase in top height over the life of the treated crop. Given an assessment of this total increase to rotation age, it is possible to assess the change in yield class over the life of the crop and hence the DR increase. Increases in growth will affect the time of certain expenditures, and hence DE. Thus in crops which are to be thinned, the age of first thinning and hence roading may well have to be brought forward. In non-thin crops the felling age may be brought forward and hence roading costs (if undertaken) may be incurred earlier. Where major increases in growth are obtained by fertilizing, there may be cases where it is profitable to thin a crop which would otherwise be left unthinned.

The benefit of fertilizer treatment is an increase in DR. This may then be compared with the discounted cost of fertilizing to measure the cost-effectiveness of the treatment. 'Check' where present is often fairly patchy, which makes appraisal very difficult, because such areas cannot usually be singled out for reatment and hence any appraisal must also take into account the benefit of the application to neighbouring trees that are growing normally. Small areas can of course be treated manually. The benefits to an area in check can be calculated simply as the DR of the crop in that area, since it is a fairly safe assumption to make that if the crop continues in check for the rest of its life, it will be of no value when it is felled; or it might have a negative value, costing more to fell than the timber is worth. Adjacent crops that are not in check receive benefit in the usual way, in terms of years saved in the rotation, and, of course, all areas have to be weighted to reflect their relative importance.

The benefit from fertilizing an area suffering from check can be summarized as: DR from the crop x percentage of the crop in check, plus increase in revenue due to shortening of rotation length. The cost is, as always, the discounted cost of fertilizing (and occasionally of any increase in harvesting costs such as roading).

Years saved is a difficult figure to estimate: judging how much earlier a crop can be felled several decades ahead, due to one application of fertilizer now, is not an easy task. The problem can be made easier sometimes by calculating the increase in top height (above what would normally be expected for that crop in the next five to ten years), that would be necessary to justify the cost of fertilizing. Taylor (1991) has updated economic appraisal.

Cleaning. This option is sometimes necessary, and can be undertaken by manual, mechanical or chemical means (see chapter 6). It is possible realistically to anticipate the cost.

Brashing. Brashing is now less frequently carried out than in the recent past. Seldom is 100% brashing undertaken. Some outlay may be recovered by reduction in the cost of marking, measuring and removal of at least the first thinning. On some occasions, brashing (as well as rack-cutting) is undertaken to ease management and to benefit the interests of shooting, fire control and management (see chapter 6). Its benefit of producing knot-free timber in the lower length is occasionally recognized. In financial terms, therefore, the decision as to whether or not to brash is a matter for weighing the costs against the benefits.

Pruning. Pruning (see chapter 6) is now rarely undertaken with the exceptions of poplar and of selected broadleaves for quality production. Since pruning is only financially worthwhile if sufficient 'premium' is eventually received for knot-free timber, it is essential to estimate how the cost compares with the expected increased price. To do this, the extra income must be discounted back to the present and compared to the pruning cost.

Options in thinning. Thinning can take different forms, be started at different times or not done at all. Its economic aspects have been noted in chapter 8. It contributes a significant share in both volume and money terms in the total financial yield. The contribution in terms of DR is usually far higher than in terms of volume – due to the effects of discounting where a smaller but earlier return can be more valuable than a larger, later return. The main purpose of thinning is to increase the discounted value of a crop. A thinning operation which does not itself produce a surplus may still be a worthwhile investment if the net result will be to increase the NPV of the crop. In all cases the DR can be calculated using Forestry Commission Yield Models for Forest Management (Booklet 48), and price-size curves for unit values (see chapters 11 and 12). The intensity of thinning, or annual amount removed per hectare, is set in the tables at 70% of the yield class.

If crops are to be thinned at all, and that may well be the issue to be resolved (see Windthrow Hazard Classification in chapter 3), there are choices to be made on the age of the first thinning, the intensity, the type (selective, namely crown or low; or systematic, namely line or chevron), the frequency at which thinning should be done, i.e. the thinning cycle, and the age at which thinning should cease. The answers to each of these will depend on such factors as the objectives, the markets available at the time and likely to be available in the future, the quality of the stand, its susceptibility to wind damage and implication of the terrain for harvesting methods and hence cost. The major decision is whether to thin and, if so, when.

Because of the variety of thinning regimes that can be applied, differences in the prices of thinnings of various sizes and locations, and differences in the costs of access and risk of windthrow, it is difficult to make general rules about the course of management which is likely to produce maximum NPV. However, given some assumptions about these factors and with the aid of the Management Tables, it is possible to calculate the most profitable course of action. Where a crop has been in check from frost, heather, nutrient deficiency or periodic browsing, there may be a delay before it starts to grow in line with the yield model. In these cases the DR should be discounted back further to allow for the delay in tree growth. For example, a five year delay in tree growth may at 5% discount rate reduce DR by 20%, which may have a substantial effect on NPV. In the extraction of thinnings, costs will largely depend on the harvesting method and roading.

Options in clear-felling and restocking. *Rotations* have been discussed in chapter 10. The *optimum felling age* is usually seen as the age when NPV from the existing and successor crops combined is maximized at a chosen discount rate. This age can be calculated from the relevant yield model and price-size curve. The implication of fixing felling age by reference to maximum NPV is that the current rate of return on the growing stock equals the chosen interest rate at that age. If land is included, maximum NPV to infinity is required. The curve of NPV against felling age is fairly flat near the point of culmination and therefore the value of NPV is insensitive to changes in felling age within a range of plus or minus five years or so.

The meaning of *optimal rotation* depends on objectives: the ecological optimum would differ from the silvicultural, while the technical optima require timber to be grown to a specification for a particular end use. The economic appraisal of financial optimal rotation entails balancing many factors, including: (i) volume growth pattern, (ii) crop price-size relationship, (iii) trends in real price of timber, (iv) discount rate, (v) value of land, and (vi) climatic, biotic and other risks to crop survival or growth (Price 1989). Two distinct approaches to combining these factors in assessing optimal rotation are: the 'total' approach aggregates predicted costs and revenues of various rotations and selects the rotation of highest NPV, IRR, or forest rent, a normal approach in planning rotations; and the 'incremental' approach regards extension of rotation as an investment decision using the indicating percent. Low discount rates usually lengthen rotations whereas high timber prices usually shorten rotations. Marketing opportunities (price-responsive cutting) should influence the actual time of the felling, higher than normal prices advancing the timing and a short-term market recession delaying it. High prices mean less volume needs to be sold to raise a given sum. (Environmental value of plantations tends to increase with their age, favouring longer rotations. The financial penalty of doing so can be calculated in terms of profit forgone.)

Private finances and the felling decision. 'Different perceptions of prices, different discount rates, different targets, mean that silvicultural decisions by individual land-owners differ from those of large organizations, even in the same circumstances and when the same criteria are employed' (Price 1989, p.140). Rotations may be influenced by the owner's age, financial circumstances (need for income or capital), capital taxation incidence, sporting, and aesthetic and environmental factors. 'The timing of these requirements is, of course, unrelated to crop age, which means major fellings are often not at the [financially] optimal time, conventionally measured' (Price 1989, p. 140).

Where the *successor crop* NPV (including allowance for land value) is 0, the felling ages for individual species and yield classes might be of one level, but rotations are likely to differ. When the successor crop is expected to make a large positive NPV then some degree of bringing forward of felling age is justified on economic grounds. Frequently the clear-felling of one stand has to be linked with that of a neighbour, whether for reasons of extraction or to reduce windthrow risk. In such cases the NPV for the whole felling coupe should be calculated for a series of felling ages in order to obtain the optimum. However, a reasonable approximation may be calculated by taking a weighted average (using area) of the individual felling ages.

Replacement of unsatisfactory crops. When deciding whether an existing crop should be replaced by another (because it is uneconomic or subject to unacceptable biotic damage), the criterion usually used is when the sum of DR of the existing crop and the NPV of the successor is at a maximum. Any delay in felling the existing crop also delays the ultimate return on the successor crop. Hence, the increase in value of the existing crop has to be sufficient to offset the fall in value of the successor crop through waiting.

It does not follow that because a crop is judged to be unsatisfactory on silvicultural grounds (for example, poorly stocked, unhealthy or of lower growth rate than an alternative species) it should be replaced at the earliest opportunity. Unless considerations of forest hygiene demand the early removal of a source of potential infection it will often be more profitable to keep an unsatisfactory crop on the ground than to clear-fell and replace it at an early stage. Cash flow may also be important. Quite poor crops will sometimes be capable of growth in value which places them at an advantage compared with replacement crops requiring new investment before they can produce any greater contributions to revenue. Further, of two stands which are growing at different rates, it will not necessarily be true that the slower-growing one should be replaced first. It may well be that the fastergrowing crop occupies a better site on which an alternative species would be relatively much more profitable than the alternative species on the poorer site.

Premature felling in anticipation of windthrow (i.e. prior to the age of economic maturity) is often undertaken to pre-empt a high risk crop becoming too vulnerable. It is necessary to compare revenue forgone from felling prematurely by a number of years, with the cost of windthrow. As windthrow is not a certainty this latter cost must be multiplied by the probability (which must lie between 0 and 100%) of windthrow actually occurring during that period.

Options in roading and extraction. Roading (discussed in chapter 10) is an investment: outlays made in the construction should reduce future costs of establishment, tending, protection and harvesting. It may result in better supervision and a higher market value for the timber. Extraction and roading costs are interdependent: each extraction method and road system has an optimal pattern of timber movement (Price 1989, pp. 89–99). Short average extraction distances and consequent low costs are achieved only through heavy road expenditure (construction costs of high specification forest roads in 1990 ranged up to £50,000/km). Conversely, minimizing road expenditure gives large average extraction distances and costs. While a general optimal density guideline can be calculated for a woodland the exact figure varies wherever the determining factors vary. Terrain has a pertinent bearing: it may mean the difference between skidder or forwarder extraction (easy ground) and cable extraction (steep, awkward slopes). Yet not all roads need to be of the highest or same standard. The economic extraction limit depends on marginal cost and marginal revenue per m^3 of timber extracted (Price 1989). If a road does not exist, volume per hectare is critical to the decision whether to construct it.

The network of roads laid down is generally chosen to minimize the combined costs of establishing, tending and harvesting. Roads may be built prior to establishing or just before the start of thinning; or, in no-thin crops, before clear-felling. The relevant costs of construction in an investment appraisal will therefore depend on the timing. So too will annual maintenance costs. The general aim is to minimize the sum of road construction and discounted maintenance costs plus the discounted costs of extracting the trees harvested over the crop's life.

One way to simplify the calculation is to determine an average cost of extraction per m^3 for the removal of all thinnings and the final felling (Busby and Grayson 1981). Then instead of calculating the relevant extraction costs for each and every cut, discounting these to the time of roading and summing them, a procedure can be adopted of multiplying the nominated extraction cost by discounted volume. Tables are available showing discounted volume for thinned stands calculated at varying discount rates to year 0. For any later year of roading up to the age of first thinning, the figures should be compounded using appropriate factors. Most woodlands contain a range of yield classes and ages and there is usually more than one species. An adequate measure of discounted volume can be found by averaging yield class and planting year and then referring to the discounted volume table for the dominant species and compounding up for the average age of the block. The cost of road maintenance must be discounted. Usage of roads fluctuates from year to year, and, given periodic maintenance outlays, may continue over several cycles of production without overall deterioration. However, heavy use, particularly during inclement weather, accelerates deterioration, and after each thinning regrading is usually required.

The increase in value of the land at the end of a rotation attributable to the provision of a road network, is termed the *residual roads value*. It can be calculated as a proportion (in the Forestry Commission, it is assumed to be 50%) of the total capital expenditure through the rotation and added as a revenue at time of felling.

Options in machinery. With progressive mechanization of forestry operations, especially harvesting, the cost of machines, plant, vehicles and equipment becomes even more important, and may dominate the economics of an operation (Price 1989, pp. 76–88). Purchase of machinery is an investment, whose pay-off is spread over several years: this complicates methods of costing. The machinery used in every forestry operation can be costed from first principles, but it is convenient if a cost for a given machine (usually per hour) is available. Price (1989) examines methods used to provide such a standard cost, and discusses whether this truly represents the cost to an enterprise of using a machine.

Machine costs conventionally comprise capital costs and running costs; and the conventional components of capital costs are depreciation and interest. Converting all these figures into a cost per unit output requires estimation of yearly work done by the machine; and this depends on working practices, and on the availability, utilization and rate of output for the machine. Once costed, it is possible to appraise whether purchase of a particular machine is justifiable, or whether an operation is worth doing, and to compare different operation methods, such as labour and capital intensive harvesting. Price (1989) in describing at length the costing of forestry machinery, emphasizes that the most sophisticated evaluation techniques are useless without sound work study data. In an investment appraisal the foregoing information should be borne in mind.

Conclusion of the appraisal. Whether or not it is possible to assign probabilities to different outcomes it may well be useful to carry out a *sensitivity analysis*.[5] This looks at the variation in the results of an appraisal when some of the assumptions are changed. Sensitivity analysis, as its name implies, indicates to which assumption the results are most sensitive. For example, it may be helpful to show the consequences of a particular course of action in terms of NPV at a variety of discount rates indicating the sensitivity of the results to the choice of discount rate.

---

5 *Sensitivity analysis* is an exercise where the assumptions made regarding events not known with certainty (such as future price changes) are altered within a plausible range. The main purpose is to discover the extent to which the outcome of an appraisal depends upon the original assumptions. Furthermore, the analysis shows the implications of various possible values and non-market costs and values to be explored. It is the process of recalculating the NPV of an option if one or more values (of cost or benefit) within that option are uncertain, and may be higher or lower than the original estimate. The range of values tested in sensitivity analyses must be based largely on the experience of the forester. 'The analysis seeks a sensible balance between presenting masses of apparently unrelated raw economic data, and characterizing each alternative by a single figure. It shows the effect of adjusting contentious figures in the appraisal' enabling decision-makers to see how robust the results are, and judge the appropriate figures themselves (Price 1989, p. 340). The analysis also shows whether uncertainty really affects the choice of project.

In the absence of any guidance on the likely range of values of the various factors the analysis has to be conducted on the basis of determining the sensitivity of the results to a given percentage change (say 10%) in each of the factors.

Budgetary constraints may force a choice between available worthwhile projects. A rational way to look at this choice is to find out which projects give the best returns for their size or cost, and choose them. In this way, the net benefit from a fixed budget can be maximized. This is where *cost-benefit ratios* noted earlier become useful. They are the ratio of NPV to cost, and may be negative if projects with negative NPVs are acceptable.

Foresters tend to use informal and sometimes conflicting rules to assign priorities among uses of limited resource needed to undertake the operation. But as, for many reasons, choices must often be made among conflicting operational priorities, Gray and Price (1990) suggest that priorities for action should be chosen in order of 'urgency index ratio', defined as the cost of delaying an operation divided by the units of limited resource needed to undertake the operation. The authors state that 'urgency index ratio can be determined by simple calculations based on readilyavailable data'.

The investment appraisal stage is only the beginning of any project or proposed course of action. It is important that the values (of cost or benefit) and quantities of inputs are monitored throughout the life of the project, enabling amendments of the programme to be made if either the original estimates prove to be wrong, or the objectives of the project change over time.

Foresters intent on undertaking an investment appraisal (or intent on calculating an expectation value – see chapter 16) will need to have access to appropriate discounting tables or a calculator or a computer as noted in section 2 *supra*. Worked examples of investment appraisals are provided by Busby and Grayson (1981) and Williams (1988), but their costs and revenues need to be up-dated.

7. Profitability of private forestry

The profitability of private forestry depends on the balance between expenditure and revenue. Expenditure (costs) is offset by grants. Revenues are controlled by growth of the crop in terms of the volume increment (see yield class in chapter 12, used by foresters to describe the maximum rate of growth of the crop measured over its rotation), and the price for which the timber can be sold (see chapter 11). The profitability both to the owner and to society as a whole can be seen as being supplemented by credit given for non-wood net benefits. The elements of profitability for the owner particularly include: capital taxation reliefs, the trade-off between cash flows especially in harvesting, and the influence of time on values, on costs of machinery and roads, and on criteria of profitable investment including investments made in conditions of risk and uncertainty. These factors and elements are the building blocks of forest economics (Price 1989).

The returns to the grower can be considered as those of an individual stand over its full rotation or be to any 'point in time' within it. A woodland usually comprises several individual stands, so its profitability at any 'point in time' is the sum of the financial results of each of its component stands. Most woodlands have a complicated structure of species, age-classes and quality, making it difficult to relate income to capital or to assess the rate of interest which the first represents on the second. Greater difficulties of assessment can relate to a forestry portfolio comprising more than one woodland or to a well-wooded extensive estate.

Scarcity of financial data. The length of time trees take to achieve merchantable dimensions makes it difficult to discuss the past, present or potential profitability of private forestry. Records of costs and revenues have seldom been maintained for the full rotation of a stand, or over many decades for a representative woodland. Many such investments stay within a family for one or more generations. Most results are treated as 'in-house'. Usually disposals are made by private treaty, or by private family arrangements such as lifetime gifts or bequests on death, whereunder the value (and hence the rate of return) is not publicly disclosed. The disposal sometimes is simply a desire to realize the asset and invest in another. Even if the results were disclosed, changes in the value of money (inflation) make it difficult to interpret the figures without careful adjustment. Records usually pay scant attention to the varying capital structure. They seldom distinguish timber income and that of Christmas trees, and land value or rental is usually excluded as are taxation reliefs. Internal estate transfers of timber, firewood, materials, labour and machinery, and joint use of fences and roads, are rarely taken into account. Shooting values are rarely credited to the woodland account, though many owners would acknowledge the critical dependence of their estates' shooting upon woodlands. Up to 1988 (April 1993 for those occupiers which have woods taxed under Schedule D), the whole financial situation was masked by the incidence of Schedule D of income tax. The incidence of capital gains tax, estate duty, capital transfer tax or inheritance tax has sometimes been relevant. Under such circumstances profitability and realistic rates of return have not been known with precision; most assessments of them must include some subjective judgements.

When calculating the expected rate of return on a stand or whole woodland, the exercise is often handicapped by the scarcity of costs of silvicultural operations and prices of timber. To help fill this gap, indications of costs have been given in chapter 5 (for establishment) and in chapter 6 (for the tending phase); indications of timber prices have been included in chapter 11.

Of the few published annual financial accounts relevant to private woodlands, the most enlightening were those of Dartington Woodlands in Devon (Hiley 1964). A few individual foresters have augmented that knowledge. Significant contributions have also been made by the annual surveys of income and expenditure on sampled private wooded estates reported by the university forestry departments at Aberdeen and Oxford. For England, Wales and Scotland they showed that between 1961 and 1980 sampled properties experienced consistent annual cash deficits, although no account was taken of changes in the value of growing-stock occasioned by felling, restocking and increment. For England and Wales a special detailed investigation of financial returns from thirty-four of the sampled properties (embodying land, growing-stock and cash flow, and assuming favourable income tax incidence and generous increases in timber volumes) over an average period of fourteen years, showed that thirty had a positive return, which was usually greater for investment plantations than for traditional woodlands.

A positive return can of course be generated in any particular year, or even over a sequence of years, particularly if timber is felled excessively or imprudently; but the result, even if it funds restocking, is not necessarily the economic optimum or the maximum profit achievable through proper silvicultural management, but simply the surplus on cash flow during the period.

Encouraging indicators of profitable forestry appear to be many well-wooded estates which have achieved notable success, chiefly by the practice of sound silviculture and marketing on a sufficiently large scale, and which occasionally have built up modest integrated woodland industries based thereon. Even so, the conversion and processing of timber has usually yielded an ampler return than has its growing.

Factors which influence profitability. These include: species, site quality (in terms of soil, climate, elevation and wind hazard, all reflected in yield class and rotation), silvicultural system, objective(s) pursued, management skill, extent of the stand or woodland, ease of removal of timber, location and proximity of markets. Additional considerations are availability of planting and management grants, and the incidence of capital taxation (favourable to forestry compared with most investments); all have an impact on cash flow.

For a particular woodland, profitability depends on the crop's species and age structure, the 'intensity of the forestry being practised', 'point of sale' of produce (standing, felled, converted or delivered), and the 'point in time' to which the result of the investment is calculated. Particularly relevant is the financial climate existing at purchase, harvesting or disposal. From the owner's viewpoint, his 'point of entry' into the investment is pertinent: i.e. whether at bare land state, established, thicket-stage, pole-stage, maturing or matured.

Comparison with State forestry profitability. The economic environment of private forestry differs in some respects from that of the Forestry Commission. The Commission has economies of scale, also bargaining power both in purchasing and marketing. Some of the side benefits of forestry (e.g. amenity, recreation, downstream employment), all important from the viewpoint of national well-being, cannot always be attained by the private grower. The Commission is set a target rate of return for its forests of 3% in real terms – of necessity to be achieved on land of mixed quality and largely in harsh upland conditions. Any of its forest investments that do not make the 3% target receive a New Planting or Restocking Subsidy, i.e. the notional amount required to raise the rate of return on plantations for which the forecast IRR is below 3% to the 3% target rate. (It is difficult to quantify and value the environmental and other social benefits that State forestry brings although large advances towards being able to do so have been made during the last five years. The value of the Commission's estate as a recreational resource may be much greater than has previously been thought.) The overall real rate of return on new planting and on restocking is now expected to rise to 2.9% (without subsidies) and 3.2% (with subsidies). The rates of return have improved progressively from 1.75% in 1977–82, 2.25% in 1982–87 and 2.5% in 1987–90. This has been achieved through the introduction of improved methods, techniques and working procedures leading to lower expenditure.

The Commission's financial returns on its plantations range from less than 1% to over 6%. Broadleaves, particularly the slow-growing species, even when grown on best sites, and conifers grown on poor sites will be at the lower end of this range; conifers grown on the best sites currently available will be at the top end. Rates of return above 5% could be achieved under the more favourable site conditions offered by better quality land. However, the benefits of higher quality sites in terms of yield class are often offset by increased establishment costs (weeds benefit from greater fertility as much as the trees). The IRRs being achieved in a wide range of circumstances are probably from −0.5% to 6% (say, 2% to 6% for conifers and

– 0.5% to 2% for broadleaves; higher for poplars and sweet chestnut coppice). The Commission's real rate of return on its commercial recreation assets has been 10% for the past three years.

In productive woodlands of the private sector, the benefit of grant-aid and fiscal advantages should add around 2% to the rates of return achieved by the Forestry Commission. However, much depends on site conditions, nearness to markets, and management skills.

Profitability of conifers. Profitability of growing conifers in Britain has generally been assumed to lie within a range of 1% to 8%, with an average around 3%. The average growth rate of the existing State forests is close to 11 m^3/ha/yr (i.e. between Yield Class 10 and 12), although this has risen consistently as silvicultural techniques have improved.

For private woodlands, a minimum real rate of return of 3% from conifers might be thought to be satisfactory. The yield from so-called 'safe investments' may well be higher in terms of constant money value, but most are subject to income tax whereas returns from forestry are not. Moreover, a woodland is something tangible which is likely to appreciate to the extent of its percentage return, whereas the capital in some other investments cannot rise. With the risks and uncertainties in forestry, the effort needed to be applied to it, and the benefit of grant-aid, a 4% to 6% return should be sought. In fact for some owners, earning as high a rate of return as possible is the criterion of efficient forestry in the economic sense. But the personal circumstances and objectives of the woodland-owner will determine what course should be adopted to make forestry most profitable and satisfying.

Comments on profitability rarely pay attention to the movements in real terms, generally upwards, in market values of woodlands. Usually they command prices well above the values of their timber and land – see chapters 11 and 16. Within the market movements there have been both losers and gainers. Moreover, the 'guillotine effect' has applied: the cost to the purchaser is that paid, the sunk costs of the vendor are ignored. It has been possible for Schedule D investors to sell at a discount because such a high percentage of their costs have been recovered by tax, rather than the capital invested. Values become more realistic as the end of the rotation gets closer. The grower besides having to fund the purchase of plantable land, has to meet a further cash flow situation, as indicated in Table 48.

Table 48. An indication of operational input (excluding land, any roading and Management Grant) and planting grant for one hectare of lowland coniferous planting throughout the first 10 years. A percentage of broadleaves would increase net costs.

Year	Item	Conifers £/ha	
		Costs	Revenue*
0	Site Preparation	200	Planting grant 616
0	Plants/Planting	300	–
0	Fencing (variable)	500	–
1	Weeding/Beating up/Management	110	–

| | | Conifers £/ha ||
Year	Item	Costs	Revenue*
2	Weeding/Maintenance/Management	90	–
3	Weeding/Maintenance/Management	70	–
4	Maintenance/Management	40	–
5	Maintenance/Management	30	Planting grant 176
6–10	Maintenance/Management: Total	60	–
9	Final Planting Grant Instalment	–	Planting grant 88
		1,400	880

* Assuming Band 2 of the FC Woodland Grant Scheme (i.e. 1.0–2.9 ha)

Thereafter, the grower's expenditure comprises annual maintenance and protection, and probably a measure of cleaning and/or brashing. In later years, income would be received from several thinnings, as well as from the final felling. However, to calculate the financial return the technique of discounting over time must be undertaken (see section 2 *supra*). Table 49 is an example of cash flow and calculation of NPV using a discount rate of 3% – for upland no-thin Sitka Spruce on a fifty-year rotation, Yield Class 14 (85% stocking).

Table 49. A cash flow (excluding land, any roading and Management Grant) and calculation of NPV using a discount rate of 3%.

| | | Cash Flow £/ha ||||
Year	Item	Expenditure	Discounted Expenditure	Revenue	Discounted Revenue
0	Plough (£100)/Fence (£125)/Drain (£50)	275	275	Planting grant 430*	417
0	Planting	188	188	–	–
0		80	80	–	–
6	Fertilize	80	67	Planting grant 123	103
14	Fertilize	35	23	Planting grant 62	45
1	Fertilize	50	49	–	–
0	Beating up	80	80	–	–
5	Weeding	23	20	–	–
1–50	Weeding Maintenance (fences, drains) protection and management	12/yr	308	–	–
50	Final Felling	–	–	11,915	2,718
	Total		1,090	12,530	3,283

* Assuming Band 4 of the FC Woodland Grant Scheme (i.e. 10 and over ha).

Results: Discounted at 3% to year 0	£/ha
Discounted Revenue	3,283
Discounted Expenditure	1,090
NPV	2,193
Annualized NPV	65/yr

Profitability of broadleaves. Profitability of growing broadleaves in Britain has traditionally been considered as much less than conifers owing to prolonged and more costly establishment (even though grant-aid is higher) and slower growth leading to poorer early thinnings and longer rotations (Busby 1982). Their average growth rate in Britain is close to 5 m^3/ha/yr (i.e. between Yield Class 4 and 6). Moreover, broadleaves generally require more fertile soils and lower elevations. On average, broadleaves (excluding poplars and sweet chestnut coppice) have financial rates of return of the order of one quarter to one half of those indicated *supra* for conifers. Their average growth rate is generally about half that of conifers, and their volume production culminates later. However, their value per unit volume in later life is often higher, and substantially so for prime sawlogs and veneer logs. Moreover, broadleaved woodlands are likely to have greater non-wood benefits such as landscape, wildlife conservation and recreation; and sporting value may be higher. In any case, not all woodlands are managed primarily for financial gain, and owners differ in their circumstances and objectives.

Until recent years the main reasons for continued planting of broadleaves have been amenity, conservation and sporting, along with tradition itself. Sites, species, locations and markets differ in most situations. Much planting of them has been an act of faith (Busby 1982). Many people have wondered why it is that production of broadleaves has often been ruled out in Britain, due to their low profitability, whereas on the Continent they are rarely considered as 'unprofitable' – due not so much to difference in silviculture, yields and markets, but to different economic criteria being applied.

An important aspect of broadleaved timber production is the value placed on high quality. The most serious defects arise from poor form, knots, shake (particularly in sweet chestnut and oak) and epicormic shoots (common in oak). Site factors may influence the occurrence of shake, while management practices can help to improve form and control the production of epicormics. Reduction of the incidence of these defects would have an appreciable effect on profitability and meet the market preference for timber of high quality.

Broadleaved rotations are rarely determined by the long-term economics of the investment. The decision on when to fell a particular stand is heavily influenced by the owner's objectives, current requirements for income and capital, and the costs of restocking that would be needed. Cutting often follows the transfer of ownership upon death of the original owner, the capital thus released being used to allay capital taxes. The prevailing state of the timber market, and the capacity of it to absorb the timber may also be relevant factors. This was most recently demonstrated after the October 1987 gale when the glut of broadleaves on the market seriously depleted prices for all but the highest quality oak and made much beech difficult to sell. Other factors may be planning authority attitudes and felling licence restrictions. Conventional ideas of maturity bear some relation to the species and its current growth rate but are largely the result of traditional attitudes and practices; as a result, the ages at which broadleaves are felled tend to be higher than purely commercial considerations would indicate as prudent. Delaying cutting beyond conventional rotation age reduces the real value of the income eventually earned as measured by discounted revenue (DR) unless timber prices are rising in real terms. The amount by which DR is reduced depends on price assumptions and other factors but in percentage terms DR probably falls by about 10% for every decade beyond the conventional rotation. The effect on NPV and hence on profitability is much more significant: a 10% decrease in DR may lead to a 50% reduction in NPV. Table 50 is an indication of the grower's cash flow.

Table 50. An indication of operational input (excluding land, any roading and Management Grant) and planting grant for one hectare of lowland broadleaved planting throughout the first 10 years. Establishment in treeshelters would significantly alter costs of many of the items.

		Broadleaves £/ha	
4.586 mm	Item	Costs	Revenue*
0	Site Preparation	200	Planting grant 963
0	Plants/Planting	400	–
0	Fencing (variable)	500	–
1	Weeding/Beating up/Management	200	–
2	Weeding/Maintenance/Management	100	–
3	Weeding/Maintenance/Management	80	–
4	Maintenance/Management	60	–
5	Maintenance/Management	50	Planting grant 275
6–10	Maintenance/Management: Total	100	–
9	Final Planting Grant Instalment	–	Planting grant 137
		1,690	1,375

* Assuming Band 2 of the FC Woodland Grant Scheme (i.e. 1.0–2.9 ha)

Profitability of broadleaved species under different management systems has been calculated in six studies during the last decade:

1   Studies of silvicultural options for broadleaved woodlands made by Pryor (1982) and Lorrain-Smith (1982) provide an economic analysis related to: acid western sessile oakwoods, sessile oakwoods, pedunculate oak-ash and standards over hazel underwood; and beech, sweet chestnut coppice, pedunculate and sessile oak over lime coppice, and birch in Scotland. Comparisons made include the economics of (i) having a tree cover most desirable for timber production by

contrast to (ii) where wildlife and non-timber characteristics take priority, (iii) a possible compromise between these views, and (iv) with an assessment of the financial burden imposed if deviation from maximum timber income should be desired. The conifer options included all the main species. All options were related to their NPV's to infinity, or expressed another way, Land Expectation Value. The level of timber price assumed had a major impact on the comparisons. The actual financial results are somewhat outdated because of changes in taxation and grant-aid, but the method is still relevant.

2   A Forestry Commission consultative paper in 1984 made comparisons of profitability (£/ha/yr) between broadleaves (oak YC 6, beech YC 6, sycamore YC 6 and 10, ash YC 6 and *Nothofagus* YC 12) and Corsican pine YC 12, based on different price levels and discount rates of 1%, 3% and 5%; also, broadleaves established by natural regeneration. The higher the discount rate the smaller were the differences in annual loss. Even in favourable circumstances, only sycamore YC 10 and *Nothofagus* YC 12 appeared to be more profitable than the Corsican pine YC 12. For almost all the other broadleaved options the cost penalty was of the order of £50 to £100 ha/yr in perpetuity.

3   Pryor and Savill (1986) made an analysis of various silvicultural systems (see chapter 7). The study is important, but subsequent removal of income tax relief and changes in grants have masked the situation of profitability. The NPVs of simple coppice, a low-input option, were relatively high. For clear-cut (and replant) systems, the NPVs of conifers were considerably higher than those for broadleaved species. The high profitability projected of 'mixed birch' (if near a turnery) showed that the intensive high-input approach is not always the most profitable, the low-input approach being particularly appropriate to small woodland owners who do not have large sums to invest. For shelterwood systems, although establishment costs were lower than for clear-cut, the overall profitability for oak was similar. Most profitable was an ash/sycamore shelterwood system. For group-felling systems, management costs are likely to be increased by the lack of uniformity. Profitability will also differ in several other ways from that of the conventional clear-cut system. The division of the forest into a number of different age-classes will give a more uniform distribution of costs and revenues through time.

The profitability of the selection system differs from that of other systems, in that the costs and revenues are constant from one year to the next. Positive cash flow is usually established. Once established, the system is thus ideal for an owner who does not have money to invest. (However, no estimates of costs or returns are available for Britain.) The annual costs would be management and tending. Harvesting would be covered by timber revenues, relatively lower due to the need for directional felling, the wide dispersion of saleable trees, and the increased care needed in extraction. Conversion of an even-aged mature high forest stand to a selection system would involve some financial sacrifice because the felling of some timber would have to be delayed until well past its financial maturity (see chapter 7). On the negative side this system might be unworkable in any practical sense in the uplands and possibly anywhere else in Britain so long as it is governed by economic criteria.

4   During 1987 a study was carried out at the Oxford Forestry Institute by Crockford et al. (1987a) on the relative profitability of a wide range of woodland management systems. 'The study considered the economics of 18 management systems ranging from pure conifers, through mixtures of conifers and broadleaves, to pure broadleaves in plantation, natural regeneration and coppice systems. The analyses used the concept of Land Expectation Values to convert Net Discounted Revenues onto an infinite time horizon for each management option and each combination of variables, in order to facilitate comparisons of the relative profitabilities of options with differing rotation lengths.' The variables included: (i) choice of species and management system; (ii) site quality and climate reflected by a choice of yield classes for the species considered; (iii) price combinations for fuelwood and timber based on national price surveys; (iv) income tax rate, for the purpose of the (now removed) Schedule D to Schedule B 'switch'; (v) the then grants; (vi) management costs and sporting rentals (if appropriate); and (vii) discount rate at a choice of rates (0, 2, 3, 4 or 5%). 'The results of the study indicated the intrinsic profitabilities and demonstrated that, for the majority, income tax relief [subsequently removed] on expenditure was a major factor in increasing the profitability at any discount rate. However, the then grants were only sufficient to offset some of the costs of establishment and did little to bridge the income gap between investment of capital and the realization of a return. This was notably conspicuous for the capital intensive plantation options' (Spilsbury and Crockford 1989). With the removal in March 1988 of income tax relief followed by increases in grants, the most relevant financial result are those which altogether exclude both reliefs and grants. The figures in Table 51 show Land Expectation Values (in £ at 1986 values) per hectare at 3% discount rate for each option under mean yield and price levels.

Table 51. Land Expectation Values at 3% in £/ha (at 1986 values) and excluding the then income tax relief and planting grants. Woodland Management Grants are not included.

Option		Yield Class	Land Expectation Values at 3% in £/ha. Excluding income tax relief and planting grants
			£/ha
1	Douglas fir	14	2,150
2	Corsican pine	12	1,590
3	Japanese larch	10	1,400
4	Sitka spruce	12	1,550
5	Douglas fir/oak mixture	14/5	930
6	Oak	5	−660
7	Corsican pine/beech mixture	12/6	540
8	Beech	6	−670
9	(i) Oak/ash/cherry mixture	5/6/8	−60
	(ii) 30 year conversion to oak/ash/cherry	5/6/8	−130
10	Coppice with standards	medium	380

Option		Yield Class	Land Expectation Values at 3% in £/ha. Excluding income tax relief and planting grants
			£/ha
11	Underplant with western red cedar	14	2,300
12	(i) Convert to pure oak coppice	4.5	700
	(ii) 30 year conversion to oak coppice	4.5	375
13	Oak/ash group felling system	5/6	150
14	Beech – natural regeneration	6	30
15	Planting simple coppice	5.5	−360
16	Convert to simple coppice	5.5	1,030
17	Birch – natural regeneration	6	290
18	Ash/sycamore – shelterwood system	6	820

Source: Crockford *et al.* (1987b).

The results illustrate that the most profitable options were the pure conifer crops. The conifer/broadleaved mixtures tend to be the next most profitable and result in a final crop which is broadleaved. New plantings of pure broadleaves were not profitable (new grants may now change this); in such situations the most profitable was planting a mixture of oak, ash and cherry but even here at 3% there was no profit. It appears more advantageous to manage broadleaved options which, at least partly, rely on natural regeneration or retention of part of the overstorey to reduce weeding and other establishment costs. Shelterwood systems for ash and sycamore natural regeneration can provide a good return in such situations.

With the subsequent improvement of grants by way of the Woodland Grant Scheme, Woodland Management Grant Scheme, Farm Woodland Scheme and the Set-aside Scheme (together with the removal of income tax incentives to forestry) woodland economics in the private sector have considerably altered. An updated study (Spilsbury and Crockford 1989) highlights the importance of these changes (other than the grants for management) in influencing profitability both within and between the management options. The results show how the profitabilities of woodland management have changed, not only for investors able to take advantage of the current incentives for planting of fertile agricultural land, but perhaps more significantly for those concerned with management of existing woodlands and new woodland planting not available for the Farm Woodland and Set-aside Schemes.

5   A study by Lorrain-Smith (1988) for TGUK, financed by the Scottish Forestry Trust, measured the economic impact of Forestry Commission 'Management Guidelines for Broadleaves' (1985) by analysing changes in profitability thereunder of five major broadleaved species, based on sampled estates throughout Britain. In essence, the Guidelines alter management

practices particularly to promote wildlife conservation and landscape. But as these are not free commodities, their cost is the profit from forestry forgone as a result of the Guidelines' restrictions. The results indicated by Lorrain-Smith, acknowledged as weakened by inadequate data on costs and prices, provide a set of price-tags for forestry conservation and landscape. Questions about whether the benefits are worth the cost, and, if so, whether and to what extent the cost should be borne by society or the woodland owners, lay beyond the scope of the study.

6   A study by Spilsbury (1990) examined the profitability of a wide range of management options ranging from intensive conifer plantations to more 'natural' systems. His comments have been noted in chapter 7.

Forestry compared with agriculture. Land for afforestation is usually taken from that being farmed, traditionally low-value grazing. Until recently there has been a tendency to treat forestry and farming as competing activities and comparison of the two has often been discussed in terms of 'Forestry versus Agriculture'. Particularly this has been the case when projecting afforestation on agricultural upland, where sheep farming has been the chief land use. Thus the social costs and benefits of agriculture and forestry are pertinent. The costs and benefits of the two enterprises have been discussed by Treasury (1972), CAS (1980) and Price (1989). Elevation and location markedly affect profitability of some timber and arable crops, but have less influence on upland grazing. There are fixed costs to each. The discount rate is all-important in comparing forestry (having a prolonged maturing period) and farming (having much faster returns). Both enterprises require their own special expertise, traditions and equipment, though some of the latter are interchangeable.

NPVs of forestry can be compared with those of agriculture, though difficulties in doing so include: cycles of operation, fiscal incidence, subsidies, farmer's own labour, rental, tenant's capital, and annual movement of sheep from lowlands and hills to uplands. In most cases cited, agriculture appears to give the better result when a 5% discount rate is used; forestry is best with up to 3% rate; and a 3% to 4% rate may favour neither activity over the other (CAS 1980). These conclusions, however, are subject to some important qualifications notably that price supports have been included as revenues in the agricultural calculations. Thus it can fairly be said that the annual cost per hectare to the Exchequer of supporting upland sheep farming for example is close to the total support cost of a complete forestry rotation. If agricultural support is substantially reduced then the comparison will change markedly. The ranges of the NPVs would widen considerably by the inclusion of specific variables, such as yield class, soil types, climate and altitude; but in general the growing conditions which favour livestock production also favour tree growth. The calculations do not take into account such benefits as landscape, shelter, and provision of employment – all of which may be in favour of forestry. Thus the economic assessments will tend to move together under a wide range of conditions, and results may often be favourable to forestry expansion in the uplands. Detailed studies would be needed for comparison of the two activities on specific sites. Furthermore, changes may occur in the relative economics of the two.

Social costs associated with forestry and agriculture vary. Forestry, introducing infrequent but concentrated use of roads by heavy vehicles, may require higher specifications; but forestry generally employs more people in the long term than extensive upland farming methods though many fewer than lowland agriculture. Agricultural use may require the maintenance of extensive communication systems, of roads, power lines and telephone lines. Either forestry or agriculture may restrict the use of the area for recreational purposes, or may detract in certain respects from the value of the area as water gathering ground or in landscape features.

8. Forests and the social good

Afforestation produces a number of joint outputs and services which can be either positive, taking the form of benefits, or negative, i.e. forests may actually reduce the provision of a service compared to the displaced land use, creating a cost (Pearce 1990). Much depends on exactly where afforestation takes place: thus an afforested area supplies trees as timber and as a source of recreational, landscape, watershed and other values including increased biological diversity. Microclimates and wildlife may be affected in one way or another. Additionally forestry acts as carbon sinks, and hence afforestation can reduce CO_2 emissions stored in the atmosphere, reducing the 'greenhouse effect', as noted *infra*. Economic security for the nation may be advanced by afforestation, and the decline of rural communities may be lessened by it. Other benefits usually attributed include the creation of protection of rural employment and the saving of imports.

Of these outputs and services only some are marketed. Typically, only timber values are reflected in the rates of return from forestry; such rates frequently fall below conventional 'discount rates' employed by private or public agencies. Hence, although afforestation may appear to be 'uneconomic', it is the whole range of outputs that is relevant to economic assessment; use of timber values alone to determine investment worth is in fact a purely commercial criterion. Hence, there is a difference between commercial rates of return and economic rates of return, the latter encompassing the value of non-marketed outputs and services.

The case for national forestry in Britain has been indicated briefly in chapter 9 section 1 and that for private forestry is discussed in chapter 17 section 1. The factors of each case and the extent that forestry should be pursued are not wholly acceptable to all relevant interests. Public debate thereon continues to be active. State involvement in forestry is substantial, and forest policy is an important subject, falling within a broad social environment and political context. Foresters should interest themselves in the social good, distinguishing objective analysis from the subjective or politically motivated ones and contributing to relevant debates by presenting the special nature of forestry as an economic enterprise, and, by helping to shift the climate of public and institutional opinion, affect the wide context of forestry (Price 1989). They should be concerned with the bases on which forest decisions can be made in pursuit of social well-being, and with the political economy of forestry, dealing with matters of regional and national significance. The need to understand the social context of forest decisions is becoming progressively clearer, and foresters now have at their disposal economic techniques for making decisions that tend to the social good (Price 1989).

Social planning is an enormous task: the variety of interests affecting government policy on any given issue is bewildering. Land use in particular is often heavily influenced by a conservationist attitude on behalf of society. Afforestation, restocking and deforestation (of the last there is now little in Britain) all represent major and dramatic transformations in the pattern of land use, and are often presented as a confrontation between economic forces on the one hand and social and environmental forces on the other. Public debates about forestry often centre on evaluation of social benefit and cost, involving such factors as: alternative investments, *cost-benefit analysis*,[6] the *social discount rate* and *shadow cost* (Price 1989). They involve the whole community, not simply timber production but also landscape, wildlife conservation, recreation and water supply. The true value of forestry should include profits generated in processing industries (see chapter 9). Society has multiple objectives and multiple products, not all reflected in profit, but which should be incorporated in decision processes. 'Courses of action which promote environmental objectives often reduce profit: for example, extending rotations or harvesting small coupes to enhance forest landscape, planting lowyielding species to promote wildlife interest, or leaving areas unplanted for informal recreation' (Price 1989, p. 247). However, 'socially responsible and responsive forestry also promotes community goodwill in a way that may ultimately enhance profits'.

Social cost and benefit relates to society as a whole, and will differ from those of the organization undertaking the forestry project. In an ideal market economy, the prevailing rate of interest in the money market should reflect the *social rate of time preference;* but in practice, the interest rate in the capital market is likely to diverge from that social rate (CAS 1980). Thus, the concept employed in defining the test discount rate for public investments (currently 6%) is inappropriate for evaluating forestry. The *social rate of return* on investment represents a miscellany of benefits in addition to revenues from marketable products. In appraisals the *social discount rate*[7] is vital: it is the key influence on the economics of State forestry.

As a result of forestry operations, some sections of the population receive significant non-market benefits while others have to bear substantial non-market costs. People in receipt of non-market benefits experience a decrease or increase in their total utility (or welfare) without any change in their money income. Where the market value of goods and services is not a true indication of their value to society, *shadow prices* have to be used: these are the prices or values attributable to goods or services to replace market valuations, taking other factors apart from market price into account. *Shadow cost* in cost-benefit analysis plays the same role as market prices in profitability calculations. It is the cost to the whole community of employing a factor of production,

---

6 *Cost-benefit analysis* can be defined as 'an economic appraisal of the costs and benefits of alternative courses of action, whether those costs and benefits are marketed or not, to whomsoever they accrue, both in present and future time, the costs and benefits being measured as far as possible in a common unit of value'. (Price 1989, p. 253).
7 *The social discount rate.* 'The social rate of return on investment represents a miscellany of benefits in addition to revenues from marketable products, [particularly] environmental and employment benefits' (Price 1989, p. 310). While all these have shadow prices, none generates reinvestable revenues. Time preference rates are implicit in consumers' choice: generally the goods or services are required *now*.

given that its employment causes some economic objectives to be better achieved, and the achievement of others to decline (Price 1989); in effect, the net loss of social benefit incurred by applying a factor of production to a particular course of action; or, the value of a factor or good so established.

Supplementary to much of the foregoing is forestry's role in reducing the 'greenhouse effect'. Anderson (1990) has considered the idea of a carbon tax on fossil fuels to reduce the rate of carbon accumulations in the atmosphere, and avers that carbon credits would provide an incentive for afforestation – suggesting credits for forestry of 8 US dollars/m^3, less the discounted costs of carbon taxes on wastes and decompositon of the products some years later. Pearce (1990) has discussed the evaluation of carbon fixing credits for UK forestry and suggests, with qualifications, a carbon credit for afforestation of £120-£300/ha, dependent on species, silvicultural system, yield class and end use mix. Discussion of global warming and the social benefits of afforestation will undoubtedly continue and appropriate action undertaken.

References

Anderson, D. (1990), communication from Prof. D. Pearce (1990). University College London. Prof. Anderson's report *The Forestry Industry and the Greenhouse Effect* has been published by the Scottish Forestry Trust and the Forestry Commission (1991).
Bateman, I. (1991), 'Recent Developments in the Evaluation of Non-Timber Forest Products: The Extended CBA method'. *Quarterly Journal of Forestry, 83* (2) and (3).
Busby, R.J.N. and Grayson, A.J. (1981), 'Investment Appraisal in Forestry'. FC Booklet 47.
Busby, R.J.N. (1982), 'Economics of Growing Broadleaves', in *Broadleaves in Britain*. ICF Symposium, Loughborough 7–8 July 1982, (eds.) Malcolm D.C. *et al*, pp. 197–201.
Centre for Agriculture Strategy (CAS) (1980), *Strategy for the UK forest industry*. Centre for Agricultural Strategy, Report 6. University of Reading.
Crockford, K.J., Corbyn, I.N. and Savill, P.S. (1987a), 'An Evaluation of the Methodology for Managing Existing Broadleaved and Coniferous Woodlands for Timber and Energy Production'. Report for the Energy Division, UK Atomic Energy Authority, ETSU B1156.
Crockford, K.J., Spilsbury, M.J. and Savill, P.S. (1987b), 'The Relative Economics of Woodland Management Systems'. Occasional Paper No. 35, Commonwealth Forestry Institute, University of Oxford.
Davidson, A.W. (1978), *Parry's Valuation Tables and Conversion Tables*. Estates Gazette, London.
Everard, J. (1974), 'Fertilizers in the Establishment of Conifers in Wales and Southern England'. FC Booklet 41.
Gane, M. and Linnard, W. (1968), 'Martin Faustmann and the Evolution of Discounted Cash Flow'. Paper No. 42, Commonwealth Forestry Institute, University of Oxford.
Gray, S.J. and Price, C. (1990), 'Urgency Index Ratios: a Simple Tool for Forest Management'. *Forestry 63* (2), pp. 161–75.
Helliwell, D.R. (1988), *Economics of Woodland Management*. Packard Publishing, Chichester.
Hiley, W.E. (1930), *The Economics of Forestry*. Clarendon Press, Oxford.
—— (1964), *A Forestry Venture*. Faber and Faber, London.
Insley, H., Harper, W.C.G. and Whiteman, A. (1987), *Investment Appraisal Handbook*. Forestry Commission. Internal publication.
Johnston, D.R., Grayson, A.J. and Bradley, R.T. (1967), *Forest Planning*. Faber and Faber, London.

Leslie, A.J. (1989), 'On the Economic Prospects for Natural Management in Temperate Hardwoods'. *Forestry 62* (2), pp. 147–66.

Lorrain-Smith, R. (1969), 'The Economy of the Private Woodland in Great Britain'. Occasional Paper No. 40, Commonwealth Forestry Institute, University of Oxford.

—— (1982), 'An Economic Analysis of Silvicultural Options for Broadleaved Woodland'. Commonwealth Forestry Institute Occasional Paper No. 19, Vol. II.

—— (1988), 'The Economic Effects of the Guidelines for the Management of Broadleaved Woodlands'. TGUK, London.

—— (1989). Memorandum 1 June 1989 to the Agriculture Committee on Land Use and Forestry, HC 16–1: Vol. II. Minutes of Evidence and Appendices, 10.1.90, pp. 444–51. HMSO.

McKillop, D.G. and Hutchinson, R.W. (1990), 'The Determination of Risk Adjusted Discount Rates for Private Sector Forestry Investment'. *Forestry, 63* (1), pp. 30–8.

Pearce, D. (1990). Personal communication, University College, London.

Price, C. (1976), 'Forestry' (Chapter 9), *Future Landscapes.* Chatto and Windus, London.

—— (1978), *Landscape Economics.* Macmillan, London.

—— (1989), *The Theory and Application of Forest Economics.* Basil Blackwell, Oxford.

Pryor, S.N. (1982), 'An Economic Analysis of Silvicultural Options for Broadleaved Woodland'. Occasional Paper No. 19. Vol. I. Commonwealth Forestry Institute, University of Oxford.

Pryor, S.N. and Savill, P.S. (1986), 'Silvicultural Systems for Broadleaved Woodland in Britain'. Occasional Paper No. 32, Commonwealth Forestry Institute, University of Oxford.

Sinden, J.A. and Worrell, A.C. (1979), *Unpriced Values.* Wiley.

Spilsbury, M.J. (1990), 'Economic Prospects for Natural Management of Woodlands in the UK'. *Forestry 63* (4) pp. 379–90.

Spilsbury, M.J., and Crockford, K.J. (1989), 'Woodland Economics and the 1988 Budget'. *Quarterly Journal of Forestry, 83* (1), pp. 25–32.

Taylor, C.M.A. (1991), 'Forest Fertilisation in Britain'. FC Bulletin 95.

Thompson, D A. (1979), 'Forest Drainage Schemes'. FC Leaflet 72.

—— (1984), 'Ploughing of Forest Soils'. FC Leaflet 71.

Treasury (1972), *Forestry in Great Britain: an interdepartmental Cost/Benefit Study.* HMSO, London.

Williams, M.R.W. (1988), *Decision-making in Forest Management.* Research Studies Press, Letchworth.

Willis, K.G. and Benson, J.F. (1989), 'Recreational Values of Forests'. *Forestry 62* (2), pp. 93–110.

Chapter Fifteen

SUPPORT FOR PRIVATE FORESTRY

1. State support and incentives

Successive governments have thought fit to encourage private forestry for reasons noted in chapter 9 and, today, particularly for the immense social benefits it generates. Support has generally been by way of a combination of direct grants for planting and management, and special tax arrangements for commercial woodlands in relation to income and capital. The long time cycle makes forestry especially responsive to changes in taxation.

Taxes are costs, and grants are income, applying to forest enterprises in diverse ways. They affect the profitability of forestry, absolutely and relative to other land uses and investments. They may affect the level of new planting, and change the kind of silviculture which is most profitable. Their effect on profitability, affect in turn optimal rotation. They also affect the post-tax and post-grant rate of return in relation to alternative investments.

Britain has not been alone in respect of grants and fiscal aids. Hart (1980) analysed the forest support systems of individual countries throughout the EC, several other countries in Europe and Scandinavia, and in the USA, and reported a wide variation in support and its effects by a mixture of legislation, grants and taxation provisions.

In Britain, financial support for private forestry throughout most of this century has been provided by the Forestry Commission as Forestry Authority chiefly through appropriate legislation and direct grants; these have been supplemented by taxation concessions. Although the support has changed in character from time to time, the general principles remained until March 1988 (Lynch 1989). They have been intended to support owners to bring their woodlands to full productivity and utility, to plant new ones, and to maintain the whole. Much success has been achieved.

Until recently, the Dedication Scheme provided the most comprehensive form of direct grant assistance, whereunder the owner entered into a Covenant or Agreement with the Forestry Commission undertaking to manage the woodlands for the main purpose of timber production in accordance with an agreed Plan of Operations, and to maintain skilled supervision. Other benefits were that no felling licences were required provided the felling accorded with the Plan of Operations, and the woodland was exempt from Tree Preservation Orders. Parallel to the Dedication Scheme was an Approved Woodland Scheme, without legally binding covenants. (Both Schemes have now been discontinued and Dedication ends compulsorily on disposal of the woodland.) Up to tax year 1987–88 a system of planting and management grants operated in parallel with special favourable taxation arrangements.

A great change in the support for private forestry was announced by the Government in the Budget of 15 March 1988 (as noted in the Introduction). The Chancellor of the Exchequer was concerned to provide a simpler and more acceptable system of support for traditional forestry and afforestation, and which avoided the previous arrangements under which top rate taxpayers in particular had been able to shelter other income from income tax by setting it against expenditure on forestry, while effectively enjoying freedom from tax on the eventual sale of timber. Subject to certain transitional provisions, commercial woodlands were removed entirely from the scope of income tax and corporation tax with effect from 15 March 1988. (The incidence of capital taxation was unchanged.) In parallel with this tax reform, the Chancellor announced that there would be increases in planting grants. In making these changes, he acknowledged that the tax system should recognize the special characteristics of forestry, with its very long time cycles between investment in planting and income from selling the felled timber.

The effect of the changes was that expenditure on the planting and maintenance of trees for timber production would no longer be allowed as a tax deduction against other income, and proceeds from the sale of timber and planting grants would not be chargeable to tax. Tax relief under the previous rules would continue to be available until 5 April 1993 for those who were already occupiers of commercial woodlands before 15 March 1988: these transitional arrangements would also apply to those who became occupiers as a result of commitments which were entered into, or of grant applications which were received by the Forestry Commission, before that date. Details of current taxation of woodlands are given in section 3 *infra*.

The then existing planting grant schemes – the Forestry Grant Scheme and the Broadleaved Woodland Grant Scheme – were closed to new applications from 15 March 1988 and were replaced on 5 April 1988 by a single scheme known as the Woodland Grant Scheme. The levels of grant (tax-free) under the new scheme are generally £375/ha higher than those under the old schemes, except that all broadleaved planting, whether on its own or in mixture with conifers, attracts the same rates of grant; and the rates for conifer planting under a new Farm Woodland Scheme remain at the old levels. A special supplement of £400/ha for conifers and £600/ha for broadleaves is also available for new planting on existing arable or improved grassland of less than ten years of age (although this is not payable on Farm Woodland Scheme applications). The Woodland Grant Scheme ensures that the best environmental standards are followed in forestry planting, and encompasses a wide range of objectives with the aim of encouraging multi-purpose woodland management and appealing to a wider range of interests. Timber production continues to be an essential objective, but need not necessarily be the primary one in every case. Other objectives, such as planting to provide or improve wildlife habitats, could be accepted as a main aim so long as timber is produced in the process. Details of the current grants and other subsidies are given in section 2 *infra*.

Results of the changes. There has been intense debate, particularly in forestry and conservation spheres, as to the effects of the foregoing changes in support. The effect on values of plantable land and stocked woodlands is discussed in chapter 16. Lorrain-Smith (1988) and Spilsbury and Crockford (1989) imply that the effect of changing the support through income tax by increased planting grant levels, was that the grants were less supportive, and inadequate particularly when related to the restraints under broadleaved policy regulations. The Report of the Agriculture Committee (10 January 1990) and the Government's response to it (9 May 1990) considered the new system of incentives provided a better means of support for private forestry. (This was before Woodland Management Grants were announced on 23 July 1990, effective from 1 April 1992.)

Plate 20. Coast redwood *Sequoia sempervirens* at Leighton near Welshpool, Powys, in Wales. A grove of thirty-three, planted 1857–58, gifted in 1958 to the Royal Forestry Society of England, Wales and Northern Ireland by C.P. Ackers OBE. The magnificent, tall, reddish-brown stems rise straight and true into the dark green canopy. Photographed *c.* 1980 [Anon.]

(a) Photographed June 1973 following a first light thinning [M. Howells]

(b) Photographed January 1991 following a second light thinning [H. Hindle]

Plate 21. Dawn redwood *Metasequoia glyptostroboides*. A one ha grove planted in 1953 on the Leighton Estate near Welshpool, Powys, in Wales.

Plate 22. Newtonhill Woodlands (306 ha), Inverness-shire, in Scotland, overlooking the Beauly Firth. Lowlying extensive mixed-age coniferous woodland P53 to P77. Mainly Sitka spruce (average YC18) and Scots pine, with small areas of larches, lodgepole pine and a few broadleaves. An example of integration of forestry and agriculture

[T. Baker]

Plate 23. Minnygryle Forest (489 ha) near Moniaive, Dumfries and Galloway, in Scotland. Established and roaded P71 on run-down grazing land. Mainly Sitka spruce (YC14–22) with some Japanese larch (YC8). Market value in 1991 *c.* £2,500/ha. A typical investment afforestation cost-relieved by the then grants and high rate taxation concessions

[A. Crow]

Plate 24. Carsphaim Forest, Dumfries and Galloway, in Scotland. Part of a 6,000 ha investment afforestation complex. Mainly Sitka spruce P71-P74. Much unplantable upper land. A remote region of run-down grazing land opened up by judicious roading. A typical investment afforestation cost-relieved by the then grants and high rate taxation concessions

[A. Crow]

Plate 25. Margree Forest (724 ha) between Dalry and Moniaive, in Scotland. Established and roaded P73–P75 on run-down grazing land. Mainly Sitka spruce with judicious planting of larches and the equivalent of 40 ha of broadleaves to provide aesthetic and ecological benefits. A typical investment afforestation cost-relieved by the then grants and high rate taxation concessions

[R.J. Smith]

Plate 26. Coed Craig Ruperra (65 ha) at Draethen, Mid Glamorgan, in Wales. A commercial woodland of exceptional content and quality planted mainly in the 1950s and 1960s. A plot of P26 Japanese larch in the older part of the woodland

[C. Hughes]

Plate 27. Ramsden Coppice and Widow's Wood (43 ha) near Holme Lacy, in Herefordshire. P71-P73 planting on lowland former broadleaved woodland using Japanese larch, western red cedar, Norway spruce, ash and red oak

[C. Hughes]

Plate 28. Llandegla Forest (652 ha) near Wrexham, Clwyd, in Wales. Established and roaded P71 and P72. Chiefly Sitka spruce with a modest attempt to include some Japanese larch for landscape purposes. Market value in 1991 *c.* £2,000/ha. A typical investment afforestation cost-relieved by the then grants and high rate taxation concessions

[C. Hughes]

Plate 29. Park Wood (20 ha) near Pen-y-Clawdd, Gwent, in Wales. Afforestation in P55 and P56 of part of a fragmented estate. Farms were sold to the sitting tenants following reservation of plantable land and rights of way for forestry. Costs reduced by forestry grants and small crops of Christmas trees

[C. Hughes]

Plate 30. Kilcot Wood (25 ha) near Newent, in Gloucestershire. Traditional oak woodland dating from 1918, with a broadleaved understorey and a wide range of plant species

[C. Hughes]

Plate 31. Hadnock Court Woods alongside the River Wye above Monmouth, Gwent, in Wales, requiring sensitive management for landscape and wildlife conservation. Such management increases costs and decreases income, and it appears equitable that financial recompense should be forthcoming for the social benefits created [C. Hughes]

Plate 32. Wyastone Leys Woods above Monmouth, Gwent, in Wales, mainly broadleaved, on rocky terrain near a lovely stretch of the River Wye, and of high nature conservation value. The owner's desire to plant a commemorative 'E.R.' was achieved with western red cedar in a matrix of European larch [Anon.]

Where the woodland owner had a low rate of tax or did not pay any tax (e.g. Pension Funds) the increase in the grants was an improved subsidy. But it is difficult to make a direct comparison between the pre-1988 grants plus income tax relief, and the new Woodland Grant Scheme without tax relief. The older grant schemes were not restricted to establishment expenditure whereas (until 1 April 1992) there is no assistance towards costs after establishment. Rates of tax have varied from year to year, from person to person, from nil to the highest marginal rate. Tax relief is a percentage of cost whereas a grant is a fixed amount irrespective of cost. Under Schedule B, grants were tax-free, whereas those under Schedule D were taxable. It appears that the Woodland Grant Scheme roughly replaced tax relief at 40%, is somewhat better than the old system for low-rate taxpayers, but is rather less attractive when tax rates are above 40% as they were immediately prior to the abolition of tax relief when the top rate was 60% (Lynch 1989). Whether or not the change in support will achieve Government's desired results remains to be seen. The effect on forestry planting, especially afforesting, remains debatable, not least because of work carried over from previous schemes.

On 23 July 1990 the Government announced that tax-free Woodland Management Grants would be available to owners of woodland – both broadleaved and coniferous – from 1 April 1992. There will be Standard Management Grants, and higher levels of grant for woodlands which are deemed to be of special environmental value. Owners will have to earn these grants, not only by managing their woodlands to a high standard, but also by taking positive measures to protect and improve their environmental quality. There will be certain other conditions, among them five-year management plans and control of grey squirrels. Provisional details are provided in section 2 *infra*.

Current grants and capital taxation reliefs are not the only encouragements to the private forestry sector: others, mostly through the Forestry Commission, are noted in section 5 *infra*. Appropriate controls and constraints on private forestry are discussed in section 4 *infra*.

2. Grants and other subsidies

Forestry Commission grants (tax-free) and Agriculture Departments' subsidies were introduced in section 1 *supra*. The supports are discussed in more detail hereunder.

Woodland Grant Scheme. This Forestry Commission grant scheme for planting and natural regeneration was introduced in April 1988, being:

Area approved for planting or natural regeneration (ha) Area band*	Rate of grant: £/ha	
	Conifers	Broadleaves
0.25–0.9	1,005	1,575
1.0–2.9	880	1,375
3.0–9.9	795	1,175
10 and over	615	975

* The grant band is determined by the total of the areas approved for planting or regeneration in each block or wood within the five-year period of the Plan of Operations.

The rate of grant for mixtures is in proportion to the area occupied by conifers and broadleaves respectively. The minimum area eligible for grant-aid is 0.25 hectare. There is no general upper limit on the area that can be planted. The extent eligible for grant (as well as for the Woodland Management Grant *infra*) can include (over and above what is required for roads and rides) modest amounts of open space and associated edge habitats which would bring sufficient benefits to the woodland environment, e.g. landscape, nature conservation, recreation and game management. Grants depend on applicants satisfying the Forestry Commission that provisions for environmental protection form an integral part of their plans. Applicants are required to set out their five-year management plans, including their proposals for timber production, landscaping and protection of archaeological sites and monuments.

Grants for *afforestation* and *restocking*, at the rates shown above, are paid in three instalments, 70% on completion of planting and further instalments of 20% and 10% at five-yearly intervals thereafter, subject to satisfactory establishment and maintenance. For *natural regeneration* the instalments are 50%, 30% and 20%. The first instalment is paid on the completion of approved work designed to encourage regeneration, the second when an adequate stocking has been achieved and the third five years later subject to satisfactory maintenance. Existing natural regeneration under twenty years of age which has not been previously grant-aided may qualify for the second and third instalment only. Provided they contain an adequate stocking of suitable species and have not previously been grant-aided, *neglected woodlands* under twenty years of age for which cleaning and other maintenance operations are required and which are not self-financing may similarly qualify for the second and third instalments of grants. These are paid at the levels applying to natural regeneration – 30% on completion of the approved operations and 20% five years later subject to satisfactory maintenance.

Coppice on traditional rotations is eligible for planting grant under the scheme for the initial establishment of new coppice stools. This also applies to establishment of *short rotation coppice* (subject to the silvicultural and environmental conditions of the Woodland Grant Scheme being met). From 1 April 1992 both types of coppice will be eligible for tax-free Woodland Management Grants *infra*.

The maximum tree spacings normally acceptable for grant payment are 2.1 m (2,250/ha) for conifers and 3 m (1,100/ha) for broadleaves. Where the Forestry Commission considers it silviculturally acceptable, however, proposals to plant at wider spacing may be permitted and grant paid on a pro-rata basis. (As to poplar, see chapter 5.) Information is available in Forestry Commission literature to be available from 4 June 1991.

Where afforestation is undertaken on existing arable land, or improved grassland which has been cultivated and reseeded within the last ten years, there is a supplement of £400/ha for conifers and £600/ha for broadleaves payable with the first instalment of grant.

Storm damage restocking supplement. Areas damaged during the storms of 16 October 1987 and 25 January 1990 qualify for restocking supplements to the usual Woodland Grant Scheme. These are paid at a rate of £150/ha for conifers and £400/ha for broadleaves. The rate for mixtures is in proportion to the area occupied by each. Restocking must take place before 31 March 1993. Supplements are similarly available for restocking of storm damaged areas done under approved Plans of Operations under all other schemes of the Forestry Commission. Information is available in Forestry Commission Leaflet 'Storm damage: Replanting support'.

Grants for new native pinewoods. In native pine wood areas of the Highlands of Scotland, planting with native Scots pine is grant-aided at the same rate as broadleaves under the Woodland Grant Scheme (rather than at the conifer levels which are lower). The aim is to emulate as far as possible the native pinewoods ecosystem, and eligibility is restricted to those areas within the former natural distribution of Scots pine dominated pine-birch forest with appropriate site characteristics. Information is available in Forestry Commission Leaflet 'Grants for new native pinewoods'.

Woodland Management Grants. These annual tax-free grants, announced on 23 July 1990 and effective from 1 April 1992, apply to both conifer and broadleaved woodlands (including traditional rotation coppice and short rotation coppice) from ten years after the establishment of the woodland (i.e. year 10) until twenty years of age for conifer woods and forty years for broadleaved woods. They are aimed at improving the management and increasing the environmental value of Britain's woodlands.

To qualify for the Woodland Management Grant woodland owners are required to agree to a five-year management plan with the Forestry Commission setting out objectives for the woodland and prescribing operations which will advance those objectives during the period of the plan. The grants will be paid annually in arrears subject to satisfactory implementation of the plan. A one-off payment of £100 is also available from the Commission for owners who draw up management plans for the first time with the benefit of professional advice for areas eligible for the Woodland Management Grants. (This assistance will not be available, however, for plans for planting grant applications.) The two types of Management Grant are:

1. *Standard Management Grants,* payable during the normal maintenance period following the initial establishment phase of the woodland for conifer woodlands between eleven and twenty years of age and for broadleaved woodlands between eleven and forty years. In return for these grants, owners will be obliged to carry out general silvicultural operations to a high standard and to take such steps as may be agreed between them and the Forestry Commission to increase the environmental value of the woodlands.

2. *Special Management Grants,* payable for woodlands of special environmental value of any age above ten years. In return for these grants, the owner will be obliged to agree to take specified action which will maintain and enhance the woodland's special character. Woodlands in this category will be those which in the Forestry Commission's view are of special value for nature conservation, landscape, public recreation or a combination of these by virtue of their nature, location or use. There will be a presumption that conifer and broadleaved woodlands properly classified as ancient and semi-natural on the Inventory drawn up by the Nature Conservancy Council will qualify, as will those of special landscape value in National Parks, Areas of Outstanding Natural Beauty and National Scenic Areas

or which are covered by woodland TPOs, but each case will, of course, be considered on its merits. Any woodland, whether or not in a nationally designated area, may qualify if the owner has proposals to establish, develop or improve free facilities for public access or for public recreation in the woodland, provided the proposals are in keeping with public demand for such facilities and are accepted by the Forestry Commission. Owners in receipt of these special management grants will not be eligible to claim the Standard Management Grant in respect of the same area.

A supplementary grant will be paid for woodlands of less than 10 hectares in either of the above categories in recognition of the higher management costs involved.

Woodlands which are currently in receipt of grants from other public bodies are not eligible for the Woodland Management Grants, except for those established under the Farm Woodland Scheme (the annual payments to farmers under that scheme compensate for agricultural income forgone, and are not provided for the purpose of defraying maintenance expenditure).

The Woodland Management Grants also apply to modest extents of open space and edge habitats (over and above what is required for roads and rides) which would bring significant benefits to the woodland environment. In areas vulnerable to grey squirrel (i.e. where the populations are high) prescriptions for their control will be an obligatory part of the approved management on which Management Grants will be paid. The annual rates of grant and their periods of eligibility are:

Type of grant	Period of eligibility (age of wood in years)	Rate of grant (£/per ha/yr)
Standard Management Grant		
Conifers	11–20	10
Broadleaves	11–40	25
Special Management Grant	11 onwards	35
Supplement for small woods (under 10 ha)		
Standard: Conifers	(as for main grant)	5
Broadleaves	(as for main grant)	10
Special grant	(as for main grant)	10

Mixed woodlands will be eligible for the broadleaved and conifer element of the grant in proportion to the area occupied by the two categories.

The foregoing extension to the Woodland Grant Scheme will come into operation on 1 April 1992, with the first grants being paid in 1993–94. Some owners are able to claim income tax relief on the cost of managing their elected Schedule D woodlands under the transitional tax arrangements which end on 5 April 1993, but tax relief will not be available on woodlands for which Woodland Management Grants are received for that final year.

Information on the Woodland Management Grants and detailed guidance for the preparation of management plans are available from the Forestry Commission. The Woodland Grant Scheme is to be extended from 1 April 1992 to include the Management Grants and the rules relating to short rotation coppice and open space.

Grants for farmland planting. The Government's aim relating to diversification of use of agricultural land is to smooth the path of change and to create the framework within which the vital objectives of a healthy rural economy and an attractive rural environment can be achieved, rather than to plan the changes in detail. All the agricultural schemes are to some extent experimental.

Farm Woodland Scheme. This scheme, run jointly by the Agriculture Departments and the Forestry Commission, is designed to encourage the establishment of new woodland on farms and came into operation on 1 October 1988 initially for a three year experimental period. It is targeted mainly on arable land and improved grassland which has been cultivated and reseeded within the last ten years. To accommodate hill farmers there is an allocation of 3,000 hectares of unimproved grassland in the Less Favoured Areas over the first three years of the scheme. Eligibility for entry to the scheme is dependent on the planting being approved by the Forestry Commission under the rules of the Woodland Grant Scheme. Tenants may apply as long as they have their landlord's consent.

Planting under the Farm Woodland Scheme attracts Forestry Commission tax-free planting and management grants as well as taxable annual payments from the Agriculture Departments as detailed below. The minimum area per holding that can be planted in the first three years of this scheme is 3 hectares and the maximum is 40 hectares. The minimum size of each planted area is one hectare. The rates of planting grant are:

Area approved for planting (ha) Area band	Conifers £/ha	Broadleaves £/ha
1.0–2.9	505	1,375
3.0–9.9	420	1,175
10.0 and over	240	975

These tax-free planting grants are paid in three instalments, 70% on completion of planting and further instalments of 20% and 10% at five-yearly intervals thereafter, subject to satisfactory establishment and maintenance. Tax-free Management Grants *supra* are also available.

The rates of annual taxable payments by the Agriculture Departments under the Farm Woodland Scheme, made as compensation for the income forgone from former farmland, are:

	Annual Payments £/ha		
Less Favoured Areas	Severely Disadvantaged Areas	Disadvantaged Areas	Elsewhere
Arable land and improved grassland which has been cultivated and reseeded within the last 10 years	100	150	190
Unimproved grassland (including rough grazing)	30	30	

These annual taxable payments are made for forty years for predominantly oak and beech planting; thirty years for other broadleaves and mixed woodland containing more than 50% broadleaves by area; twenty years for other woodland; and ten years for traditional coppice. The first payment is made the year after planting. Information is available from the Agriculture Departments, ADAS and NFU.

Set-aside Scheme. This is a voluntary scheme designed to reduce surpluses of arable crops. In return for taking out of production an area equivalent to at least 20% of their land growing certain arable crops in the base year 1987–88, farmers receive annual taxable payments of up to £200/ha. Those who wish to plant trees on set-aside land have a choice between: direct set-aside to woodland, and set-aside through the Farm Woodland Scheme. Apart from areas of less than 0.25 hectares and short rotation coppice, planting under either of these options will involve approval under the terms of the Woodland Grant Scheme. Fruit orchards, hardy nursery stock production and Christmas tree production are excluded from the Set-aside Scheme. Direct set-aside is eligible for the Woodland Grant Scheme grants plus set-aside annual taxable payments of £180/ha in Less Favoured Areas, £200/ha elsewhere. Annual payments will continue for the duration of the set-aside agreement only (up to five years). For set-aside through the Farm Woodland Scheme, planting grants, annual payments and the other rules of the Farm Woodland Scheme apply, but the area planted can be counted towards the 20% set-aside total. Farmers wishing to plant more than the 40 hectares limit allowed under the Farm Woodland Scheme may apply to plant the balance under direct set-aside. Additional to the foregoing are annual grants of £100 to £200/ha for growing trees to protect wildlife and the countryside.

Under a Farm Diversification Grant Scheme, grants are available to assist an existing agricultural business and to develop alternative commercial uses for land and buildings. It does not cover forestry, but would for example cover facilities for processing timber. Information is available from the Agriculture Departments, ADAS and NFU. Financial assistance is also available under Environmentally Sensitive Area Schemes for certain farming areas designated for their outstanding ecological or landscape importance, and which are under threat from actual or potential changes in farming practice. Grants for shelterbelts and for hedges and hedgerow trees have been noted in chapter 5.

Grants for amenity planting. Grants are available from the two Countryside Commissions for the planting and management of amenity woodlands of up to 0.25 hectares. Where the primary objective of the work is nature conservation, discretionary grants of up to 50% may also be available from the successors to the Nature Conservancy Council. Other grants for amenity and/or conservation planting may be obtainable from the Department of the Environment, National Park Authorities, the Woodland Trust, the Tree Council and some Local Authorities. Information on these and other grants is set out by Lorrain-Smith (1989) and can also be obtained from the organizations concerned.

3. Taxation of woodlands

Governments have thought fit to ensure that occupiers of commercial woodlands are treated encouragingly and sympathetically in regard to the incidence of both income and capital taxation. The whole taxation system is recognition of the long-term nature of forestry and that woodlands are a great benefit to the nation, that forestry is a business where the return on capital is generally modest, and the constraint of illiquidity a burden. The Finance Act 1988 reduced some of the taxation concessions (subject to transitional arrangements until 5 April 1993) but cash support continued by way of increased grants for planting (and, from 1 April 1992, for management). The incidence of capital taxation was not altered.

Income tax. The Finance Act 1988 made significant changes in the tax treatment of commercial woodlands (Hart 1988; Forestry Commission 1989; Lynch 1989). (Previously, an occupier of commercial woodlands was liable to income tax under Schedule B on one-third of the annual value of the woodland, and there was no separate tax charge on the profits from forestry; alternatively, the occupier could elect to be assessed on the profits of forestry under Schedule D.) The Act abolished the charge to tax under Schedule B (with effect from 6 April 1988). So income and corporation tax is no longer chargeable on the occupation of commercial woodlands, and expenditure on establishing and maintaining commercial woodlands is not allowable as a tax deduction. Tax relief is no longer available for any interest paid in connection with forestry activity, on a loan, for example, to buy woodlands or to invest in a company or partnership engaged in forestry activity. Non-commercial woodlands were already outside income tax incidence (since 1963). However, the taxpayer already in occupation of commercial woodlands within Schedule B retains the right to elect for Schedule D treatment under the transitional arrangements noted below.

Schedule D transitional arrangements until 5 April 1993. There are transitional arrangements for those who were occupiers of commercial woodlands before 15 March 1988. They can elect to be assessed and charged to tax under Schedule D for the tax years up to and including 1992–93, and if they do so, they can continue to offset the costs of establishing and maintaining such elected woodlands against taxable income for the same period. The arrangements are also available to anyone who became an occupier of commercial woodlands after 15 March 1988 as a result of a firm commitment evidenced in writing and entered into before that date; or to anyone who had made a grant application to the Forestry Commission for commercial woodlands before 15 March 1988. The arrangements are not available to an occupier who receives a grant under the Woodland Grant Scheme. (Grants paid under this scheme are free from income and corporation tax, and are set at levels which reflect the fact that occupiers can no longer claim tax relief for the costs of establishing and maintaining.)

An occupier who had an election for Schedule D treatment in force at 15 March 1988, will automatically qualify for transitional relief for the woodlands covered by the election. Indeed the transitional arrangements are mandatory and this can be particularly unfortunate if large insurance claims are received from, for example storm damage as such receipts are taxable under Schedule D. Others who wish to benefit from the arrangements must inform the Inspector of Taxes for the district in which the woodlands are situated in writing not later than two years after the end of the chargeable period to which the election relates. The election much extend to all woodlands on the same estate managed on a commercial basis. But an occupier can treat woodlands which have been afforested or restocked as being on a separate estate if notice is given to the Inspector within two years of the afforesting or restocking. If the occupier does not wish this 'notice' to apply to all the woodlands planted within the last two years, particulars should be given. In this way, if an occupier wishes to clear and restock storm damaged woodland in stages a series of elections can be made, one for each area as it is restocked under the Woodland Grant Scheme: they will be treated as separate estates outside the income tax system and tax relief can continue for the clearance of the remainder of the woodland. An election to be assessed under Schedule D is binding for the remainder of the transitional period as long as the woodlands are occupied by the person making the election.

Schedule D assessment during the transitional period. Once a Schedule D election has been made, profits from the occupation of commercial woodlands will be calculated as if they were profits of a trade and should include: receipts from the sale of timber and other produce, grants from the Forestry Commission (other than those made under the Woodland Grant Scheme), grants made by other government departments, and any insurance money received as a result of damage to, or destruction of, trees. (Insurance receipts of a capital nature should be left out of the computation but an occupier may be liable to capital gains tax on them yet only in so far as they relate to the land itself as capital insurance receipts in respect of trees and underwood are exempt from capital gains tax.) The foregoing rules also apply where commercial woodlands are occupied by a company chargeable to corporation tax.

The normal basis of assessment under Schedule D is the 'profit' for the year immediately preceding the year of assessment. But in the case of afforested or restocked woodlands treated as a separate estate the computation of the profits should be made as if a new business had been set up, without taking into account any profits or losses arising from the land in question before the date of afforesting or restocking, as the case may be. The income tax year begins on 6 April and ends on the following 5 April. A woodland occupier should therefore make up accounts for tax purposes for a period of twelve months ending on a day within the tax year preceding the year of assessment. The Board of Inland Revenue is prepared to accept accounts which are made up each year by reference to receipts and payments in the year (the 'cash' basis), instead of on the more accurate and correct basis of accrued receipts and expenses (the 'earnings' basis). The profits of the first three years from the date of commencement must be computed on the 'earnings' basis. A form of account for occupiers who desire to change to a cash basis (Form 10W) can be obtained from the Inspector of Taxes. The accounts do not include growth increment or other changes in stock, and hence are simply income and expenditure accounts, not trading accounts.

An occupier who has elected for the transitional arrangements and who makes a loss in any year from the occupation of woodlands may claim to set that loss against other taxable income for the year of claim (or the following year). For this purpose, a loss may include an excess of capital allowances; usually those due for the year following a year of loss. During the transitional period an owner or tenant of forestry land who has elected to be assessed for income tax under Schedule D and who has incurred capital expenditure on the construction or reconstruction of forestry buildings, dwellings, fences, or other work for the purposes of forestry on the land, may be able to claim capital allowances. Since 1 April 1986 this kind of capital expenditure qualifies for annual writing down allowance of 4%. A system of balancing adjustments operates, at the taxpayer's request, when an asset is destroyed, demolished or sold. This enables the allowances to be brought into line with actual depreciation. Allowances for capital expenditure on plant and machinery used in the woodlands may be claimed on the same basis as for a trade.

Capital gains tax on land following the disposal of woodlands. Special rules apply to the computation of gains and losses arising on the disposal of commercial woodlands. ('Disposal' includes any occasion on which the ownership of the asset is transferred, whether in whole or in part from one person to another, except on death, for example by sale, exchange or gift.) The effect of the rules is to exclude growing timber (as distinct from the land on which it stands) from the computation. Thus the computation does not take any account of the price paid for the trees growing on the land at the time of the disposal or the cost of acquiring or establishing and maintaining those trees. In the case of the land, the gain is the increase in value since it was acquired (or since 31 March 1982, whichever is the later). An exemption is given for index-linking, i.e. increased annually in line with inflation. Capital expenditure on the construction, reconstruction, extension or adaptation of forestry buildings, dwellings and fences and other capital improvements is deductible in arriving at the gain or loss on disposal. During the transitional period to 5 April 1993 any income tax allowances for such capital expenditure are left out of account in the capital gains tax computation, unless the land is disposed of at a loss. If it is disposed of for less than the owner paid for it, the loss is not allowable for capital gains tax purposes to the extent that it has been covered by capital allowances.

Commercial woodlands (but not growing timber) is a qualifying business asset for the purposes of capital gains tax *rollover relief*. This means that where all or part of the proceeds from disposal of other qualifying assets are reinvested in commercial woodland (or vice versa) within certain time limits, rollover relief may be available to defer all or part of the tax which would otherwise be payable immediately. However, gains arising on commercial woodlands (i.e. excluding the growing timber) do not qualify for Retirement Relief.

Woodlands which are not run on a commercial basis are subject to the normal capital gains tax rules. Felled trees are treated as chattels for capital gains tax purposes. A chargeable gain can arise if an individual tree is sold for more than £6,000 (1990).

Inheritance tax on woodlands. The Finance Act 1986 introduced inheritance tax for transfers made on or after 18 March 1986. The tax applies to 'transfers of value' – broadly, transfers which reduce the value of the transferor's estate. Most outright gifts become exempt from tax if the transferor survives seven years from the date of the gift. During the seven-year period they are called 'Potentially Exempt Transfers', and may be made by one individual to another individual, by an individual into an accumulation and maintenance trust, or by an individual into a trust for the disabled; but cannot be a 'Potentially Exempt Transfer' so long as the transferor enjoys a benefit from the gift.

The amount chargeable to inheritance tax on a lifetime gift is measured by the fall in the value of the estate as the result of the transfer. If the transferor, rather than the transferee, pays the tax, the amount of tax has to be included in the fall in value. On death, tax is charged on the estate as valued immediately before that event. The estate includes settled property in which the deceased has an interest in possession.

The rate of inheritance tax on any transfer depends on the cumulative total of transfers (other than exempt transfers) made in the previous seven years. This applies to transfers made on death as well as to lifetime transfers which are immediately chargeable (for example, transfers into discretionary trusts) or which become chargeable because the transferor dies within seven years of making them. No tax is payable below a threshold of the cumulative total, which threshold (£150,000 in 1994) normally is increased from 6 April every year in line with the increase in the retail price index for the year to the previous December. Above the threshold, tax on transfers is charged at a single rate (40% in 1994). Lifetime transfers which are chargeable when made are charged at half the rate of tax which would apply on death. Where such a transfer is made within seven years of death, tax will become due at the full rate but will be tapered if that transfer occurred more than three years before the death. There are a number of reliefs and exemptions: in particular, most transfers between spouses are exempt.

Where woodlands subject to a deferred *estate duty* charge are transferred as *part* of a lifetime gift involving other property then inheritance tax is not immediately chargeable on the other property – instead the other property is regarded as a 'Potentially Exempt Transfer', i.e. no inheritance tax will arise unless the original owner dies within seven years of making the transfer.

Woodlands relief – deferment of tax on trees. Woodlands, both trees and the land, are a chargeable asset but in addition to general exemptions already noted there are other reliefs which can be claimed in appropriate circumstances. When a woodland is transferred on death an election may be made within two years of the death to have the value of standing trees ('trees' include 'underwood'), but not the land on which they are growing, left out of account in determining the value of the estate for inheritance tax purposes. When such an election is made the tax on the trees is deferred until they are disposed of by sale or otherwise. If, however, the beneficiary dies before all the timber is disposed of, the remaining deferred charge on the earlier death is cancelled, and a fresh election may be made to cover timber which forms part of the beneficiary's estate at his or her death. The woodlands must be in the United Kingdom and must not constitute agricultural property; for example small areas of woodland, the occupation of which is ancillary to that of agricultural land or pasture, do not qualify for the special deferment relief for woodlands (although they may qualify for the separate relief given for agricultural property). If the woodlands were acquired by the deceased other than by gift or inheritance, he or she must have been beneficially entitled to them throughout the five years before his/her death. When woodlands transferred on death can be divided into clearly distinct geographical areas, election may be made for each area separately. If woodlands are held jointly a joint election is necessary. An election once made, is final.

Careful consideration must be given to the possibility of electing to defer the tax on the tree crop in relation to a transfer on death; particularly in the light of the woodland structure (age and content) and the circumstances both of the deceased and of potential beneficiaries. The increase in the value of the tree crop over what might well be many years since that death, could involve a charge to tax substantially higher than that whch would have been payable at the date of death. Even with the Business Relief *infra,* the ultimate burden of tax remains unquantifiable. The eventual timber value will of course include volume increment and unit price increases, and usually will result in a much higher value than that at the death. With a twentyto thirty-year-old conifer plantation, one can expect to make payments of tax every five years as and when the thinnings are marketed.

Problems connected with deferment which may arise where woodlands pass on death to more than one beneficiary, include: the decision of one beneficiary not to elect for deferment is likely to affect the inheritance tax liability of another beneficiary, or other beneficiaries, who do so elect; and where an election for deferment has been made by more than one beneficiary, the subsequent decision of any one beneficiary to dispose of his timber is likely to affect the tax liabilities of the other(s).

However, if a beneficiary dies before the deferred tax becomes due in whole or in part, the liability in so far as it is still outstanding is cancelled. The timber is then treated in respect of the second death in the same way as on the first death, i.e. the inheritance tax liability on the second death may either be paid or deferred. With a long rotation, for example most broadleaves, this can happen several times in relation to a single stand of timber. The tax is cancelled on each successive death to ensure as far as possible that inheritance tax is paid only once on the timber when it is finally disposed of (and then deduction is allowed of selling expenses and restocking costs).

Tax payable on disposal of timber. When the value of trees has been left out of account on a death and they are subsequently disposed of (other than to a spouse) tax becomes chargeable on the net proceeds of sale if that disposal is for full consideration. In other cases, the tax is charged on the net open market value of the trees at the time of disposal. The rate of tax is found by adding the amount chargeable to the value of the estate of the last person on whose death the trees were left out of account and treating it as the highest part of that total. The scale in force at the date of disposal is then applied to the total. If there is more than one disposal the amount chargeable on a later disposal is added to the deceased's estate as increased by the amounts chargeable on earlier disposal(s). The disposal which triggers the deferred charge may itself be an occasion on which inheritance tax (or capital transfer tax where still applicable[1]) is payable, for example if the timber is given away by the beneficiary. In that case, the value transferred is calculated as if the value of the trees had been reduced by the deferred tax chargeable on them by reference to the preceding death.

Admissible deductions. In calculating the net proceeds of sale or the net value of trees allowance is made for the expenses of the disposal, and for the costs incurred in restocking, within three years of disposal, the trees disposed of so long as they have not been previously allowed. The Board of Inland Revenue has discretion to extend the three-year period if the delay in restocking has been caused by circumstances outside the owner's control.

Business Relief. The value of certain business property included in chargeable transfers during lifetime or on death may be reduced if certain conditions are satisfied. The business must be run on a profit-making basis and the 'relevant business

---

1. Capital transfer tax was charged on a sliding scale depending on the total value of chargeable lifetime transfers cumulated during the previous ten years. There were separate scales for transfers made on or within three years of death and for all other lifetime transfers. The reliefs and exemptions from capital transfer tax were broadly similar to those now ruling for inheritance tax.

property' must have been owned by the transferor for a minimum of two full years immediately before the transfer. In the case of a woodlands business, Business Relief extends to the value attributable to the timber itself as well as to the residual value of the business or the interest in it. The relief also applies to the deferred charge triggered by a later disposal. It will apply to reduce the net proceeds or value of the timber in calculating the deferred tax charge if the conditions for Business Relief would have been satisfied at the time of death on which the woodlands relief was granted. Where an election is made for woodlands relief for the timber itself, Business Relief may be due on the value attributable to the rest of the business including the land on which the trees are growing. The current (1991) rates of Business Relief are either 50% or 30%. The high relief rate is for: a solely owned business or an interest in a business such as a share in a partnership, shares, whether quoted or unquoted, where the transferor had control of the company immediately before the transfer, unquoted shareholdings of over 25%, and a life tenant's business or interest in a business. The low relief rate is for: other unquoted minority shareholdings, land, buildings, machinery or plant owned by the transferor and used in his or her business.

Heritage relief. Woodlands may qualify for the conditional exemption from inheritance tax (or capital transfer tax where still applicable) which is available for assets of national heritage quality. The exemption applies to land of outstanding scenic, historic or scientific interest as well as to certain other property. Trees may share in the exemption if they contribute to the qualifying interest. Ancient semi-natural woodlands which are, or could be, properly included on the Nature Conservancy Council's Inventory of Ancient Woodland will be eligible for consideration for exemption on scientific as well as scenic or historic grounds but each case will need to be considered on its merits. In return for exemption undertakings have to be given to preserve and maintain, and to provide public access to, the qualifying property. Tax becomes payable if the undertakings are broken or the property is sold or ownership changes (by lifetime gift or on death) and the relevant undertakings are not then given or renewed. Business Relief is not available in these circumstances. Consideration for heritage status can only be given after a death or after a life-time transfer has actually taken place; and hence there cannot be long-term management to make the most of the very values that would put the woodland into heritage category. The heritage relief provisions are described in booklet IR67 *Capital taxation and the national heritage* available from the Board of Inland Revenue.

Payment of inheritance tax. Inheritance tax (or capital transfer tax where still applicable) is not charged on timber cut for use on the estate. (When an election is made to defer tax on underwood passing on death, a charge to tax is made only when the first subsequent cutting of underwood takes place.) Tax is due and payable within six months of the end of the month in which the transfer or event giving rise to the tax liability occurs, but in the case of lifetime gifts made between 6 April and 30 September inclusive, by the end of April in the following year. When an election for woodland relief has been made, the person liable to tax is the person who is entitled to the proceeds of sale (or would be so entitled if the disposal were a sale). The deferred tax is payable within six months of the month in which the disposal occurs. Except in the case of a lifetime gift where the donor is paying the tax (subject to the qualification below), or in the case of a deferred charge arising on a dispersal of timber, the tax attributable to the value of the woodland business (including the land and timber comprised in it), or to the value of the land and timber on their own, may be paid by ten annual instalments. The first instalment is payable on the date the tax falls due. If any of the timber is sold, however, the balance of the tax attributable to it becomes payable immediately. Generally, interest (11% in 1990, but variable from time to time by Statutory Instrument) runs from the date when payment of the tax falls due. But instalments of tax on a woodland business are interest-free provided they are paid by the due date. When a life-time gift also triggers a deferred charge, the conditions relating to payment are relaxed for the tax payable on the gift. Tax may then be paid by instalments and the facility continues regardless of any sale. Also the instalments, including those for land and timber on their own, are interest-free if paid by the due date.

Estate duty. Estate duty continues to be payable on the net proceeds of the sale of timber which was left out of account on a death before 13 March 1975 until the first transfer of value of the woodland (unless it is a transfer to a spouse). There are also special transitional arrangements for gifts made before 27 March 1974 where the transferor dies within seven years (four years in Northern Ireland) of the date of the gift. Comprehensive information on Estate duty and its phasing out is provided by Hart (1988), the Forestry Commission (1989) and Lynch (1989).

Sporting in woodlands. Net receipts from sporting are assessable to income tax under Schedule A (s.75). The expenses, such as keepering, beating, stocking with game, management and advertising for guns, will be deductible. If the owner himself exercises the shooting rights as well as letting them, then only a proportion of the expenses will be deductible from shooting rents. Where the expenses cannot be directly allocated to let, or in-hand days, an apportionment should be made. Rents received under sporting leases are subject to VAT at the standard rate.

Recreation in woodlands. Charges for recreation in woodlands fall under normal rules of taxation. Wayleaves and rents in amenity woodlands are assessed under Schedule A. Allowance relating to costs of recreation (maintenance, repairs, insurance, management and other non-capital expenditure) are given under Schedule A.

Estate forest tree nurseries. Value resulting from the raising of seedlings, transplants and whips is not taxable if the stock is used in the owner's own woodlands. However, profit from sales made to others (e.g. from sales of more than a reasonably small quantity of surplus plants) is taxable as for a trade. In other words, beyond normal raising of forest-nursery stock for the owner's use there is a point where commercial trading is reached and profits become taxable.

Christmas trees. When Christmas trees fully occupy a substantial area, profits on growing them are taxable. But the taxation position is different when: (i) the trees are planted in mixture with other species to be grown as timber; or when (ii) the grower ensures that he always leaves sufficient trees growing after each part 'removal' to comply with correct silvicultural espacement and ultimately to provide a final crop of timber; or when (iii) tops only of thinnings are sold. In all three cases the net income therefrom is virtually tax-free.

Estate sawmills. The sale of trees standing or felled constitute part of the occupation or management of commercial woodlands; so does felled timber converted in order to make it most commercially marketable, i.e. into sawlogs, industrial small diameter roundwood, pitprops, fencing material, firewood and the like. Where a woodland owner has an estate sawmill, a measure of primary processing beyond the foregoing can also be undertaken, i.e. to the extent of simple planking and boarding, without the profits being taxable under the general fiscal rules *(Christie v. Davies,* 1945, 26TC398; *CIR v. Williamson Brothers,* 1950, 31TC370; *Collins v. Fraser,* 1969, 46TC143).

When the transitional Schedule D tax arrangements for occupation of woodlands end on 5 April 1993, the taxation incidence of estate sawmills will be determined by the measure of simple conversion (profit being tax-free) and that of further processing/conversion (profits being taxable). Obviously, profits arising from conversion of bought-in timber will be taxable. The relevant circumstances of an estate sawmill will need to be clarified with the Inspector of Taxes.

Grants. Forestry Commission planting and management grants are not taxable; but the annual payments under the Farm Woodland Scheme and the Set-aside Scheme are taxable. Some owners are able to claim income tax relief on the cost of managing their elected Schedule D woodlands until 5 April 1993, but such tax relief will not be available on woodlands for which Woodland Management Grants are received for that final year.

Rates. Woodlands are derated except those designated as 'amenity'. Land-drainage rates are chargeable on the gross annual value of woodlands. Sporting rights are only rateable when severed from the occupation of the land on which the right is exercisable (Country Landowners Association 1989a). Letting the sporting rights does not create rateability. When there *is* severance from the occupation of the land, e.g. when sold or gifted to another separately from the land, or when reserved by the owner from the letting of the land, there is rating liability. The same would apply if the tenant on a sub-letting of the land reserved the sporting rights. In instances where severance has already taken place, the sporting rights will, of course, remain rateable if they are let. There is no rating liability where an owner-occupier of land holds the sporting rights, or (unless they are already severed) he lets the sporting rights, or lets the land with the sporting rights.

Value Added Tax (17.5% in 1991). Forestry operations constitute taxable supplies and purchases, and thus attract respectively Output VAT and Input VAT, subject to appropriate limits (£35,000 in 1991). For a woodland owner who compulsorily or voluntarily has registered for VAT purposes, the incidence of the tax is neutral: Input VAT borne is recoverable; Output VAT collected is passed to H.M. Customs and Excise. Detailed information on VAT for landowners is available (Country Landowners Association 1989b).

Where timber or other woodland produce (including Christmas trees) is sold standing or felled *without the land,* Output VAT is chargeable. Where the sale includes the land, no VAT is applicable. Wood-logs, fire-lighters and wood charcoal (for fuel) are zero-rated; so too are sweet chestnut and walnut seeds and plants. Cordwood is not always zero-rated.

Certain land transactions are standard rated, including: (i) sales of land where the owner has opted to tax the estate by election, (ii) surrenders and assignments to the landlord of tenancies, (iii) sales of land which include valuable sporting rights (VAT is levied on the value of the sporting element subject to a *de minimus* exemption where this is less than 2% of the total value), and (iv) grants of leases and tenancies where sporting rights are transferred (again VAT is levied on the value of the sporting element but the *de minimus* exemption applies where this is less than 5% of the total value).

Woodland owners embarking upon afforestation programmes are advised to voluntarily register for VAT. H.M. Customs and Excise accept that in such cases a long period may elapse before timber sales bearing Output VAT are made, but if certain undertakings are given concerning the long term commercial nature of the enterprise, they will normaly accept voluntary registration and allow Input VAT in respect of forestry operations to be recovered. (Discretionary registration may be permitted where there will be future taxable supplies.)

The economic benefits and uncertainties of current capital taxation incidence have been noted in chapter 14. Even with an appropriate system of taxation coupled with adequate grants, forestry and woodland ownership may be unattractive and unprofitable unless sound silviculture, harvesting and marketing are practised. No business can survive healthily without prudent and efficient management. The skill of foresters is of paramount importance. Sound planning and foresight, often opportunism, play an important part in successful forestry practice. Correct choice of species, economy in operation, prudent planning, choice of the best rotations, intensive marketing, all have their place in sound woodland management.

4. Controls and restraints

The grant-aid noted in section 2 *supra,* understandably is accompanied by appropriate control and surveillance by the Forestry Commission as Forestry Authority. Its hand is light and sensibly applied; and many welcome a sympathetic control which regulates silviculture, particularly multi-purpose forestry, along with appropriate constraints as to felling trees.

Environmental aspects. Both the Forestry Commission and the Agriculture Departments have a statutory duty to pay regard to environmental considerations. In most planting grant schemes, and particularly under the Woodland Management Scheme and the Farm Woodland Scheme, emphasis is given to achieving environmental benefits from them, and steps are taken to prevent environmental damage. Applications are scrutinized with regard to landscape, nature conservation, recreation and public access, archaeology and hydrology, as well as timber production. Environmentally sensitive areas, SSSIs and the like, need special attention: forestry operations in them are sometimes listed as potentially damaging. Watercourses must be safeguarded. Scheduled monuments or historic buildings, as well as all archaeological sites must be protected. The same applies to public rights of way: discussions may have to be made by applicants with the Forestry Commission and with local authorities, if they so request, about possible new public access provisions.

Environmental assessment of afforestation projects. The purpose of an Environmental Assessment (EA) is to determine so far as practical the environmental effects of a proposed afforestation, to identify the actions that should be taken in the design of the proposals to mitigate any adverse effects, and to assist the Forestry Commission when deciding on proposals for which grant-aid is being sought. In general, an EA will only be required for afforestation proposals which are likely to have significant effects on the environment and which may lead to adverse ecological changes by reason of their size, nature or location. In 1988 the Commission published guidelines Leaflet 'Environmental Assessment of Afforestation Projects', to assist applicants for forestry grants on the circumstances in which an EA may be required and the procedures to be followed.

Indicative forestry strategies have been introduced to determine the appropriate locations for new forests in Scotland. Regional authorities have been encouraged to prepare these strategies as part of their structure plan wherever there is potential for major forestry expansion. An example of a type of identification of three categories of land for forestry development are: 'Preferred' areas where forestry would be positively promoted, 'Potential' areas with only limited sensitivity in relation to afforestation, and 'Sensitive' areas where the impact on the environment would require careful consideration before planting grant applications could be approved. Supplementary to the strategies are the forestry capability maps for Scotland (noted in chapter 3). Kennedy (1990) comments on the subject. Consideration is likely to be given to the extension of such a system to England and Wales in the light of experience gained in Scotland and of studies being carried out by the Countryside Commission.

Consultation with statutory authorities. Since 1974, the Forestry Commission has consulted as appropriate with the Agriculture Departments, local planning authorities and other statutory authorities on applications for grant-aid or felling licences and on draft Plans of Operations containing planting or felling proposals. These arrangements, which apply in England, Scotland and Wales, but not in Northern Ireland, are formally set out in Ministerial directions issued to the Forestry Commissioners under the terms of the Forestry Act 1967. Their purpose is to ensure that the requirements of land use, agriculture, amenity, recreation and nature conservation are taken into account before decisions are reached on grant-aid for planting and on permissions for felling. They are also designed to reconcile any conflict of view that might arise between the authorities consulted or between the applicant and such authorities. The administration of the consultation procedures was reviewed with the authorities concerned on 1 December 1983. Forestry Commission Leaflet 'Consultation Procedures for Forestry Grants and Felling Permissions' (1987) describes how the present procedures operate and the circumstances in which consultation is undertaken.

Control of tree felling. With certain exceptions, it is an offence to fell growing trees without first having obtained a licence from the Forestry Commission. Full information is available in the Commission's Leaflet 'Control of Tree Felling' (1987). The felling of trees in Great Britain (but not in Northern Ireland or the inner London Boroughs) is controlled by the Commission in exercise of its powers under the Forestry Act 1967. A licence from the Commission is normally required to fell growing trees (though not for topping or lopping) but in any calendar quarter up to 5 m^3 may be felled by an occupier without a licence provided not more than 2 m^3 are sold. A licence is not required if any of the following conditions apply:

1   The felling is in accordance with an approved Plan of Operations under one of the Forestry Commission's grant schemes;

2   The trees are in a garden, orchard, churchyard or public open space;

3   The trees are all below 8 cm in diameter, measured 1.3 m from the ground; or in the case of thinnings, below 10 cm in diameter; or in the case of coppice or underwood, below 15 cm in diameter;

4   The trees are interfering with permitted development or statutory works by public bodies;

5   The trees are dead, dangerous, causing a nuisance or are badly affected by Dutch elm disease; and

6   The felling is in compliance with an Act of Parliament.

In certain circumstances – whether or not a felling licence is needed – special permission may be required from another body for any proposed felling. This can occur where an area is designated as a Conservation Area or as a Site of Special Scientific Interest (SSSI) or the trees are protected by a Tree Preservation Order (TPO). Recent developments regarding control of felling are discussed by Griffin and Watkins (1988).

Tree Preservation Orders. Local planning authorities have powers to protect trees by the imposition of a Tree Preservation Order (TPO). Guidance on the circumstances in which a TPO should be imposed has been given by the Department of the Environment in Circular 36/1978. The DoE recognize that 'an Order may not be appropriate where owners are known to manage their estates acceptably with proper regard for amenity where such management is expected to continue'. In woodland which has been grant-aided by the Forestry Commission under one of its schemes, a TPO may only be made if the Commission gives its specific consent to the making of the Order. Such consent will usually be withheld where woodlands are under satisfactory management.

The procedure for making a TPO is set out in the Town and Country Planning (Tree Preservation Order) Regulations, 1969, S.I. 1969, No. 17, as amended by the Tree Preservation Order (Amendment) Regulations, 1981, No. 14. TPOs do not extend to prohibit the cutting of trees which are dead, dying or dangerous. Other trees also exempt from the prohibition include those cut down at the request of a Statutory Undertaking, an electricity company, a local authority, or the Ministry of Defence.

Penalties for non-compliance with a TPO are set out in S.102 of the Town and Country Planning Act 1971; also in S.62(1 A) of that Act as inserted by the Town and Country Planning (Amendment) Act 1985. The Acts deal with regulations as to applications to fell, refusal of consent, and compensation payable. General information on TPOs is provided by the DoE (1988) and the Country Landowners Association (1990). Recent developments regarding TPOs are discussed by Griffin and Watkins (1988). The TPO procedures are currently under review by the DoE and the Welsh Office. Management grants from the Forestry Commission are available for woodlands under TPOs as from 1 April 1992.

5. Other aid and encouragement

The support given by the Forestry Commission to private forestry in the provision of planting and management grants noted in section 2, and by the capital taxation concessions outlined in section 3, is supplemented by significant assistance in other ways noted hereafter.

As Forestry Authority the Commission stresses to Government the importance of a viable and profitable private sector. It administers the Woodland Grant Schemes, operates the licensing of felling, and undertakes census of private woodlands. It encourages groups for control of grey squirrel, and provides invaluable help relating to plant health and insect pests and diseases. It provides guidelines and examples of management for wood production as well as for non-wood values – landscape, recreation, wildlife conservation and to safeguard water resources. Fine examples are given of provision of recreation sites and nature trails. The private sector have valued being able to deal with staff who have practical experience of day to day forestry operations, problems and opportunities, and who are able to provide examples of high standards of multi-purpose forestry.

The Commission produces abundant and diverse literature, regularly up-dated, on aspects of silviculture from seed collection onwards. Work study help is made available. Harvesting and marketing are helped by example and information, and particularly by stimulating investment in the wood processing industries. Much help is given about marketing and replacement of storm damaged trees. Price-size curves are provided for both conifers and broadleaves, as well as a range of management and measurement tables. Classifications of soils, terrain, and windthrow hazard have been made available. Fire protection measures are advised upon, and some local help given with forest fire control. New forest machines and equipment are tested and reported upon. Advice is given relating to herbicides, pesticides and fertilizers.

The Commission's efforts to educate the public about its Forestry Enterprise objectives and management equally benefit the private forestry sector. Procedure is by way of talks, literature, agricultural shows, schools, and information and visitor centres in many State forests, notably at Grizedale Forest, Forest of Dean, New Forest, and Kielder Forest. Noteworthy, too, are the maintenance and management of the Pinetum at Bedgebury in Kent and the Arboretum at Westonbirt in Gloucestershire.

The Commission helps in the integration of advisory services (Coed Cymru is a useful illustration of the effective integration which has been successful in delivering advice to woodland owners). It is seeking to increase the help which its own foresters can give to the private sector, particularly in the co-ordination of advice, the training of advisers and the better utility of the produce from broadleaved woodlands. Particularly important is the Commission's role in the provision of forestry education and training noted *infra*. Of equal importance are the Commission's research and development – functions noted in section 6 *infra*. In the wider sphere, the Commission attends to the Regulations for forestry within the EC, among them those relating to plant health, pollution, provenance, and farm forestry.

Education and training. Forestry education and training are evolving dramatically to meet the changing needs and demands of both government and the industry. Qualifications and statements of competence are now as important as the training course itself. The Forestry Commission through its Education and Training Branch, and support of the Forestry Training Council (FTC), operating from its headquarters in Edinburgh, plays a notable role in today's forestry and industry.

The FTC was reorganized and reconstituted on 1 April 1987, being recognized by the Training Agency as the Industry Lead Body representing the many components of the forestry industry. Its main functions are to identify and advise on training needs and to make proposals regarding the implementation of Government policy on training. The FTC maintains a 'Register of Approved Instructors' who have been assessed and adjudged competent to standards required by the Council. Many of the instructors are self-employed people working in the forestry industry, others are based at colleges of agriculture or further education and some are employed by the Forestry Commission; geographical cover is more or less complete.

The Commission, through its Education and Training Branch, besides training its own staff arrange courses for other organizations on request – and encourage basic craft training in practical forestry skills, forest management and a whole range of subjects which can be tailored specifically to the needs of the organization concerned. Specific courses are arranged for the practical forestry examinations of the Royal Institution of Chartered Surveyors – courses held in the Management Training Centres located in the Forest of Dean and Ae Forest.

Vocational training in forestry is available at Holme Lacy College in Herefordshire (and through the college at work sites organized throughout south-west England and in Wales), the Cumbria College at Newton Rigg, Sparsholt College near Winchester, and the Scottish School of Forestry, Inverness. The Forestry Commission and private employers provide practical experience for potential forestry students. Block release college based courses at craftsman level are City and Guilds Phases I and II in England and Wales and equivalent SCOTVEC modules in Scotland; similar courses are available at the highest levels for City and Guilds Phases III and IV and equivalent SCOTVEC modules. Merrist Wood College at Worplesdon in Surrey specializes in training for arboriculture and nursery practices. Specialist training is available at Hooke Park College for Advanced Manufactury in Wood (Parnham Trust). Practical training in forestry is offered by The Barony College at Parkgate near Dumfries. The Buckinghamshire College in High Wycombe offers training in timber technology, utilization, sawmilling, management and marketing.

Higher National Diploma courses are available at Newton Rigg and at Inverness. BSc courses at honours level in forestry or a forestry-related subject are available at the universities of Aberdeen, Edinburgh and the University College of North Wales at Bangor. Post-graduate qualifications (MSc, MPhil, and PhD or D.Phil) are available at all these universities and also at the University of Oxford. The National Diploma in Forestry (NDF), considered to be an ordinary degree level qualification, is offered by the Central Forestry Examination Board. The Institute of Chartered Foresters confers chartered status to professional foresters by examination.

Information on all the foregoing courses and qualifications are obtainable from the schools, colleges or universities concerned, from the Forestry Commission, the Forestry Training Council, the Institute of Chartered Foresters, and the Association of Professional Foresters.

Training in the new technological skills needed to take on management as part of a farming system is available to farmers and landowners under a Woodland Training Project supported by the EC Social Fund. There is available a two-year YTS Scheme in forestry, entitled 'Preferred National Training Pattern for Forestry'. Grants may be available from the Training Agency to employers who are seeking to improve the efficiency of their woodland operations. A notable organization aiding forestry education and training is the Forestry Trust for Conservation and Education at Theale in Berkshire.

There is an increasing need for updating training in forestry at all levels – most foresters were trained under circumstances much removed from those of today. Lorrain-Smith (1986) discussed the needs and the subjects for training courses. Implementation unfortunately has been dogged by logistical and financial problems.

Contents of current forestry courses are designed to provide an adequate background for a very wide range of activities. The changes in curricula and the methods of delivery of vocational training will continue into the foreseeable future. The National Council for Vocational Qualifications (NCVQ) and in Scotland SCOTVEC, and in England BTEC, will increasingly exercise influence by accrediting only those qualifications which enjoy a large measure of support from industry. The first National Vocational Qualification (NVQ) for forestry workers is to be administered by the National Proficiency Tests Council (NPTC). The Foundation Level 1 NVQ has nine Units which cover basic competences – health and safety, planting/beating up, hand weeding, fencing, repair and maintenance of hand tools, cleaning, brashing/pruning, draining and basic tree identification. It has been drawn up by the Forestry Training Council, the National Examinations Board for Agriculture, Horticulture and Allied Industries and the NPTC in consultation with the industry. Potential students as well as employers should be cautioned to ensure that they ascertain for themselves what the contents of courses are – rather than assume they know: a rich source of disillusionment.

Health and safety at work. A legal obligation is placed on employers, by virtue of the Health and Safety at Work etc Act 1974, to provide adequate instruction, training and supervision as is necessary to ensure, so far as is reasonably practicable, the health and safety at work of their employees. This requirement to train, has been extended by the requirements for skill and competence to be displayed by all users of pesticides (The Control of Pesticide Regulations 1986). Important, too, are COSHH regulations ('Control of substances hazardous to health' regulations) made under the 1974 Act noted above. They deal particularly with toxic or other hazardous effects.

The Forestry Safety Council (FSC) was set up by the Forestry Commission to promote all aspects of safety, particularly safe working practices throughout the forestry industry. As an aid to maintaining the safe working standards of operators, the FSC publish a series of Safety Guides each of which gives advice on safety in a particular forest operation; each is accompanied by a Safety Check List intended for use by supervisors and safety representatives. The FSC Safety Guides currently available, free of charge, cover the following operations (some have been noted in earlier chapters):

N	Noise and Hearing Conservation
1	Clearing Saw
2	Dipping Plants in Insecticides
4	Pre-Planting Spraying of Container grown Seedlings
6	Tractor Mounted Weeding Machines

7	Planting
8	Hand Weeding
9	Brashing and Pruning with Handsaw
10	The Chainsaw
11	Felling by Chainsaw
12	Chainsaw Snedding (Delimbing)
13	Chainsaw – Crosscutting and Stacking
14	Chainsaw – Take Down of Hung-up Trees
15	Chainsaw – Clearance of Windblow
17	Chainsaw – Felling Large Hardwoods
18	Tree Climbing and Pruning
21	Forest Tractors
22	Extraction by Skidder
23	Extraction by Forwarder
24	Mechanical Harvesting
25	Extraction by Cable-crane
26	Use of Tractors with Winches in Directional Felling and Take-Down
30	Mobile Saw Bench
31	Mobile Peeling Machine
32	Fencing
33	Hand Held Power Posthole Borer (Rock Drill)
34	First Aid
35	All-Terrain Cycles

There is also a Joint FSC/Electricity Supply Industry Code of Practice: The Avoidance of Danger from Overhead Electric Lines and Underground Electric Cables in Forests and Plantations.

The FSC Safety Guides have no particular legal status, nonetheless they are used by the enforcement authority (HSE) as one of several criteria in deciding whether a breach has been committed. Their content has been successfully used to secure conviction during legal proceedings. Training should not, however, be undertaken merely on the grounds of fulfilling a legal obligation which in most cases is concerned with the requirements to work safely: the aims and benefits of training must be in the context of improving professional competence and productivity.

6. Forestry research and development

It is important for the private forestry sector to be aware of, and benefit from, relevant forestry research and development. This as well as most of the applied research is chiefly carried out by the Forestry Commission which acts as contractor for both the State and private sectors (Grayson 1987; Burdekin 1989). Forestry research is broadly divided into nine subject areas:

1. Genetics and tree improvement, including seed, selection, breeding, progeny and clonal testing;
2. Tree biology, including tree physiology and mycorrhizae;

3   Silviculture, including cultivation, establishment, tree stability, soils and nutrients;

4   Biotic damage, including fungal and viral pathogens, vertebrate and insect pests;

5   Biomass production, including distribution of dry matter within and between trees, competition studies, and basic physical properties of tissues;

6   Harvesting techniques, including harvesting of stems, roots, branches and leaves;

7   Wood science and processing, including technical properties of wood utilization development;

8   Environmental effects, including wildlife conservation, recreation, landscape, water and air pollution; and

9   Forest planning, including application of technical forestry knowledge for management of forests, and relations with other land users.

The current organization of the Research Division is explicit in Figure 31.

Other agencies are also engaged in forest research. Projects carried out by the Institute of Terrestrial Ecology which bear partly or wholly on forestry range from studies of the susceptibility of trees to autumn frost and nutrient cycling under different tree species to the biology of deer and the concentration of air pollutants in mists. Important work of value to forestry is undertaken on viral control of various insect pests by the Institute of Virology and Environmental Microbiology, and on water yield of catchments under forest cover by the Institute of Hydrology. All three institutes are members of the Natural Environment Research Council. Agricultural Departments support an increasing volume of research, both north and south of the Border, concerning farm woodlands and agro-forestry, the latter covering the establishment and maintenance of widely spaced trees combined with grazing by sheep or cattle. The Department of the Environment supports significant amounts of research bearing on forestry in a variety of institutes and universities. About 60% of forest research at universities is undertaken by Aberdeen, Bangor, Edinburgh and Oxford; a significant proportion is therefore undertaken at other universities. For a general view of the inter-relationships among the various bodies foresters are recommended to refer to the annual summary of expenditures on forestry research prepared by the secretary of the Forestry Commission Forest Research Station at Alice Holt Lodge.

The growing of large quantities of wood is by itself no guarantee of success in forestry; it is essential to have regard to quality of the product that can be made from wood, in two ways: by designing silvicultural techniques in the broadest sense to produce more uniform and high quality material, and by developing utilization processes which are appropriate to the raw material supplied, at realistic cost. *Wood research* is undertaken, partly under Forestry Commission contract, by the Building Research Establishment of the Department of the Environment at Garston near Watford. Its work covers a broad spectrum of projects ranging from studies of wood as a material to preservation, finishing, sawmilling and wood structures. The Timber Research and Development Association, located at Hughenden Valley near High Wycombe, is supported by the timber industry with a modest contribution from the Forestry Commission. Its work is mainly of an applied and developmental nature. Three of the four university forestry faculties and in particular those of Bangor and Oxford also undertake some wood science research.

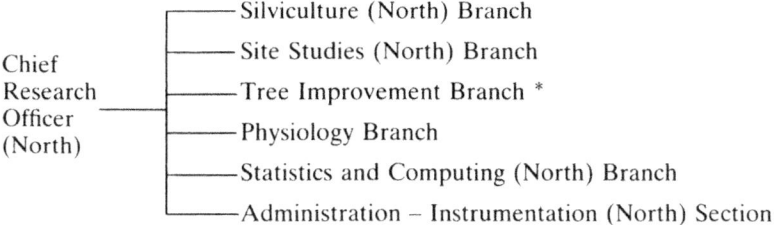

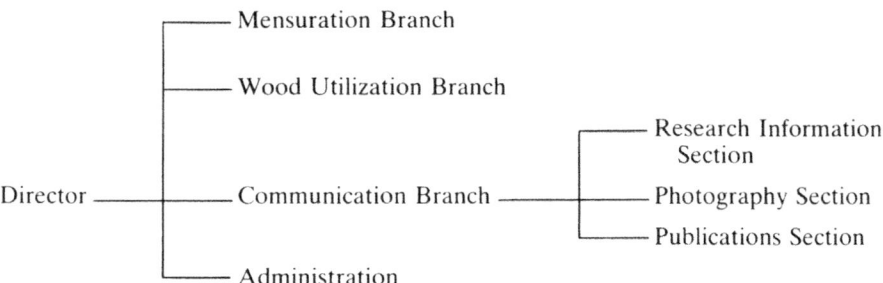

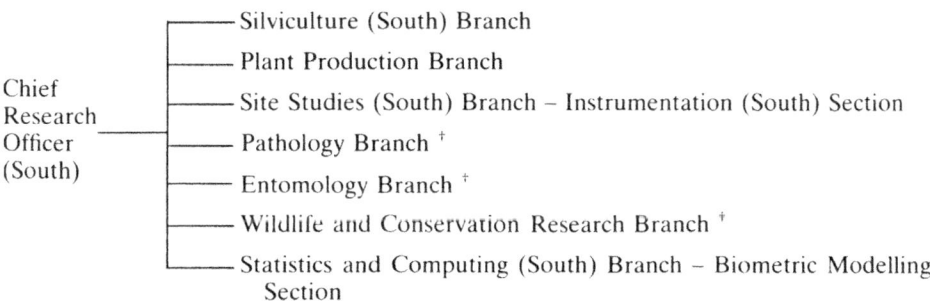

* Branch with Section at Alice Holt.
† Branches with Sections at the Northern Research Station.

Figure 30: Forestry Commission Research Division Organization.

Forestry Commission liaison with the private sector as regards research and development is maintained by two principal means. The first is directly with the Home Grown Timber Advisory Committee and its Technical Sub-Committee on which are represented all the private forestry and forest products interests. The second is through the Technical Sub-Committee of TGUK.

Following the 1980 report on scientific aspects of forestry by Lord Sherfield's sub-committee of the Lords Select Committee on Science and Technology, the government established the Forestry Research Coordination Committee under the chairmanship of the Forestry Commissioner responsible for research. This has an important function in assessing research requirements, in ensuring good liaison among the many organisations undertaking forestry research and in co-ordinating particular programmes such as that begun in 1988 on farm woodlands.

The development and application of new techniques in forestry has been rapid and broadly successful. British forestry practice and most of the forests now existing in Britain are direct extensions of research, results of which have been applied over the last seventy years. The Forestry Commission has been compelled by circumstances to develop and apply new techniques of afforestation. The technical difficulties of creating large productive new forests on bare degraded uplands were formidable; mainly it was a case of establishing exotic conifers. The accomplishments were remarkable. The productivity of the afforestation is now being reflected in the rapidly rising annual output of wood including sawlogs.

The Commission's Research Division, as part of the Forestry Authority, serves the needs of the forest industry as a whole. Foresters already have a wide range of options, from species choice through to felling regime, that they deploy in managing woodlands and these have different effects on the wood they grow and habitats created. New demands and different effects on the values of products call for more options and accordingly the principal aim of forest research is to improve understanding of the way in which management influences the quantity and quality both of the wood produced and of the environmental benefits supplied, including wildlife conservation. A major swing in research during recent years has concerned conservation of birds, mammals and plants, while on protection aspects significant advances have been made in providing guides to managing red deer populations. In addition the Commission has undertaken a two-pronged attack on the question of air pollution effects on trees, using countrywide surveys and controlled experiments in specially designed chambers. A major concern is that the operations carried out in forestry should be done at low cost consistent with safety and with minimum impact on the environment. These requirements provide challenges for research and development in both biological and engineering areas.

Biological research within forestry is of its nature closely related to the sorts of trees and conditions found in different regions. Accordingly such research is conducted by Branches within the Research Division. The office and main Research Station are at Alice Holt Lodge, Wrecclesham, Farnham, Surrey GU10 4LH (Tel. 0420–22255), with the Northern Research Station at the Bush Estate, Roslin, Midlothian EH25 9SY (Tel. 031–445–2176). The eighty scientists and professional foresters engaged in research are supported by technical staff at the two stations as well as at out-stations located throughout England, Wales and Scotland. The Director of Research is advised by an Advisory Committee for Forest Research consisting of leading academic and other figures in plant sciences and related fields. Close contact is maintained with other research organisations concerned with silviculture, forest products, agriculture, wildlife conservation and the environment.

Each of the two Research Stations, as well as the Branches within them, has a main area to cover, defined by the function and site, thus Silviculture (North) staff work primarily in Scotland, northern England and Wales. Silviculture (South) staff work mainly in the southern half of England. In the North, project leaders are responsible for those aspects of silviculture especially concerned with upland commercial forestry. This includes research into all methods of coniferous plant production and the large scale production of genetically improved coniferous stock; site preparation and drainage; nutrition in plantations at all stages, along with investigations of the effects and control of heather in coniferous crops; tree stability investigations; regeneration of existing coniferous crops; and species choice and seed origin in relation to site and managerial objectives. In the South, research is concerned with silviculture in the lowlands and the emphasis is on broadleaved species. Nursery work, weed control and treeshelters are topics aimed at improving establishment technique. Control of epicormic growth in oak, work on vegetative propagation and the support of investigations into timber quality of oak reflect the current concern over Britain's major native trees.

Close linkage and a degree of interchange between personnel engaged in applied research and management is most desirable. About half of the research project leaders in the Forestry Commission are forest officers transferred to the Research Division from forest management duties. The other project leaders are career scientists permanently engaged in research. There is a movement of forest officers into the Division – generally to those Branches that are most directly concerned with general forest management. After five years or more they return to field duties. This process of exchange of forestry officers greatly facilitates the flow of technical information into, and out of, research. Similar exchanges have been effective at the technical level.

Current developments in forest research are set out in the Forestry Commission's annual *Report on Forest Research,* published up to 31 March each year. Reports of review groups set up by the Forestry Research Coordination Committee are also available from the head office of the Research Division; these cover all the main subjects of biological and economic research in forestry.

A feature in recent years is the work being carried out for other Government Departments on a contract basis. The Department of the Environment supports an *Arboricultural Advisory and Information Service,* work on urban tree health and problems of tree establishment. The Department of Energy through its Energy Technical Support Unit supports work on short rotation coppice.

The work of the *Tree Improvement Branch* of the Northern Research Station is noted in chapter 2. The *Physiology Branch* has as its main purpose the provision of basic information for the more applied research Branches. Much of the research is carried out on plants in controlled environments. Investigations have been made into root growth and development, mycorrhizae, induction of flowering, and micropropagation by way of subculture *in vitro.* The work of the *Pathology Branch* and the *Entomology Branch* is referred to in chapter 6. The main functions of the *Seed Branch* are noted in chapter 2. The principal concerns of the *Wildlife and Conservation Branch* are noted in chapter 13.

The *Mensuration Branch* of the Research Division is responsible for research into growth and yield in forest crops, for the construction and development of yield models, the preparation of various management guides and mensurational aids for field staff, and the development of electronic field data collection equipment. It is also responsible for the management of approximately 750 permanent sample plots from which primary data are obtained, and for the development of the sample plot computer data base.

Research by other Divisions of the Forestry Commission. The bulk of the Development Division's work is concerned with the economics of tree crops and their yield of wood as well as environmental effects. The Division is also responsible for Forest Survey and Work Study.

The *Forest Survey Branch's* function is to collect, process and present data concerning forest crops, for operational and corporate planning and to provide technical support and advice. The branch also investigates applications of remote sensing, including aerial photography, to forest surveys, the development of the forest (sub-compartment) data base and processing of these data for forecast of production and general management needs, and the development of inventory methods. All these activities are dependent upon electronic data handling and computational procedures involving close co-operation with the Statistics and Computing Branch of the Research Division.

The *Work Study Branch* is responsible for operational efficiency, and studies cover all aspects from crop establishment to harvesting. Its services are noted in chapter 16. In addition the Branch provides a back-up service to the Safety Officer in safety research, techniques and evaluation of safety clothing and equipment.

In the field of *economics,* the many subjects of study include: the demand for goods and services supplied by forests, the impact of new technical developments on supply, and appraisals of more general social and economic trends bearing on forest policy and management. While certain of these studies are of a policy nature, the remainder, as with work carried out by or for the Research Division, are directly geared to the provision of guidance for foresters.

Information obtained by research is only of use if it is communicated, and hence both the Research and the Development Divisions are concerned with providing practical advice to those concerned with forestry and tree growing. The scope of research required to serve these aims is very wide, covering all aspects of the formation, regeneration, protection, management and harvesting of forest crops. The private forestry sector is fortunate to receive the benefit of the Forestry Commission's continuing research and development.

References

Agriculture Committee (January 1990), Second Report for the Session 1989–90 on 'Land Use and Forestry' (HC 16–1) 10 January 1990. HMSO.
—— (May 1990), Response to the Second Report from the Agriculture Committee, Session 1989–90, 'Land Use and Forestry' (HC 16–1) 9 May 1990. HMSO.
Burdekin, D.A. (1989), 'Trends in Forestry Research 1982–1988'. FC Occasional Paper 23.
Country Landowners Association (1989a), 'Rating'. CLA Publication R1/89. London.
—— (1989b), 'VAT and Landowners'. CLA Publication T4/89. London.
—— (1990), 'Trees and the Landowner'. CLA Publication F/90. London.

Department of the Environment (1988), *Protected Trees: a Guide to Tree Preservation Procedures.*
Forestry Commission (1989), 'Taxation of Woodlands'. FC Bulletin 84.
Grayson, A.J. (1987), 'Evaluation of Forestry Research'. FC Occasional Paper 15.
Griffin, N. and Watkins, C. (1988), 'The Control of Tree Felling: Recent Developments in Statute and Case Law'. *Quarterly Journal of Forestry, 82* (1), pp. 26–32.
Hart, C.E. (1980), 'Effect of Taxation on Forest Management and Roundwood Supply: A Report for United Nations, EEC and FAO'. Supplement 4 to Vol. 33 of the 'Timber Bulletin for Europe'. Geneva, September 1980.
—— (1988), *Taxation of Woodlands.* Chenies, Coleford.
Kennedy, D. (1990), 'Where Will Scotland's Future Forests Be?' *Forestry and British Timber,* March 1990.
Lorrain-Smith, R. (1986), *Training Needs of Professional Foresters in Mid-career.* A Study undertaken for the Institute of Chartered Foresters. Edinburgh.
—— (1988), 'The Economic Impact of the Guidelines for the Management of Broadleaved Woodland'. TGUK, London.
—— (1989), 'Grants and Other Financial Assistance for Trees'. Calderdale Metropolitan Borough Council, Halifax.
Lynch, T.D. (1989), *The Taxation of Woodlands in the United Kingdom.* Green, Edinburgh.
Spilsbury, M.J. and Crockford, K.J. (1989), 'Woodland Economics and the 1988 Budget'. *Quarterly Journal of Forestry, 83* (1), pp. 25–32.

Chapter Sixteen

WOODLAND ACCOUNTS AND FORESTRY VALUATIONS

1. Woodland accounts

For forestry to be critically assessed and compared with alternative financial investments, and for management to be efficient, it is necessary to have an adequate system of accountancy and costing. Woodland accounts have two main purposes. First, financial accounting – as for any business and required by law; second, cost accounting.

Financial accounting is required primarily to record income and expenditure, and, with the addition of depreciation, interest, valuation of assets and liabilities, to build up a trading account, a profit and loss account, and a balance sheet. These results relate to the whole enterprise but can be analysed to relate to individual stands and to whole woodlands. Financial accounting gives a current and periodic view of the woodland enterprise and its component parts, indicating at any point in time whether in the whole or in any part a profit is being achieved or a loss is being sustained. It helps towards profit maximization, when this is the objective, or loss avoidance, or any other financial objective.

Some woodland accountancy systems lead only to a receipts and expenditure account: when income in any year exceeds expenditure a 'profit' is assumed; or a 'loss' when expenditure exceeds income. These assumptions of course are wrong, most notably because no account is taken of the increase or decrease in the value of the growing stock, nor of changes in the value of equipment, materials, and prepared produce. None the less, such accounts, if analysed can provide much helpful cost information.

Cost accounting is likewise an important 'tool' in forest management – aiding it by way of costings, records and so forth in operating, producing and marketing. Unless information is recorded and analysed, it is difficult to practise silviculture and marketing in the most advantageous manner. Without appropriate costs, budgeting and planning may be meaningless, and if expected returns are not calculated commercial forestry can enter into the realms of faith (Openshaw 1980) as well as fantasy.

Once the main financial accounts are analysed, individual operational costs are revealed, and may indicate ways of reducing them and of increasing efficiency. Costing of products can distinguish the profitable from the unprofitable.

Costs of individual operations or products accurately ascertained, can be used for comparisons, fixing of piece-work and bonus rates, productivity measurements, budgeting and control. By aggregating the individual costs and overheads, and recording revenues, an income and expenditure account for a stand, a woodland, or a whole forestry enterprise can be ascertained. However, such accounting shows only the year-to-year running without measure being taken of the change in capital value.

Methods of accounting vary – ranging from the simplest suitable for a single woodland to extensive detailed systems used by owners of major well-wooded estates or large afforestations. (Accounts for the latter are often prepared by computer consultants, by one of the forest management or accountancy companies, or by a firm of chartered surveyors.) Basic accounting principles are described in standard accountancy textbooks. Whatever the system used it should enable cash, journal and ledger entries to be collected to form eventually a receipts and expenditure account, a trading account, a profit and loss acount, and a balance sheet. Much of the work is now done on computers (see section 5 *infra*).

Any accounting system is difficult to describe briefly. Of those relating specifically to forestry, that advanced by Openshaw (1980) is particularly suitable. The first requisite is adequate records supplemented by a map/plan showing for each stand its extent, species and year of planting. Openshaw provides suitable standard account heads for financial accounting (in broad terms, land, capital and labour) and for cost accounting (in detail, individual operations and products). The second requisite is primary records (time-sheets, and wage, machinery, and materials analyses).

Comprehensive accountancy systems lead to detailed information for any required period, for the woodlands as a whole, for individual sub-compartments, and for departments such as a forest tree nursery, sawmill, wood-yard or preservative plant. Detailed analyses of the accounts are undertaken supplemented by depreciation of machinery, overheads, materials and prepared products. For the profit and loss account and balance sheet, there will be opening and closing valuations of growing stock, the closing one reflecting the increase by growth increment, afforestation and restocking, as well as any decrease due to removals by clear-felling. (Removals by thinning are unlikely to decrease the capital value, and should eventually increase it. However, in a sale, value may fall because a recent thinning has taken place.) Shooting and/or recreational rents or revenues generated by the woodlands should be credited. Difficulties arise in assessing the non-wood values generated by the woodlands, and in crediting their monetary value to the forestry enterprise – difficulties discussed in chapters 12, 13 and 14. Other complications in woodland accounting to be borne in mind include:

1 Until 5 April 1993 some parts of the woodlands may be assessed under Schedule D of income tax, in which case an income and expenditure account (without changes in stocks) must be submitted annually to the Inland Revenue (see chapter 15);

2 There may be internal trading (transfers of fencing materials, machinery, labour);

3 The forestry enterprise may include a sawmill or wood-yard whose profits may be taxable and compulsorily accountable if attained from the conversion of timber from the Schedule D areas (until 5 April 1993) or from bought-in timber (see chapter 15).

4 Part of an estate forest nursery may be taxable and compulsorily accountable (see chapter 15).

Woodland accounts are important not only as records of historic costs and revenues but for what they may suggest about future cash flows. The extraordinary length of forestry's production cycle means that annual accounts are of interest only if they are sufficiently detailed to indicate operational and product costs. Changes in stock values are important; and revaluation to allow for inflation over long periods is appropriate (Price 1989). This is again referred to in section 4 *infra*.

2. Woodland records

Supplementary to financial and cost accounting, foresters should make records of their management both for themselves and for the time when their office falls to another. An adequate history can be built up each year when any operation is carried out or when anything else arises worth recording. This can be based on sub-compartment or stand records discussed in chapter 12. Foresters can also make use of an *Annual Sub-Compartment Financial Record* of receipts and expenditure, which can be aggregated to provide at any time a *Cumulative Financial Record'*, and if all of them are condensed into one they can provide a *Consolidated Record* for the whole of the woodlands. Computers will obviate the use of much manual recording and analysis (see section 5 *infra*).

3. Forestry costings

Labour. Analysis of wages, machinery, material and overheads can determine costs of operations and products, as well as of each sub-compartment at any point in time. The basis of most costings is labour time-sheets (supplemented where relevant by costings of machinery). They are usually ruled and headed for weekly or fortnightly use, but differ in form on almost all estates. Some are simple, with or without columns for overtime and piece-work, and cater for machinery used; others are comprehensive. The total cost of labour for the work done may be easily ascertained from a summary made at the foot of the time-sheet, calculated at an hourly or weekly rate. Analysis to calculate the cost of the operations concerned can be undertaken either on the back of the time-sheet or on a separate sheet. The analysis of each sheet will then be posted to a columned wage-book or wage-sheet, and the total operations of the week or fortnight can easily be ascertained by aggregating.

Supplementary to time-sheets must be some means of recording all materials (purchased or sold) moved to, from or within the woodlands. For this purpose some kind of forester's day book or large diary may suffice. Sales invoices may be prepared from it and rendered for goods supplied other than by cash sales. To the costings of labour must be added relevant items of purchase or transfer inward and of sale or transfer outward, besides cost of transport and machinery; all must have been analysed according to the system in use. Analysing will indicate costs of, and returns from, any operation, as well as outputs; the last can be related to productivity. Without costings, losses on individual operations, such as conversion, may remain hidden for long periods.

Direct costs include wages of forest workers, outlay on materials/chemicals, and running costs of machinery. *Oncosts* are mostly associated with labour – holidays, wet time, machine failure, contributions to pension and welfare funds, and transport of workers. It is usual to sum these costs, calculate them as a proportion of direct costs, and add this percentage to the direct labour cost of operations. *Overheads* are enterprise costs which cannot easily be assigned to any particular operation, or to any particular stand. They are often those costs recorded centrally for convention or convenience. Many would be allocated to jobs (as labour oncosts are), and if there are labour oncosts then there are probably materials and machinery oncosts too (Lorrain-Smith 1988). They vary greatly from one enterprise to another, and are among the major costs of forestry; though usually easy to ascertain, they are often difficult to allocate appropriately. Examples include the costs of research and development, administrative and clerical services, and supervision at all levels. Like oncosts, overheads are usually added as a percentage to direct costs. Allocation must be to the operation or product on which the overheads are spent. Most overheads are associated either with supervision of labour or with securing the best possible prices for products; thus a system may be to apply one percentage of the total overheads to wages and another to sales. The following overheads need appropriate allocation according to circumstances: tools and other equipment (new, renewals, repairs, maintenance); maintenance of woodland boundary fences, roads, rides and drains (see chapter 6); protection: fire and storm insurance premiums; fire precautions and equipment; control of pests and diseases; and costs of workers' rented dwellings.

Forest machinery. With progressive mechanization of forest operations, particularly cultivating and harvesting, the cost of machinery becomes ever more important, and may dominate the economics of an operation or product or the profitability of a crop. Purchase of machinery, an investment whose pay-off is spread over several years, complicates methods of costing. The machinery used in every forest operation can be costed from first principles, but it is convenient if a cost for a given machine (usually per hour) is ascertained. Price (1989) examines methods used to provide such a standard cost; and discusses whether this standard cost truly represents the cost to an enterprise of using a machine.

Machinery costs conventionally comprise capital and running expenses. Since machine and operator form an inseparable unit as far as useful output is concerned, the running cost of large machines often includes the labour cost of operating them. Oncost is added to direct payment. Tax and insurance recur infrequently, but can none the less be allocated hourly. The conventional components of capital cost are depreciation and interest; they are allocated over the units of work – hours, or output achieved (Openshaw 1980; Price 1989). Applications of machine costings include setting a machine-hire charge (with a mark-up for profit, unless the loan of the machinery is internal), costing an operation to decide if it is worth doing, and comparing operational methods, such as labour and capital intensive harvesting. It is usual to calculate the cost of each machine and eventually allocate it to operations undertaken. If a particular machine or vehicle has only one operator or driver, the log-book indicates the usage, but often it is necessary to have a specific log-book or log-sheet for each machine.

Work Study. Work Study, consisting of work measurement and method study, including machine assessment, plays an important role in silvicultural and harvesting operations. Knowledge of it is essential to foresters.

For labour, the aim is usually to determine a 'Standard Time', that is, the time in which a trained and motivated worker of average ability would be expected to complete a specified job in specified circumstances: in effect, a combination of three factors to complete a job – direct work, indirect work, and rest. To the 'direct time' taken to complete a single working cycle (such as felling and delimbing a tree), an 'Overhead Time' must be added for repairs/maintenance of working tools, together with a percentage rest allowance which varies with arduousness of the job. Tables of 'Standard Times' (output guides) are important sources of information in planning requirements for and costing an operation. In all cases a 'Standard Time' is multiplied by a price per standard minute (PSM) to give a piece-work price. They can be used to calculate output, determine requirements of labour and machines, productivity increases, cost an operation, assess performances, and form the basis of an incentive payment scheme. For machines, output is usually measured as the units of output achieved in a given working time, normally an hour. Costing of machinery is comprehensively described by Price (1989).

Work study has several objectives relevant to economics: (i) defining optimum working methods; (ii) improving health and safety at work and, thereby, raising productivity; (iii) defining a rate of work (time) that when taken with an appropriate price per standard minute (PSM) gives workers a realistic incentive to increase productivity; and (iv) especially to decision-making, measurement of output rates of workers and machines.

The Forestry Commission's Work Study Branch 'Output Guides and Standard Timetables' have always been generally available to the private sector, and its method study work/reports are now also available. Foresters requiring information on an operation can contact the Chief Work Study Officer, 231 Corstorphine Road, Edinburgh EH12 7AT. The Branch, while it investigates to satisfy the Commission's own needs, disseminates the information for the good of the forest industry, but offers no guarantee: foresters can accept and use its information to add to their decision making. The 'Index of Output Guides and Standard Timetables by Forest Operation', available on application, lists all those currently available covering a particular operation or method. As examples, relevant to harvesting, they include thinning and felling for various conifers; for machine harvesting, some processors and harvesters; for conversion, cross-cutting and peeling; and for extraction, skidding, forwarding and cable-crane extraction. For planting stock supply, they include cone collection and transplant lifting by machine. For silviculture, operations covered include scarification, new planting, restocking, fencing, drain maintenance, weeding and brashing.

The Commission's earlier Booklet 45 'Standard Time Tables and Output Guides' (1978) has been updated and reissued in loose-leaf format with an introduction along with the output guides themselves. This enables the holder to keep a copy up to date by obtaining new output guides as they are issued by the Work Study Branch. If an output guide is not available for a particular operation, the Branch's Information Note 13 ('Setting Piece-work Rates') will provide advice. As noted above, a 'Standard Time' is multiplied by a price per standard minute (PSM) to give a piece-work price. Piece-work gives an incentive to get on with the job and make more money; and reduces unit costs by spreading fixed oncost charges over more units of work. But the incentive is greatly diminished if there is not further piece-work to follow.

Scale economies/diseconomies. These comprise the phenomenon by which output increases/decreases more than proportionately with given increases in the scale of factors of production employed. Scale economies are cost advantages from working on a large scale, e.g. in relation to technology, labour, machines, bulk purchases and bulk marketing. Diseconomies of scale may arise.

4. The growing-stock

The preparation of a balance sheet for a forest enterprise requires decisions on how to classify the growing-stock among assets and how to estimate its value. The growing-stock is a hybrid between current and fixed assets (Openshaw 1980); it is partly stock (though standing and often not ready for harvesting) and therefore partly a current asset, and partly a factor of production and hence part of fixed assets. It is not merely a matter of terminology since fixed assets do not appear in the conventional profit and loss account but stocks (current assets) do, and are traditionally valued at cost or market price, whichever is the lower. The asset of the growing-stock should be recorded separately from either fixed or current assets provided that their value is assessed in a way which is practical and indicative of changes in the growing-stock, changes which can then be incorporated in the profit and loss account. The growing-stock would be recorded in the balance sheet after fixed assets and before current assets.

Hiley (1954), recognizing the need to include in a woodland's profit and loss account the appreciation in value of the growing-stock by annual growth, adopted a method that used increment for the management of the Dartington woodlands in Devon. His method for merchantable timber stands in each sub-compartment was to increase yearly the assessed value of the stand (at inventory), not by physically measuring increment but by a percentage which was subjectively decided as a proportion of the true increment to allow for removals by thinnings. It was considered to be too complicated to include thinnings in the adjustment (Osmaston 1968). Afforestation and restocking were valued at cost, values which were then increased annually by a subjectively decided percentage of not more than 3.5%, so that their book value roughly equalled their market value when the plantations became merchantable.

Gane (1966) showed that increment must be used in assessing appreciation of value, proposing that market prices should be applied to the increment with due allowance for difference in prices obtained for thinnings and final crops. The proportion of increment removed in thinnings is valued at the unit rate obtained for thinnings of the size or age concerned but the value of the increment retained (i.e. remaining in the stand after thinning) is assessed at the unit value of the mature timber that it is destined to produce. The total value of the increment of a stand is thus calculable. For the calculations of the value of the increment, stands are classified into suitable categories of species, yield and age classes. By using yield tables (see chapter 12) a practical procedure of calculating the value of the increment of a stand is possible. This incremental value is added to the initial assessed value (at inventory) of the growing-stock and the value of all removals subtracted to give the yearly estimated value. In calculating the value of removals, Gane proposed that the same book values used for the increment should be used for the removals rather than actual market prices received each year.

Foresters sometimes are opposed to analysis forms, multiple columns and the like, asserting that the cost of account books and the amount of 'dead' paper work are considerable, that there is always some expenditure that does not find a ready place in any column provided, that if there are sufficient columns to allow full analysis and direct costing from them, the books and their constant writing up or posting are expensive beyond the value of the result, and that if the books are kept simple with few columns further working up of the figures is required and there is then no advantage over the unanalysed account. However, computers (see section 5 *infra*) have removed most of the tedious work of recording and analysis.

The numerous figures obtainable from woodland detailed accountancy may entail so much extra booking that the analysis may be too expensive and may in the long run obscure the main account that the extra work fails to fulfil its chief purposes. The detailed costing of *every* operation and of *every* product can be tedious, and may not be worthwhile if only a few of the numerous facts and figures obtained are acted upon. Calculators and computers have greatly eased the burden, but it may still be desirable to select and set only the most worthwhile objectives and to reduce the number of analyses as far as is reasonably possible. The least to be aimed at may be the ascertainment of either the profit or the loss on each principal product, and determination of the cost of each major operation. Whatever the view, some form of accountancy and costing is a necessary adjunct to successful management.

An important factor, generally overlooked, is the demonstration in financial and cost accounts of the expense of unwise or uneconomic 'whims and fancies' on the part of the owner or his supervisory staff. The time is ripe, too, for a comprehensive treatment of costs for woodland operations (Lorrain-Smith 1989), as well as for an estate sawmill, a tree nursery or preservative plant, along with overheads. Other matters with important implications include the calculation of the best roading layout economically justifiable for each woodland or group of woodlands (see chapter 10).

Foresters intending either to begin or to improve a system of woodland accounts can obtain guidance from Openshaw (1980). If a computerized system is to be pursued, professional assistance is readily available, as noted in the next section.

5. Computers in forestry

Computers can be used in all types of business – in general accounting and in management. They are invaluable for invoicing, credit control, paying, budgeting and financial accounting, also in summarizing time-sheets into wages/salaries/ payrolls/ payslips, and costing therefrom.

Computers are important 'tools' in forest management. They are able to make numerous calculations speedily, based on data and instructions fed in as programs, which can be stored on tape or disc. Some computers are relatively simple, and can be set on a desk; others can be hand-held and used in the forest. They have removed much drudgery from forest calculations and recording.

The techniques of discounting (and of investment appraisal as discussed in chapter 14) are relatively easy to apply to a simple investment with a few cash flows and a fairly short time span. With a long time span such as that of forest investment, identifying the cash flows can be more difficult. Judgements must be made, for example, on the yields and value of timber which will be produced, twenty years or more in the future, from a crop about to be planted. Costs and benefits can occur as single payments or annually over a range of years, so calculations of NPV may prove complicated. Unless the forester has access to a computer, calculations have to be made by hand with the help of ready-reckoners, for example tables of DR.

Forestry tasks which a computer can undertake much faster than any forester using a manual system include: compartment records, stock control (materials, trees, machinery), volume estimation, timber valuation, growth prediction, financial forecasting, cash flows, and investment appraisal (analysis). Growing-stock can be recorded by species, yield class, age, dimension and volume; and it can be up-dated by recording volume increment, and changes due to felling, restocking and afforestation (Lorrain-Smith 1990). Once data have been entered they can be extracted in any combination to provide the report that the user requires, and this is one of the main advantages over a manual system.

Computers are much in demand for production and income forecasts, and they provide invaluable help in calculations for investment appraisal, decision-making, NDR and IRR. They can be used for control of establishment, tending, harvesting, and departmental enterprises.

There are many ready-made software programs on the market which have been written for specific forestry applications. Computing services offered by specialist firms[1] have greatly reduced the time spent on calculating, recording and accounting. Some firms offer a series of interrelated computer programs which can be purchased separately or combined to form a comprehensive management tool to suit the individual requirements of different forestry businesses. It is not always necessary to purchase a computer as several firms offer a bureau service.

Computerization in the private sector has proceeded very quickly since the conference held by the Institute of Chartered Foresters in Edinburgh in 1984 (Mason and Muetzelfeldt 1986). The major forestry management companies and several of the larger firms of chartered surveyors and chartered foresters have 'in-house' computer services for themselves and their clients. Some of the larger forestry estates have also realized the potential of computers in controlling their woodlands. Lorrain-Smith (1990) provides detailed guidance on the use of spreadsheets for computerized management applications.

The Forestry Commission uses a wide variety of computers in its headquarters and Forest districts (Arthurs 1987). Through its Forest Investment Appraisal Package (FIAP) the Commission provides its local foresters access to a wide range of information and gives answers to calculations quickly (Harper 1986). It helps forest managers to manage their crops. By specifying crop information and the pattern of cash flows for a model hectare, it is possible to compare options using the discounted cash flow approach. Other uses of computers in the Commission include portable microcomputers to help to record measurements of their sample plots to predict timber outputs (Gay 1988). Calculating and checking tariff data is also computerized. The Commission is working on further projects, among them to

---

1 Badenoch Land Management Ltd ('Woodplan'), Alvie Estate Office, Kincraig, Kingussie, Inverness-shire PH21 1NE.
British Forest Surveys (H. Malyon), Snowdon House, Academy Street, Ecclefechan, Lockerbie, Dumfries DG11 3DE.
Simon Masters, ('Masterplan Forestry'), New Barn, Runnington, Wellington, near Taunton, Somerset TA21 0QJ.
'Cleggbase', c/o John Clegg & Co., 2 Rutland Square, Edinburgh EH1 2AS.

integrate Operational Planning and Control computer systems with its growingstock database and business systems.

Foresters should be aware of the benefit from the basic uses of computers. There is neither the prospect nor the need for them to understand the mechanisms by which complicated mathematical problems are solved, but it is desirable that they know how to formulate a problem in terms intelligible to the computer. However, the proper interpretation and application of results provided by a computer depend on foresters' experience and judgement.

6. Forestry valuations

Forestry valuations are necessary for sale/purchase, accountancy/management, capital taxation, probate, insurance, compensation (after windthrow, fire, or severance through such as road construction and expropriation), and wayleaves (for water, oil and gas pipes, and for telegraph or power lines). Chiefly involved is the application of the market value to the land (hectares) and of the worth of volumes (m^3) of trees measured by methods discussed in chapter 1 or, for young plantations, calculations of the (discounted) value of their future net revenues.

The choice of method of valuation is largely determined by the objectives of the vendor/purchaser; or sometimes by statute to satisfy the District Valuer. Valuation may well differ between vendor and purchaser, or between owner and acquirer. Their objectives and ideas of value may be at variance. Thus, for example, a woodland crop, though it has only one market value, may appear to have two: its current worth to a roundwood merchant or sawmiller for harvesting and processing (bearing in mind restocking commitments), and its worth to the owner (or a new investor) as a source of future income and/or capital gain above the (discounted) inevitable outgoings.

Valuation of woodlands. Current market values of woodlands are indicated in chapter 17 – the important factors influencing values being location, species, yield class, stability, past management and proximity to markets. Valuations, according to their purpose, may suffice as realistic guesstimates, but where a high degree of accuracy is necessary, individual stands and whole woodlands must be valued in detail. Where only a reasonably sound estimate is required, quicker and less costly methods can be used, some being described below. In valuations involving discounting of future costs and revenues, the appropriate discount rate will be that of the owner or intending purchaser or acquiring party – determined by them and not by the crop's performance. It is usually assumed that future stumpage (standing) prices will not change.

The owner's or intending purchaser's circumstances and motives will have a great influence on their respective notions of value. Of an estimated value, it is always uncertain how much a prospective buyer might offer above, or how much a vendor might accept below. Both would be aware of the capital taxation incidence and concessions, and of the grants available towards management and restocking.

Capital valuation has a special meaning because the tree is regarded as both the (standing) capital and the (felled) product. Therefore, the term 'capital valuation' refers chiefly to the valuation of the growing-stock. At any point in time there is a high ratio of growing-stock (inventory) to annual increment. Account should therefore be taken of this naturally appreciating capital as well as of the inanimate capital such as land, infrastructure, machinery and equipment.

Valuation for accountancy and management. The need for these valuations has been noted in section 1 *supra*. An effective plan for the management of woodlands can be assisted by a realistic knowledge of their capital value. Valuations can be costly, but an appropriate value can be compiled by methods noted below. Once an opening book-value (inventory) has been established, annual capital values, particularly for stands of measurable volume, can be updated annually by: (i) adding the notional value of timber increment, say 3–6% for conifers and 0.5–2% for broadleaves (higher for poplar and sweet chestnut coppice); (ii) adding costs, less grants, of establishing new plantations or of restocking (increment need not be applied to young plantations until they are fully established); and (iii) deducting the book-value of clear-felled stands (those thinned during the year need not result in a reduction in their book-values as the potential of the remainder of the stand should be more valuable, and the net income generated has produced a cash reserve). However, a valuation made on book-value alone can only be approximate; adequate inspection and appraisal of the crops at every relevant date are essential.

Valuation for compensation. Woodlands in part or whole can be compulsorily acquired under statute. Valuations relative to them would be made under the usual methods of valuing woodlands, noted in this section. In the case of wayleaves and easements (for roads, gas, oil, water and transmission lines), compensation may fall under three categories: (i) the crops removed (often immaturely); (ii) the potential damage to adjacent stands; and (iii) the 'sterilizing' for forestry, yet the land has to be kept clean, fences and drains maintained, fire guarded against, and pests controlled. Compensation for (i) and (ii) can be valued by conventional methods, but compensation for (iii) is generally an annual payment (probably re-negotiated periodically) which may or may not be commuted later to a lump sum. (The Forestry Commission agrees standard values for crops taken from its properties by Electricity or Water companies.)

Valuations to assess compensation for the effects of windthrow and of harm to wildlife habitats or sporting interests on the remaining components or woodlands which have been bisected (e.g. by motorways) are separate considerations. Sheate and Taylor (1990) discuss the wildlife conservation aspects. Another type of compensation is that under the Countryside and Wildlife Act 1981.

Valuation for insurance. For woodlands up to twenty-five years old, insurers and brokers use a schedule of crop values based on a cost-plus-notional increment scale – representing an owner's expenditure at current average costs, net of grants, increasing annually to reflect additional maintenance expenditure plus compound interest. The premium may vary according to individual circumstances and the locality, species and structure of the woodland. This basis of valuation is applied for the period until merchantable timber is produced, and thereafter the woodlands are often insured at market value or at devastation value. The latter is the value of the timber net of harvesting costs (Insley (ed.) 1988). Although market or devastation value is commonly used in woodland insurance, for investors this value will be much lower than expectation value up until shortly before the economic rotation age. Thus once the crop has grown beyond the establishment stage it may become appropriate to insure the crop on agreed expectation value; or at market value; or.

for storm damage, to cover the difference between what the owner actually receives for his windthrown timber and expectation value; the latter should in addition take into account the additional cost of clearance. One insurer's scale of plantation values is included later in I(c) and II(c) *infra*. Arriving at an independent value for insurance can be undertaken by any relevant method indicated in / and // *infra*. Older woodlands can be insured at figures based on the market values of the timber at the time of loss, which values may be only a fraction of their potential as a final crop.

Valuation for capital taxation. In a disposal of a woodland where capital gains tax is relevant (only applicable to the land) an apportioned arms length value between land and crop must be agreed by the disposer/vendor and the beneficiary/ purchaser. Pertinent to the valuation of the land is the fact that it is effectively 'sterilized' during the remainder of the rotation. There can (at least in theory) be a wide variation in value between land where the crop is nearing the clear-fell stage and land that is newly planted and hence subject to a crop's whole rotation. For capital gains tax, the base value of the land is either the cost or the market value at 31 March 1982, whichever is the later, the value being index-linked. For inheritance tax, both land and crop must be separately valued. (However, a valuation of the crop is unnecessary if an election for deferment on it is to be made.)

Types of forestry valuations. A woodland usually represents a substantial store of wealth, even a young plantation without merchantable timber having a market value reflecting the expected future production. However, a woodland does not have the same value to everyone. Objectives differ (only aesthetically sensitive buyers will pay a premium for attractive woodland; a roundwood merchant or wood processor will pay only for timber and land values). Perceptions differ in expectations of timber prices, also in circumstances (for example, the expected incidence of capital taxation). The desired value depends also on the purpose of the valuation. Owners may wish as high a valuation as can be justified when negotiating with potential purchasers, or establishing value as loan collateral, or seeking compensation for woodland compulsorily acquired or destroyed by third parties. By contrast, as low a valuation as can be justified is wanted for probate, when inheritance tax liability is to be assessed, or when negotiating with potential sellers. An objective valuation is especially desired when deciding whether to sell or buy, or when insurance is being arranged.

Valuation of a wood may be based on various principles according to circumstances and objectives (Long 1984; Price 1989), chiefly on (i) historic cost; (ii) expected contribution to future revenues and benefits; or (iii) present market value. Only in a perfectly functioning market do all three coincide (Price 1989). Variability is inherent in interpretation of such matters as the appropriate discount rate in (ii) above.

Historic cost. Current owners of a wood usually have either inherited it with some idea of what it was then worth or paid cash for it. At the simplest, they will not wish to sell the wood for less, after adding the cost of any improvements they have effected and some mark-up for inflation. If the wood is simply an amenity (for example, used mainly for shooting or hunting) there may be no further considerations, but if the wood is an investment, the owner may add compound interest on net costs incurred, deducting from the base price any revenue accruing since acquisition, also with compound interest. The figures used should be either actual cash transactions adjusted upward for inflation (or compounded at a monetary rate of interest) or, in the absence of sufficiently detailed records, a present-day cash flow for an equivalent operation. Price (1989) gives an example of a historic cost valuation. Historic cost, as far as purchasers are concerned, is an irredeemable sunk cost; and if sellers are rational they too should regard past costs as irrelevant. (A role remaining for historic cost is to ensure fairness when a wood is compulsorily acquired, or an inheritance shared.) Woodland accounts (as outlined in section 1 *supra*) can be an important part of forest valuation, not so much as records of historic costs and revenues, but for what they suggest about future cash flows.

Expectation value. This value only applies to future income and expenditure. Particularly for purchasers, attention focuses on future expectations of profit or benefit that the wood affords: they want a discounted predicted cash flow from existing or successor crops. A factor in the cash flow is periodic costs of protection, maintenance and management. Expectation value (see I(d) and III(a) *infra*) should be calculated separately for stands of different species, age and productivity (yield class), and aggregated in a combined valuation. The further ahead that costs/prices are predicted the more inaccurately are they likely to be determined.

Present market value. Where there are frequent transactions involving rather similar woods in similar locations, an indication of market price can be apparent. Specialist woodland agents such as Bell-Ingram, Bidwells, John Clegg & Co., Savills and Strutt & Parker, and the major forestry management companies, know that a particular type of wood, adjusted for extent, in a given location, has a basic price range, with credits or debits given for special attractive or detrimental features. However, when transactions are infrequent, or a wood has unique characteristics, no clear market value is apparent: intuition and experience come into play. Important factors include location, species, yield class, stability, rotation, level of past management, and proximity to markets; likewise are sporting and non-wood benefits. Vendors usually keep in mind the past cash flow, while buyers are only concerned with future cash flow.

Where valuations between parties differ, market price may be difficult to determine, and in any case depends on attitudes of sellers and buyers. The market price lies somewhere between the lowest price the seller will accept and the highest price any potential buyer is prepared to pay. (When the lowest acceptable price exceeds the highest offered price, no transaction occurs.) Otherwise, sale price may depend on the relative bargaining strengths and circumstances of the seller and buyer: if there are many potential keen buyers the price is likely to be high; if there is only one buyer and the seller urgently needs cash the price is likely to be low. Eventually an acceptable compromise usually is reached.

Infrastructure. The value of infrastructure (access, internal roads, bridges, buildings, permanent fences) lies partly in contributing to enjoyment of satisfactions but chiefly in its ability to contribute to future profitability, a value subsumed in the expectation value of growing-stock (the value of standing timber depends largely on its accessibility). However, if the necessary infrastructure does not exist, the discounted future cost of providing it is a debit against expectation value. When a wood contains areas of young, medium and old growing-stock, the infrastructure has greater value for the thinning of the medium age crops and the clear-felling of the older crops, than for the young plantations; for the latter the infrastructure will provide most benefits only in the distant future. The effect overall of infrastructure in relation to protection, maintenance and management is also important. The (discounted) cost of any such infrastructure still to be built is a debit against woodland value.

Recreational and sporting values. Existing commercial recreational facilities, as well as let sporting, should be valued at the discounted stream of net revenues expected from their rental. Potential for such rentals, and the cost of planning and installing them, may also enter valuations.

Methods of forestry valuations. Methods of valuation are noted below, along with a few examples. Results obtained from each method can be compared. The examples relate mainly to conifers and exclude land. Broadleaves in the young to early pole-stage – other than poplar and sweet chestnut coppice – might be expected to have values of one half to three-quarters of those for conifers, although their establishment and tending costs may have been much higher. Their value would increase with age.

I. *Young plantations up to 5 years old.*

(a) *Value based on historic cost,* i.e. actual past net cost plus compound interest. Where historical cost is unknown, an indication can be obtained from University Surveys (Oxford and Aberdeen) and from Hart (1987).

(b) *Value based on replacement cost,* i.e. the current expected net cost plus compound interest. Costs can be extrapolated from the sources of information noted in *I* (a) *supra,* and particularly in chapters 5 and 6.

(c) *Insurance value,* based on values used by some insurance companies, of which the following are examples (excluding land): (£/hectare):

Age of crop Years	England and S. Wales £/ha	Scotland and Mid Wales £/ha
1	704	352
2	792	440
3	880	528
4	968	616
5	1,056	704

The insurance values exclude re-cultivation of ground and renewal of fences. They are higher than historic costs or current market values, and hence are more likely to be used for insurance claims than for sale/ purchase, or inheritance tax assessments.

(d) *Expectation value.* This value only applies to future income and expenditure. Forestry Commission Management Tables can be used to give the expected volume yield from thinnings and final crop, at the rotation age.

The expected yield class should be estimated from neighbouring tree crops. Unit values should be applied to the volumes at the relevant point in time, and discounted to the present (one to five years), using an appropriate discount rate. Constant costs and prices are used. (Difficulties lie in determining future costs and prices: the assumption of costs/prices over such a long period limit the usefulness of expectation value.) In the simplest calculations, net thinnings are excluded, their value being assumed (arbitrarily) to equate to interim costs of protection, maintenance, management, insurance, tending, brashing and cleaning. Examples at the simplest are given below (excluding land): (£/hectare):

Species	Expected Yield Class	Rotation age	Final Crop			Approximate NPV at 5%
			Vol. m^3	Approximate Nominal Value £/m^3	Value £/ha	£
Scots pine	10	80	356	37	13,172	265
Japanese larch	10	50	232	35	8,120	700
Oak	6	150	238	41	9,758	10
Beech	6	100	268	26	6,968	55

The NPVs are lower than historic costs, and well below replacement costs – understandably as the excluded volume of thinnings would amount to, respectively, 294, 185, 397 and 241 m^3/ha. Obviously such calculations of NPVs are inappropriate for young plantations. They are more likely to be used for inheritance tax assessments than for compensation claims and sale/purchase. The examples indicate the impact of species, age, yield class and rotation (as well as discount rate) on expectation values of young plantations, and the effect (arbitrary) of excluding thinnings.

(e) *Comparable sales.* Comparison can be made with any known recent relevant sales. Such information is available to relatively few foresters in the private sector, whereas District Valuers have a well established and highly efficient system for the exchange of information/levels of valuation – ostensibly ensuring some uniformity. Values are indicated in chapter 17.

(f) *Present market value.* Placing a present market value on young plantations requires a high degree of experience. (Values are indicated in chapter 17.) Only a sale can confirm the market value, but some indication may be obtained by the methods noted above.

II. Thicket-stage plantations 8 to 15 years old. Plantations of these ages rarely contain merchantable timber and are difficult to measure or estimate for volume. Management Tables can be used to calculate the expected volume yield from thinnings and final crop, to the rotation age. The yield class should be apparent from neighbouring tree crops. Unit values should be applied to the volumes at the relevant points in time, and discounted to the present (eight to fifteen years), using an appropriate discount rate. Difficulties lie in determining future costs and prices: the older the crop, the easier becomes the possibility of allowing for realistic interim costs and prices. Several methods of valuation are suggested below, along with some examples:

(a) *Value based on historic cost.* As for I (a) *supra*.

(b) *Value based on replacement cost.* As for I (b) *supra*.

(c) *Insurance value*, based on values used by some insurance companies, of which the following are examples (excluding land): (£/hectare):

Age of crop Years	England and S. Wales £/ha	Scotland and Mid Wales £/ha
8	1,320	968
9	1,408	1,056
10	1,496	1,144
11	1,584	1,232
12	1,672	1,320
13	1,760	1,408
14	1,848	1,496
15	1,936	1,584

The insurance values exclude re-cultivation of ground and renewal of fences. They are higher than historic costs or current market values, and hence are more likely to be used for compensation claims than for sale/purchase, or inheritance tax assessments.

(d) *Expectation value.* Reference should be made to I (d) *supra*. Many assumptions again have to be made as to the future, but the older the crop, the easier becomes the possibility of allowing for realistic interim costs and prices. The crops are unlikely to be measurable. Reference should also be made to III(a) *infra*.

(e) *Comparable sales.* Comparison can be made with any known recent relevant sales – see /(e) *supra*. Values are indicated in chapter 17.

(f) *Present market value.* Placing a present market value on thicket-stage plantations requires a high degree of experience. The expectation value – I(d) *supra* and III(a) *infra* – will be the main investment criterion. Where circumstances are such that the valuation has to be 'the market value standing, if cut now', it will be 'the devastation value'. (Values are indicated in chapter 17.) The trees are unlikely to be measurable, in which case one or more of the methods II(a) to (e) *supra* can be used. However, where the trees are measurable (though it may be unreasonable to attempt this) two methods of valuation at the simplest are:

(i) Decide by the average espacement, how many trees per hectare are thriving, for example only 2,000 of 2,500 originally planted. If the estimated volume per tree is 0.025 m^3, the volume per hectare is 50 m^3 which at, say, £5 m^3 amounts to £250/ha.

(ii) Use Management Tables and apply unit prices (£/m³) as in the wide-ranging examples below (excluding land): (£/hectare):

Species	Yield Class	Age Years	Vol. m³	Approximate Nominal Value £/m³	Value £/ha
European larch	10	15	51	5	255
Japanese larch	10	15	64	5	320
Sitka spruce	20	15	73	6	438
Ash/Sycamore	8	15	32	4	128

Values could be increased if some trees are considered to have special value (e.g. larch for rustic work; spruce/fir as Christmas trees), since the tables disregard for volume stems with a base diameter under 7 cm, as well as tops above the 7 cm diameter of the stem. The above values are much lower than historic cost, and very much below replacement cost; and hence are unlikely to be acceptable even for inheritance tax assessments and certainly not for compensation claims and sale/purchase. The examples of the use of this method indicate how low are the values arrived at for thicket-stage plantations with little if any merchantable produce.

The values in (f) (ii) would be greatly increased if expectation values are calculated as in *III*(a) *infra*.

III. *Pole-stage to mid-age crops.* At these stages, measurement of merchantable timber is possible – by the relevant methods noted in chapter 1. Valuation based on historic cost or replacement cost, or on insurance values is rarely appropriate. Three methods of valuation are:

(a) *Expectation value.* Reference should be made to *I*(d) *supra*. Management Tables can be used to give the expected volume yield from thinnings and final crop, to the rotation age. The yield class should now be apparent. Unit values should be applied to the volumes at the relevant points in time, and discounted to the present, using an appropriate discount rate. The older the crop, the easier becomes the ability to use realistic interim costs and prices.

The example below assumes that net income is only obtained at the final felling (thinnings are excluded) and that the four chosen species are of age 20. (Excluding land): (£/hectare):

Species	Yield Class	Rotation of maximum MAI	Final Crop			Approximate NPV	
			Vol. m³	Nominal Value £/m³	Value £/ha	at 5% £	at 10% £
Scots pine	10	75	337	37	12,469	860	70
Japanese larch	12	45	250	36	9,000	2,680	840
Oak	6	80	222	24	5,328	280	20
Beech	8	85	310	27	8,370	350	20

The above NPVs illustrate that the chosen discount rate affects value: high discount rates give lower values, low discount rates give higher values. Rotation length is also important: the shorter the rotation the higher is usually the value. Of less importance are average price/m³ and yield class, although the latter determines the rotation of maximum mean annual increment (MAI): the higher the yield class the shorter is the rotation.

Depending on the discount rate chosen, expectation value is likely to be used for inheritance tax assessment and purchases (using higher discount rates) or for compensation claims and sales (using lower discount rates). Valuations which include net income from the thinnings would of course be higher.

(b) *Comparable sales.* Comparison can be made with any known recent comparable sales – see *I*(e) and *II*(e) above. Values are indicated in chapter 17.

(c) *Present market value.* The value of the crops being mostly relevant to their standing timber value, measurement is likely to be necessary. (Values are indicated in chapter 17). The result can be compared with the expectation value in *III*(a) above.

IV. Nearly or quite mature crops. Valuations as noted under expectation value in *III*(a) *supra* above are unlikely to be used, and although resort can be made to comparable sales, the valuation is based almost invariably on actual measurement of the standing timber. Values are indicated in chapter 17. In valuing a woodland or stand/parcel of timber, a sawmiller is likely to 'calculate back' from his sawnwood sale price minus his harvesting/transporting/conversion costs. A roundwood merchant trading in sawlogs and/or small diameter roundwood, is likely to 'calculate back' from the prices he knows he can obtain from, respectively, sawmillers and other wood processors.

V. Hazel underwood. Productive hazel underwood should be valued where possible by expected out-turn of produce per hectare, if grown on a regular rotation of, say, eight years for wattle hurdles, pea sticks, bean rods, thatching spars and speekes. The standing value is unlikely to exceed £100 per hectare, according to quality, local demand, and availability of workers skilled in the conversion of underwood. More often, hazel is of poor quality especially when grown under standards, has little commercial value and, where it is to be replaced by restocking with other species, it may cost a substantial sum to remove. However, its retention for nature conservation can be important.

VI. Coppice. The value of coppice depends substantially on location, species, age, stocking, quality, access and markets. Much comprises mixed broadleaves, but the most valued is pure sweet chestnut. The volumes can be ascertained by measuring sample plots; or counting the number of poles (or number of stools and multiplying by the average of poles thereon) and measuring in m³; or estimating the weight per pole, multiplying as relevant and pricing per tonne. Some coppices of low quality may have 'a net minus value', costing more to remove than the income therefrom. Yields and prices/ha are indicated in chapter 7. Coppice with standards will be valued as coppice, plus the timber value of the standards.

VII. Forestry land. Current market values of forestry land are indicated in chapter 17 – the main factors influencing values being particularly locality, productivity, infrastructure and proximity to expected future market. Methods of calculating values of bare land are noted in chapter 14. There is often a difference in value between bare land and land underlying a forest crop.

VIII. Forest properties. The value of a forest property can most meaningfully be ascertained by a properly conducted sale. A valuation is usually undertaken for sale, purchase, probate or capital taxation. The incidence of the latter is referred to in *IX infra* and in chapter 15. A valuation will chiefly involve setting a price on the land and the crops it carries; the principles followed will be those noted in *I* to *VII supra*. In addition will be the value of dwellings, buildings, plant, machinery, other equipment, materials and prepared products. There may also be an estate sawmill, preservative plant, and a forest tree nursery. Value will be influenced by location, topography, access, the infrastructure (roads, bridges, buildings, permanent fencing), the level of past management, proximity to markets, and sporting possibilities. Other considerations are public rights of way, easements and wayleaves. There may be TPOs and SSSI regulations to comply with. The terms under which forestry grant-aid has been received have to be honoured, also any restocking commitments. There may be 'hope value' for land and minerals, all subject to planning incidence.

IX. Effect of grants, taxation and other personal circumstances. The personal capital tax position of the owner of a woodland has a considerable bearing on its value to him or her. Income tax plays no part in forestry (at least after 5 April 1993). There is no capital gains tax on crops; and inheritance tax can benefit from 50% or 30% Business Relief and other concessions. Planting and management grants are available for afforestation, restocking, and appropriate natural regeneration. Where, as at present (1991), capital taxation arrangements and grant-aid assist forestry, a valuation specific to a given person or agency should include the influence on cash flows. Investors with the highest marginal rate (under Schedule D until 5 April 1993) claim more income tax relief on costs, and pay more income tax on revenues than those with lowest marginal tax. Depending on the precise arrangements (again until 5 April 1993), plantable land may be more or less valuable to them. Similarly, individuals with a low time preference rate give more emphasis to long-delayed revenues, so may value a young woodland relatively highly. These differences of circumstance and perception allow mutual advantage in sales of land and young plantations. Perhaps the most striking example of how capital taxation concessions affect prices is in the current market value of most forest properties either large or small. Many change hands at prices in excess of any orthodox values based on present values of land and merchantable crops. This is partly due also to non-wood benefits and satisfactions. Thus woodlands appear to pass not at their worth as land and timber but at values enhanced by concessions and benefits applicable to their ownership.

X. Valuation of amenity. A system for the valuation of amenity trees suggested by Helliwell (1967) and adopted by the Tree Council (1974), was designed to assess the amenity value afforded to society in general. A comparable system for assessing the amenity value of woodlands was included in Helliwell's 1967 paper, and has been commented upon by Campbell (1985). A modified version of the system was developed by Helliwell assisted by members of the Arboricultural Association (1986), to assess the amenity value of woodlands. These two systems were brought together again in a booklet published by the Arboricultural Association (1990), after consultation with the Tree Council.

Amenity value of a woodland is of course in addition to net returns from timber production and the value for nature conservation, and perhaps for sporting. The system enables the assessor to ascribe an amenity value to a woodland on a points scale. This figure is then multiplied by a conversion factor in order to arrive at an appropriate monetary value for planning, compensation or other purposes. The conversion factor is amended from time to time to take account of changes in the value of the £. Where the amenity to a private owner is concerned, a slightly different approach is needed, and the maximum values obtained are not so great.

XI. *Valuation of non-wood benefits.* Non-wood benefits are difficult to value by conventional methods. The Forestry Commission will continue with its efforts to quantify these benefits, although it should be noted that for some, such as the values of forests as wildlife habitats, there are no established techniques for valuation. Many of the benefits provided by woods cannot be sold for money; often the public can freely enjoy landscape, climatic and environmental benefits, much wildlife conservation and even recreation.

Non-wood benefits have been discussed in chapter 13, and the difficulties of evaluating them noted; and again in chapters 5 and 14. Most such benefits (and their values) depend on the woodland's location, access, age and content. Chapter 14 provides examples of benefits that can be evaluated – relevant to agriculture, those of grazing, water catchments, sporting and charged-for recreation. The level of value of the benefit of carbon-fixing is suggested in chapter 15.

Charges for valuations. Fees and commissions charged for forestry valuations vary throughout the country. As a general guide, the charges for valuation of timber and forestry properties are likely to be: 2.5% on the first £10,000, 1% on the next £10,000, 0.5% on the next £30,000, and 0.25% thereafter. Where detailed timber measurements or unusually protracted negotiations are involved, these may be the subject of an additional and specified time charge, or an additional fee may be agreed between the parties.

Some of the professional 'pitfalls' affecting the making of valuations are pointed out by Taylor (1990) along with his suggestions as to the treatment of 'disclaimers' related to them.

References

Arboricultural Association (1986), *An Evaluation Method for Amenity Woodlands.*
—— (1990), *Amenity Valuation of Trees and Woodlands.*
Arthurs, E.K. (1987), 'Strong Start to FC's Computer Strategy'. *Forestry and British Timber,* December 1987.
Campbell, M.J.J. (1985), 'A Woodland Ranking System'. *Arboricultural Journal, 9,* pp. 183–91.
Gane, M. (1966), 'The Valuation of Growing-stock Changes'. *Quarterly Journal of Forestry, 60* (2), pp. 110–120.

Gane, M. and Linnard, W. (1968), 'Martin Faustmann and the Evolution of Discounted Cash Flow'. Occasional Paper No. 42, Commonwealth Forestry Institute, University of Oxford.
Gay, J. (1988), 'Computer Recording in the Forest'. *Timber Grower,* Summer 1988, pp. 32/3.
Harper, W.C.G. (1986), 'Forest Investment Appraisal Package (FIAP)'. FC Report on Forest Research, 1986, p. 56.
Hart, C.E. (1987), *Private Woodlands: a Guide to British Timber Prices and Forestry Costings.* Chenies, Coleford.
Helliwell, D.R. (1967), 'The Amenity Value of Trees and Woodlands'. *Arboricultural Journal, 3,* pp. 128–31.
Hiley, W.E. (1954), *Woodland Management.* Faber and Faber, London.
Insley, H. (ed). 1988, 'Farm Woodland Planning'. FC Bulletin 80.
Long, H.A.R. (1984), 'The Valuation of Woodlands'. *Journal of Valuation,* 1984 (2), pp. 161–6.
Lorrain-Smith, R. (1988), *The Economic Effects of the Guidelines for the Management of Broadleaved Woodland.* TGUK, London.
—— (1989). Memorandum 1.6.89 to the Agriculture Committee on Land Use and Forestry: Vol. II: Minutes of Evidence and Appendices, pp. 444–51. 10.1.90. HMSO.
—— (1990), 'Using a Spreadsheet in Forest Management'. *Forestry and British Timber,* April 1990 and following five issues.
Mason, W.L. and Muetzelfeldt, R. (1986), *Computers in Forestry.* Edinburgh Conference of Institute of Chartered Foresters, 1984.
Mitchell *et al*. (1944), 'The Private Woodlands Survey', FC Technical Paper 5.
Openshaw, K. (1980), *Cost and Financial Accounting in Forestry.* Pergamon Press, Oxford.
Osmaston, F.C. (1968), *The Management of Forests.* George Allen and Unwin, London.
Price, C. (1989), *The Theory and Application of Forest Economics.* Basil Blackwell, Oxford.
Sheate, W.R. and Taylor, R.M. (1990), 'The Effect of Motorway Development on Adjacent Woodland'. *Journal of Environmental Management, 31,* pp. 261–7.
Taylor, D.W.G. (1990), 'Disclaimers'. ICF News '90, Issue 4, pp. 14–16.
Tree Council, The (1974), *An Evaluation Method for Amenity Trees.* The Tree Council, London.

Chapter Seventeen

INVESTMENT IN PRIVATE FORESTRY

1. The case for investment

Forestry investment is the sharing in the development of one of nature's few renewable raw materials for man's use. It takes several forms, and can offer many satisfactions to a wide range of investors. Success therein depends upon bringing together silvicultural and financial skills in a specialized way. The case for national forestry noted in chapter 9, section 1, substantially also relates to the private sector. Woodlands have productive, protective and social functions. The productive function is mainly to sustain supplies of timber and minor forest products. The protective functions are concerned with the environment and include: protecting the physical features of the land; regulating the quantity and quality of water supplies; conserving wild plants and animals through management of their habitats; protecting cherished landscapes; and providing shelter for crops, livestock and buildings. Social functions include recreation and amenity.

Currently there is much debate about the advantages of multi-purpose and integrated land use, relating particularly to agriculture, forestry and agroforestry (to which chapter 13, section 8 is addressed). Besides the need for wood production, there are demands on woodlands for aesthetic and environmental requirements, for which woodland values increase with age. All aspects of environmental quality are much influenced by forest management decisions seeking specific objectives. Silvicultural system, species mixture, plantation layout, age and rotation length are all important. In every aspect of forestry, the private sector plays a valued role. Factors which favour investment in it, and conditions it must face, are noted hereafter.

Favourable factors. Government declared policy (noted in chapters 9, 10 and 15), is that forestry has an important role to play in the well-being of this country and the industry has its full support and an assured future. There is a supplementary successful 'partnership' between the Forestry Commission as the Forestry Enterprise and the private sector. The Commission as the Forest Authority encourages and benefits private forestry in many ways as noted in chapter 15. Britain annually imports over £7 billion of wood and wood products (only food, fuel and motor vehicles cost more in imports). Self-sufficiency is only about 10% and is unlikely to exceed 25% in the foreseeable future. Britain is one of the least forested countries in the whole of Europe – about 10%, as opposed to a European average of 25%. The continuing growth in consumption in both developed and less developed countries, taken with the depletion of forest resources in many regions of the world, provides the assurance that, even over the very long time-scale involved, the demand for wood will remain buoyant.

Britain has a very favourable environment for forestry: not surprisingly for a once densely wooded island, it has excellent conditions for growing trees. For afforestation, marginal agricultural land is available at fluctuating prices. Farmland withdrawn from arable use also is contributing to afforestation but is reducing the average size of commercial plantations. Planting and management grants (and, on farmland, other payments) are available; and capital taxation of commercial woodlands attracts relief. Financial returns from investment in forestry over the longer term compare well with other investments (Francis 1989). Real rates of return of 2%-6% can be obtained (see chapter 14) according to species, site factors and location; the average with grant-aid may be at least 4%. Adequate efficient wood processing industries are extant, seeking immense continuing supplies of small diameter roundwood and sawlogs.

Woodland owners have assurances that their investment is welcomed and encouraged by the State; and that Government, as appropriate, will continue to encourage investment by wood processors and wood users. The processing industry requires assurance of adequate supplies of realistically-priced small diameter roundwood and sawlogs, commensurate with the financial risks inherent in the immense investment in buildings, plant and machinery. The whole forestry industry seeks a stable political and monetary background; as well as a long-term forest policy.

Other factors. Forestry being a long-term investment, it is not for those seeking a quick return. Moreover, the financial rate of return may be modest in comparison with some investments, although the non-wood benefits and personal satisfactions usually compensate for this. Woodlands can be at risk to damage by wind, fire, biotic agencies, wild and domestic animals; also by man – through harmful trespass, vandalism, or *laissez-faire* management. Future timber prices remain uncertain. Forestry experiences several social pressures, having wide responsibilities to the community at large relative to the environment and, increasingly, to access. Complaints from the public may arise when afforestation is undertaken, particularly on hills and uplands, for example: (i) landscape changes (plough furrows, planting lines, straight fences); (ii) ecological changes; (iii) when thinning lines appear in the landscape; (iv) when clear-felling occurs; and (v) when restocking is delayed. Private investment must take note of these factors and of the multipurpose land use concept.

Past investments. Until the late 1940s, private forestry was mainly a traditional activity of most owners of wooded estates. It largely comprised tending, harvesting and restocking in the ordinary course of prudent estate management in which amenity and game management played a substantial part. Many thousands of small woods on farms, as well as numerous isolated woods, received little if any management. (Many woods had supplied invaluable timber during two world wars, but had not been rehabilitated.) To most owners, the return on capital was less important than the amenity, sporting and other benefits and personal satisfactions. Yet they fully appreciated the advantage of possessing standing timber, often for their own use, and particularly as a means of meeting estate duty or funding other requirements. (This 'savings box' aspect of forestry still remains a function which attracts many traditional and new private investors; often the effect is prolonged rotations.) Many woods were fragmented by the break-up of estates. Seldom did individuals or companies purchase woodlands for investment purposes. Acquisitions by sawmillers were with a view to promptly harvest the produce. Most purchases were to transfer potentially estate duty-vulnerable capital funds into woodlands in order to qualify for the considerable savings in estate duty while at the same time achieving a hedge against inflation.

The emergence in the early 1950s of forestry companies specializing in commercial management and contracting, gave an impetus to new investment in both traditional woodlands and afforestations by usually high-rate payers of income tax and surtax. The Schedule D/B income tax 'switch' attracted perhaps far more importance than did the growth of the trees themselves.

Throughout the 1950s and 1960s, with the continuing support of grants and relief from income taxes, investments were made in either stocked woodland or bare plantable land. Afforestation was often effected at a net cost of around £150/ha (land plus planting). Most investments were by 'absentee' owners, usually under the joint guidance of silviculturists and accountants, starting with Economic Forestry Ltd. – later EFG plc – the chief instigator being the entrepreneurial chartered accountant Kenneth Rankin. He was one of the pioneers, since 1952, of three major forestry management companies, a specialist firm of chartered accountants, six smaller forestry organizations, and, latterly, a forest tree nursery. Even small investors entered forestry, generally through co-ownership schemes/ syndicates. Thus arose new classes of woodland ownership. Concurrently, many traditional wooded estates began to rehabilitate their war-ravaged woodlands, encouraged by the Forestry Commission. Investment progressed steadily, usually with taxation incentives foremost, but gradually more emphasis was placed on the fundamental benefits of woodlands, particularly the growth performance of the tree in volume and value.

From the mid-1970s, financial institutions (Pension Funds and Life Assurance Companies) became attracted to forest investment opportunities, mostly in thicketstage to mid-aged woodlands; also occasionally in the purchase and afforestation of bare plantable land. They were mainly tax-exempt funds, encouraged only by planting grants. They thought the concept of forestry to be prudent, involving the two sound commodities of land and timber. The collapse of the private woodlands market, in the wake of capital transfer tax provisions and high inflation rates, resulted in bargain forestry investments being acquired by institutions.

Some of the woodland investments disposed of in the 1970s showed handsome profits, the gain on the crop being exempt from capital gains tax. However, not all disposals were successful. There was a preponderance of young plantations from about five to fifteen years old. Some investors (not the traditional owners) found themselves financially over-committed, and became forced sellers even at low prices. During 1974–75 it was possible to acquire substantial blocks of good quality plantations about five years old at less than £250/ha including the land. However, by 1979–80, similar properties had increased in value to about two to three times that figure.

Towards the end of the 1970s, demands for woodlands began to run ahead of availability. Investors' earlier policy of careful and selective acquisition of high quality woodlands (i.e. fertile land, fenced and roaded, holding sound full-stocked stands) gave way to investment in lesser quality woodland where the chief characteristic was availability rather than value. Almost any woodland of reasonable extent attracted very competitive bids from several interested institutions and wealthy individuals. Potential yields on those prices fell to around 3% instead of the 4%–5% previously sought. From around 1980, institutional investment became stable again; they, along with potentially high taxed (capital-wise) individuals, competed in the search for high quality woodlands.

Institutional investors. UK pension funds started investing in forestry in 1975 (Campbell 1983), and by 1983 some 25–30 funds were directly involved, making a major impact on the investment market for woodlands. Each year thereafter, two or three new funds made their first forestry investment. Their total value of annual investment in forestry, approximately £15 million, increased steadily. Their investment probably equated with about 10% of their property portfolio or perhaps 1%-2% of their total available fund. By 1983 funds owned at least 50,000 ha of commercial plantations, equivalent to about 5% of all privately owned woodland, but nearer 20% of the new commercial afforestations.

The institutional fund manager's aim is to keep pace with or beat inflation. But the growth rate of 4%-5% per annum at constant prices that might be achieved in a conifer rotation is not the sole conern. Equal interest is in arguments to support the expectation that timber prices will increase in the medium to long term faster than prices of other products and of inflation generally. Note is taken of such factors as consistently-rising demand for roundwood, the importing by the UK of more than 90% of its wood requirements (currently costing over £7 billion annually) and the importing by the EC of about 50% of its wood requirements. Furthermore, consideration is given to the possibility of a world shortage of timber within the first quarter of the coming century, and of a wide range of alternative uses of timber as energy costs increase.

On the other hand, the fund manager gives due consideration to such factors as: (i) the essential long-term nature of the investment; (ii) the negative or modest cash flow performance until such times as thinnings are removed or the final crop is felled (yet the necessity of restocking remains); (iii) the fact that institutions cannot take full advantage of the capital tax reliefs that are available to the individual investor; (vi) the rises and falls in demand and price of the industrial market for roundwood; and (v) fluctuations in plantation prices and poor liquidity compared with some other investments.

Fund managers have become increasingly more selective, undertaking deeper research into factors affecting performance, such as costs, revenues, internal rates of return, windthrow hazard, as well as monitoring trends and market forces. They dislike negative cash flow. They have shown concern regarding Pine beauty moth on lodgepole pine in Scotland, beetle attack on spruces in parts of England and Wales, and Green spruce aphids on Sitka spruce; as well as problems of fire and windthrow. They ponder the eventual commitment to second rotations, particularly in the uplands. The current relatively favourable prices for small diameter roundwood and sawlogs, and the level of grant-aid, generate only modest interest in oneto five-year-old plantations or new planting A further irritation is the relative difficulties of managing and forecasting profits compared with a portfolio of equities or fixed interest stock. Indeed, this proves a significant factor in any decision to reverse investment policy and sell the woodlands. There is also the lack of hard and regular data on financial returns of forestry.

Forestry can be the ideal investment for an institution that does not need running yield now, as opposed to capital appreciation. Institutions ponder whether timber demand in the next century will outstrip supply. They are aware that they need not necessarily have to stay the full rotation, but can realize at any stage. Their interest has had a beneficial effect on other private investment in forestry (likewise on State forestry when its woodlands have been available for purchase) by indicating that there is, and is likely to be, a ready market for woodlands, especially extensive plantations. This is a major incentive and motivation for new investors (knowing that a forestry investment is relatively illiquid compared with stocks and shares).

Wood-processing mills as investors. At least one paper mill and, occasionally, some sawmills, in recent years have invested in stocked woodlands, with the intention of ensuring a proportion of future supplies of, respectively, small diameter roundwood and sawlogs. The same paper mill is specializing in a scheme of woodland management and harvesting which provides an added financial benefit to any private grower.

Selection of a forestry investment. Investment in woodlands is generally accepted as being a financially sound and capital tax-efficient method of building up an appreciating asset, as well as providing a hedge against inflation. It can offer a tax-free income as well as several non-wood benefits and personal satisfactions. It may be a method both of increasing and conserving a family's wealth; and has other roles in estate and family financial planning.

Individuals who recognize the attraction of woodland ownership may not know how it can be acquired. They do not always appreciate that investment can be undertaken on a relatively small scale; that although the life of a tree crop is long they can 'buy into' a woodland investment of almost any age, species, type, structure and condition; that existing objectives can be changed if desired; that loans can be obtained from various sources; and that the investment can be disposed of at any stage in its life.

Selection of a forestry investment should be undertaken with the benefit of professional advice from silvicultural, financial and legal viewpoints.[1] The silvicultural, and some of the financial advice, can be obtained from chartered foresters (either self-employed or within a firm of chartered surveyors) or from the five major forestry management companies (Tilhill Forestry, Fountain Forestry, Scottish Woodlands, English Woodlands, and the Forestry Investment Management). Names and addresses are available from the Royal Institution of Chartered Surveyors, the Institute of Chartered Foresters, or the Association of Professional Foresters. A wide range of woodland investments and independent advice is always on offer. An investment is also obtainable through a co-operative scheme, but this means simply a share in a single wood or a group of woods without any defined areas being in the sole name of the individual participant.[2] There are several individuals and firms who can provide relevant valuations; or competent soil and other surveyors; others can offer computer services aiding investment appraisals and management.

When considering a particular investment on offer, the individual should inspect the bare plantable land, or the stocked woodland, and consider what advisers say about it. Then, study the cash flow projections including grant-aid, the basis of the

---

1 The Financial Services Act 1986, Section 76, appears to preclude any person other than an 'authorised person' from advising or marketing any collective investment scheme, other than to 'another authorised person'. The section needs review and amendment to permit the marketing of forestry and forestry investment schemes.
2 Syndicate forestry was a form of joint-ownership which enjoyed modest success until the above Act inadvertently cast doubts upon the tax status and legality of such holdings. Changes would enable the possibility of co-ownerships being brought together as a means of broadening the base of forestry investment.

computed rate of return and the discounted present worth. Thereafter, consider the capital taxation implications, the relevant constraints and responsibilities, and the risks and uncertainties. Special attention must be given to rights of way, and to access, fencing provisions, management agreements, way-leaves, restrictive covenants, and sporting and mineral rights. Finally, the method of management should be decided upon – a matter discussed later in this chapter.

The suitability of any woodland investment will be affected by the age, financial circumstances and dependants of the individual. A projected investment should fit into the family scheme of living as well as into family financial planning; otherwise the costs and responsibilities may outweigh the returns and personal satisfactions. Much depends on whether the commuting distance makes possible occasional residential occupancy; if it does, the owner has the possibility of reducing custodial expense and maximizing satisfactions. A prospective investor should, therefore, consider his objectives keeping in mind the needs of his family, posing to himself such questions as: would the family share his interest and enthusiasm; would he, and his family, have the time to spend on their property to manage it adequately and to realize their satisfactions commensurate with the cost, effort and risks involved; and would the property on his decease become a serious financial burden or source of anxiety to the family. Emphasis should be placed on how to reap the greatest satisfaction from ownership and how to avoid disappointments. Foresters can often help an owner's understanding of silviculture and wildlife, and give an introduction to responsible management practices.

A new owner of woodland, whether acquired by purchase, gift or inheritance, will ignore the sunk costs of establishing and tending borne by the previous owner; being solely concerned with the woodland's worth in money and other benefits or satisfactions; also, perhaps, with the net income possibly attainable through a reasonably sustained yield, or with the potential sum he might obtain above the cost of harvesting when this occurs, and above the cost of consequential restocking.

Types of forestry investment. Woodlands exist in great variety, from the lowland broadleaves so admired for their beauty, colour and utility, to upland uniform monoculture conifer plantations, which, however, can contain an often surprising diversity of fauna and flora and attractive microscale landscape. In between are woodlands of small, medium or large extent; and virtually of all shapes, ages and conditions. Many are run with the main objective of profit-maximizing wood production, but this does not exclude pursuit of other compatible beneficial and satisfying objectives, particularly sporting.

The first decision to be made is whether to invest in bare plantable land and afforest it or instead to purchase stocked woodlands. Then, to decide whether to invest in conifers or broadleaves, or a mixture. Finally, to choose according to silvicultural system, i.e. high forest, coppice (e.g. sweet chestnut) or coppice with standards; or special crops such as poplar, or short rotation coppice. Types of woodlands available for investment are indicated below.

Bare plantable land. Investment in afforestation, geared to tax savings, has usually been a case of acquiring plantable land in late winter and afforesting it during the following spring – a typical time-scale. This procedure may continue, under grants Appropriate plantable land (usually agricultural grazings) is to be found in the uplands, or may comprise bare forest land ready for restocking. (Sometimes woodland sites are made available after clear-felling by timber merchants to avoid them undertaking the compulsory restocking.) Land costs are dependent on location, fertility, and terrain, and whether fenced and/or roaded. Moreover, the Forestry Commission and consulted authorities should have approved the land for planting and hence for eligibility of grant-aid. Sometimes it may be feasible to purchase low-priced land and enhance its suitability by cultivation, drainage and fertilizing. Further value can be added by appropriate fencing and roading. There is an interaction between land quality and establishment costs.

When completely established (generally in three to five years), the land and young plantations can be disposed of to another investor; or be held for appreciation and personal satisfactions, eventually to be sold, gifted or bequeathed as desired. Gifts may be particularly appropriate in family financial planning when values are at a low stage, in view of the favourable incidence of inheritance tax relating to woodlands.

Stocked woodlands. Here the primary aims may be an acceptable tax-free return on capital, and capital appreciation (favourably relieved as to capital gains tax and inheritance tax), supplemented by other benefits and personal satisfactions. Stocked woodlands fall broadly into four categories:

Young plantations, up to ten years old for conifers and up to fifteen years for broadleaves, where all costs (except perhaps roading) have been incurred, although continued maintenance, protection and tending will be required. At this young stage, depending on them being reasonably extensive, they are of interest to institutions and substantial individuals seeking, respectively, mediumto longterm supplies of roundwood, or capital appreciation. They are a particularly appropriate investment when 'rolling over' a capital gain into forestry.

Mid-age woodlands, from twenty to forty years old for conifers and thirty to fifty years for broadleaves. These may have been cleaned, and partly brashed and pruned, and may have had their first commercial thinning. As they age, and the time of final harvesting becomes closer, realization of the final timber income comes within the span of an owner's lifetime. For instance, a man of forty could purchase a twenty-five-year-old conifer crop, retain it for twenty-five years (receiving interim tax-free income from thinnings) and then clear-fell, reaping a tax-free sum as part of his personal pension planning. Throughout he could benefit from the wood's amenity, wildlife and recreational facilities.

Maturing or mature woodlands, from thirty-five to fifty-five years for conifers and fifty-five to one hundred years for broadleaves. The woodland holding a substantial volume of merchantable timber, may have received its final thinning. As it approaches maturity, the value of the timber becomes increasingly higher and more marketable. There is a degree of flexibility, as harvesting rarely has to take place at any particular time, but there is an optimum economic rotation for both volume growth and value. Such woods provide ideal purchases for inheritance tax relief, allowing high values to be gifted or bequeathed, leaving a relatively liquid asset in the hands of the donee, which can be realized as necessary.

Special types of woodland include poplar plantations and stands of coppice (particularly sweet chestnut). Their attributes are discussed, respectively, in chapters 5 and 7. Both can produce attractive financial yields under most circumstances.

There are woodland investments appropriate for particular sets of estate planning circumstances. However, a single woodland, correctly chosen and managed, can answer more than one requirement. The location is always paramount, and will certainly affect profitability. Some extremes of north of Scotland and West Cornwall have a poor record for stability and profitability. Internal and external access and closeness to markets are vitally important. In the uplands economic considerations generally dictate investment to be in conifers. But such locations are not as convenient as might be desired by owners wanting to enjoy the satisfactions offered by their woodland. In many of the otherwise desirable parts of Britain, the grey squirrel can hinder the economic establishment of broadleaved woodlands. Rabbits and deer can create major problems.

The commercial case for investment in private forestry. The four main arguments for investment in commercial private forestry are noted below.

Land. Land, although relatively illiquid, may be the ultimate hedge against inflation. It is unlikely that anyone purchasing forest property at current prices will not reap a real profit on the underlying land. Since the last world war, the rise in land prices has significantly exceeded the rate of inflation. However, there has been no significant rise in forestry land values since 1982–83. Even marginal land suitable for forestry is now scarce: in Wales, land is virtually unobtainable at any realistic price; in Scotland, upland tracts are still available, but only occasionally is there sign of land for planting 'coming down the hill'.

Timber. The immense cost of imports of wood and wood products, also the UK's adverse situation as regards domestic production compared with other EC countries, have already been noted. World supplies appear to be in some doubt. Traditional suppliers prefer to market value added products. The UK paper, panel and sawnwood processors, heavily committed in investment, express some concern as to whether their requirements of roundwood will be adequately met. The balance between demand and supply of roundwood is likely to remain tight. It is difficult to envisage circumstances in which demand for roundwood will not lead to higher prices.

Grant-aid. Planting and management grants are available. They are higher for broadleaves than conifers, and rates for planting are progressively higher for areas under 10 hectares, which are considered to be of high environmental importance. For some farmland planting, special grants, as well as annual taxable payments, apply.

Taxation. Commercial forestry is relieved from income tax and corporation tax. It enjoys a favourable capital tax regime, recognizing the very long-term commitment involved in maintaining what is an important strategic resource as well as providing many invaluable social benefits; and it may be eligible for Heritage status (see chapter 15).

The foregoing arguments for investment in commercial private forestry are of course often supplemented by invaluable environmental benefits and personal satisfactions.

Specific investment objectives. Woodlands at different stages of growth/maturity appeal to different types of investor which aids the marketability of what is fundamentally a long-term commodity investment. People rarely outlive the crops they plant. Though the investor may benefit from grants, and generally from at least one thinning (supplemented by personal satisfactions, amenity, recreation and sporting) his investment is likely to be realized at any stage in the rotation either by sale or gift during lifetime or bequest by will. In all these cases the disposal value and rate of return achieved can be ascertained. At any stage in the rotation, the investment will have performed in a measurable way, by producing a redemption yield – which can be compared with other known rates from investments; and it will be evident how far the specific objectives of the original investment have been achieved.

A prospective investor may consider a shortor a medium-term investment, with a view to the entire property being disposed of, either following complete establishment (say, at three to five years) or before first thinning (say, twenty to thirty years). Consideration of such projects should take into account the shorter length of time involved, as well as the favourable capital tax incidence whereunder on disposal during lifetime of the entire property (i.e. the land and the crop) the only tax will be capital gains tax on any increase in value of the land. Three e xamples of specific investment objectives are:

Creating capital out of income. Capital can be created by way of investment in afforestation. This largely depends on the effectiveness of grant-aid; and on whether the land is in hand or has to be purchased. A particular advantage is in its flexibility: fluctuations in personal income from year to year can be accommodated by regulating the planting programme; but sufficient income must be available in each year to maintain and protect earlier plantations. (Such investments have been found satisfactory by joint professional partners, and by groups of directors of private companies. In some cases they have been related to private pension fund arrangements.)

Protecting capital. Commercially managed woodlands can attract relief from capital gains tax and inheritance tax. Freedom from capital gains tax upon timber is a considerable attraction for those seeking to protect funds from inflation over a medium to long term; and the Business Relief at 50% under inheritance tax regulations is clearly attractive: potential inheritance tax liabilities can be significantly reduced by transferring into forestry. The middle years of the rotation are those in which the highest rates of annual volume growth are achieved (for conifers, generally in the region of 10%–15%), and the net results of this growth in monetary terms will not be eroded by capital taxation. (During this growth period, the plantations are generally most attractive to financial institutions. Schemes are available whereby an institution can not only enjoy a rental yield from individual investors, but also secure a share in the timber crop once it has been established.) Similar arrangements help to provide solutions to potential inheritance tax problems facing landowners whose estates include plantable land. The introduction of an extraneous investor can result in a joint venture which will produce either capital or income to the landowner, whilst providing an opportunity for the investor to undertake the costs of afforestation.

Obtaining income from timber. The purchase of productive mid-age or maturing woodlands will provide the opportunity of receiving tax-free income. Investment of this type can offer opportunities for estate planning by spanning the generations. For example, an arrangement can be made between a parent who has retired from active business, and has children, whereby a maturing woodland is acquired, thereby providing a tax-free income for the elderly parent. Subsequently, after an area has been felled, it is transferred, either by gift or by lease, to the children for them to restock, with grant-aid, for the ultimate benefit of any grandchildren.

Management of forestry investments. Successful forest management requires professional skill and specialized knowledge. The investor can have as much, or as little, involvement in management as he desires. Owners who intend their woodland to be run on a commercial basis need to select an operating system from a range of available options. Much will depend on the extent, location and structure of the woodland, as well as on the intensity of the forestry to be practised (see chapter 13). Where the owner does not wish to be burdened with management, either because of time constraints or limited expertise, he can engage the services of a forestry consultant or a management company. Over 225,000 ha are managed by companies, and probably about 100,000 ha by individual consultants, chartered surveyors and independent foresters. Hence, adequate technical and specialist management is available for all types of woodland investment. Appropriate insurance cover, where prudent, can be arranged for natural hazards including fire, windthrow and (occasionally) biotic damage.

Provided the woodland is sufficiently large (though much depends more on factors such as its age, structure and prescription of work) it may justify the direct full-time employment of an appropriate woodland staff, and perhaps of a qualified forester. This has been the traditional approach on well-wooded estates. However, this operating system is becoming rarer, partly as a result of woodland ownerships having been divorced from the other enterprises of many landed estates. (In consequence, the opportunities and benefits of integration with agriculture or other enterprises are becoming fewer; the remaining integration is usually between forestry and sporting.) The long production cycle and the uneven pattern of requirements of labour and machinery from season to season, and year to year, make financially difficult the employment of a full-time forestry staff by a wooded estate or an individual woodland owner.

Several forestry co-operatives (for instance, Western Woodland Owners Ltd.), associations and agencies have arisen, one of their objectives being to even out imbalances by sharing labour and machinery between a number of different woodland ownerships. They extend to their members or associates services ranging from contracting for a single silvicultural operation to complete management, including contractors capable of undertaking management and contracting work, without the involvement of other categories of manager. (Names and addresses are available from the Association of Professional Foresters.)

Forestry is not an enterprise to be looked at once a year as purely a capital tax-efficient investment, but as an important, and continuing, component of financial planning. The best way for an owner to protect his investment is either to undertake management himself, or to make aware whoever is responsible for it that the owner cares about what is being undertaken and how the woodland is progressing. Sound managers welcome this interest.

2. The market for woodlands and plantable land

Woodland properties reach the market for many reasons, by various methods of sale and in a wide range of location, extent, age, structure and condition. Their values fluctuate according to those factors and to the favourable incidence of capital taxation, the level of grant-aid, the economic, financial and political climate, and the varied motivations, objectives and circumstances of investors.

Wide variations in values are also evident when one compares: (i) conifers with broadleaves, (ii) fast-growing species with slow species, (iii) young plantations with mid-age, maturing or mature, and (iv) diverse locations (e.g. South Scotland, North England and North Wales with either North Scotland or lowland regions). Similarly, when one considers: (v) economics (dependent on species, yield class, windthrow hazard, terrain, infrastructure, timber markets, and intensity and quality of past management), (vi) the influence of interests such as landscape, amenity, recreation and sporting, and (vii) public access. Often woodlands having high conservation features, particularly if broadleaved, generally command premium prices well above the values of their timber and land. The same applies to most woodlands having substantial sporting value, or a holiday dwelling.

The relevant silvicultural systems are mainly high forest, less frequently coppice and coppice with standards, the group selection systems being quite rare. The types of woodland may be: (i) 'traditional' (chiefly coniferous in Scotland, broadleaved in England), (ii) 'plantations' (chiefly conifer monocultures, predominantly in the uplands), (iii) 'sporting woodlands' and (iv) 'conservation woodlands' (amenity, wildlife, scientific) – perhaps subject to SSSIs, TPOs or registration as 'ancient semi-natural woodland'. Within the four types, timber production will be one of the primary objectives; but various interests and objectives will often overlap, supplementing each other. Some will have public access, others will not. Disposals of woodlands by both the State and the private sector during the last decade are noted below.

Forestry Commission disposals of woodlands and plantable land: 1981–90. Since 1981 the Forestry Commission has been disposing of some of its plantations and plantable land. Up to 1984, the main objective was to reduce the Commission's call on public funds. In that year, the main emphasis was changed to rationalizing the Commission's forestry estate in order to improve efficiency. In June 1989, the Secretary of State for Scotland announced that the disposal programme would continue and that the Commission had been asked to dispose of some 100,000 hectares of forestry land and properties by the end of the century so as to consolidate their forestry estate in a rational and orderly manner. At times, the disposal programme attracted criticism from the media, trade unions, countryside interests concerned with conservation and recreation (especially the Ramblers' Association), and the general public. The woodlands sold have comprised a wide variety of ages and species, in four main extent categories: (i) small areas (from a few hectares) mainly of interest to neighbouring landowners and local residents; (ii) blocks of up to about 100 hectares which, when sold singly, have attracted private investors or, when grouped into large 'packages', financial institutions commensurate with their preferences – outside their core portfolio; and (iii) large commercial plantations.

The Commission, in its early days, sold a number of woodlands themselves; but the majority of sales are now effected through specialist firms of chartered surveyors. Some purchasers have been lessors of the underlying land.[3] Generally there has been a ready market for the woodlands offered, but not all have been easy to sell. Institutional interest has fluctuated, usually depending on competing forms of investment.

Prices have varied considerably between different parts of the country and even within a particular locality. Small woodlands have achieved prices above the value of their timber and land, due to potential sporting value or other inherent satisfactions, as well as reliefs associated with capital taxation. Amenity and conservation values have also enhanced prices. Highest prices paid were for mixed woods. Larger blocks, particularly coniferous plantations, achieved prices that were solely dependent upon the economic return from forestry as an investment. They sold well in southern Scotland, northern England, and Wales, mainly to institutions, wood processors, or substantial individuals. Much information on the Commission's sales during 1981–87 is provided by Hart (1987). However, for woodlands of such variation, generalizations on prices are difficult to suggest: they ranged from around £1,000 per hectare to £4,500 per hectare. 'Average' figures, or even specific figures of value, can be misleading as a comparison with other woodlands.

During 1988–89 the Forestry Commission programme of disposal amounted to some 180 blocks of forest land totalling 4,328 hectares. Most of the areas sold were under 25 hectares. The total proceeds were £4,993,200, and the average per hectare was £1,182. Sales of forest areas under trees during 1989–90 totalled £6,972,000 and comprised:

Conservancy		ha	Plantable land:ha	Average £/ha for forest areas under trees and plantable land*
England	N	743	15	1,145
	E	509	0	1,957
	W	341	0	2,698
Wales		1,038	0	1,721
Scotland	N	550	34	1,161
	Mid	715	407	957
	S	613	20	1,027
		4,509	476	

* The average prices above can be deceptive as forest areas under trees and plantable land are not separated.

The total cash receipts since the disposal programme started in 1981 amounted to £123.3 million. In all, about 100,000 hectares of woodlands were sold

---

3 Most Forestry Commission leases are for 999 years, at a low fixed rent. Some have provision for lessors to resume and many restrict the right to assign, or limit the use to which the land may be put. In England and Wales, about one quarter of the Commission's forestry is on leasehold land. During recent years the Commission has afforded opportunities to former owners to repurchase land sold or leased to them. The repurchases made an important contribution in the early years of disposal. The opportunity arises only in relation to woodlands selected for sale by the Commission, and then usually under the Governmental offer-back procedure.

Future disposals of woodlands by the Forestry Commission are scheduled to comprise another 100,000 hectares – around 11% of the woodlands – during the 1990s (50,000 hectares to occur during the first five years), some 40% of which is expected to be in England and Wales and 60% in Scotland. This recognizes the Forestry Commission's need to be able to plan ahead without the uncertainty caused by frequent reviews. The Government is concerned that the general public should continue to enjoy access to the woodlands sold, in a way which is compatible with management for forestry. This will probably reduce values and restrict markets. The Forestry Commission will give advance notice to local authorities of intended disposals of woodlands under its management, and offer to enter into legal agreements with them which will provide for continued public access to the woodlands after sale. Such agreements will be compatible with the management of the woodlands for forestry and other purposes and will be binding on subsequent owners. These arrangements will apply to those woodlands where there are no existing legal or other constraints on public access to the land. Guidelines setting out the details of these arrangements are available from the Forestry Commission.

Private sector sales of woodlands: 1981–90. There has been an active market in the sale and acquisition of private woodlands of a wide variety. The agents for most of the transactions have been John Clegg & Co.,[4] supplemented by other chartered surveyors (among them Bell-Ingram,[5] Bidwells,[6] Savills[7] and Strutt & Parker[8]) and five main forestry management companies.

Results of the transactions are rarely disclosed by vendors or purchasers, but some values become patent through auctions, or are indicated by an unaccepted tender. The most forthcoming partial disclosure of results has been contained in annual reports by John Clegg & Co. which have given many indications of price levels. The disclosed results are only of limited use as comparisons or guides, because rarely is it possible to compare dike' with 'like' – so varied are the location, extent and quality of sites and crops. The woodlands have comprised a mixture of a wide range of age-classes; or simply young plantations; or older stands. Varying widely too have been the species, stocking, growth (actual or potential), access, fencing, roading, and proximity to markets; also the intensity and quality of past management. Thus comparison of values of woodlands sold or on offer is often invidious.

Effects of the 1988 Budget on woodland sales. The 1988 Budget Statement (and the resultant Finance Act 1988), as noted in the Introduction, removed commercial forestry from income tax and corporation tax (subject to transitional conditional arrangements for Schedules B and D until 5 April 1993). This marked the conversion of

---

4 John Clegg & Co., The Bury Estate Office, Church Street, Chesham, Buckinghamshire HP5 1JF (0494 784711); 2 Rutland Street, Edinburgh EH1 2AS (031 229 8800); Apex House, Wonaston Road, Monmouth, Gwent NP5 4YE (0600 715311).
5 Bell-Ingram, 7 Walker Street, Edinburgh EH3 7JY (031 225 3271).
6 Bidwells, Forestry Department, 5 Atholl Place, Perth PH1 5NE (0738 30666). Head Office: Trumpington Road, Cambridge CB2 2LD (0223 841841).
7 Savills, 20 Grosvenor Hill, Berkeley Square, London W1X 0HQ (071 499 8644).
8 Strutt & Parker, 26 Walker Street, Edinburgh EH3 7JY (031 226 2500; 13 Hill Street, Berkeley Square, London W1X 8DL (071 629 7282).

support for private forestry from a mix of fiscal incentives and grants to an entirely direct grant (tax-free) regime. (Inheritance tax and capital gains tax concessions for forestry remained unchanged.) In conjunction with the 1988 Budget changes, the Forestry Commission's new Woodland Grant Scheme was announced. Restrictions were introduced on further conifer afforestation in the English uplands. The target on broadleaved planting and management was prominent. The current (1991) grants and taxation incidence have been discussed in chapter 15. New planting was significantly curtailed.

By contrast to the effect on afforestation, the traditionally stable market for stocked woodlands experienced little change, despite inflation and high interest rates, and remained relatively active, demonstrating their customary resilience in the face of economic and political difficulties.

Market values of woodlands and plantable land. During and since the late 1980s, woodland values in many parts of the country were increased by the demands for small diameter roundwood needed for the paper and board industries, and for sawlogs for the expanded efficient sawmills.

There is much difference between values of low quality and better quality woodlands; especially as the ages increase. Quality and intensity of past management are very important (for instance, 'check' or inadequate drain maintenance in upland plantations depress values). Restrictions and constraints also can depress market values. Those relating to roads and bridges (as well as terrain or access) can reduce values of isolated and otherwise affected woodlands. 'Registration' as ancient seminatural woodlands, if subject to any management constraints, tends to reduce their values; yet conversely it might increase the value if a purchaser was keen to acquire a woodland to be conserved instead of commercially managed. Much will depend on the degree of silviculture and harvesting that will be permitted. Some 'Registered' woodlands are likely to qualify for the conditional exemption from inheritance tax for assets of national heritage quality. (However, at present consideration for heritage status can only be given after death. There cannot therefore be long-term management to make the most of the very values that would put the woodland into heritage category.) It has been suggested that the exemption could properly be extended to broadleaved and native woodland where the essential character of the woodland is maintained.

Among the most difficult exercises in forestry is to estimate what figure a particular plantation, woodland or wooded estate may command in the open market. The land value can be fairly closely appraised. So too can the value of the crop – (i) up to the thicket-stage, either on historical or current costs, or (more usually) on discounted expectation value; (ii) in the poleor thinning-stages, either on current timber value or on discounted expectation value; or (iii) much older, on current timber value. Additionally, a realistic value can be assumed for sporting (especially if let) and something can be added for aesthetic values. However, despite summing the land, the crop, the sporting and other benefits, a higher figure is likely to be effected in a sale/purchase. Thus calculating a 'value' using a variety of conventional mensurational methods and undertaking an investment appraisal (helped by the many calculation aids now available) do little more than establish a *base* for a likely value. The difficulty lies in assessing the other factors which influence the market value, particulary supply and demand, inherent and surrounding sporting values, accessibility (for management and visits), and the role that the woodland can play in abating capital taxes. The premium paid is really another kind of return on capital. A market value can never be known with precision until a transaction has been agreed between a willing seller and a willing buyer.

Obviously, many woodlands change hands at sums above their strict investment value or the commercial values of their timber and land. The main reasons are the generally favourable incidence of capital taxation, grant-aid, the many non-wood benefits, and personal satisfactions. Location is a major factor.

Market values of plantable land. Land currently (1991) is in rather short supply, as vendors and buyers still adjust to the post-1988 situation. Land prices are one of the major obstacles to afforestation. Bare land for afforestation, 'cleared' for change of land use and hence potentially eligible for planting and management grants, currently commands prices of £350-£750/ha, dependent chiefly on location, terrain, access and potential for tree growth. Land prices in Scotland are still below pre-1988 Budget, but upland farmers are seeking £600-£750/ha whereas investors, under the prevailing economic conditions, rarely appear interested at more than £300-£500/ha (even though grant-aid is sometimes sufficient to cover at least half of initial costs of establishment, but not of roading). The effect of support to upland farming is pertinent. In the south of England, prices up to £2,500/ha have been paid for bare land – obviously unrepresentative for forestry – sometimes reflecting hope of development value.

The Forestry Commission's land acquisitions in 1987–88 totalled 1,638/ha, average cost £673/ha, all in Scotland. In 1988–89, purchases totalled 1,851 ha, average cost £410/ha, mainly in Scotland, the low price reflecting the high proportion of acquisitions made in their North Scotland Conservancy. In 1989–90 444 ha were acquired, the average cost being £343/ha.

Land underlying stocked woodland, or bare ready for restocking, generally has a value within a range of £250-£1,000/ha, dependent on location and quality. The value would increase, in theory at least, according to the length of time before the land again becomes bare and available for restocking.

Market values of stocked woodlands. Market values of stocked woodlands currently (1991) appear to lie within the following ranges:

Land and trees'. £/ha:

(A) Small areas of upland conifers:

	Age	Crop £/ha	Land £/ha	Total £/ha
Young plantations	up to 10	350–500		600–1,250
Thicket-stage	10–15	750–1,200		1,000–1,950
Pole-stage	15–20	1,100–1,700		1,350–2,450
Thinning stages to	20–25	1,600–3,000	250–750	1,850–3,750
maturity	25–30	2,000–3,250		2,250–4,000
	30–50+	2,750–5,000+		3,000–5,750+

(B) Extensive areas of upland conifers containing a substantial proportion of Sitka spruce and an overall weighted yield class of 14 or above:

	Age	Crop £/ha	Land £/ha	Total £/ha
Young plantations	up to 10	500–700		750–1,450
Thicket-stage	10–15	1,000–1,300		1,250–2,050
Pole-stage	15–20	1,200–2,000		1,450–2,750
Thinning stages to	20–25	1,700–3,000	250–750	1,950–3,750
maturity	25–30	2,000–3,500		2,250–4,250
	30–50+	3,000–6,000+		3,250–6,750+

Main factors affecting value of (A) and (B) above:

1. In general: Extent, location, access, terrain, internal roading, any weight restrictions on classified roads, fencing, species, yield class, Windthrow Hazard Class, markets, expectation value, past management (good or bad) not least as it affects future costs of management (e.g. need for heather control or fertilizer application). TPOs, SSSIs, or Registration as ancient semi-natural woodland may affect values. Harvesting constraints may deter sawmillers and timber merchants, but generally do not deter woodland and conservation trusts and the like. General constraints may include public pressures, public access and environmental considerations.

2. For investment: Either current merchantable value of timber, or the expected *real* rate of return (a purchaser's expectation often depends on those available from alternative investments).

(C) Lowland broadleaves and/or mixed conifers:

	Age	Crop £/ha	Land £/ha	Total £/ha
Young plantations	up to 10	500–750		1,000–1,750
Thicket-stage	10–15	800–1,200		1,300–2,200
	15–20	1,300–1,600		1,800–2,600
Pole-stage	20–25	1,800–2,000	500–1,050	2,300–3,000
Thinning stages to	25–30	2,100–2,500		2,600–3,500
maturity	30–50	2,300–3,000		2,800–4,000
	50–100	2,500–5,000+		3,000–6,000+

Main factors affecting value of (C) above: Similar to 1 and 2 above, where relevant, but in the older ages, the values of merchantable timber, and whether high forest or coppice are important.

In all the above categories there has to be an awareness that licences are necessary for thinning and felling and that, following clear-felling, restocking is compulsory, but for which tax-free grant-aid is available.

Except for best quality and well-stocked woodlands in siviculturally and otherwise 'attractive' areas within reach of generally adequate markets, the weighted average values are likely to lie in the middle of the ranges indicated above. The lower part of the ranges might reflect poor quality understocked crops, remote locations, bad access, or poor past management. Best prices for extensive forestry properties have been where the regime of management has been that likely to result in the highest financial yield over the longer term. Values are generally enhanced if there are important sporting possibilities, e.g. deer stalking, and pheasant shooting.

Types of investors in private woodlands. Little information is available on the woodlands owned by different types of private sector interests (traditional estate, personal investor, institution, farm), or on their respective shares of new planting – information which would seem to be essential to enable Government to target its support to the private sector. (From the statistics on woodlands within the Forestry Commission's grant systems, broad estimates on ownership types are: traditional estate 35%, personal investor 25%, corporate investor 25%, farmer 5%, and other 10%.)

Longstanding owners of traditional woodlands include estate owners, individuals, farmers, universities, the National Trust, the Royal Society for the Protection of Birds, conservation and charitable trusts, water companies and local or national authorities. During recent decades, new investors in extensive 'plantation afforestations' and stocked woodlands have mainly comprised high income tax payers, individuals within syndicates, and (more recently) financial institutions, paper mills, and a few overseas investors. A simplified classification of current types of new investors in woodlands are those who seek: (i) a sound financial return from silviculture, or (ii) personal satisfactions, or (iii) to promote conservation interests, or (iv) of necessity, to ensure supplies of small diameter roundwood or sawlogs for the processing enterprises. Sawmillers and timber merchants are not looking for investments *per se* but mainly for stands of mature timber. The eight main types of investor currently are:

1 Individuals. Personal ownership of woodlands holds unique attractions, their motives to invest, or to continue with a woodland holding, appearing to be as varied as the occupations of the owners themselves. Many desire that their woodlands provide personal satisfactions of ownership (seclusion, responses to owner's care, trees of unusual size or character), amenity, nature conservation and sporting; and to be a source of tax-free income, and capital appreciation relieved from taxation. Some desire them to be socially useful and aesthetically pleasing to the community at large. The inherent economical and personal advantages of woodlands have led many business and other professional people to acquire them during recent decades. By contrast to farming, which requires close attention on a day-to-day basis, woodlands need infrequent but intensive care (dependent on their age and structure, and on the set objectives), usually supplemented by some latitude in timing of silvicultural operations and a substantial degree of flexibility in harvesting. (Generally, the owner can dispose of his timber, or the whole woodland, at any stage of a rotation.) A supplementary satisfaction of ownership comes from personal participation in management decisions and, less frequently, in assisting with some of the less physically exacting silvicultural operations.

Individual owners are rarely dependent for income from their woodlands; yet some may be dependent on them for capital commitments when needed. Many are particularly interested in the sporting values. Some buyers are landowners accepting the opportunity to purchase a neighbouring woodland. In addition there are relatively rich individual buyers competing with the types of substantial investors noted in 3 to 7 below, being particularly interested in long-term real rates of return as well as the reliefs related to capital taxation. They may possess capital gains tax roll-over availability. Some woodlands change hands as a component in the sale of a farm or a mixed landed property; others by gift or inheritance. Personal investors may seek some or all of the following financial benefits either by purchasing stocked woodlands, or by afforesting bare land: a *real* rate of return, grant-aid (tax-free) and payments relative to farmland, income (tax-free), freedom from capital gains tax on trees, inheritance tax relief, and use of roll-over relief on land.

The financial and fixed benefits are likely to be supplemented by personal satisfactions. Small investors usually prefer broadleaved or mixed woodlands in mid-age and upwards. Substantial individuals look for extensive productive woodlands. Whereas numerous individuals are interested in purchasing small woodlands, the buying and afforesting of bare land is now usually contemplated only by those who possess the necessary cash or roll-over relief for the land and at least half of the costs of establishment; and who can dispense with any return for a long way ahead.

2 Co-ownership involves schemes formed to meet the desires of many small investors for a modest holding in woodlands, for varying reasons. A large number of syndicates have been set up, by at least two of the forestry management companies, operating at a level which retains the economies of scale present in silvicultural operations, harvesting and marketing. Participators usually require diversified woodlands of varied sizes, ages, structure, tree species, and locations. They can usually enjoy all the woodlands comprised in the scheme, but individually do not own any specific part. Some of the schemes, as well as personal investments noted in 1 above, are targeted to estate or educational planning. Part of the current system of government support seems to discriminate against 'unitization'. There is, however, government support for wider share ownership in UK forestry, and ministers have promised any necessary changes in legislation to ensure that syndicate members are treated no differently than individual woodland owners. (The Financial Services Act 1986 refers.)

3 The National Trust, woodland and conservation trusts and charities, including the Royal Society for the Protection of Birds, seeking to satisfy and promote membership objectives; supplemented by some of the financial and other benefits noted in 1 above. They usually prefer broadleaved or mixed woodlands of mid-age and upwards.

4 Financial institutions (Pension Funds and Life Assurance Companies) seeking in the medium to long term a real rate of return of 4%-5% (and, very occasionally, grant-aid towards afforestation); supplemented, where relevant, by some of the financial and other benefits noted in 1 above. Important to them is the grant-aid available when the time for restocking arises. Institutions generally seek extensive conifer plantations in the thicket-, pole-, or early thinning stages, treating the underlying land as a property investment expected to increase in value in its own right or to at least keep pace with inflation.

5 Industrial consumers of small diameter roundwood, and associated harvesting companies, seeking of necessity plantations to secure supplies in the short or medium term (selling-on any sawlogs); supplemented, where relevant, as in 4 above. They usually seek conifer plantations in the poleor early thinning stages.

6 Sawmillers or timber merchants, seeking of necessity sawlogs mainly in the quite short term, rarely for long investment (selling-on the inevitable small diameter roundwood, and any veneer logs). They usually seek maturing or mature woodlands.

7 A few overseas investors, notably from Scandinavia (denied the right to buy woodlands in their own country; and encouraged by the abolition of exchange control). They seek high quality and otherwise attractive extensive woodlands and sporting estates, mainly for long-term investment.

8 Farmers, afforesting surplus arable and improved grassland or 'Set-aside' land, for which annual taxable payments are available – this financial benefit (along with other benefits) being additional to those noted in 1 above. Many participating farmers are likely to be particularly concerned with amenity and sporting.

The future. The supply and demand levels of stocked woodlands (also of bare land) will depend largely on the financial, economic and political climate, not least the rate of inflation and the cost of borrowing.

The present immense and efficient wood processing industry has, along with forestry, established a vital integrated industry. It has helped to encourage investment in stocked woodlands, not only by some of the processors themselves but also by individuals, institutions and others who consider the investment to be sound and prudent. However, to date, even the presence of the immense potential market has proved somewhat insufficient encouragement for adequate new planting when considered in the light of the 1988 Budget changes. For bare land, relatively few vendors are forthcoming even at a price above a realistic level for viable afforestation. Most sought after is better quality land, i.e. of higher growth potential. However, all new planters have the satisfaction at least of knowing that should they desire for any reason to dispose of their plantation there is, from the early thicket-stage onwards, a likely 'secondary market', i.e. wood processors, institutions, and substantial individuals.

Turning to stocked woodlands, average plantations up to five years old are often difficult to sell at realistic prices, sometimes achieving little above bare land value. Other plantations difficult to market include those on degraded heather land, or above Windthrow Hazard Class 4 while having no special aesthetic or sporting appeal. However, thriving plantations of almost any age will continue to sell well. They are in short supply, and some potential buyers remain unaccommodated. Yet the demand is rather more opportunist than active. There is also a demand for small woodlands throughout Britain, as well as for parcels of timber maturing or mature, and woodlands offering opportunities to develop a wide range of recreational, sporting and conservation activities. Particularly required are good quality plantations of medium to large extent, with adequate access, in favourable locations for markets. There is no abatement of demand in the south and south-east of England; also elsewhere if the woodlands are attractive in one form or another, and well situated. Especially is this so if they have a pleasing appearance and acceptable structure, or have good sporting potential, and preferaby where they will cost little to maintain, protect and tend. An imminent thinning may provide an added attraction. However, if a wood is remote or the access poor, the only likely purchaser may be a neighbouring landowner, local resident, or a nearby sawmill. Maturing or mature small to large woods are in demand mainly by sawmillers or timber merchants intent on early harvesting. Price will generally depend on species, stocking, quality and access. Where strong competition from individuals is present, woods will command a premium.

Institutional enthusiasm for woodlands may vary considerably, much dependent on alternative investments. In recent years some have sold their holdings, some have increased their acquisitions, and a few newcomers have emerged. Steady buying of extensive plantations by wood processors and major sawmillers is likely to continue, particularly to ensure in the short or medium term a proportion of necessary supplies of small diameter roundwood and sawlogs. This is encouraged by the knowledge that premature felling of stands is an economic decision solely for the woodland owner, and cannot be decreed by the Forestry Authority. Most such buyers are becoming increasingly selective, and in particular seek well managed properties. Foreign buyers will continue to enter the market.

Factors which may affect the stocked woodlands market, include (besides the economic and political climate): (i) the Forestry Commission projected sale of 100,000 hectares by the year 2000; and (ii) sales by private owners (especially those needing, before 5 April 1993, to reduce their heavily burdened borrowings) of upland plantations of various extents, ages and conditions. The two types of disposal may result in some values falling, especially where public access to them is relevant. Other vendors will include: those whose woodland investments outlast them (for whatever reason); those who see their original objective in purchasing or establishing woodlands achieved in one way or another; and those whose financial or personal affairs alter in emphasis, and need to realize capital for development or diversification in their enterprises.

One of the principal factors of the woodland market will continue to be a shortage of good woodlands on offer for sale, rather than a lack of potential purchasers. British investors are now acquiring woodlands in France, as those in Southern England are either very expensive or simply not available for sale. Investment looks likely to become more international in nature, as 1992 approaches.

The forest management companies will continue to be among the prime movers in promoting forestry as an investment under the new conditions, although sophisticated purchasers may be more willing to listen to advice from independent consultants who do not profit from contract turnover. However, investment in forestry will need to perform as good as other types of available investment and take into account the long-term nature of forestry and the illiquidity of the investment at certain times of its life. Conditions are now awaited to favour a substantial increase in afforestation. Particular promoters of it, in addition to the Forestry Commission, are: FICGB, TGUK, ICF, APF, NFU, forestry trade unions and wood-processors.

3. Loans for forestry

Loans for commercial forestry – chiefly for the purchase of land and the growing of trees – are obtainable from several institutions, notably: the Royal Bank of Scotland (in association with Fountain Forestry Ltd.), the Bank of Scotland, Barclays Scotland (with Bell-Ingram Forestry), Barclays Bank plc (through Barclays Mercantile Highland Finance Ltd.), the Agricultural Mortgage Corporation, and the Scottish Agricultural Securities Corporation plc.

Banks without a specific loan scheme for forestry but willing to consider financial forestry propositions, include Midland Bank plc and National Westminster Bank plc. Loans or other forms of financial assistance are also available for harvesting, processing or marketing of timber. Brief accounts of the main sources, with addresses, are provided by Lorrain-Smith (1989).

4. Forestry in the European Community

In world terms the European Community (EC) has a relatively small area of forest representing 1.2% of the world's forest. International comparisons of land are given in Table 52. The EC has 43 million ha of closed forest, about 20% of the total land area, whereas the total woodland amounts to 53 million hectares, or 24% of the land surface. The proportion of private woodlands, averaging 58%, varies greatly throughout Member States; in the UK it is about 60%. Forestry employs over two million people in wood production, harvesting, processing and marketing. A total of about 125 million m^3 of wood is produced each year, but the EC harvests only 55% of its forest product requirements. The annual trade deficit is of the order of £15 billion.

The consumption of wood and wood products in the EC is likely to increase by around 75 million m^3 by the end of the century – about 30% of the current figures. The projected increase in harvesting in EC forests by the year 2000 will produce only an extra 30 million m^3. Thus, in order to ensure its security of supply, the EC is faced with a need to develop more strongly its own wood production where it makes economic sense to do so, as well as helping to ensure that exporting countries, particularly developing ones, achieve sustainable production. The EC climate and soil conditions are often better suited to forest production than those in the countries from which for instance temperate imports are supplied, such as Canada, Scandinavia and the USSR.

There is not a Common Forestry Policy within the EC chiefly because of the wish to maintain national sovereignty in the context of very diverse forestry conditions between the Member States. The diversity of EC forestry is clear, but some distinctions can be drawn on a north/south basis as described in *Europe's Green Mantle:*[9]

> What might be termed the 'Northern type' of forest can be typified as being high forest, often even-aged plantation, consisting of one or few species, managed largely or exclusively by maintaining a high growing-stock volume. Its chief purpose is to maintain a supply of tangible forest products, mainly in the form of timber, especially for sawnwood, paper and board products. Though not in physical isolation from it, the management of this forest type is usually a separate activity from agriculture.

The 'Southern forest type' differs in form, purpose and management. It occurs in parts of France, much of Italy and most of Greece and becomes more important

---

9 *Europe's Green Mantle – the heritage and future of our forests*, Green Europe No. 204, European Communities 1985.

with the accession of Spain and Portugal to the Community. Seldom consisting of high forest in the accepted sense, except where expressly managed or in plantations, the forest trees are often stunted to a remnant 'Maquis' vegetation when grazed or burned. It usually serves as protection forest, conserving water and preventing soil erosion. It provides forage and shelter to grazing herds and throughout its extent suffers greatly from the risk of fire. In total contrast to the 'Northern type', it is inextricably bound up in the agricultural life of the Mediterranean zone which is very dependent on it, so is seldom viewed as a distinct exploitable resource in the traditional northern sense, except where cork is concerned although cork production is limited to Portugal, Andalucia, Extramadura and Cataluna.

Table 52. Land use – International comparisons. Forestry areas include unproductive woodland. Other land includes mountains, tundra, desert etc.

Country	Total Land Area (million ha)	Percentage of total area		
		Forestry	Agriculture	Urban or Other
Great Britain	22.7	10	77	13
United Kingdom	24.1	10	77	13
Belgium/Luxembourg	3.3	21	46	33
Denmark	4.2	12	66	22
France	55.0	27	57	16
West Germany	24.4	30	49	21
Greece	31.1	20	70	10
Ireland	6.9	5	82	13
Italy	29.4	23	58	19
Netherlands	3.4	9	59	32
Portugal	9.2	40	36	24
Spain	49.9	31	62	7
EEC Countries	222.9	25	60	15
Norway	30.7	27	3	70
Sweden	41.2	68	9	23
Finland	30.5	76	8	16
USA	916.7	29	47	24
Canada	922.1	38	8	54
USSR	2,227.2	42	27	31
Japan	37.7	67	14	19
World	13,076.5	31	36	33

Source: Forestry Commission 'Forestry Facts & Figures' 1989–90.

The current Forestry Action Programme for the Community, the first in a series, runs from 1989 to 1992, and was proposed in conjunction with a global Community Forestry Strategy which has yet to be developed in the international context. The main elements of the Programme are:

- Aid to meet part of Member States' own assistance for afforestation of agricultural land, granted under Council Regulation (EEC) No 1609/89;

- Under the Reform of the EC Structural Funds the aid for rural development is concentrated on schemes in those areas which the Community has designated as qualifying for special assistance, under the so-called Objective 1 and Objective 5b areas. Areas covered in the UK are Northern Ireland (Objective 1) the Highlands and Islands and parts of Dumfries and Galloway in Scotland, parts of rural Wales, and areas of Devon and Cornwall (all Objective 5b). The UK gains up to £6 million annually by contributions towards the cost of its Grant Schemes. The measures cover afforestation, the construction and improvement of roading, the extension and restoration of forests in areas susceptible to erosion, the reconstitution of forests destroyed by fires or other natural catastrophes, the rationalization of holdings, pilot projects for fighting forest fires, aid to forestry associations, aid to awareness campaigns and extension work (this aid is under Council Regulation (EC) No 1610/89);

- Small-scale investment aid for the primary processing and marketing of cork and forestry products, under Council Regulations (EC) Nos 1611 and 1612/89, respectively;

- Extension and reinforcement of the forestry protection Regulations (3528/86 and 3529/86) against atmospheric pollution and forest fires, under Council Regulations (EC) Nos 1613 and 1614/89, respectively;

- The establishment of a European Forestry Information and Communication System (EFICS) under Council Regulation (EC) No 1615/89, so as to improve the exchange of forestry sector information within the Community; and

- The setting up of a Standing Forestry Committee under Council Decision No 89/367/EC. This Committee, as well as its specialist sub-groups, may be consulted by the Commission (of the Communities) on matters concerning the forestry sector. It is composed of Member State representatives – two per country – who may be accompanied by their experts for given subjects.

The Programme envisages forestry playing a greater role in rural development, particularly within the context of the reform of the Common Agricultural Policy, the development of rural services and infrastructures in the Community, environmental improvement, and social and structural policies. The proposals were based on the EC Commission's view that there is a need for the expansion of forestry in the Community, which is only some 55% self-sufficient in forest products. Most of the provisions of the Programme are optional and do not constrain the UK Government's freedom to pursue forestry policies appropriate to the UK.

Separately from the Forestry Action Programme, the UK follows other EC legislative provisions concerning forestry, including that for Environmental Impact Assessments on new developments. In the UK's case such EIAs are only carried out on a discretionary basis for forestry projects felt likely to promote environmental concern.

Further important EC provisions apply to forest reproductive material (FRM) where thirteen major commercial species are covered as to their origin and external quality. These arrangements are likely to be extended to ornamental forest species towards 1992, by which time the EC Plant Health Directives will have been revised to comply with the internal market. This will mean that instead of border checks on phytosanitary certificates, plant material will be issued with a 'plant passport' at source, which will be verifiable at destination and at points in between, regardless of national boundaries. (The present arrangements are incorporated into the Plant Health (Forestry) (Great Britain) Order 1989 and a separate Order for Northern Ireland, incorporating and bringing up to date legislation on tree pest control and the import and export of trees, wood and bark.)

During this decade widespread changes will be made to the national standards and regulations throughout the Community to assist in the removal of technical barriers to trade between Member States from within the single internal market commencing in 1993. Eurocode 5 and other interpretive documents will be produced to cover solid timber, panel products and preservative treatment. Otherwise, the largely unrestricted entry and circulation of forestry products into and within the EC will remain mostly unchanged.

J. Wall is Administrator, Forestry, Division DGV1 – F-II–2, Commission of the European Communities, at Rue de la Loi, 12010/198, 1040 Brussels, Belgium. Timber Growers United Kingdom (TGUK) is a member of the Central Committee of Forest Ownership (CCPF); and TGUK/FICGB have one representative on the Commite du Filiere Bois (The Wood Chain Committee).

5. European forestry and timber trends

Britain's private forestry sector will benefit from understanding something about European forestry and timber trends.

Forests cover one third of Europe, and benefit millions of Europeans through ownership, through income from employment in forestry, the forest industries and ancillary activities, and through recreation, landscape values, nature conservation and environmental protection. Wood is a vital raw material for the construction, furniture, packaging and printing sectors. It is also renewable. While wood production is, and will remain, the single most important function of forestry in many countries, the generating of revenue-environmental and social benefits of the forest are gaining in relative importance. Forestry plays an active part in the protection of the environment and the stabilization of soils.

Trends in the demand and supply situation in Europe for wood and wood products are studied periodically by the Food and Agriculture Organization (FAO) and the Economic Commission for Europe (ECE) in Geneva and published by the United Nations. The most recent Study (FAO/ECE 1986) is 'European timber trends and prospects to the year 2000 and beyond' (known as ETTS IV). The growing importance of forests and their products to the well-being of society, and the policies needed to ensure that their essential role will be fulfilled in the future, were discussed at a joint session of the Timber Committee of the ECE and the FAO European Forestry Commission, in Geneva from 12 to 16 October 1987[10] taking as a basis the findings of ETTS IV noted above.

ETTS IV forecasts that removals from European forests, the use of industrial wood residues and waste paper and imports from other regions, will be able to keep pace with the expansion in the consumption of sawnwood, wood-based panels, paper and paperboard and energy wood to the end of the present century. Negative effects of air pollution could, however, disturb this balance (oversupply through sanitation fellings followed by reduced levels of increment and removals). The conclusion that there will be a balanced supply/demand situation differs from those of earlier Studies.

ETTS IV foresees a continuation to the end of the century of the keen competition between suppliers that has been a feature of the European markets for most forest products over the past decade. However, there are wide differences between the supply/demand situation in different regions and the broad conclusions above do not necessarily apply to all countries. In some countries, the needs of society for wood in a number of economic branches (construction, pulp and paper industry, furniture, etc.) are constantly growing. This trend will remain unchanged in the near future. Therefore one of the main tasks in the field of forestry and forest industries in these countries is the wider use of forest resources and the fuller and more rational utilization of the forest biomass, including low quality timber and secondary wood wastes.

The changed outlook for the wood supply/demand balance, the ever-increasing demand on the forest for environmental and other non-wood goods and services, the threat to forests from fires and air pollution, and the marked increase in public interest in forest policy in many countries have stimulated basic reappraisals of national policies in some countries and added new dimensions to policy discussions in others. The joint session of the Timber Committee and the European Forestry Commission issued a Declaration intended to stimulate and provide an international basis for these reappraisals, and thus contribute to the evolution of policies. It selected a number of key areas meriting special attention and adopted the conclusions and recommendations of ETTS IV, stating that action is particularly called for in the following areas:

1   The growing relative importance attached to the *non-wood functions of the forest* demands in some countries the adoption of new and integrated approaches to the formulation and implementation of forest policy. Attention should be given to the question as to whether existing sources of revenue – mainly from the production of wood – provide sufficient incentive to generate the necessary changes in direction of forest management and/or how far eventual financial compensation has to be paid for the obligation to manage for the public;

---

10 The joint session was attended by representatives of: Austria, Belgium, Bulgaria, Byelorussian SSR, Canada, Czechoslovakia, Finland, France, German Democratic Republic, Germany, Federal Republic of, Greece, Hungary, Italy, Luxembourg, The Netherlands, Norway, Poland, Portugal, Romania, Spain, Sweden, Switzerland, Turkey, Ukrainian SSR, USSR, United Kingdom of Great Britain and Northern Ireland, United States of America and Yugoslavia.

2   Damage caused to the forest by fire, air pollution and other agents such as game or grazing results in the long-term deterioration of the environment and losses in many other ways. Recent international surveys confirm that in large parts of the ECE region the number of trees affected by air pollution, both coniferous and non-coniferous, remains intolerably high. Measures for the *protection of the forest* deserve very strong support, especially those directed to the reduction of the causes of damage, notably emission of air pollutants. International co-ordination of research should be strengthened in order to avoid duplication and to stimulate complementary programmes;

3   Trends in *agriculture* in certain regions could lead to the transfer of some agricultural land to other uses. Provided the land's potential is respected and a balance in rural land use maintained, this offers a major opportunity for the forestry sector to develop a long-term strategy that will ensure that afforestation blends into an overall rural land-use policy and, through careful planning, to provide economic benefits to local communities and the forestryforest industry chain. To encourage conversion of agricultural land to forestry, it may be necessary to keep open the possibility for this land to revert to non-forestry uses;

4   In some countries, it was considered that the growing need of society for forest products should encourage the governments to promote the *wider utilization of forest resources* and also the fuller and more effective utilization of all extracted biomass, including low quality timber and wood wastes;

5   Governments need to define the role that the large number of *small forest holdings* could and should play in contributing to society's needs in general and rural development in particular. Having defined that role, they should provide active support to the strengthening of management, especially through extension services and other forms of support;

6   The forest and forest industry sector should take up the challenge to *improve the marketability of its products,* by improving information on structural developments for markets and end-uses, by investing in research and development, by more aggressive promotion and marketing strategies and by greatly extending education and training in the use of forest products;

7   Policies *for wood-based energy development* should be pursued, since it can be expected that changes from the present energy situation will, sooner or later, result in renewed interest in wood as a source of energy. Market information should be made available to enable suppliers and users of wood to make the best use of available resources;

8   In some countries, governments have an important role to play in promoting the dynamic development of the forest and forest industry sector by encouraging better *communication and understanding* amongst its various components and promoting greater co-operation in furthering common interests; and

9   Governments should actively encourage greater well-balanced *public participation* in the policy-making process for the forest and forest products sector and take steps to strengthen the public's and legislators' understanding of the complex issues involved.

Finally, it is not sufficient to make policy: it must also be implemented. This implies a strong institutional framework in the member countries, backed by full political support. Given the long-term nature of forestry, stability should be encouraged and unnecessary changes avoided.

Work has begun on the next ETTS (i.e. V) due for publication in the mid-1990s. Because of the increasing importance of non-wood benefits, they will be treated in a separate Study, although the inter-play between them and wood supply and demand, on which ETTS V will concentrate, will be considered. It will also be necessary to restructure ETTS V, as compared with earlier studies, to take account of the remarkable changes in Eastern Europe.

The FAO/ECE publish bi-annually the *Timber Bulletin: Monthly prices for forest products*. This is in response to the expressed need for improved coverage and dissemination of information on prices for forest products. Following the work by the Joint FAO/ECE Working Party on Forest Economics and Statistics of assembling price series available in member countries, the publication *Forest Products Statistics, Price series, 1950–1976* was issued as supplement 6 to volume XXX of the *FAO/ECE Timber Bulletin for Europe* (Geneva, December 1977). A selection was made from that publication of the series of prices which appeared to be representative of the product groups in question and to have some importance for the international market. These series are published on a regular basis. Monthly data, for the most recent three years, are published in numbers 3 and 7 of the *Timber Bulletin*. In addition, annual data for the same series for the most recent six years are published in number 7 of the *Timber Bulletin,* which also presents data on a range of economic indicators (index numbers of industrial production, producers, wholesale and consumer prices, gross domestic product, population, exchange rates, dwelling completions). Long-term series of forest products prices are also regularly published by FAO as Forestry Papers.

T.J. Peck is Director of the ECE/FAO Agriculture and Timber Division at Palais des Nations, CH-1211, Geneva 10, Switzerland. Hummel (1989) has given an analysis of the policies by which the countries of Europe intend to take forestry into the 21st Century.

References

Campbell, J. (1983), 'Ecology in the Eighties: Excellence in British Forestry'. *Commonwealth Forestry Review, 62* (4) 1983, pp. 243–9.
FAO/ECE (1986), *European Timber Trends and Prospects to the year 2000 and beyond* (ETTS IV), in two volumes, ref. ECE/TIM/30. United Nations, Geneva.
Francis, G.J. (1989), 'The Case for Investment in Forestry' in *Proceedings of a Discussion Meeting on UK Forest Policy into the 1990s,* held at Bath University 31 March to 2 April 1989. Institute of Chartered Foresters, pp. 87–108.
Hart, C.E. (1987), *Private Woodlands: a Guide to British Timber Prices and Forestry Costings.* Chenies, Coleford.
Hummel, F.C. (1989), *FAO Forestry Paper 92. Forestry Policies in Europe: An Analysis.* FAO, Rome.
Lorrain-Smith, R. (1989), *Grants and Other Financial Assistance for Trees.* Calderdale Metropolitan Borough Council, Halifax.

INDEX

Grouped-headings include: Acts of Parliament (Statutes), Animals (mammals), Birds, Climate, Diseases, Economics, Forestry, Forestry Commission, Forestry Operations, Insect pests. Measurement (Forest), Nursery, Plants, Poplar, Taxation, Timber, Trees, Utilization/uses, Vegetation (flowers, grasses, ground flora, shrubs), and Wood.

Aaronson Bros, plc 369, 438; *see also* Fig. 22
Aberdeen University, Forest Economics Department 583, 586
 forestry projects and research 271, 383, 586
 survey of Private Forestry Costs and Returns 537, 549
access, woodland 624
accounts/accounting, woodland 5, 592, 593
 computers 594, 598–600
 costs 592–6; *see also* costs
 financial 592
 records 594
 labour 537. 594
'acid rain' *see* air pollution
Ackers, C.P., OBE 256
Acts of Parliament (Statutes)
 Control of Pollution Act (1974) 504
 Deer Acts (1963 and 1980) 509
 Financial Services Act (1986) 316, 616, 629
 Firearms Act (1968) 232
 Food and Environmental Protection Act (1985) 504
 Forestry Act (1967) 580
 Ground Game Act (1880) 232
 Health and Safety at Work etc. Act (1974) 584
 Pests Act (1954) 232
 Water Act (1974) 504
 Wildlife and Countryside Act (1981) 7, 232, 486
 see also legislation
Ae Forest, Scotland 583
aerial spraying of chemicals 182, 222
afforestation (new planting) 144, 149, 180–2, 321
 assumption against conifers in English uplands 10, 625
 recent extent 322
age-classes 253
 broadleaved 144
agriculture
 compared with forestry 558, 559
 integration with silviculture 390, 514, 515, 516
 interaction with silviculture 514, 515, 516 *see also* farm forestry
Agricultural Development and Advisory Service (ADAS) 7, 233, 518, 520, 570
Agriculture and Fisheries for Scotland, Department of (DAFS) 1, 146
Agriculture Committee, House of Commons 5, 564
agriculture departments 569, 580
Agriculture, Fisheries and Food, Ministry of (MAFF) 1, 146, 514
agro-forestry, agro-silviculture, intercropping 516
air pollution and tree health 76–9
Alder 13, 14, 28, 114
 common *Alnus glutinosa* (L.) Gaertn. 13, 114
 grey *A. incana* (L.) Moench. 13, 114
 italian *A. cordata* Desf. 14, 114
 red *A. rubra* Bong. 14, 114, 125
 rotation 303
 silvicultural characteristics 114
 thinning 303
 timber: identification 28; markets 417; price *see* prices – timber and turnery, turneries; properties 343; uses 343, 354
 value for nitrogen fixing 91
Alice Holt Lodge, Forestry Commission Research Station 224, 586
altitude *see* climate and 'tree line'
amenity woodlands 520, 521, 570
 grants 570
 taxation *see* taxation of woodlands

valuation 609, 610
ancient semi-natural woodland 146, 437, 489, 625; see also Plate 3 and Register of Ancient Woodland
Anderson, M.L., MC 248, 256
animals, mammals 231–8, 487
  badger *Meles meles* 238
  bats, woodland 494
  deer 236–8, 509: bark stripping, browsing, fraying 236; control 236–8; damage by 236–8; fallow *Dama dama* 238; fences against 159, 161; management 509, 510; muntjac *Muntiacus muntjak (reevesi)* 238; red *Cervus elaphus* 237 and Plate 15; roc *Capreolus capreolus* 237; sika *Cervus nippon* 238
  domestic (cattle, goat, horse, sheep) 160
  dormouse *Glis glis* 234
  fox *Vulpes vulpes* 235
  grey squirrel *Sciurus carolinensis* 122, 136, 234
  hare: brown *Lepus capensis* 233; mountain *Lepus timidus* 233
  polecat *Mustela putorius* 233
  pine marten *Martes martes* 233
  rabbit: *Oryctolagus cuniculus* 231–3; fences against 159, 160; myxomatosis 231
  red squirrel *Sciurus vulgaris* 235
  stoat *Mustela erminea* 508
  voles: bank *Clethrionomys glareo-lus* 234; field *Microtus agrestis* 234
  weasel *Mustela nivalis* 508
arboreta *see* Westonbirt

Arboricultural Advisory & Information Service 512, 589
Arboricultural Association 610
Area of Outstanding Natural Beauty (AONB) 483
Ash, common *Fraxinus excelsior* L. 13, 113, 136
  'dieback' 113
  diseases and insect pests 113
  rotation 301
  silvicultural characteristics 113
  suitable conditions 136
  thinning 301, 309
  timber: identification 28; markets 415; price 433, 434, 436; properties 342; uses 342, 352, 353, 354, 355, 356, 359
Ash, Mountain (Rowan) *see* Mountain ash (Rowan) *Sorbus aucuparia* L.
Aspen *see* Poplar
Association of Professional Foresters (APF) 6, 389, 631
Atlas cedar *Cedrus atlantica* 14, 27, 111, 341
  f. *glauca* 111
Austrian pine *Pinus nigra* var. *nigra* Harrison 14, 110

Bangor University (University College of North Wales)
  forestry projects and research 537, 586, 587
  School of Agriculture & Forest Sciences 583
bark
  allowance 33, 37
  characteristics 20, 22
  debarking *see* peeling
  identification 19
  oak, tanbark 356, 439–43

products 356, 357, 360
  overbark (ob) 37
  underbark (ub) 37
  bark stripping, oak 357, 439, 441
  see also peeling and tanbark
basket willow coppice 286, 287
Bath, Lord, of Longleat Estate 2, 283, 284
Bathurst, Lord, of Cirencester Park 2
bats in woodland 494
Bedgebury pinetum 582
Beech, Common *Fagus sylvatica* L. 13, 14, 112, 136
  diseases and insect pests 112
  rotation 301
  silvicultural characteristics 112
  suitable conditions 136
  thinning 301, 309
  timber: identification 28; markets 415; price 434, 436; properties 342; uses 342, 352, 353, 354, 355
Beech, Copper *Fagus sylvatica purpurea* 112
Bidwells Forestry 603, 624
biomass for energy, 270–3, 382, 383
Birch
  Downy *Betula pubescens* Ehrh. 13, 114
  Silver *B. pendula* Roth. 13, 114
  rotation 302
  silvicultural characteristics 114
  thinning 302, 309
  timber: identification 28; markets 417; price *see* prices – timber and turnery, turneries; properties 343; uses 343, 352, 354, 356
Bird cherry *Prunus padus* L. 13, 119
birds, woodland 238, 499–502, 565

blackgame, black grouse *Lyrurus tetrix* 238
capercaillie *Tetrao urogallus* 238, 499
crow *Corvus corone* 499
gamebirds *see* shooting
goshawk *Accipter gentilis* 499
insectivorous 238, 500
jackdaw *Corvus monedula* 499
jay *Garrulus glandarius* 499
magpie *Pica pica* 499
nestboxes 499
of prey 499
owl, various spp. 499
partridge *Perdix perdix* 501
pheasant *Phasianus colchicus* 238, 501, 505, 506
sparrow-hawk *Accipter nisus* 499, Plate 13
starling *Sturnus vulgaris* 238, 499
woodpecker, various spp. 500
wood-pigeon *Columba palumbus* 505, 508
see also Royal Society for the Protection of Birds (RSPB)
Blackthorn, sloe *Prunus spinosa* 119
boardmill factories *see* fibreboard and particleboard
botanical features, trees 12, 15–20
Boughton Estate 485
Bourne, R. 256, 257
Bowhill, Selkirk, Scotland 256
Box *Buxus sempervirens* 13, 30, 120, 346
bracken 89, 152, 176
Bradford, Lord 2, 258
Bradford–Hutt Silvicultural Plan, Tavistock 258–62; *see also* Plate 6
branches, tree 38, 452, 469
  epicormic 219
  habit 17, 19, 111

persistency 105, 112, 216
Brazier, J.D. 348
breeding of trees *see* genetics in silviculture
Britain, forestry *passim*
  broadleaved resource 143, 144
  case for national forestry 322–4, 612
  case for private forestry 323, 613
  cost of imports 322, 377, 612
  effect of wars 321, 613
  forest history 321: uneven-aged, irregular forestry 255
  forest policy 2: afforestation 321, 322; benefits 322, 323. 324; broadleaved 143–7
  forest profitability 323
  forest resource 143, 144, 321, 322, 612: extent compared with other EC Member States 1, 322; extent of productive and unproductive woodland 321; private sector 1, 2, 5, 6, 143–7; State 1, 2
  growth rates compared with Europe and Scandinavia 323
  investment in timber processing 324, 326
  'knock-on' effects of forestry 324; *see also* Fig. 21
  rate of return compared with Europe and Scandinavia 323
  restocking *see* restocking
  self-sufficiency, timber 612
  sylvan heritage 1, 11. 12, 15
  silvicultural systems *see* silvicultural systems
  timber consumption 322, 326, 327; demand 322, 326, 327, *see also* timber;

supply/production 322, 326, 327, *see also* Forestry Commission; yield class compared with Europe and Scandinavia 323
British Association for Shooting and Conservation *see* Appendix
British Christmas Tree Growers' Association 202
British Coal Corporation 362, 366;
British Columbia, province of Canada 131, 132
British Deer Society 509
Britial Rail *see* transport of timber
British Standards 75
  fencing 365
  pallets 366
  plants 41, 165
  preservatives 407
  timber 348, 365
British Timber Merchants Association (England and Wales) 8
British Trust for Ornithology 501
British Wood Preserving Association (BWPA) 409
broadleaves *see* individual species guidelines for management 147, 473, 474, 475, 476, 557
  in uplands 185
Bryant and May (Forestry) Ltd 193
Buccleuch, Duke of 2, 485
Building Research Establishment (BRE) 332, 335, 348, 369, 406, 586

Caberboard Ltd 369, 437; *see also* Fig. 22
Caledonian Paper plc 369, 436; *see also* Fig. 22
Caledonian Scots pine

(indigenous) 13, 130, 496, 567; see also Plate 13
California, State of, U.S.A. 68
Canada see British Columbia and Queen Charlotte Islands
canopy, closure 213
carbon-fixing see global warming
careers in forestry see education and training
Caucasian fir *Abies nordmanniana* 14, 198
Cedar see Atlas, Deodar, Japanese red, Lebanon, and Western red
Cell Grown Plant Producers Association 8
census of woodland and trees (1979–82) 321
Central Scottish Woodland Project 514
chainsaw 379, 400
charcoal 357, 358, 439
Chartered foresters/ Chartered surveyors (firms)
  Bell-Ingram 603, 624, 631
  Bidwells Forestry 603, 624
  John Clegg & Co. 603, 624
  Savills 603, 624
  Strutt & Parker 603, 624
'check' of spruce in heather 140, 181, 543
Checkendon, Oxon 254, 256
chemicals see fertilizing, legislation, repellants, site preparation, thinning and weeding
Cherry, Bird see Bird cherry Wild (Gean) see Wild cherry
Chestnut, Horse *Aesculus hippocastanum* L. 13, 29, 118, 344, 354
Chestnut, Sweet *Castanea sativa* Mill. 13. 113, 136, 266–8

coppice 266–8, 356
diseases and insect pests 113
rotation 302
'shake' in 435
silvicultural characteristics 113
suitable conditions 136
thinning 302
timber: identification 28; market 416. 417; price 356; properties 343. 434, 435; uses 343, 352, 353, 354, 356
yield table 267, 268
Chilterns 92, 146, 254, 256, 275
chipboard see fibreboard
chipping machine, portable, 383
chlorosis, lime-induced 92, 110, 223
choice of species see species choice
Christmas trees 198–202
  costs 200, 201
  insect pests 199
  profitability 199, 201
  rotation 199
  shearing 199
  species 198
  taxation see taxation of woodlands
Cirencester Park, Glos. 257, 258
classifications
  ground vegetation 87–9
  land capability for forestry 129
  site 68
  soils 91–3, 95–7, 129
  terrain 390
  topex value 85
  tree 12, 293
  windthrow hazard 83
  see also silvicultural systems
clear-felling 544; see also silvicultural systems
cleaving 356
Clegg, John, & Co. 603, 624
climate 1, 69–87
  drought 69, 70
  exposure 85
  factors 69–87

frost/cold 69, 70, 112, 165
ice, rime 74, 75
lightning 75
pollution 76–9
rainfall (precipitation/ evaporation; wetness/ dryness) 71–3
snow 74
sunscorch, heat 70
temperature (warmth/cold; drought/frost) 69–70
see also global warming, microclimate and wind
Clinton, Lord 2
Coast redwood *Sequoia sempervirens* (D. Don) Endl. 14, 110
  one of the tallest species in the world 110
  timber: identification 27; markets 415; properties 341; uses 341, 352; see also Plates 17 and 20
Coed Cymru 486, 582
cold storage of plants and seeds 56, 57, 60, 61, 167
community forests 512, 513
compartment, financial record 456–69, 594; see also subcompartment
computer, computing 598–600; see also accounts/accounting, woodlands
conifers see individual species
conifer heart rot *Heterobasidion annosum* see diseases
conservation, 7, 121, 485–501
  management guidelines 239, 489–94
  wildlife 121, 239
  see also animals and birds
constraints on forestry see restraints

contract, management 465
  terms 389
contractors, woodland 389, 465
control, felling 9
  Forestry Commission 9, 579, 581
  planning 9
  see also thinning and Tree Preservation Orders
conversion of silvicultural systems 273
conversion of timber 356, 397–402, 420, 439–43
  prices 439–43
  products 398, 399, 439–43
  sawmill see sawmills
  timber 420, 439–43
  underwood see underwood
  woodland 397–9
  wood-yard 401
co-operation, co-operative associations, wooodland 413, 486, 582, 621
  Western Woodland Owners Ltd. 621
co-ownership of woodlands 629
coppice 248, 364–8, 458, 469, 487
  conversion to high forest 273
  extent 144
  markets 356
  short rotation 270–3
  soil deterioration under 98
  stored 269, 270
  sweet chestnut see Chestnut, Sweet
  coppice with standards 144, 249, 268–70, 488
cordwood, 357, 380, 420
Corrour, Scotland 256
Corsican pine *Pinus nigra* var. *maritima* (Aiton) Melville. 14, 106
  diseases and insect pests 106, 130
  rotation 298
  silvicultural characteristics 106
  suitable conditions 130
  thinning 298, 308
  timber: identification 26; markets 414; price 432; properties 336; uses 336, 351, 352, 353, 361
COSHH regulations 584
costs 537, 594, 595, 598
  plantable land 622–31
  stocked woodland 622–31
Cotswolds 92, 245, 285, 483
Country Landowners Association (CLA) 6, 578
Countryside Commission (CC) 7, 145, 485, 512, 513
Countryside Commission for Scotland (CCS) 7, 485
Countryside Council for Wales (CCW) 7
County Trusts 7, 491
Crab-apple *Malus sylvestris* 13, 30, 119
Cricket-bat willow *Salix alba* cv. 'Coerulea' 13, 202–4
  nursery practice 202
  price of sets 202
  profitability 203
  pruning (disbudding) 202
  rotation 203
  timber 30, 204, 353: markets 204; price 204
  watermark disease *Erwinia salicis* 203
Crowe, Dame Sylvia 477
Culbin plantations, Scotland 130
cutting, market-responsive, priceresponsive 376
Cypress see Lawson, Leyland, and Monterey

damage to trees and woodland see air pollution, animals, birds, climate, diseases, fire, insect pests and wind
Dartington Woodlands Estate, Devon 254, 255, 547, 597
Dawn redwood *Metasequoia glyptostroboides* 14, 111; see also Plate 21
Dean, forest of 110, 139, 141, 255, 275, 280, 514, 582, 583
decision-making see economics
dedication scheme see Forestry Commission
deer see animals
deforestation 1, 80
degraded sites 149
dendrochronology 21
Deodar cedar *Cedrus deodora* 14, 27, 111, 341
density, natural regeneration 282
  stand 163, 164, 542
  see also nursery and wood
Devonshire, Duke of 2
direct seeding, sowing 150
Direct Worktops 369, 438; see also Fig. 22
diseases, canker, fungal 228–31
  forest: beech bark disease associated with *Nectria coccinea* 227, 230; biological control 229; blister rust on Weymouth pine *Cronartium ribicola* 229; butt and root rot fungus *Heterobasidion annosum* 106, 107, 108, 109, 110, 125, 229, 298, 380, 381; *Coryneum cardinale* canker on Leyland cypress 110; dieback of ash 113, 230; Dutch elm disease fungus *Ceratocystis ulmi* 118, 230; *Gremmeniella abietina* Dieback fungus

on pines 106, 130, 230; honey fungus *Armillaria mellea* 108, 109, 229; larch canker *Lachnellula wilkommii* 230; Lophodermella sulcigena 230; *Lophodermium* spp. 54, 105; *Peniophora gigantea*, a competing fungus 229, 381; *Peridermium pini* 105, 230; *Phaeolus schweinitzii* 230; poplar canker *Xanthomonas (Aplanobacter) populi* 193; *Ramichloridium pini* 106, 230; *Rhizina undulata* 230; *Stereum gausapatum* 218; *Stereum sanguinolentum* 229; watermark disease of Cricket-bat willow *Erwinia salicis* 203 nursery: *Brunchorstia pinea* 230; damping-off in nurseries *Pythium* spp. 53; *Dydymascella thujina* on western red cedar 54, 229; Grey mould *Botrytis cinerea* 53; *Meria laricis* on larch in nurseries 54; Oak mildew *Microsphaera alphitoides* 54; root rot *Phytophthora* spp. 54, 55
'distress' flowering of trees 209
Dogwood *Cornus sanguinea* 119
Douglas fir *Pseudotsuga menziesii* (Mirb.) Franco. 14, 107, 132
diseases and insect pests 62, 108
rotation 299
silvicultural characteristics 107
suitable conditions 132
thinning 299, 309
timber: identification 26; markets 414; price 432; properties 337; uses 351, 352, 353, 354, 360, 363

see also seed
Downy birch *see* Birch
drainage 154–6, 541
cost 189, 242, 243
woodlands 154–6, 189
drying, seasoning *see* timber
Dulverton, Lord 2
Dutch elm disease *see* diseases Dyer, R. 262

Ebworth Estate, Glos. 280, 582; *see also* Plate 7
Economic Forestry Group plc 614
economics, forestry 8, 524–61, 590
analysis, appraisal 547, 548
annuities, annuitizing 531, 534
bargaining strength in marketing and purchasing 524
benefit – cost ratio 531, 534, 548
cash flow 531, 533, 554
compound interest 527, 528
conservation 491, 536
cost-benefit 534, 535, 560
cost-effective 377
costs 537
data analysis 455, 456
decision-making 525, 526
discount factors 527
discount rate 527
discounted cash flow 527
discounting 526, 534
diseconomies of scale 597
economic appraisal 533–48
economies of scale in labour, machinery and marketing 597
Faustmann formula 539
forest rent 528
Forestry Commission 590

game management, sporting 509
Gross Domestic Product (GDP) 531, 532
increment percent 529
indicating percent 529
inflation 531, 532, 534, 538
interest 527
internal rate of return (IRR) 529
irregular silviculture 262–4
investment appraisal 533–48
land expectation value (LEV) 539, 540, 556, 557
landscape 483, 536
literature 524, 533, 534
maximization 525
natural regeneration 284
net annual income 528
net present value (NPV) 530
non-discounting criteria 528
non-wood benefits 536
objectives 525, 534, 535
opportunity cost 525, 535
options 544–8
protection 542
pruning 218
recreation 536, 630
Retail Price Index (RPI) 532
revenues 537
risk 532
roading 390–3
sawmilling 403
sensitivity analysis 534, 536, 547
shadow cost 560
shadow prices 560
social costs 560
social discount rate 560
social rate of return 560
social rate of time preference 560
sporting 509, 510
sunk costs 535, 537
sustained yield 273, 274
thinning 310–12

time preference 526
time preference rate 526
time scales 526
uncertainty 532, 533, 534
volume increment percent 529
see also Forestry Commission and Aberdeen, Bangor, Edinburgh and Oxford universities
ecosystem, forest 99, 100
Edinburgh University, forestry department within the Institute of Ecology and Resource Management 583, 586
 forestry projects and research 583, 586
education and training 582–5
 approved instructors 583
 Barony College, Parkgate 583
 Buckinghamshire College 21, 22, 583
 courses: forestry 582–4; nursery practice 583; timber, wood, preservation 583; underwood conversion 583
 Cumbria College, Newton Rigg 583
 examinations: BTEC 584; Central Forestry Examination Board 583; City and Guilds 583; Institute of Chartered Foresters 583; National Diploma of Forestry 583; Royal Institution of Chartered Surveyors 583; SCOTVEC 583
 Forestry Commission *see* Forestry Commission
 Forestry Training Council (FTC) 7, 583, 584
 Forestry Trust for Conservation and Education 486, 584
 Holme Lacy College, Herefordshire 583
 Hooke Park College (The Parnham Trust) 583
 Merrist Wood College, Worplesdon 583
 National Council for Vocational Qualifications (NCVQ) 584
 Scottish School of Forestry, Inverness 583
 Sparsholt College, Winchester 583
 see also Aberdeen, Bangor, Edinburgh and Oxford universities
Egger (UK) Ltd 369, 438; *see also* Fig. 22
Eildon, Melrose 256
Elderberry *Sambucus nigra* 119
electricity supply poles *see* transmission poles
Elm, English *Ulmus procera* Salisb. 13, 118
 diseases and insect pests 118
 Dutch elm disease *see* diseases
 hybrid elm Sappora Autumn Gold' 118
 silvicultural characteristics 118
 timber: identification 29; markets 416; price 434; properties 344; uses 344, 352, 353, 354
Elm, Wych *Ulmus glabra* Huds. 13, 344
employment in forestry and forest industries 1, 322, 324
energy crops 270, 517, 589
Energy, Department of 270, 271, 377, 589
English Nature (NCCE) 7
enrichment 285, 286
Environment, Department of (DoE) 512, 514, 581, 586
environmental impacts, considerations and interests 579–80
 assessments 579–80
 clear-felling 395–7
 see also afforestation, roading and thinning
epicormic shoots 219, 303, 553
epiphytes 113, 489
equipment *see* chainsaw, machinery
establishment of plantations 149
 age when complete 186, 213
 cost 186, 187
 failures in 185, 186
 operational costs 188–90
 see also forestry – operations
establishment of special tree crops 190–204
Eucalypts *Eucalyptus* spp. 14, 116
European Community (EC) 1, 514, 532–5, 582, 632–5; *see also* plants and seed
European forestry 1, 248, 253, 275, 322, 632–8
European Forestry Commission of the United Nations Food and Agriculture Organisation (FAO) 383, 426, 635
European timber trends 1, 68, 635–8
European larch *Larix decidua* Mill. 14, 106, 131
 diseases and insect pests 106, 131
 rotation 298
 silvicultural characteristics 106
 suitable conditions 131
 thinning 298, 309
 timber: 'boat-skin' larch, 432; identification 26; markets 414; price 432; properties 338; uses 338, 353, 354, 360, 363
European silver fir *Abies*

alba Mill. 14, 254
aphids *Adelges nusslini* 254
even-aged (regular) crops *see* silvicultural systems
exotic (introduced) trees 13, 14
export
  roundwood 322, 327, 369
  veneer logs 359
extraction of woodland produce 384–6, 546
  cable-cranes 384
  costs 386
  forwarders 384, 386
  harvesters 385
  horse 384, 399
  loaders 385
  processors 385, 386
  product mix 385
  skidders 384

farm forestry 514–20
  agro-forestry 516
  change to forestry 514–20
  Farm Woodland Grants 564, 569, 570
  hedgerows 206–8
  integration with forestry 515, 516
  reversion to agriculture 519
  shelterbelts 204–6
  short rotation coppice crops 517
  woodlands 517
  see also Set-aside
farmers 628, 630; *see also* farm forestry
Farmers' unions 6, 631
Farming and Wildlife Trust Ltd 7
Farming, Forestry and Wildlife Advisory Groups (FFWAGs) 7
Farm Woodland Scheme 507, 518; *see also* Set-aside
fauna *see* animals and birds
Faustmann formula *see* economics
felling 378–80

  costs 380
  cutter (feller) selection 216, 297
  decision 545, 553, 554
  delimbing (snedding) 380
  harvesters 385
  licences 396, 580, 581
  optimum felling age 544
  optimum rotation 544
  piece-work rate 380
  premature in anticipation of windthrow 546
  processors 385
  quality 379
  replacement of unsatisfactory crops 545
  season 378, 379
  tools 379; *see also* chainsaws
fencing 156–62, 353, 356, 363, 365, 441–3, 541
  'bars' 363, 432
  costs 189, 242, 243
  motorway 365
  netting 158
  posts, stakes, struts 158
  specification 156–62; deer 159, 161; rabbit 159, 160; stock 160
  terms 158
  timber 356, 441–3
  tools 159
  see also treeshelters
fertilizing 179, 182–4, 219–25
  application 179, 182–4, 219–25; aerial 179, 182, 222
  broadleaves 179, 223
  conifers 179, 219–22
  costs 189, 245
  deficiencies 179
  foliar analysis and sampling 220
  initial 179, 189, 541
  lowland 179
  plantation 179
  top-dressing 219–25, 542, 543: broadleaves 223; conifers 219–22
  types 179

  upland 179
fibreboard, 324, 327, 353, 358, 367, 369, 370, 371, 437, 438; *see also* Fig. 22
  imports 327
  timber 353:
    consumption 327, 331;
    potential supply 331;
    prices 437, 438;
    species 353, 437, 438;
    specification 437, 438;
    supply troughs and peaks 326, 368
Field maple *Acer campestre* L. 13, 119
finance, forestry *see* economics and loans for forestry
financial accounts *see* accounts/ accounting, woodlands
financial maturity *see* rotation
Financial Plan *see* management, forest – management scheme
Finland, forestry in 68, 88
  Cajander, A.J. 88
Fir *see* individual species
fire, breaks, belts 239–41
  control 239–41
  hazard 239
  insurance 241, 242
  interactions 240
  prevention 239–41
  risk 239
firewood, fire-blocks, fuelwood *see* timber and utilization/uses of timber/wood
flora *see* vegetation
flowers of trees 17, 18, 113
foliage *see* greenery for florists' use
  juvenile *see* juvenile foliage, nursery and plants
foliar analysis and sampling *see* fertilizing
Food and Agriculture Organisation (FAO) 383
forest(s) *see* Ae, Dean,

Glentress, Grizedale, New, Thetford and Windsor
community 512
functions, social 559–61, 612
management companies 4, 614, 616; EFG plc 614; employment 322; Forestry Investment Management 616; Fountain Forestry 616, 631; Scottish Woodlands Ltd 616; Tilhill Forestry 616; *see also* Chartered foresters/Chartered surveyors (firms)
new national 514
State *see* Forestry Commission
Forest Enterprise 1, 321, 584, 612
Forest Reproductive Material Regulations 57, 59
forester *see* Chartered foresters/ Chartered surveyors (firms)
forestry, *passim*
 case for national forestry 323, 324, 559–61
 case for private forestry 323, 559–61
 compared with agriculture 558, 559
 costings 594, 595
 'knock-on' effects 324, Fig. 21
 land reclamation 208, 209
 loans 631, 632
 operations: after-care (tending) 243, 244, 245, 542, brashing 215, 216, 243, 244, 543, cleaning 214, 243, 244, maintenance of fences and roads 190, 242, 546, protection 190, 223–41, 542, rackcutting 215, 543, *see also* pruning; establishing 186, 187, beating up (gapping) 178, 190, 541, 542, *see also* drainage, fencing, planting, preparation of ground, weeding; harvesting, *see* conversion, extraction, felling, harvesting, thinning, transport of timber
 private sector *see* private sector
Forestry Acts *see* Acts of Parliament (Statutes)
Forestry and British Timber 413, 430
Forestry Commission 1, 5, 311, 563, 579, 585–9, 612; *see also* Fig. 1
 achievements 9, 62, 322, 323, 484, 582, 588
 acquisition of plantable land 626
 Arboricultural Advisory & Information Service, 512, 589
 Branches: Education and Training 381, 582–4; Entomology 224, 587, 589; Mensuration 587, 589; Pathology 228, 587, 589; Photography 587; Physiology 587, 589; Plant production 587; Seed 58, 59, 589; Statistics and Computing 582, 590, 599; Survey 590; Tree Improvement 65, 587, 589; Wildlife and Conservation 499, 587, 589; Wood Utilization 587; Work Study 377, 379, 381, 590, 596
 broadleaved policy 143–7, 475, 557
 census of woodlands and trees (1979–82) 321
 classifications *see* classifications
 chairman (J.R. Johnstone CBE) 1
 commissioner for private forestry (R.T. Bradley) 1
 commissioners 1
 consultation procedures 586
 dedication scheme 563
 deputy chairman and director general (T.R. Cutler) 1
 disposal of some State woodlands 622–4, 631
 education and training courses 582–4
 employees 322
 encouragement of private forestry *see* private sector forestry and chapter 15
 encouragement of processing industries 326
 estimates of financial yield 323
 extent of financial yield 323
 extent of woodland/forests managed 1, 321, 322
 felling licences 396, 580, 581
 forest parks, national 9
 Forestry Authority 2, 582, 612
 Forestry Safety Guides 584, 585
 Forestry Safety Council 584, 585
 Forestry Training Council 583, 584
 formation (1919) 1, 21
 future timber price assumptions 323
 grant-aid to private forestry *see* private sector forestry and chapter 15
 head of private forestry and environment department (A.H.A. Scott) 2
 history (from 1919) 321
 literature, publications *passim* main offices 2, 3

Management and Training Centres 583
nature reserves 489
organization, changes in 2, 3
partnership with private sector 1, 612
policy 2, 324:
  broadleaved 328, 475–6; marketing 418; private woodlands *see* private sector forestry; public access *see* access, woodland
profitability of State forestry 323, 550, 551
research and development 585–90: Research Station, Alice Holt Lodge 224, 586, 587, 588; Research Station, Roslin, 224, 587, 588
role 1, 2, 322
sample plots 590
sawlog (conifer) grades 328, 349–51
supplier of tree seed 58, 127
support to private sector 563–70, 582–90
tables *see* measurement, forest
Forestry Industry Committee of Great Britain (FICGB) 6, 367, 631
Forestart (seed firm) 58, 127
'Forestor' sawmilling machinery 405
Formwood Ltd 439
Fountain Forestry 616, 631
Fraser of Allander Institute 9
free growth 304
frost *see* climate
fuelwood *see* timber – firewood, utilization/uses of timber/wood – firewood

gale/storm damage 80–7; *see also* wind, windthrow
gamebirds, game management *see* shooting
Game Conservancy 505
Garfitt, J.E. 256, 257, 258, 279
Gean *see* Wild cherry
General Post Office *see* transmission poles
genetics in silviculture, tree breeding 23, 62–6
  choice of seed trees 64
  genotype 62, 63
  phenotype 62, 63
  potential gain 66
geology 90; *see also* maps and soil
Glentress Forest, Scotland 256
global warming 1, 79, 80, 323, 559, 561
government, forest policy *see* Britain and Forestry Commission
grading *see* nursery, roundwood, sawlogs, sawnwood
Grand fir *Abies grandis* Lindl. 14, 109, 134
  diseases and insect pests 109
  silvicultural characteristics 109
  suitable conditions 134
  tallest tree in Britain 109
thinning 299, 309
timber: identification 27; price 432; properties 340; uses 340, 351, 355, 360
grants, forestry
  see agriculture departments, Countryside Commissions, Forestry Commission and Nature Conservancy Councils
grants
  management, planting *see* woodland – Management Grants
  see also coppice, coppice – short rotation, Set-aside, shelterbelt, shelter, woodland – neglected, grants for
greenery for florists' use 191, 360, 441
grey squirrel *see* animals
Grizedale Forest, 536, 582
ground vegetation, flora 20, 87–9
growing-stock 445, 597, 598, 600
growth, tree 68, 180. 347, 348, 375
  'check' 140, 141
  improving 347, 348
  in Britain 323, 347, 348
  in Europe 323
  in Scandinavia 323
  rate 68. 469, 551, 554–7
  season 70, 75
  see also increment
Guelder rose *Viburnum opulus* 119
guidelines for management *see* broadleaves, landscape, recreation, water and wildlife
Guiting, Glos. 257

hardwoods *see* individual species
harvesting 373–89, 391, 394, 395–7
  after windthrow 82, 380–2
  contracts 388, 389
  costs 386, 387
  environmental considerations 395–7
  interaction with restocking 394, 395
  machines, equipment 384–6, 395
  systems: full length 377, 378; full tree 377, 382–3; shortwood 377, 378
  techniques 376–87, 391: conversion 397–402;

equipment 384, 385; extracting 384–6; felling 378–80; whole tree, forest biomass 382–3
see also roads, thinning and transport of timber
Hawthorn, Common *Crataegus monogyna* Jacq. 13, 120
Midland *C. laevigata* 13, 120
Hazel *Corylus avellana* L. 13, 118, 268, 346, 352, 353, 355, 417
health and safety 584, 585
see also Forestry Commission and Forestry Safety Council
health factors (trees) see trees – health
health of trees, surveys see plant health
heathland, heath, heather 181; see also 'check' of spruce in heather
hedgerows, farm 206–8
hedgerow trees 206–8
heeling-in (sheughing) see planting
helicopter, use 182, 222, 241
herbicides 175–8
high forest, extent 144; see also silvicultural systems
Highland Forest Products plc 369, 437; see also Fig. 22
Hiley, W.E., CBE 255, 256, 597
history of forestry in Britain 321
Hockridge, Herts. 257
Holly *Ilex acquifolium* L. 13, 30, 120, 346
Holm oak see Oak
Home Grown Timber Advisory Committee 331, 588
Hoppus system of measuring timber, see measurement, forest

Hornbeam *Carpinus betulus* L. 13, 29, 115, 355, 417
Horse chestnut see Chestnut, Horse
Horticultural Trades Association 8
hunting 507; see also animals and sporting
Huntley Woodlands 110 and Plate 18
Hutt, P., MBE see Bradford-Hutt Silvicultural Plan (Tavistock)
hybrid trees see Hybrid larch, Leyland cypress, and Poplar,
hybridization 12, 107, 111, 112
Hybrid larch *Larix* x *eurolepis* Henry. 12, 107, 132
diseases and insect pests 107
rotation 299
silvicultural characteristics 107
suitable conditions 132
thinning 299, 309
timber: identification 26; markets 414; price 432; properties 338; uses 338, 353, 354, 360, 363

identification
sawnwood 25
tree species 15–20
timber/wood 20–30
various 20
Iggesund Paperboard Ltd 369, 437; see also Fig. 22
imports of timber (1989) 327
impregnation of timber see preservation of timber
improvement of forest trees see genetics
income from forestry see economics, profitability and taxation of woodlands

increment, timber 446, 447, 597, 601
current (CAI) 445, 446, 447
mean annual (MAI) 445, 446, 447
periodic (PAI) 445, 446, 447
percent 529
volume 446
see also measurement, forest
Indicative forestry strategies 9, 580
industrial areas, planting in 208, 209
industries, timber/wood 324–32, 367–71, 436–9; see also Figs. 21 and 22
employment see employment
forestry *passim*
processing of small diameter round wood 368: investment 324; wood-based panel boards 369–71, 439; see also fibreboard, particleboard (chipboard), pulpmills and sawmills, sawmilling, estate
inflation see economics
infrastructure, forest 603
insect pests, damage, control
forest: ash scale *Pseudochermes fraxini* 227; ash bud moth *Prays fraxinella* 113; black pine beetles *Hylastes* spp. 56, 125, 167, 183, 225; clay-coloured weevil *Otiorhynchus singularis* 56, 225; Douglas fir seed wasp *Megastigmus spermotrophus* 62; Dutch elm beetle, large *Scolytus scolytus* 230, small *S. multistrictus* 230; felted beech coccus *Cryptococcus fagisuga* 227, 230; fox-coloured

sawfly *Neodiprion sertifer* 225; great spruce dark beetle *Dendroctonus micans* 226, 227, 228; green spruce aphid *Elatobium abietinum* 108, 199, 225; larch woolly aphid *Adelges laricis* 56, 225; large larch bark beetle *Ips cembrae* 226; large pine sawfly *Diprion pini* 225; large pine weevil *Hylobius abietis* 56, 125, 167, 183, 225; nut-leaved weevil *Strophosomus melanogrammus* 225; oak leaf roller moth *Tortix viridana* 62, 112, 226; osier weevil *Cryptorhyncus lapathi* 287; pineapple gall woolly aphids *Adelges* spp. 199; pine beauty moth *Panolis flammea* 225; pine looper moth *Bupalus piniaria* 225; pine shoot beetle *Tomicus piniperda* 226; pine shoot moth *Rhyacionia buoliana* 226; *rhizophagus grandis,* a predatory beetle 227, 228; spruce sawfly *Gilpinia hercyniae* 226; web-spinning larch sawfly *Cephalcia lariciphila* 226; weevils *Pissodes* spp. 226; woolly aphid *Adelges cooleyi* 56, 108
nursery: cockchafer *Melolontha melolontha* 55; cutworm 55; springtail *Bourletiella hortensis* 56; *see also* seed
insecticides 175, 227
Institute of Agricultural Engineers 377
Institute of Chartered Foresters (ICF) 6, 583, 599, 631

Institute of Terrestrial Ecology (ITE) 586
Institute of Wood Science (IWS) *see* Appendix
Institute of Virology and Environmental Microbiology 226, 586
Institution, Royal of Chartered Surveyors 7
institutions, financial 5, 614, 615
insurance premium 190, 242 valuation *see* valuations
internal trading, transfers 593
investment in forestry 524, 525, 612–31
appraisal 533–48, 547, 548
commercial case for 619
future 630–1
management 621
objectives 525
selection 616, 617
specific objectives 620
syndicates 616
taxation *see* taxation of woodlands
types 617, 618: bare plantable land 617, 619; stocked woodland 618
types of investor 629–30
investment in processing industries 324, 326
investors 628–30
co-ownership 629
individuals 628
institutions, financial 614, 615, 629
National Trust 629
overseas 630
RSPB 629
wood processors 616, 629, 630
Woodland Trust 486
in-vitro propagation 589
Ipsden Estate, Oxon. *see* Reade, M.G. irregular (uneven-aged) structure silviculture *see* silvicultural systems

Japanese larch *Larix kaempferi* (Lambert) Carr. 14, 107, 132
diseases and insect pests 107
rotation 298
silvicultural characteristics 107
suitable conditions 132
thinning 298, 309
timber: identification 26; markets 414; price 432; properties 338; uses 338, 353, 354, 360, 363
Japanese paper pots (JPPs) *see* plants
Japanese red cedar *Cryptomeria japonica* (L.f.) Don. III
Jones, E.W. 248
Juniper, Common *Juniperus communis* 13
juvenile foliage 16
wood 3

kiln, drying 328
knots *see* pruning and wood
Kronospan Ltd. 369, 438; *see also* Fig. 22

labour 465, 537, 595
contract 465
cost 537
in nurseries *see* nursery
in sawmills *see* sawmills
oncosts 537, 594
piece-work 596
productivity 377
rates, piece-work *see* work study
wet weather work 64, 70
Lake District 73; *see also* Grizedale Forest and Thirlmere, Lake
Lammas shoots 54, 226
land, forestry
expectation value *see* land expectation value
market price 8, 625
plantable 622–31
reclamation and forestry

208, 209
use, capability for forestry *see* agriculture and Set-aside
land expectation value (LEV) *see* economics Faustmann formula
landscape, economics of 483, 536
 guidelines: management 476–83; planning 479–83
 Landscape Institute 483
Larch *see* European, Hybrid, and Japanese
Lawson cypress *Chamaecyparis Lawsonia* (A. Murr) Parl. 14, 109
 silvicultural characteristics 109
 thinning 299, 309
 timber: identification 27; markets 415; price 432; properties 340; uses 340, 355
lease, feu, woodland, 623
Lebanon, cedar of, *Cedrus libani* 14, 27, 111, 341
legislation 175, 227, 230, 232
 chemicals 227
 Control of Pesticide Regulations (1986) 175, 227
 Control of Substances Hazardous to Health (COSHH) 175
 herbicides 175, 227
 insecticides 175, 227
 pesticides 175, 227
 plant health 51, 59, 224
 see also Acts of Parliament (Statutes) and felling
legumes 209
Leighton estate, Powys 110, 111; *see also* Plates 20 and 21
Leyland cypress x *Cupressocyparis leylandii* (Jacks and Dallimore) 12, 111
licence, felling *see* Forestry Commission
light requirement of trees 75, 99, 104, 105, 112
lightning damage 75
Lime, Common *Tilia* x *europaea* L. 13, 30, 118, 344, 354, 356, 417
 large-leaved *T. platyphyllos* Scop. 13, 30, 118, 344, 354, 356, 417
 small-leaved *T. cordata* Mill. 13, 30, 118, 344, 354, 356, 417
Lines, R., OBE 126–9
lining out *see* transplanting
loans for forestry 631, 632
local authorities 7, 581
Lodgepole pine *Pinus contorta* Douglas 14, 106, 131, 299, 309
 diseases and insect pests 106
 silvicultural characteristics 106
 suitable conditions 131
 timber: identification 26; markets 414; prices 432; properties 336; uses 336, 351, 352, 353
logs *see* sawlogs and timber – firewood
Long Ashton Research Station 270, 287
Longleat Woodlands Estate, Wiltshire 110, 283–5: *see also* Plates 8 and 9
lop and top 358, 380, 420
Loughgall Horticultural Centre 270

Macaulay Land Use Research Institute, Aberdeen 95–7, 129
machinery 386, 395, 399, 400, 465, 466, 457, 595
sawmilling *see* sawmills, sawmilling, estate
mammals *see* animals
management companies *see* forest(s) – management companies
management, forest 461–521
 broadleaves 468, 469, 473, 474
 charges 466, 467
 choice of forest intensity 461–3
 deer management 509
 forest investments 621
 grants *see* grants
 guidelines *see* broadleaves, landscape, recreation, water and wildlife
 including non-wood benefits 471–6
 landscape 477–84
 Management Scheme 463, 464: Financial Plan 466; Operations Plan 464; Organization Plan 464–6
 multi-purpose 471
 primarily for wood production 467–71
 recreation 485, 486
 sporting, game, 504–10
 water 502–4
 wildlife conservation 485–501
Management Plan *see* management, forest
management tables *see* measurement, forest
Maple *see* Norway maple and Field maple
maps, aerial 455
 drift 90
 geological 90
 land capability for forestry 95–7
 ordnance survey 90, 455, 458
 rainfall 72
 soil 90
 stock 455
 wind zonation 84
 woodland 455
Maritime pine *Pinus pinaster* Ait. 14
marketing, timber 355–8, 360–3, 368, 411–25

agreement for sale and purchase
broadleaves 355, 356, 415, 416
charges 424
conifers 414, 415
continuity 412
contracts for sale and purchase 424
coppice 356, 417
determinants of sale value 411
Forestry Commission assistance to private sector 389, 411
Forestry Commission marketing policy, 418
grading of roundwood market-responsive cutting 376
methods 423, 424
particulars and conditions of sale 423, 424
price-responsive cutting 376
sales of timber 419: converted 420; felled 420; standing 419
sawlogs 413–17
seeking markets 413–17
small diameter roundwood 355–8, 360–3, 414–17
small thinnings 355, 356, 414–17
underwood 355, 356; *see also* prices, price-size relationship, price-size curves and sales markets for home-grown timber 356, 412–17; *see also* marketing and individual species
market values *see* valuations, forestry
Matthews, J.D., CBE 248
McHardy, J. (Longleat Woodlands) 283–5
mensuration *see* measurement, forest
measurement, forest 31–9, 380
area 31
bark allowance 37
basal area 32, 33, 295
branchwood 38
breast height 32
conversion factors 31, 37, 38
cord 38, 357
diameter 32
form factor 34
height, tree 32
Hoppus 39, 403, 431
increment, annual *see* increment
relascope 32, 295
sample plots 32, 36, 590
stacked timber 38
taper 34
tariffing 30
tariff tables 37
timber height 32
top height 32
total height 32
volume, felled 37: standing 35
volume/weight 31, 33, 34, 38, 39
weight 31, 33, 34, 432
Weise's rule 32
see also management tables, yield class and yield tables
mechanization in forestry *see* forestry and machinery
in nurseries *see* nursery
merchants, roundwood *see* timber
Méthode du Controle 253, 256
microclimate, forest 69
Mitchell, A.F. 11, 13, 14, 15
mixed *v.* pure crops 138–43
mixtures 138–43
broadleaved 142
broadleaved with conifer 142
components 141
conifer 141
patterns 140
self-thinning 139, 319, 320
treatment 139, 143
types 141
moisture *see* soil
money, changes in value *see* inflation
monocultures 180
Monterey cypress *Cupressus macrocarpa* 14
Monterey pine *Pinus radiata* D. Don 14, 68
Moore, Major General D.G. 318
Mountain ash (Rowan) *Sorbus aucuparia* L. 13, 119
mulching *see* weeding of plantations
myxomatosis *see* rabbits
mycorrhizae 45, 91

National Agricultural Centre 519
National Committees, Forestry *see* Forestry Commission
national forests *see* Forestry Commission
national nature reserves 147
national parks 147
National Trust 256, 491, 521, 628, 629
native (indigenous) trees 12, 13, 14, 121, 143
natural pruning *see* pruning
natural regeneration 257, 258, 275–85, 488
advance 277
broadleaves 276, 277, 281
conifers 281–4
density 282
economics 284
examples in Britain 279–80
grants 566
management 257, 257
overstorey 278
preparation 276
respacing 279
tending 279 uplands: broadleaves 281; conifers 281–4
natural woodland 100
management 262–4
'naturalness' 100, 147, 262–4

Natural Environment Research Council (NERC) 586
nature conservation *see* conservation – wildlife
Nature Conservancy Council for England (NCCE) 7, 121, 146, 499, 567, 570
Nature Conservancy Council for Scotland (NCCS) 7
nature reserves 9, 147, 489
New Forest, England 275, 582
New Midlands Forest 514
New National Forests 514
new planting *see* afforestation
nitrogen
 deficiency 76
 fixing 91, 209
Noble fir *Abies procera*, Rehd. 14, 110, 134, 309
 silvicultural characteristics 110
 suitable conditions 134
 timber: identification 27; markets 414; price 432; properties 341; uses 341, 351, 354
non-wood, non-market costs and benefits 534, 536, 548
normal forest 273, 528, 529
normality 273, 274
 the private owner 274
Norway maple *Acer platanoides* L. 13, 29, 115, 354
Norway spruce *Picea abies* (L.) Karst. 14, 108, 133
 Christmas trees 133
 diseases and insect pests 108
 rotation 299
 silvicultural characteristics 108
 suitable conditions 133
 thinning 299, 309

timber: identification 27; markets 414; price 432; properties 338; uses 338, 351, 352, 353, 354
Nothofagus spp. *see* Southern beech
nurse, trees, crops 139
nursery, forest tree 41–57
 costs 42
 diseases 53–5
 estate 41
 fallows 43, 53
 insect pests 55–7
 irrigation 45, 47
 labour requirement 42
 literature 42
 mechanization 46
 mycorrhizae 45
 nutrients 42, 43
 practice 41–7
 precision sowing with undercutting 48
 preparation 44
 problems 53–6
 pros and *cons* 41, 57
 protection 45, 48, 53–6
 seedbeds 44
 site 43
 size 42, 43
 soils 43
 stock, of plants taxation *see* taxation of woodlands
 trade 5, 41
 transplanting (lining-out) 46, 47
 undercutting 42, 46
 water supply 45, 47
 weeding 44, 52, 53
 whether to start a nursery 41
 see also cold storage of plants and seeds, plants, poplar, seed, seedbeds and willow
nutrients, nutrition 76

Oak, Holm *Q. ilex* L. 13, 29, 117
Pedunculate *Quercus robur* L. 13, 112, 135, 342
Red *Q. rubra* du Roi. 13, 29, 114, 346

Sessile *Q. petraea* (Mattuschka) Liebl. 13, 112, 135, 342: barkstripping, tanbark; 'Brown' oak timber 25, 435; diseases and insect pests 62, 112, 218; hybridization 112; natural regeneration 280; rotation 300; 'shake' in 334, 435; silvicultural characteristics 112; stem rot 218, 229; suitable conditions 135; thinning 300, 309; timber: identification 28; markets 415; price 435, 436; properties 342; uses 342, 351, 352, 353, 354, 355, 356, 359
Turkey *Q. Cerris* L. 13, 29, 115, 346
objectives in woodland management 525; *see also* Britain and Forestry Commission
'oceanic' forestry 317, 318
open-cast mining areas, planting of 208, 209
operational costs *see* forestry
Organization Plan *see* management, forest
origin/provenance, tree 63, 68, 116, 126–9, 131–6
osiers *see* basket willow coppice
output guide *see* work study
over-crop, overwood, overstorey 278
overheads 537, 595
 labour 537, 595
 machinery 595
 materials 595
owners of woodlands 1, 5, 6, 321, 613, 616, 620
 management by 621
 types 628–31
Oxford University 256
 Commonwealth

Forestry Institute,
Forest Economics
Department 136, 583:
forestry projects and
research 136, 556,
557, 558, 586, 587;
survey of Private
Forestry Costs and
Returns 537, 549

paper and paperboard
324, 352, 353, 358,
368, 369; *see also* Fig.
22
imports 327
timber: consumption
327, 331; potential
supply 331; prices
436, 437; processes
368; species 353, 436,
437; specification 436,
437; supply troughs
and peaks 326, 368
see also pulpmills and
pulpwood
Parmoor Estate, Oxon.
256, 257
particleboard (chipboard)
324, 353, 358, 369,
370, 371, 376, 437,
438; *see also* Fig. 22
imports 327
timber: consumption
327, 331; potential
supply 331; prices
4437, 438; species
353, 437, 438;
specification 437, 438;
supply troughs and
peaks 326, 368
Patten, J., water colours
by 18
Peck, T.J. (ECE/FAO)
638
Pedunculate oak *see* Oak
– Pedunculate
peeling (of bark) 335,
356, 399
Penistan, M. 256
pesticides 175, 227, 584
nursery 55
pheasants and forestry *see*
shooting
Pine *see* individual
species

pinetum *see* Bedgebury,
pinetum
pioneer species 139
pit-props (round mining
timber) *see* timber
pitwood *see* timber
plan of operations *see*
management, forest
Plane, London *Platanus*
x *hispanica* Muenchh.
(P. x *acerifolia)* 13, 29,
117, 344, 352
planning, forest 445–59
data processing and
collection 455, 456
labour *see* labour
machinery *see*
machinery
sub-compartment notes
456–8
plantations 149, 150,
180–2, 213–45
after-care (tending)
213–45
costs 186, 187, 243,
244, 245
establishment 149, 150,
213–45
plants (forest trees)
165–8, 541
age 165, 166
bare-rooted 166
British Standard 41,
165
container grown
seedlings 49, 106,
166, 168
Cell Grown Plant
Producers Association 8
'Ecopots' 49
Japanese paper pots 49,
166
costs 167, 168, 189
covert 20, 506
damage *see* animals,
birds, diseases and
insect pests
despatching 166
dormancy 170
dipping in sodium
alginate 167
'EC Species' 57, 59
export regulations 224
Forest Reproductive
Material Regulations
57, 59

genetically-improved
Sitka spruce 49, 166
handling 166, 169
import regulations 224
poplar 165, 168
prices 165–8
production *see*
nursery propagation *see*
propagation, vegetative
provenance/origin 165
root growth potential
(RGP) 165
rootrainers 49, 166
sheughing (heeling-in)
170
storage 167; *see also*
cold storage of plants
and seeds
stumped back 169
treatment against insect
pests 167
type 166
undercut 166
'whips' 168
willow 165
see also nursery and
individual species
plant health 76–9, 224
directives *see* European
Community
planting (forest trees)
168–71, 541
cost 189
espacement 162–4, 541
grant-aid *see* grants
upper limit 448
manual 169, 170
mechanical 170
plant handling 170
'Pottiputki' planting tool
170
quality 171
sheughing (heeling in)
170
stock *see* plants
time of planting 170,
171
upper limit 448
ploughing *see* preparation
of ground for
planting
policy, forest *see*
Britain and Forestry
Commission
pollarding 286
pollination *see* genetics

in silviculture
pollution, air 76–9
  water 73, 74
Poplar 20, 116, 137, 191–8
  Aspen *Populus tremula* L. 13
  'Balsam spire' *P. trichocarpa* x *balsamifera* (TT32) 14, 192
  Belgian varieties, new 192
  Black *P. nigra* L. 13, 192
  Black cottonwood *P. trichocarpa* 14, 192
  British Standard 165
  canker 51, 117, 192, 193
  costs 197
  damage by roots 156
  diseases and insect pests 51, 117, 193, 230
  Grey *P.* x *canescens* 13
  Lombardy *P. nigra* 'Italica' 13
  nursery practice 50, 51
  pruning 193, 194
  Robusta *P.* x *euramericana* 14, 192
  roots 156
  rotation 193
  Serotina *P.* x *euramericana* 13, 192
  silvicultural characteristics 116
  spacing 192, 193
  suitable conditions 116, 137, 191
  timber 196, 197: identification 30; markets 196, 416; prices 197, 435; properties 345; uses 196, 345, 352, 353, 354
  White *P. alba* 13
  yield class 193, 195, 196
preparation of ground for planting 150–4, 541
  chemical 152
  costs 188
  cultivation 151
  mounding 146, 153
  ploughing 151–4, 541; see also Plate 5
  ripping 154
  scarifying 154
  subsoiling 154
preservation of timber/wood 405–9
  copper chrome arsenic (CCA) 407
  creosote 407
  decay in service 405
  hot and cold treatment with creosote 407
  plant 407
  pressure 407
  selection of treatment 408
prices Christmas tree 200
  cricket-bat willow 204
  land 8, 622–31
  minor woodland products 439–43
  plants, forest tree 167, 168
  roundwood see roundwood
  stumpage 430
  timber 425–43: boardwood 437, 438; broadleaved 433–43; conifer 432–43; future 425, 426, 427; pulpwood 436, 437; roundwood 433–43; sawlogs 432–4; world 425, 426
  veneer logs 434, 435
  woodland, stocked 8, 622–31
  see also individual species
price–size relationship, roundwood 427–30
price–size curves, roundwood 427–30
  broadleaves 429, 430
  conifers 428, 429
private sector forestry *passim*
  achievements 2, 613
  case for 323, 612–31
  changes in 4, 563, 564
  controls and restraints 579–83, 612
  employment 1, 322
  extent of woodlands 1, 321, 322
  future roundwood production 330: achieving 330; forecasting 330; see also forestry
  government policy 4, 5, 582
  grant-aid 565–70
  investment 612–31
  ownership 1, 5, 6, 321, 613, 616, 620
  partnership with State forestry 1, 612
  productive woodlands 321
  profitability 524, 525, 548–59: scarcity of data 549
  role 2, 613
  sales of woodlands 624
  support received 4, 322, 563–70, 582–90, 612
  taxation 4, 565, 570–9
  unproductive woodlands 321
processing industry 436–9
  investment 324, 326
  'knock-on' effects 321; see also Fig. 21
production class 448, 449
  forecast tables 455
productivity see labour and work study
profitability of forestry 469, 521, 528–33
  broadleaves 469, 551, 553, 554
  conifers 550, 551–3
  forest rent 528
  Forestry Commission see Forestry Commission
  maximization see economics
  measures 528–33
  net annual income 528
profitability of sawmilling 364
Project Sylvanus 479, 487, 488
propagation, vegetative 50, 51, 52, 55, 66, 589; see also genetics

in silviculture and nursery
provenance, tree *see* origin/ provenance
pruning 137, 217–19, 470, 543
 artificial 137, 217–19
 broadleaved ornamental timber trees 190, 470
 broadleaves 137, 218, 470
 conifers 217
 costs 244
 cricket-bat willow 219
 economic, evaluation 218
 formative 213
 natural 217–19
 poplar 137, 219
public access *see* access, woodland
pulpmills 324, 326, 327, 367, 368, 369, 436–7; *see also* Fig. 22
 imports 327
 processes 352, 368
 timber: consumption 327, 331; potential supply 331; prices 436, 437; species 352, 436, 437; specification 436, 437; supply troughs and peaks 326, 367, 368
pulpwood 324, 322, 327, 352, 353, 358, 436–7
 exports 322, 327, 369
 imports 327
 processes 352, 368
 timber: consumption 327, 331; potential supply 331; prices 436, 437; species 436, 437; specification 436, 437; supply troughs and peaks 326, 367, 368
pure crops 138–43
pure *v.* mixed crops 138–43
Pusey Wood, Oxon. 254
Pyrok Manufacturing Ltd 369, 438; *see also* Fig. 22

Queen Charlotte Islands, British Columbia, Canada 65, 133, 134, 139

rabbits *see* animals
railways *see* transport of timber
rainfall *see* climate
Ramblers' Association 485, 622, 627
Rankin, K.N. 614
rates 578
 derated (woodlands) 578
 sporting 578
Reade, M.G., Ipsden Estate, Oxon. 254, 256
reafforestation *see* restocking
reclamation, land, planting of 91
recreation, forest 121, 485, 486
 economics 536, 636
 guidelines for management 485, 486
 valuation 536
Red oak *see* Oak
red squirrel *see* animals
Redwood *see* Coast redwood
 Dawn *see* Dawn redwood
regeneration
 artificial *see* planting and direct seeding
 natural *see* natural regeneration
Register of Ancient Woodlands (NC 489, 567, 625
regular (e enaged) silviculture *see* silvicultural systems
repellant, ch mical 231
replacement of unsatisfactory crops 545
research, forest 585–90; *see also* Building Research Establishment, Forestry Commission, Institute of Terrestrial Ecology and universities

research, wood 586, 587
residues, wood *see* wood
respacing of natural regeneration 279, 282
 upland conifers 316–18
restocking, reafforestation 125, 126, 149, 155, 182–5, 322, 394
 interaction with harvesting 394, 395
 operational costs 188–90
 recent extent 322
 restocking after storm damage grant 566
restraints on forestry 9, 579–83
revenues 537
Rhododendron ponticum 152, 177, 506
roads, forest 242, 243, 245, 389–94, 546
 costs 392, 546
 residual value 547
 specification 392
 traffic on highways 387, 393, 394
roll-over relief of capital gains on land *see* taxation of woodlands
roots, tree 156, 163, 165, 182
rootrainers *see* plants
Roslin, Forestry Commission Northern Research Station 224, 587
Rossie Priory near Dundee 254
rotation 373–6, 544, 545, 553, 554
 constraints, restraints 376
 financial 374
 highest net annual 373
 length 373
 maximum mean annual volume 374
 market-responsive cutting 376
 optimum 544
 physical 373
 price-responsive cutting 376
 second 373
 silvicultural 373
 technical 373

see also basket
willow, coppice
– short rotation,
felling decision and
individual species
roundwood
achieving production
331, 332
conversion 439–43
demand forecast 331
export 327; *see also*
pulpwood and veneer
logs
production forecast 329,
330, 331
supply forecast 329,
330, 332
see also individual
species and timber
Rowan *see* Mountain
ash (Rowan) *Sorbus
aucuparia* L.
Royal Agricultural
College, Cirencester 18
water colour tree
paintings *see* Patten, J.
Royal Forestry Society
of England, Wales and
Northern Ireland 6
Royal Institution of
Chartered Surveyors
(RICS) 7, 583
Royal Scottish Forestry
Society 6
Royal Society for Nature
Conservation 7
Royal Society for the
Protection of Birds
(RSPB) 7, 491, 502,
628, 629
rural crafts 268, 355,
398, 399, 461

safe working methods 7,
400, 401, 584, 585
Forestry Safety Council
(FSC) 7, 584, 585
Health and Safety at
Work etc. Act (1974)
584
Safety Guides, Forestry
584, 585
St. Regis Paper Co. Ltd
369, 437; *see also* Fig.
22

sales
timber 411–20: charges
424; conditions
423–5; contracts
424; converted 420;
felled 420; grading
see grading; lotting
421; methods 422–5,
auction, public 414,
422, 423, negotiation,
private treaty 422,
423, tender 422,
423; particulars
423–5; points 419–21;
standing 419, 420;
weight 420
woodlands 622–4
Sallow (Goat) *see* Willow
salt-laden wind damage
48, 77
sample plots *see*
measurement,
forest and Forestry
Commission
sand dunes *see* soil
sawlogs 326, 363, 432–4
broadleaved 328, 358,
363, 433, 434
conifer 328, 363, 432,
433
Forestry Commission
grading of conifers
328
price 434, 435
production 327
specification 432–4
sawmills, sawmilling,
estate 7, 8, 326, 328,
349, 359, 363–7,
402–5
cost 324
economics *see*
economics employment
322
equipment 403, 404
labour 404
portable 403
power 403
profitability 364
recovery percentage
328, 364
residues *see* wood
residues
sawnwood 359
taxation *see* taxation of
woodlands

trade 7, 8
wastage, 359, 403
sawnwood
grading 348, 349
hardwood 359, 360
imports 327
mining 366
softwood 328, 363–7
Scandinavia 425
scarifying, scarification *see*
preparation of ground
Schlich, Sir W. 255, 256
Scotland 1, 7, 321, 499
afforestation restraints
580
forestry 321; *see also*
Caledonian Scots pine
land capability for
forestry 95–7
new national forests
514
Secretary of State 1
soil survey 90
Scots pine *Pinus sylvestris*
L. 13, 105, 130
diseases and insect pests
105
rotation 298
silvicultural charac-
teristics 105
suitable conditions 130
thinning 298, 308
timber: identification
26; markets 414; price
432; properties 335;
uses 335, 351, 352,
353, 354, 361
see also Caledonian Scots
pine (indigenous)
Scottish Forestry Trust
557
Scottish Landowners
Federation (SLF) *see*
Appendix
Scottish Natural Heritage
(SNH) 7
Scottish Wildlife Trust 7
Scottish Woodlands Ltd
413, 616
scrub 144, 149
seasoning of wood *see*
timber and wood
seed (tree) 18, 57–62,
126
bank 88, 487
collection 59, 60

cost *see* prices
damaging agents 59, 62
dispersal, dissemination 18
dormancy 61, 62
Douglas fir seed wasp 62
'EC Species' 57, 59
girdling of trees 65
insects attacking 62
Knopper gall wasp 62
legislation 57, 59
masts, mast years 277
orchards 60, 63, 65
periodicity 277
pretreatment 61
prices 44, 58, 126
provenance/origin 44, 63
processing 61
sowing *see* seedbeds
stands 64, 65
storage 56, 60
stratification 62
suppliers, sources 44, 58: certificate 59; Forestry Commission 44, 58, 127; Forestart (seed firm) 58, 127
testing 59
treatment 61
see also individual species and seedbeds
seedbeds 44, 45, 52
fertilizing 45
gritting 45
pests and diseases 53, 55, 59, 62
post-emergence spraying 53
precision sowing and undercutting 48
pre-emergence spraying 52
preparation 44
protection 45, 46, 48
sterilization 44, 52
sowing methods 45: density 45
undercutting 42
weeding 52, 53
see also seed
seeding, sowing, direct *see* direct seeding, sowing
seedlings *see* plants

selection felling *see* felling
selection system *see* silvicultural systems
selection woodland *see* silvicultural systems
semi-natural woodland *see* ancient semi-natural woodland
Sequoia see Coast redwood
Sequoiadendron see Wellingtonia
Serbian spruce *Picea omorika* Purkyne. 110
Service tree *see* Wild service tree
Sessile oak *see* Oak – Sessile
Set-aside (farmland) 517, 570
sewage sludge 209
shake in oak 334, 435, 553
sweet chestnut 435, 553
sheep *see* animals
shelterbelt, shelter 121, 149, 150, 204–6
cost 205
farmland 204
grant-aid 206
plantation 205
shelterwood system *see* silvicultural systems
shooting 144, 497, 504–10, 593
cooperation with gamekeeper 505, 506
covert plants 60, 506
economics 509
flushing points 506
gamebirds 504, 505
pheasant 505, 506
rough 507
values 505, 506, 509, 510
see also Game Conservancy, sporting, and taxation of woodlands Shotton Paper Company plc 367, 369, 436; *see also* Fig. 22
Silver birch *see* Birch
Silver fir *see* Grand fir,

European silver fir, and Noble fir
silvicultural characteristics 103–20
definition 103
silvics 103
see also individual species
silvicultural systems 9, 248–73, 479, 487, 488
classification 248, 249
conversion 273
coppice systems 249, 264–73: coppice 248, 264–8; coppice with standards 249, 268–70
even-aged, regular, uniform 249, 250, 251
high forest systems 144, 248, 250–5: accessory 249, 255, high forest with reserves 249, 255, two-storied high forest 249, 255; clear cutting (regular) 248, 249, 250; group regeneration 252; group 249, 252; group selection 252, 253; irregular shelterwood 9, 249, 250, 251, 252; selection (irregular) 249, 252, 253; strip 249, 250, 251; uniform 249, 250, 251; wedge 249, 250, 251
single tree system 249, 272
uneven-aged, irregular 9, 249, 251, 255, 262–4; *see also* Bradford-Hutt Silvicultural Plan, Tavistock and Méthôde du Contrôle
silviculture *passim*
single stem plantations 272
singling, formative pruning 213; *see also* natural regeneration
site factors, woodland 68–7

aspect 69
classification 68
climate 69–87
  cold 69, 70
  conservation of drought 69, 70
  elevation/altitude 69
  exposure (Topex) 85
  frost 69, 70
  geology 90
  heat 69, 70
  microclimate 69
  moisture 71
  rainfall *see* climate
  rewetting 155
  temperature 69, 70
  topography 69
  vegetation 87–90
Site of Special Scientific Interest (SSSI) 7, 14, 147, 489
Sitka spruce *Picea sitchensis* (Bong.) Carr. 108, 133
  diseases and insects pests 108
  genetically improved 49, 66, 108
  natural regeneration 282
  rotation 299
  silvicultural characteristics 108
  suitable conditions 133
  thinning 299, 309
  timber: identification 27; markets 414; price 432; properties 337; uses 337, 351, 352, 353, 354, 364
small diameter roundwood *see* roundwood
smoke, damage *see* air pollution
softwoods *see* individual species
soil 88–98, 138, 139, 181, 257
  acidification 92
  brown forest 91, 94
  calcareous 88, 92, 94
  chalkland 88, 92, 94
  clay 92, 94
  classifications/groups 91–3, 95–7
  depth 90, 104

deterioration 98, 138, 139
drift 90
fertile 89, 90, 92
gley 92, 94
greensand 98, 267, 283
heathland, heath 92, 179, 184
horizons, profiles 90, 92
humus 89
infertile 89
ironpan 90, 92, 94, 152
land capability for forestry 95, 129
lime, limestone 88, 92, 94
littoral 95
map 90, 95–7
modification 98
moisture 88, 90
moorland 93
mor 92
mull 91
mycorrhizae 91
nutrient status 90
pan *see* soil – ironpan
peatland 89, 92, 94
pH values 92, 136, 137
plants indicative of 87–9
podzol 92, 94
properties 90
protection 99
reclaimed 90
rendzina 92
sand, sand dunes 89
surveys 90, 97
tree relationships 97, 98
undeveloped 93, 95
wet 92
see also geology and nursery
soil requirements of trees *see* individual species
South Africa 68
Southern beech, Roble *Nothofagus obliqua* (Mirb.) Bl. 14, 29, 115, 116, 303, 345
Raoul *N. procera* (Poepp. & Endl.) Oerst. 14, 29, 115, 116, 303, 345

spacing, espacement 87, 162–4, 541
  final crop
  narrow (close), initial 162
  plantation (initial) 162–4
  wide, (initial) 162
  see also nursery and respacing
species choice 87–9, 91–3, 120–37, 540
  biological constraints 120, 137
  broadleaves 135–7
  conditions justifying selection 130–7
  conditions unsuitable 130–7
  conifers 130–4
  grey squirrel restraints 121
  lowlands 123
  mixtures *see* mixtures
  objectives 120, 126–9, 137
  origins/provenances 63, 68, 116, 126–9, 131–6
  owner's restraints 120, 122
  pure crops
  site factors influencing 120, 123
  special crops 190–204: agroforestry 516; Christmas trees 198–202; decorative quality broadleaved timber 190; ornamental (decorative foliage) 191; poplar *see* poplar
  timber production 122
  uplands 123
  see also individual species, recreation, shelter, sporting and conservation – wildlife
Spindle tree *Euonymus europaeus* 118
spoil-heap, colliery, planting of 208, 209
sporting in woodland 121, 504–10
  crediting to woodland accounts 510

economics 509
management 144, 504–10
rates on *see* rates
taxation *see* taxation of woodlands
value 505, 506, 509, 510
see also shooting
Spruce *see* Norway, Sitka and Serbian
squirrels *see* animals
standards, tree *see* coppice with standard; and British Standards
State forestry *see* Forestry Commission
Stenner of Tiverton, Ltd 405
stocking density of crops 305; *see also* maps and growing-stock
stools, stoolbeds 50, 51; *see also* coppice
storage *see* cold storage of plants and seeds
storm damage *see* gale/storm damage and wind, windthrow
strip system of regeneration *see* silvicultural systems
stumpage (standing price) 430, 538, 600
sub-compartment, woodland 456–8, 594
subsidies *see* grants
successor/succession crop 545
sunscorch *see* climate
sustained yield *see* yield
Sweet chestnut *see* Chestnut, Sweet
Switzerland 253
Sycamore *Acer pseudoplatanus* L. 13, 113, 136
 diseases and insect pests 113
 rotation 302
 silvicultural characteristics 113
 suitable conditions 136
 thinning 302
 timber: identification 28; markets 415; price 434; properties 242; uses 324, 352, 353, 354, 359
sylvan heritage 1, 11, 12
syndicated forestry 2, 616, 629
 see also County, National and Woodland Trusts

tanbark, tannin 356: *see also* bark stripping
tanneries 357
Tavistock Woodlands Estate *see* Bradford–Hutt Silvicultural Plan and Plate 6
taxation of woodlands 4, 538, 565, 570–9, 609
 amenity woodland 520
 budget 1988, 563, 571
 capital expenditure allowance 572
 capital gains tax on land 573: rollover relief 573
 capital transfer tax 575
 Christmas trees 577
 corporation tax 571
 effect on values of woodlands 609
 estate duty 577
 income tax 571–73
 inheritance tax 573–7: Business Relief 575; Heritage Relief/ status 11, 12, 576
 nurseries 577
 recreation 577
 sawmills 402, 578
 sporting 577
 see also rates and Value Added Tax (VAT)
telegraph poles *see* transmission poles
tending (after-care) of plantations 213–45
terrain 386, 390
 classification 390
Thetford Forest England 536
thinning(s) 291–320, 544
 broadleaves 300–3, 304–7
 chemical 318, 319
 chevron 308
 classes 291
 classification of trees 292, 293, Fig. 20
 conifer 298, 299, 307
 control 315, 316
 cost of 319, 380
 crown 292, 294, 296, 298
 cutter selection 319
 cycle 294, 297
 economics 292, 310–12
 effect on growth 292
 first commercial 297, 307, 308, 309
 Forestry Commission practice 312–14
 influence: on girth 291; on timber quality 292
 intensity 294, 544: marginal 451
 line 308, 310
 low 292, 294, 296, 298, 307, 308
 markets for, *see* marketing marking 297, 319
 mixed plantations 304–7
 non-thin regime 313
 options 544
 periodicity 310
 practice 312–14
 regimes 291, 292–6
 respacing 316–18
 self-thinning mixtures 319, 320
 systematic 308
 timing of first thinning 307, 308
 traditional 296–303
 type 294
 volumes *see* management, forest and yield tables
 wind, effect 312, 314
 see also individual species
Thonock, Lincs. 257
Thirlmere, Lake 73
Tilhill Economic Forestry 616
Tilhill Forestry 616
timber, including properties 332–46
 achieving production 331

'bars' 363, 432
biomass 270–3, 382, 383
boardwood *see* fibreboard and particleboard (chipboard)
branchwood 38, 452, 469
broadleaves (hardwoods) 28–30
cement-bonded panels 353, 367, 370, 371, 439
conifers (softwoods) 26, 27
consumption, demand, 1, 326, 327
conversion *see* conversion
decorative 347
density 34
diseases, rot 25, 347
drying, seasoning 432
durability, natural 334, 347
exports: pulpwood *see* pulpwood; veneer logs *see* veneer
features 26–30
firewood 326, 327, 357, 439
fluting 334, 347
forecasting production 331, 332
future 326, 327, 329
grading: broadleaves 363, 412, 421; conifers 349–51, 422
hedgerow 207, 518
identification 20–30
improving quality 347
knots *see* wood
marketing *see* marketing
markets *see* individual species and marketing
measurement *see* measurement. forest
mining 366; *see also* pitprops
moisture content 33, 332
moulded-board 353, 367, 370, 371, 439
permeability, preservative treatment *see* wood

prices: roundwood 430–3; sawlogs 433, 434; veneer logs 434, 435
price-size curves *see* price-size curves
production 326, 327, 331, 332
properties *see* wood
pulpwood *see* pulpwood
sales *see* sales sawlogs 25
sawnwood 25, 359
shake 334, 347
straightness 347, 350
strength 347, 348
supply 1, 326, 327, 331
uses, utilization *see* utilization/uses of timber/wood
weight 347
wet storage 381
windblown 380–2
wood-based panel board *see* fibreboard and particleboard (chipboard)
wood wool 353, 370, 371
world prices, trends 425, 426
valuation *see* valuations *see also* biomass, individual species and veneer logs
Timber Growers U.K. Ltd (TGUK) 6, 330, 360, 413, 430, 499, 557, 631
Code 471, 485
Timber Grower (TGUK) 430
Timber Research and Development Association, Ltd (TRADA) 8, 21, 22, 332, 348, 349, 350, 409, 586
Timber Trade Federation 361
Timber Trades Journal 413
tissue culture 589
tools 379, 398
topex values 83, 85
Torvale Building Products

Ltd 439
training *see* education and training
transmission poles 360–2
electricity 361
preservation treatment 361
telegraph 361
transplanting (lining-out) *see* nursery
transplants *see* plants
transport of timber 387–9
costs 388, 389
road 387
rail 387
sea 387, 388
traffic on highways 387, 393, 394
'tree line' 69
upper planting limit 448
trees
angiosperms 12
biomass 382–3
branching habit 17, 19, 111
branch persistency 105, 112
breeding 23. 62–6; *see also* genetics in silviculture champion, monarchs 11, 15, 283
Christmas 198–202
classification, types 12, 293
damage to *see* air pollution, animals, birds, climate, diseases, fire, insect pests and wind
dangerous 207
defects *see* wood
diseases *see* diseases
exotic (introduced) 13, 14
farm 207, 518
form and outline 19
frost-hardy 105, 112
frost-tender 105, 112
growth rates 68, 445, 451
gymnosperms 12
hazardous 207
health 76–9
identification 15–20

improvement 62–6
insects harmful to *see* insect pests
light requirements 99, 104, 105, 112
'line' 69
native (indigenous) 12, 13
nurse 139
pioneer 139
'plus' 64, 65
poisonous *see* yew
relationship with soils 97, 98
ring dating *see* dendrochronology
silvicultural characteristics 103–20
silvics 103
shade, tolerance of 75
soil depth required 104
surgery 61, 207
surveys 207
taper 38
utilization *see* utilization/uses of timber/wood
valuation *see* valuation
wind-firm 105, 112
windthrow, harvesting of 382–3
wind-weak 105, 112
'wolf' 293, 306; *see also* Fig. 20
see also individual species, plants, seed, seedbeds, timber and wood
Tree Council 609, 610
Tree Preservation Order (TPO) 7, 145, 563, 581
treeshelters 171, 172, 187–9
Troup, R.S. 248, 255
Trusts *see* County, National and Woodland
Turkey oak *see* Oak – Turkey
turnery, turneries 357, 433
two-storied high forest *see* silvicultural systems

underplanting 139
see also enrichment
understorey 301, 302
underwood 268
hazel 268
utilization 268
workers (cutters) 268, 398
valuation 608
uneven-aged (irregular) structure silviculture *see* silvicultural systems
uniform system *see* silvicultural systems
United Kingdom and Ireland Particle Board Association 7
United Kingdom Softwood Sawmillers Association (UKSSA) 8
United Kingdom Wood Processors Association (UKWPA) 7, 367
University *see* Aberdeen, Bangor, Cambridge, Edinburgh, Oxford, and University College London
United Nations Economic Commission for Europe 383
upland forestry 180–5
afforestation 180, 181
broadleaves 185
restocking 125, 126, 155, 182–4
urban and urban fringe woods 510–12
urban forestry 511, 512
USA, forestry in 68, 127, 128, 131–5, 425
utilization/uses of timber/wood 335–47, 357–71, 439–43
bark *see* bark beams 351
bean rods 355, 356, 440
bee-hives 344
besoms, birch 356
boats 353
boxes 353
broadleaves 355–9
building construction 351, 365
by-products 334, 335

cable-drums 353
carving 346, 354
charcoal 354, 356, 357, 358, 439
chip-baskets 352, 358
chipboard *see* particleboard (chipboard)
chips, wood 359, 366, 383
coffin boards 354
conifers 360–3
cooperage 354
coppice 417
crates 353, 356
electricity supply poles *see* transmission poles
faggots, wood 356
farm buildings 365
fascines 356
fencing *see* fencing
fibreboard *see* fibreboard
firewood, fire-blocks, fuelwood 354, 357, 439
flag poles 352, 362
furniture 352, 356
gates 354, 356, 365
handles, striking, ash 355
hedging materials 353, 355, 356, 440
hop-poles 344, 356, 440
hurdles, bar, ash 356, 440
hurdles, wattle, hazel 353, 355, 356, 440
joinery 351, 365
jumps, birch 343, 356, 440
ladder poles 354, 362
ladder rungs 354
medium density fibreboard (MDF) 437
metal-refining poles 356
mining-timber 352, 363, 366
motor vehicles 354
musical instruments 354
oriented strand board (OSB) 437
packaging 353, 359
pallets 353, 366
paper *see* pulpwood

particleboard *see*
 particleboard
 (chipboard)
pea sticks 355, 356,
 440
piling 354
pitprops 352, 362, 363
pitwood *see*
 mining-timber and
 pitprops
plywood 352
poles, barn 365
poles, various 360, 362
posts, rails, stakes, struts
 158, 440–43
power poles *see*
 transmission poles
pulpwood *see* pulpwood
residues, wood *see*
 wood – chips and
 wood – residues
roundwood, 327, 328,
 329, 351–63, 430–3:
 broadleaved 328, 329,
 355–9; conifer 328,
 329, 360–3; mining
 timber 327; sawlogs,
 broadleaved 358, 363,
 conifer 363; small
 diameter roundwood,
 broadleaved 328,
 355–8, conifer 328,
 360–3; small thinnings,
 broadleaved 355,
 356, conifer 360–3;
 underwood 355, 356,
 358; veneer logs *see*
 veneer timbers; *see
 also* roundwood
rugby poles 352
rustic poles 360, 440
sawlogs *see* sawlogs
scaffold poles 352, 362
shingles, roof, oak 351
sleepers 353
small thinnings 332–46,
 360–2: broadleaves
 355; conifers 360–2
sports goods, ash 353
tanbark *see* tanbark
telegraph poles *see*
 transmission poles
tent-pegs, ash 355
thatching speakes/spars
 356
toys 417

transmission poles 352,
 360–2
turnery 354, 357, 417
vats *see* cooperage
veneer *see* veneer
walking sticks 439, 446
wood wool *see*
 wood-wool
wood-cement
 board *see* timber –
 cement-bonded panels
see also individual
 species

valuations, forestry
 600–10
accountancy and
 management 601
amenity 609
capital taxation 602
charges 610
comparable sales 605,
 606, 608
compensation 601
coppice 608
devastation 606
effect of taxation
 concessions 609, 624;
 see also taxation of
 woodland
effect of grants 609
expectation value 603,
 604, 606, 607
forest properties 609
forestry land 609
historical cost 602, 604,
 606
infrastructure 603
insurance 601, 604, 606
nearly or quite mature
 crops 608
non-wood benefits 610
'pitfalls' 610
plantable land 626
plantations 8 to 15
 years old 605–7
plantations up to 5
 years old 604, 605
pole-stage and mid-age
 crops 607, 608
present market value
 603, 605, 606, 608
recreation 604
replacement cost 604,
 606

sale or purchase 600
sporting 604
thicket-stage 605–7
types 600–2
underwood 608
woodlands 600, 626–8
Value Added Tax (VAT)
 538, 578, 579
Vancouver Island, British
 Columbia, Canada
 131–4
vegetation (flowers,
 grasses, ground flora,
 shrubs) 20, 87–9, 136
 barbary *Berberis vulgaris*
  506
 bilberry, blaeberry
  Vaccinium myrtillus 89
 bluebell *Endymion
  non-scriptus* 90
 bog myrtle *Myrica gale*
  89
 bracken *Pteris aquilinum*
  L. 89, 176
 blackberry, bramble
  Rubus fruticosus 89
 broom *Sytisus scoparius*
  177
 clematis, old man's
  beard, traveller's joy
  Clematis vitalba 88
 Cotoneaster spp. 506
 cotton grass *Eriophorum
  vaginatum* 88, 94
 dog's mercury *Mercurialis
  perrenis* 88
 ferns, various 88
 foxglove *Digitalis
  purpurea* L. 89
 garlic, Ransoms *Allium
  ursinum* 88
 gorse, whin *Ulex
  europaeus* 89, 177
 heather, ling *Calluna
  vulgaris* L. 89, 94;
  bell *Erica cinerea* 89;
  cross-leaved *Erica
  tetralix* 89
 honeysuckle *Lonicera*
  spp. 506
 horsetail, *Equisetum
  sylvaticum* 88
 laurel, cherry *Prunus
  laurocerasus* 177
 marram grass *Psamma
  arenaria* 89

mat grass *Nardus stricta* 89
moss, *sphagnum* 88, 94
nettle, stinging *Urtica dioica* L. 89
privet *Ligustrum vulgare* 506
purple moor grass *Molinia caerulea* L. 88, 94
raspberry *Rubus idaeus* L. 506
rhododendron *see* *Rhododendron ponticum*
rock rose *Helanthemum* spp. 88
rosebay willow-herb *Epilobium angustifolium* 89
rushes *Juncus* spp. 88, 94
sedges *Carex* spp. 88, 94
snowberry *Symphoricarpus rivularis* 506
strawberry, wild *Fragaria vesca* 88
tufted hair grass *Deschampsia caespitos* 88
thyme, wild *Thymus* spp. 88
guide to species choice in forestry 87–9, 136
vegetative propagation 49, 52, 63, 65
Venables, Henry, Hardwoods Ltd 436
veneer timber 352, 358, 359, 434, 435
export 359
merchants 359
prices 434, 435
species (including yew) 359, 434, 435
Vyrnwy, Lake, in Wales 73

Wakefield, Norfolk 254
Wales 7
Secretary of State 1
Wall, J. (EC) 634
Walnut, Black *Juglans nigra* L. 13, 29, 118, 344, 351, 352, 354, 359, 435

Common *J. regia* L. 13, 29, 118, 344, 351, 352, 354, 359, 435
wars, effect of 2, 321
water 73, 74, 502–54
catchment areas, watersheds 73, 74
guidelines for management 502–4
management of forests for 74, 502–4
nursery *see* nursery
quality 74
Wayfaring tree *Viburnum lantana* 119
Weasenham, Norfolk 254
weeding of plantations 172–8, 541
chemical 175–8
costs 178, 190
manual 174
mechanical 175
mulching, mulch mats 177, 178
see also herbicides
weeding of nurseries *see* nursery
weeds *see* weeding of plantations
welfare *see* health and safety
Wellingtonia (Giant Sequoia) *Sequoiadendron giganteum* (Lindl.) Buchholz. 14, 27, 111
largest species in the world 110
silvicultural characteristics 110
timber properties 341
timber uses 341, 352
Welsh Office Agriculture Department 7
Western hemlock *Tsuga heterophylla* (Raf.) Sarg. 14, 109, 134
diseases and insect pests 109
silvicultural characteristics 109
suitable conditions 134
thinning 299, 309
timber: identification 27; markets 414; price 432; properties 340; uses 340, 351, 356

Western red cedar *Thuja plicata* D. Don. 14, 109, 134
diseases and insect pests 54, 109
silvicultural characteristics 109
suitable conditions 134
thinning 299, 309
timber: identification 27; markets 414; price 432; properties 340; uses 354, 355
Westonbirt arboretum 582; *see also* Plate 4
Weymouth, Viscount, Longleat Estate 283, 284
'whips' *see* plants
Whitebeam, English *Sorbus aria* (L.) Crantz 13, 119
Swedish *S. intermedia* (Ehrh.) Pers. 13, 119
Wild cherry, Gean *Prunus avium* L. 13, 29, 113, 137, 143, 303, 343, 351, 352, 354, 435
Wild service tree *Sorbus torminalis* (L.) Crantz 13, 119
wildlife
assessment of damage 239
economics 491, 536
guidelines for conservation management 121, 490–9
habitat 489
see also Acts of Parliament (Statutes), animals and birds
wildwood 479
Willow 13, 30, 51, 116, 345, 352, 353, 354
Almond *Salix triandra* L. 286
Basket willow coppice 286, 287
Bay *Salix pentandra* L. 13
British Standard 165
Common osier *S. viminalis* 270, 286

Crack *S. fragilis* L. 13
cricket-bat *see*
   Cricket-bat willow
   diseases and insect
   pests 287
Purple *S. purpurea* L. 286
rods 51
Sallow (Goat) *S. caprea* L. 13
sets 51
timber properties 345
timber uses 345
wind, windthrow, 80–7, 381
  breaks 149, 150
  catastrophic 81
  critical heights 86
  damage 80–7
  endemic 82
  reducing risk of windthrow 87
  restocking after windthrow 82, 566
  salt-laden 77
  stability of species 105, 112
  storms (of 1987 and 1990) 82, 381
  terminal heights 86
  Windthrow Hazard Classification 83–7
  see also climate
Windsor estate, Berks. 250, 285
wood preservation *see* preservation of timber/wood
wood processing 436–9
  moulded-board 439
  paper and paperboard 436, 437
  sawmills *see* sawmills, sawmilling, estate
  turneries 357, 433
  wood-based panel board 437, 438
  wood wool 439
wood (timber) including properties 335–46
  annual growth rings 20
  ash (elements) 334
  by-products 355, 357
  cellulose 334
  chemistry 20, 334, 335, 357
  chips 359, 366, 383
  cleavability 356
  colour 24, 347
  compression wood 334
  decorative 347
  defects 347
  density 34, 332, 347
  diffuse-porous 23
  distillation 334, 335, 357
  drought cracks 347
  drying, seasoning 432
  durability, natural 334, 347, 406
  earlywood, springwood 21
  elasticity 347
  extractives 334, 335
  features 26–30, 347
  fibre 22, 368
  figure 24, 347
  grain 23, 24, 333, 347
    spiral 23
  heartwood 20
  hemicellulose 334
  improving quality 347, 348
  insect attack 334
  juvenile 23, 292, 333, 347
  knots 21, 22, 303, 333, 350
  latewood, summerwood 21, 22
  lignin 23, 334
  lustre 347
  moisture content 24, 33, 334
  movement 332
  permeability 334, 347, 407
  preservative treatment 334
  processing 436–9
  properties 332
  quality 163, 292, 347–9
  reaction wood 334
  research 586, 587
  residues 326, 327, 328, 329, 358, 359, 366
  resin pockets 334
  ring-porous 24
  rot 334, 347
  sapwood 20
  sapstain 334, 350
  shake 334, 347
  shrinkage 332
  smell 25
  straightness 347
  strength 347, 348
  structure 20–30; *see also* Figs. 2 and 3
  tannin 334
  tension wood 334
  texture 24, 332
  weight 24, 347
  wool 439
  see also individual species and utilization/uses of timber/wood
wood residues *see* wood
woodland
  ancient *see* ancient semi-natural
  census *see* census of woodland and trees (1979–82)
  dedication *see* Forestry Commission
  ecosystem 99, 100
  Grant schemes 518
  heritage status *see* taxation of woodland
  Management Grants 147, 564–8
  markets 622–31
  measurement *see* measurement, forest
  neglected, grants for 566
  ownership, private, *see* owners of woodlands
  prices 625–7
  public access *see* access, woodlands
  sales of 622–4
  taxation *see* taxation of woodlands
  valuation *see* valuations
Woodland Trust 486
woods, small 476, 506, 511, 515
woods, urban 510–12
wood-yard 401
working plan 463
Workman, J., OBE 280
Workmans Wood National Nature Reserve 280; *see also* Plate 7
work study 377, 379, 381, 595, 596

output guides and standard time tables 377, 379, 381, 595, 596
piece-work 596
standard minute, price per (PSM) 596
see also Forestry Commission Work Study Branch
world timber prices 1, 425, 426
Wright, Carleton 204
Wright, J.S., and Son Ltd 203

Wych elm *see* Elm, Wych *Ulmus glabra* Huds.

Yew, common *Taxus baccata* 13, 27, 111
timber properties 341
timber uses 341, 352, 354
veneer quality 341, 435
yield, timber 538
forecasting 455
models 451–5
sustained 273, 274

yield class 105, 112, 180, 445–9
broadleaves 553
conifers 551
general (GYC) 447, 448, 450, 451
local (LYC) 451
of bare land 448
upper planting limit 448
see also individual species
yield tables 451–5